Element	Symbol	Atomic No.	Atomic Weight	Element	Symbol	Atomic No.	Atomic Weight
Gold	Au	79	196.967	Praseodymium	Pr	59	140.907
Hafnium	Hf	72	178.49	Promethium	Pm	61	[147]*
Helium	He	2	4.0026	Protactinium	Pa	91	[231]*
Holmium	Ho	67	164.930	Radium	Ra	88	[226]*
Hydrogen	H	1	1.00797[a]	Radon	Rn	86	[222]*
Indium	In	49	114.82	Rhenium	Re	75	186.2
Iodine	I	53	126.9044	Rhodium	Rh	45	102.905
Iridium	Ir	77	192.2	Rubidium	Rb	37	85.47
Iron	Fe	26	55.847[b]	Ruthenium	Ru	44	101.07
Krypton	Kr	36	83.80	Samarium	Sm	62	150.35
Lanthanum	La	57	138.91	Scandium	Sc	21	44.956
Lead	Pb	82	207.19	Selenium	Se	34	78.96
Lithium	Li	3	6.939	Silicon	Si	14	28.086[a]
Lutetium	Lu	71	174.97	Silver	Ag	47	107.870[b]
Magnesium	Mg	12	24.312	Sodium	Na	11	22.9898
Manganese	Mn	25	54.9380	Strontium	Sr	38	87.62
Mendelevium	Md	101	[256]*	Sulfur	S	16	32.064[a]
Mercury	Hg	80	200.59	Tantalum	Ta	73	180.948
Molybdenum	Mo	42	95.94	Technetium	Tc	43	[99]*
Neodymium	Nd	60	144.24	Tellurium	Te	52	127.60
Neon	Ne	10	20.183	Terbium	Tb	65	158.924
Neptunium	Np	93	[237]*	Thallium	Tl	81	204.37
Nickel	Ni	28	58.71	Thorium	Th	90	232.038
Niobium	Nb	41	92.906	Thulium	Tm	69	168.934
Nitrogen	N	7	14.0067	Tin	Sn	50	118.69
Nobelium	No	102	...	Titanium	Ti	22	47.90
Osmium	Os	76	190.2	Tungsten	W	74	183.85
Oxygen	O	8	15.9994[a]	Uranium	U	92	238.03
Palladium	Pd	46	106.4	Vanadium	V	23	50.942
Phosphorus	P	15	30.9738	Xenon	Xe	54	131.30
Platinum	Pt	78	195.09	Ytterbium	Yb	70	173.04
Plutonium	Pu	94	[242]*	Yttrium	Y	39	88.905
Polonium	Po	84	[210]*	Zinc	Zn	30	65.37
Potassium	K	19	39.102	Zirconium	Zr	40	91.22

QUANTITATIVE ANALYTICAL CHEMISTRY

Fifth Edition

QUANTITATIVE ANALYTICAL CHEMISTRY

JAMES S. FRITZ
Iowa State University

GEORGE H. SCHENK
Wayne State University

PRENTICE HALL, Englewood Cliffs, New Jersey 07632

Cover Administrator: Linda Dickinson
Cover Designer: Marcia Boeing
Production Administrator: Lorraine Perrotta
Manufacturing Buyer: William J. Alberti
Editorial-Production Services: York Production Services

© 1987 , 1979 , 1974 , 1969 , 1966 by Prentice-Hall, Inc.
A Simon & Schuster Company
Englewood Cliffs, New Jersey 07632

Library of Congress Cataloging-in-Publication Data

Fritz, James S. (James Sherwood), 1924–
   Quantitative analytical chemistry.

   Bibliography: p.
   Includes index.
   1. Chemistry, Analytic—Quantitative. I. Schenk,
George H. II. Title.
QD101.2.F74 1987   544   86-26598

ISBN 0-205-10480-0
ISBN 0-205-10554-8 (International)

10 9 8 7 6 5

Printed in the United States of America.

Contents _____

32 Electronanalytical Procedures   628

33 Separation Procedures   641

Appendixes   655

Answers to Selected Problems   677

Index   685

Preface

       The analytical chemistry practiced in professional laboratories is clearly changing. Increasingly, devices and sensors are being used that will perform analytical measurements as quickly and automatically as possible. Robots are coming into use to perform repetitive tasks such as weighing, dissolving, and pretreating analytical samples. Sacrifices in the accuracy of analyses are often made to obtain greater speed and convenience.

       Such rapid change makes it difficult to decide what should be included in a basic textbook on analytical chemistry. We have concluded that a balanced treatment is called for. Any intelligent approach to chemical analysis requires an understanding of the basic *principles* and a thorough knowledge of the associated *chemistry*. After all, analyses are done on *chemical* samples. We need to understand the chemistry on which analytical procedures is based and how changes in various parameters can affect the equilibrium of the chemical system. At the same time, it is essential to provide a thoroughly modern treatment of the various forms of spectroscopy, chromatography, and other analytical methods that are now being used so frequently and extensively.

       This fifth edition represents the most extensive revision the book has undergone since publication of the first edition in 1966. Our commitment to produce a book that is written in a clear and understandable manner remains unchanged. However, many changes have been made in the content of this edition. First of all, the order of chapters has been changed somewhat. The three chapters on spectroscopy (Chapters 17–19) are now placed together just after the chapters on predominantly chemical and electrochemical methods of analysis. Chapter 20 on liquid-liquid extraction and four chapters on chromatography (Chapters 21–24) make up the remainder of Part I.

       Virtually every chapter in Part I contains important revisions, together with many new questions and problems. Chapter 3 on treatment of analytical data has been rewritten and expanded to emphasize the use of a simple statistical approach.

The Henderson-Hasselbach equation for calculating pH is now discussed in Chapter 7 on acid-base equilibrium. A new and more logical method for determining the shape of titration curves is given in Chapter 8. Chapter 12 on oxidation-reduction theory has been extensively revised to provide a clearer and more logical treatment of this important subject. Chapter 13 on oxidation-reduction titrations has been partially rewritten, and some older factual material has been eliminated. Much of Chapter 14 has been rewritten to give a clearer understanding of the role of reaction rates in chemical analysis. New material on ion-selective electrodes has been added to Chapter 16.

The three chapters on spectroscopy are now placed together. Chapter 17 is on the principles of spectrophotometry, Chapter 18 covers spectrophotometric methods for organic compounds, and Chapter 19 deals with atomic spectroscopy. Chapter 19 now covers the plasma and graphite furnace, but discussion of older arc-spark methods has been eliminated.

Separations are now used extensively in chemical analysis. In many laboratories, some form of chromatography is the overwhelming choice for quantitative analysis. Chapter 20 on liquid-liquid extraction has been extensively revised, and the four chapters on chromatography have been completely rewritten. Chapter 21 covers the general principles and theory of chromatography. Chapter 22 provides a modern treatment of gas chromatography with emphasis on the use of capillary columns. Chapter 23 covers modern liquid chromatography (HPLC). Chapter 24 is on ion-exchange chromatography and features "ion chromatography," which is one of the fastest growing areas of analytical chemistry.

Part II has also been greatly rewritten, with a special section in the first chapter (Chapter 25) on laboratory safety. Several new experiments have been added. In addition, the spectrophotometry experiment chapter (Chapter 31) has been rewritten to include directions for the latest student-level spectrophotometer, the LKB Novaspec microprocessor-spectrophotometer. This instrument has some powerful advantages for routine acquisition of absorption spectra, as demonstrated in Experiment 22.

The authors wish to thank the following reviewers for their many excellent suggestions: Larry R. Faulkner, University of Illinois at Urbana-Champaign; Quintus Fernando, The University of Arizona-Tucson; M. Dale Hawley, Kansas State University; Jerome W. O'Laughlin, University of Missouri-Columbia; Su-Moon Park, The University of New Mexico; Lloyd R. Parker, Jr., Vassar College; William L. Switzer, North Carolina State University; John P. Walters, St. Olaf College; James D. Winefordner, University of Florida.

James S. Fritz
George H. Schenk

PART I

PRINCIPLES
AND
THEORY

1

Introduction

1–1. THE NATURE OF ANALYTICAL CHEMISTRY

What Is Analytical Chemistry?   Analytical chemistry is the branch of chemistry dealing with the separation and analysis of chemical substances. Traditionally, analysis has been concerned largely with chemical *composition*, but it is coming more and more to include the determination of chemical *structure* and the measurement of physical properties. Analytical chemistry includes both qualitative and quantitative analysis. Qualitative analysis is concerned with *what* is present, quantitative analysis with *how much*.

The qualitative identification of the substances present in an analytical sample is of great importance. Often the answer to a vexing industrial problem lies in the identification of something that is adulterating a product or fouling up a chemical process. Sometimes a chemical "spot test," based on a distinctive color-forming reaction, can be used for identification purposes. Emission spectrometry (Chapter 19) is a fast and useful way of detecting the elements present in inorganic samples. Chromatography is an extremely valuable way to identify organic compounds, especially when coupled with a mass-spectrograph (Chapter 22). Thin-layer chromatography (Chapter 23) is also very useful for qualitative identifications. It has been used by police laboratories, for example, to distinguish between the many inks used in ballpoint pens. Infrared spectra (Chapter 18) serve as excellent "fingerprints" for identifying organic and many inorganic compounds. These and other ingenious methods are the techniques of modern qualitative analysis.

Courses in quantitative analysis have traditionally dealt almost exclusively with the analysis of inorganic material. Nevertheless, analytical chemistry properly includes the analysis of organic material too. Analytical chemistry finds extensive application in the analysis of organic compounds, pharmaceuticals, biochemicals, body fluids, hair, the atmosphere, polluted water, foods, soils, and many other types of substances.

What Is an Analytical Chemist?   A true analyst, or analytical chemist, has several characteristics. He or she has a knowledge of the methods and instruments used for analysis. He understands the principles of analysis, so that he can apply and, if necessary, modify analytical methods to solve a particular problem; frequently, he is a research chemist who studies the theory of analytical processes or develops completely new methods of analysis. He can evaluate and interpret the results of a quantitative analysis.

Above all, an analytical chemist is a problem solver. It has been said that if you can state a problem clearly, it can be solved. An analytical chemist must do just this. By asking questions and gathering information, he or she determines what the actual problem *is*, then uses experience and intelligence to map out a scheme for solving it.

Thus, an analytical chemist is a skilled, well-trained chemist—in sharp contrast with the more numerous technicians or "determinators," who simply twist the dials of an instrument or follow "cookbook" analytical procedures.

What Information Does Chemical Analysis Provide?   *Qualitative analysis* may be used to indicate the presence or absence of certain elements, ions, or molecules. For example, the first step in "screening" a suspicious solid sample for lysergic acid diethylamide (LSD) is to examine it under ultraviolet light [1]. Most hallucinogens such as LSD show up as fluorescent or discolored areas that can then be dissolved and tested further. A *structural determination* may be used to define the entire structure of a new drug or to verify only the structure or stereochemistry of a certain part of a newly synthesized molecule.

The most important aspect of analysis is still *quantitative analysis*, with which this book is mainly concerned. A quantitative analysis provides data regarding the chemical composition of matter. These data may be quite detailed, or they may be incomplete and general. The types of quantitative analysis may be classified as follows:

Complete Analysis.   The amount of each constituent of the sample is determined quantitatively. For example, a complete analysis of a gasoline sample would tell the percentage of each compound present (hydrocarbons, tetraethyllead, tricresyl phosphate, etc.). In many samples, such an analysis would be a waste of time; instead a "complete" analysis is run for a select number of species. For example, in clinical laboratories, a "complete" blood analysis may involve the determination of eight or twelve species: glucose, Na^+, K^+, bilirubin, alkaline phosphates, etc. [2].

Ultimate Analysis.   The amount of each element in a sample is determined without regard to the actual compounds or ions present. An ultimate analysis of a gasoline sample would tell the percentage of carbon, hydrogen, oxygen, lead, phosphorus, etc.

Partial Analysis.   The amount of a certain selected constituent in a sample is determined. A partial analysis of gasoline might tell the percentage of tetraethyllead

[1] Scientific Methods of Crime Investigation, *Chemistry*, *43*, 12 (1969).
[2] L. T. Skeggs, *Anal. Chem.*, *38*, 31A (May 1966).

or the percentage of aromatic hydrocarbons. In the routine assay of commercial aspirin tablets, the amount of salicylic acid impurity usually gives the best indication of purity. Often a partial analysis will provide all of the information that is needed.

Where Is Analytical Chemistry Used? The need for quantitative analysis is widespread—not only in chemistry, but also in commerce and in other fields of science and technology. Some of its important uses are:

Relating Chemical Composition to Physical Properties. The efficiency of a catalyst, the mechanical properties of a metal, the performance of a fuel, etc., may depend largely on chemical composition.

Quality Control. Chemical analysis is vital in maintaining good quality in the air we breathe and the water we drink. Standards must be set and frequent analyses performed to see that they are met. In industry, analysis is needed to see that raw materials meet specifications and to check the purity of the final product.

Determining the Amount of a Valuable Constituent. The determination of the amount of butterfat in cream, the amount of uranium in an ore, and the protein content of food are but a few of many examples.

Diagnosis. Chemical analysis is finding increasing use in diagnosing disease. For example, the presence of a measurable amount of bilirubin and more than 12 mg% of alkaline phosphatase (an enzyme) in a patient's blood serum is indicative of an impaired liver function [2].

Research. Analytical chemistry is of major importance in connection with many research projects. Some examples are determining traces of metal that pass into solution (corrosion studies), finding out about a competitor's product, analyzing reaction mixtures to see which reaction conditions give the best product yield, and measuring extraction distribution ratios to find the best conditions for a large-scale extraction process.

What Methods Are Used in Quantitative Analysis? The scope of analysis is tremendous. Many varied and ingenious methods of analysis have been invented, and research on new methods is proceeding rapidly. Some of the major ways of measuring the amount of a given substance in a sample are listed in the next chapter; it is not an exhaustive list, nor does it include methods of separating complex mixtures.

Methods of quantitative analysis are based on chemical reactions, on the measurement of certain chemical or physical properties (such as spectra), or on the measurement of a combination of chemical and physical properties (photometric titration, for example). In some areas of analysis, instrumental methods based largely on the measurement of some physical property have preempted other analytical methods. Instrumental methods are often fast and lend themselves well to automation. For instance, a special emission spectrograph analyzes metal samples for several constituents in a few minutes, and the results are immediately transmitted to the production part of the plant.

Analysis may be carried out on discrete samples, or it may be done on a *continuous* basis whereby the concentration of a particular ion or molecule is *monitored* continuously. This is done when a change in concentration can affect the environment, the safety of a substance, or the quality of a product. This type of analysis is particularly important in protecting our health and well-being. For example, the fluoride-ion concentration in public water supplies must be checked continuously so that the fluoride concentration does not become dangerously high or so low that it is ineffective at protecting teeth from decay. This is done by continuous colorimetric measurement (see Chapter 17) or by using a fluoride-selective ion electrode (see Chapter 16). It is also important in combating air pollution. In cities such as Los Angeles, New York, and Chicago, the level of contaminants such as sulfur dioxide in the air are automatically and continuously measured and recorded. Keyed in with the analysis data are multistage alert plans that go into effect if the SO_2 readings rise beyond specified limits. The heart of these cities' protection against danger from air pollution is continuous analysis developed by use of analytical chemistry.

Although a number of purely instrumental methods of analysis have been developed, chemical methods are still vital and widely used, for several reasons. Many instruments require extensive calibration for each type of sample to be analyzed, whereas chemical methods can be quickly adapted to analyzing new types of samples. Few instrumental methods can match the accuracy and precision of chemical methods. For example, the emission spectrograph is superb for the analysis of trace constituents in metal alloys, but its accuracy with respect to major constituents is greatly inferior to that achieved by chemical methods. Properly used, instrumental and chemical methods supplement each other, and the best analytical laboratories make extensive use of both. Finally, a knowledge of chemical methods provides the background needed for a real understanding of analysis and analytical problems.

1–2. SOME FUNDAMENTAL CONCEPTS

In the remainder of this chapter, we review a few of the concepts upon which quantitative chemistry is based, and we explain some of the terms and conventions used in this book. Other important concepts, with which it is hoped the student is already familiar, are reviewed briefly in later chapters. For example, chemical equilibrium is considered in Chapter 8, and the balancing of oxidation-reduction equations in Appendix 4.

Moles. Chemistry is concerned with the reactions of elements and of combinations of elements (compounds). The smallest particle of an element is the atom; the smallest particle of a compound is the molecule.

Unfortunately, chemists cannot *count* (except indirectly) the number of molecules reacting; they must know the relative *weights* of various molecules. Atomic weights are the relative weights of the atoms and are based on the carbon isotope of mass number 12 (formerly, atomic weights were based on oxygen as 16.0000). A molecular weight is the sum of the atomic weights of the atoms that make up a molecular compound. A formula weight is the sum of atomic weights in the formula of an element, ionic substance, or molecular compound.

Atomic weights take into account the "abundance" of the various naturally occurring isotopes of each element. Isotopes are atoms of the same element (same atomic number) that have different atomic weights. For example, nitrogen as it occurs naturally contains primarily ^{14}N and a little ^{15}N; lithium is mostly ^{7}Li but has some ^{6}Li. The atomic weight of nitrogen thus is a little greater than 14 (14.007), and the atomic weight of lithium is less than 7 (6.939). The fact that isotopes of some elements can be partially separated can complicate the job of the analytical chemist. For example, in a gravimetric analysis, the percentage of lithium is calculated from the weight of lithium sulfate precipitate and the ratio of the atomic weight of lithium to the molecular weight of lithium sulfate. A sample in which the normal ratio of ^{6}Li to ^{7}Li had been changed by a partial isotope separation would give erroneous results if the usual atomic weight for lithium were used.

The fundamental quantity a chemist works with is the mole. A mole is 6.023×10^{23} molecules of a substance (this fantastically large number is called Avogadro's number). We shall use the term *mole* in a broad sense to describe amounts of molecular compounds, free elements, and ions. A mole is the *formula weight* of a molecule, element, or ion, expressed in grams. Thus,

$$\begin{aligned}
1 \text{ mole of } H_2O &= 18.01 \text{ grams} \\
1 \text{ mole of } Na_2SO_4 &= 142.04 \text{ grams} \\
1 \text{ mole of } Na &= 22.99 \text{ grams} \\
1 \text{ mole of } Cl_2 &= 70.90 \text{ grams} \\
1 \text{ mole of } Cl^- &= 35.45 \text{ grams}
\end{aligned}$$

The number of moles equals the number of grams divided by the formula weight:

$$\text{Moles of urea } (H_2NCONH_2) = \frac{\text{grams}}{\text{mol wt}} = \frac{\text{grams}}{60.06}$$

$$\text{Moles of sulfate } (SO_4{}^{2-}) = \frac{\text{grams}}{\text{ionic wt}} = \frac{\text{grams}}{96.06}$$

$$\text{Moles of silver } (Ag) = \frac{\text{grams}}{\text{atomic wt}} = \frac{\text{grams}}{107.87}$$

Concentration of Solutions. Any one of several different methods can be used to express the concentration of solutions. The systems commonly used are listed in Table 1–1.

In analytical chemistry, the molar and normal systems are the most frequently employed. Some chemists use the formal system extensively. Analytical calculations with these systems are discussed in detail later in the book.

It is important to distinguish between the *analytical* concentration and the *equilibrium* concentration of a solution.

The analytical concentration is the total number of grams, moles, etc., of solute present in a given volume of solution. Nothing is said about whether or not the solute ionizes in solution. For example, $0.1M$ KCl (also written $C_{KCl} = 0.1$) means that the solution contains 7.45 g (0.1 mole) of potassium chloride per liter of solution. The fact that potassium chloride exists in aqueous solution entirely as K^+ and Cl^- is not taken into account in stating the analytical concentration of KCl. As

Table 1–1. Systems for Expressing Concentrations of Solutions

Name of System	Symbol	Definition
Molar	M	$\dfrac{\text{moles of solute}}{\text{liters of solution}}$
Molal	m	$\dfrac{\text{moles of solute}}{\text{kilograms of solvent}}$
Formal	F	$\dfrac{\text{gram–formula-weights of solute}}{\text{liters of solution}}$
Mole fraction	$\mathcal{N}$	$\dfrac{\text{moles of solute}}{\text{moles of solvent} + \text{moles of solute}}$
Normal	N	$\dfrac{\text{equivalents of solute}}{\text{liters of solution}}$
Grams per volume	——	$\dfrac{\text{grams of solute}}{\text{liters of solution}}$
Weight percent	wt %	$\dfrac{100 \times \text{grams of solute}}{\text{grams of solvent} + \text{grams of solute}}$
Volume percent	vol %	$\dfrac{100 \times \text{liters of solute}}{\text{liters of solution}}$
Parts per million	ppm	$\dfrac{\text{milligrams of solute}}{\text{kilograms of solution}}$ or $\dfrac{\text{milligrams}}{\text{liter}}$
Parts per billion	ppb	$\dfrac{\text{micrograms}}{\text{liter}}$

another example, $1.0M$ CH_3COOH (or $C_{CH_3COOH} = 1.0$) means that the solution contains 1.0 mole (60 g) of acetic acid dissolved in 1.0 liter of solution.

Equilibrium concentration is the concentration of ions or molecules actually present in solution and takes into account the possible dissociation of the solute into ions. The equilibrium concentration is indicated by enclosing the ion or molecule in brackets. Brackets around a species also mean that we are dealing with its *molar* concentration: $[K^+] = 0.1$ means that the solution contains 0.1 mole of potassium ion per liter; $[CH_3COOH] = 1.0$ means that the equilibrium concentration is 1.0 mole per liter of molecular (nonionized) acetic acid. Because acetic acid in water is partially ionized, the analytical concentration of acetic acid must be slightly greater than $1.0M$ to give an equilibrium concentration of molecular acetic acid equal to $1.0M$.

The parts per million (ppm) unit is very common in trace analysis. This unit was devised by Lord Kelvin during the British beer poisoning epidemic of 1900. Kelvin headed a Royal Commission that attempted to set the world's first tolerance on a poison (arsenic). A 1-ppm solution contains 1 mg of solute per 10^6 mg of a solvent. Since 1 liter of water is about one million milligrams, a 1-ppm solution also contains about 1 mg of solute per liter of solution. Often ppm is defined as mg/liter, even though a liter of the solution may weigh somewhat more, or less, than 1 kg. The sensitivity of analytical methods has improved so much that parts per billion (ppb) has also become a common unit of concentration.

When dealing with solids, the ppm unit must be used in terms of mg of constituent per kg of solid. For example, in 1969–1970, fish were found in the Lake Erie area that were contaminated with mercury in excess of the FDA tolerance of 0.5 ppm. Samples of walleye pike containing 1.40–3.57 ppm of mercury were found; this meant that they contained from 1.40 to 3.57 mg of mercury per kilogram of fish.

Activity and Activity Coefficients. Solutes may be roughly classified according to their ability to ionize and conduct an electric current: nonelectrolytes, such as sugar and urea; weak electrolytes, such as weakly ionized acids and bases; and strong electrolytes, such as HCl or KCl, which are highly or completely ionized in aqueous solution (potassium chloride is known to be completely ionized in the solid state).

Except in very dilute solution, the effective concentration of ions in solution (determined by the lowering of the freezing point of water, by electrical conductivity, or by other means) is usually less than the actual concentration of ions known to be present. The term *activity* is used to denote the active or effective concentration of an ion or molecule in solution. Activity may be related to molar concentration through the use of an *activity coefficient*; in the equation

$$a_i = f_i[i]$$

a_i is the activity of an ion, f_i is the activity coefficient of that ion, and $[i]$ is the molar concentration of the ion. In very dilute solution, f_i approaches 1; that is, $a_i \approx [i]$. As the concentration increases, the activity coefficient becomes smaller and the values of a_i and $[i]$ become more divergent. The activity coefficient of an ion of charge greater than 1 is smaller at any given concentration than that of an ion whose charge is unity. Activity coefficients of nonionic substances are approximately 1 except in very concentrated solutions.

Differences between concentration and activity arise because of ionic interaction between positively and negatively charged ions moving about in a solution. Ions of like charge will repel each other, and ions of unlike charge will attract each other. Attraction and repulsion are not as strong in water as in solvents having lower dielectric constants, but they do exist. Most of the time, the space around a positive ion will have an excess negative charge, and the space around a negative ion will, on the average, have an excess positive charge. Thus the motion of the average ion (either plus or minus) is impeded to some extent, and it is not as active as an entirely free ion. As the solution becomes more dilute, the ions in solution are farther apart and have less effect on one another.

From the laws of electrical attraction and repulsion and from the Boltzmann distribution law, which describes the tendency of thermal agitation to counteract electrostatic effects, Debye and Hückel derived an equation that enables activity coefficients to be calculated theoretically. According to Debye and Hückel, the activity coefficient of any ion depends on the *ionic strength* of the solution. The ionic strength of a solution is *not* equivalent to the total ionic concentration, but is defined by the equation

$$\mu = \tfrac{1}{2}\sum [i]Z_i{}^2$$

where μ is the ionic strength, $[i]$ is the molar concentration of an ion, and Z_i is the charge ($+$ or $-$) of that ion.

Example: Calculate the ionic strength of a $0.1M$ KCl solution. $[K^+] = 0.1$, and $[Cl^-] = 0.1$; then

$$\mu = \tfrac{1}{2}(0.1 + 0.1) = 0.1$$

Example: Calculate the ionic strength of a $0.1M$ Na_2SO_4 solution. $[Na^+] = 0.2$, and $[SO_4{}^{2-}] = 0.1$; then

$$\mu = \tfrac{1}{2}[0.2 + 0.1(4)] = 0.3$$

In calculations of the ionic strength of a solution, the contribution of weakly ionized substances, such as weak acids, may be ignored.

The Debye–Hückel equation, which relates the activity coefficient of an ion to the ionic strength of the solution, is

$$-\log f_i = 0.5Z_i{}^2\sqrt{\mu}$$

where f_i is the activity coefficient of an ion, Z_i is the charge ($+$ or $-$) of that ion, and μ is the ionic strength of the solution. This equation is useful in estimating activity coefficients in a solution of fairly low ionic strength. The activity coefficients in Table 1–2 have been calculated from a form of the Debye–Hückel equation that takes ionic size into account.

Experimentally, only *mean* activity coefficients ($f_{\pm}$) can be measured (the mean of f_+ and f_-); no experimental method is available for measuring single-ion activity coefficients. However, the $f_{\pm}$ values calculated for various electrolytes from the theoretical individual-ion activity coefficients are in satisfactory agreement with the experimentally determined mean activity coefficients, up to an ionic strength of approximately 0.1.

The activity of solvents, solids, and gases is defined differently from that of solutes in solutions. A pure liquid (water, for example) is assigned the value of unit activity ($a = 1$). Similarly, a pure solid has unit activity. A gas has unit activity at 1 atm pressure.

An acquaintance with the concept of activity is important to understanding several aspects of analytical chemistry. Activity coefficients must be used for rigorous calculations in which equilibrium constants are used. Even though activity coefficients are often ignored in such calculations, the student should be aware of the approximations involved and should know under what conditions such approximations are valid. The potentials of several electrodes of analytical importance depend on the activities of certain ions in solution. For example, a pH meter measures the activity of hydrogen ions in solution, not the concentration.

Chemical Equations. When a solute dissolves in water or another solvent, the crystal structure is destroyed and there is an accompanying temperature change. The solute may go into solution as molecules, as ion pairs, or as simple or complex ions. The solute molecules or ions become *solvated* by interacting with the solvent.

Table 1–2. Individual Ion Activity Coefficients as a Function of Ionic Strength (μ)

Ion	Activity Coefficient							
	$\mu = 0.001$	$\mu = 0.002$	$\mu = 0.005$	$\mu = 0.01$	$\mu = 0.02$	$\mu = 0.05$	$\mu = 0.1$	$\mu = 0.2$
H^+	0.975	0.967	0.950	0.933	0.914	0.88	0.86	0.83
Li^+	0.975	0.965	0.948	0.929	0.907	0.87	0.835	0.80
Na^+, IO_3^-, HSO_4^-	0.975	0.964	0.947	0.928	0.902	0.86	0.82	0.775
$K^+, F^-, Cl^-, Br^-, I^-$	0.975	0.964	0.945	0.925	0.899	0.85	0.805	0.755
NH_4^+, Ag^+	0.975	0.964	0.945	0.924	0.898	0.85	0.80	0.75
Mg^{2+}, Be^{2+}	0.906	0.872	0.813	0.755	0.69	0.595	0.52	0.45
$Ca^{2+}, Cu^{2+}, Zn^{2+}, Mn^{2+}, Ni^{2+}, Co^{2+}$	0.905	0.870	0.809	0.749	0.675	0.57	0.485	0.405
Ba^{2+}, Cd^{2+}	0.903	0.868	0.805	0.744	0.67	0.555	0.465	0.38
Pb^{2+}	0.903	0.868	0.805	0.742	0.665	0.55	0.455	0.37
SO_4^{2-}, HPO_4^{2-}	0.903	0.867	0.803	0.740	0.660	0.545	0.445	0.355
$Al^{3+}, Fe^{3+}, Cr^{3+}$	0.802	0.738	0.632	0.54	0.445	0.325	0.245	0.18
PO_4^{3-}	0.796	0.725	0.612	0.505	0.395	0.25	0.16	0.095
$Th^{4+}, Zr^{4+}, Ce^{4+}$	0.678	0.588	0.455	0.35	0.255	0.155	0.10	0.065

Source: J. Kielland, *J. Am. Chem. Soc.*, **59**, 1675 (1937).

Solvation is seldom taken into account in writing the formulas of ions or molecules except in the case of the hydrogen ion, which has an unusually high solvation energy and cannot exist in solution unsolvated. Thus, in aqueous solution it exists as H_3O^+, the hydronium ion, but is conveniently written H^+ and referred to as the hydrogen ion.

In writing a chemical equation involving substances that are highly ionized in solution, it is more accurate (and usually more convenient) to indicate the reaction of the ions than to write the stoichiometric formulas of the solutes that supply the reactive ions. Consider what happens when an aqueous solution of silver nitrate is added to an aqueous solution of sodium chloride. The one solution contains Ag^+ and NO_3^-; the other contains Na^+ and Cl^-. When mixed, the Ag^+ and Cl^- combine to form a white precipitate of silver chloride:

$$Ag^+ + Cl^- \rightarrow AgCl(s)$$

This represents the reaction taking place very well. But what of the Na^+ and NO_3^- in solution? Shouldn't we write $NaNO_3$ or $Na^+NO_3^-$ as another product of the reaction? For an aqueous solution, the answer is "no," because the Na^+ and NO_3^- remain simply as free ions in solution and are not combined with one another. An equation that accounted for all ions would be

$$Ag^+ + NO_3^- + Na^+ + Cl^- \rightarrow AgCl(s) + Na^+ + NO_3^-$$

The actual chemical reaction, however, is simply the combination of silver ions and chloride ions forming a precipitate.

For clarity we shall sometimes write in parentheses the compound dissolved in solution, to show what supplied the reactive ion; for example,

$$Ag^+ + Cl^- \rightarrow AgCl(s)$$
$$\text{(AgNO}_3\text{)}\quad\text{(NaCl)}$$

$$2H^+ + CO_3^{2-} \rightarrow H_2O + CO_2$$
$$\text{(HCl)}\quad\text{(Na}_2\text{CO}_3\text{)}$$

Many substances exist in solution as mixtures of several species. In some cases, it becomes difficult to write a chemical reaction and be precise about the reacting species. We shall write weak electrolytes in the molecular form, even though they may be ionized to the extent of a few ions per hundred molecules; for example,

$$OH^- + CH_3COOH \rightarrow H_2O + CH_3COO^-$$
$$\text{(NaOH)}$$

Acetic acid (CH_3COOH) is predominately in the molecular form, although it is partly dissociated into H^+ and CH_3COO^-.

Sometimes it is difficult to find a single, predominating form of a chemical substance. In such cases, it is often convenient to write the symbol of the reactive element with a Roman numeral to indicate its oxidation state. For example, in sulfuric acid solution tetravalent cerium is present as a mixture of forms, such as Ce^{4+}, $Ce(OH)^{3+}$, $Ce(SO_4)^{2+}$, $Ce(SO_4)_2$ and $Ce(SO_4)_3^{2-}$. The notation Ce(IV) or cerium(IV) simply indicates that we have tetravalent cerium in solution, without specifying the ionic or molecular form it takes.

QUESTIONS AND PROBLEMS

Concentration Problems

1. For each of the following, give the combining ratio of the two reactants in moles and in grams.

   (a) $2CH_3OH + 2Na \rightarrow 2CH_3ONa + H_2(g)$
   (b) $2Fe^{3+} + Zn \rightarrow 2Fe^{2+} + Zn^{2+}$
   (c) $H_3PO_4 + 2NH_3 \rightarrow 2NH_4^+ + HPO_4^{2-}$
   (d) $Fe^{2+} + 3C_{10}H_8N_2 \rightarrow Fe(C_{10}H_8N_2)_3^{2+}$
   (e) $HIO_4 + 7I^-(+7H^+) \rightarrow 4I_2 + 4H_2O$

2. What is the molarity of a urea solution containing 10 g of urea (NH_2CONH_2) in 750 mL of solution?

3. What is the molarity of a potassium sulfate solution containing 1.74 g/liter of the anhydrous salt? If the salt is completely ionized in solution, calculate the molar equilibrium concentration of potassium ions and sulfate ions.

4. Calculate the number of moles of ammonium chloride in each of the following:

   (a) 9 mL of a $2.0M$ solution
   (b) 500 mL of a $0.2M$ solution
   (c) 45 mL of a $0.6M$ solution
   (d) 100 mL of a solution containing 10.7 g/liter.

5. To what volume must 10 mL of $13M$ HCl be diluted to give a $0.5M$ solution?

6. The analytical concentration of several solutions follows. For each solution, write the formulas of the ions or molecules involved and give their equilibrium concentrations (assume complete ionization for strong electrolytes).

   (a) $0.1M$ KCl   (d) $0.5M$ $CuSO_4$
   (b) $0.1M$ H_2SO_4 (e) $0.15M$ $MgCl_2$
   (c) $0.5M$ glucose $(C_6H_{12}O_6)$ (f) $0.15M$ methyl alcohol (CH_3OH)

7. A $1.0M$ solution of acetic acid is about 0.4% ionized. Calculate the equilibrium concentration of hydrogen ion, acetate ion, and molecular acetic acid in such a solution.

8. Write an equation for the reaction of hydrogen bromide with water. Which ion is written as a solvated ion?

Ionic Strength and Activity

9. Calculate the ionic strength μ for each of the following:

   (a) $0.5M$ $NaClO_4$ (c) $0.5M$ $Al(ClO_4)_3$
   (b) $0.5M$ $Mg(ClO_4)_2$ (d) $0.5M$ $Th(ClO_4)_4$

10. Using the simple Debye–Hückel equation, estimate the following activity coefficients:

    (a) H^+ at $\mu = 0.05$ (c) Ca^2 at $\mu = 0.0005$
    (b) Ca^{2+} at $\mu = 0.01$ (d) PO_4^{3-} at $\mu = 0.02$

11. Explain qualitatively how the total ionic strength will change when $0.01M$ sodium hydroxide is added to a $0.01M$ solution of a weak acid, HA.

12. A pH meter measures the activity, a_{H^+}, of solvated hydrogen ion in solution. The measured pH of a certain hydrochloric acid solution is 2.00, which corresponds to

$a_{H^+} = 0.01$. If the activity coefficient, f_{H^+}, is 0.91, what is $[H^+]$, the concentration of H^+ in solution?

13. What would be the qualitative effect on lithium-ion activity of adding 0.1 mole per liter of solid potassium chloride to a 0.01M solution of lithium nitrate? Explain. (Both lithium nitrate and potassium chloride are strong electrolytes.)

14. For each of the following mixtures, write the specified element with a Roman numeral to indicate its oxidation state.

Mixture	Element	Element and Oxidation State
$Cr_2O_7^{2-}$, CrO_4^{2-}	Cr	Cr(VI), *example*
$SnCl_2$, $SnCl_4^{2-}$	Sn	
TiO^{2+}, $Ti(OH)_3^+$	Ti	
SbO^+, $H_2SbO_3^-$	Sb	
HIO_4, H_5IO_6	I	

2

Steps in a
Chemical Analysis

Although the methods of analytical chemistry vary tremendously, several operations are common to most quantitative analyses. These will be discussed in turn, so that the reader may gain a perspective on the entire process.

2–1. PLAN OF ANALYSIS

It is fairly common for a sample to be brought in with the request that it be "analyzed." It frequently turns out, on questioning, that the person with the sample has only a vague idea of the information really needed. It is the analyst's job to ascertain the actual problem and to plan appropriate analyses to provide the answer needed. Before doing any quantitative analysis, the following points should be considered.

1. What analytical information is needed?
2. What analyses are necessary to provide this information? In this connection, the accuracy needed should be specified. It is well to consider also the information added by each determination proposed.
3. Which analytical methods should be used? The merits of various analytical methods should be compared and the substances that interfere with each procedure carefully noted. The number of samples to be analyzed, the accuracy required, and the type of analytical equipment available have an important bearing on the final choice of methods. The analyst obviously needs experience and a knowledge of the literature of analytical chemistry to make a wise choice.

2–2. SAMPLING

The major problem in sampling is to obtain a "laboratory-sized" sample that is representative of the larger whole. Often it is difficult to obtain such a sample because many substances are not homogeneous with regard to particle size or composition, especially in large amounts.

> One famous example is the analysis of the lunar rocks brought back by the Apollo 11 flight to the moon. The hope was that organic compounds might be found in the lunar samples and indicate something about possible life on the moon, or the beginning of life on earth [1]. Total inorganic and organic carbon content was found by analysis to be less than 200 ppm, with the organic carbon content reported as 40 ppm by one analytic group and 1 ppm by another.
>
> After the data were reported, many doubts were expressed about the meaning of the results. For example, the samples had been scooped up from the *surface* of the moon, not the *subsurface*. Since the surface is exposed to temperatures well above the boiling point or sublimation point of many organic compounds during the prolonged lunar day, it is highly questionable that a surface sample is representative of the chemical composition of the moon's crust. Also, the samples were taken on a few pounds of material removed from a very restricted area around the landing spot. It is unlikely that this small sample would be representative of the entire surface of the moon.

The importance of obtaining a good sample can hardly be overemphasized. Suppose that the chemical analysis of a laboratory-sized sample is accurate to ± 1 ppt (part per thousand), but the composition of the laboratory sample and the average composition of the substance sampled differ by 10 ppt. Obviously, the final answer is accurate to only ± 10 ppt instead of ± 1 ppt. Since an analysis can be only as good as the sample, an accurate analysis of a poor sample is largely wasted.

> A good example in the health sciences is the analysis of pharmaceutical tablets or capsules. The analysis of a single capsule or tablet, such as an aspirin tablet, is hardly an accurate indication of the purity of a bottle of 100 tablets or a case of such bottles. It is quite possible that one such aspirin tablet, or any other tablet, could be of much higher purity or of much lower purity than the average tablet. It is known that aspirin decomposes in the presence of moisture to salicylic acid and volatile acetic acid. The procedure used by the FDA (Food and Drug Administration) is to analyze 20 aspirin tablets for percentage of salicylic acid. The 20 tablets are ground together to form a homogeneous powder, and a sample weight equivalent to the average weight of one tablet is used for the determination. In this way, the composition of the laboratory sample has a much higher probability of accurately representing the average composition of the entire lot. If the salicylic acid content exceeds 0.15% for unbuffered aspirin, or 0.75% for buffered aspirin, it is judged to exceed the standard limit for salicylic acid. (It is important to keep the salicylic acid content low, because it irritates the stomach lining much more than aspirin does.)

Solutions of homogeneous liquids are usually easy to sample because any portion will give a representative sample. Obtaining a proper sample of a non-homogeneous solid presents much more of a problem, especially if the material contains some large particles. Coal is an example. Coal is largely organic, but the composition of the organic part varies from piece to piece, and there are apt to be layers of inorganic salts and small inclusions of foreign matter such as iron py-

[1] *Chem. and Eng. News*, pp 42–47, Jan. 12, 1970.

Figure 2–1. Sampling a liquid flowing in a pipe

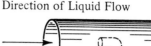

Direction of Liquid Flow

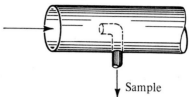

↓ Sample

rites (FeS$_2$). Thus, analysis of two chunks of coal would probably give different analytical results, and neither would be representative of a large load of coal. The sampling problem may be further complicated by segregation of some materials; for example, coal dust may be richer in some components than larger chunks.

To properly sample bulky and inhomogeneous substances such as ore or coal, a large gross sample must first be taken. A good way to do this is to grab pieces at random as the substance is being unloaded from a conveyor belt. Then the gross sample is crushed to a smaller particle size. Portions of the crushed sample are taken, and the rest is discarded. One technique is to pile up the sample repeatedly, each time retaining opposite quadrants of the pile and discarding the other two quadrants. Further reduction in particle size is accomplished by passing it through a disk pulverizer and grinding it in a ball mill or with a mortar and pestle. Sieves may be used to see whether the final sample is of the proper mesh size.

Exact procedures are specified by various professional societies for sampling many types of materials. These methods are based on careful studies and long experience. In other cases, the analyst follows known principles but often must use considerable ingenuity in sampling. Frequently, special knowledge is needed to know where and when to take a sample to best serve a particular purpose. Thus, for proper sampling of blood, the sample must be taken at a time when the body is functioning normally and when the ingestion of food is not disturbing the level of chemicals in the blood. Frequently, blood samples are taken in the morning before the patient has had breakfast.

In general, large quantities of materials are best sampled when they are moved. Bulky solids are sampled by grabbing random samples during unloading. A liquid moving in a full pipe can be sampled through a smaller-diameter pipe as shown in Figure 2–1. Metals and solid materials are sampled by drilling, sawing, or milling (the sawdust, shavings, etc., constitute the sample). A "thief" (a sampling device with a removable stopper) can be used to sample liquids containing suspended matter. The thief is lowered to the desired depth and opened temporarily to catch the sample. Inhomogeneous liquids or soft solids can be sampled by sticking a pipe into the full depth of the substance and lifting out the core inside the pipe. Gas samples are usually obtained by allowing the gas to displace a confining liquid in a bottle equipped with suitable stopcocks.

Bicking [2] gives an excellent discussion of sampling, including descriptions and drawings of many of the sampling devices used.

[2] C. A. Bicking in *Treatise on Analytical Chemistry*, Part I, Sec. B, Chap. 6, I. M. Kolthoff and P. J. Elving, eds, New York: Wiley-Interscience.

2-3. DRYING THE SAMPLE

Samples often contain water, either as chemically combined hydrates or as occluded or surface-adsorbed moisture. The water content of different materials varies within wide limits. Water is an inherent part of most biological substances and constitutes more than 90% of the fresh weight of some plant materials. By contrast, the amount of water adsorbed on the surface of a metal sample is normally only a few parts per million. Unfortunately, the moisture content of most samples is not constant but varies with changes in the atmospheric humidity. The amount of moisture adsorbed also varies with particle size, since samples of smaller particle size offer more surface for adsorption. To afford reproducible analytical results, samples are usually dried before analysis, and the percentage composition of the sample is then calculated on a dry basis.

The goal of drying is either to make the sample anhydrous or to remove adsorbed moisture but retain chemically combined water. Ordinarily, samples are dried in an oven at 100°–110°C for one or two hours. However, some samples are partially decomposed under these conditions and should be dried at a lower temperature or not at all. A weight-temperature curve obtained by means of a thermobalance (see Chapter 4, Section 4–5) is often a good means of determining the proper drying conditions for a difficult sample. Surface moisture is usually lost at a low temperature, and the temperature range at which a definite hydrate is stable is indicated by a plateau on the curve. Unfortunately, the weight-temperature curves of some substances have no plateau regions but show a more or less continuous weight loss with increased temperature. For such cases, the purpose of drying may have to be limited to that of drying at some arbitrary temperature for a fixed time so that different laboratories can at least obtain reproducible results. Drying of heat-sensitive samples may be done in a desiccator containing a drying agent or in a vacuum desiccator.

Sometimes it is advantageous to analyze samples on an "as received" basis. While a sample is being weighed out and analyzed, the moisture content of another portion is determined accurately by drying or by some other method for determining water (see the following example). The analysis of the sample on a dry basis can be calculated from the data obtained.

Example: A food sample analyzed "as received" is found to contain 14.57% protein and 10.4% water. Calculate the percentage of protein on a dry basis.

Before drying there is 14.57 mg of protein per 100 mg of sample. But since 10.4 mg of the sample is moisture, the percentage of protein on a dry basis is

$$\frac{14.57 \text{ mg}}{89.6 \text{ mg}} \times 100 = 16.26\% \text{ protein}$$

Although it involves an additional measurement, analysis of samples by the method just described is recommended when results are needed as soon as possible or for samples that are difficult to weigh out or handle when completely dry. For example, ion-exchange resins are quite easy to weigh when kept in a normal atmosphere, but they are quite hygroscopic after being oven dried. For distribution coefficient experiments (see Chapter 20) in which many samples may be needed, "air dry" samples are normally taken; the water content is determined by measuring carefully the loss of weight of two or three samples on drying.

The loss in weight on oven drying is not the only way to determine the amount of water in samples. Another method is to heat the sample and sweep the vapors given off through a weighed absorption tube by means of a stream of dry gas. Anhydrous magnesium perchlorate is commonly used as an absorbent for water. The amount of water is calculated by reweighing the absorption tube and noting the gain in weight.

The Karl Fischer method (Chapter 13) is yet another way to determine water content. It is used primarily for determining the water content in organic compounds.

2–4. MEASURING THE SAMPLE

The results of a quantitative analysis are usually reported in weight percentage, that is, the number of grams of the desired species per 100 g of sample. However, volume percentage (milliliters of species per 100 mL of sample), mole percentage (moles of species per 100 moles of sample), or other relative terms may be used.

In a quantitative analysis, the purpose of all steps after weighing is to determine the quantity of a species or of several species relative to the known weight or volume of a sample. Weighing samples on an analytical balance is usually a very accurate operation (this is discussed in Chapter 26). Measuring liquid samples by volume with a pipet or buret is less accurate than weighing but is very rapid and convenient. Sometimes samples are weighed, dissolved, and diluted to a definite volume, after which measured aliquots of the sample solution are pipetted for analysis. This method may be used when a sample is too light to be weighed accurately on an ordinary analytical balance.

2–5. DISSOLVING THE SAMPLE

A solid sample usually must be put into solution before an analysis can be completed. Often dissolution is one of the most time-consuming steps in a quantitative analysis. Two excellent reviews of methods used to dissolve analytical samples have been published [3, 4].

Decomposition Under Pressure in a Closed System. Often the increased pressure and higher temperature that can be realized in a closed system will greatly speed the dissolution process. For example, by placing the sample and solvent in a hard glass tube, which is then put inside a closed steel cylinder, temperatures up to 300°C and

[3] W. F. Pickering, *C.R.C. Crit. Rev. in Anal. Chem.*, 3, 271 (1973).
[4] Z. Sulcek, P. Povondra, and J. Dolezal, *C.R.C. Crit. Rev. in Anal. Chem.*, 6, 255 (1977).

pressures above 100 atmospheres are possible. Another device is a small steel autoclave that has a Teflon lining to prevent attack of the container itself. Ordinarily corrosive acids, such as HF, can be used, but 150°C is the approximate temperature limit.

Decomposition in Open Systems.   This is the more commonly used dissolution method. The sample is covered with a suitable solvent and usually is gently heated to make it dissolve faster. The solvent used should dissolve the sample completely in as short a time as possible. A solvent should be chosen that will not interfere in the subsequent steps of the analysis. The solvents used to dissolve most samples can be classified as follows:

Water.   Many inorganic salts and some organic compounds dissolve readily in ordinary distilled water. Occasionally, a small amount of acid is added to prevent hydrolysis and partial precipitation of certain metal cations.

Organic Solvents.   These are alcohols, chlorinated hydrocarbons, ketones, etc. They are ordinarily used to dissolve organic compounds prior to analysis.

Mineral Acids.   Concentrated or slightly diluted acids will dissolve most metals and metal alloys and many oxides, carbonates, sulfides, etc. Nitric acid, hydrochloric acid, aqua regia (nitric plus hydrochloric acid), or sulfuric acid are most commonly used, although perchloric acid ($HClO_4$) or phosphoric acid may be used in some cases. Hydrofluoric acid, either alone or mixed with another acid, readily dissolves metals that form strong fluoride complexes in aqueous solution. Some of these metals (niobium and tantalum, for example) are virtually insoluble in other solvents.

Fusion.   Samples that fail to dissolve in any other way can usually be brought into solution by fusing them with a high-temperature acid such as potassium pyrosulfate ($K_2S_2O_7$) a base such as sodium carbonate, or an oxidant such as sodium peroxide (Na_2O_2). A finely ground sample is mixed intimately with the granular solid flux, and the mixture is melted in a crucible. The melted flux attacks and dissolves the sample. Then the crucible is cooled, and the solidified melt is dissolved in dilute, aqueous acid or in water alone.

Table 2–1 lists some commonly used solvents for metals and a few other types of inorganic samples.

2–6. SEPARATING INTERFERING SUBSTANCES

The ideal quantitative method for a substance would be a *specific* method; that is, a method that would measure the desired substance accurately in the presence of any possible combination of foreign substances. Unfortunately, while many methods are *selective*, few analytical methods come very near to being specific. A selective method can be used to determine any of a small group of ions or compounds in the presence of certain foreign ions or compounds. Usually, there are some foreign substances that prevent the direct measurement of a species; these are called *interferences*.

Table 2–1. Some Solvents Used to Dissolve Inorganic Substances

Metals Element	Solvent	Element	Solvent
Ag	HNO_3	Ni	acids
Al	HCl; NaOH	Pb	HNO_3
		Rare	
As	HNO_3; aq. regia; H_2SO_4	earths	HNO_3; HCl; $HClO_4$
Bi	HNO_3; aq. regia; H_2SO_4	Sb	H_2SO_4; HNO_3 + tartrate
Cd	HNO_3	Sn	HCl; aq. regia
Co	acids	Ta	HF + HNO_3
Cr	$HClO_4$; HCl; dil. H_2SO_4	Th	HNO_3; HCl
Cu	HNO_3; HCl + H_2O_2	Ti	HF; H_2SO_4
Fe	acids	U	HNO_3
Hg	HNO_3; H_2SO_4	V	HNO_3; H_2SO_4
Mg	acids	W	HF + HNO_3; H_3PO_4 + $HClO_4$
Mo	HNO_3	Zn	acids; NaOH
Nb	HF + HNO_3	Zr	HF

Carbonates, Oxides, Sulfides

Usually soluble in acids; a few require fusion.

Phosphates

Some are soluble in acids; many require fusion with an alkaline flux.

Silicates

Most samples dissolve in HF; silica is volatilized as H_2SiF_6; fusion is used extensively.

Separating interferences from the species to be determined is an important step in many quantitative analytical procedures. Quantitative precipitation of either the interfering substances or the species to be analyzed is a common separation method (Chapter 4). Separation may be accomplished by electrodeposition (Chapter 15) or by means of ion exchange (Chapter 24). Solvent extraction is another useful technique. Even highly complex mixtures can often be separated by chromatography (Chapters 21–23).

2–7. MEASURING THE DESIRED SUBSTANCE

This is the step in which the quantity of the substance being determined is actually measured. Much of this book, and indeed much of the literature of analytical chemistry, is concerned with theories and practices connected with quantitative analytical measurements. A wide variety of methods have been developed for quantitative measurements, as can be seen from Table 2–2.

The measurement step usually requires the adjustment of several conditions. For example, many procedures require the pH of the solution to be within a given range before precipitation, or before a color is developed for spectrophotometry. Sometimes the oxidation state of certain elements has to be adjusted before a

Table 2-2. Classification of Analytical Measurement Methods

1. *Gravimetric.* Isolate and weigh precipitate.
    (a) Inorganic precipitating reagent.
    (b) Organic precipitating reagent.
    (c) Electrodeposition.
2. *Titrimetric.* Measure volume of standard solution required to react with substance to be determined.
    (a) Precipitate formation.
    (b) Acid-base.
    (c) Complex formation.
    (d) Oxidation-reduction.
3. *Absorption of radiant energy.* Measure amount of energy absorbed at a particular wavelength.
    (a) Visible spectrophotometry (colorimetry).
    (b) Ultraviolet spectrophotometry.
    (c) Infrared spectrophotometry.
    (d) X-rays.
    (e) Nuclear magnetic resonance. Interaction of radio waves with atomic nuclei in a strong magnetic field.
4. *Emission of radiant energy.* Sample bombarded with a large amount of energy (electricity, heat, etc.), which raises substances to an excited state and causes them to emit energy in some form. Measure amount of energy emitted at a particular wavelength.
    (a) Emission spectroscopy. Electric arc or spark excitation; light emitted.
    (b) Flame photometry. Sample sprayed into flame; light emitted.
    (c) X-ray fluorescence. X-rays strike sample; sample emits other X-rays of a type peculiar to the sample.
5. *Gas analysis*
    (a) Volumetry. Measure change in volume upon the absorption or evolution of a gas.
    (b) Manometry. Measure gas pressure or change in pressure.
6. *Electrical*
    (a) Polarography. Measure electric current, which is proportional to the concentration of substance reduced or oxidized in an electrochemical reaction at a microelectrode.
    (b) Coulometry. Measure quantity of current needed to cause a quantitative electrochemical reaction.
    (c) Potentiometry. Measure potential of an electrode in equilibrium with substance to be determined.
    (d) Conductimetry. Measure conductance of a solution.
7. *Miscellaneous*
    (a) Polarimetry. Measure rotation of polarized light by solution of substance.
    (b) Refractometry. Measure refractive index, which determines composition of simple mixture.
    (c) Mass spectrometry. Measure amount of ions of various charge relative to mass ratios.
    (d) Activation analysis. Induce radioactivity and count radioactive particles.
    (e) Thermal conductimetry. Measure thermal conductance.
    (f) Kinetic methods. Measure rate of a chemical reaction.

titration or other quantitative measurement. Furthermore, the dilution of the sample solution must be such that the species to be measured are in the proper concentration range.

2–8. CALCULATING AND EVALUATING THE RESULTS

The measurement step provides the data needed to calculate the quantity (usually in weight units) of each species in the sample. The percentage of each species is then obtained by dividing this number by the weight of the sample and multiplying by 100, paying attention to the proper use of significant figures in calculations. Analyses are ordinarily carried out in triplicate (on three samples simultaneously), or at least in duplicate. If the results of three samples agree closely, the *precision* of the analyses is good. Usually (but not invariably), good precision means that the accuracy of the analyses is also good. Precision, accuracy, and significant figures are discussed in Chapter 3.

When the results have been calculated, it is then the task of the analyst to indicate how reliable they probably are. This is done by considering the inherent accuracy of the method used and by subjecting the numerical data to simple statistical treatment (Chapter 3).

3

Treatment of Analytical Data

3–1. INTRODUCTION

The task of the analytical chemist goes beyond that of correctly performing the manipulations and readings required in a procedure. In the quantitative analysis laboratory the analyst must also do the following:

1. Properly record and correctly calculate the results of each analysis.
2. Determine the *best value* to report. This is usually the arithmetic mean, or average, of the individual results, but there is often the question of whether to include a result that seems out of line with the others.
3. Estimate how good the results are with respect to precision, or scatter.

Simple statistical tests or quantities are useful in connection with the second and third steps.

The ability of an analyst to evaluate larger numbers of analytical data is of major importance. In industrial laboratories, medical centers, and other places where chemical analyses are performed, a huge number of determinations are made every day. With increasing automation of analyses and greater use of computers, the amount of data produced can be staggering. Laboratories need to know what limits of error can be placed on the analytical results, whether the analyses are providing adequate control for a process, whether medical analyses are painting a true picture of the patient's condition, etc.

Increasingly, laboratories are turning more to statistical methods for evaluating analytical data. In this chapter, a number of useful statistical methods are described. In addition, some elementary but basic statistical principles are discussed so that the student will have some background to build on.

Some Common Terms. A series of repeated analytical measurements is often represented by X_1, X_2, X_3, etc. The *mean* (which is also called the average) is simply the sum of these measurements divided by the number of measurements.

$$\bar{X} = \frac{\sum X_i}{n} \tag{3-1}$$

where $\bar{X}$ is the mean, n is the number of measurements, and $\sum X_i$ represents the summation of all the values of X.

Accuracy is the nearness of a single result (X_i) or of a mean ($\bar{X}$) of a set of results to the true value (μ). Accuracy is usually expressed in terms of error, $X_i - \mu$ or $\bar{X} - \mu$.

Precision is the agreement of a set of results among themselves. Precision is usually expressed in terms of the deviation of a set of experimental results from the arithmetic mean of the set. Precision is really a measure of the ability to reproduce a result. Although good precision is usually an indication of good accuracy, it is entirely possible to obtain good precision but poor accuracy, and vice versa.

The *average deviation* ($\bar{d}$) is sometimes used as a measure of precision. It is defined as the sum of the absolute deviations from the mean, divided by the number of measurements.

$$\bar{d} = \frac{\sum |X_i - \bar{X}|}{n} \tag{3-2}$$

Absolute deviation is represented by $|X_i - \bar{X}|$ in this equation and means that the deviations are taken without regard to sign.

Example: A pH measurement, repeated five times, gave the following values: 5.42, 5.45, 5.46, 5.40, 5.46. Calculate the mean and the average deviation.

$$\bar{X} = \frac{27.19}{5} = 5.44$$

$$\bar{d} = \frac{0.02 + 0.01 + 0.02 + 0.04 + 0.02}{5} = 0.02$$

Average deviation has no real statistical significance, and its use is not recommended. It is much better to express precision in terms of *standard deviation* or *variance*. These will be discussed in some detail later in this chapter.

3-2. UNCERTAINTY; SIGNIFICANT FIGURES

Any analytical measurement, no matter how carefully made, is subject to some uncertainty. For example, the analytical balance is normally read to the fourth decimal place; it follows that the uncertainty of a weight such as 1.2345 g is uncertain by at least ± 0.0001 g. This implies that the weight could range from 1.2344 to 1.2346 g.

Measurements made with any analytical instrument, whether it has a scale or digital readout, are also subject to some uncertainty. There is a difference between the approach for scale readouts and that for digital readouts, so we will consider them separately.

Digital Readouts. It is fairly common for the final digit in a digital readout instrument such as an analytical balance or spectrophotometer to be slightly unsteady and to fluctuate between two or more numbers. In the absence of qualifying information, the uncertainty of such a single measurement is taken to be ± 1 in the final digit of the readout.

Scale Readouts. Often the approximate uncertainty of a scale readout will be easily apparent. If the closest scale markings are quite close together, a reasonable uncertainty of the reading might be half the distance between the markings. For example, if the marking represents 0.1 of a unit, then a reasonable uncertainty might be 0.1/2 or 0.05 unit. Burets are a special case of scale readout. A standard 50-mL buret has graduation marks every 0.1 mL. The level of liquid between the marks can be estimated to no better than the nearest 0.01 mL for an experienced analyst and perhaps 0.02 mL for a student. It follows that the uncertainty of a single buret reading is perhaps ± 0.01 mL for the analyst and ± 0.02 mL for the student.

     The uncertainty expressed in the units measured (g, mL, scale division, etc.) is called the *absolute uncertainty*. Another convenient term is the *relative uncertainty*, which is the absolute uncertainty divided by the number measured and multiplied by 1000 to give the uncertainty in parts per thousand.

Example: A spectrophotometer gives a reading for a certain colored solution of 0.476 absorbance unit with an uncertainty of 0.003 absorbance unit. Calculate the relative uncertainty.

$$\text{Rel. uncertainty} = \frac{0.003 \text{ AU}}{0.476 \text{ AU}}(1000) = 6 \text{ ppt}$$

Significant Figures. The uncertainty of an analytical measurement or a result that has been calculated from a measurement should be reflected in the number of significant figures used to record the result or measurement. Significant figures, by definition, are those digits in a number that are known with certainty plus the first uncertain digit. The last digit of a number is generally considered uncertain by ± 1 in the absence of qualifying information. (Qualifying information would indicate a larger uncertainty; for example, the age of some moon rocks was reported as 3.86 ± 0.04 billion years.)

     To illustrate the concept of significant figures, consider the following numbers, all of which have three significant figures: 0.104, 1.04, 104, and 1.04×10^4. The 1 and the middle 0 are certain, and the 4 is uncertain but significant. Note that an exponential number has no effect on the number of significant figures.

Treatment of Zeroes. Special attention should be paid to whether zeroes are significant or not. The following cases are important:

1A. *Initial zeroes in numbers,* such as in 0.104 or 0.0014, are never significant, since they serve only to locate the decimal point.

1B. *Initial zeroes to the right of the decimal point in logs* are always significant.
Thus, 0.004, the log of 1.01, has three significant figures; all three would be necessary to distinguish it from 0.114, the log of 1.30.

2. Zeroes that appear between other digits in numbers or logs are always significant.

3. Terminal zeroes are generally considered to be significant.

If a terminal zero is not significant but is used only to fix a decimal point, it should be eliminated and the number written in scientific notation. For example, 10,100 as written is considered to have five significant figures. If only three digits are meant to be significant, then it should be written as 1.01×10^4.

Rounding Off Versus Nonsignificant Figures. In any calculation, such as when data must be reentered on a calculator, it is desirable to carry along at least two nonsignificant figures and eliminate them only when reporting the final result.

The final result may be achieved by rounding off (*recommended procedure*) or by reporting one nonsignificant figure (alternative procedure):

1. *Recommended procedure.* Round off by eliminating nonsignificant figures. Increase the last retained digit by 1 if the first discarded digit is 5 or greater. If the latter is less than 5, do not change the last retained digit.

2. *Alternative procedure.* Report the first nonsignificant figure as a subscript and drop all other nonsignificant figures. For example, if 1.1149 and 1.1050 are both to be reported with three significant figures, write them as 1.11_4 and 1.10_5.

Reporting a nonsignificant figure has been recommended by Kaiser [1] as showing the trend in the last significant figure, which rounding off does not. In the preceding example, 1.1149 and 1.1050 would be both rounded off to 1.11, yet there is a difference of almost 0.01 between them. Reporting 1.1149 as 1.11_4 shows the trend is between 1.11 and 1.12; reporting 1.1050 as 1.10_5 shows the trend is between 1.10 and 1.11.

Addition and Subtraction. The proper use of significant figures in addition and subtraction is to compare only the *absolute* uncertainties of the numbers used in the calculation. This usually means that you should retain only as many digits to the right of the decimal in the answer as the number with the fewest digits to the right of the decimal. This number of course has the largest absolute uncertainty. It will be necessary to round the last retained digit up if the next discarded digit is 5 or greater. An alternative method is to adjust the number of significant figures in each number before adding or subtracting (see the following example).

If numbers with positive or negative exponents are involved, adjust the exponents so that they are all the same before adding or subtracting, as shown in this example.

Example: Suppose the following concentrations of OH^- must be added: $4.00 \times 10^{-2}M$ OH^-, $5.55 \times 10^{-3}M$ OH^-, and $1 \times 10^{-6}M$ OH^-. Adjust all the exponents to be

[1] H. Kaiser, *Anal. Chem.*, *43*, 26A (Apr. 1970).

10^{-2} and note that $4.00 \times 10^{-2}M$ has the largest uncertainty, $\pm 0.01 \times 10^{-2}M$. Add by either of the following methods:

Adjust answer only	*Adjust each number*
$4.00 \times 10^{-2}M$	$4.00 \times 10^{-2}M$
$0.555 \times 10^{-2}M$	$0.56 \times 10^{-2}M$
$0.0001 \times 10^{-2}M$	$0.00 \times 10^{-2}M$ (negligible)
$\overline{4.5551}$	$\overline{4.56 \times 10^{-2}M}$

$\vdots$

(round up)

$\downarrow$

$4.56 \times 10^{-2}M$

Clearly, it is meaningless to express the answer beyond the second decimal when the uncertainty of the total must be at least $\pm 0.01 \times 10^{-2}M$. Also note that since $1 \times 10^{-6}M$ OH^- is negligible in comparison to the other numbers, the fact that it has only one significant figure is of no consequence.

Multiplication and Division.   The proper use of significant figures in multiplication and division is to compare only the *relative* uncertainties of the numbers used in the calculation. The rule is that the relative uncertainty of the answer should be of the same order of magnitude as the number with the largest relative uncertainty. In practice this means that it should fall between 0.2 to 2 times the largest relative uncertainty in the data. (A less accurate rule is to use the same number of significant figures in a product or quotient as in the value with the fewest number of significant figures.)

The following example will illustrate proper use of significant figures in both multiplication and division *as well as give you a general guide for most of your titration calculations.*

Example: A sample of 2.0000 g (weighing by difference) of impure potassium acid phthalate (form wt 204.228) requires 40.00 mL of $0.1000M$ sodium hydroxide titrant for neutralization. When the data as set up in the following equation are fed into the calculator, the exact six-digit answer is obtained:

$$(40.00 \text{ mL})(0.1000M)(204.228 \text{ mg/mmole})(100)/2{,}000.0 \text{ mg} = 40.8456\%$$

How many significant figures should be reported for the percent purity?

Usually, all the relative uncertainties would be calculated and compared. However, in most titrations, the relative uncertainties of the formula weight and sample weight (if >200 mg) are relatively small, so that only the relative uncertainties of the concentration and titrant volume (using an absolute uncertainty of 2×0.02 mL for buret readings) need be calculated:

$$\frac{0.0001M}{0.1000M}(1000) = 1 \text{ ppt}$$

$$\frac{0.04 \text{ mL}}{40.00 \text{ mL}}(1000) = 1 \text{ ppt}$$

The percent purity should have a relative uncertainty *between* 0.2 and 2 times 1 ppt. The two possibilities are 40.8% and 40.84%:

$$\text{Rel. un. of } 40.8\%: \quad (0.1/40.8)1000 = 2.4 \text{ ppt} \quad (>2 \text{ times } 1 \text{ ppt})$$

$$\text{Rel. un. of } 40.84\%: \quad (0.01/40.84)100 = 0.24 \text{ ppt} \quad (>0.2 \text{ times } 1 \text{ ppt})$$

The percent purity is thus correctly expressed as 40.85% after rounding up, or as $40.84_5\%$ with one nonsignificant figure.

Log Terms. The *general rule* for log K, pH, etc. is that the log term should be expressed with the same number of significant figures to the *right* of the decimal as the number of significant figures in the number. The exponent does not affect digits to the right of the decimal because it indicates zeroes, not measured numbers.

Example: The equilibrium constant for a reaction is 1.70×10^4. Calculate log K. After K is entered on an eight-digit calculator and the log is taken, the calculator readout is 4.2304489, which should be expressed to a total of $3 + 1$, or 4, significant figures:

$$\log(1.70 \times 10^4) = 4.2304489 = 4.230 \quad (\text{or } 4.230_4)$$

Calculation of pH from $[H^+]$. For pH calculations, the general rule for log terms is accurate except when the first digit in the $[H^+]$ is 8 or 9. In those cases, at least one additional digit written as a subscript should be used instead of rounding. Consider the following three cases:

$[H^+]$	pH (general)	pH ($+1$ subscript)
$9.2 \times 10^{-2} M$	1.04	1.03_6
$9.1 \times 10^{-2} M$	1.04	1.04_1
$9.0 \times 10^{-2} M$	1.05	1.04_5

Note that in all three cases, the pH is close to the same value, even though the $[H^+]$ varies from 9.2 to $9.4 \times 10^{-2} M$. Using a one-digit subscript will differentiate among all three of the concentrations.

3–3. ERROR AND DEVIATION

Systematic and Random Errors. Errors in analysis may be classified as systematic (determinate) or random (indeterminate). Systematic errors are caused by a defect in the analytical method or by an improperly functioning instrument or analyst. For example, if the indicator used in a titration changes before the equivalence point, a systematic error will result. Similarly, a titration with a dirty buret will cause a somewhat systematic error (high results). The only way to deal with this type of error is to rectify the cause.

Random errors are unavoidable because there is some uncertainty in every physical measurement. The most careful analyst can only read a 50-mL buret accurately to the nearest 0.01 or 0.02 mL, for example. However, a truly random error is just as likely to be positive as negative. This fact makes the average of several replicate measurements more reliable than any individual measurement. Unfortunately, random errors do set a definite limit on accuracy, even when the measurement is repeated many times.

Absolute and Relative Error.   Absolute error is the difference between a measured value and the true value, μ, expressed in the units of the two numbers. For example:

Type of measurement	x_i	μ	Error
Weighing	0.1009 g	0.1007 g	$+0.0002$ g
Buret reading	24.97 mL	25.00 mL	-0.03 mL
Peak area	50.5 mm^2	50.9 mm^2	-0.4 mm^2

Relative error is the absolute error divided by the true value. Relative error, therefore, has no dimensions. However, the relative error is usually multiplied by 1000 so that it can be expressed in parts per thousand (ppt) or by 100 so that it can be expressed in parts per hundred (pph). Usually the sign of the error ($+$ or $-$) is dropped. The relative errors of the measurement examples listed above are:

$$\text{Weighing:}\quad \text{Rel. error} = \frac{0.0002 \text{ g}}{0.1007 \text{ g}}(1000) = 0.2 \text{ ppt}$$
$$(0.02 \text{ pph})$$

$$\text{Buret reading:}\quad \frac{0.03 \text{ mL}}{25.00 \text{ mL}}(1000) = 1.2 \text{ ppt}$$
$$(0.12 \text{ pph})$$

$$\text{Peak area:}\quad \frac{0.4 \text{ mm}^2}{50.9 \text{ mm}^2}(1000) = 7.9 \text{ ppt}$$
$$(0.79 \text{ pph})$$

Sometimes relative error is given in percent instead of in pph. This practice can be confusing, especially when the error in an analytical determination is being discussed. For example, if an analytical result is 14.05% chloride and the true value is 14.00%, the absolute error is $+0.05\%$ but the relative error is

$$\frac{0.05\%}{14.00\%}(100) \quad \text{or} \quad 0.36 \text{ pph}$$

3–4. DISTRIBUTION OF EXPERIMENTAL RESULTS

Random Variables.   When a measurement or determination is repeated several times, the results tend to be similar, but they are seldom *exactly* the same. For example, three titrations of the same sample will rarely yield exactly the same answer. Even if the results should turn out to be the same, measurement of the titration volume to one or two additional digits will certainly produce variations in the results. This does not mean that there is not a single correct answer. What it means is that there is a certain unavoidable variation in the results that can be obtained with the equipment that one is using. Better equipment may reduce this variation, but it cannot be eliminated entirely. Nor is there any way of predicting ahead of time just what the results of the next trial will be. This variation in our measurements is *random*. Thus, the outcome of a measurement or determination is called a *random variable*. This is a term that has some rather technical meanings in the study of probability and statistics (from which we have taken it), but here we deal with the concept in a loose intuitive fashion.

Each *type* of measurement can be designated by a random variable. For example, measurement of the pH of a given solution might be designated by the random variable X. As we shall see, X actually represents a *distribution* of results such as a *normal distribution curve*, discussed in Section 3–5. By measuring the pH (either once or several times) we are actually sampling this distribution of X. A sequence of repeated pH measurements of the same sample is written X_1, X_2, X_3, etc. Specific numerical results for such a sequence are written in lowercase: x_1, x_2, x_3, etc.

Each *type* of analytical measurement is, of course, different, and each will have its own distribution. Letters such as X, Y, and Z are used to distinguish different random variables. For example, in a titration, X might represent the volume of titrant used, Y the concentration of titrant, and Z the weight of analytical sample taken.

When we are concerned with analytical measurements the usual situation is that X will be a continuous random variable, which means that it can take on any value. The other type of random variable is a *discrete* random variable, which can assume only certain values such as 1, 2, 3, ...; 0.5, 1.0, 1.5, ...; etc. The discussion in this chapter will be concerned with continuous random variables.

Random variables can be combined mathematically (added, multiplied, etc.), as we shall see in Section 3–7.

Histograms. Clearly, an experimental measurement cannot be a totally random quantity, or else experimental results would mean nothing. Generally speaking, we should get pretty much the same value each time the measurement is repeated, subject to only a relatively small variation. Experimental results close to the "true" value should occur fairly often, while values far from the "true" value should occur very seldom. The nature of the particular random variable in question is described by the relative frequencies with which the various results are obtained as the measurement is repeated over and over. Such results can be plotted graphically in a histogram, as shown in Figure 3–1. To construct a histogram, the results are divided into various class intervals. The number of results in each interval is plotted as the frequency (on the vertical axis) against the range of each class interval on the horizontal axis.

Alternatively, the *relative frequency* can be plotted on the vertical axis. The relative frequency is obtained simply by dividing the frequency of each class interval by the total number of results, n.

Figure 3–1.   A histogram.

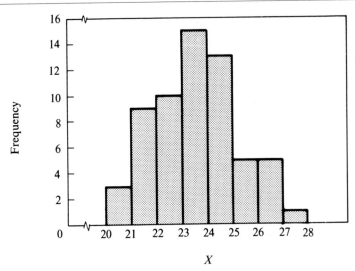

Unless n is quite large, the histogram obtained will be imperfect but still will give some idea about the distribution of results. If a very large number of repeated measurements were made and the vertical and horizontal axes rescaled appropriately, the histogram would become more perfect and would be converted into a smooth curve called a *density function*. The density function for random variable X is a plot of density, $f(x)$, versus X. This plot can be represented by a mathematical equation. Assuming that X is a continuous random variable, the function $f(x)$ must satisfy the conditions that $f(x)$ is equal to or greater than 0 and that the integrated area under the curve is equal to 1.

Several types of density functions are known: exponential, binomial, Poisson, and normal (or Gaussian), to name a few. Analytical data are generally considered to follow a *normal distribution* (see Section 3–6).

A density function gives the *probability* that any single result associated with the density function lies within certain limits. The values for probability lie between 0 and 1, a value of 1 indicating that something is completely probable (a certainty). If a and b are real numbers with $a < b$, the probability that an experimental result between a and b will be obtained is given by the integrated area under the curve from a to b:

$$P\{a < X < b\} = \int_a^b f(x)\, dx \qquad\qquad (3\text{–}3)$$

Example: Suppose the underlying distribution for X is exponential with parameter λ. For this, the density function, $f(x)$, is given by the equation $f(x) = \lambda e^{-\lambda x}$. Calculate the probability that X will lie between a and b when $a = 0$, $b = \frac{1}{2}$, and $\lambda = 3.0$. (See Figure 3–2.)

Figure 3–2. Exponential density function, $f(x) = \lambda e^{-\lambda x}$, when $\lambda = 3.0$.

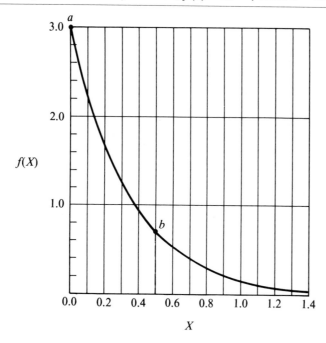

This probability is expressed as follows:

$$P\{a < X < b\} = \int_a^b f(x)\,dx = \int_a^b \lambda e^{-\lambda x}$$

Integrating between a and b:

$$P\{a < X < b\} = -e^{-\lambda b} + e^{-\lambda a}$$

when $a = 0$ and $b = \frac{1}{2}$.

$$P = -0.223 + 1 = 0.777$$

To obtain the probability that a value of X lies within a certain area of the density function curve, it is necessary to integrate the mathematical equation between the desired limits of X. Sometimes algebraic integration is difficult or impossible, as in the case of a normal distribution. However, for a normal distribution curve, numerical integration has been used to calculate the areas under various portions of the curve. These results are given in tables, available in various books (see Table 3–1), which can be used to calculate various probabilities.

Mean and Standard Deviation. Were we to repeat any given measurement (represented by random variable X) a *very* large number of times, the values would follow some kind of a distribution. It would be possible to measure the mean of the distribution, μ, and the scatter of results about the mean, which is given by the standard deviation, σ. Such a procedure is totally impractical, and the precise values of μ and σ are in a sense theoretical. However, it is possible to make a sufficiently large number of measurements to construct a histogram that will approximate the underlying density function. The histogram will be imperfect, but it should be sufficiently well defined to recognize its mathematical distribution form (such as a normal distribution). It should also provide a close estimate of the mean (which locates the distribution on the X-axis) and the variance (or standard deviation) that determines the width and height of the distribution.

In statistics, the mean (μ) of a distribution of random variable X is often referred to as the *expected value of X* and is written as $E(X)$. The expected value of a histogram is the summation of X_i times the relative frequency for all values of i:

$$E(X) = \sum_i X_i(\text{rel. freq.})_i \qquad\qquad (3\text{--}4)$$

A more precise value for the mean is given in terms of the density function, $f(x)$:

$$E(X) = \int_{-\infty}^{\infty} Xf(x)\,dx \qquad\qquad (3\text{--}5)$$

Equation 3–4 becomes a better and better approximation for Equation 3–5 as the interval length in a histogram gets smaller and the number of results (i) becomes larger.

Next we consider the dispersion (scatter) of values about the mean. This is measured by the standard deviation, σ, or the variance, σ^2. As with the mean, the value of the standard deviation or variance can be estimated from a histogram, but it is determined more precisely from the underlying density function. Note that the variance of a random variable X may be written as $\text{Var}(X)$ instead of σ^2. For a histogram:

$$\text{Var}(X) = \sum_i (X_i - \mu)^2(\text{rel. freq.})_i \qquad\qquad (3\text{--}6)$$

This equation is approximately equal to the variance, which is defined more precisely by the following equation.

$$\text{Var}(X) = \int_{-\infty}^{\infty} (X_i - \mu)^2 f(x)\,dx \qquad\qquad (3\text{--}7)$$

The standard deviation, σ, is simply the square root of the variance.

Mean and Standard Deviation of a Sample.   Sometimes we are actually presented with data in the form of a density function. An example is a chromatographic peak, which approximates a normal distribution curve and can be considered as the statistical result of a very large number of individual samples passing through a chromatographic column. The mean and variance can be measured directly from the recorded peak, as described in Chapter 21.

The more common case is when we have a relatively small number of data and must use these to make inferences as to the mean and standard deviation of the underlying distribution. In essence, we make a few measurements and use these as a *statistical sample* to estimate the mean and standard deviation of the underlying density function. The mean of this statistical sample is called $\bar{X}$ (X bar), and the standard deviation is indicated by S.

$\bar{X}$ and S are also random variables and have distributions of their own. Specific numerical values are written in lowercase: $\bar{x}, s$.

The sample mean, $\bar{X}$, and the standard deviation, S, are said to be estimators for the mean and standard deviation of the underlying distribution, μ and σ, respectively. As the number of measurements in the statistical sample, n, becomes larger, $\bar{X}$ and S provide estimates that are closer and closer to μ and σ, respectively.

The *mean* of a statistical sample is merely the average of the values:

$$\bar{X} = \frac{X_1 + X_2 + X_3 + \cdots + X_n}{n} \quad \text{or} \quad \bar{X} = \frac{\sum_1^n X_i}{n} \tag{3-8}$$

The *variance* is

$$S^2 = \frac{(X_1 - \bar{X})^2 + (X_2 - \bar{X})^2 + \cdots + (X_n - \bar{X})^2}{n - 1}$$

or

$$S^2 = \frac{\sum_1^n (X_i - \bar{X})^2}{n - 1} \tag{3-9}$$

The *standard deviation* is the square root of the variance:

$$S = \left[\frac{\sum_1^n (X_i - \bar{X})^2}{n - 1} \right]^{1/2} \tag{3-10}$$

The quantity $n - 1$ is the "degrees of freedom" of a set of n results in which S has been calculated from the results. After $\bar{X}$ has been calculated and the values of the results are listed, the value of the last, or nth, result is already fixed. In effect, one of the n values has been used up in calculating the mean, leaving $n - 1$ degrees of freedom in the system.

The units of standard deviation are the same as the data used in its calculation (g, mL, %, etc.). *Relative standard deviation* (Rel. S) may be calculated by dividing the standard deviation by the mean and multiplying by 1000 so that relative S is in ppt. A similar term, the *coefficient of variation* (CV), is obtained by dividing the standard deviation by the mean and multiplying by 100:

$$\text{Rel. } S = \frac{S}{\bar{X}}(1000) \tag{3-11a}$$

$$CV = \frac{S}{\bar{X}}(100) \tag{3-11b}$$

Example: Calculate the standard deviation of the following measurements: 10.1, 19.5, 9.9, 9.5, 10.6, 9.4, 11.5, 9.5, 10.0, 9.5. $\bar{x} = 10.0$.

x_1	$x_1 - \bar{x}$	$(x_1 - \bar{x})^2$
10.1	0.1	0.01
10.5	0.5	0.25
9.9	−0.1	0.01
9.5	−0.5	0.25
10.6	0.6	0.36
9.4	−0.6	0.36
11.5	1.5	2.25
9.5	−0.5	0.25
9.5	−0.5	0.25
10.0	0	0
9.5	−0.5	0.25
		4.24

$$S^2 = \frac{4.24}{10} = 0.42; \quad S = \sqrt{0.42} = 0.65$$

$$\text{Rel. } S = \frac{0.65}{10.0}(1000) = 65 \text{ ppt}$$

Sometimes analytical data from several different samples or sets of measurements are pooled to establish the standard deviation of the method used. The reason for doing this is that S for a larger number of results becomes a better estimator for σ. However, pooling assumes that the various analytical samples provide equally good results for estimating the underlying standard deviation, σ. For this to be true, the samples must be similar in chemical composition.

Example:

Example No.	Results	Mean	$\sum(X_i - X)^2$
1	19.76%	19.72%	0.0312
	19.58		
	19.82		
2	31.56	31.68	0.1593
	31.85		
	31.90		
	31.42		
3	24.17	24.07	0.0201
	24.08		
	23.97		
4	30.01	29.96	0.2538
	29.66		
	30.34		
	29.83		
			0.4644

$$\text{Pooled } S = \left[\frac{0.4644}{14 - 4}\right]^{1/2} = 0.216$$

$$\text{Rel. } S = \frac{0.216\%}{27.00\%}(1000) = 8.0 \text{ ppt}$$

$$(\text{aver. }\%)$$

Note that there are 10 degrees of freedom in this calculation. This is because 4 of the 14 results are "used up" in finding mean of the 4 samples.

3–5. NORMAL DISTRIBUTION CURVE

Earlier, we stated that analytical measurements or results that are subject only to random error are assumed to have a normal distribution.

The density function of the normal random variable X, with mean μ and variance σ^2, is given by the equation

$$f(x) = \frac{1}{\sigma\sqrt{2\pi}}e^{-(1/2)[(X - \mu)/\sigma]^2} \qquad (3–12)$$

where σ is the standard deviation and e is the natural base of logarithms $= 2.718$. The resulting plot is bell-shaped. As the standard deviation, σ, becomes larger, the curve flattens out (see Figure 3–3). The location of the curve on the X-axis is determined by its mean, μ.

A normal distribution curve is easy to calculate with the aid of a programmable calculator. The mathematical steps in Equation 3–12 are entered into the program, and the desired values of μ and σ are put into two memories. Then it is necessary only to enter the various values of X and obtain for each the corresponding value of $f(x)$.

Sometimes analytical data are obtained directly in the form of a normal distribution. For example, chromatographic peaks are essentially Gaussian, or normal, in nature. In such cases, the mean and variance (or standard deviation) can be obtained from the curve itself. The mean is of course the value of X at the peak maximum. The value of σ is obtained by measuring the peak width at one-half the peak height where peak width $= 2.355\sigma$. At the base of the peak, the peak width $= 4\sigma$, as will be explained in Chapter 21.

Instead of dealing with a family of distribution curves of differing shape and mean, a single distribution curve can be obtained by transforming any normal random variable X to a new normal random variable Z with mean 0 and variance 1:

$$f(Z) = \frac{1}{\sqrt{2\pi}}e^{-Z^2/2} \qquad (3–13)$$

where

$$Z = \frac{X - \mu}{\sigma}$$

Figure 3–3. Normal distribution curves for $\mu = 25.0$, and $\sigma = 0.1$ (higher peak) and $\sigma = 0.2$ (lower peak).

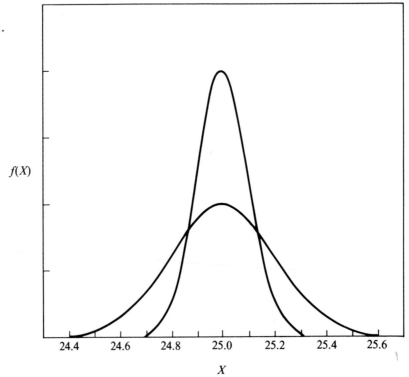

A curve plotted in this manner is called a *standard normal distribution* curve. Such a curve is shown in Figure 3–4. This gives us some interesting information about the probabilities of individual analytical measurements.

1. Because the curve is symmetrical, there tends to be a negative error for each positive error of the same value. The expected value of X corresponds with the mean, μ.
2. Integration of the area under the normal distribution curve within definite limits gives the probability that a measurement will lie within that area. For example, 68.3% of the results are between -1σ and $+1\sigma$ from the mean.
3. The relative frequency of measurements having a large deviation from the mean is very small. Because 99.7% of the measurements fall within ± 3 from the mean, only 0.3% of the measurements will fall outside these limits.

Areas of the standard normal distribution, $A(Z)$, between 0 and Z are given in statistics textbooks. An abbreviated list is given in Table 3–1.

Table 3–1. Areas of the standard normal distribution. $A(Z)$ is the area under the curve between 0 and Z. (Note that the area must be doubled to give the area under the curve from $-Z$ to $+Z$.)

Z	A(Z)	Z	A(Z)
0.1	0.0398	1.5	0.4334
0.2	0.0793	1.64	0.4495
0.5	0.1915	1.8	0.4641
0.9	0.3159	1.96	0.4750
1.0	0.3413	2.0	0.4773
1.1	0.3643	2.5	0.4938
1.2	0.3849	3.0	0.4987

Figure 3–4. Standard normal distribution curve.

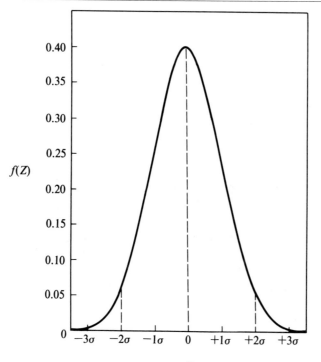

3-6. DISTRIBUTION OF MEANS

Usually, an analytical measurement or determination is performed in replicate, and the mean of the several measurements is computed. We generally report the mean as the "best" value of the replicate measurements. Thus, the mean is our most vital concern. We would like to know how the number of replicate measurements affects the goodness of the mean and how much confidence can be placed in the value we report for a determination.

Because a single analytical measurement is a random variable (X), the mean of n measurements is a random variable $(\bar{X})$. The result of a single determination, such as a titration, is also a random variable that comes from mathematically combining several measurements. Therefore, the mean of several titrations or other analytical results is a random variable. We observed that any type of repeated measurement gives results that are not exactly the same, but instead follow some type of underlying distribution. When several measurements are taken and the mean is calculated, and then several more are taken and a second mean is calculated, we usually find that the means are not exactly the same. If we follow this procedure many times, *we can plot a frequency distribution of means in the same way that distribution of individual measurements was handled earlier. In statistics this is called a sampling distribution.*

Perhaps *the* most important question in chemical measurements is how accurate the *mean* of several repeated measurements is likely to be. This is not an easy question to answer with absolute certainty; but by using simple statistical principles, we can get a numerical answer that is very likely to be correct.

It can be shown that the variance of the distribution of sample means is equal to the variance of the underlying distribution, divided by the number of individual measurements making up the sample means, n.

$$\text{Var}(\bar{X}) = \frac{\text{Var}(X)}{n} \tag{3-14}$$

or

$$\sigma_{\bar{X}} = \frac{\sigma_X}{\sqrt{n}} \tag{3-15}$$

In Equation 3-15, $\sigma_{\bar{X}}$ is logically called the "standard deviation of the mean," although another name, the "standard error of the mean," is sometimes used. This is a very significant equation because it tells us that the standard deviation of the mean becomes smaller and smaller as the number of individual measurements, n, increases. For example, the standard deviation of the mean is cut in half by a fourfold increase in n.

Confidence Limits.   The standard deviation of the mean is a *point* estimate of μ. However, a point estimate does not indicate the *confidence* that can be placed in such an estimate. When an objective measure of reliability is required, we report a *range of values* rather than a single value. These interval estimates are called *confidence intervals* or *confidence limits*.

The confidence limit of a mean, $\bar{X}$, is calculated by using the following equation:

$$\text{Confidence limit} = \bar{X} \pm \frac{Z\sigma_X}{\sqrt{n}} \tag{3–16}$$

The confidence limit gives the range that the mean should be in a certain percentage of the time. Thus, if we wish to know within what limits the mean of n chloride determinations will fall 95% of the time, we use 1.96 as the value of Z to insert into Equation 3–16. This is because 95% of the area under the standard normal distribution falls within ± 1.96 standard deviations. Areas under the standard normal distribution are given as a function of Z in Table 3–1.

Example: Calculate the confidence limit at the 95% probability level for four replicate analyses in which the mean is 21.70% chloride and the known standard deviation, σ, for such a determination is 0.08%. From Table 3–1, Z for 95% probability is 1.96.

$$\text{Confidence limit} = 21.70 \pm \frac{1.96 \times 0.08}{\sqrt{4}} = 21.70 \pm 0.08\% \text{ Cl}$$

At the 90% probability level, $Z = 1.64$, and the confidence limit is

$$21.70 \pm \frac{1.64 \times 0.08}{\sqrt{4}} = 21.70 \pm 0.06\% \text{ Cl}^-$$

When σ is not known from previous experience, it becomes necessary to calculate the standard deviation, S, for a small set of measurements. Thus, S is a less precise value than σ, and the confidence limit for the mean will be larger when S is used. W. S. Gosset, writing under the pseudonym of "Student," worked out a "t" distribution that is used for calculating confidence limits of the mean when S, instead of σ, is known.

To formulate the confidence limits, the standard deviation S of the sample is calculated, and a value for t is found by consulting Table 3–2. The constant t depends on the probability level and on n, the number of measurements in the sample. (Many statisticians prefer $n - 1$, which is termed the degrees of freedom.) The confidence limits are calculated from the equation

$$\text{Confidence limit} = \bar{X} \pm \frac{tS_X}{\sqrt{n}} \tag{3–17}$$

Example: Calculate the confidence limit (95% probability) for four replicate chloride analyses in which the mean is 21.70% and the standard deviation, s, is 0.08% Cl$^-$. (This value of s was chosen to permit a direct comparison with the previous example, in which σ was 0.08%.)

From Table 3–2, $t = 3.182$ for $n = 4$.

$$\text{Confidence limit} = 21.70 \pm \frac{3.18 \times 0.08}{\sqrt{4}} = 21.70 \pm 0.13\%$$

At the 90% probability level, the confidence limit is 0.09%.

Table 3–2. Values of t for Calculating Confidence Limits

Number of Measurements n	Degrees of Freedom $n-1$	Risk and Probability Level		
		0.10 90%	0.05 95%	0.01 99%
2	1	6.314	12.706	63.657
3	2	2.920	4.303	9.925
4	3	2.353	3.182	5.841
5	4	2.132	2.776	4.604
6	5	2.015	2.571	4.032
7	6	1.943	2.447	3.707
8	7	1.895	2.365	3.499
9	8	1.860	2.306	3.355
10	9	1.833	2.262	3.250
11	10	1.812	2.228	3.169
12	11	1.796	2.201	3.106
13	12	1.782	2.179	3.055
14	13	1.771	2.160	3.012
15	14	1.761	2.145	2.977
16	15	1.753	2.131	2.947
21	20	1.725	2.086	2.845
26	25	1.708	2.060	2.787
31	30	1.697	2.042	2.750
41	40	1.684	2.021	2.704
61	60	1.671	2.000	2.660
$\infty + 1$	∞	1.645	1.960	2.576

Table 3–3. Ratio of confidence limits at the 90% level. The ratio given is the confidence limit using S_X to that using σ_X.

Number of measurements, n	Ratio
2	3.84
3	1.78
4	1.43
5	1.30
6	1.22
8	1.15
10	1.11
15	1.07
21	1.05
41	1.02

By comparing the ratio of the confidence limit when S is used to that when σ is used, we can get an idea of how good S is as an estimator for σ as n changes. This is done at the 90% confidence level in Table 3–3. The results show that S is a poor estimator for σ when $n = 2$ or 3. Things improve as n increases, and S becomes a very good estimator for σ at $n = 15$ or 20.

3–7. MATHEMATICAL OPERATIONS WITH RANDOM VARIABLES

Suppose we add, subtract, multiply, or divide two or more random variables and want to know the variance or standard deviation of the result. This is a useful thing to do in following the propagation of errors in analytical determinations that involve several steps or measurements. For example, the weight (in mg) of a titrated substance is obtained by multiplying the milliliters of titrant (a random variable) by the molarity of titrant (another random variable) and then multiplying by the molecular weight.

Addition and Subtraction of Random Variables.   Recall first that a random variable, such as X, represents a distribution. The mean of such a distribution is μ, but it is often called the *expected value*. Thus, $E(X)$ denotes the expected value of the random variable, X. In adding or subtracting two random variables the expected value of the sum or difference is obtained by adding or subtracting the expected values:

$$E(X + Y) = E(X) + E(Y) \qquad\qquad \textbf{(3–18)}$$

$$E(X - Y) = E(X) - E(Y) \qquad\qquad \textbf{(3–19)}$$

The variance of the sum of two independent random variables is the *sum of the variances*. However, the variance of the difference of two independent random variables is also the *sum* of their variances:

$$Z = X + Y$$

$$\text{Var}(Z) = \text{Var}(X) + \text{Var}(Y), \quad \text{or} \quad \sigma_Z^2 = \sigma_X^2 + \sigma_Y^2 \qquad \textbf{(3–20)}$$

$$Z = X - Y$$

$$\text{Var}(Z) = \text{Var}(X) + \text{Var}(Y), \quad \text{or} \quad \sigma_Z^2 = \sigma_X^2 + \sigma_Y^2 \qquad \textbf{(3–21)}$$

Example: Suppose that each buret reading has a standard deviation of ± 0.02 mL. A titration volume is the difference between the final buret reading and the initial, or zero, reading. Estimate the standard deviation of the titration volume:

$$Z = X - Y$$

$$\sigma_Z^2 = \sigma_X^2 + \sigma_Y^2 = (0.02)^2 + (0.02)^2 = 0.0008$$

$$\sigma_Z = \sqrt{0.0008} = 0.028 \text{ mL}$$

If random variables are normally distributed, the entire distribution can be added or subtracted. For example, suppose that in spectrophotometry a colored solution has an absorbance of 0.608 with a standard deviation of 0.007 and from this a blank of absorbance 0.105 with a standard deviation of 0.007 is to be subtracted. The expected value of the difference is $0.608 - 0.105 = 0.503$. The variance of the difference is $(0.007)^2 + (0.007)^2 = 98 \times 10^{-6}$. The standard deviation of the difference is the square root of the variance, or 0.010. The distributions of the two absorbance readings and for their difference are obtained by substituting the appropriate values of μ and σ into Equation 3–12 and plotting points for various values of the random variable. This is shown graphically in Figure 3–5.

Multiplication and Division of Random Variables. The expected value of the product of two normally distributed random variables is the product of the expected values:

$$E(XY) = E(X)E(Y) \qquad (3-22)$$

The variance of the product of two independent random variables is more complicated. From statistics, the variance of $Z = XY$ is given by the equation

$$\text{Var}(Z) = Y^2\text{Var}(X) + X^2\text{Var}(Y) + \text{Var}(X)\text{Var}(Y) \qquad (3-23)$$

Dividing by the square of the initial equation, $Z^2 = X^2Y^2$,

$$\frac{\text{Var}(Z)}{Z^2} = \frac{\text{Var}(X)}{X^2} + \frac{\text{Var}(Y)}{Y^2} + \frac{\text{Var}(X)\text{Var}(Y)}{X^2Y^2} \qquad (3-24)$$

For analytical measurements, the variance of a random variable is usually quite small in comparison to the random variable itself. Therefore, the third term in Equation 3–22 will be negligibly small, and we can state the following rule: The

Figure 3–5. Absorbance measurements in spectrophotometry as random variables. S = sample, B = blank.

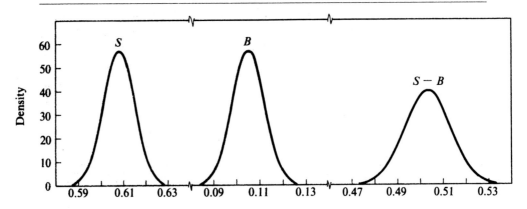

Absorbance

relative variance of a product of random variables is the sum of the relative variances:

$$\frac{\mathrm{Var}(Z)}{Z^2} = \frac{\mathrm{Var}(X)}{X^2} + \frac{\mathrm{Var}(Y)}{Y^2} \qquad (3\text{-}25)$$

This equation can also be written:

$$\left(\frac{\sigma_Z}{Z}\right)^2 = \left(\frac{\sigma_X}{X}\right)^2 + \left(\frac{\sigma_Y}{Y}\right)^2 \qquad (3\text{-}26)$$

It can also be shown that the relative variance of the quotient of two independent random variables is the *sum* of their relative variances. Thus for an equation involving both multiplication and division of random variables: $W = \dfrac{XY}{Z}$

$$\left(\frac{\sigma_W}{W}\right)^2 = \left(\frac{\sigma_X}{X}\right)^2 + \left(\frac{\sigma_Y}{Y}\right)^2 + \left(\frac{\sigma_Z}{Z}\right)^2 \qquad (3\text{-}27)$$

3-8. HANDLING SMALL SETS OF DATA

The material thus far has been mainly concerned with a discussion of statistical principles and with statistical evaluation of medium to large sets of analytical data. However, a student in the laboratory usually does an analysis in triplicate on three carefully weighed samples or on three aliquots of a sample solution. After calculating the results, the student would like any available statistical help in answering two basic questions: What is the "best value to report" and how good do my results appear to be? Unfortunately, it is impossible to obtain a very meaningful statistical evaluation of only three results. Nevertheless, we will examine these two questions to see what might be done.

Usually, the best value to report for an analysis is the mean, or average, of the individual results. However, it often happens that two results will agree rather well and the third value will be appreciably higher or lower. In this case, there is a natural tendency to reject the third value and to report the mean of the remaining two values. Unless there is a good experimental reason to do so, the questionable value should not be rejected, since it often happens that the close agreement of the two results is fortuitous; the third result should then be used to calculate the mean. A method to test whether such a result should be rejected will now be described.

The Q Test. The Q test uses the range to determine whether a questionable result should be rejected when $n = 3$ to 10 [4]. This test is most commonly conducted at the 90% confidence level, which means that the questionable result may be rejected with 90% statistical confidence that it is significantly different from the other results. The rejection quotient is labeled $Q_{0.90}$; at the 96% confidence level, it is designated $Q_{0.96}$,

[4] R. B. Dean and W. J. Dixon, *Anal. Chem.*, 23, 636 (1951).

Table 3-4. The Q Test

Questionable Result	Formula for Testing	n	$Q_{0.90}$	$Q_{0.96}$	$Q_{0.99}$
Smallest value (X_1)	$Q = \dfrac{X_2 - X_1}{X_n - X_1}$	3	0.94	0.98	0.99
		4	0.76	0.85	0.93
		5	0.64	0.73	0.82
Largest value (X_n)	$Q = \dfrac{X_n - X_{n-1}}{X_n - X_1}$	6	0.56	0.64	0.74
		7	0.51	0.59	0.68
		8	0.47	0.54	0.63
		9	0.44	0.51	0.60
		10	0.41	0.48	0.57

Note: If $Q \geq Q_{0.90}$, reject $X_?$. Arrange sample in order of increasing magnitude.

etc. Values for rejection quotients have been compiled by Dixon [5] and are given in Table 3-4.

To perform the Q test, the results are arranged in increasing order of magnitude and labeled $X_1, X_2, \ldots, X_n$. The difference between a questionable result and its nearest neighbor is divided by the range to obtain a quotient, Q. If this Q is equal to or greater than the quotient given in Table 3-4 for n results, the questionable one is rejected.

Example: Use the Q test to determine whether any of the following results should be rejected: 40.12, 40.15, 40.55.

The close agreement of the first two results makes it tempting to discard the third result. However, the Q test will not quite permit this.

$$Q = \frac{X_3 - X_2}{X_3 - X_1} = \frac{0.40}{0.43} = 0.93 \qquad (0.94 \text{ needed for rejection})$$

Unconvinced by the failure of the Q test to reject a result that "obviously" was not as good as the first two, the analyst decided to do two additional analyses. The new results were: 40.20, 40.28. The average of the five results is 40.26, which is much closer to the average of the original three results (40.27) than to the average of the two lower original results (40.14).

In this case the additional analyses supported the retention of *all* of the first three results. Of course, the additional analyses could have been lower and showed that the 40.14 average was more correct. But unless additional analyses are made, it is better statistically to average all three unless one can be rejected by the Q test.

Evaluation of Results. When the mean of individual results has been properly computed, the problem of evaluating the analysis remains. Do the results appear to be satisfactory or not? Although this question cannot be answered definitely when a truly unknown chemical sample is involved, the precision of the individual results usually indicates the correctness of the value reported. (This assumes, of course, that

[5] W. J. Dixon, *Ann. Math. Stat.*, 22, 68 (1951).

an appropriate method of analysis has been selected and that no systematic error has been made.) Thus, if the precision is good, the analysis probably is satisfactory. But what precision should be considered as "good"? This can only be answered from past experience. If we know that a given analytical procedure should give a relative standard deviation of 5 ppt or less and that this precision has been obtained in the past on similar samples, then it is reasonable to assume that a single analysis in triplicate should give precision of this order. By estimating the standard deviation (Equation 3-10), it is convenient and easy to evaluate the results of an analysis by comparing the precision with that expected. (The Q test should be applied before estimating s.)

Example: A certain analytical method should give a relative standard deviation of 5 ppt or better. A sample is first analyzed three times by using this method to give the following *set* of results: 40.12%, 40.15%, and 40.55%. Because the 40.55% result appears questionable, two additional results are obtained as follows: 40.20% and 40.39%. What is the standard deviation of the first set ($n = 3$) and of the complete set ($n = 5$)?

The range of the first set is 0.43% (note that the Q test does not reject 40.55%); the range of the second set is 0.43%. The *absolute* value of each standard deviation is calculated as follows:

$$s \text{ (first set)} = 0.59 \times 0.43 = 0.25 \qquad \text{(Rel. } s = 6.3 \text{ ppt)}$$

$$s \text{ (compl. set)} = 0.43 \times 0.43 = 0.18 \qquad \text{(Rel. } s = 4.6 \text{ ppt)}$$

Thus the first set, which contains the questionable results, does not meet the precision requirements, but the complete set does have a relative standard deviation within the expected limits for the method used.

3-9.   ANALYTICAL CONTROL

Chemical analyses are often done for the purpose of controlling some industrial process or for assuring the quality of a product. Sometimes, quality control must be done exceptionally well to protect our health and safety. For example, human and veterinary drugs are tested routinely for purity, potency, and dosage (the amount in a tablet). To ensure adequate control, we need to know with some certainty that the analytical methods being used are valid and that they continue to give correct results over a period of time.

The use of "control charts" is perhaps the simplest and most common way of ascertaining that the quality of chemical analyses is being maintained. A control chart is simply a chronological plot of results from the periodic analysis of some reference material. Control charts can be used to indicate any significant trends and, more important, to point out any gradual deterioration in the analytical results so that a difficulty can be corrected before any serious consequences occur. The central value (the long-term mean, for example) is indicated by a solid line on the chart. Statistical limits, such as ± 2 standard deviations from the central value, are shown by dotted lines above and below the solid line. As long as the results lie within the established limits, the analytical program is considered to be satisfactory and under control.

Although many types of control charts are used, the most common are those in which the mean of several replicate analyses of the reference material is plotted as a function of time. Such a chart shows whether the accuracy of analyses is under control statistically. A second chart of the standard deviation or range of each set of results is often plotted concurrently to indicate whether the repeatability of the results remains under control.

To set up a control chart, several replicate analyses of a standard reference material are made periodically, perhaps once each week. If possible, the true value of the determination made on the reference material should be known. If it is not already known, the correct value can be estimated very accurately by numerous analyses performed over a period of time. The reference material selected should be very stable and should approximate as closely as possible the composition of actual samples to be analyzed. Each week (or other time period) the mean of several replicate analyses is plotted on a chart. After several weeks, enough data will be available to calculate the mean of all the data and to establish upper and lower control limits.

This process is best illustrated by working with some actual data. Table 3–5 gives results for weekly determinations of sulfur in a reference material. (In actual practice, data for at least 12–15 weeks would be needed to set up a good control chart.) The mean of all individual results (16.40%) gives a good estimate of the true value. The standard deviation of all results (0.059%) provides a reasonable estimate of the standard deviation of the population, σ. Since three replicate analyses were done each week, the standard deviation of the mean of each subset ($s_{\bar{x}}$) is given by Equation 3–15:

$$s_{\bar{x}} = \frac{0.059}{\sqrt{3}} = 0.034$$

The data are plotted as shown in Figure 3–6. "Warning" control limits are set at $\pm 2\sigma$, and "action" limits are set at $\pm 3\sigma$. The results for the first eight weeks show

Table 3–5. Analysis of a Reference Material for Sulfur

| Week | Results, % Sulfur | | | |
	x_1	x_2	x_3	$\bar{x}$
1	16.40	16.48	16.48	16.45
2	16.42	16.35	16.35	16.37
3	16.45	16.35	16.47	16.42
4	16.37	16.44	16.30	16.37
5	16.48	16.32	16.40	16.40
6	16.29	16.36	16.39	16.35
7	16.44	16.40	16.39	16.41
8	16.50	16.40	16.45	16.45

Figure 3–6. A control chart.

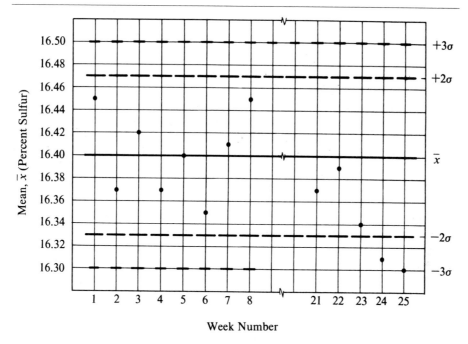

Week Number

no discernible trend, and all of the values are within the limits. Some weeks later, the results showed a sudden trend toward low values; the warning limits were exceeded, and the action limit was reached. At this point, rather extensive "troubleshooting" was undertaken to locate and correct the source of the difficulty.

3–10. CALIBRATION CURVES

In chemical analysis, it is often necessary to prepare calibration plots for use in calculating the results of analytical measurements. These are usually prepared by taking standards of known concentration, measuring the desired property of each standard, and plotting the property measured against the concentration. For example, in spectrophotometry (Chapter 17), the absorbances of sample solutions that absorb visible or ultraviolet light are plotted against the concentration of the standards. In chromatography, the height or area of peaks are plotted against the known concentrations of standards. When such a plot has been prepared, the concentration of desired substance in an unknown sample is obtained by measuring the absorbance, peak height, etc. under the same conditions as were used for the standards and reading the unknown concentration from the horizontal axis of the plot.

Whenever possible, parameters are selected that will give a linear calibration plot. Sometimes the form of a term has to be adjusted so that a linear plot will be obtained. Thus, the potential of an ion-selective electrode must be plotted against the *logarithm* of concentration (or activity) in order to obtain a linear plot (see Figure 16–6).

In preparing a calibration plot that is expected to be linear, it is usually best to plot the points graphically to be sure that the plot is linear over the entire concentration range and does not level off at the higher or lower end. Although unknown concentrations can be read directly from the graphical plot, significantly better accuracy is possible by using linear regression analysis. This is often called a "least squares" analysis because of the mathematical procedure that is used. Of the possible straight lines that can be drawn through or near the data points, the one chosen minimizes the sum of the squared deviations. The deviation for each point is the difference between the actual data point and a point with the same X-axis value that lies exactly on the straight line. Linear regression calculates the slope of the "best" straight line and also gives the intercept of the line on the Y-axis. The intercept is useful because it tells whether there is a significant blank measurement even when the concentration of the standard is zero.

The least squares method of fitting a straight line to a set of data and calculating the slope and Y-intercept of the line is discussed in Appendix 5. From the example given, it will be seen that calculation from the mathematical equation is somewhat tedious. However, the computation is very easy with a calculator that has a built-in linear regression program. One has only to punch in the X- and Y-coordinates of the data points and then press appropriate keys to obtain the slope and intercept. Another key gives the correlation coefficient, r, which shows how well the data points fit the straight line. A perfect fit is indicated by an r value of 1.000 (or $r = -1.000$ if the plot has a negative slope).

Example: In spectrophotometry (Chapter 17) the absorbance (A) of a colored solution is proportional to the concentration of the colored chemical species. From the data given, calculate the equation of the calibration plot by linear regression. Also calculate the concentration of an unknown that has an absorbance of 0.467.

A	C
0.203	$1.20 \times 10^{-5} M$
0.411	$2.40 \times 10^{-5} M$
0.600	$3.60 \times 10^{-5} M$
0.819	$4.80 \times 10^{-5} M$
0.985	$6.00 \times 10^{-5} M$

The equation takes the form $A = \epsilon c + \text{blank}$, where ϵ is the slope and blank is any blank absorbance. By using the linear regression keys on a hand calculator, the intercept (blank) = 0.012, and the slope (ϵ) = 16,430. The correlation coefficient (r) = 0.9993, which indicates a very good fit of the data points. Thus, our equation is $A = 16430c + 0.012$. The concentration of an unknown where $A = 0.467$ is

$$c = \frac{0.467 - 0.012}{16,430} = 2.77 \times 10^{-5} M$$

QUESTIONS AND PROBLEMS

Uncertainty, Significant Figures

1. Assuming that the absolute uncertainty in each of the following is ± 1 in the last digit, calculate the relative uncertainty in ppt.

   (a) 0.8084

   (b) 115.03

   (c) 0.00020

   (d) 4.94×10^6

2. Assuming that the relative uncertainty of each of the following is ± 2 ppt, rewrite each number, using the correct number of significant figures.

   (a) 9.97193

   (b) 0.062851

   (c) 0.0079

   (d) 15,000,000

   (e) 3.295×10^{-4}

3. Compute the correct answer for each set of figures, using the correct number of significant figures.

   (a) $204.2 + 19.15 + 3.035$

   (b) $0.456 - 0.029$

   (c) $14.13 + 3.13 - 5.2$

   (d) $1.9 \times 10^{-4} + 3.7 \times 10^{-6}$

4. pH is defined as the negative logarithm of hydrogen ion concentration. Calculate the pH, using the proper number of significant figures, for each of the following hydrogen-ion concentrations. Use a subscript when the first digit of the $[H^+]$ makes it necessary.

   (a) $3.8 \times 10^{-6} M$

   (b) $1 \times 10^{-12} M$

   (c) $9.5 \times 10^{-2} M$

   (d) $8.6 \times 10^{-2} M$

   (e) $0.796 M$

   (f) $0.870 M$

5. Calculate the correct $[H^+]$ of pure water in each case, assuming that K_w has been measured only to the number of significant figures given. $[H^+] = \sqrt{K_w}$.

   (a) $K_w = 1.0 \times 10^{-14}$

   (b) $K_w = 1.01 \times 10^{-14}$

   (c) $K_w = 1.008 \times 10^{-14}$

6. Calculate the value of K in scientific notation from the log of K, using the correct number of significant figures.

   (a) $\log K = 2.40$

   (b) $\log K = 0.04$

   (c) $\log K = 10.3$

   (d) $\log K = -0.04$

7. Various authors list different values for the pH of the gastric fluid in the stomach. Convert the values below to $[H^+]$ of the gastric fluid and indicate which is probably safest to report as a typical value. $pH = -\log[H^+]$.

   (a) $pH = 2$

   (b) $pH = 1.7$

   (c) $pH = 1.698$

Basic Definitions

8. Distinguish between random and systematic error. Which type is more amenable to treatment using statistical concepts?

9 Distinguish between absolute error and relative error. If a weighing of 5 g can be made on a top-loading balance with an expected absolute error of ± 0.01 g, calculate the relative error in ppt.

10. Distinguish between precision and accuracy. List some commonly used measures of precision.

11. Distinguish between a histogram and a density function. For each, tell what is plotted on the X- and Y-axes.

12. Explain what is meant by a random variable. Draw a rough graph of a distribution of results that might be represented by a random variable, X. How are individual numerical values of this random variable designated?

13. A second estimator for the mean of the underlying distribution of an analytical sample is the *mode*. It is defined as the value that occurs most frequently in a sample. Indicate the likelihood of a mode existing for data where (a) all 100 measurements have one significant figure, (b) all five measurements have one significant figure, and (c) all 20 measurements have four significant figures.

14. A third estimator for the mean of the underlying distribution of an analytical sample is the *median*, M. It is defined as the middle value of a sample of an odd number of results arranged in order of magnitude, or the average of the two middle results of a sample of even number. The advantage of M over the mean is that a gross error in one result in a small sample will cause a large error in only the mean, not the median. Calculate the mean and the median of the pH of gastric juice from two pH measurements of 1.213 and 1.29. Explain why there is no difference.

15. The median of an even number of results is always more efficient than that of an odd number of results because the two middle values of an even-numbered sample are averaged. Calculate the median of each of these samples and compare the efficiency of each median in terms of the number of values used and not used.

(a) 1.1, 1.2, 1.4, and 3.3   (b) 1.1, 1.2, 1.4, 1.5, and 3.3

Distribution of Results

16. A quantitative analysis class obtained the following results for percentage chloride in an unknown, rounded off to the nearest 0.1%. (The number in parentheses is the number of results for each percentage.) 18.9% (2), 19.0% (0), 19.1% (2), 19.2% (7), 19.3% (15), 19.4% (14), 19.5% (21), 19.6% (19), 19.7% (4), 19.8% (2), 19.9% (1), 20.1% (1), 20.5% (1).

(a) Plot a histogram for these data.
(b) Calculate the mean and standard deviation.

17. Using the data from the previous problem, discard any results that differ from the mean by more than three standard deviations, then prepare a histogram of the remaining results, plotting relative frequency on the vertical axis. If a smooth curve is drawn through the center of each vertical bar, estimate the area under such a curve. Explain how relative frequency (the vertical axis) could be converted to density (on the vertical axis) so that the area under the curve would be equal to 1.0, as in a density function plot.

18. If a density function plot is skewed (not symmetric), will the mean be at the highest vertical part of the curve? Explain.

19. Explain how probability can be defined in terms of a density function. To illustrate this concept, calculate the probability that any given result in the histogram in Problem 16 will be either 19.3% or 19.4%.

20. If we have a density function plot of X that is symmetric, how can we graphically find the mean? The expected value, $E(X)$?

21. A histogram can be used to approximate a density function plot. This is done by plotting relative frequency versus the average value of X for each class interval. Plot such a curve from the data given below and estimate the mean and standard deviation.

The following lists the average value for X and then the relative frequency.

10.0, 0.002	11.4, 0.079	12.8, 0.060
10.2, 0.004	11.6, 0.097	13.0, 0.042
10.4, 0.009	11.8, 0.109	13.2, 0.027
10.6, 0.016	12.0, 0.113	13.4, 0.016
10.8, 0.027	12.2, 0.109	13.6, 0.009
11.0, 0.042	12.4, 0.097	13.8, 0.004
11.2, 0.060	12.6, 0.079	14.0, 0.002

Mean and Standard Deviation of a Statistical Sample

22. What is a statistical sample? Explain the relationship of $\bar{X}$ to μ and of S to σ.

23. For the following set of spectrophotometric measurements (each in absorbance units), calculate the sample mean, variance, and standard deviation: 0.496, 0.489, 0.495, 0.501, 0.502, 0.490.

24. Estimate the absolute and relative values of the standard deviation of the following data: 40.02, 40.11, 40.16, 40.18, 40.19.

25. The following data were obtained with a method that should give a relative standard deviation of 5 ppt or better: 30.15, 30.55, 30.12.

   (a) Estimate the absolute value of the standard deviation.
   (b) Evaluate the precision of these results, comparing it to the known relative standard deviation.

26. A certain analytical method should give a relative standard deviation of 6 ppt or better. A sample is first analyzed three times using this method to give the following set of results: 43.22%, 43.25%, and 43.65%. Because the 43.65% result appears questionable, two additional results are obtained as follows: 43.30% and 43.49%.

   (a) What is the standard deviation of the first set ($n = 3$) and of the complete set ($n = 5$)?
   (b) Evaluate the precision of each of the two sets of results.

27. Calculate the standard deviation of a method used to measure pH from the pooled data given below for an industrial effluent:

Date	pH values
Nov. 8	3.45, 3.48, 3.48
Nov. 9	3.46, 3.43, 3.45
Nov. 10	3.37, 3.33, 3.33
Nov. 11	3.21, 3.18, 3.22
Nov. 12	3.18, 3.21, 3.17

28. Wells supplying drinking water to a midwestern city were contaminated with traces of methyl naphthalene. Samples from each of four wells were analyzed for methyl naphthalene in triplicate. Estimate the standard deviation of the analytical method used from the following data: Well No. 1: 0.21, 0.18, 0.17 ppb; No. 2: 0.15, 0.18, 0.20 ppb; No. 3: 0.26, 0.24, 0.18 ppb; No. 4: 0.21, 0.23, 0.20 ppb.

Handling small sets of data

29. Calculate the mean of each of the percent nitrogen results *after* testing any questionable values at the 90% confidence level.

   (a) 11.11%, 11.15%, and 12.09%   (b) 4.97, 5.23, 5.20, and 5.17

30. After obtaining arsenic percentages of 10.00, 10.10, and 11.00%, a student decides that 11.00% is a questionable result and desires to reject it.

    (a) Can he do so with 90% confidence?
    (b) He then obtains a fourth result of 10.20%. What is the mean of his results after testing 11.00% again?

31. The median is often used if the precision is poor. Suppose the fourth result in the previous problem were 10.30% instead of 10.20% and suppose the precision were indeed poorer than past precision. Test 11.00% and decide what to calculate as the best value.

32. Questionable results most often arise from two determinate (systematic) errors: incorrect transcription of results and sample identification. Three pH measurements were recorded as 1.601, 1.607, and 1.064.

    (a) Which of these results may have been incorrectly recorded and what might the correct value have been?
    (b) Assume that the values were correct; calculate the mean.

33. After measuring many blood samples in the 7.40–7.50 range, a technician notes that she has recorded a pH of 7.93. Suspicious, she measures it twice more, obtaining values of 7.42 and 7.46.

    (a) Can she reject 7.93 using the Q test?
    (b) Should she conclude there is a determinate (systematic) instrumental error?
    (c) What other possible error could have been made if the instrument is not at fault?

Normal Distribution Curve

34. If a programmable calculator or computer is available, calculate points and plot a normal distribution curve (Equation 3–12) for:

    (a) $\mu = 20.0, \sigma = 1.5$ (b) $\mu = 24.0, \sigma = 1.2$

35. Explain how a standard normal distribution differs from a normal distribution curve.

36. Repeated analysis of a sample used as a student "unknown" has given a "true" value of 22.05% with a standard deviation of $\pm 0.06\%$. Assuming a normal distribution, calculate the probability that the value of your first analysis will fall within the limits given.

    (a) 22.05% and 22.11%   (c) 21.96% and 22.05%
    (b) 22.04% and 22.06%   (d) $>22.17\%$

Confidence Limits

37. Calculate the confidence limits at the 95% level for the mean obtained in Problem 21.

38. The confidence limit for a mean, $\bar{X}$, is always smaller when σ_X is known than when S_X must be calculated from the individual measurements.

    (a) Assuming that $S_X = \sigma_X$, compare the confidence interval at the 95% level for S_X and σ_X for $n = 3, 4, 5, 6, 8,$ and 10.
    (b) The "efficiency" of S_X as an estimator for σ_X (at a given confidence level) is the ratio of the confidence interval using σ_X to that using S_X at each value of n. Calculate this efficiency for each value of n.

39. A company that sells platinum catalysts to the petrochemical industry needs very accurate chemical assays because of the very high price of platinum. For a catalyst

containing 60% Pt, the standard deviation of the analytical method is known to be 0.20%. What will be the standard deviation of the mean if each sample is analyzed in triplicate? If each sample is analyzed six times?

40. Modern microprocessors permit some analytical measurements to be made and averaged many times in a short time period. Suppose a spectrophotometric measurement can be made with a standard deviation of ± 0.001 absorbance unit. How many times must the measurement be made so that the reading of the mean can be made with a standard deviation of ± 0.0001 absorbance unit.

41. A dairy wants to keep the butterfat content of its milk as close as possible to 2.00% so that it will not be "giving away" valuable excess butterfat. If σ for the analytical method used is $\pm 0.10\%$ and each sample is analyzed in triplicate, calculate the butterfat content to which the milk should be blended so that the analysis mean will show *at least* 2.00% butterfat at the 90% confidence level.

Random Variables (Mathematical Operations)

42. The standard deviation for each of two buret readings in a titration is 0.02 mL, and the standard deviation of locating the titration end point is 0.04 mL. Calculate the expected standard deviation of the titrant volume reading. Also calculate the relative standard deviation of the reading for a 25.00-mL titration.

43. The absorbance of a colored complex is 0.836 with a standard deviation of ± 0.005. From this is subtracted a blank of 0.134, which has a standard deviation of ± 0.004. Estimate the standard deviation and relative standard deviation of the difference between the sample and blank readings.

44. Calculation of the result of a titration can be stated as $Z = WXY$, where each capital letter represents a random variable. Calculate the numerical value for Z and also its expected standard deviation from the following data: $W = 25.00$ mL, $\sigma = 0.03$ mL; $X = 0.1000$ mmole/mL, $\sigma = 0.0005$ mmole/mL; $Y = 204.2$ mg/mmole, $\sigma = 0.05$ mg/mmole.

Analytical Control

45. Indicate the purpose of a control chart and outline briefly how such a chart is set up and used.

46. An agent from a government regulatory agency is checking to see whether the waste water discharged from your factory is "under control." The agent notes that results of your daily analyses for specified pollutants are within legal limits but questions the validity of the analytical procedures you used. What kind of information should you have to answer this objection?

Regression Analysis

47. In spectrophotometry the molar absorptivity, ϵ, is often calculated from the slope of a plot of absorbance versus concentration. ($A = \epsilon C$.) From the data given, use linear regression to calculate the slope, ϵ, of the best straight line through the data points. The first value given for each data point is the absorbance, and the second in the molar concentration. 0.108, 1.00×10^{-5}; 0.210, 2.00×10^{-5}; 0.325, 3.00×10^{-5}; 0.425, 4.00×10^{-5}; 0.535, 5.00×10^{-5}.

48. A calibration plot of quantity measured (Y-axis) versus concentration (X-axis) should pass through (0, 0) unless there is a blank due to reagent impurities. Calculate the equation for the best straight line through the data points given below. From the appropriate

intercept, calculate the blank in concentration units. Data (X, Y): 202, 2.00; 350, 4.00; 489, 6.00; 624, 8.00; 741, 10.00.

Theory of the Q Test

49. The calculation of the rejection quotient, such as $Q_{0.90}$, for $n = 3$ is done by using a formula of Dixon [*Annals of Math. Statistics 22*, 68 (1951)] for R_a, a general rejection quotient:

$$R_a = 0.500 \times (\sqrt{3}/2)\tan[\pi/3(0.500 - \alpha)]$$

where α is the fractional probability that a questionable value is *larger* than R_a; for $Q_{0.90}$, α thus is 0.05.
   (a) Calculate $Q_{0.90}$ for $n = 3$ to three significant figures.
   (b) Calculate $Q_{0.96}$ for $n = 3$ to three significant figures.
   (c) Calculate $Q_{0.99}$ for $n = 3$ to three significant figures.

4

Gravimetric Methods of Analysis

The quantitative determination of a substance by precipitation followed by isolation and weighing of the precipitate is called *gravimetric analysis*. Gravimetric methods are widely employed in quantitative analysis, although most chemists prefer to use the more rapid titrimetric, spectrometric, or chromatographic methods whenever possible. Precipitation can be a very selective process; for that reason, precipitation methods are valuable for quantitative analytical separations.

The general scheme of a gravimetric analysis is fairly simple. A weighed sample is dissolved, after which an excess of precipitating reagent is added. The precipitate is filtered, washed, dried or ignited, and weighed. From the weight and known composition of the precipitate, the amount of the ion can be calculated. From this and the weight of sample taken, the percentage of the desired substance in the original sample can be calculated.

For a successful determination, the following requirements must be met:

1. The desired substance must be completely precipitated. Most analytical precipitates are low enough in solubility that the solubility loss is negligible. The common ion from the excess of precipitating reagent reduces the solubility of the precipitate. For example, silver chloride is slightly soluble, but the excess silver nitrate added to precipitate the chloride shifts the equilibrium and represses the solubility of the precipitate:

$$AgCl(s) \rightleftharpoons Ag^+ + Cl^-$$

$$\leftarrow \boxed{\begin{array}{c} \text{excess} \\ Ag^+ \end{array}}$$

2. The weighed form of the precipitate should be a compound of known composition. Calculations with a gravimetric factor (Section 4–6) have this as their basis.

3. The precipitate must be pure and easily filtered. Sometimes it is very difficult to obtain a precipitate that is free from impurities.

4–1. MECHANISM OF PRECIPITATION

The first step in precipitation is the formation of very tiny particles of precipitate, called nuclei. The process of forming these particles is called *nucleation*. Following nucleation, particle growth takes place in three dimensions, so that the tiny nuclei become relatively large (macro-sized) precipitate particles.

When the precipitating reagents are mixed in solution, there is an induction period before nucleation occurs. This induction period varies with different precipitates; it is very short for silver chloride but is unusually long for barium sulfate. (With very dilute solutions, the induction period for barium sulfate nucleation is several minutes long.) In most cases, however, nucleation occurs spontaneously almost as soon as the precipitating reagents are mixed.

After the first burst of nucleation, it is easier for the nuclei to grow into macro-sized particles than it is to form more nuclei. The cations and anions in solution collide with the small particles and attach themselves to the surface by chemical bonding. This results in the growth of a three-dimensional crystal lattice.

> The first nuclei formed are far too small to be seen. In one case studied, it was found that each nucleus contains only about four molecules (remember that 1 mole = 10^{23} molecules) and that 10^9 to 10^{12} nuclei are formed per mole of the precipitate ions present [1].

Precipitates always have some ions adsorbed on their surfaces. Either the lattice cation or the lattice anion will be adsorbed during precipitation, depending on which is in excess. For example, if silver chloride is precipitated by the slow addition of silver nitrate to excess sodium chloride, chloride ions will be adsorbed on the precipitate surface and await the arrival of more silver ions to continue crystal growth. The lattice ion adsorbed (in this example, chloride) is called the *primary adsorbed ion*. There is, of course, the possibility that one of the other ions in solution (NO_3^- or Na^+ in this example) will be adsorbed, but in general the lattice ion that is in excess strongly predominates.

Because of primary adsorption, the surface of a precipitate will have a plus or minus charge, depending on whether the lattice cation or anion is in excess. To balance this charge, ions of the opposite charge are attracted to the portions of solution immediately surrounding the precipitate particles. These ions are called *counter ions*. Counter ions are less tightly held than are primary adsorbed ions. The counter ion layer is somewhat diffuse and contains some other cations and anions in addition to the counter ions.

Primary Adsorbed Ion	Counter Ion
The lattice ion that is in excess	Opposite in charge to primary adsorbed ion
Held by chemical bond	Held by electrostatic attraction
Fixed on precipitate surface	Loosely held in solution surrounding precipitate

[1] F. R. Duke and L. M. Brown, *J. Am. Chem. Soc.*, *71*, 1443 (1954).

Symbolism for Primary Adsorbed Ions and Counter Ions. We have just seen that precipitates tend to have an electrical double layer of adsorbed ions—the primary absorbed ions on the outer surfaces of the precipitate lattice and the counter ions in the solution layer surrounding the precipitate particles. The primary adsorbed ion will be indicated by writing the formula of the precipitate followed by two vertical dots and the absorbed ion. This will symbolize that an electron pair is being shared at the surface of the precipitate. For a silver halide precipitate, $AgX(s)$, two cases are possible

$$\frac{\text{Excess Ag}^+}{AgX:Ag^+} \qquad \frac{\text{Excess } X^-}{X Ag:X^-}$$

In the first case, the excess Ag^+ is attracted to the halides in the precipitate lattice. In the second case, the excess X^- is attracted to the silver in the lattice.

The electrical double layer of a precipitate will be represented as before, but with the counter ion also added. Either two or six dots separate the counter ion from the primary adsorbed ion. The six dots symbolize that the counter-ion layer is very diffuse, as in tiny colloidal precipitate particles. The two dots represent a closer approach of the counter ion to the primary adsorbed ion, as in a coagulated precipitate.

Nitrate counter ion in AgX precipitate:

Nitrate counter ion farther away (colloidal $AgX:Ag^+$):

$$AgX:Ag^+ \cdot \cdot NO_3^- \qquad AgX:Ag^+ \cdot \cdot \cdot \cdot \cdot \cdot NO_3^-$$

Where there are two or more ions that can act as counter ions, then the ion that forms the least soluble compound will have the strongest attraction to the primary adsorbed ion and will be the counter ion. For example, if both nitrate and perchlorate ions (ClO_4^-) are present when a silver halide is precipitated, the nitrate ion will be the counter ion because silver nitrate is about one-fifth as soluble as silver perchlorate.

4-2. CONDITIONS FOR ANALYTICAL PRECIPITATION

Ideally, an analytical precipitate for gravimetric analysis should consist of perfect crystals large enough to be readily washed and filtered. The perfect crystals would be free from impurities within and would present a minimum of surface area for the adsorption of impurities without. The precipitate should also be insoluble enough that loss of precipitate due to solubility would be negligible.

The way in which pure, filterable crystals are obtained depends on the type of precipitate formed. At this point, we will describe the three types briefly; we will discuss them in detail later.

Types of Precipitates.   The three types of analytical precipitates are the curdy precipitates, the gelatinous precipitates, and the crystalline precipitates. Curdy and gelatinous precipitates both form the same way: The cation and anion react to form a soluble *colloid* (described in items 1 and 2), and then grow (coagulate) to filterable-sized particles. Crystalline precipitates form by precipitating out as small, imperfect crystals that grow to purer, larger crystals.

1. *Curdy precipitates.*   The most useful curdy precipitates are the silver halide compounds. These initially precipitate as suspended, colloidal particles that do not build up to a large enough size to precipitate. For example, if NaCl is to be precipitated by adding an excess of $AgNO_3$, the first step is the formation of colloidal silver chloride:

$$Cl^- + 2AgNO_3 \rightarrow AgCl:Ag^+ \cdots\cdots NO_3^- \text{ (colloidal)}$$

Because the nitrate ions are too far away to neutralize the Ag^+ ions, the positively charged particles repel each other and remain colloidal. Heating is necessary to produce a precipitate.

2. *Gelatinous precipitates.*   The best examples of the gelatinous precipitates are the $+3$ metal hydroxides (hydrous oxides), such as $Fe(OH)_3$. In precipitating these compounds, a colloid is also formed first, and heating again is necessary to form a precipitate. In contrast to the curdy precipitates, the gelatinous precipitates trap a great deal of water and many more impurities.

3. *Crystalline precipitates.*   The best example of the crystalline precipitates are the alkaline earth sulfates such as barium sulfate. They precipitate as regularly shaped, discrete particles. For example, if the sulfate ion is to be precipitated by adding an excess of $BaCl_2$, the first step is the formation of small imperfect crystals:

$$nSO_4^{2-} + 2nBaCl_2 \rightarrow nBaSO_4:Ba^{2+} \cdots 2Cl^-\text{(s)}$$

Here the chloride counter ions approach the primary adsorbed barium ion closely enough to prevent repulsion by barium ions. Hence, these species can group together and precipitate. Since these crystals are not pure and are often too small to filter effectively, they are heated to achieve larger, purer crystals:

$$x[nBaSO_4:Ba^{2+} \cdots 2Cl^-\text{(s)}] \rightarrow xnBaSO_4:Ba^{2+} \cdots 2Cl^-\text{(s)}$$

$$\quad\text{(small crystals)} \qquad\qquad\qquad\qquad \text{(large crystals)}$$

Von Weimarn's Study of Particle Size.   Von Weimarn found that the particle size of precipitates is *inversely proportional* to the relative supersaturation of the solution during precipitation [2]:

$$\text{relative supersaturation} = \frac{Q - S}{S}$$

where Q is the molar concentration of the mixed reagents *before* any precipitation

[2] P. P. Von Weimarn, *Chem. Rev.*, 2, 217 (1925).

Figure 4–1. Particle size as a function of supersaturation.

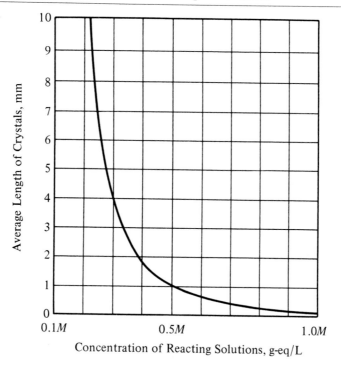

occurs and S is the molar solubility of the precipitate when the system has come to equilibrium. This effect is illustrated in Figure 4–1.

The *rate* of precipitation increases as the relative supersaturation increases. To give a pure precipitate of large particle size, then, the relative supersaturation should be low so that the rate of precipitation will be slow.

Actually, the rate of precipitation depends on the rates of two processes, nucleation and particle growth. The rates of both processes depend on the supersaturation, $Q - S$: the nucleation rate k is $(Q - S)^n$, where n may be about 4, and the growth rate is $k'A(Q - S)$, where A is the area of the precipitate. The values of k and k' are such that, after the first nucleation, particle growth takes place rather than further nucleation, provided that $Q - S$ is low. If $Q - S$ becomes too great, nucleation may predominate over growth, with the result that the precipitate will be colloidal.

For the best possible precipitate, conditions should be adjusted so that Q will be as low as possible and S will be relatively large.

Obviously, S must not be too great, or the precipitation will not be quantitative. However, a "quantitative" precipitation requires that less than 1 part per 1000 of the desired ion remain unprecipitated. Increasing S from 1 part per 1,000,000 to 1 part per 10,000 does not affect the quantitativeness of the precipitation because the solubility loss is so slight as to be negligible in either case.

The following techniques are commonly used for precipitation.

Precipitation from Dilute Solution.   This keeps Q low.

Slow Addition of Precipitating Reagent with Effective Stirring.   Adding the precipitating reagent slowly also keeps Q low; stirring avoids locally high concentrations of precipitating reagent.

Precipitation at a pH Near the Acidic End of the pH Range in Which Precipitation Is Quantitative.   Many precipitates are more soluble at the lower, or more acidic, pH values because the anion tends to protonate and dissolve to a greater extent. Hence the rate of precipitation is slowed by the lower concentration of the unprotonated anion.

Precipitation from Hot Solution.   The solubility S of precipitates increases with temperature; hence, an increase in S decreases the supersaturation.

Digestion of Precipitates.   Digestion of any precipitate involves heating the precipitate in contact with the solution from which it precipitated. However, digestion affects crystalline precipitates somewhat differently than curdy (and gelatinous) precipitates.

1. *Digestion of Crystalline Precipitates.*   During the digestion of crystalline precipitates, the larger crystals grow at the expense of the smaller ones. The result is a marked improvement in the filterability of the precipitate. To some extent, digestion causes the small particles to dissolve and reprecipitate onto the larger, more perfectly formed crystals. Finally, digestion of crystalline precipitates achieves some internal perfection, or "ripening," of the precipitation in which the amount of entrapped impurities often decreases.

2. *Digestion of Curdy Precipitates.*   Many precipitates are so insoluble that they form colloids rather than small crystalline particles. Colloids have a particle size between 1 and 100 nm. They appear to be soluble but actually form a suspension and pass through ordinary filter media. Such a colloidal suspension scatters light, making it appear cloudy. At room temperature, colloidal particles of $AgCl:Ag^+ \cdots\cdots NO_3^-$ repeal each other because the nitrate counter ions are too far away to reduce the repulsion of the positive charges. Digestion just below the boiling point enables coagulation to occur; the heat supplies energy to reduce the repulsion and to effectively "shrink" the layers of primary ions and counter ions. This produces a curdy precipitate that is stable even at room temperature.

The digestion of colloidal particles of silver chloride in excess silver nitrate may be shown symbolically as follows:

$$
\begin{array}{ccc}
\begin{array}{l} ClAgCl:Ag^+ \cdots\cdots NO_3^- \\ AgClAg \updownarrow \\ ClAgCl:Ag^+ \cdots\cdots NO_3^- \end{array}
& \xrightarrow{\text{digestion}} &
\begin{array}{l} -ClAgCl:Ag^+ \cdot\cdot NO_3^- \\ AgClAg \\ ClAgCl:Ag^+ \cdot\cdot NO_3^- \\ \quad\quad\quad\quad| \end{array}
\end{array}
$$

(colloidal suspension
with repulsion (arrow))   (corner of a curd)

Note that the symbolism shows that the distance between each primary absorbed ion and its counter ion is reduced from six dots to two dots. This allows the three colloidal particles to "shrink" together with no repulsion of the positive silver ions.

4–3.   IMPURITIES IN PRECIPITATES

The term *coprecipitation* refers to the carrying down of normally soluble impurities during the precipitation of an insoluble compound. Coprecipitation occurs to some extent in any analytical precipitation, but it is especially marked with barium sulfate and with colloidal precipitates such as the hydrous oxides. By careful precipitation and thorough washing, the effects of coprecipitation can be minimized—but not always eliminated.

Surface Adsorption.   Adsorption of ions at the surface occurs with all precipitates. Contamination by adsorption causes a significant error only in cases in which the precipitate has a very large surface area. As was explained earlier, the primary adsorbed ion of a precipitate is the excess lattice ion. Thus, when all chloride has been precipitated by the addition of excess silver nitrate, Ag^+ will be the primary adsorbed ion on the AgCl precipitate. In this case, the counter ion will be a foreign anion, such as $NO_3{}^-$. The net effect, then, is a silver chloride precipitate with a layer of silver nitrate adsorbed on its surface.

In the case of metal hydroxides (hydrous oxides), the primary adsorbed ion depends on the pH of the solution. For example, the isoelectric point (neutral point) of hydrous aluminum oxide is about pH 8.0. Below this value, the surface of the precipitate has a positive charge and attracts anionic counter ions; at higher pH values, the precipitate surface is negatively charged and attracts cations more strongly as counter ions. Thus, coprecipitation of foreign metal cations can be reduced by precipitating a hydrous oxide at as low a pH as possible. Even so, the precipitate is so adsorptive that some foreign cations contaminate it.

> Cation coprecipitation is about the same when metal cations are added to a solution just *after* precipitation of a hydrous oxide as it is when a hydrous oxide is precipitated in their presence. This is evidence that we are dealing with surface adsorption.

Digestion of a precipitate reduces the amount of surface and makes it more dense. This often reduces greatly the surface adsorption of foreign ions. Thorough *washing* of a precipitate removes surface impurities or replaces them with adsorbed substances that are volatile on ignition. Gelatinous precipitates, such as hydrous oxides, usually cannot be washed free from adsorbed impurities.

Occlusion.   Occlusion is a type of coprecipitation in which impurities are trapped within a growing crystal. The impurities are distributed unevenly through the precipitate, mostly occupying places where the crystal structure of the precipitate is imperfect.

One type of occlusion may be thought of as gross mechanical entrapment. This occurs when a precipitate grows in such a manner that there are holes or pockets in the crystals. These contain solvent (water) with the dissolved impurities that were present as the crystal formed.

In another type of occlusion, surface adsorption during precipitation plays an important role. This type is illustrated by the precipitation of sulfate by the slow addition of barium chloride. During the precipitation, the sulfate is in excess and is the primary adsorbed ion. A positive ion, for example Na^+, is adsorbed as the counter ion. As more precipitant is added, Ba^{2+} replaces the adsorbed Na^+, and the growth of precipitate particles continues. If the particles grow too fast, however, not all of the Na^+ will be replaced by Ba^{2+}, and the precipitate may grow around the adsorbed Na^+, enclosing it.

Because sulfate is in excess during its precipitation by the slow addition of barium chloride, there is a tendency for cation coprecipitation to occur through the occlusion of cations adsorbed as counter ions. A reverse addition of reagents (slow addition of sulfate to excess barium chloride) would be expected to decrease cation coprecipitation, because Ba^{2+} would be in excess during precipitation and the counter ion would be an anion. Experiment shows that reverse addition does decrease cation coprecipitation (and increases anion coprecipitation). Some cation coprecipitation still takes place through occlusion because *local excesses* of precipitant cause a primary adsorption of sulfate and adsorption of cations as counter ions.

The best way to deal with occlusion is to avoid it in the first place. Techniques for precipitation from homogeneous solution have been worked out for many analytical precipitates in which the precipitating ion is slowly generated throughout the solution instead of being added all at once. Digestion also often reduces the amount of occluded impurities in a precipitate, but usually some impurities remain. Washing is ineffective for the removal of occluded substances. If the precipitate dissolves readily in acids, purification by reprecipitation may be useful.

Isomorphous Inclusion. Compounds with the same type of formula crystallizing in similar geometric forms are said to be *isomorphous*. When the lattice dimensions of two isomorphous compounds are about the same, one compound can replace part of the other in a crystal. The result is the formation of mixed crystals. For example, $MgNH_4PO_4$ and $MgKPO_4$ are isomorphous; the ionic radii of K^+ and NH_4^+ are virtually the same. During precipitation of Mg^{2+} as $MgNH_4PO_4$, the K^+ replaces some of the NH_4^+ in the precipitate crystals. Thus, the precipitate contains some $MgKPO_4$, even though the solubility product for this compound is not exceeded. This causes an error in the gravimetric determination of magnesium, because $MgNH_4PO_4$ and $MgKPO_4$ are ignited to compounds of different molecular weights.

The error incurred through coprecipitation by isomorphous inclusion is usually very serious. Furthermore, little can be done to avoid it unless the offending ion can be removed before precipitation. It is fortunate that isomorphous replacement in analytical precipitates is fairly rare. The partial replacement of NH_4^+ by K^+ in $MgNH_4PO_4$, of Br^- by Cl^- in $AgBr$, of SO_4^{2-} by CrO_4^{2-} in $BaSO_4$, and of Ba^{2+} by Pb^{2+} in $BaSO_4$ are examples of isomorphous inclusion.

Gathering. Several pages have now been devoted to a discussion of the troubles caused by coprecipitation, but it should be pointed out that coprecipitation can be used to advantage. Suppose that we wish to precipitate an ion that is present in such a low concentration that adding a precipitating reagent fails to give a precipitate (that is, the solubility product of the desired precipitate is not exceeded). In such cases, the trace ion can often be removed quantitatively from solution by coprecipitation onto a macro quantity of another precipitate. The precipitate used to coprecipitate the trace ion is called a *gathering agent*, or *carrier*.

A method for the isolation of traces of lead from urine illustrates the use of gathering [3]. A calcium salt and a phosphate salt are added to the urine, causing calcium phosphate to precipitate. The lead is removed from solution by coprecipitation, even though the solubility product of lead phosphate has not been exceeded. The precipitate can then be dissolved in a small volume of aqueous acid and the lead determined by a colorimetric method.

Postprecipitation.   Sometimes a precipitate standing in contact with the mother liquor becomes contaminated by a foreign compound that precipitates on top of it. This is called *postprecipitation* because the foreign precipitate comes down *after* the desired precipitate has formed. For example, postprecipitation of magnesium oxalate occurs if a precipitate of calcium oxalate is allowed to stand too long before being filtered. Zinc sulfide postprecipitates onto cadmium, copper, or mercuric sulfide, which precipitates from acidic solution.

4–4.   WASHING AND FILTERING PRECIPITATES

Coprecipitation, especially surface adsorption, cannot be avoided, but the resulting error can be avoided or minimized by washing the precipitate. A precipitate that settles rapidly can be washed by decantation (see Chapter 27, Section 27–3). After the precipitate has been filtered, it can be washed in the filter funnel or crucible with several small portions of wash liquid.

Washing Curdy Precipitates with Pure Water.   The choice of a wash solution is particularly important for curdy precipitates. Intuitively, one might use pure water to wash a curdy precipitate, but pure water will *peptize* the curdy precipitate back to the colloidal form and wash it through the filter. As an example of peptization, consider washing curdy silver chloride (as $AgCl:Ag^+ \cdot \cdot NO_3^-$) with water. The water contains no nitrate ions, so it gradually lowers the concentration of nitrate counter ions on the surface; that is, the nitrate ions get farther and farther away. Then the positive silver counter ions repel each other and revert back to the colloidal form.

Consider the corner of the curd of silver chloride discussed in Section 4–2. In the first step, the nitrates are pulled farther away and allow repulsion of the primary

[3] L. T. Fairhall and R. G. Keenan, *J. Am. Chem. Soc., 63,* 3076 (1941).

ions to occur:

$$
\begin{array}{ccc}
\text{—ClAgCl:Ag}^+ \cdots \text{NO}_3{}^- & & \text{ClAgCl:Ag}^+ \cdots\cdots \text{NO}_3{}^- \\
\text{AgClAg} & \xrightarrow[\text{washing}]{\text{pure water}} & \text{AgClAg} \updownarrow \\
\text{ClAgCl:Ag}^+ \cdots \text{NO}_3{}^- & & \text{ClAgCl:Ag}^+ \cdots\cdots \text{NO}_3{}^- \\
\mid & & \\
\text{(corner of a} & & \text{(colloidal suspension} \\
\text{curd)} & & \text{with repulsion)}
\end{array}
$$

In the second step, the individual $\text{ClAgCl:Ag}^+ \cdots\cdots \text{NO}_3{}^-$ species migrate away from the surface of the curdy precipitate and are washed through the filter.

Washing Curdy Precipitates with Electrolyte Solution. To avoid peptization, one should use a solution of an electrolyte containing the counter ion. Such a solution will not wash the counter ion away from the surface. In addition, the electrolyte must be volatile so that it is vaporized when the precipitate is dried. For example, for washing curdy precipitates of the type $\text{AgCl:Ag}^+ \cdots \text{NO}_3{}^-$, a volatile nitrate compound must be used. This eliminates salts such as sodium and potassium nitrate; the only common electrolytes that could be used are nitric acid and ammonium nitrate. Each vaporizes as follows during drying:

$$\text{AgCl:H}^+ \cdots \text{NO}_3{}^- \xrightarrow{\text{heat}} \text{AgCl(s)} + \text{HNO}_3\text{(g)} \rightarrow \text{ further decomposition}$$

$$\text{AgCl:NH}_4{}^+ \cdots \text{NO}_3{}^- \xrightarrow{\text{heat}} \text{AgCl(s)} + \text{NH}_3\text{(g)} + \text{HNO}_3\text{(g)}$$

$$\rightarrow \text{ further decomposition}$$

Washing with such an electrolyte also replaces some of the primary adsorbed ions, such as the Ag^+ described previously, but not all of such ions. Often this is of the order of one silver nitrate for every 10^3 silver chloride molecules (1 ppt).

Filtering Precipitates. The choice of a filter medium depends on the type of precipitate and on the temperature at which the precipitate is to be heated. Curdy precipitates are usually filtered with a crucible that has a built-in porous sintered-glass disk or mat; hence, such crucibles are called *sintered-glass filtering crucibles.* Crystalline precipitates are often filtered through filter paper. The selection of a filter and the technique of filtration are discussed in Chapter 27, Section 27–3.

4–5. HEATING THE PRECIPITATE

After a precipitate has been filtered and washed, it must be heated and weighed. Heating serves several purposes. One is to remove water from the precipitate. Another is to volatilize the adsorbed electrolyte from the wash liquid and any other volatile impurities that may be present. In some cases, heating converts the precipitate to another compound more suitable for weighing than the original precipitate. For example, a precipitate of calcium oxalate (CaC_2O_4) can be converted either to calcium carbonate (CaCO_3) or to calcium oxide (CaO) for weighing.

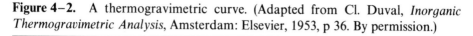

Figure 4–2. A thermogravimetric curve. (Adapted from Cl. Duval, *Inorganic Thermogravimetric Analysis*, Amsterdam: Elsevier, 1953, p 36. By permission.)

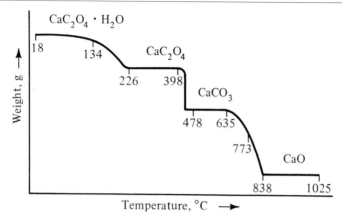

Heating temperatures vary greatly, depending on the precipitate. Silver chloride is dried in an oven at about 120°C. A precipitate of magnesium ammonium phosphate ($MgNH_4PO_4$) must be ignited and thus decomposed into magnesium pyrophosphate ($Mg_2P_2O_7$) in an electric muffle furnace at about 900°C. The heating temperatures for various precipitates have been determined largely by trial and error through the years. A more scientific way of determining a correct ignition temperature is through the use of a *thermobalance*. This instrument measures the weight of a precipitate as the temperature of the precipitate is slowly increased from room temperature to about 1000°C. The weight of precipitate as a function of temperature is plotted manually or recorded automatically (see Figure 4–2). A suitable heating temperature is at a plateau on the curve.

4–6. CALCULATING THE RESULTS

The final step in a gravimetric analysis is the calculation of results. Usually, the purpose of a quantitative analysis is to determine the percentage of a certain element or ion in a sample. The data obtained include the weight of the sample and the weight of the precipitate containing the substance to be determined.

The weight of the element or ion to be determined is calculated from the weight of the precipitate with the aid of a *gravimetric factor*. The gravimetric factor is the ratio of the formula weight of substance sought to that of substance weighed:

$$\text{Factor} = \frac{\text{formula wt (subst. sought)}}{\text{formula wt (subst. weighed)}}$$

The expression for a gravimetric factor may be derived from simple proportion. For example, the ratio of sulfur to barium sulfate is

$$\frac{S \text{ (grams)}}{BaSO_4 \text{ (grams)}} = \frac{S \text{ (form wt)}}{BaSO_4 \text{ (form wt)}}$$

If we know $BaSO_4$ (grams) and wish to calculate S (grams), solving this equation gives

$$S \text{ (grams)} = BaSO_4 \text{ (grams)} \times \frac{S \text{ (form wt)}}{BaSO_4 \text{ (form wt)}}$$

The last term is the gravimetric factor.

The same number of key atoms must be present in the numerator of the gravimetric factor as in the denominator. Thus, it may be necessary to multiply one or both formula weights by suitable integers to balance the key atoms. This is illustrated by the following example.

Example: Derive an expression for the gravimetric factor required to calculate the weight of magnesium in a precipitate of magnesium pyrophosphate, $Mg_2P_2O_7$.

The weight proportion of magnesium in magnesium pyrophosphate is equal to two atomic weights of magnesium divided by the formula weight of magnesium pyrophosphate.

$$\frac{Mg \text{ (grams)}}{Mg_2P_2O_7 \text{ (grams)}} = \frac{2Mg \text{ (form wt)}}{Mg_2P_2O_7 \text{ (form wt)}}$$

Solving for the grams of magnesium,

$$Mg \text{ (grams)} = Mg_2P_2O_7 \text{ (grams)} \times \frac{2Mg \text{ (form wt)}}{Mg_2P_2O_7 \text{ (form wt)}}$$

Additional examples of gravimetric factors are as follows:

Sought	Weighed	Gravimetric Factor
K	$KClO_4$	$\dfrac{K}{KClO_4}$
K_2O	$KClO_4$	$\dfrac{K_2O}{2KClO_4}$
Fe	Fe_2O_3	$\dfrac{2Fe}{Fe_2O_3}$
Fe_3O_4	Fe_2O_3	$\dfrac{2Fe_3O_4}{3Fe_2O_3}$
$KAlSi_3O_8$	SiO_2	$\dfrac{KAlSi_3O_8}{3SiO_2}$

In each case, the weight of substance sought is obtained by multiplying the weight of the precipitate by the gravimetric factor:

$$\text{weighed (grams)} \times \frac{\text{sought (form wt)}}{\text{weighed (form wt)}} = \text{sought (grams)}$$

Note that the species and units of the right-hand side of the equation can be checked by dimensional analysis; for example,

$$\text{BaSO}_4 \text{ (grams)} \times \frac{\text{SO}_4 \text{ (form wt)}}{\text{BaSO}_4 \text{ (form wt)}} = \text{SO}_4 \text{ (grams)}$$

Once the weight of the substance to be determined is known, the percentage is calculated by dividing by the weight of the sample and multiplying by 100:

$$\frac{\text{sought (grams)}}{\text{sample (grams)}} \times 100 = \% \text{ sought}$$

Calculation of the results of a gravimetric analysis is summarized in the following equation:

$$\frac{\text{weighed (grams)} \times \text{factor} \times 100}{\text{sample (grams)}} = \% \text{ sought}$$

4–7. EXAMPLES OF PRECIPITATION METHODS

Determination of Chloride. Chloride is precipitated from a slightly acidic solution as silver chloride:

$$\text{Cl}^- + \text{Ag}^+ \rightarrow \text{AgCl(s)}$$

Silver chloride is an example of a curdy precipitate; at first the precipitated silver chloride particles are colloidal, but these coagulate to form a curd. This type of precipitate varies, depending on whether an excess of $AgNO_3$ is added to precipitate the chloride ion or whether an excess of KCl is added to precipitate the silver ion. If an excess of $AgNO_3$ is used to precipitate chloride, then a colloidal suspension is formed as follows:

$$\text{Cl}^- + 2\text{AgNO}_3 \rightarrow \text{AgCl:Ag}^+ \cdots \cdots \text{NO}_3^-$$

During digestion, a curd is formed, as discussed in Section 4–2. A suitable wash solution must contain the nitrate as counter ion (Section 4–4), and this is usually added as nitric acid (or NH_4NO_3). If an excess of KCl is used to precipitate the silver ion, then a colloidal suspension is formed as follows:

$$\text{Ag}^+ + 2\text{KCl} \rightarrow \text{ClAg:Cl}^- \cdots \cdots \text{K}^+$$

During digestion, a curd will again form, but the primary adsorbed ion will always be the chloride ion, and the counter ion will be the potassium ion. The wash solution cannot be a potassium salt, since such salts are not volatile; hydrochloric acid or ammonium chloride solutions can be used instead.

Exposure to even laboratory light will cause a small degree of photodecomposition of silver chloride:

$$\text{AgCl(s)} \xrightarrow{\text{light}} \text{Ag(s)} + \tfrac{1}{2}\text{Cl}_2\text{(g)}$$

If the precipitate has been washed immediately after digestion, the results for chloride will be slightly low (usually $<0.1\%$). If the precipitate cannot be washed immediately after digestion and stands in strong light in the presence of excess silver nitrate, the chlorine reacts to form more silver chloride, giving slightly high results.

Other Applications.   The precipitation of chloride from solutions of sodium chloride, hydrochloric acid, etc. is an extremely accurate way of standardizing such solutions. Hydrochloric acid can usually be standardized to $4-5$ significant figures by using careful technique. Since the mmoles of NaCl or HCl are equal to the mmoles of AgCl, the molarity of such solutions are calculated as follows:

$$\text{M of NaCl or HCl} = \frac{\text{mg AgCl}/143.3 \text{ mg/mmole of AgCl}}{\text{mL NaCl or HCl}}$$

A number of other anions can also be determined by gravimetric precipitation using silver nitrate. Thus bromide and iodide are precipitated quantitatively as AgBr and AgI. Hypochlorite (OCl^-), chlorite ($ClO_2{}^-$), and chlorate ($ClO_3{}^-$) can be reduced to chloride and precipitated as AgCl.

Determination of Sulfate.   Sulfate is determined gravimetrically by precipating as barium sulfate:

$$Ba^{2+} + SO_4{}^{2-} \rightarrow BaSO_4(s)$$

The barium sulfate is precipitated from acidic solution so that barium salts of other anions will not coprecipitate as much. However, if the acidity is too high, part of the barium sulfate precipitate will dissolve.

Coprecipitation is especially serious in the sulfate determination. Anions coprecipitate along with the barium ions and cause high results. Among monovalent anions, nitrate coprecipitation is among the worst; at equivalent molarity, it is four or five times that of chloride. The coprecipitated barium anion salt is usually converted to barium oxide before being weighed by being ignited with the barium sulfate precipitate.

Cation coprecipitation causes the results to be low instead of high, because most coprecipitated metal sulfates have a lower formula weight than does barium sulfate. An idea of the error caused can be obtained from Table 4–1. Besides those listed, the error due to coprecipitation of iron(II) or calcium(II) is especially serious.

Coprecipitation is so serious in the sulfate determination that accurate results can be had only when anion coprecipitation, which causes high results, and cation coprecipitation, which causes low results, exactly cancel each other.

Barium sulfate is ignited at high temperature and weighed as $BaSO_4$. The precipitate is a fine crystalline precipitate and is usually filtered through paper designed to retain such a precipitate. In this case, there is no danger of peptization, so the precipitate can be washed with pure water. The precipitate must be ignited carefully with plenty of air to avoid partial reduction of the barium sulfate:

$$BaSO_4 + 2C \rightarrow BaS + 2CO_2$$

This method can be used in determining sulfides, thiosulfate, and sulfur of

Table 4-1. Coprecipitation of Metal Sulfates

Metal Sulfate	Sulfate Copptd., mmoles/g $BaSO_4$	Error in Grav. Sulfate Detn., Assuming Stoich. Form. for Copptd. Sulfate, pph
Al^{3+}	0.017	-0.34[a]
Mg^{2+}	0.024	-0.27
Na^+	0.029	-0.26
K^+	0.033	-0.19
Ni^{2+}	0.031	-0.26
Cu^{2+}	0.041	-0.30
Mn^{2+}	0.064	-0.52

a. Barium chloride solution added to metal sulfate in dilute hydrochloric acid, 18 hours standing before filtration. Coprecipitated sulfate weighed as Al_2O_3.

Source: Data from J. Johnson and L. H. Adams, *J. Am. Chem. Soc.*, **33**, 829 (1911).

lower oxidation states, provided these ions are first oxidized to sulfate. Persulfate $(S_2O_8{}^{2-})$ can be determined by precipitating it as barium sulfate, if it is first reduced to sulfate.

Determination of Iron. Ferric ion is precipitated by the addition of aqueous ammonia:

$$Fe^{3+} + 3NH_3 + 3H_2O \rightarrow Fe(OH)_3(s) + 3NH_4{}^+$$

This is an example of a gelatinous precipitate. It is very bulky, highly adsorptive, and is almost completely lacking in any ordered crystal structure.

In principle, iron(III) can be separated from copper(II) or from any other metal ion that forms an ammonia complex or does not precipitate in the presence of aqueous ammonia. However, coprecipitation is so bad that the separation is never complete. Precipitation from homogeneous solution by the addition of urea and boiling greatly reduces, but does not eliminate, coprecipitation:

$$NH_2CONH_2 + H_2O \xrightarrow{\text{boil}} 2NH_3 + CO_2$$

Reprecipitation, by dissolving the filtered precipitate in acid and carrying out a second precipitation, also reduces the impurities in the final precipitate.

The hydrous oxide precipitate is filtered through a hardened filter paper, to avoid clogging the pores of the paper. Filter paper pulp added to the precipitate facilitates filtration and washing. Since the precipitate peptizes if washed with pure water, dilute ammonium nitrate is the recommended wash solution for hydrous ferric oxide. (Ammonium chloride wash solution volatilizes $FeCl_3$ on ignition, which will cause low results.)

After being filtered and washed, the precipitate is wrapped in the filter paper and placed in a crucible of exactly known weight. The paper is carefully smoked off; then the precipitate is ignited at a high temperature and weighed as Fe_2O_3.

This method can also be used to precipitate and determine aluminum, chromium, manganese, or titanium as the hydrous oxide. It should be emphasized that this is a *poor* way to determine iron or any of these metals; quicker and more accurate analytical methods are available for these metals. The hydrous oxide precipitation method is valuable primarily for the separation (and sometimes for the determination) of *small amounts* of these metals.

Silica. Determination of silica is one of the most common uses of gravimetric methods. Silica occurs extensively in nature and must therefore be determined or removed by precipitation so that it will not interfere with other analyses. The method used is based on the insolubility of silica in acidic, aqueous solution. The freshly precipitated silica is highly hydrated and colloidal. It would be virtually impossible to handle it as an analytical precipitate at this stage. What is done is to *dehydrate* the silica by adding an acid and heating to evaporate the water and acid.

Although "acid" silicates and some other minerals require fusion to decompose the sample, many minerals and most alloys will dissolve in a suitable acid. For example, limestone is usually dissolved in hydrochloric acid. As the sample dissolves, the silicates are converted into insoluble, hydrated silica:

$$(Ca, Mg, CO_3, SiO_3) + H^+ \rightarrow Ca^{2+} + Mg^{2+} + SiO_2 \cdot xH_2O(s) + CO_2$$
$$\text{(HCl)}$$

The best way to dehydrate the silica is to add perchloric acid and evaporate the acid to white fumes. This converts the silica to a granular solid not unlike fine sand. Metal ions in solution are converted to anhydrous metal perchlorates, most of which are unusually soluble in water. After evaporation to dehydrate the silica, the beaker and contents are cooled and water added to dissolve the metal perchlorates and reduce the viscosity of the perchloric acid. Then the silica precipitate is filtered, washed, dried, and weighed as SiO_2.

An alternative procedure uses hydrochloric acid instead of perchloric acid. However, since hydrochloric acid does not have the dehydrating abilities of perchloric acid, it is necessary to evaporate entirely to dryness and to bake the precipitate for a while. Then more hydrochloric acid is added, and the evaporation and baking are repeated. If the silica is not sufficiently dehydrated, some of the particles may pass through the filter medium and be lost. Finally, the precipitate is diluted, filtered, dried, and weighed as before.

Precipitation of silica results in the coprecipitation of metal ions, some of which remain even after dehydration. A "corrected" silica determination is often made to compensate for these impurities. This is based on the volatility of the silicon tetrafluoride that is formed when hydrofluoric acid is added to silica and heat is applied:

$$4HF + SiO_2 \rightarrow SiF_4(g) + 2H_2O$$

(Naturally, a platinum crucible must be used because glass contains silica, which would react with hydrofluoric acid.) A small amount of sulfuric acid is also added to assist in the volatilization of the hydrofluoric acid. The residue, containing oxides of coprecipitated metals, is ignited and weighed. The difference in weight between the original silica precipitate and the impurity residue is used to calculate the corrected percentage of silica.

Table 4–2. Gravimetric Analytical Methods for Selected Elements

Element	Precipitation Form	Weighing Form	Important Interfering Elements
K	$KClO_4$	$KClO_4$	NH_4^+, Rb, Cs
	$KB(C_6H_5)_4$[a]	$KB(C_6H_5)_4$	NH_4^+, Rb, Cs
Mg	$MgNH_4PO_4$	$Mg_2P_2O_7$	All metals except Na and K
Ca	CaC_2O_4	$CaCO_3$ or CaO	All metals except Mg, Na, K
Ba	$BaCrO_4$	$BaCrO_4$	Pb
Zr	Zr Mandelate[b]	ZrO_2	Hf, F^-, PO_4^{3-}
Th	$Th(C_2O_4)_2$	ThO_2	Rare earths, Zr, F^-
Fe	$Fe(OH)_3$	Fe_2O_3	Al, Cr, Ti, many others
	Fe Cupferrate	Fe_2O_3	Tetravalent metals
Ni	$Ni(dmg)_2$[c]	$Ni(dmg)_2$	Pd
Cu	Cu (electrodeposition)	Cu	Ag, Bi, As, Sb, Sn
Ag	AgCl	AgCl	Hg(I)
Zn	$ZnNH_4PO_4$	$Zn_2P_2O_7$	Alkalis, all metals except Mg
Al	$Al(OH)_3$	Al_2O_3	Fe, Cr, Ti, and others
	Al (oxine)$_3$[d]	Al (oxine)$_3$	Alkalis, most metals except Mg
Sn	$SnO_2 \cdot xH_2O$	SnO_2	Sb, Si
Pb	$PbSO_4$	$PbSO_4$	Ca, Sr, Ba
Si	$SiO_2 \cdot xH_2O$	SiO_2	Sn
P	$MgNH_4PO_4$	$Mg_2P_2O_7$	MoO_4^{2-}
S	$BaSO_4$	$BaSO_4$	NO_3^-, ClO_3^-, PO_4^{3-}
F	$PbClF$[e]	PbClF	SO_4^{2-}, PO_4^{3-}
Cl	AgCl	AgCl	Br^-, I^-, CN^-, SCN^-

a. See review of A. J. Barnard, Jr., *Chemist-Analyst*, **44**, 104 (1955) and **45**, 110 (1956).
b. C. A. Kumins, *Anal. Chem*, **19**, 376 (1947).
c. Here dmg = dimethylglyoxime.
d. See review by J. I. Hoffman, *Chemist-Analyst*, **49**, 126 (1960). Here oxine = 8-hydroxyquinoline.
e. R. A. Bournique and L. H. Dahmer, *Anal. Chem.* **36**, 1786 (1964).

Other Gravimetric Determinations. A large number of gravimetric methods have been published. Probably the best single source of information is a three-volume work by L. Erdey [4]. The three volumes run well over 1000 pages and include detailed procedures, theory, and separation methods for virtually all of the elements. Table 4–2 gives the precipitation and weighing forms for just a few typical gravimetric methods.

The *order* in which various substances are precipitated is frequently quite important. This is true whether the goal is primarily separation or the substances are to be first precipitated and then measured by weighing the precipitate, i.e., gravimetric analysis. This point is well illustrated by an analysis of a sample containing both calcium(II) and magnesium(II), which frequently occur together. Calcium must precipitate first (as calcium oxalate) because both calcium and magnesium form insoluble phosphates.

[4] L. Erdey, *Gravimetric Analysis*, Oxford Pergamon, 1965.

Table 4–3. Some Useful Precipitates for Separations

Type of Precipitate	Elements Precipitated Quantitatively	Partially Precipitated
Hydrous oxide (pcpt. by NH_3)	Al_2O_3, Fe_2O_3, $La_2O_3^a$, TiO_2, ThO_2, U_3O_8, ZrO_2	Cr_2O_3
Hydrous oxide (pcpt. by acid)	Nb_2O_5, SiO_2, SnO_2, Ta_2O_5, WO_3	——
Chloride	$AgCl$, Hg_2Cl_2, $BiOCl$	$PbCl_2$, $SbOCl$
Sulfate	$BaSO_4$, $PbSO_4$	$CaSO_4$, $SrSO_3$
Oxalate (acid soln)	$Th(C_2O_4)_2$	——
Oxalate (basic soln)	CaC_2O_4	Many metals
Dimethylglyoxime	$Ni(dmg)_2$, $Pd(dmg)_2$	——
Cupferron (acid soln)	Fe(III), Mo(VI), Sn(IV), Ti(IV), U(VI), Zr(IV)	Bi(III), Cu(II), Th(IV)

a. All other rare earths also precipitated.

Separations.   Precipitation methods are valuable just for separating ions from one another. For this purpose, it is not necessary to have a precipitate of exact stoichiometric composition suitable for weighing. After filtration and washing, the precipitate may be redissolved (usually by adding acid) and the isolated element(s) determined by titration or other means.

In Table 4–3, a partial list is given of insoluble compounds that may be used for separation. Some types, such as hydrous oxides, are highly adsorbent precipitates and are good only for isolating rather small quantities of the precipitated elements. Many separations depend on control of acidity. Thus thorium(IV) may be separated as the oxalate from most metal ions in acidic solution; in basic solution, calcium(II) is separated from magnesium by oxalate precipitation, but most other metal ions interfere.

Notice that organic reagents are frequently used to precipitate inorganic ions. For example, sodium tetraphenyl boron, $NaB(C_6H_5)_4$, is used to precipitate potassium(I). Cupferron, $C_8H_5N(NO)O^-NH_4^+$, precipitates iron(III) and tetravalent metal ions such as tin(IV), titanium(IV), thorium(IV), and zirconium(IV) from strongly acidic aqueous solutions [5].

4–8. PRECIPITATION FROM HOMOGENEOUS SOLUTION

A special technique, called precipitation from homogeneous solution, is the ultimate refinement for many of the methods in Section 4–7. In this technique, the precipitating ion is not added as such to the solution but is slowly generated

[5] K. L. Cheng, *Chemist-Analyst*, 50, 126 (1961).

Table 4–4.   Some Reagents Used for Precipitations from Homogeneous Solution

Precipitating Ion	Hydrolytic Reagent
$C_2O_4^{2-}$	Diethyloxalate, $(C_2H_5)_2C_2O_4$
PO_4^{3-}	Trimethylphosphate, $(CH_3)_3PO_4$
SO_4^{2-}	Sulfamic acid, NH_2SO_3H
S^{2-}	Thioacetamide, CH_3CSNH_2
Oxinate$^-$	8-Hydroxyquinoline acetate

throughout the solution by a homogeneous chemical reaction. Local excesses of precipitant, which are inevitable in the conventional precipitation process, are avoided. In precipitation from homogeneous solution, the supersaturation, $Q - S$, is kept extremely low at all times, with the result that a very pure, dense precipitate is formed. Substances that ordinarily precipitate only as amorphous solids frequently precipitate from homogeneous solution as well-formed crystals (see Figure 4–3).

Analytical methods involving precipitation from homogeneous solution have been reviewed [6]. The major techniques can be classified as follows:

1. *Increase in pH.*   Usually the pH is made more alkaline by hydrolysis of urea, NH_2CONH_2, in boiling, aqueous solution. The ammonia slowly liberated raises

$$NH_2CONH_2 + H_2O \xrightarrow{\text{heat}} 2NH_3 + CO_2$$

the pH of the solution homogeneously, causing metal ions that form insoluble hydroxides (or hydrous oxides) to precipitate.

In the precipitation of aluminum from homogeneous solution, urea is added to an acidic solution of aluminum containing some sulfuric or succinic acid. No precipitation occurs until the solution has been boiled long enough (about 1 hour) for the ammonia to raise the pH to the necessary value. In this procedure, aluminum precipitates as the basic sulfate or the basic succinate, not as aluminum hydroxide. The precipitate obtained is much more dense and free from impurities than are aluminum precipitates formed by the conventional addition of ammonia.

Another example of precipitation from homogeneous solution is the precipitation of barium chromate. Chromate is added to barium in a solution acidic enough to prevent precipitate formation. Then urea is added, and the solution is heated to boiling. The ammonia released slowly raises the pH of the solution. When the pH reaches a certain value, precipitation of barium chromate begins. The precipitate of barium chromate is more perfectly formed and free from impurities than are precipitates obtained by the best conventional techniques.

2. *Anion Release.*   An organic ester or some other reagent is added to a sample solution containing metal ions to be precipitated. When the solution is heated, the reagent slowly hydrolyzes to form an anion that precipitates metal ions. Several examples of this technique are given in Table 4–4.

[6] P. F. S. Cartwright, E. J. Newman, and D. W. Wilson, *Analyst*, 92, 663 (1967).

Figure 4–3. Crystals precipitated from homogeneous solutions: (top) copper 8-hydroxyquinolate produced by direct addition; (bottom) copper 8-hydroxy-quinolate produced by solvent evaporation. Distance shown in 0.1 mm. [From L. Howick and J. Jones, *Talanta*, *10*, 197 (1963). By permission of Microform International Marketing Corporation.]

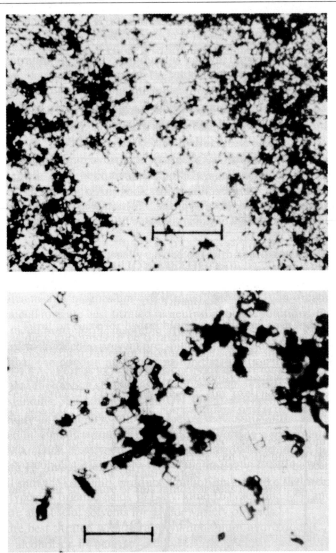

3. *Cation Release.*   This is illustrated by the precipitation of tungstic oxide from homogeneous solution. Tungsten(VI) in the sample is complexed as peroxy-tungstate by addition of hydrogen peroxide. The complex is then decomposed slowly by boiling it with 1:1 nitric acid, causing WO_3 to precipitate.

4. *Precipitation from Mixed Solvents.* An example is illustrated in Figure 4–3. A metal-organic complex such as copper 8-hydroxyquinolate is formed in water containing enough of a volatile organic solvent (acetone) to keep the complex dissolved. As the acetone evaporates, the solution becomes predominately aqueous, and the copper-organic complex precipitates as well-formed crystals.

5. *Valency Change.* This is less common. One example is the slow oxidation of cerium(III) iodate, which is soluble, to cerium(IV) iodate, which is insoluble.

QUESTIONS AND PROBLEMS

Principles and Methods

1. An aqueous solution contains dissolved lead nitrate and sodium nitrate. The lead(II) is precipitated by the dropwise addition of sodium chromate, Na_2CrO_4. When an excess of sodium chromate has been added, what will be (a) the primary adsorbed ion and (b) the counter ion?

2. Which is easier to wash away or replace by an ion from the wash solution: the primary adsorbed ion or the counter ion? Explain briefly.

3. Define (a) supersaturation and (b) relative supersaturation in terms of Q and S. What precipitation conditions will minimize Q? What precipitation conditions can be used to decrease the supersaturation by increasing S?

4. What is digestion? In what ways can digestion improve the quality of an analytical precipitate?

5. Explain how precipitation from homogeneous solution results in a precipitate with better purity and particle size than can be obtained by conventional precipitation.

6. Suggest two different methods of precipitating calcium oxalate from homogeneous solution for the gravimetric determination of calcium.

7. Briefly define the following terms, giving one example of each: (a) occlusion, (b) isomorphous replacement, (c) postprecipitation.

8. A radiochemist has a solution containing only 1 μmole (10^{-6} mole) of a radioactive metal ion per liter. How can he quickly isolate about 1 μmole of this metal as a solid compound for purposes of counting the radioactivity?

9. What is peptization and how is it caused? With which types of precipitate is peptization likely to occur?

10. Although many procedures state that a temperature of approximately 900°C is needed to ignite a magnesium ammonium phosphate precipitate to form magnesium pyrophosphate, one authority claims that a dull red heat (about 500°C) is sufficient. Suggest an experimental way of determining who is right.

11. Explain how sunlight can cause the results of gravimetric chloride determination to be either too high or too low, depending on the circumstances.

12. If barium sulfate is precipitated so that Ba^{2+} is the primary absorbed ion, which of the following anions would you expect to be the predominating counter ion: bromide, chlorate, or chloride? (Consult a chemical handbook for solubility data to help answer this question.)

13. In the gravimetric sulfate determination, how does the coprecipitation of potassium sulfate affect the results? In the same determination, how does the coprecipitation of barium nitrate affect the results?

14. Explain why a silica precipitate must be dehydrated and how dehydration is accomplished.
15. What is the purpose of performing a *corrected* silica determination? Outline the scheme for this determination.
16. Suggest a gravimetric method for the separate determination of both constituents in each of the following mixtures (note that the order in which the constituents are precipitated is frequently important): (a) Ca^{2+}, Mg^{2+}; (b) Mg^{2+}, Ni^{2+}; (c) Pb^{2+}, Zn^{2+}; (d) Cl^-, SO_4^{2-}.
17. Limestone is primarily $MgCa(CO_3)_2$ but also contains some silicates and small amounts of iron and aluminum. Outline a sequential procedure for determining calcium, magnesium, aluminum plus iron, and silica in limestone.
18. Consult Tables 4–2 and 4–3 and outline in the proper order a separation scheme for isolating each element in a brass containing Cu, Fe ($\sim 1\%$), Pb, Sn, and Zn.
19. Hydrous tin(IV) oxide is precipitated by addition of nitric acid in a manner somewhat analogous to the precipitation of silica by acid. Consult a handbook for a volatile tin(IV) compound with which the weight of the tin oxide could be corrected for coprecipitated impurities.

Gravimetric Factor and Percentage Problems

20. For each of the following, indicate (but do not calculate) the gravimetric factor.

	Weighed	Sought	Factor
(a)	$AgBr$	Br	
(b)	$AgBr$	C_6H_5Br	
(c)	$BaSO_4$	S	
(d)	$BaSO_4$	K_2SO_4	
(e)	$BaSO_4$	FeS_2	
(f)	$Mg_2P_2O_7$	MgO	
(g)	$Mg_2P_2O_7$	P	
(h)	CO_2	$CaMg(CO_3)_2$	
(i)	CeF_3	F	

21. Calculate the weight of CaO that should be obtained from 100 mg of pure tooth enamel, $Ca_5(PO_4)_3(OH)$, after dissolving the enamel sample, removing the phosphate, precipitating calcium as CaC_2O_4, and igniting the precipitate to CaO for weighing.
22. Calculate the percentage of uranium in U_3O_8.
23. Calculate the weight of silver chloride that will be produced by precipitating the chloride from 1 g of pure potassium chloride.
24. Compute the volume of solution containing 40 mg/mL of silver nitrate needed to completely precipitate the chloride from 1 g of pure potassium chloride.
25. Calculate the volume of solution containing 60 mg/mL of barium chloride dihydrate needed to completely precipitate the sulfate from 1 g of pure potassium sulfate.
26. What volume of a solution containing 18.0 g/L of oxalic acid ($H_2C_2O_4$) is needed to precipitate the lanthanum as lanthanum oxalate, $La_2(C_2O_4)_3$, from 1.000 g sample containing 20% lanthanum?
27. Calculate the weight of silica (SiO_2) precipitate that would be obtained from a 1 g sample of pure $NaAl(SiO_3)_2$.
28. A 1.05-g steel sample was fused in an oxygen atmosphere. The carbon in the steel was converted to carbon dioxide and swept by a stream of oxygen into a weighed absorption tube. The carbon dioxide absorbed caused a gain of 0.040 g in the weight of the absorption tube. Calculate the percentage of carbon in the steel.

29. Phosphate can be determined gravimetrically by precipitation as ammonium phosphomolybdate and weighing as $(NH_4)_3PMo_{12}O_{40}$. If 10.0 mg of precipitate is the smallest that can be collected and weighed to within $\pm 1.0\%$, calculate the smallest weight of phosphate (PO_4) that can be determined by this procedure.

30. Suppose that hydrous ferric oxide coprecipitates 1% of its ignited weight of calcium. If a limestone sample is 30% calcium, what is the maximum percentage of Fe_2O_3 that can be present in the sample and cause no more than 1/1000 of the calcium in solution to be coprecipitated?

Environmental and Health Analysis Problems

31. The FDA tolerance for mercury in foods is 0.5 ppm (0.5 mg/kg). A 113.0-g sample of fish taken from Lake Erie is found by indirect analysis to contain 0.11 mg of mercury. (a) Calculate the percentage of mercury in the fish. (b) Show by calculation whether the mercury content exceeds FDA standards.

32. The range of the toxic metal cadmium in a typical 1.12-g cigarette has been found to be 1.14–1.90 micrograms. (Up to 50% of the cadmium is exhausted in smoke not inhaled.) (a) Calculate the *percentage* range of cadmium in a typical cigarette. (b) A particular cigarette weighs 1.08 g and contains 0.00015% Cd; assuming that only three-fourths of the cigarette is smoked, calculate the minimum amount of cadmium that enters the lungs from it.

33. Mercury ingested by fish exists in lakes primarily as soluble CH_3Hg^+ or insoluble $(CH_3)_2Hg$. A certain sample of pike contained 3.5 ppm of mercury. Use the appropriate gravimetric factor to calculate the ppm of (a) CH_3Hg^+ in the sample and (b) $(CH_3)_2$ Hg in the sample.

34. The phosphate in a 1.000-g fertilizer sample was precipitated as magnesium ammonium phosphate, $MgNH_4PO_4$. Ignition at 900°C converted the precipitate to magnesium pyrophosphate, $Mg_2P_2O_7$, which weighed 0.2550 g. Calculate the percentage of phosphorus in the fertilizer.

35. An organic insecticide was decomposed by an oxygen combustion procedure to convert the chlorine in the molecule to water-soluble chloride. The chloride was then precipitated as silver chloride. A 0.7715-g precipitate of silver chloride was obtained from a 0.500-g sample of the insecticide. Calculate the percentage of chlorine in the insecticide.

Moon Rock and Earth Rock Analyses

36. The calcium from a 0.6000-g limestone sample was precipitated as calcium oxalate and ignited to calcium carbonate for weighing. The ignited precipitate weighed 0.2820 g. Calculate the percentage of calcium in the limestone.

37. In the determination of silica (weighed as SiO_2 after precipitation and dehydration), metal ion impurities are often coprecipitated. In a "corrected" silica determination, the impure precipitate is weighed, the silica is volatilized by reaction with HF to form volatile SiF_4 or H_2SiF_6, and the impurities remaining are weighed. From the floowing data, calculate the correct percentage of silica in a rock sample.

$$\text{Wt sample} = 1.000 \text{ g}$$

$$\text{Silica} + \text{Impurities} = 0.1262 \text{ g}$$

$$\text{Pcpt. after HF treatment} = 0.0012 \text{ g}$$

38. Samples of anorthositic rock from the moon contain about 46% SiO_2 and 29% Al_2O_3. (a) Calculate the percentages of Si and Al in this category of lunar rock. (b) A specific

mineral with the formula $CaAl_2Si_2O_8$ has been found to occur in the anorthositic rock. If this mineral were the major constituent in the rock, could its composition account for the high aluminum and silicon content of the anorthosite?

39. An unknown mineral containing calcium, iron, and the metasilicate ion, SiO_3^{2-}, was discovered in the Apollo 11 lunar samples. It contained 16.2% Ca, 22.5% Fe, 22.6% Si, and 38.7% O. Calculate the empirical formula of this mineral.

40. Lunar samples have contained from 6% to 11% titanium, much more than most terrestrial rocks. The mineral ilmenite, $FeTiO_3$, has been found in some lunar samples. If it were the only titanium mineral present, what percentage of it would have to be present to account for the 11% titanium?

Calculation of Empirical Formulas and Molecular Formulas

41. A 0.3999-g sample of reagent grade $Al_2(SO_4)_3 \cdot xH_2O$ was dissolved and analyzed for aluminum by the aluminum oxinate method in Experiment 4. If the precipitate of aluminum oxinate, $Al(C_9H_6NO)_3$, weighed 0.4185 g, calculate the value of x, the number of water molecules in the reagent.

42. An organic compound has the formula $C_6H_{6-x}OCl_x$. A sample of the pure compound weighing 0.1500 g was decomposed and the chloride precipitated and weighed as silver chloride. The precipitate weighed 0.4040 g. Calculate the value of x and give the correct formula for the compound.

43. A sample of the pure compound $(CH_3)_4NBr_x$ weighing 0.0962 g was dissolved and treated with a reducing agent to ensure that all bromine was present as Br^-. A precipitate of silver bromide weighing 0.1730 g was obtained. Calculate the value of x in the formula of the compound.

44. An organic compound containing only carbon and oxygen is found by analysis to contain 50.0% C and 50.0% O. Its molecular weight is 289 ± 2. Calculate its empirical formula.

45. Anhydrous potassium hydrogen sulfate, $KHSO_4$, loses water when heated. If a 135.9-g sample of $KHSO_4$ loses 9.0008 g of water, calculate the formula of the chemical compound remaining.

Special Problems

46. Write equations for the formation of the corner of a curd of the digested silver halide from:

   (a) Adding excess $AgNO_3$ to NaBr   (b) Adding excess $AgClO_4$ to NaCl

47. Write equations for the formation of the corner of a curd of the digested silver bromide from:

   (a) Adding excess NaBr to $AgNO_3$ (b) Adding excess KBr to $AgClO_4$

48. State whether each solution can be used to wash insoluble silver chloride after precipitation of chloride with excess silver nitrate and also after precipitation of silver ion with excess potassium chloride.
   (a) Dilute HNO_3 (c) Dilute HCl
   (b) Dilute NH_4NO_3 (d) Dilute H_2SO_4

49. Which of the following solutions would not be a good wash solution for washing silver chloride precipitated with excess silver nitrate?

   (a) Dilute HNO_3 (c) Dilute HBr
   (b) Dilute acetic acid   (d) Dilute H_2SO_4

5

Titrimetric (Volumetric) Methods of Analysis

Titrimetric methods (methods that involve a titration) constitute some of the most important procedures used in quantitative analysis. The purpose of this chapter is to indicate the elementary principles and scope of titration and, especially to acquaint the student with the calculations involved. Detailed theoretical treatments and actual methods of titration are given in later chapters.

5–1. GENERAL PRINCIPLES

Principle of Titration. Titration is a quick, accurate, and widely used way of measuring the amount of a substance in solution. A titration is performed by adding exactly the volume of a *standard* solution (a solution of exactly known concentration) needed to react with an unknown quantity of a second substance. The standard solution is called the *titrant*; the volume of titrant needed for the titration is carefully measured by means of a *buret*. If the volume and concentration of the titrant are known, the unknown quantity of the substance titrated can be calculated.

A titration is based on a chemical reaction that may be represented as

$$a\text{A} + b\text{B} \rightarrow \text{products}$$

where A is the titrant, B is the substance titrated, and a and b are the numbers of moles of each.

In a titration, there are a few chief requirements, as follows.

1. The reaction should be *stoichiometric*; that is, there should be a definite whole-number ratio between a and b in the reaction.
2. The rate of the chemical reaction should be rapid so that the titration can be carried out quickly.

3. The reaction should be quantitative. For the usual analytical accuracy, it must be at least 99.9% complete when a stoichiometric amount of titrant has been added.

4. Some method must be available for determining the point in the titration at which a stoichiometric amount of titrant has been added and the reaction is complete. When this point is determined experimentally by an indicator color change or by some change in the electrochemical or physical properties of the solution, it is called the *end point* of the titration. The point at which the theoretical amount of titrant has been added is called the *equivalence point* of the titration. Ideally, the end point and the equivalence point coincide, but for various reasons (such as the indicator changing color a little too early or too late), there is frequently some difference between the two.

Standard Solutions.   One way to prepare a standard solution is to accurately weigh a portion of some highly pure chemical (a *primary standard*) and then to dilute the solution carefully to a known volume in a volumetric flask. Another way is to prepare a stock solution of approximately known concentration and then to *standardize* it by titrating a known amount of some primary standard with this solution. For example, a stock solution of purified sodium hydroxide is prepared, and this is standardized by titrating accurately weighed portions of a primary standard acid, such as potassium acid phthalate.

A chemical must fulfill several requirements in order to serve as a satisfactory primary standard.

1. The material should be of known composition and highly pure; preferably 100% pure, although a slightly lower purity is acceptable provided this is accurately known.

2. It should undergo a rapid and stoichiometric chemical reaction with the solution being standardized. The equilibrium constant for the standardization reaction should be favorable. For example, in acid-base titrimetry, the primary standard should be as strong an acid or base as possible.

3. The material should stay stable indefinitely at room temperature and should withstand drying in an open oven without change. It should not absorb water or carbon dioxide from the atmosphere.

4. It should, if possible, have a high equivalent weight because weighing out a greater weight of material tends to minimize the relative error caused by weighing.

Scope of Titration Methods.   Any of the following types of chemical reaction can serve as the basis of a titration.

Precipitation.   An example is the determination of chloride by titration with a standard solution of a silver salt:

$$Ag^+ + Cl^- \rightarrow AgCl(s)$$
$$\text{(titrant)}$$

The end point in this titration can be detected by any of several visual-indicator methods or by a potentiometric titration with a silver indicator electrode. Precipitate-formation titrations are discussed in Chapter 9.

Acid-Base. Hundreds of compounds, both organic and inorganic, can be determined by a titration based on their acidic or basic properties. Acids are determined by titrating them with a standard solution of a strong base, such as sodium hydroxide:

$$OH^- \quad + \quad HA \quad \rightarrow A^- + H_2O$$

(NaOH titrant)   (the acid titrated)

Bases are titrated with a standard solution of a strong acid, such as hydrochloric or perchloric acid:

$$H^+ \quad + \quad B \quad \rightarrow BH^+$$

(HCl titrant)   (the acid titrated)

Organic acids and bases are frequently titrated in a nonaqueous solvent instead of in aqueous solution.

The end point of an acid-base titration is usually detected by adding a small amount of an indicator, which changes color abruptly when the last bit of the acid or base being titrated is neutralized. Another way is to monitor hydrogen-ion concentration during the titration by means of a pH meter. Acid-base titration methods are discussed in Chapter 8.

Complex Formation. Most metal ions can be accurately determined by titration with a standard solution of an organic complexing agent such as EDTA (ethylenediaminetetraacetic acid). This reagent reacts with most metal cations to form a very stable, water-soluble complex. The reacting ratio of EDTA to metal ion is almost always 1 to 1. A drop or two of an indicator solution is added, which forms a highly colored complex with the metal ion. The color remains as long as some metal ion remains untitrated. When a stoichiometric amount of EDTA titrant has been added, the complex of metal ion and indicator dissociates, causing a color change and marking the end point. The theory and practice of complex-formation titrations are considered in Chapter 11.

Oxidation-Reduction. Several elements that have more than one oxidation state can be determined by titration with a standard oxidizing or reducing agent. Perhaps the most common example is the titration of ferrous iron with permanganate:

$$MnO_4^- + 5Fe^{2+} + 8H^+ \rightarrow 5Fe^{3+} + Mn^{2+} + 4H_2O$$

The end point of this titration is marked by the appearance of a permanent violet color caused by the first excess of highly colored permanganate. The end point in other oxidation-reduction titrations is determined by means of visual indicators or by a potentiometric titration. Oxidation-reduction titrations are taken up in Chapter 13. The conditions for some typical titrations are summarized in Table 5–1.

Methods of Performing Titrations. The classical method of doing a titration is to carefully add titrant from a buret until the end point is indicated by an indicator's changing color or by some change in an instrumental reading. The titrant can be

Table 5–1.   Conditions for some Typical Titrations

Type of Titration	Titrant	Primary Standard	Substance Determined	Indicator
Precipitate Formation	$AgNO_3$	KCl	Cl^- or Br^-	Dichlorofluorescein
Acid-Base	NaOH	KHP[a]	Acids	Phenolphthalein
Acid-Base	HCl	APyr.[b]	Bases	Methyl Orange
Complex Formation	EDTA	Zn	Ca^{2+}, Mg^{2+}, etc.	Eriochrome Black T
Oxidation-Reduction	$KMnO_4$	$Na_2C_2O_4$	Fe^{2+}, etc.	MnO_4^- Color
Oxidation-Reduction	$KMnO_4$	As_2O_3	Fe^{2+}, etc.	Ferroin

a. Potassium acid phthalate.
b. 4-Aminopyridine.

added rapidly at first, but later it must be added in small increments so that the end point can be accurately located. Ordinarily, a 25- or 50-mL buret is used, although a smaller 5- or 10-mL buret can be used for titration of a few milligrams of a substance.

Manual titrations can be done quickly, but they require manipulation by the operator. Burets are available that deliver titrant by depressing a piston that is attached to a motor-driven worm gear. The volume of titrant is obtained from a digital readout of the piston position at the start and completion of the titration.

In laboratories where a large number of samples must be titrated every day, a completely automated titration apparatus is often used. One such device will titrate fifteen samples in succession completely automatically. The sample solutions to be titrated are placed in beakers inside cups on a turntable. A visual indicator is added to each, and then the instrument takes over. Titrant is added from a buret by means of a motor-driven plunger that displaces the liquid titrant. The plunger is driven by a screw, so the volume of liquid displaced from the buret can be calculated by mechanically counting the number of revolutions of the screw. The end point of the titration is marked by an indicator color change that is observed electrically by the change in output current of a simple spectrophotometer. Temporary indicator color changes just before the end point may cause the buret delivery to be temporarily interrupted, so titrant increments are added more slowly near the end point, just as would be done in a manual titration. When a permanent color change is attained, the volume of titrant is printed out digitally on a paper tape. Then the light pipe and stirrer assembly are lifted out and rinsed automatically, the turntable moves the next sample in line to be titrated, and the whole process is repeated.

In a given situation, industry wants the most rapid and automatic method of analysis available. Continuous analysis is the ideal method for monitoring the chemical composition of plant streams. One technique for continuous analysis is based on a variation of the titration principle. A small portion of the chemical plant stream is bled off continuously at a constant flow rate. This is mixed with a titrant stream so that a quantitative chemical reaction will take place between the titrant and the substance being measured in the sample stream (Figure 5–1). The *flow rate* of the titrant stream is controlled continuously so that just the right amount of

Figure 5–1. Schematic diagram of a continuous analyzer. The rate of adding the titration reagent is controlled by the detector, which acts as a null-point device. The concentration of titratable substance in the sample is proportional to the speed of the reagent pump.

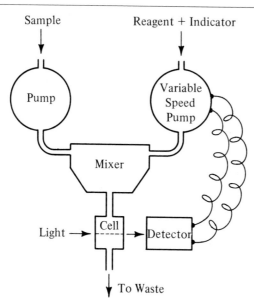

titrant enters the mixing chamber to react with the substance being measured. The titrant flow rate is adjusted by means of a precision, variable-speed pump or by varying the height of pressure head on the titrant stream. A sensing device such as a photometer or a pair of electrodes is used to indicate when the flow rates are balanced so that the mixing of the two streams will produce a stoichiometric reaction. When there is an imbalance, the rate of the titrant stream is either increased or decreased until balance is again attained. A recorded plot of the titrant flow rate against time gives a record of the substance's concentration in the sample stream.

An extensive compilation of automated and continuous methods of analysis is given by Blaedel and Laessig [1].

5–2. CALCULATIONS WITH MOLARITY

Definitions and Relationships. To perform calculations in titrimetric analysis, the student must become adept at manipulating expressions involving moles and molar

[1] W. J. Blaedel and R. H. Laessig, "Automation of the analytical process through continuous analysis" in *Advances in Analytical Chemistry and Instrumentation*, Vol. 5, New York: Interscience, 1966.

solutions. In chemical analysis, it is frequently convenient to use the term *millimole* (abbreviation mmole) instead of mole. Definitions of these fundamental terms follow.

Mole. A mole is the formula weight of a substance expressed in *grams*:

$$\text{moles} = \frac{g}{\text{form wt}}$$

Millimole. A millimole is the formula weight of a substance expressed in *milligrams*:

$$\text{mmoles} = \frac{mg}{\text{form wt}}$$

Molar concentration. The molar concentration (or molarity) M of a solution is the number of moles of solute present in one liter of solution:

$$M = \frac{\text{moles}}{\text{liters}}$$

Molar concentration. The molar concentration (or molarity) M of a solution is the number of millimoles of solute present in one milliliter of solution:

$$M = \frac{\text{mmoles}}{\text{mL}}$$

Several expressions (simply rearrangements or combinations of the preceding) are useful in volumetric calculations:

$$(\text{liters})(M) = \text{moles}$$
$$(\text{moles})(\text{form wt}) = g$$
$$(\text{liters})(M)(\text{form wt}) = g$$

$$(\text{mL})(M) = \text{mmoles}$$
$$(\text{mmoles})(\text{form wt}) = mg$$
$$(\text{mL})(M)(\text{form wt}) = mg$$

The expressions in the right-hand column (mmoles, milliliters, milligrams) are preferred to those in the left-hand column (moles, liters, grams) because most analytical titrations are concerned with rather small quantities of solutions and chemicals. It is easier to say "thirty-six milliliters" and write "36 mL" than it is to say "thirty-six thousandths of a liter" and write "0.036 L." There also is less chance of error in writing 2.1 mmoles than there is in writing 0.0021 mole.

Preparation of a Titrant. Titrants are usually prepared by dissolving a weighed amount of a primary standard in a known volume or by dissolving an impure chemical and standardizing it later. Sometimes, however, a titrant is prepared by dilution of a more concentrated reagent solution. The *amount* of reagent (in millimoles) will not be changed by dilution, but the *concentration* (molarity) will, of course, change. The new concentration, or the desired volume, can be calculated from the simple relationship:

$$(mL_1)(M_1) = (mL_2)(M_2) \qquad (5-1)$$

where 1 refers to the initial solution and 2 refers to the diluted solution.

Example: What volume of concentrated ($12.0M$) hydrochloric acid should be used to prepare 500 mL of $0.100M$ HCl solution?

Let the $12.0M$ HCl be designated as solution 1 and the $0.100M$ HCl as solution 2. Then substituting into Equation 5–1,

$$(mL_1)(12.0M) = (500 \text{ mL})(0.100M)$$

$$mL_1 = 4.17 \text{ mL (conc. HCl)}$$

Calculating the Results of a Titration.   To calculate the results of a titration, it is necessary to know the volume and molarity of titrant used and also the *combining ratio* of substance titrated to titrant. The combining ratio can be easily obtained from the balanced chemical equation, as follows:

$$aA \quad + \quad bB \quad \longrightarrow \text{ products} \qquad (5-2)$$
$$\text{(titrant)} \quad \text{(substance titrated)}$$

The mmoles of titrant A is easily calculated simply by multiplying the number of milliliters of A used in the titration by the molarity of A.

$$(mL_A)(M_A) = \text{mmoles}_A \qquad (5-3)$$

The mmoles of substance titrated B is obtained by multiplying the mmoles of A by the combining ratio b/a.

$$(\text{mmoles}_A)\left(\frac{b}{a}\right) = \text{mmoles}_B \qquad (5-4)$$

The validity of Equation 5–4 is made clearer by some examples.

Example: A given sample containing sodium carbonates requires 26.20 mL of $0.1000M$ hydrochloric acid for titration according to the reaction, $2HCl + Na_2CO_3 \rightarrow$ products. Calculate the weight in milligrams of sodium carbonate.

$$26.20 \times 0.1000 = 2.620 \text{ mmoles HCl}$$

According to the reaction, 2 mmoles of HCl react with 1 mmole of Na_2CO_3. Therefore, the amount of Na_2CO_3 is one-half the millimoles of HCl:

$$(\tfrac{1}{2})(2.620) = 1.320 \text{ mmoles Na}_2CO_3$$

This same calculation according to Equations 5–3 and 5–4 is

$$(26.20)(0.1000)(\tfrac{1}{2}) = 1.320 \text{ mmoles Na}_2CO_3$$

Example: Dichromate is used to titrate iron(II) according to the question $Cr_2O_7^{2-} + 6Fe^{2+} \rightarrow$ products. If 0.743 mmole of dichromate is used in the titration, calculate the mmoles of iron(II).

One dichromate combines with *six* iron(II); therefore,

$$0.743 \times 6 = 4.458 \text{ mmoles}_{Fe}$$

We can combine Equations 5–3 and 5–4:

$$(mL_A)(M_A)\left(\frac{b}{a}\right) = \text{mmoles}_B \qquad (5\text{--}5)$$

Usually, we want the *weight* of substance titrated. Weight (in mg) is obtained by multiplying mmoles by formula weight (form wt). Multiplying both sides of Equation 5–5 by the formula weight of **B**:

$$\boxed{(mL_A)(M_A)\left(\frac{b}{a}\right)(\text{form wt}_B) = mg_B} \qquad (5\text{--}6)$$

Equation 5–6 can be used to calculate the results of any single titration. If percentage of **B** is desired, the weight of **B** is divided by the weight of the sample and multiplied by 100:

$$\frac{mg_B}{mg_S} \times 100 = \%B \qquad (5\text{--}7)$$

Note that the weight of sample must be in the same units as **B**. Finally, Equations 5–6 and 5–7 may be combined to give a general equation for obtaining percentage **B**.

$$\boxed{\dfrac{(mL_A)(M_A)\left(\dfrac{b}{a}\right)(\text{form wt}_B)(100)}{mg_S} = \%B} \qquad (5\text{--}8)$$

Several examples will now be given to illustrate the use of Equations 5–6 and 5–8 in calculating the results of titrations.

Example: What weight of acetic acid (CH_3COOH) is in a 5.00-mL sample of vinegar that requires 35.00 mL of 0.1000M sodium hydroxide for titration?

$$A + B \rightarrow \text{products}$$

$$\underset{\text{(NaOH)}}{OH^-} + CH_3COOH \rightarrow CH_3COO^- + H_2O$$

The combining ratio b/a is 1:

$$(mL_{OH^-})(M_{OH^-})(\text{form wt}_{CH_3COOH}) = mg_{CH_3COOH}$$

$$(35.00)(0.1000)(60.03) = 210.1 \text{ mg of acetic acid}$$

Example: Calculate the percentage of sodium carbonate in a sample of known weight that has been titrated with standard hydrochloric acid to a methyl orange end point.

$$2H^+ + CO_3^{2-} \rightarrow CO_2(g) + H_2O$$
$$(2HCl) \quad (Na_2CO_3)$$

The combining ratio is $\frac{1}{2}$:

$$\frac{(mL_{HCl})(M_{HCl})(\frac{1}{2})(\text{form wt}_{Na_2CO_3})(100)}{mg_{sample}} = \%Na_2CO_3$$

Example: Calculate the percentage of fluoride in a 92.5-mg sample that requires 19.80 mL of 0.0500M calcium perchlorate for titration.

$$Ca^{2+} + 2F^- \rightarrow CaF_2(s)$$

$b/a = 2$; formula weight of $F^- = 19.00$:

$$\frac{(mL_{Ca})(M_{Ca})(2)(\text{form wt}_F)(100)}{mg_{sample}} = \%F^-$$

$$\frac{(19.80)(0.0500)(2)(19.00)(100)}{92.5} = 40.7\%F^-$$

Calculating the Molarity of a Solution from a Standardizing Titration. If an accurately weighed sample of a primary standard B is dissolved and titrated with A, the molarity of A is calculated as follows:

$$aA + bB \rightarrow \text{products}$$
$$(mL_A)(M_A)(b/a)(\text{form wt}_B) = mg_B$$

$$M_A = \frac{mg_B}{(mL_A)(b/a)(\text{form wt}_B)}$$

Frequently, $b/a = 1$, and this factor becomes trivial.

Example: Exactly 410.4 mg of primary standard potassium acid phthalate (KHP), formula weight 204.2, is weighed out and dissolved in water. If titrating the KHP requires 36.70 mL of sodium hydroxide solution, what is the molarity of the sodium hydroxide?

$$OH^- + HP^- \rightarrow P^{2-} + H_2O$$
$$(NaOH) \quad (KHP)$$

$$(36.70)(M)(204.2) = 410.4$$

$$M = \frac{410.4}{(36.70)(204.2)} = 0.0548$$

Back-Titrations. Sometimes the rate of a chemical reaction is too low for a titration to be carried out directly by adding titrant A from a buret to combine exactly with the amount of B in solution. Instead, a measured amount of A is added that is in *excess* of that theoretically needed to react with B. If necessary, the solution is heated to drive the slow reaction to completion. Then the excess A is *back-titrated* with a reagent C that will react rapidly with A. This scheme may be summarized:

$$aA + bB \rightarrow \text{products} + \text{excess A} \qquad (5\text{--}9)$$

(titrant) (substance titrated)

$$cC + dA \rightarrow \text{products} \qquad (5\text{--}10)$$

(backtitrant) (excess)

The calculation method is to obtain the *net* number of millimeters of A used to combine with B, and then to calculate the results as was done for a direct titration.

$$A_{added} - A_{excess} = A_{net} \qquad (5\text{--}11)$$

$$[(mL_A)(M_A)] - \left[(mL_C)(M_C)\left(\frac{d}{c}\right)\right] = mmoles_A \qquad (5\text{--}12)$$

$$(mmoles_A)\left(\frac{b}{a}\right)(\text{form wt}_B) = mg_B \qquad (5\text{--}13)$$

Example: Zirconium(IV) reacts rather slowly with EDTA and so must be determined by using a back-titration. Exactly 10.00 mL of 0.0502M EDTA is added to a solution containing zirconium(IV). Back-titration of the excess EDTA requires 2.08 mL of 0.0540M bismuth nitrate. Calculate the weight of zirconium in the solution.

$$EDTA + Zr(IV) \rightarrow Zr\text{-EDTA} + \text{excess EDTA}$$

$$Bi^{3+} + EDTA_{excess} \rightarrow Bi\text{-EDTA}$$

In both reactions, the combining ratio is 1:1. Substituting into Equations 5–12 and 5–13:

$$[(10.00)(0.0502)] - [(2.08)(0.0504)](91.22) = 36.0 \ mg_{Zr}$$

Example: Manganese dioxide can be determined by reduction with excess ferrous sulfate. The excess iron(II) is then back-titrated with potassium dichromate

$$2Fe^{2+} + MnO_2(s) + 4H^+ \rightarrow 2Fe^{3+} + Mn^{2+} + 2H_2O + \text{excess } Fe^{2+}$$

$$Cr_2O_7^{2-} + 6Fe^{2+} + 14H^+ \rightarrow 2Cr^{3+} + 6Fe^{3+} + 7H_2O$$

Sample = 200.0 mg. Iron(II) added = 50.00 mL of 0.1000M. Dichromate for back-titration = 16.07 mL of 0.0230M. Calculate % MnO_2. (Form wt of MnO_2 = 86.94.)
Substituting into Equations 5–12 and 5–13:

$$[(50.00 \times 0.1000) - (16.07 \times 0.023 \times 6)] \times \tfrac{1}{2} \times 86.94 = 120.9 \ mg \ MnO_2$$

$$\frac{120.9}{200} \times 100 = 60.47\% \ MnO_2$$

5–3. CALCULATIONS WITH NORMALITY

Definitions and Relationships. If the reacting ratio of reactants A and B is not unity, calculations are sometimes made easier if the concentrations of solutions are expressed in terms of *normality* instead of molarity and the *equivalent weight* of the substance titrated is used instead of the formula weight. In the "normal system," the concentration of solutions is adjusted, or normalized, to affect differences in reacting ratios in chemical reactions. A $1N$ (1 normal) solution contains one *equivalent* per liter, or one *milliequivalent* per milliliter:

$$N = \frac{eq}{L}, \qquad N = \frac{meq}{mL}$$

One equivalent of titrant will react with one equivalent of substance titrated, and one milliequivalent of titrant will react with one milliequivalent of substance titrated.

The definition of an equivalent in terms of molarity depends on the reaction a substance undergoes. Keep in mind that 1 mole = 1 Avogadro number (6.023×10^{23}) of atoms or molecules. In acid-base reactions, one equivalent is the number of grams of a substance that supplies, or combines with, one Avogadro number of hydrogen ions. In oxidation-reduction, one equivalent is the number of grams of a substance that supplies, or combines with, one Avogadro number of electrons. In precipitation and in complex-formation reactions, one equivalent is the number of grams of a substance that supplies one Avogadro number of $1+$ ions (or one-half Avogadro number of $2+$ ions, etc.) or the amount that reacts with one Avogadro number of $1+$ ions:

1 eq of HCl = 1 mole, or 36.5 g of HCl

1 eq of $H_2SO_4 = \frac{1}{2}$ mole, or 49 g, of H_2SO_4

1 eq of Fe^{2+} = 1 mole (assuming the redox reaction $Fe^{2+} \rightarrow Fe^{3+} + e^-$)

1 eq of $KMnO_4 = \frac{1}{5}$ mole (assuming the reaction $KMnO_4 + 5e^- + \cdots \rightarrow Mn^{2+} + \cdots$, etc.)

From these examples, it is seen that the equivalent weight of a substance is related to its formula weight:

$$eq \; wt \; HCl \; (acid\text{-}base) = form \; wt$$

$$eq \; wt \; H_2SO_4 \; (acid\text{-}base) = \tfrac{1}{2} \, form \; wt$$

$$eq \; wt \; Fe^{2+} \; (redox) = form \; wt$$

$$eq \; wt \; KMnO_4 \; (redox) = \tfrac{1}{5} \, form \; wt$$

The general rules for computing an equivalent weight follow.

Acid-Base Titrations. Equivalent weight is the formula weight divided by the number of reactive hydrogen atoms in an acid. In a base, it is the number of hydrogen atoms required to neutralize each molecule of base:

$$\text{eq wt} = \frac{\text{form wt}}{\text{number of } H^+}$$

Precipitate-Formation and Complex-Formation Titrations. The equivalent weight of a metal ion is the formula weight of the metal ion divided by the charge on the ion:

$$\text{eq wt} = \frac{\text{form wt}}{\text{ion charge}}$$

The equivalent weight of an anion is the formula weight of the anion divided by the number of metal-ion equivalents it reacts with:

$$Ba^{2+} + SO_4^{2-} \rightarrow BaSO_4 \begin{cases} \text{eq wt } Ba^{2+} = \text{form wt} \div 2 \\ \text{eq wt } SO_4^{2-} = \text{form wt} \div 2 \end{cases}$$

$$3Ag^+ + PO_4^{3-} \rightarrow Ag_3PO_4 \begin{cases} \text{eq wt } Ag^+ = \text{form wt} \\ \text{eq wt } PO_4^{3-} = \text{form wt} \div 3 \end{cases}$$

$$Ag^+ + 2CN^- \rightarrow Ag(CN)_2^- \begin{cases} \text{eq wt } Ag^+ = \text{form wt} \\ \text{eq wt } CN^- = \text{form wt} \div \frac{1}{2} = \text{form wt} \times 2 \end{cases}$$

Oxidation-Reduction Titrations. Here the equivalent weight of any substance is the formula weight of substance weighed out (or to be calculated) divided by the number of electrons gained or lost in the reaction:

$$MnO_4^- + 5Fe^{2+} + 8H^+ \rightarrow 5Fe^{3+} + Mn^{2+} + 4H_2O$$

$$\underset{(7+)}{MnO_4^-} \rightarrow \underset{(2+)}{Mn^{2+}} \qquad \text{eq wt } MnO_4^- = \text{form wt} \div 5$$

$$Fe^{2+} \rightarrow Fe^{3+} \qquad \text{eq wt Fe} = \text{form wt}$$

$$Cr_2O_7^{2-} + 3Sn^{2+} + 14H^+ \rightarrow 3Sn^{4+} + 2Cr^3 + 7H_2O$$

$$\underset{(12+)}{Cr_2O_7^{2-}} \rightarrow \underset{(6+)}{2Cr^{3+}} \qquad \text{eq wt } Cr_2O_7^{2-} = \text{form wt} \div 6$$

$$Sn^{2+} \rightarrow Sn^{4+} \qquad \text{eq wt Sn} = \text{form wt} \div 2$$

Once equivalent weight is understood, the other relationships in the normal system can be stated:

$$\text{eq} = \frac{g}{\text{eq wt}}$$

$$N = \frac{\text{eq}}{L}$$

In quantitative analysis, the units milliequivalents and milliliters are commonly used in place of equivalents and liters to avoid the frequent use of small numbers with one or more zeros after the decimal point. For reference, the corresponding relationships in the normal and molar systems are given side by side:

$$meq = \frac{mg}{eq\ wt}$$

$$N = \frac{meq}{mL}$$

$$mmole = \frac{mg}{form\ wt}$$

$$M = \frac{mmoles}{mL}$$

Calculating the Results of a Titration.   Calculations with normality are similar to those with molarity. The difference is that at the stoichiometric point in a titration, the milliequivalents of titrant A always equal the milliequivalents of titrated substance B. The combining ratio of B to A used in the molar system has been taken into account in preparing normal solutions. It follows that a given volume of a normal solution will react with an equal volume of another solution having the same normality:

$$mL_A N_A = mL_B N_B$$

Example: A hydrochloric acid solution is standardized by titration with standard sodium hydroxide. If 25.00 mL of HCl requires 32.20 mL of $0.0950N$ NaOH for titration, what is the normality of the HCl solution?

$$(32.20)(0.0950) = (25.00)(N_{HCl})$$

$$N_{HCl} = 0.1224$$

Example: What value of $12.1N$ perchloric acid ($HClO_4$) must be diluted to 1.000 L to give a $0.100N$ solution?

$$(mL_{HClO_4})(12.1) = (1000)(0.100)$$

$$mL_{HClO_4} = 8.26$$

In calculating the results of a titration of a substance B with titrant A, one of the following equations is used:

$$(mL_A)(N_A)(eq\ wt_B) = mg_B$$

$$\frac{(mL_A)(N_A)(eq\ wt_B)(100)}{mg_{sample}} = \%B$$

Example: A 0.2000-g sample of a metal alloy is dissolved, and the tin is reduced to tin(II). Titration of the tin(II) requires 22.20 mL of $0.1000N$ $K_2Cr_2O_7$. Calculate the percentage of tin in the alloy.

To solve this problem, we must know that in the titration, tin is oxidized from tin(II) to tin(IV). Thus tin undergoes a two-electron change, and the equivalent weight of tin is its atomic weight divided by 2, or 59.35. Then

$$\frac{(mL_A)(N_A)(eq\ wt_B)(100)}{mg_{sample}} = \%B$$

$$\frac{(22.20)(0.1000)(59.35)(100)}{200.0} = 65.9\%\ tin$$

Example: A 1.0000-g sample requires 28.16 mL of 0.1000N sodium hydroxide for titration to a phenolphthalein-indicator end point. Calculate the percentage of phosphoric acid (H_3PO_4) in the sample.

This problem illustrates an important point—that the balanced chemical reaction should always be considered in calculating the results of a titration. In this instance, the formula H_3PO_4 might imply that the equivalent weight of phosphoric acid is the formula weight divided by 3. However, the third hydrogen is too weakly acidic to titrate. By using phenolphthalein indicator, two hydrogens are titrated, and

$$eq\ wt_{H_3PO_4} = \frac{form\ wt}{2} = 49.00$$

With methyl orange indicator, only *one* hydrogen of phosphoric acid is titrated with sodium hydroxide, and the equivalent weight is the formula weight of phosphoric acid.

$$\frac{(mL_A)(N_A)(form\ wt_B)(100)}{mg_{sample}} = \%B$$

$$\frac{(28.16)(0.1000)(49.00)(100)}{1000} = 13.80\%\ H_3PO_4$$

If an accurately weighed primary standard B is dissolved and titrated with A, the normality of A is calculated as follows:

$$(mL_A)(N_A)(eq\ wt_B) = mg_B$$

$$N_A = \frac{mg_B}{(mL_A)(eq\ wt_B)}$$

Example: A 150.0-mg sample of pure sodium carbonate (Na_2CO_3) requires 30.06 mL of hydrochloric acid solution for titration:

$$2H^+ + CO_3^{2-} \rightarrow CO_2(g) + H_2O$$
$$\text{(2HCl)}\quad \text{(Na}_2\text{CO}_3)$$

Calculate the normality of the hydrochloric acid:

$$eq\ wt_{Na_2CO_3} = \frac{form\ wt}{2} = 52.99$$

$$(30.06)(N_{HCl})(52.99) = 150.0$$

$$N_{HCl} = 0.09416$$

QUESTIONS AND PROBLEMS

Titration Concepts

1. Define briefly each of the following terms: (a) titrant, (b) standard solution, (c) end point, (d) equivalence point.
2. What is a primary standard? List the requirements of a satisfactory primary standard.
3. What requirements must a chemical reaction fulfill to be used for a titration? List the four types of reactions that can be used as titration reactions.
4. The hardness of water $(Ca^{2+} + Mg^{2+})$ can be determined by a complex-formation titration with EDTA. Outline a technique by which the hardness of water used in an industrial plant can be measured continuously.
5. Explain how a flowing sample can be titrated continuously.
6. For each of the following, calculate the equivalent weight of both reactants in terms of formula weight. (For example, eq wt = form wt/2.)

   (a) $2NaOH + H_2C_2O_4 \rightarrow Na_2C_2O_4 + 2H_2O$
   (b) $2HCl + Ba(OH)_2 \rightarrow BaCl_2 + 2H_2O$
   (c) $I_2 + H_2S \rightarrow S + 2HI$
   (d) $2FeCl_3 + SnCl_2 \rightarrow 2FeCl_2 + SnCl_4$
   (e) $Pb(NO_3)_2 + H_2SO_4 \rightarrow PbSO_4(s) + 2HNO_3$
   (f) $AgNO_3 + 2NaCN \rightarrow Ag(CN)_2^- + {}^2Na^+ + NO_3^-$

Sample Molarity Problems

7. Calculate the molarity of each of the following solutions.

   (a) $AgNO_3$, 117.4 g/L
   (b) KSCN, 0.972 g/100 mL
   (c) $BaCl_2 \cdot 2H_2O$, 200 mg/L
   (d) Na_2SO_4, 72.0 mg/72 mL

8. Calculate the total amount of compound (in milligrams) in each of the following solutions.

   (a) 100 mL of 0.500M NaOH
   (b) 10.0 mL of 0.100M Br_2
   (c) 24.7 mL of 0.100M KSCN
   (d) 5.00 mL of 0.010M $KMnO_4$

9. Calculate the volume of 12.0M hydrochloric acid needed to prepare 1.0 L of approximately 0.25M hydrochloric acid.
10. Calculate the volume of 50% (16M) sodium hydroxide needed to prepare 2.0 L of approximately 0.1M sodium hydroxide.
11. Calculate the molarity of concentrated phosphoric acid (85% by weight), specific gravity 1.69.
12. Calculate the molarity of concentrated hydrobromic acid (48% by weight), specific gravity 1.486.

Standardization

13. A solution of hydrochloric acid is standardized and found to be 1.183M. Calculate the volume of this solution that, diluted to 1.000 L in a volumetric flask, produces a 0.1000M solution of hydrochloric acid.
14. Exactly 46.32 mL of sodium hydroxide is used to titrate a 1,200.0-mg sample of primary standard potassium acid phthalate, or KHP (form wt = 204.2). Calculate the molarity of the sodium hydroxide to four significant figures.

15. Exactly 24.69 mL of hydrochloric acid is required to titrate a 278.0-mg sample of the primary standard *tris*(hydroxymethyl)aminomethane according to the following reaction:

$$HCl + (CH_2OH)_3CNH_2 \rightarrow (CH_2OH)_3CNH_3{}^+Cl^-$$

Calculate the molarity of the HCl.

16. A 8.5332-g portion of primary standard silver nitrate (form wt = 169.9) is weighed into a 500-mL volumetric flask and diluted to volume. Calculate its molarity. In deciding on significant figures, recall that the tolerance for a 500-mL volumetric flask is of the order of 0.1 mL.

17. A 10.00-mL sample of sodium chloride solution is diluted to 50.00 mL. A 20.00-mL aliquot is then withdrawn and titrated with 3.923 mL of 0.0110M silver nitrate. Calculate the molarity of the 10.00 mL of NaCl.

18. A 25.00-mL sample of calcium chloride solution is diluted to 50.00 mL. A 20.00-mL aliquot is then withdrawn and titrated with 3.923 mL of 0.0110M silver nitrate. Calculate the molarity of the 25.00 mL of $CaCl_2$.

19. A 395.6-mg sample of primary standard arsenic(III) oxide is dissolved in 25.00 mL of acid, forming two molecules of H_3AsO_3 for each molecule of As_2O_3 (form wt = 197.84).

    (a) Calculate the molarity of H_3AsO_3.
    (b) Calculate the normality of the H_3AsO_3 (for oxidation to H_3AsO_4).

20. A 25.00-mL sample of 0.0300N (0.0150M) H_3AsO_3 is oxidized to H_3AsO_4 by 24.10 mL of iodine (I_2) titrant.

    (a) Calculate the molarity of the I_2.   (b) Calculate the normality of the I_2.

21. A 93.0-mg sample of primary standard As_2O_3 is dissolved to give two molecules of H_3AsO_3 per one molecule of As_2O_3. The resulting H_3AsO_3 is oxidized to H_3AsO_4 in a reaction in which two ions of Ce^{4+} react with one molecule of H_3AsO_3. If 18.40 mL of Ce^{4+} is used and the formula weight of As_2O_3 is 197.84, calculate:

    (a) The molarity of the Ce^{4+} (b) The normality of the Ce^{4+}

Calculation of Percentage or Concentration

22. A 0.500-g sample containing sodium dihydrogen phosphate is titrated with sodium hydroxide:

$$OH^- + H_2PO_4{}^- \rightarrow HPO_4{}^{2-} + H_2O$$

If 23.06 mL of 0.0985M sodium hydroxide is required for the titration, what is the percentage of NaH_2PO_4 in the sample?

23. Tin(II) is titrated with dichromate according to the following balanced equation:

$$Cr_2O_7{}^{2-} + 3Sn^{2+} + 14H^+ \rightarrow 3Sn^{4+} + 2Cr^{3+} + 7H_2O$$

Calculate the weight of tin(II) in a sample that requires 20.00 mL of 0.1000M $Cr_2O_7{}^{2-}$ for titration. (The atomic weight of tin is 118.7.)

24. Fluoride in a uranium salt is determined by reacting a 1.037-g sample with water vapor at 1000°C, distilling the fluoride as HF, and titrating the fluoride in the distillate with 3.14 mL of 0.1000M thorium nitrate:

$$Th^{4+} + 4F^- \rightarrow ThF_4(s)$$

Calculate the percentage of fluoride in the sample.

25. A 300.0-mg sample of acid contains either impure H_3PO_4 or impure NaH_2PO_4. It is titrated with 21.00 mL of 0.1000M sodium hydroxide to the phenolphthalein end point to give the HPO_4^{2-} ion. Calculate the %H_3PO_4 (form wt = 98.00) and the %NaH_2PO_4 (form wt = 119.98) and decide whether the results are reasonable for either or both.

26. Calculate the percentage purity of a 500.0-mg sample of impure sodium carbonate that requires 22.00 mL of 0.1800M HCl for complete neutralization.

27. A 300.0-mg sample of impure $MgCl_2$ is titrated with 45.00 mL of 0.1000M $AgNO_3$ to $2AgCl + Mg(NO_3)_2$.

    (a) Calculate the percentage chloride (at wt = 35.45) in the sample.

    (b) Calculate the percentage $MgCl_2$ (form wt = 95.23) in the sample.

28. A 500.0-mg sample of chloride requires 15.50 mL of 0.1100M $AgNO_3$ for tiration by an accurate end point. Calculate the percentage chloride (at wt = 35.45).

29. A 1.000-g portion of the same sample in the previous problem requires 31.50 mL of 0.1050M $AgNO_3$ with a questionable end point. Calculate the percentage Cl^- and compare results.

30. An impure 1.0000-g sample of arsenious acid (H_3AsO_3) is oxidized to H_3AsO_4 by titrating with 45.00 mL of 0.0800N (0.0400M) iodine (I_2). Calculate the percentage H_3AsO_3 (form wt = 125.9) and percentage As (at wt = 74.92).

31. A 377.0-mg sample of As_2O_3 is dissolved to give two molecules of H_3AsO_3 per one molecule of As_2O_3 and is oxidized to H_3AsO_4 with 31.48 mL of 0.0502M (0.1040N) iodine (I_2). Calculate the percentage of As_2O_3 (form wt = 197.84) in the sample.

Back-Titration Problems

32. Iron(III) is best determined by addition of excess EDTA, followed by back-titration with a metal ion that reacts rapidly with EDTA. A 700.0-mg sample is dissolved, 20.00 mL of 0.0500M EDTA is added, and the excess EDTA is titrated with 5.08 mL of 0.0420M copper(II). Calculate the percentage of Fe_2O_3 in the sample.

33. The rate of diffusion of a volatile organic acid from container A into container B, which contains 2.00 mL of 0.8040M potassium hydroxide, is being measured. After 2 hours, the acid in A requires 1.53 mL of 0.0100M sodium hydroxide for titration; the excess potassium hydroxide in B requires 1.90 mL of of 0.2000M hydrochloric acid for titration. Calculate (a) the amount of organic acid in each container and (b) the percentage of organic acid that has diffused into B.

34. Aluminum(III) and zinc(II) both react with EDTA to form a 1:1 soluble complex. A 550.0-mg sample is analyzed for aluminum(III) by adding 50.00 mL of 0.0510M EDTA and back-titrating the excess EDTA with 14.40 mL of 0.0480M zinc(II). Calculate the percentage of aluminum in the sample.

35. A 50.00-mL aliquot of 0.1000M calcium nitrate is added to a 1.0000-g sample containing sodium fluoride. After the calcium fluoride precipitate has been filtered and collected, the excess calcium(II) is titrated with EDTA. This titration requires 24.20 mL of 0.0500M EDTA. Calculate the percentage of NaF in the sample.

36. Calculate the percentage iodide (at wt = 126.9) in a 1.0000-g sample to which is added an excess of 50.00 mL of 0.1000M $AgNO_3$. The unreacted $AgNO_3$ is then back-titrated with 16.00 mL of 0.0800M KSCN, producing AgSCN(s).

37. Calculate the percentage of sodium carbonate (form wt = 106.0) in a 510.0-mg impure sample that requires 50.00 mL of 0.1111M excess $AgNO_3$ for the precipitation of Ag_2CO_3. The unreacted $AgNO_3$ is back-titrated with 11.00 mL of 0.1050M KSCN, producing AgSCN(s).

38. Calculate the percentage of potassium arsenate (form wt = 256.2) in a 620.0-mg impure sample that requires 50.00 mL of $0.1111M$ excess $AgNO_3$ for the precipitation of Ag_3AsO_4. The unreacted $AgNO_3$ is back-titrated with 12.00 mL of $0.1010M$ KSCN, producing AgSCN(s).

39. The epoxy group, which is found in organic chemicals used to make resins and other polymers, may be determined analytically by reacting it with HBr to form a nonacidic bromohydrin:

$$-CH-CH- + HBr \rightarrow -CH-CH-$$
$$\underset{O}{\diagdown\diagup} \qquad\qquad \underset{OH\ \ Br}{|\ \ \ |}$$

A 0.4000-g sample of a pure epoxy compound is allowed to react with 20.00 mL of $0.1000M$ HBr. The excess HBr then requires 6.15 mL of $0.1080M$ sodium hydroxide for titration. Calculate the formula weight of the organic compound, assuming that only one epoxy group is present in the molecule.

Miscellaneous Calculations

40. A newly synthesized organic reagent is found to form a complex with calcium(II). To learn more about the nature of the complex, a 50.0-mg sample of the reagent (form wt = 181.2) is titrated with calcium(II) using a calcium-ion electrode to detect the titration end point. This titration requires 11.46 mL of $0.0120M$ calcium chloride. What is the combining ratio of reagent to calcium in the complex?

41. A complex containing bismuth and iodide is decomposed, after which the iodide ion is titrated with silver(I) and the bismuth(III) with EDTA. A 550-mg sample requires 14.50 mL of $0.0500M$ EDTA for titration of the bismuth, and a 440-mg sample requires 23.25 mL of $0.1000M$ silver(I) for titration of the iodide. Calculate the ratio of iodide to bismuth in the original complex.

42. A 10.00-mL aliquot of sulfuric acid solution requires 28.16 mL of $0.1000M$ sodium hydroxide for titration. What volume of $0.1000M$ $BaCl_2$ will be required to titrate a second 10.00-mL aliquot of the H_2SO_4 solution if $BaSO_4$ is precipitated?

43. A 345.0-mg sample of a pure unknown monoprotic acid is dissolved and titrated with 27.40 mL of $0.1000M$ NaOH. Calculate the formula weight of the monoprotic acid.

6

Chemical Equilibrium

Chemical equilibrium plays a major role in chemistry, particularly in the quantitative aspects of chemistry. The analytical chemist needs to know which reactions are quantitative and how various reaction conditions will affect the completeness of a reaction. Rather simple calculations can be made using chemical *equilibrium constants* that will answer these questions.

Another use of equilibrium constants is in calculating the change in the concentration of a particular ion, such as H^+, during a titration. Calculations of this type make it possible to predict the course of the titration and how sharp the end point will be.

In this chapter, the general principles of chemical equilibrium and the concept of equilibrium constants will be discussed. Calculations made with equilibrium constants will be illustrated, and some of the uses of these calculations in quantitative analytical chemistry will be pointed out.

6–1. EQUILIBRIUM AND EQUILIBRIUM CONSTANTS

Suppose we have a chemical reaction of the type

$$A + B + \cdots \rightleftharpoons C + \cdots$$

where the three dots indicate that the equation may be expanded, if necessary, to include more species. Let us assume that the reaction is reversible; that is, the reaction can proceed in the reverse as well as in the forward direction. When A and B are mixed, the reaction that forms the product C proceeds such that the concentrations of A and B become smaller and the concentration of the product C becomes larger. *Equilibrium* is reached when the concentrations of A, B, and C have achieved definite values that do not change with time.

The time required for equilibrium to be attained varies from one reaction to another. The time may be anything from a fraction of a second to many days. Most of the reactions that we will consider reach equilibrium within a few minutes.

Note that equilibrium is a dynamic condition rather than a static one. At equilibrium, the reactions continue in both the forward and the reverse directions. However, the concentrations of A, B, and C in solution remain constant because at equilibrium the forward and reverse reaction rates are equal.

The equilibrium point of a chemical reaction may be shifted by varying the concentrations of reactants or products. If the concentration of A or B is increased or the concentration of C is decreased (by precipitation or volatilization, for example), the equilibrium point is shifted farther to the right. Conversely, an increase in the concentration of C will shift it farther to the left. This is simply a qualitative statement of the *law of mass action.*

The law of mass action is a special case of the principle of le Chatelier, which states that when a stress is applied to a chemical system, the system acts so as to counteract the stress.

A more quantitative treatment of equilibrium is possible through the use of the *equilibrium constant.* The equilibrium constant expresses the concentration of species in solution at equilibrium and is defined as the product of the activities (Chapter 1, Section 1–2) of species on the right-hand side of the equation, divided by the product of the activities of those on the left-hand side. Each species in the equilibrium-constant expression is raised to the power of its coefficient in the chemical equation. For the reaction

$$aA + bB + \cdots \rightleftharpoons cC + \cdots$$

where a is moles of A, etc., the equilibrium-constant expression is

$$K = \frac{{}^aC^c}{{}^aA^a \, {}^aB^b}$$

The equilibrium constant for a given reaction has a specific value at a stated temperature. Unless otherwise indicated, the value given for an equilibrium constant in a table is for an equilibrium at room temperature, 25°C.

Because activities complicate calculations, differences between activity and concentrations are often ignored, and equilibrium constants are written with concentration. Thus, for the reaction

$$aA + bB \rightleftharpoons cC$$

the equilibrium-constant expression may be written

$$K = \frac{[C]^c}{[A]^a[B]^b}$$

where the brackets represent the *molar concentration* of the species enclosed. An equilibrium constant written only in terms of concentration is *not an exact expression*—but as a first approximation, it gives an answer that in most cases is accurate enough to be useful to the analytical chemist. For this reason, concentrations will be used in most of the equilibrium calculations in this book. The validity of replacing activities with concentrations is discussed further in Section 6–7.

Most reactions used in chemical analysis fall into one of the following categories. An example of each major type of reaction is given together with the equilibrium constant for the reaction.

1. Dissociation:

$$HA \rightleftharpoons H^+ + A^- \qquad K = \frac{[H^+][A^-]}{[HA]}$$

$$MA_2(s) \rightleftharpoons M^2 + 2A^- \qquad K = [M^{2+}][A^-]^2$$

2. Formation:

$$M^{2+} + A^{2-} \rightleftharpoons MA \qquad K = \frac{[MA]}{[M^{2+}][A^{2-}]}$$

3. Oxidation-Reduction:

$$M_1(ox) + M_2(red) \rightleftharpoons M_1(red) + M_2(ox) \qquad K = \frac{[M_1(red)][M_2(ox)]}{[M_1(ox)][M_2(red)]}$$

where (ox) indicates an oxidized form and (red) indicates a reduced form.

It is not uncommon for reactions to be more complicated than in the preceding examples; but in such cases, formation and dissociation reactions (except for precipitates) usually occur in *steps*. There is a separate equilibrium constant for each step in the reaction. Dissociation (ionization) of an acid with two acidic hydrogens is an example:

$$H_2A \rightleftharpoons H^+ + HA^- \qquad K_1 = \frac{[H^+][HA^-]}{[H_2A]}$$

$$HA^- \rightleftharpoons H^+ + A^{2-} \qquad K_2 = \frac{[H^+][A^{2-}]}{[HA^-]}$$

As we shall see in Chapter 7, many analytical problems can be solved rather easily by employing stepwise equilibrium constants.

Many equilibrium constants are available in the chemical literature. By using the equilibrium constant, it is possible to calculate the equilibrium concentration of each substance present in a chemical reaction, or of a single substance at various stages during a titration. Such calculations, together with a sound knowledge of chemical equilibrium in general, are helpful in understanding how various conditions affect an analytical procedure. For example, one can calculate the effect of a common ion, changes in acidity, or the formation of a complex on the solubility of a precipitate. The pH at which a buffer functions or the feasibility of performing a given acid-base titration is made clear by simple calculations involving equilibrium constants. The completeness of a color-forming reaction used in spectrophotometry can be calculated. In solvent extraction or chromatography, the distribution of a solute between two phases is governed by the laws of equilibrium, and suitable calculations can be made with the aid of equilibrium constants.

The use of equilibrium-constant calculations will be illustrated in the following sections. It should be kept in mind that the purpose of such calculations is to *estimate* the equilibrium concentrations of various species, sometimes to only one or two significant figures. However, a reasonable approximation of equilibrium concentrations is often all that is required by the analyst.

6–2. CALCULATIONS WITH EQUILIBRIUM CONSTANTS

The types of calculations and the reasoning involved in making the calculations are perhaps best illustrated by numerical examples.

Example: For the reversible chemical reaction

$$A + B \rightleftharpoons C$$

the equilibrium constant, K, is shown to be 13.0. If we mix 0.1 mmole of A with 0.3 mmole of B and dilute to 10.0 mL, calculate how much A, B, and C will be in solution when the reaction has attained equilibrium.

Let x = mmoles of C at equilibrium. Then A = $(0.1 - x)$ mmoles and B = $(0.3 - x)$ mmoles. The *concentration* of each species is the mmoles divided by the volume (10.0 mL). Substituting into the equilibrium constant expression,

$$K = \frac{[C]}{[A][B]} = 13 = \frac{(x/10)}{\left(\dfrac{0.1 - x}{10}\right)\left(\dfrac{0.3 - x}{10}\right)}$$

This gives the quadratic equation $13x^2 - 15.2x + 0.39 = 0$. Solving by the quadratic formula in Appendix 6,

$$x = 1.14 \quad \text{or} \quad 0.0263$$

Only the second root makes sense here; thus, $[A] = 0.0737/10 = 0.00737M$, $[B] = (0.3 - 0.026)10 = 0.0274M$, $[C] = 0.0263/10 = 0.00263M$.

Example: For the reversible reaction

$$A + B \rightleftharpoons C$$

the equilibrium constant is 2.5×10^6. If we mix 1.0 mmole of A with 1.2 mmole of B and dilute to 10.0 mL, calculate the concentration of A remaining when equilibrium has been reached.

Because the equilibrium constant is so large, we can see that *almost all* of the A will react with an equivalent amount of B to form C. In such cases, we should let x = the concentration of the most dilute species, namely, A. Thus,

$$[A] = x, \quad [B] = \frac{1.2 - 1.0}{10} = 0.020, \quad [C] = \frac{1.0}{10} = 0.100.$$

Substituting into the equilibrium constant expression,

$$2.5 \times 10^6 = \frac{(0.100)}{(x)(0.020)}$$

$$x = 2.0 \times 10^{-6} = [A]$$

6–3. IONIZATION OF WEAK ACIDS

Although the subject of acid-base equilibrium is considered in some detail in Chapter 7, we shall now consider briefly how an equilibrium constant can be used to calculate the hydrogen-ion concentration of a slightly ionized acid.

Since strong acids, such as HCl and HNO_3, are completely ionized in aqueous solution, the hydrogen-ion concentration [1] can be calculated easily from the concentration of acid known to be in solution. Weak acids, however, are ionized only to a small extent. The hydrogen-ion concentration can be calculated if the equilibrium constant and the total concentration of acid are known.

The dissociation of a weak acid into ions may be represented as

$$HA \rightleftharpoons H^+ + A^-$$

where HA is the acid, H^+ is the hydrogen from dissociation of the acid, and A^- is the anion from dissociation of the acid. The equilibrium-constant expression for the ionization of this acid is

$$K = \frac{[H^+][A^-]}{[HA]}$$

For the ionization of weak acids or bases, the equilibrium constant is frequently called the *ionization constant*. This constant is written K_a (denoting the ionization of an acid) and K_b (denoting the ionization of a base). Numerical values for many weak acids and bases can be found in handbooks and in textbooks on analytical chemistry. From the ionization constant and the concentration of acid, the hydrogen-ion concentration can be calculated.

Example: Calculate the hydrogen-ion concentration of a 0.10M solution of acetic acid in water; for acetic acid, $K_a = 1.74 \times 10^{-5}$.

$$HA \rightleftharpoons H^+ + A^-$$

$$K_a = \frac{[H^+][A^-]}{[HA]}$$

The concentrations H^+ and A^- are unknown but equal. Thus, the

---

[1] The symbol H^+ and the term "hydrogen ion" are used as a matter of convenience. Actually, hydrogen ions in solution are always solvated. Solvated hydrogen ions are probably $H(H_2O)_4^+$, but they are often written simply as the hydronium ion, H_3O^+.

Figure 6–1. Error caused by assuming $[HA] = C_{HA}$ instead of $[HA] = C_{HA} - [H^+]$, where $[HA]$ is the equilibrium concentration and C_{HA} is the original, or analytical, concentration of HA.

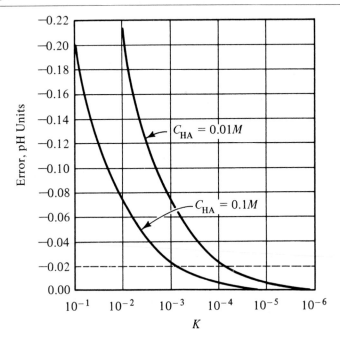

numerator of the K_a expression is $[H^+]^2$. The concentration of HA at equilibrium equals the initial concentration minus the concentration that ionizes:

$$[HA] = 0.10 - [H^+]$$

Substituting into the K_a expression,

$$1.74 \times 10^{-5} = \frac{[H^+]^2}{0.10 - [H^+]}$$

Note that this is a quadratic equation. Rearrangement gives

$$[H^+]^2 + 1.74 \times 10^{-5}[H^+] - 1.74 \times 10^{-6} = 0$$

Solving for $[H^+]$ using the solution to a quadratic equation in Appendix 5,

$$[H^+] = 1.3_1 \times 10^{-3}\,M$$

In this example, the denominator $(0.10 - [H^+])$ turns out to be 0.099, which is very close indeed to the original concentration of HA, $0.10M$. Computation is simplified if we neglect the $[H^+]$ term in the denominator:

$$1.74 \times 10^{-5} = \frac{[H^+]^2}{0.10}$$

$$[H^+] = \sqrt{1.74 \times 10^{-6}} = 1.3_2 \times 10^{-3}\,M$$

The error resulting from this simpler calculation (neglecting the $[H^+]$ term in the denominator) is plotted in Figure 6–1. An error of ± 0.02 pH units is usually considered negligible, since experimental pH measurements are seldom more accurate than this. Thus, the simpler calculation method can be used when the concentration of HA is at least 100 times greater than the ionization constant, K_a. If $[HA]$ is not 100 times greater than K_a, the $[H^+]$ term must be included in the denominator and $[H^+]$ obtained by solving a quadratic equation, as illustrated previously.

6–4. FORMATION OF COMPLEXES

A substance that forms a complex with a metal ion is called a *ligand*. Complex formation often occurs in steps, one ligand being added in each step (see Chapter 11):

$$M + L \rightleftharpoons ML$$

$$ML + L \rightleftharpoons ML_2, \quad \text{etc.}$$

In these equations, the charges on the various species have been omitted for simplicity.

In the case of metal complexes, the equilibrium constant is usually written for the *formation* of the complexes rather than for dissociation. The stepwise equilibrium constants for the preceding reactions may be written

$$K_1 = \frac{[ML]}{[M][L]}$$

$$K_2 = \frac{[ML_2]}{[ML][L]}$$

These are called *formation constants*. It will be seen that, as a complex becomes more stable, the formation constant becomes larger. In this chapter, we shall consider the simplest case, in which only a 1:1 metal-ligand complex is formed.

In a solution containing a complex (ML), it is frequently of interest to calculate the concentration of free metal ion (M) in solution. This calculation is easy if the concentration of the complex and its formation constant are known.

Example: Zinc(II) forms a 1:1 complex with a ligand called triethylenetetramine, $(H_2-NCH_2CH_2NHCH_2-)_2$. The formation constant for this complex is $10^{12.0}$. Calculate the zinc-ion concentration of a $0.01M$ solution of the complex.

$$Zn^{2+} + L \rightleftharpoons ZnL^{2+}$$

$$K = \frac{[ZnL^{2+}]}{[Zn^{2+}][L]}$$

The large value for K indicates that the equilibrium lies far to the right, but a small amount of the complex dissociates. The $[Zn^{2+}]$ and $[L]$ from this dissociation are unknown but equal. Substituting into the K expression,

$$10^{12.0} = \frac{0.01}{[Zn^{2+}][L]} = \frac{10^{-2.0}}{[Zn^{2+}]^2}$$

$$[Zn^{2+}] = \sqrt{10^{-14.0}} = 10^{-7.0}M$$

The common-ion effect applies to complexation reactions as well as the ionization of weak acids.

Example: Calculate the zinc-ion concentration in a solution prepared by mixing 20.0 mL of 0.01M zinc nitrate with 30.0 mL of 0.01M triethylenetetramine.

The *initial* concentration of each reactant is calculated as follows:

$$[Zn^{2+}] = \frac{20.0 \text{ mL} \times 0.01M}{50.0 \text{ mL}} = 0.004M$$

$$[L] = \frac{30.0 \text{ mL} \times 0.01M}{50.0 \text{ mL}} = 0.006M$$

Practically all of the Zn^{2+} reacts with an equivalent amount of L, so at equilibrium,

$$[Zn^{2+}] \text{ is unknown,} \quad [L] = 0.002M, \quad [ZnL^{2+}] = 0.004M$$

Substituting into the K expression,

$$10^{12.0} = \frac{0.004}{[Zn^{2+}][0.002]}$$

$$[Zn^{2+}] = \frac{2.00}{10^{12.0}} = 2 \times 10^{-12}$$

6-5. SOLUBILITY OF PRECIPITATES; CALCULATION WITH THE SOLUBILITY PRODUCT

Solubility in Water. When a soluble compound MA dissociates into M and A, the concentration of MA influences the concentrations of M and A. Here, M stands for a metal cation and A for an anion; for simplicity, the charges on M and A are omitted. The greater the concentration of MA, the greater the concentrations of M and A. However, when MA is a precipitate, the concentrations of M and A in solution do *not* depend on the amount of MA precipitated. Of course, there must be some MA precipitated, and the precipitate must have reached equilibrium with the solution. Therefore, for precipitates, the equilibrium constant for the reaction

$$MA(s) \rightleftharpoons M + A$$

is written

$$K = \frac{[M][A]}{1}, \quad \text{or simply} \quad K = [M][A]$$

(Recall from Chapter 1 that $a = 1$ for a solid.) This equilibrium constant is given the special name *solubility product* and may be written

$$K_{sp} = [M][A]$$

The solubility products of many precipitates are given in handbooks and textbooks. From this constant, the solubility of a precipitate under various conditions may be calculated.

Example: Calculate the solubility of silver chloride in water at 20°C, both in moles per liter and in grams per hundred milliliters. The solubility product at 20°C is 1.0×10^{-10}.

$$AgCl(s) \rightleftharpoons Ag^+ + Cl^-$$

$$K_{sp,AgCl} = [Ag^+][Cl^-]$$

Each mole of dissolved AgCl gives 1 mole of Ag^+ and 1 mole of Cl^-. Thus, the solubility of AgCl is $[Ag^+] = [Cl^-]$. Substituting into the solubility-product expression,

$$1.0 \times 10^{-10} = [Ag^+]^2$$

$$[Ag^+] = 1.0 \times 10^{-5} M$$

$$\text{form wt}_{AgCl} \times 10^{-5} = 14_3 \times 10^{-5} = 1.4_3 \times 10^{-3} \text{ g/L}$$

$$\text{solubility} = 1.4_3 \times 10^{-4} \text{ g/100 mL}$$

Example: Calculate the solubility of silver chromate in water at 25°C; the K_{sp} of Ag_2CrO_4 at this temperature is 1.1×10^{-12}.

$$Ag_2CrO_4 \rightleftharpoons 2Ag^+ + CrO_4^{2-}$$

$$K_{sp} = [Ag^+]^2[CrO_4^{2-}]$$

Each mole of silver chromate that dissolves gives two moles of silver ion and one mole of chromate ion. Note that the solubility of silver chromate, therefore, is $[CrO_4^{2-}]$; note also that the concentration of Ag^+ is twice that of CrO_4^{2-}.

Let x be number of moles per liter of silver chromate that dissolve:

$$[CrO_4^{2-}] = x$$

$$[Ag^+] = 2[CrO_4^{2-}] = 2x$$

Substituting into the solubility-product expression,

$$1.1 \times 10^{-12} = (2x)^2(x) = 4x^3$$

$$x^3 = \frac{1.1}{4} \times 10^{-12} = 0.275 \times 10^{-12} = 275 \times 10^{-15}$$

$$x = \sqrt[3]{275 \times 10^{-15}} = 6.5 \times 10^{-5} M$$

Common-Ion Effect. A soluble compound that readily ionizes to give an ion common to the precipitate will reduce the concentration of the other ion from the precipitate in solution:

$$AB(s) \rightleftharpoons A + B$$

$$\leftarrow \boxed{\text{more B}}$$

If the concentration of added compound is known, the concentration of ion from the precipitate that is *not* common to the added compound can be calculated.

Example: Calculate the concentration of silver ion in the aqueous solution in equilibrium with a precipitate of silver chromate, when the solution contains sodium chromate so that $[CrO_4{}^{2-}] = 0.010M$.

$$Ag_2CrO_4(s) \rightleftharpoons 2Ag^+ + CrO_4{}^{2-}$$

$$K_{sp} = [Ag^+]^2[CrO_4{}^{2-}]$$

The silver-ion concentration is unknown and is to be calculated; the chromate-ion concentration is 0.010M. Substituting into the solubility product expression,

$$1.1 \times 10^{-12} = [Ag^+]^2[1.0^{-2}]$$

$$[Ag^+]^2 = 1.1 \times 10^{-10}$$

$$[Ag^+] = 1.0_4 \times 10^{-5}M$$

6–6. SIMULTANEOUS EQUILIBRIA: USE OF CONDITIONAL CONSTANTS

Effect of Acidity on Solubility. In many cases, the acidity of the solution needs to be considered in solubility product calculations involving a precipitate. Consider a precipitate MA(s) that dissolves to a slight extent to give M^+ and A^-, but the A^- can combine with H^+ in solution to give HA. Qualitatively, it will be seen that H^+ in solution will shift the solubility product equilibrium to make the precipitate more soluble:

$$MA(s) \rightleftharpoons M^+ + A^-$$

$$\Updownarrow H^+ \qquad\qquad (6\text{–}1)$$

$$HA$$

The quantitative effect of acidity on the solubility of a precipitate can be calculated by using the hydrogen-ion concentration of the solution and the value of K_a for HA. A convenient approach is first to define a term α_A, which is the fraction of the A *in solution* that exists as A^-.

$$\alpha_A = \frac{[A^-]}{[HA] + [A^-]} \qquad (6-2)$$

$$\frac{1}{\alpha_A} = \frac{[HA]}{[A^-]} + \frac{[A^-]}{[A^-]} \qquad (6-3)$$

The value of α_A as a function of $[H^+]$ can readily be evaluated from the K_a expression for HA.

$$K_a = \frac{[H^+][A^-]}{[HA]}$$

$$\frac{[HA]}{[A^-]} = \frac{H^+}{K_a}$$

Substituting into Equation 6–3,

$$\frac{1}{\alpha_A} = \frac{[H^+]}{K_a} + 1 \qquad (6-4)$$

Example: An acid HA having an ionization constant K_a of 6.30×10^{-5} is buffered at pH 4.0 ($[H^+] = 10^{-4.0}$). Calculate α_A, the fraction of HA present as A^- under these conditions.

$$\frac{1}{\alpha_A} = \frac{[H^+]}{K_a} + 1$$

$$\frac{1}{\alpha_A} = \frac{10^{-4.0}}{6.30 \times 10^{-5}} + 1 = 1.59 + 1 = 2.59$$

$$\alpha_A = \frac{1}{2.59} = 0.386$$

Returning to the solubility equilibrium (Equation 6–1), we see that if x moles of MA(s) dissolve, $[M^+] = x$ and the *sum* of $[HA] + [A^-] = x$. The solubility product expression for MA(s) requires $[A^-]$ rather than the sum of $[HA] + [A]$.

$$K_{sp,MA} = [M^+][A^-]$$

An expression for $[A^-]$ can be obtained from the α_A definition (Equation 6–2):

$$[A^-] = \alpha_A([HA] + [A^-]) \qquad (6-5)$$

Substituting into the K_{sp} expression,

$$K_{sp,MA} = [M^+]([HA] + [A^-])\alpha_A = x^2\alpha_A \qquad (6-6)$$

This equation can be used to calculate the precipitate's solubility in acidic solution.

Example: Calculate the M^+ concentration of a pH 4.0 solution in equilibrium with MA(s) when $K_{sp} = 2.06 \times 10^{-9}$ and K_a for HA = 6.30×10^{-5}.

In the preceding example, α_A was found to be 0.386 under the same conditions. Substituting into Equation 6–6,

$$2.06 \times 10^{-9} = x^2(0.386)$$

$$x = [M^+] = \sqrt{53.4 \times 10^{-10}} = 7.30 \times 10^{-5}M$$

This compares with $[M^+] = 4.54 \times 10^{-5}M$ in neutral or slightly alkaline solution.

Effect of Complex Formation on Solubility. Complex formation frequently needs to be considered in solubility-product calculations. Often a complexing agent is added that will permit the precipitation of one metal ion by a precipitating agent but will prevent that of a second metal ion. A practical example is the addition of tartrate to prevent the precipitation of hydrous ferric oxide at the slightly alkaline conditions needed for precipitating nickel with dimethylglyoxime. In this case, tartrate is called a *masking agent* because it masks or prevents the precipitation of iron(III). If the necessary constants are known, the success or failure of a separation employing a given masking agent can be predicted.

The effect of a masking agent or complexing ligand on the solubility of a precipitate may be calculated with the help of an alpha coefficient similar to that used above for acidity. A complexing ligand, L, may react with the metal ion from the precipitate to form a single metal-ligand complex or a series of complexes containing various L:M ratios. If a single complex is formed,

$$M + L \rightleftharpoons ML$$

$$K_{ML} = \frac{[ML]}{[M][L]} \tag{6–7}$$

we define α_M as follows:

$$\alpha_M = \frac{[M]}{[M] + [ML]} \tag{6–8}$$

$$\frac{1}{\alpha_M} = 1 + \frac{[ML]}{[M]} \tag{6–9}$$

Combining Equations 6–7 and 6–9,

$$\frac{1}{\alpha_M} = 1 + K_{ML}[L] \tag{6–10}$$

Example: The formation constant of a complex, ML, is 5.37×10^3. Calculate α_M, the fraction of metal ion in solution present as M, in a solution containing $0.10M$ free ligand, L.

Substituting into Equation 6–10,

$$\frac{1}{\alpha_M} = 1 + 5.37 \times 10^3 \times 10^{-1} = 538; \qquad \alpha_M = \frac{1}{538} = 1.86 \times 10^{-3}$$

Once α_M is known, the effect of the complexing ligand on the solubility of a precipitate can be calculated in a manner analogous to that used for the effect of hydrogen ions on solubility.

Example: Calculate the solubility of a precipitate, MA ($K_{sp} = 2.06 \times 10^{-9}$), in a solution containing $0.10M$ free ligand ($\alpha_M = 1.86 \times 10^{-3}$).

If x moles of the precipitate dissolves, $[A^-] = x$ and $[M] + [ML] = x$. From the α_M definition (Equation 6–8): $[M] = \alpha_M([M] + [ML]) = \alpha_M x$.

The K_{sp} expression is

$$K_{sp} = [M][A]$$

Substituting into this expression,

$$2.06 \times 10^{-9} = (\alpha_M x)(x) = (1.86 \times 10^{-3})(x^2)$$

$$x = \sqrt{1.11 \times 10^{-6}} = 1.05 \times 10^{-3}M$$

The solubility of the precipitate in the absence of a complexing ligand is $4.54 \times 10^{-5}M$.

Complex formation can also occur when an excess of the precipitating ion is added. For example, some metal ions form an insoluble oxalate precipitate (MC_2O_4), but they can add another ion of oxalate to form a metal oxalate complex:

$$M^{2+} + C_2O_4^{2-} \rightleftharpoons MC_2O_4(s)$$

$$MC_2O_4(s) + C_2O_4^{2-} \rightleftharpoons M(C_2O_4)_2^{2-}$$

Since the complex is soluble, adding more oxalate ion will actually increase the solubility of the precipitate by forming more of the complex, instead of reducing the solubility of the precipitate by the common-ion effect of the added oxalate.

Actually, complex formation is extremely common. If the formation of soluble complexes is neglected, solubility-product calculations sometimes lead to ridiculous answers. Ringbom cites the calculation of the solubility of mercuric sulfide in a $0.1M$ solution of sodium sulfide [2]. The solubility product for mercuric sulfide is 10^{-52}. Substituting a sulfide-ion concentration of $0.1M$ from the sodium sulfide leads to a mercuric-ion concentration of 10^{-51}:

$$10^{-52} = [Hg^{2+}][S^{2-}]$$

$$[S^{2-}] = 10^{-1} \quad \text{from } Na_2S$$

$$[Hg^{2+}] = \frac{10^{-52}}{10^{-1}} = 10^{-51}M$$

This value is so small that it would take more water than is on the earth to find even one dissolved mercuric ion. The actual solubility of mercuric sulfide under these conditions is about $10^{-3}M$, which is considerably different from $10^{-51}M$. The explanation of the discrepancy is that soluble mercuric sulfide complexes are present in solution and greatly enhance the solubility.

[2] A. Ringbom, *J. Chem. Ed.*, *35*, 282 (1958).

6-7. ACTIVITY COEFFICIENTS AND CHEMICAL EQUILIBRIUM

For more accurate calculations with equilibrium constants, we should use the *activities* of ions and molecules instead of their concentrations. For the simple equilibrium

$$A + B \rightleftharpoons C$$

the equilibrium-constant expression is

$$K = \frac{a_C}{a_A a_B}$$

Since activity equals molar concentration times an activity coefficient (see Chapter 1, Section 1-2),

$$a_i = f_i[i]$$

where the subscript refers to the ion. The equilibrium-constant expression can then be written as

$$K = \frac{[C]}{[A][B]} \times \frac{f_C}{f_A f_B}$$

The use of activity coefficients complicates calculations with equilibrium constants. Whenever possible, the analytical chemist prefers to ignore them and work with molar concentrations only. Frequently, the ionic strength of a solution (Chapter 1, Section 1-2) can be held fairly constant throughout a series of experiments, in which case the term containing the activity coefficients becomes a constant:

$$\frac{f_C}{f_A f_B} = k$$

$$\frac{K}{k} = \frac{[C]}{[A][B]} = K'$$

By using the modified constant K', it is possible to work with concentrations instead of activities.

The equilibrium constants K' are often determined for several ionic strengths. From a plot of the equilibrium constant versus ionic strength, the value of K at zero ionic strength may be calculated by extrapolating the curve. Frequently, this is the value of K that is given in the chemical literature.

QUESTIONS AND PROBLEMS

Equilibrium Constants, General

1. Equilibrium is a dynamic rather than a static condition. Suggest an experiment that would prove this for the equilibrium between a precipitate and its ions in solution. (Discuss this with your instructor if necessary.)

2. The ability of a chemical reaction to proceed is measured by the value of its equilibrium constant. For each of the following, write the expression for the equilibrium constant and calculate its numerical value. Use the following constants: K_{sp} for AgCl is 1.8×10^{-10}; K_{sp} for AgBr is 4.9×10^{-13}; K_a for HCN is 4.9×10^{-10}; K_a for NH_4^+ is 5.62×10^{-10}.

(a) $Ag^+ + Cl^- \rightleftharpoons AgCl(s)$ (c) $H^+ + CN^- \rightleftharpoons HCN$

(b) $Ag^+ + Br^- \rightleftharpoons AgBr(s)$ (d) $H^+ + NH_3 \rightleftharpoons NH_4^+$

3. The formation constant for the reaction $A + B \rightleftharpoons AB$ is 4.0. (All of these species are soluble in water.) If equal volumes of $0.2M$ A and $0.4M$ B are mixed, calculate the concentrations of A, B, and AB when equilibrium has been reached.

4. The equilibrium constant for the reaction $CA \rightleftharpoons C^+ + A^-$ is $10^{-6.0}$. Calculate the equilibrium concentration of C^+ in a $0.01M$ solution of CA that is also $0.04M$ in A^-.

Ionization Constants

5. The ionization constant of dichloroacetic acid is 5.50×10^{-2}. Calculate the hydrogen-ion concentration of a $0.10M$ solution of dichloroacetic acid in two ways: (a) assuming that the equilibrium concentration of dichloroacetic acid is 0.10 and (b) taking into account the dichloroacetic acid ionization when calculating the equilibrium concentration.

6. Calculate the hydrogen-ion concentration of a solution containing $0.01M$ dichloroacetic acid and $0.01M$ dichloroacetate in two ways: (a) assuming that the equilibrium concentrations of dichloroacetic acid and dichloroacetate are both $0.01M$ and (b) taking into account the effect of dichloroacetic acid ionization on the equilibrium concentrations of both dichloroacetic acid and dichloroacetate ($K_a = 5.50 \times 10^{-2}$).

7. Calculate the hydrogen-ion concentration of a solution that contains 2.44 g/L of benzoic acid, C_6H_5COOH ($K_a = 6.30 \times 10^{-5}$).

8. Calculate the hydrogen-ion concentration of a solution that contains 1.22 g/L of benzoic acid and 2.88 g/L of sodium benzoate, C_6H_5COONa. K_a for benzoic acid is 6.30×10^{-5}.

9. In what proportion should a solution of formic acid, HCOOH, and sodium formate, $HCOO^-Na^+$, be mixed to give a solution of pH 4.0 (hydrogen-ion concentration, $10^{-4.0}$)? (K_a for formic acid is 1.77×10^{-4}.)

Complex Formation

10. Copper(II) forms a soluble complex with ethylenediamine (en) as follows:

$$Cu^{2+} + 2en \rightarrow Cu(en)_2^{2+} \qquad K = 10^{19.60}$$

Calculate the copper ion concentration of a solution containing $0.010M$ $Cu(en)_2^{2+}$ and $0.010M$ excess en.

11. The formation constant of a soluble magnesium tartrate complex is $K_{Mg\,tart} = 22.9$. If 20.0 mL of $0.02M$ Mg^{2+} is mixed with 20.0 mL of $0.05M$ $tart^{2-}$, calculate the concentrations of Mg^{2+}, Mg tart, and $tart^{2-}$ in the resulting solution at equilibrium.

12. Calcium(II) forms a weak, 1:1 complex (CaL) with the lactate ion ($K = 10.0$). If 10.0 mL of $0.10M$ Ca^{2+} is mixed with 15.0 mL of $0.10M$ lactate, calculate the calcium-ion concentration of the resulting solution.

Solubility Product

13. Calculate the solubility of a saturated solution of each of the following compounds in moles per liter.

(a) $PbMoO_4$, $K_{sp} = 10^{-13.0}$ (b) PbF_2, $K_{sp} = 10^{-7.57}$

14. Calculate (a) the solubility in moles per liter and (b) the silver-ion concentration of a saturated aqueous solution of silver oxalate, $Ag_2C_2O_4$ $(K_{sp} = 1.3 \times 10^{-11})$.

15. Calculate which gives the lower concentration of Ag^+ in solution at equilibrium: (a) AgCl $(K_{sp} = 1.8 \times 10^{-10})$ in equilibrium with pure H_2O or (b) Ag_2CrO_4 $(K_{sp} = 1.1 \times 10^{-12})$ in equilibrium with pure H_2O.

16. Calculate the hydroxide-ion concentration and the pH of a saturated aqueous solution of calcium hydroxide, $Ca(OH)_2$ $(K_{sp} = 5.50 \times 10^{-6})$.

17. Sodium hydroxide is added to a solution containing magnesium(II) so that a precipitate of $Mg(OH)_2$ forms, and the pH of the aqueous solution is 11.0 $([OH^-] = 10^{-3.0})$. From the solubility product of magnesium hydroxide $(K_{sp} = 1.82 \times 10^{-11})$, calculate the molar concentration of the magnesium ion in solution.

18. Silver nitrate is added to a very dilute (0.0001M) solution of sodium chloride so that the silver-ion concentration is 0.01M. Under these conditions, is the precipitation of the chloride quantitative (i.e., at least 99.9% of the chloride precipitates)? K_{sp} for silver chloride is 1.8×10^{-10}.

Solubility Product, Activity Coefficients

19. Calculate the concentration of both Ba^{2+} and IO_3^- in a saturated solution of barium iodate. K_{sp} for $Ba(IO_3)_2 = 1.5 \times 10^{-9}$. Also calculate the ionic strength of this saturated solution. (See Chapter 1.)

20. Using data calculated in the previous problem, calculate the value of the modified solubility product constant, K', for barium iodate.

21. The solubility product constant for AgCl at $\mu = 0$ is 1.80×10^{-10}. Calculate the value of the modified solubility product constant, K', for silver chloride at each of the following ionic strengths.

(a) $\mu = 0.001$ (b) $\mu = 0.01$ (c) $\mu = 0.10$

22. The solubility product of thorium oxalate, $ThOx_2$, at $\mu = 0$ is $10^{-22.0}$. (a) Estimate the solubility product at $\mu = 0.1$ if $f_{Th^{4+}} = 0.10$ and $f_{Ox^{2-}} = 0.44$. (b) Calculate and compare the $[Th^{4+}]$ in a saturated solution of thorium oxalate at $\mu = 0$ and $\mu = 0.1$.

Simultaneous Equilibria

23. The ionization constant for hydrofluoric acid at $\mu = 0.1$ is $k_A = 6.75 \times 10^{-4}$. Calculate α_F, the fraction of HF that exists as the fluoride ion (F^-) at each of the following pH values.

(a) pH = 2.0   (c) pH = 4.0
(b) pH = 3.0   (d) pH = 5.0

24. The solubility product of PbF_2 at $\mu = 0.1$ is 1.3×10^{-7}. Using the data calculated in the previous problem, calculate the concentration of Pb^{2+} in solution at equilibrium when a solution containing 0.10M dissolved fluoride $(F^- + HF)$ is buffered at each of the following pH values.

(a) pH = 2.0   (c) pH = 4.0
(b) pH = 3.0   (d) pH = 5.0

25. Calculate the molar concentration of Pb^{2+} in equilibrium with a precipitate of PbI_2 in a solution containing 0.10M excess I^-. The solubility product of $PbI_2 = 3.16 \times 10^{-8}$ for $\mu = 0.1$.

26. Repeat the calculation of the concentration of lead(II) in solution in the previous problem, but taking into account the formation of soluble lead(II)-iodide complexes. From the formation constants of the various lead(II)-iodide complexes, $\alpha_{Pb} = 0.08$. (*Note:* In this case the total lead(II) in solution, $[Pb']$, should be calculated instead of $[Pb^{2+}]$.)

27. Calculate whether it will be possible to precipitate $BaSO_4$ without precipitating any $PbSO_4$ if the solution contains $0.01M$ excess sulfate ion (after any precipitation has occurred) and $0.01M$ EDTA is present as a masking agent. The solution is buffered at pH 5.0. (Necessary constants, etc: K_{sp} for $BaSO_4 = 6.31 \times 10^{-10}$, K_{sp} for $PbSO_4 = 1.0 \times 10^{-7}$. α_{Ba} (for EDTA complexing at pH 5.0) $= 10^{-0.1}$; $\alpha_{Pb} = 10^{-9.4}$.)

Miscellaneous

28. A colored acid, HA, has an ionization constant of 2.0×10^{-5}. The acid has a spectrophotometric absorption peak at 487 nm with no absorbance of the conjugate base, NaA, at this wavelength. The conjugate base has an absorption peak at 610 nm with a slight absorbance of the acid at this wavelength. If a dilute solution of the acid is titrated with standard sodium hydroxide, sketch the titration curve (a) when the titration is followed by measuring the absorbance at 610 nm with a spectrophotometer and (b) when the titration is followed by measuring the absorbance at 487 nm.

29. A precipitate of a silver dye of stoichiometric but unknown concentration is mixed with varying, known concentrations of excess silver(I). The dye remaining in solution is then measured spectrophotometrically. What should a plot of $\log[Ag^+]_{added}$ vs. $\log[dye]$ look like? From this plot, tell how you would (a) find the composition of the precipitate, and (b) estimate the value of K_{sp}.

7

Acid-Base Equilibria

7-1. ACID-BASE THEORY

Acid-Base Definition. According to Brönsted, an acid is a substance that can give up protons. A base is any compound or ion that can accept protons.

This simple definition omits types of acids, often called Lewis acids, that do not give up protons yet have the properties and undergo the reactions of acids. Lewis acids are substances, such as $AlCl_3$ and BF_3, that have unfilled electron shells and react by accepting an electron pair from a base. For example,

$$\text{F---B} + \text{:N---H} \rightarrow \text{F---B:N---H}$$

The Brönsted definition of acids and bases leads to the simple relationship

$$\text{acid}_1 \rightarrow H^+ + \text{base}_1$$

$$\text{base}_2 + H^+ \rightarrow \text{acid}_2$$

Combining these two equations, we have the general equation for the reaction of acid_1 with base_2:

$$\text{acid}_1 + \text{base}_2 \rightarrow \text{acid}_2 + \text{base}_1$$

Obviously, the products must be more weakly acidic and basic than the reactants. Acids and bases that differ only in number of protons (H^+) are called *conjugate pairs.* Here are some examples:

Acid	Conjugate Base
HCN	CN$^-$
HCl	Cl$^-$
CH_3COOH	CH_3COO^-
$NH_4{}^+$	NH_3
⟨NH$^+$⟩	⟨N⟩
H_2CO_3	$HCO_3{}^-$
$HCO_3{}^-$	$CO_3{}^{2-}$

Many acids and bases are organic. The most common type of organic acid is a *carboxylic acid.* This has the type formula RCOOH in which R represents some organic group such as methyl (CH_3—), ethyl (CH_3CH_2—) or butyl (C_4H_9—) and —COOH is the carboxyl group, in which the hydrogen is acidic. Carboxylic acids are named with an –*ic* suffix, and the corresponding salt of a carboxylic acid has an –*ate* suffix. For example, CH_3COOH is acetic acid, and CH_3COONa is sodium acetate, the conjugate base of acetic acid; $CH_3CH_2CH_2COOH$ is butyric acid, and $CH_3CH_2CH_2COO^-$ is the butyrate ion, the conjugate base.

An *amine* is the most common class of organic base. Amines may be thought of as organic derivatives of ammonia in which one, two, or three of the ammonia hydrogens are replaced by organic R groups. Thus, $C_4H_9NH_2$ is butylamine, $(CH_3)_2NH$ is dimethylamine, and $(CH_3CH_2)_3$ is triethylamine. The conjugate acid of an amine is the parent amine with an added proton. The name has an –*ium* suffix. For example, $C_4H_9NH_3{}^+$ is the butylammonium ion, $C_4H_9NH_3Cl$ is butylammonium chloride, $C_5H_5NH^+$ (the conjugate acid of pyridine) is the pyridinium ion.

Ionization of Acids and Bases. A substance that can act either as an acid or as a base is said to be *amphiprotic.* Many solvents are amphiprotic. For example, water acts as a base toward acids and as an acid toward bases. When an acid HA is dissolved in an amphiprotic solvent SH, the resulting ionization is actually an acid-base reaction.

$acid_1$	$base_2$		$acid_2$	$base_1$	
HA	+ SH	$\rightleftharpoons$ $SH_2{}^+$		+ A$^-$	*General Case, in Solvent* SH
HA	+ H_2O	$\rightleftharpoons$ H_3O^+		+ A$^-$	*In Water*
HA	+ CH_3OH	$\rightleftharpoons$ $CH_3OH_2{}^+$		+ A$^-$	*In Methyl Alcohol*
HA	+ CH_3COOH	$\rightleftharpoons$ $CH_3COOH_2{}^+$		+ A$^-$	*In Glacial Acetic Acid*

Note that, in each case, the proton (H^+) in solution is solvated.

The extent of ionization depends on several things. One is the inherent acidity of HA. A strong acid, such as hydrochloric acid, is completely ionized in water, whereas a weak acid, such as acetic acid, is only slightly ionized. Another is the basic strength of the solvent. A basic solvent will promote ionization of an acid by virtue of an acid-base reaction with the dissolved acid. All the solvents listed in the previous paragraph have some basic properties, although glacial acetic acid is a much weaker base than water, for example. Finally, the dielectric constant of the solvent—a measure of the electrical insulating ability of the solvent—has an effect on ionization. Since water has an unusually high dielectric constant, ions are relatively free from attraction and repulsion effects. (Some effects do exist, as will be recalled from the difference between ionic concentration and activity in water.) In most organic solvents, however, ions have a tendency to be present as *ion pairs* (cation-anion pairs). Thus, in glacial acetic acid solution, even strong acids exist mostly as ion pairs, very few *free* $CH_3COOH_2^+$ or A^- ions being present.

When a base is dissolved in an amphiprotic solvent (SH), the solvent acts as an acid, and the resulting ionization increases the solvent anion concentration (S^-):

$$\underset{base_1}{B} + \underset{acid_2}{SH} \quad \rightleftharpoons \quad \underset{acid_1}{BH^+} + \underset{base_2}{S^-} \qquad \textit{General Case, in Solvent } SH$$

$$B + H_2O \quad \rightleftharpoons \quad BH^+ + OH^- \qquad \textit{In Water}$$

$$B + CH_3OH \quad \rightleftharpoons \quad BH^+ + CH_3O^- \qquad \textit{In Methyl Alcohol}$$

$$B + CH_3COOH \rightleftharpoons BH^+ + CH_3COO^- \qquad \textit{In Glacial Acetic Acid}$$

The reaction of an acid and base in solution may well take place from the combining of solvated proton (from ionization of the acid) and solvent anion (from ionization of the base):

$$HA + SH \rightleftharpoons SH_2^+ + A^-$$

$$B + SH \rightleftharpoons BH^+ + S^-$$

$$SH_2^+ + S^- \rightleftharpoons 2SH$$

The sum of these three equations, however, is the simple reaction

$$\underset{acid_1}{HA} + \underset{base_2}{B} \quad \rightleftharpoons \underset{acid_2}{BH^+} + \underset{base_1}{A^-}$$

7–2. ACIDITY OF SOLUTIONS: pH

Amphiprotic solvents ionize to a very slight extent and form a cation and an anion:

$$2SH \rightleftharpoons SH_2^+ + S^- \qquad \textit{General Case}$$

$$2H_2O \rightleftharpoons H_3O^+ + OH^- \qquad \textit{Water}$$

$$2CH_3OH \rightleftharpoons CH_3OH_2^+ + CH_3O^- \qquad \textit{Methyl Alcohol}$$

$$2CH_3COOH \rightleftharpoons CH_3COOH_2^+ + Ac^- \qquad \textit{Glacial Acetic Acid}$$

The degree to which a solvent ionizes may be expressed by the *autoprotolysis* (self-ionization) *constant* K_a. For the general case,

$$K_a = a_{SH_2^+} a_{S^-}$$

Electroneutrality demands that the following be true:

$$a_{SH^+} = a_{S^-} = \sqrt{K_a}$$

(The activity of the solvent SH is constant and does not appear in the K_s expression.) Values of K_s for several solvents are listed in Table 7–1.

The autoprotolysis constant for water is usually designated K_w:

$$K_w = a_{H_3O^+} a_{OH^-}$$

For convenience, we shall hereinafter write simply H^+ instead of H_3O^+; thus,

$$K_w = a_{H^+} a_{OH^-}$$

At room temperature, $K_w = 10^{-14}$, and the hydrogen ion and hydroxyl ion activity of pure water are equal to $10^{-7} M$. When an acid is dissolved in water, the hydrogen-ion activity is increased to an extent depending on the concentration and degree of ionization of the acid. Any increase in hydrogen-ion activity results in a corresponding decrease in hydroxyl-ion activity because the product of the hydrogen- and hydroxyl-ion activities must always be a constant, 10^{-14}.

The acidity of an aqueous solution is frequently expressed in terms of pH. The pH is strictly defined as follows:

$$pH = \log \frac{1}{a_{H^+}} = -\log a_{H^+}$$

In analogous fashion, pOH and pK_w are defined:

$$pOH = -\log a_{OH^-}$$

$$pK_w = -\log K_w$$

Note that this "p" notation avoids the use of exponents.

Table 7–1. Autoprotolysis Constants of Some Solvents

Solvent	$-\log K_s$
Sulfuric Acid (100%)	3.6
Formic Acid	6.2
Water	14.0
Acetic Acid	14.5
Deuterium Oxide (heavy water)	14.7
Ethylenediamine	15.3
Methyl Alcohol	16.7
Ethyl Alcohol	19.1

When one or more solutes are dissolved in water, the *activity* of H^+ is not exactly equal to the *concentration* of H^+. (It might be pointed out here that a pH meter (see Chapter 16, Section 16–1) measures a_{H^+} and not $[H^+]$.) Because of the chemist's desire to simplify calculations as much as possible, however, activity coefficients are often ignored, and the following *approximate* definitions of pH and pOH are used:

$$pH = -\log[H^+]$$

$$pOH = -\log[OH^-]$$

Because the error introduced is not very great in dilute solutions, we shall use these approximations in all calculations involving pH. Using either the exact or approximate definitions of pH and pOH, the expression for the autoprotolysis constant for water becomes:

$$pK_w = pH + pOH$$

$$pH = 14 - pOH$$

$$pOH = 14 - pH$$

In water, $pK_w = 14$ and a pH of 7 is neutral. Higher pH values are increasingly basic and lower values are increasingly acidic.

Negative pH values are possible (if $[H^+] = 10M$, $pH = -1$), as are pH values greater than 14 in strongly alkaline media. However, measurement of pH is very inaccurate in these regions, so the molar concentration of strong acid or base is generally used instead of pH.

Note that a change of one pH unit represents a tenfold change in acidity. For example, in going from pH 8 to pH 5 (a change of 3 pH units), the H^+ concentration of the solution increased 1000 times.

The pH scale in H_2O encompasses a tremendous change in H^+ concentration. In going from $1M$ HCl (pH = 0) to $1M$ NaOH (pOH = 0, pH = 14), the H^+ concentration of the solution has been reduced by a factor of 10^{14}. This number is so large that it dwarfs even our national debt! Nevertheless, a pH meter is able to measure H^+ concentration (pH) throughout most of this range.

By using the appropriate value of the autoprotolysis constant, K_s, it should be possible to establish pH scales for solvents other than water. For example, the pH scale for ethyl alcohol ($pK_s = 19.1$) would be from 0 to 19.1 with a neutral pH value of 9.55.

7-3. CALCULATION OF pH: SOLUTIONS OF STRONG ACIDS AND BASES

Acids and bases are commonly classified as *strong* or *weak*, to indicate the approximate degree of ionization. The strong acids include hydrochloric, hydrobromic, hydriodic, nitric, perchloric ($HClO_4$), sulfuric, and organic sulfonic acids (RSO_3H). Acids in this group are assumed to be 100% ionized in dilute aqueous solution. If the concentration of a strong acid is known, the pH can be easily calculated.

Example: Calculate the pH of a 0.01M solution of hydrochloric acid. Since the hydrochloric acid is completely ionized, the hydrogen-ion concentration [H^+] is 0.01M, or $1 \times 10^{-2}M$, and

$$pH = -\log 1 \times 10^{-2} = 2.0$$

Example: Calculate the pH of 0.0125M hydrochloric acid.

$$[H^+] = 0.0125 = 1.25 \times 10^{-2}$$

$$pH = 2 - \log 1.25 = 2 - 0.097 = 1.903$$

The log of 1.25 is obtained from a log table or preferably from a calculator. If [H^+] is written in the form above (1.25×10^{-2}), with the decimal after the first digit, the log of the number is easily found by placing a decimal point before the first number (zero included) of the log in a table.

Strong bases are also considered to be 100% ionized in dilute aqueous solution. They are designated MOH, and they include alkali metal hydroxides, certain alkaline earth hydroxides, and quaternary ammonium hydroxides (R_4NOH):

$$M^+OH^- \rightarrow M^+ + OH^-$$

The pH of aqueous solutions is found after first calculating the pOH.

Example: Calculate the pH of 0.025M sodium hydroxide:

$$[OH^-] = 2.5 \times 10^{-2}$$

$$pOH = 2 - \log 2.5 = 2 - 0.40 = 1.60$$

$$pH = 14 - pOH = 12.40$$

7-4. CALCULATION OF pH: SOLUTIONS OF WEAK ACIDS AND BASES

Ionization Constants.   Hundreds of acids and bases are classified as weak acids or bases because they are only slightly ionized in solution. For convenience, we shall classify common types of weak acids and bases as follows:

Acids

1. HA = acid without a charge that ionizes to give H^+ plus a basic anion, A^-. This class includes a great many organic acids that contain a carboxyl group, —COOH, in which the H atom is acidic. For example, propionic acid, CH_3CH_2COOH, ionizes to give $H^+ + CH_3CH_2COO^-$.
2. BH^+ = a charged acid that is the conjugate acid of an uncharged base, B. Acids of this class ionize to give $H^+ + B$.

Bases

1. A^- = an anionic base that is the conjugate base of the acid HA. An alkali metal salt of a carboxylic acid ($CH_3COO^-Na^+$, for instance), or the salt of a weak inorganic acid such as Na^+F^-, is an example of this class.
2. B = an uncharged base, usually containing nitrogen. Examples include ammonia (NH_3) and amines such as butylamine ($C_4H_9NH_2$) and pyridine (C_5H_5N).

The following equations are given for the ionization reaction and the ionization-constant expression for each type of weak acid and base.

Acids	*Bases*

(1)
$$HA \rightleftharpoons H^+ + A^- \qquad\qquad A^- + H_2O \rightleftharpoons HA + OH^-$$

$$K_a = \frac{[H^+][A^-]}{[HA]} \qquad\qquad K_b = \frac{[HA][OH^-]}{[A^-]}$$

(2)
$$BH^+ \rightleftharpoons H^+ + B \qquad\qquad B + H_2O \rightleftharpoons BH^+ + OH^-$$

$$K_a = \frac{[H^+][B]}{[BH^+]} \qquad\qquad K_b = \frac{[BH^+][OH^-]}{[B]}$$

Note that both the (1) acid and base and the (2) acid and base constitute conjugate acid-base pairs. For any conjugate acid-base pair, $K_a \times K_b = K_w$

Pair (1)

$$K_aK_b = \frac{[H^+][A^-]}{[HA]} \times \frac{[HA][OH^-]}{[A^-]} = [H^+][OH^-] = K_w$$

Pair (2)

$$K_aK_b = \frac{[H^+][B]}{[BH^+]} \times \frac{[BH^+][OH^-]}{[B]} = [H^+][OH^-] = K_w$$

From this, it follows that $pK_a + pK_b = 14.00$.

Values of K_a for acids of the HA type are found in handbooks, textbooks, and other compilations and are commonly given in semiexponential form or as pK_a, which is the negative logarithm of K_a. For example, a value might be given as $K_a = 2.0 \times 10^{-5}$, $K_a = 10^{-4.70}$, or $pK_a = 4.70$. K_b or pK_b values for bases are often given, but in some compilations only the K_a (or pK_a) of the conjugate acid is listed.

Example: A table lists the acid ionization-constant of triethanolammonium ion, BH^+, as $pK_a = 7.8$. Calculate the pK_b of triethanolamine B.

$$pK_b = 14 - pK_a = 6.2$$

In similar fashion, the pK_b of an anionic base can be calculated from the pK_a of the acid.

Example: The pK_a for hydrocyanic acid (HCN) is 9.31. Calculate the pK_b for sodium cyanide (NaCN).

$$pK_b = 14 - 9.31 = 4.69$$

Calculation of pH. The pH of a weak acid can be calculated from the ionization constant, K_a. Similarly, the pH of a weak base can be calculated from its K_b value. A weak acid of initial concentration C ionizes as follows:

$$HA \rightleftharpoons H^+ + A^- \tag{7-1}$$

At equilibrium, the concentrations of H^+ and A^- are equal but unknown; the concentration of HA is C minus the amount that dissociates, or $C - H^+$. Inserting this into the K_a expression, we get

$$K_a = \frac{[H^+]^2}{C - [H^+]} \tag{7-2}$$

This takes the form of a quadratic equation,

$$H^2 + KH - KC = 0 \tag{7-3}$$

for which the positive root is

$$H = \frac{-K + (K^2 + 4KC)^{1/2}}{2} \tag{7-4}$$

$$pH = -\log H = -\log\left[\frac{-K + (K^2 + 4KC)^{1/2}}{2}\right] \tag{7-5}$$

This rather formidable-looking equation can be solved easily by using a calculator and even more quickly with a programmable calculator. For Hewlett-Packard calculators, the following program can be used: ENTER, STO 1, X^2, RCL 1, RCL 2, X, 4, X, +, $\sqrt{}$, RCL 1, $-$, 2, $\div$, log, CHS. In the run mode, the value of C is stored in memory 2, and the desired value of K is then entered to calculate pH.

Example: Calculate the pH of a 0.10M solution of chloroacetic acid, $K_a = 1.38 \times 10^{-3}$. Substituting into Equation 7–5 for $C = 0.10$ and $K = 1.38 \times 10^{-3}$, pH = 1.96.

A widely used and simpler method of calculation uses Equation 7–2 but makes the approximation that the denominator is C rather than $C - [H^+]$. This causes almost no error for the usual case in which $[H^+]$ is small in comparison to C.

By making this approximation, the calculation of pH is greatly simplified:

$$K_a = \frac{[H^+]^2}{C} \tag{7-6}$$

$$[H^+] = (K_a C)^{1/2} \tag{7-7}$$

$$pH = -\log H = -\log (K_a C)^{1/2} \tag{7-8}$$

Example: Calculate the pH of a 0.10M solution of chloroacetic acid, $K_a = 1.38 \times 10^{-3}$. Substituting into Equation 7–8 for $C = 0.10$ and $K_a = 1.38 \times 10^{-3}$, pH = 1.93.

Note the small difference (0.03 pH unit) obtained in the two calculation methods. This difference *increases* as K_a becomes larger ($\sim 10^{-2}$, $\sim 10^{-1}$, etc.) or as the concentration, C, becomes smaller.

In Section 6–3, we concluded that Equation 7–6 can be used if $C \geq 100\, K_a$; otherwise, Equation 7–5 should be used.

Concentrations of a weak acid given in grams per volume must be converted to molar concentrations before the pH can be calculated.

Example: A solution contains 6.1 g/L of benzoic acid, C_6H_5COOH; the K_a for benzoic acid is $6.3 \times 10^{-5} = 10^{-4.2}$. Calculate the pH.

$$HA \rightleftharpoons H^+ + A^-$$

$$K_a = \frac{[H^+][A^-]}{[HA]}$$

Because $[H^+] = [A^-]$, the numerator of the K_a expression is $[H^+]^2$. The concentration of free benzoic acid HA must be expressed as molarity. To do this, divide the number of grams per liter by the formula weight:

$$[HA] = \frac{6.1}{122} = 0.050M$$

Substituting into the K_a expression,

$$10^{-4.20} = \frac{[H^+]^2}{5.0 \times 10^{-2}}$$

$$[H^+] = \sqrt{5.0 \times 10^{-6.20}}$$

$$[H^+] = 2.24 \times 10^{-3.10}$$

$$pH = 3.10 - \log 2.24 = 2.75$$

The pH of a solution of a weak base can be calculated from the initial concentration, C, and from the value of K_b in the same way as for the pH of a weak acid. The only difference is that $[OH^-]$ and pOH is calculated first. The pH is then obtained by subtracting the pOH from 14:

$$B + H_2O \rightleftharpoons BH^+ + OH^- \tag{7-9}$$

$$K_b = \frac{[BH^+][OH^-]}{[B]} = \frac{[OH^-]^2}{C - [OH]} \tag{7–10}$$

If K_b is 10^{-4} or smaller, a simpler equation can be used:

$$K_b = \frac{[OH^-]^2}{C} \tag{7–11}$$

Example: Calculate the pH of a 0.20M solution of pyridine, $K_b = 1.5 \times 10^{-9} = 10^{-8.82}$. Substituting into Equation 7–11,

$$1.5 \times 10^{-9} = \frac{[OH^-]^2}{0.20}$$

$$[OH^-] = 1.7 \times 10^{-5}; \qquad pOH = 4.76$$

$$pH = 14 - 4.76 = 9.24$$

The pH of a solution containing both an acid and a base as the conjugate pair is most easily calculated by using the acid-ionization constant, K_a.

$$K_a = \frac{[H^+][A^-]}{[HA]} \tag{7–12}$$

The value for K_a and the concentrations of A^- and HA are inserted; then $[H^+]$ and the pH can be calculated. Alternatively, Equation 7–12 can be converted to log form:

$$\log K_a = \log[H^+] + \log\left(\frac{[A^-]}{[HA]}\right) \tag{7–13}$$

Rearranging terms, we get

$$-\log[H^+] = -\log K_a + \log\left(\frac{[A^-]}{[HA]}\right)$$

$$pH = pK_a + \log\left(\frac{[A^-]}{[HA]}\right) \tag{7–14}$$

Equation 7–14 is often used in biochemical and life science texts, where it is known as the Henderson–Hasselbach equation.

Example: Calculate the pH of a solution that contains 0.01M o-nitrophenol (HA) and 0.02M sodium o-nitrophenolate (A^-). pK_a for o-nitrophenol = 7.21. Substituting into Equation 7–14,

$$pH = 7.21 + \log(0.02/0.01) = 7.21 + 0.30 = 7.51$$

Example: Calculate the pH of a 0.50M solution of hydroxylammonium chloride, $(NH_3OH)^+ Cl^-$, that has been 25% neutralized with sodium hydroxide. The pK_b for hydroxylamine is 7.91.

$$[NH_3OH^+] = [BH^+] = 0.375 \qquad \textit{(Equivalent to HA in Equation 7–14)}$$

$$[NH_2OH] = [B] = 0.125 \qquad \textit{(Equivalent to A}^- \textit{ in Equation 7–14)}$$

$$pK_a = 14 - pK_b = 6.09$$

Substituting into Equation 7–14, we get

$$pH = 6.09 + \log(0.125/0.375) = 6.09 - 0.48 = 5.61$$

7–5. IONIZATION OF POLYPROTIC ACIDS

Calculation of pH. Acids with more than one acidic hydrogen ionize in steps. An ionization-constant expression may be written for each step. The stepwise-ionization and ionization-constant expressions for a diprotic acid, H_2A are as follows:

$$H_2A \rightleftharpoons H^+ + HA^- \qquad K_1 = \frac{[H^+][HA^-]}{[H_2A]}$$

$$HA^- \rightleftharpoons H^+ + A^{2-} \qquad K_2 = \frac{[H^+][A^{2-}]}{[HA^-]}$$

The methods used for calculating the pH of solutions containing various combinations of the species H_2A, HA^-, and A^{2-} from the ionization constants are summarized as follows:

1. *A Solution Containing* H_2A, *or* $H_2A + HA^-$. If K_1 is a hundred times or so
   greater than K_2, the second ionization will have very little effect and can be
ignored. The pH of the solution is calculated from the K_1 expression.

Example: Calculate the pH of a 0.15M solution of malonic acid, $CH_2(COOH)_2$. The
ionization constants for malonic acid are $K_1 = 1.40 \times 10^{-3} = 10^{-2.85}$ and
$K_2 = 2.2 \times 10^{-6} = 10^{-5.66}$.

   K_1 is enough larger than K_2 that the second ionization can be safely
ignored and the pH calculated only from the K_1 expression. Note that K_1 is large
enough that the equilibrium concentration of malonic acid (H_2A) must be taken
as $0.15 - [H^+]$.

$$K_1 = \frac{[H^+][HA^-]}{[H_2A]} = \frac{[H^+]^2}{0.15 - [H^+]}$$

$$[H^+]^2 + 1.40 \times 10^{-3}[H^+] - 2.1 \times 10^{-4} = 0$$

Solving this equation by the quadratic formula,

$$[H^+] = 1.38 \times 10^{-2}$$

$$pH = 1.86$$

2. *A Solution Containing* HA^-. Here *both* ionizations affect the composition of the solution and must be considered. In the second ionization, $[H^+]$ is *not* equal to $[A^{2-}]$ because some H^+ combines with HA^- to form H_2A (the reverse reaction of the first ionization). Therefore,

$$[A^{2-}] = [H^+] + [H_2A]$$

From the K_2 expression, we have

$$[A^{2-}] = \frac{K_2[HA^-]}{[H^+]}$$

Equating the right sides of these two equations,

$$[H^+] + [H_2A] = \frac{K_2[HA^-]}{[H^+]}$$

In this equation, we can replace $[H_2A]$ with a quantity obtained from the K_1 expression:

$$[H^+] + \frac{[H^+][HA^-]}{K_1} = \frac{K_2[HA^-]}{[H^+]}$$

Rearranging this equation gives the following:

$$[H^+]^2(K_1 + [HA^-]) = K_1 K_2[HA^-]$$

$$[H^+]^2 = \frac{K_1 K_2[HA^-]}{K_1 + [HA^-]}$$

At usual concentrations, $[HA^-]$ will generally be much larger than K_1, so the denominator will be approximately equal to $[HA^-]$:

$$[H^+]^2 \approx \frac{K_1 K_2[HA^-]}{[HA^-]} \approx K_1 K_2$$

$$\boxed{[H^+] \approx \sqrt{K_1 K_2}} \quad \text{or} \quad \boxed{pH \approx \frac{pK_1 + pK_2}{2}}$$

Example: Calculate the pH of a solution of sodium hydrogen malonate. The ionization constants for malonic acid are $pK_1 = 2.85$ and $pK_2 = 5.66$.

Unless the solution is very dilute, the pH is independent of concentration and is calculated from the simple expression previously given:

$$pH \approx \frac{2.85 + 5.66}{2} \approx 4.26$$

3. *A Solution Containing* $HA^- + A^{2-}$. If K_1 is 100 times or more greater than K_2, there will be very little H_2A in the solution at equilibrium, and the first

ionization-constant need not be used. The pH is calculated very easily by using the K_2 expression.

Example: Calculate the pH of a solution having, at equilibrium, a hydrogen malonate ion (HA^-) concentration of $0.15M$ and a malonate ion (A^{2-}) concentration of $0.05M$.

$$K_2 = \frac{[H^+][A^{2-}]}{[HA^-]}$$

$$2.2 \times 10^{-6} = \frac{[H^+](0.05)}{(0.15)}$$

$$[H^+] = 6.6 \times 10^{-6}$$

$$pH = 5.18$$

Calculation of Species Present.   As a strong base neutralizes a diprotic acid, the pH increases and the proportions of H_2A, HA^-, and A^{2-} in solution change. It is often useful to know the composition of the solution as a function of pH. A good example is in complex-formation reactions because substances that can form complexes with metal ions usually have acid-base properties. Consider, for example, the acid form of a complexing agent H_2L, which ionizes as follows:

$$H_2L \rightleftharpoons H^+ + HL^-$$

$$HL^- \rightleftharpoons H^+ + L^{2-}$$

The unprotonated anion, L^{2-}, can react with a metal ion to form a complex or a series of complexes:

$$M^{2+} + L^{2-} \rightleftharpoons ML$$

$$ML + L^{2-} \rightleftharpoons ML_2^{2-}$$

As the hydrogen-ion concentration increases (more acidic pH), the concentration of L^{2-} available for reaction with M^{2+} will decrease through reaction with H^+ and will form HL^- and H_2L. For calculations with complex-formation constants, the fraction of L that exists as L^{2-} is needed. This fraction is defined as α_L (see p. 192).

$$\alpha_L = \frac{[L^{2-}]}{[H_2L] + [HL^-] + [L^{2-}]} \tag{7-15}$$

An expression for the calculation of α_L at any pH may be derived from the ionization constants of H_2L. The inverse of Equation 7-15 may be written:

$$\frac{1}{\alpha_L} = \frac{[H_2L]}{[L^{2-}]} + \frac{[HL^-]}{[L^{2-}]} + \frac{[L^{2-}]}{[L^{2-}]} \tag{7-16}$$

From the two ionization-constant expressions for H_2L,

$$[H_2L] = \frac{[H^+]^2[L^{2-}]}{K_1 K_2} \tag{7-17}$$

$$[HL^-] = \frac{[H^+][L^{2-}]}{K_2} \qquad (7-18)$$

Substituting Equations 7–17 and 7–18 into Equation 7–16,

$$\frac{1}{\alpha_L} = \frac{[H^+]^2}{K_1 K_2} + \frac{[H^+]}{K_2} + 1 \qquad (7-19)$$

A similar expression can be derived for cases in which the ligand forms an acid having more than two hydrogen atoms. For example, the calculation of α_L for H_4L is as follows:

$$\frac{1}{\alpha_L} = \frac{[H^+]^4}{K_1 K_2 K_3 K_4} + \frac{[H^+]^3}{K_2 K_3 K_4} + \frac{[H^+]^2}{K_3 K_4} + \frac{[H^+]}{K_4} + 1 \qquad (7-20)$$

In calculations of α_L made with an equation like Equation 7–19 or 7–20, it frequently happens that one or more terms are not significant and may be disregarded.

Example: Calculate α_L for a tartrate solution at pH 5.0. The ionization constants for tartaric acid are $K_1 = 9.2 \times 10^{-4}$ and $K_2 = 4.3 \times 10^{-5}$. (α_L is the fraction of tartrate that is present as the unprotonated anion, $Tart^{2-}$.)

Substituting into Equation 7–19,

$$\frac{1}{\alpha_L} = \frac{10^{-10}}{4.0 \times 10^{-8}} + \frac{10^{-5}}{4.3 \times 10^{-5}} + 1$$

$$= 0.0025 + 0.23 + 1$$

The first term is too small to be significant:

$$\frac{1}{\alpha_L} = 1.23$$

$$\alpha_L = \frac{1}{1.23} = 0.81$$

Expressions similar to that used for calculating the fraction of the unprotonated anion may be derived for other species in solution. In Figure 7–1, the fractions of various tartrate species are graphed as a function of the pH of the solution.

7-6. BUFFERS

A buffer is a compound or mixture that, when added to a solution, helps maintain a specific pH. A buffered solution resists large changes in pH that otherwise would occur if the solution were diluted or if *either* a strong acid or a strong base were added to the solution. Buffers are very important in many chemical and biochemical systems.

Figure 7–1.   Tartaric acid species as a function of pH.

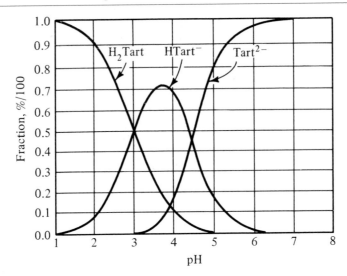

A buffer is composed of a weak acid and the salt of that acid, a weak base and the salt of that base, or an acid salt such as potassium acid phthalate. A buffer is formed during the titration of a weak acid with a strong base or during the titration of a weak base with a strong acid.

The formation of buffers and their ability to resist large changes in pH is perhaps best illustrated by an acid-base titration curve. A titration curve is a graphical plot of pH versus milliliters of titrant added (or vs. percentage titration). The buffer region of a titration curve is shown in Figure 7–2. Here a weak acid, HA, is titrated with a strong base, NaOH, to form the conjugate base, A^-.

$$HA + OH^- \rightarrow A^- + H_2O$$
$$\text{(NaOH)}$$

Between the initial and final points, the solution contains a mixture of HA and A^-. If we start with a solution of A^- and titrate with a strong acid such as HCl, the same titration curve is obtained but in the reverse direction (right to left).

Figure 7–2 shows the buffer region of a titration curve in which the ionization constant of the acid, HA, is $10^{-5.20}$. A similar titration curve involving an acid of the BH^+ type ($K_a = 10^{-8.00}$) and a base of the B type is shown in Figure 7–3 for $0.10M$ solutions, neglecting any dilution effects.

Too much acid or base may be added so that the pH gets outside the buffer region and rises or falls more rapidly. The amount of acid or base that a buffered solution can handle with only a certain small pH change is called its *buffer capacity*. In more quantitative terms, buffer capacity has been defined as the number of moles of a strong base needed to raise the pH of one liter of buffer by one pH unit. Obviously, the buffer capacity is greater when larger amounts of buffer are in

Figure 7–2. Buffer region (consisting of HA and A⁻) obtained in a titration curve.

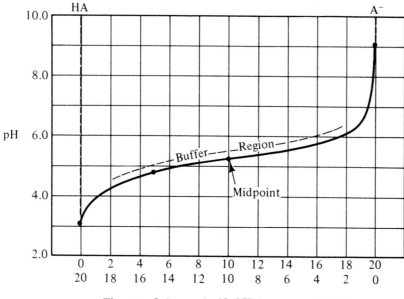

Titrant, mL (top scale, NaOH; bottom scale, HCl)

Figure 7–3. Buffer region (consisting of BH⁺ and B) obtained in a titration curve.

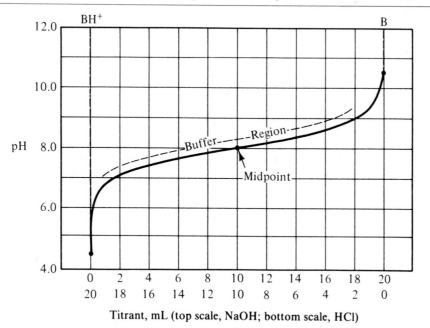

Titrant, mL (top scale, NaOH; bottom scale, HCl)

solution. For a *given concentration* of buffer, the greatest buffer capacity is obtained when the conjugate acid and base of the buffer are present in equimolar concentrations. This is the situation at the midpoint of a titration (see Figure 7–3).

The pH at which a buffer operates depends on the ratio of conjugate acid and base in the buffer and on the ionization constant of the weak acid or base involved. The pK_a of an acid in a buffer (with its conjugate base) should be as near as possible to the pH at which the buffer is to operate. The ratio of acid to base (or base to acid) required for buffering a solution at a given pH can be calculated from the pK_a of the acid.

Example: Calculate the ratio in which sodium acetate and acetic acid must be mixed to give a solution that is buffered at pH 5.00.

The ratio is calculated by substituting into the K_a expression for acetic acid. K_a and [H^+] are known; the ratio of acetate to acetic acid is to be calculated.

$$K_a = \frac{[H^+][Ac^-]}{[HAc]}$$

$$1.8 \times 10^{-5} = \frac{(1.0 \times 10^{-5})[Ac^-]}{[HAc]}$$

$$\frac{[Ac^-]}{[HAc]} = 1.8$$

Thus, sodium acetate and acetic acid should be mixed together in a molar ratio of 1.8 to 1. The pH is not appreciably affected by the dilution.

Example: In what molar ratio should butylamine and hydrochloric acid be mixed to prepare a buffer of pH 10.0? At $\mu = 0.1$, the pK_b for butylamine (B) is 3.30, and pK_a for the conjugate acid (BH$^+$) is 10.70.

The ratio of B to BH$^+$ needed to give a pH 10.0 solution is calculated by substituting into the K_a expression for BH$^+$:

$$10^{-10.70} = \frac{[H^+][B]}{[BH^+]} = \frac{10^{-10.0}[B]}{[BH^+]}$$

$$\frac{[B]}{[BH^+]} = 10^{-0.70} = 0.20$$

The calculated ratio of 0.20 is equivalent to a B/BH$^+$ ratio of 1:5. This may be obtained by adding 0.5 mole of hydrochloric acid to 0.6 mole of B, thus forming 0.5 mole of BH$^+$ and leaving 0.1 mole of B.

While calculations of this type are very useful, they are only approximations. If a solution must be buffered at an exact pH, the pH of the solution should be checked with a pH meter after the buffer has been added. Frequently, a little more acid or base will have to be added. Directions are given in handbooks for preparing buffer solutions with a precisely known pH (Clark and Lubs buffers and MacIlvaine buffers). These solutions serve as pH standards; they are used to standardize pH meters and for various other purposes.

QUESTIONS AND PROBLEMS

Acid-Base Theory

1. Write the formula for the conjugate base of each of the following acids. Give two conjugate bases for diprotic acids.

   (a) Nitrous acid, HNO_2
   (b) Fumaric acid, $C_2H_2(COOH)_2$

   (c) Ammonium chloride, NH_4Cl
   (d) Glycinium chloride, $\underset{\underset{NH_3Cl}{|}}{CH_2COOH}$

2. Write the formula for the conjugate acid of each of the following bases.

   (a) Sodium cyanide (NaCN)
   (b) Sodium hydrazoate (NaN_3)

   (c) Hydrazine (NH_2NH_2)
   (d) Pyridine (C_5H_5N)

3. Write equations to show how each of the amphiprotic solvents below reacts with an acid (HA) or a base (B).

   (a) Ethyl alcohol (C_2H_5OH)
   (b) Glacial acetic acid (CH_3CO_2H)

   (c) Deuterium oxide (D_2O)
   (d) Formic acid (HCO_2H)

4. Calculate the pH of each of the pure solvents, given the K_s for self-ionization.

   (a) Ethyl alcohol, $K_s = 10^{-19.1}$
   (b) Deuterium oxide, $K_s = 10^{-14.7}$

   (c) Glacial acetic acid, $K_s = 10^{-14.5}$

5. In anhydrous ethyl alcohol ($K_s = 10^{-19.1}$), an acid HA has a K_a value of $10^{-4.2}$. Calculate the K_b value for the conjugate base A^- in anhydrous ethyl alcohol.

pH of Acids and Bases

6. Calculate the pH of each of the following aqueous solutions.

   (a) $0.057M$ HCl
   (b) $0.200M$ H_2SO_4 (assume complete ionization of only the first H^+ and report the pH to only one significant figure)
   (c) $0.200M$ H_2SO_4 (assume that the first H^+ ionizes completely and use $K_2 = 1.0 \times 10^{-2}$ for the second H^+; report pH to two significant figures)
   (d) $0.025M$ NaOH
   (e) $0.0020M$ KOH

7. In aqueous solution, the ionization constant for dichloroacetic acid is 5.0×10^{-2}. Calculate the $[H^+]$ of each concentration of this acid using both Equation 7–4 (or 7–5) and Equation 7–7; state which is accurate if there is a discrepancy.

   (a) $10.0M$ dichloroacetic acid
   (b) $0.168M$ dichloroacetic acid

   (c) $0.050M$ dichloroacetic acid

8. For each of the following, calculate the pH of a $0.10M$ aqueous solution. Check the accuracy of Equation 7–7 by using Equation 7–4 (or 7–5).

   (a) Hydrogen cyanide (HCN), $K_a = 4.9 \times 10^{-10}$
   (b) Hydrazoic acid (HN_3), $K_a = 1.9 \times 10^{-5}$
   (c) Ammonium chloride (NH_4Cl), $K_a = 5.62 \times 10^{-10}$
   (d) Methylammonium chloride (CH_3NH_3Cl), $K_a = 2.30 \times 10^{-11}$

9. For each of the following, calculate the pH of a $0.10M$ aqueous solution (see Problem 8 for the ionization constant of the conjugate acid or base).

(a) Sodium cyanide (NaCN)
(b) Sodium hydrazoate (NaN$_3$)
(c) Ammonia (NH$_3$ or NH$_4$OH)
(d) Methylamine (CH$_3$NH$_2$)

10. Calculate the pH of each of the following solutions. The K_a is given for the acidic member of the conjugate acid-base pair in Problem 8.

(a) $0.010M$ hydrogen cyanide and $0.050M$ sodium cyanide
(b) $0.010M$ hydrazoic acid and $0.001M$ sodium hydrazoate
(c) $0.50M$ ammonium chloride and $0.50M$ ammonia
(d) $0.001M$ methylammonium chloride and $0.082M$ methyl amine

11. Calculate the difference in pH of a $0.10M$ solution of acetic acid at $25°C$ and at $60°C$. K_a at $25° = 1.772 \times 10^{-5}$; K_a at $60°C = 1.551 \times 10^{-5}$.

12. Calculate the pH of $0.10M$ iodic acid, $K_a = 1.67 \times 10^{-1}$.

13. Calculate the pH of $0.10M$ urea. The K_a of the conjugate acid is 7.1×10^{-1}.

14. Calculate the pH of a solution prepared by diluting 5.71 mL of concentrated acetic acid to exactly 1 liter. Concentrated acetic acid is 100% CH$_3$COOH and has a density of 1.05 g/mL.

15. Weak acids are more completely ionized as their solutions become more dilute. Calculate the fraction of benzoic acid ($K_a = 6.25 \times 10^{-5}$) that is ionized for each of the following solutions.

(a) $1.0 \times 10^{-1}M$
(b) $1.0 \times 10^{-2}M$
(c) $1.0 \times 10^{-3}M$
(d) $1.0 \times 10^{-4}M$

16. 2-Furoic acid (see Section 8–1) has an equivalent weight of 112.08 and a $K_a = 8.63 \times 10^{-4}$. A 224.2-mg sample of 2-furoic acid is weighed out and dissolved in 250 mL of water.

(a) Calculate the pH.
(b) Calculate the pH when 8.09 mL of $0.1000M$ sodium hydroxide has been added.
(c) Calculate the pH when 15.04 mL of $0.1000M$ sodium hydroxide has been added.
(d) Calculate the pH when 20.00 mL of $0.1000M$ sodium hydroxide has been added.

17. Calculate the pH of a solution when 25.0 mL of $0.10M$ hydrazine hydrochloride (NH$_2$NH$_3$$^+Cl^-$) is mixed with 25.0 mL of $0.078M$ sodium hydroxide. The K_a for hydrazine hydrochloride is 1.0×10^{-8}.

Polyprotic Acids

18. The ionization constants of o-phthalic acid are $K_{a1} = 1.20 \times 10^{-3}$, $K_{a2} = 3.9 \times 10^{-6}$.

(a) Calculate the pH of a $0.010M$ solution of phthalic acid.
(b) Calculate the pH of a $0.010M$ solution of potassium acid phthalate.
(c) Calculate the pH of a solution containing $0.010M$ potassium acid phthalate and $0.010M$ potassium phthalate.

19. Calculate the pH of a $0.01M$ solution of disodium EDTA. (See Appendix 2 for the ionization constants.)

20. Calculate the pH of a $0.10M$ solution of sodium bicarbonate. (See Appendix 2 for the ionization constants of carbonic acid at $\mu = 0.1$.)

21. Calculate the pH of a solution resulting from mixing 10.0 mL of $0.20M$ sodium hydrogen sulfite with 5.0 mL of $0.20M$ sodium hydroxide. (See Appendix 2 for the ionization constants of sulfurous acid at $\mu = 0.1$.)

22. Calculate the concentration of oxalic acid, hydrogen oxalate ions, and oxalate ions in a solution by dissolving 10.0 mmoles of oxalic acid in 100 mL of water and adjusting the pH to 4.0. (See Appendix 2 for the ionization constants of oxalic acid at $\mu = 0.1$.)

23. m–Amino benzoic acid (MAB) has a basic amino group and a carboxyl group, each attached to a benzene ring (represented by a hexagon with a circle inside):

The ionization constant for the protonated amino group (K_1) is 8.85×10^{-4} and the ionization constant for the carboxyl K_2) is 2.52×10^{-5}.

(a) Calculate the pH of a $0.100M$ solution at the isoelectric point (neutral MAB).
(b) Calculate the pH of a $0.100M$ solution of the sodium salt of MAB ($-COO^- Na^+$).

24. (a) Write or derive expressions for calculating the fraction of malonic acid that exists as the free acid (H_2M), the monovalent anion (HM^-), and the divalent anion (M^{2-}).
(b) The acid ionization constants for malonic acid are $K_1 = 1.43 \times 10^{-3}$, $K_2 = 2.2 \times 10^{-6}$. Calculate and plot the fraction of H_2M, HM^-, and M^{2-} as a function of pH. (See Figure 7–2 for an example.) A programmed calculator or computer will speed the calculation but is not essential.

25. Calculate the fraction of EDTA present as the 4– anion at pH 5.5. See Appendix 2 for ionization constants. Check your answer with the graph in Figure 11–3.

26. Calculate the fraction of nitrilotriacetic acid (NTA) present as the 3– anion at pH 5.5. See Appendix 2 for ionization constants.

Buffers

27. What is a buffer? What is meant by "buffer capacity"? At what ratio of salt to acid or salt to base is the capacity of a buffer the greatest?

28. Tell how you would experimentally determine the pK of an unknown water-soluble organic base.

29. Consult the table of ionization constants in Appendix 2 and pick out a suitable acid or base to be used in preparing a buffer for each of the following pH values.

(a) pH 3.0   (c) pH 5.2   (e) pH 9.0
(b) pH 4.5   (d) pH 7.5

30. (a) From the list given in the following table, select a suitable buffer to obtain a pH of:

(1) 8.4   (2) 5.0

(b) Calculate the mole ratio of BH^+ to B to obtain the buffer of pH 8.4. [See part (a).]

Base	pK_a of BH^+
Ammonia	9.25
Butylamine	10.61
Hydrazine	8.00
Ethylenediamine	7.30
Pyridine	5.20
Urea	0.20

31. A buffer contains $0.10M$ glycine and $0.20M$ sodium glycinate. K_a for glycine = 1.66×10^{-10}.

    (a) Calculate the pH.
    (b) Calculate the buffer capacity.
    (c) Calculate the buffer capacity of a buffer containing $0.10M$ glycine and $0.10M$ sodium glycinate.

32. At what molar ratio should ammonium hydroxide and ammonium chloride be mixed to buffer a solution at pH 9.8? The K_a for ammonium hydroxide is 1.8×10^{-5}.

33. Calculate the molar ratio of pyridine and hydrochloric acid necessary to produce a buffer of 6.0. The K_b for pyridine is 1.5×10^{-9}.

34. Exactly 20.0 mL of a $0.10M$ solution of 4-aminopyridine ($K_b = 2.34 \times 10^{-5}$) is titrated with $0.10M$ hydrochloric acid. Neglecting dilution effects, calculate and plot the pH versus milliliters of hydrochloric acid when 0, 10.0, 18.0, and 20.0 mL of hydrochloric acid have been added.

Miscellaneous

35. Spectrophotometry is often a good means of measuring the pK of a very weak acid or base. The ratio of free acid (HA) to anion (A$^-$) in an acid is measured spectrophotometrically in a series of solutions buffered at known pH values. For each solution, the pK_a is calculated from the following equation, which can be simply derived from the K_a expression

$$pK_a = pH + \log \frac{[HA]}{[A^-]}$$

The best pK_a value is the average of the several values obtained.

   From the data [1] given in this table, complete the table and calculate the best pK_a value for 8-hydroxyquinoline. Each solution contains $3.3 \times 10^{-5}M$ 8-hydroxyquinoline buffered at various pH values. Spectrophotometric measurements were made at 335 nm. The absorbance at 335 nm when the solution is completely in the anion (A$^-$) form is 0.558; the absorbance at 335 nm when completely in neutral (HA) form is 0.045

pH	Absorbance	[HA]/[A$^-$]/[A$^-$]	pK_a
9.12	0.123		
9.52	0.216		
9.89	0.310		
10.12	0.370		
10.53	0.465		

Average pK_a =

(*Hint:* In solving for the ratio of HA to A$^-$, remember that the solution contains both HA and A$^-$, and that the absorbances are additive. Let x be the fraction of HA present and $1 - x$ the fraction of A$^-$ present.)

[1] Data from A. Albert, *Biochem. J.*, *54*, 646 (1953).

8

Acid-Base Titrations

Acid-base titration is a quick and accurate means of determining acidic or basic substances in analytical samples. Several inorganic acids and bases and hundreds of organic compounds are sufficiently acidic or basic to be determined by acid-base titration. Organic compounds are often titrated in a nonaqueous solvent rather than in water (Chapter 9). A number of useful analytical methods also depend indirectly on acid-base titration. For example, a salt such as potassium chloride can be determined by converting it to hydrochloric acid by means of an ion-exchange column, and then titrating the acid with a standard base.

A standard solution of a strongly basic titrant, such as sodium hydroxide, is used to titrate acids. Bases are titrated with a standard solution of hydrochloric acid or some other strongly acidic titrant. Most commonly, the end point of an acid-base titration is detected by observing the color change of an indicator. However, a pH meter (with appropriate electrodes) can be used to follow the pH changes that occur during an acid-base titration and to locate the end point.

In this chapter, methods for preparing and standardizing titrants will be discussed first. Then titration curves will be considered in order to give a better understanding of various types of acid-base titrations and to correlate equilibrium theory in Chapters 6 and 7 with practical acid-base titrations. The theory and use of acid-base indicators will then be discussed, followed by a section describing some valuable analytical methods employing acid-base titrations.

8–1. PREPARATION AND STANDARDIZATION OF TITRANTS

Sodium Hydroxide. Sodium hydroxide is *not* a primary standard; it always contains some water and some sodium carbonate. To make it a suitable titrant, the carbonate must first be removed, then a sodium hydroxide stock solution of approximately the desired concentration prepared. The exact molarity of the solution is determined by standardizing it against a suitable primary-standard acid.

The sodium carbonate impurities must be removed because they react with the acid to form a buffer, which greatly diminishes the sharpness of the end point in titrating a weak acid. A common way to purify sodium hydroxide is to prepare a nearly saturated aqueous solution of it. Sodium carbonate, being less soluble than sodium hydroxide, settles out of the concentrated solution as a precipitate. After the solution has been allowed to stand, the clear supernatant liquid is carefully decanted and diluted with distilled water (often the distilled water is boiled to remove dissolved carbon dioxide). While this method is useful for sodium hydroxide, it does not work for potassium hydroxide.

Another method of removing carbonate is to precipitate it by adding an excess of a barium salt to the sodium or potassium hydroxide solution:

$$Ba^{2+} + CO_3^{2-} \rightarrow BaCO_3(s)$$

This introduces barium ions as an impurity but usually this does not matter. A method has been proposed in which barium hydroxide is added to precipitate the carbonate. The excess barium is then replaced by Na^+ or K^+ when the hydroxide solution is passed through a cation-exchange column in the sodium or potassium form (see Chapter 21).

Being a strong base, sodium hydroxide will react with carbon dioxide from the atmosphere to form sodium carbonate:

$$2NaOH + CO_2 \rightarrow Na_2CO_3 + H_2O$$

So that carbonate will not form on storage, sodium hydroxide solutions should be protected from atmospheric carbon dioxide. When a large siphon bottle is used, an Ascarite (NaOH on asbestos) tube is used to absorb carbon dioxide from the air that enters the bottle.

Sodium or potassium hydroxide solutions may be standardized by weighing out any of several primary standard acids and titrating them with the base to be standardized. Potassium hydrogen phthalate (KHP), the mono potassium salt of phthalic acid, is very pure and has a high equivalent weight (204.2). It is a moderately weak acid ($K_a = 3.9 \times 10^{-6}$, $pK_a = 5.4$), but will give a satisfactory end point when titrated with sodium hydroxide. 2-Furoic acid [1] is also very pure. It has a lower equivalent weight (112.08) but is a moderately strong acid ($K_a = 8.63 \times 10^{-4}$, $pK_a = 3.06$) and thus gives a very sharp end point when titrated with sodium hydroxide.

Potassium acid phthalate

2-Furoic acid

[1] W. F. Koch, W. C. Hoyle, and H. Diehl, *Talanta, 22,* 717 (1975); *23,* 509 (1976).

A hydrochloric acid stock solution of approximately the desired concentration can be prepared simply by diluting concentrated hydrochloric acid with distilled water. Since concentrated hydrochloric acid is not a primary standard, the diluted titrant must be standardized. 4-Aminopyridine, $C_5H_4N(NH_2)$, is probably the best primary standard available [2, 3]. Its formula weight is slightly low (94.12), but its purity and stability are excellent. A base known as THAM, tris-(hydroxymethyl)aminomethane, has also been used as a primary standard, but its purity is often undependable [4].

Anhydrous sodium carbonate (Na_2CO_3) is frequently used to standardize hydrochloric acid solutions. However, unless the hydrochloric acid is to be used to titrate samples containing carbonate, the use of sodium carbonate is not highly recommended. The equivalent weight of sodium carbonate is quite low (53 with the usual methyl orange or methyl red end point), and the carbon dioxide released complicates end-point detection somewhat.

Very often, hydrochloric acid solutions are standardized by being compared with a *secondary standard*—a standard solution of sodium hydroxide. If the sodium hydroxide solution has been carefully standardized, the concentration of the hydrochloric acid solution can be determined accurately.

8–2. TITRATION CURVES

Experimental Method for Titration Curves. The best way to follow the course of an acid-base titration is to measure the pH as the titration progresses, and to plot a titration curve of pH versus milliliters of titrant (or pH versus percent neutralization). During the greater part of an acid-base titration, the pH changes gradually as titrant is added. Near the equivalence point, however, the pH changes abruptly. The rate of change (Δ pH per Δ mL of titrant) is greatest at the equivalence point.

Data for an acid-base titration curve may be obtained experimentally with a pH meter. The instrument measures the pH of a solution from the difference in potential between two electrodes dipping into the solution. A glass electrode is used as the *indicator electrode* because its potential varies according to the pH of the solution. A calomel electrode is used as the *reference electrode* because its potential does not change, even in solutions of widely varying pH. The difference in potential of these electrodes, measured in volts or millivolts, is a linear function of the pH of the solution. The scale of a pH meter is so designed that the voltage can be read directly in terms of pH.

A glass electrode usually consists of a silver and silver chloride electrode in contact with dilute aqueous hydrochloric acid, surrounded by a glass bulb that acts as a

[2] W. F. Koch, W. C. Hoyle, and H. Diehl, *Talanta*, *22*, 717 (1975).

[3] W. F. Koch and H. Diehl, *Talanta*, *23*, 509 (1976).

[4] W. F. Koch, D. L. Biggs, and H. Diehl, *Talanta*, *22*, 637 (1975).

conducting membrane. A potential difference develops at the interface between the solution and the glass membrane and depends on the *difference* between the hydrogen-ion concentrations on each side of the glass. The hydrogen-ion concentration of the acid solution inside the bulb is constant; therefore, the potential of the glass electrode depends on the hydrogen concentration *outside* of the bulb—that is, in the solution to be measured.

A calomel electrode contains elemental mercury and a paste made of calomel (Hg_2Cl_2) and mercury metal. This paste is in contact with an aqueous solution of potassium chloride. The electrode is so arranged that the potassium chloride solution serves as a salt bridge between the electrode and the solution into which the electrode is dipping.

Strong Acid Titrated with Strong Base; Strong Base Titrated with Strong Acid.   In a titration involving a strong acid and a strong base, the pH changes abruptly at the equivalence point. The titration curves for hydrochloric acid titrated with sodium hydroxide, and for sodium hydroxide titrated with hydrochloric acid, are shown in Figures 8–1 and 8–2. At the equivalence point, a small increment of titrant causes a pH change of several units. Any indicator that has a color transition range within the nearly vertical portion of the titration curve is suitable for the titration. Methyl red, phenolphthalein, and several other indicators may be used for this type of titration.

Acid-base titrations are usually carried out with approximately $0.1M$ to $0.5M$ titrant. Frequently the acid or base to be titrated is diluted until it is about $0.01M$. As shown in Figure 8–1, the effect of dilution is to shorten the pH break at the

Figure 8–1.   Titration curve for hydrochloric acid titrated with sodium hydroxide at different concentrations.

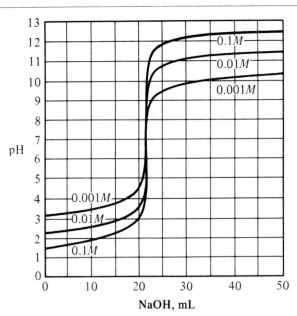

NaOH, mL

Figure 8–2. Titration curve for sodium hydroxide titrated with 0.1M hydrochloric acid.

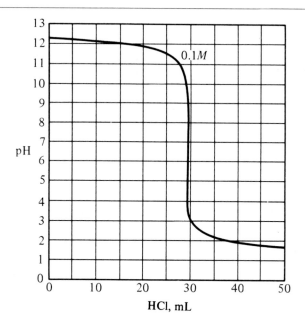

equivalence point. While some reduction in concentration can be tolerated, titration with extremely dilute solutions must be avoided.

Weak Acid Titrated with Strong Base; Weak Base Titrated with Strong Acid. A weak acid of the neutral type (HA) or of the charged type (BH^+) reacts with a strongly basic titrant, such as sodium hydroxide, to form the conjugate base, A^- or B:

$$HA + OH^- \rightarrow A^- + H_2O$$

$$BH^+ + OH^- \rightarrow B + H_2O$$

The course of such a titration curve will depend on the value of the ionization constant and to a lesser extent on the concentration of the acid titrated. Points on the curve can be calculated as follows:

0% Titration. At this point, we have a solution of a weak acid of concentration, C. (In the illustrations, the acid will be of the HA type; however, calculations are identical for an acid of the BH^+ type.) Substituting into the K_a expression,

$$K_a = \frac{[H^+][A^-]}{[HA]} = \frac{[H^+]^2}{C}$$

$$[H^+] = (K_a C)^{1/2}$$

$$pH = -\log[H^+]$$

Example: For a solution of benzoic acid, $C = 0.10M$, $K_a = 10^{-4.20}$:

$$[H^+] = (10^{-4.20} \cdot 10^{-1.00})^{1/2} = 10^{-2.60}, \qquad pH = 2.60$$

5–95% Titration. Each addition of sodium hydroxide converts an equivalent amount of HA to the conjugate base, A^-. In this buffer region, the pH is largely independent of concentration and depends only on the ratio of A^- to HA and on the value of K_a.

$$K_a = \frac{[H^+][A^-]}{[HA]}$$

If P is the percent titration, the ratio of $[A^-]$ to $[HA]$ is given by $P/(100 - P)$, and $[H^+]$ can be easily calculated:

$$K_a = \frac{[H^+]P}{100 - P}$$

$$[H^+] = \frac{K_a(100 - P)}{P}$$

$$pH = -\log[H^+]$$

Example: For a solution of benzoic acid ($K_a = 10^{-4.20}$) at 20%, 50%, and 90% titration:

20%: $\quad [H^+] = \dfrac{(10^{-4.20})(80)}{(20)} = 4.0 \times 10^{-4.20}; pH = 3.60$

50%: $\quad [H^+] = \dfrac{(10^{-4.20})(50)}{(50)} = 1 \times 10^{-4.20}; pH = 4.20$

90%: $\quad [H^+] = \dfrac{(10^{-4.20})(10)}{(90)} = 0.11 \times 10^{-4.20}; pH = 5.15$

Calculations of this type can be carried out very quickly for any percent titration in the buffer region using a programmable calculator or computer. Combining the equations for calculating $[H^+]$ and for converting $[H^+]$ to pH gives

$$pH = -\log K_a - \log(100 - P) + \log P$$

The exact program depends on the calculator used but would probably involve putting the value of $-\log K_a$ (pK_a) in a memory. A value for P would go into a second memory. In calculating pH for the next value of P, that value would replace the previous value in the second memory.

For Hewlett-Packard calculators, the following program could be used: STO 1, log, RCL 1, CHS, 100, +, log, −, RCL 2, +. In the run mode, store the value of pK_a in memory 2, then punch the desired value of P and then R/S to calculate the pH automatically.

100% Titration. At this point, all of the original acid has been converted to the conjugate base, A^- or B. If we neglect dilution during the titration and consider the concentration of the base to be the same as the starting concentration, C, the pH can be calculated from the K_b value of the conjugate base:

$$K_b = \frac{10^{-14}}{K_a} = \frac{[BH^+][OH^-]}{[B]} = \frac{[OH^-]^2}{C}$$

$$[OH^-] = (K_b C)^{1/2}$$

$$pOH = -\log[OH^-], \qquad pH = 14 - pOH$$

Example: For a solution of benzoic acid ($K_a = 10^{-4.20}$), titrated completely to benzoate, $C = 0.10M$:

$$[OH^-] = (10^{-9.80} \cdot 10^{-1.00})^{1/2} = 10^{-5.40}$$

$$pOH = 5.40, \qquad pH = 14 - 5.40 = 8.60$$

A plot of the titration curve given in the examples (0.10M benzoic acid, titrated with sodium hydroxide) from 0% to 100% titration is given in Figure 8–3. If additional sodium hydroxide is added beyond the stoichiometric point in the titration, the pH is calculated from the concentration of excess sodium hydroxide. The reason for this is that the weak base, A^-, has an inconsequential effect on the pH in comparison to the strong base, sodium hydroxide. Note in Figure 8–3 that the pH gradually levels off as more sodium hydroxide is added.

Suppose we reverse the direction of titration and titrate a weak base with a strong acid such as hydrochloric acid. The titration of sodium benzoate with

Figure 8–3. Titration of 0.1M benzoic acid with 0.1M sodium hydroxide.

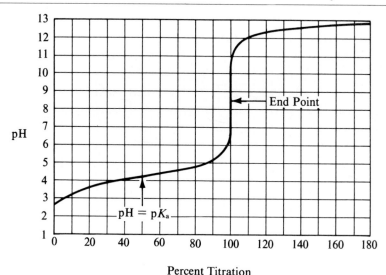

Percent Titration

hydrochloric acid will serve as an example:

$$A^- + H^+ \rightarrow HA$$
$$\text{(HCl)}$$

The course of this titration curve is obtained by starting at the pH of A^- in Figure 8–3 and retracing the curve from right to left to HA in the figure. If we now add additional hydrochloric acid, the pH will drop further and then level off as shown by the part of the curve in Figure 8–3 to the left of HA. In this example, note that the end point in the titration of A^- with hydrochloric acid is not very sharp. This is because A^- is such a weak base ($K_b = 10^{-9.80}$).

A family of titration curves for acids with varying K_a values can easily be calculated. These curves, shown in Figure 8–4, cover virtually any situation in which a monofunctional acid or base is titrated. The curve for an acid is obtained by tracing the curve at the selected K_a value from HA or BH^+ to the point at the right at which excess sodium hydroxide has been added. For the reverse case, in which a base is titrated with hydrochloric acid, start at A^- or B and follow the curve to the left out to where excess hydrochloric acid has been added.

Several things should be noted about these titration curves. One is that at the midpoint in any titration (halfway between HA and A^- or between BH^+ and B), the pH is always equal to the pK_a of the acid. In the titration of any *acid*, the magnitude (and sharpness) of the pH break at the end point decreases as the K_a of the acid becomes smaller. For acids as weak as $K_a = 10^{-9}$, the end point break is so slight that titration in aqueous solution is not considered to be feasible. In the titration of a

Figure 8–4. Titration curves as a function of K_a.

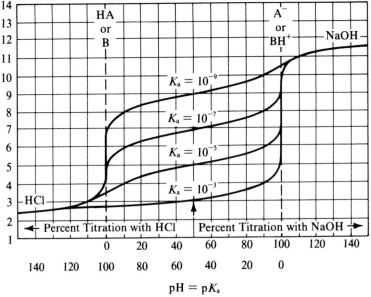

base, the titration becomes *better* as the K_a of the conjugate acid becomes smaller. This is because the K_b is becoming larger.

Titration of Two Acids or Two Bases of Different Strengths. If there is an appreciable difference in acidic strength, the stronger acid in a mixture will be titrated first and give a pH break at its equivalence point. Then the weaker acid will be titrated and give a second pH break at its equivalence point. This is illustrated in Figure 8–5 for hydrochloric and acetic acids, where the curves of the individual acids are shown with the curve of the acid mixture. A titration in which a separate end point is obtained for each of two or more constituents in a mixture is called a *differentiating titration*.

A differentiating titration of two acids is possibly only if the ratio of the ionization constants is approximately 10^3 or greater—that is, a difference of 3 or more pK units. In Figure 8–6, however, the ratio of the K_a values for lactic and acetic acids is only $10^{-3.88}/10^{-4.86}$, or $10^{0.98}$. The individual titration curves are thus too similar, and the curve for titration of the mixture shows only one pH break when the *sum* of the two acids has been titrated.

Many compounds have two or more acidic or basic groups in the same molecule. Acids of this type are called diprotic acids (two groups), triprotic acids (three groups) or, in general, polyprotic acids. Bases with two groups are called diacid bases, etc. Usually, the neutralization of one acidic or basic group causes an electronic rearrangement within the molecule that makes it more difficult to neutralize the next acidic or basic group. For this reason, the compound ionizes in

Figure 8–5. Curve for the differentiating titration of hydrochloric and acetic acids with sodium hydroxide.

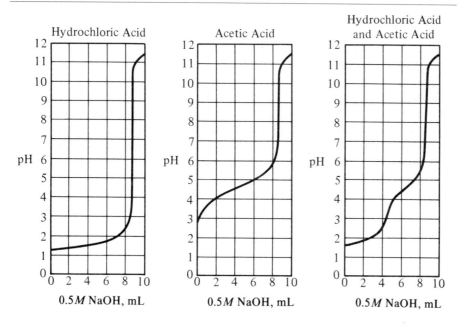

Figure 8–6. Curve for the titration of lactic and acetic acids with sodium hydroxide.

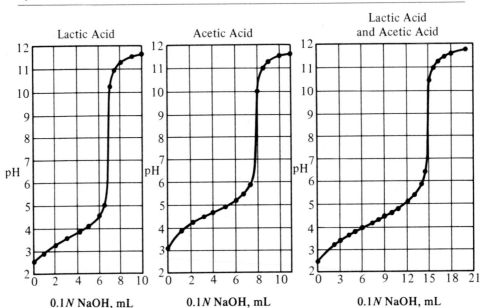

steps. An ionization constant may be written for each step; for example:

$$H_2A \rightleftharpoons H^+ + HA^-, \qquad K_1 = \frac{[H^+][HA^-]}{[H_2A]}$$

$$HA^- \rightleftharpoons H^+ + A^{2-}, \qquad K_2 = \frac{[H^+][A^{2-}]}{[HA^-]}$$

If the ratio K_1/K_2 is greater than approximately 10^3, a diprotic acid will give two pH breaks when titrated with a strong base, just as the titration of two *different* acids may produce two pH breaks. The sharpness of the first break increases as the ratio of K_1/K_2 becomes greater.

The method for calculating pH at various stages in the titration of a diprotic acid was illustrated in Section 8–2. This may be summarized as follows: (a) The pH during the titration of the first acidic hydrogen is calculated from K_1 using the ratio of $[HA^-]$ to $[H_2A]$. (b) The pH at the first equivalence point is calculated from the approximate expression, pH $= (pK_1 + pK_2)/2$. (c) The pH during the titration of the second acidic hydrogen is calculated from K_2 using the ratio of $[A^{2-}]$ to $[HA^-]$. (d) At the second equivalence point, the pH is calculated from the ionization of A^{2-}, the conjugate base of HA^-.

Example: In the titration of 20 mL of $0.1M$ malonic acid (H_2A) with $0.10M$ sodium hydroxide, calculate the pH when 10, 20, 30, and 40 mL of sodium hydroxide have been added. Also sketch the approximate titration curve for the entire titration. For malonic acid, $pK_1 = 2.85$ and $pK_2 = 5.66$.

10 mL of NaOH *Added:* The first hydrogen of the malonic acid is now 50% neutralized, so $[HA^-] = [H_2A]$. Substituting into the K_1 expression,

$$10^{-2.85} = \frac{[H^+][\cancel{HA^-}]}{[\cancel{H_2A}]} = [H^+]$$

$$pH = 2.85$$

20 mL of NaOH *Added:* We are now at the first equivalence point, and the pH is calculated from the approximate expression

$$pH \approx \frac{pK_1 + pK_2}{2} \approx \frac{2.85 + 5.66}{2} \approx 4.26$$

30 mL of NaOH *Added:* The first 20 mL of NaOH has neutralized the first acidic hydrogen of malonic acid, and the next 10 mL of NaOH has neutralized 50% of the second acidic hydrogen. Therefore, $[HA^-] = [A^{2-}]$. Substituting into the K_2 expression,

$$10^{-5.66} = \frac{[H^+][\cancel{A^{2-}}]}{[\cancel{HA^-}]} = [H^+]$$

$$pH = 5.66$$

40 mL of NaOH *Added:* The malonic acid has been titrated to A^{2-}, which we shall estimate as $0.01M$. The pH is calculated from pK_b for A^{2-}, which equals $14 - pK_2$, or 8.34.

$$A^{2-} + H_2O \rightleftharpoons HA^- + OH^-$$

$$10^{-8.34} = \frac{[HA][OH^-]}{[A^{2-}]} = \frac{[OH^-]^2}{10^{-2.0}}$$

$$[OH^-] = \sqrt{10^{-10.34}}$$

$$pOH = 5.17, pH = 8.83$$

Using these key points, the titration curve can be sketched as in Figure 8–7.

8–3. ACID-BASE INDICATORS

Acid-base indicators are highly colored weak acids or bases. Most are two-color indicators in which the acidic and basic forms have contrasting colors. There are also a few one-color indicators such as phenolphthalein, in which the acidic form is colorless and the basic form is magenta.

In titrating a single acid or base, the indicator acts as a second acid or base. Thus, in titrating an acid with sodium hydroxide, the second acid (the indicator) is *weaker* than the main acid and titrates after it. The indicator is also present in much lower concentration. A titration curve with a very large amount of indicator present might look like curve *a* in Figure 8–8. The main acid is titrated first, followed by the

Figure 8–7.   Curve for the titration of 20 mL of $0.10M$ malonic acid with $0.10M$ sodium hydroxide.

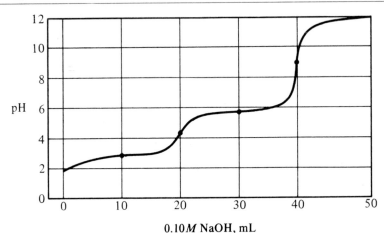

weaker indicator acid. In actual practice, however, so little of the highly colored indicator is used that the titration curve approximates curve b in the figure.

It will be seen from Figure 8–8 that an indicator changes color over a pH range and not at a single pH. The transition range of an indicator depends on the ability of the observer to detect subtle color changes. For a two-color indicator, the transition range covers approximately two pH units.

Suppose we have an indicator acid, HIndic. Since it is a weak acid, we may write an equation for its ionization and an ionization-constant expression:

$$\text{HIndic} \rightleftharpoons \text{H}^+ + \text{Indic}^-, \qquad K_a = \frac{[\text{H}^+][\text{Indic}^-]}{[\text{HIndic}]}$$

The acidic form of the indicator is HIndic, which (we shall assume) is colored. When the indicator is neutralized with a strong base, the indicator is present as Indic^-, which exhibits a color different from that of the acidic form. Starting with the acidic form of a two-color indicator, most people cannot distinguish any color change until more than a tenth of the indicator has been converted to its basic form, Indic^-. By substituting into the ionization-constant expression, the pH at the acid end of the indicator can be calculated:

$$K_a = \frac{[\text{H}^+]1}{10}$$

$$\text{H}^+ = K_a(10)$$

$$\text{pH} = pK_a - 1$$

As neutralization continues, the indicator appears visually to have its full basic color

Figure 8–8. Titration of an acid with a visual indicator added: Curve (*a*) with an abnormally large amount of indicator; curve (*b*) with a normal amount of indicator. The shaded area shows the approximate pH transition range of the indicator.

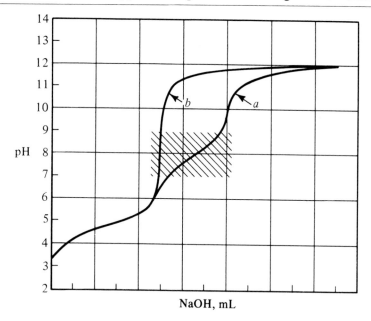

NaOH, mL

when the solution contains 1 part of the acidic indicator color to about 10 parts of the basic color. The pH at the basic end of the transition is calculated by substituting into the K_a expression:

$$K_a = \frac{[H^+]10}{1}$$

$$H^+ = \frac{K_a}{10}$$

$$pH = pK_a + 1$$

The difference in pH between the acidic and basic ends of the indicator transition is

$$pH_{basic} - pH_{acidic} = (pK_a + 1) - (pK_a - 1) = 2$$

The transition ranges of some common indicators are shown in Figure 8–9. The wide variety in the properties of different indicators makes it possible to select one with a transition range that comes at the steepest part of an acid-base titration curve. From the titration curves illustrated in this chapter, it will be observed that the equivalent-point pH is usually not 7.0. For instance, in the titration of the weak acid in Figure 8–3, the pH at the equivalence point is 9.0, which is within the transition range of phenolphthalein. In the titration of a weak base with

Figure 8–9.   Transition ranges of some acid-base indicators.

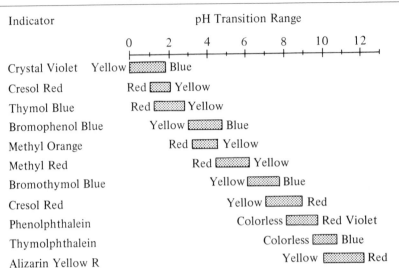

Indicator	pH Transition Range

hydrochloric acid which gives an equivalence-point pH of 4.2, methyl orange or bromophenol blue is a suitable indicator (see Figure 8–9.)

Since an indicator changes color over a pH *range*, it is often difficult to know just which color to take as the end point. One answer to this question is first to do an acid-base titration (with indicator present) using a pH meter and then to determine the color of the indicator at the equivalence point. Subsequent samples can be titrated to exactly this color. Sometimes a color standard is prepared, and the titration is carried out until the indicator color matches that of the standard.

8–4.   SOME ACID-BASE METHODS

Molecular Weight of a Weak Acid.   If the ionization constant and concentration of a weak acid are approximately known, a suitable indicator can be selected, and the acid can be titrated directly with standard sodium hydroxide. Ordinarily, phenolphthalein is a satisfactory indicator. In research, analytical information is often needed about a pure, unknown acid. Acid-base titration can be used to determine the molecular weight of the acid or, if the number of acidic groups in the molecule is unknown, the equivalent weight. If the titration is carried out potentiometrically with a pH meter, the pK_a of the acid can also be measured. This is determined from the pH at the midpoint (50% neutralization) of the titration, where [HA], the concentration of unneutralized acid, is equal to [A$^-$], the concentration of salt from neutralization of the acid. Substituting into the ionization-constant expression of the acid,

$$K_a = \frac{[H^+][A^-]}{[HA]}$$

$$K_a = [H^+]$$

$$pK_a = pH$$

Titration of Sodium Carbonate, and Mixtures Containing Carbonate. Sodium carbonate is the conjugate base of the bicarbonate ion, which in turn is the conjugate base of carbonic acid. The pK_a values listed in Appendix 2 for carbonic acid are interpreted as follows: pK_{a_1} governs the ionization of carbonic acid to the bicarbonate ion, and pK_{a_2} governs the ionization of the bicarbonate ion to the carbonate ion.

When sodium carbonate is titrated with a strong acid (such as hydrochloric acid), the carbonate ion is first converted to the bicarbonate ion, and then to carbonic acid:

$$CO_3^{2-} + H^+ \rightarrow HCO_3^-$$
$$\text{(Na}_2\text{CO}_3\text{)} \quad \text{(HCl)}$$

$$HCO_3^- + H^+ \rightarrow H_2CO_3$$
$$\text{(HCl)}$$

The general shape of the titration curve can be deduced from the two acid-ionization constants of carbonic acid: $pK_1 = 6.37$, $pK_2 = 10.32$. Thus, if carbonic acid were titrated with sodium hydroxide, the pH halfway to the first equivalence point (i.e., where the solution contains 50% H_2CO_3 and 50% HCO_3^-) is 6.37 (pH = pK_1). At the first equivalence point, HCO_3^- is present, and the pH is $(pK_1 + pK_2)/2$, or 8.35. Halfway from the first to the second equivalence point, the solution contains 50% HCO_3^- and 50% CO_3^{2-}, and the pH is 10.32 (pH = pK_2).

When sodium carbonate is titrated with standard hydrochloric acid, the titration curve will of course be reversed. The experimental curve for titration of sodium carbonate (Figure 8–10) shows fairly good agreement with the pH values calculated above. In Figure 8–10, the second end point is sharpened by boiling to remove most of the carbonic acid as carbon dioxide gas.

$$H_2CO_3 \xrightarrow{\text{heat}} H_2O + CO_2(g)$$

At the first end point (when carbonate has been titrated to bicarbonate) the phenolphthalein indicator changes from magenta to colorless. This change is gradual, and the accuracy is mediocre. After this end point, conversion of bicarbonate to carbonic acid begins. This titration is complete at the second end point. The methyl orange changes from yellow to pink at the second end point. This end point is sharper than the phenolphthalein end point, but the change is still somewhat gradual.

The best procedure is to use methyl red indicator and titrate until it becomes red; the color change will be very gradual. At this point, the solution is then boiled

Figure 8–10.   Titration curve for sodium carbonate titrated with hydrochloric acid and methyl red indicator.

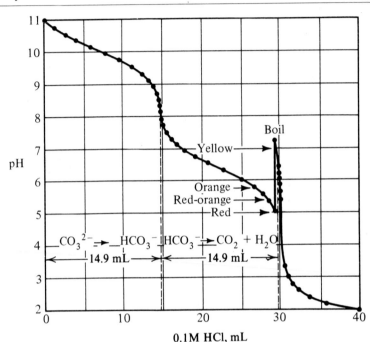

0.1M HCl, mL

for a minute, so that the dissolved carbon dioxide is volatilized. Then it is cooled, and the titration is continued until there is a sharp methyl red indicator change from yellow to pink. The pH jump in Figure 8–10 shows the effect the boiling has on the titration curve.

Sodium carbonate and sodium bicarbonate often occur together. The amount of each in a mixture may be determined by a differentiating acid-base titration (see Figure 8–11). Sodium carbonate is the stronger base ($pK_{b_1} = 3.68$, $pK_{b_2} = 7.63$). Sodium bicarbonate is a weaker base ($pK_b = 7.63$), and its titration gives only a single end point. The mixture may be analyzed in the following manner:

1. Titrate with standard hydrochloric acid to the phenolphthalein end point. *Only the CO_3^{2-} is titrated.* Note that at the end point, CO_3^{2-} is neutralized to HCO_3^-.
2. Continue the titration to the methyl orange or methyl red end point. *All* of the HCO_3^- present will be neutralized—that is, the HCO_3^- originally present and the HCO_3^- from the partial neutralization of CO_3^{2-}. Thus a larger volume of the titrant will be required to go from the first to the second end point than from the start of the titration to the first end point. The amounts of CO_3^{2-} and HCO_3^- in the original sample can be calculated from the buret readings at the phenolphthalein and methyl orange (or methyl red) end points.

Figure 8–11. Titration curves for (a) sodium carbonate, (b) sodium bicarbonate, and (c) a mixture of the two titrated with hydrochloric acid. Curves are moved horizontally for clarity.

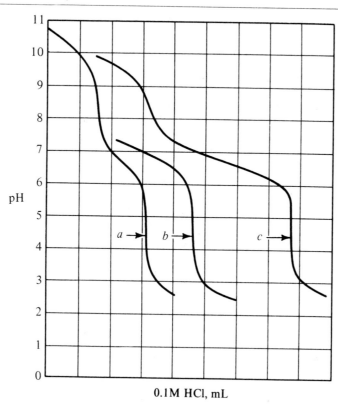

0.1M HCl, mL

Example: A mixture of sodium carbonate and sodium bicarbonate is dissolved in water and titrated with 0.1M HCl. The buret reading at the phenolphthalein end point is 12 mL, and the buret readings at the methyl orange end point is 34 mL. How many millimoles of carbonate and of bicarbonate are in the sample?

12 mL of HCl to titrate CO_3^{2-} to HCO_3^- (phenolphthalein end point)

+12 mL of HCl to continue titration of *this* HCO_3^- to methyl orange end point

24 mL of HCl to neutralize completely the CO_3^{2-}

$34 - 24 = 10$ mL to titrate HCO_3^- *originally present*

$12 \times 0.1 = 1.2$ mmoles of CO_3^{2-}

$10 \times 0.1 = 1.0$ mmoles of HCO_3^-

Mixtures of sodium hydroxide and sodium carbonate can also be analyzed by titration with hydrochloric acid to two different end points. Sodium hydroxide is a stronger base than sodium carbonate, but not enough stronger to result in a third end point in the titration. At the first (phenolphthalein) end point, NaOH plus CO_3^{2-} is titrated; between the first and second end points, the HCO_3^- from the CO_3^{2-} in the original sample is titrated. The titration curve for a sodium hydroxide and sodium carbonate mixture is shown in Figure 8–12. If the buret reading is 30 mL at the first end point and 42 mL at the second, then the titration of HCO_3^- required $42 - 30 = 12$ mL of HCl. An *additional* 12 mL, therefore, was required to titrate the original CO_3^{2-} to HCO_3^-, and titration of the OH^- in the original sample required $42 - 24 = 18$ mL of HCl.

Mixtures of sodium hydroxide and sodium bicarbonate cannot exist in solution because they react and form carbonate:

$$OH^- + HCO_3^- \rightarrow CO_3^{2-} + H_2O$$
$$\text{(NaOH)}$$

Figure 8–12. Titration curve for a mixture of sodium hydroxide and sodium carbonate titrated with hydrochloric acid.

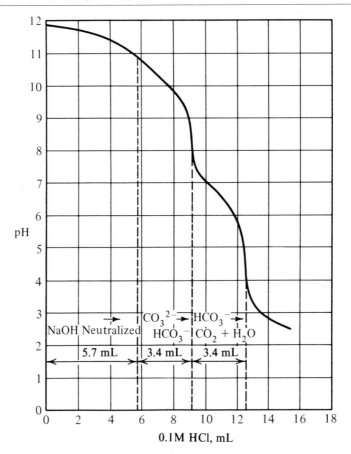

Kjeldahl Nitrogen Determination. The Kjeldahl method is very important for analyzing organic samples containing nitrogen, such as pure organic compounds, foods, fertilizers, etc. The protein content of food and animal feeds is estimated from a Kjeldahl nitrogen determination.

The method consists of several steps, each of which will be discussed briefly.

Step 1, Prereduction. The Kjeldahl method determines amine or amide nitrogen. Prior reduction is required for inorganic nitrates and for organic nitro and azo compounds and certain other compounds.

Step 2, Digestion. The sample is decomposed by digestion with hot, concentrated sulfuric acid. The organic matter is oxidized to carbon dioxide and water; the nitrogen is converted to ammonium hydrogen sulfate:

$$\text{Organic C, H, N} \xrightarrow[\text{H}_2\text{SO}_4]{\text{O}} CO_2 + H_2O + NH_4HSO_4$$

Potassium hydrogen sulfate added to the digestion mixture raises the boiling point. A mercuric, cupric, or selenium compound is added as a catalyst to speed the decomposition.

Step 3, Distillation. When digestion is complete, the solution is cooled, and a concentrated aqueous solution of sodium hydroxide is added carefully, so that it forms a separate layer on top of the sulfuric acid. The flask is connected to a distillation apparatus and is agitated until the layers mix. Sodium hydroxide neutralizes the sulfuric acid and evolves ammonia from the ammonium salt:

$$2OH^- + NH_4HSO_4 \rightarrow NH_3(g) + 2H_2O + SO_4^{2-}$$
$$\text{(2NaOH)}$$

The flask is heated so that ammonia, together with some water, is distilled. The distillate is collected in a receiver that contains standard hydrochloric acid or saturated boric acid, to neutralize the ammonia and prevent loss by volatilization.

Step 4, Titration. In the ordinary method, an accurately measured volume of hydrochloric acid is added to the receiver before the distillation. The amount of hydrochloric acid must be in excess of that required for neutralizing the ammonia. The ammonia distillate is neutralized by the acid as follows:

$$H^+ + NH_3 \rightarrow NH_4^+$$
$$\text{(HCl)}$$

Next, the excess hydrochloric acid is titrated with standard sodium hydroxide. The amount of ammonia (and hence the amount of nitrogen in the sample) is computed from the *difference* (in millimoles) between HCl in the receiver and of NaOH needed for back-titrating the excess HCl.

The boric acid modification of this method requires only one standard solution and is more direct. Ammonia is distilled and collected in a saturated solution of boric acid, H_3BO_3. Boric acid is a very weak acid ($K_a = 10^{-9}$), and the exact amount of it does not need to be known. The reaction forms ammonium borate, $NH_4H_2BO_3$:

$$NH_3 + H_3BO_3 \rightarrow NH_4^+ + H_2BO_3^-$$

Figure 8–13. Cation-exchange column.

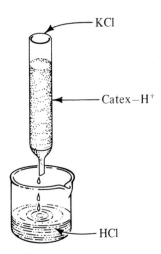

Borate is the conjugate base of boric acid and has a pK_b of 5. The borate is titrated with standard hydrochloric acid without interference from the excess of boric acid:

$$H^+ + H_2BO_3{}^- \rightarrow H_3BO_3$$

(HCl)

The digestion and distillation steps in a classical Kjeldahl determination are time-consuming, and a complete Kjeldahl determination may require one to two hours. A new method greatly speeds up a Kjeldahl nitrogen determination [5]. A powdered sample is weighed into a volumetric flask, a mixture of sulfuric acid and hydrogen peroxide is added, and the flask and contents are heated on an electric heater that also has a fume hood operated by a water aspirator. After a short heating period (2–3 min), a second portion of sulfuric acid–hydrogen peroxide is added, and a brief additional heating completes the digestion. The flask is removed from the heat, cooled, and diluted to volume with water. Then a measured portion of the solution is pipetted into a second volumetric flask, Nessler's Reagent is added to form a color with the ammonia, and the flask is diluted to volume. The amount of nitrogen in the original sample is calculated from a spectrophotometric measurement of the color. (See Chapter 17 on spectrophotometry.) A complete nitrogen determination can be completed in less than 10 min on most samples.

Determination of Salts Using Ion Exchange. The concentration of a salt in solution can usually be determined by an acid-base titration following passage of the solution through a cation-exchange column (See Chapter 24.) For example, if a potassium chloride solution is passed through a cation-exchange column in the hydrogen form, the potassium ions are taken up by the column and an equivalent amount of hydrogen ions goes into solution (Figure 8–13). The net result is that potassium chloride is quantitatively converted to hydrochloric acid. The acid can be easily determined with standard sodium hydroxide.

[5] Hach Co., Ames, Jowa, 1984.

If the salt of a divalent cation is passed through the cation-exchange column, two moles of hydrogen ions go into solution for each mole of cations:

$$M^{2+} + 2\,catex-H^+ \rightarrow catex_2-M^{2+} + 2H^+$$

The uses of ion exchange in analytical chemistry are discussed in Chapter 24.

QUESTIONS AND PROBLEMS

Definitions and General Concepts

1. Why must the sodium hydroxide solutions used in acid-base titrations be free from carbonate? Describe how a carbonate-free sodium hydroxide solution is prepared.
2. What is a primary standard? List the properties of an ideal primary standard for acid-base titrations (see Chapter 5). Compare potassium acid phthalate and 2-furoic acid as primary standards for sodium hydroxide solutions.
3. Give an example of a secondary standard used in acid-base titrations.
4. Tell what an indicator electrode and a reference electrode are. What electrodes are used for acid-base titrations with a pH meter?

Titration Curves, Indicators

5. Calculate the pH at several points and sketch carefully on graph paper each of the following titration curves.

   (a) Titration of $0.10M$ lactic acid with $0.10M$ sodium hydroxide. K_a for lactic acid $= 1.32 \times 10^{-4}$.
   (b) Titration of $0.01M$ o-nitrophenol with $0.01M$ sodium hydroxide. K_a for o-nitrophenol $= 6.20 \times 10^{-8}$.
   (c) Titration of $0.10M$ sodium lactate with $0.10M$ hydrochloric acid. [See part (a).]
   (d) Titration of $0.01M$ sodium cyanide with $0.01M$ hydrochloric acid. K_a for HCN $= 4.90 \times 10^{-10}$.

6. An analyst attempted to titrate 500 mL of a solution estimated to contain about 10 mg/L of sulfuric acid with $0.01M$ sodium hydroxide. Sulfuric acid is a strong acid, and its titration with methyl orange indicator (transition range pH 3.0–4.8) is usually an easy matter, but in this case no definite end point was obtained. Explain what the difficulty is and what might be done about it.
7. 4-Aminopyridine ($K_b = 2.34 \times 10^{-5}$) is used as a primary standard for a hydrochloric acid solution. Calculate the pH when 50% and when 100% of the 4-aminopyridine has been neutralized by $0.10M$ hydrochloric acid. (Assume that the concentration of neutralization product is $0.01M$ at 100% titration.) Sketch the entire curve and select a suitable indicator.
8. Draw an approximate curve for the titration of 40 mL of $0.03M$ triethanolamine ($K_b = 6.6 \times 10^{-7}$) with $0.06M$ hydrochloric acid. Calculate the pH at 50% and at 100% neutralization (BH$^+$ will be $0.02M$ at the equivalence point). Consult Figure 8–9 and select a suitable indicator for the titration.
9. Calculate the pH at the stoichiometric point for each of the following titrations. In each case, assume the product of the titration has a concentration of $0.01M$ at the stoichiometric point.

(a) Hydrazine ($K_b = 1.00 \times 10^{-6}$), titrated with HCl.
(b) Ethylamine ($K_b = 4.68 \times 10^{-4}$), titrated with HCl.
(c) Cyanoacetic acid ($K_a = 3.50 \times 10^{-3}$), titrated with NaOH.
(d) o-Nitrophenol ($K_a = 6.20 \times 10^{-8}$), titrated with NaOH.

10. The indicator methyl red ($K_a = 7.9 \times 10^{-6}$) is added to a buffered aqueous solution of unknown pH. By spectrophotometric measurements, the ratio of the basic to acidic form of the indicator is found to be 2.45 to 1 in this solution. Calculate the pH of the solution.
11. Explain why a two-color indicator changes over a pH range (usually about 2 pH units).
12. The indicator bromothymol blue changes from yellow (acidic) to blue (basic). In a particular titration, the color changes from yellow to yellow-green and finally to blue as 0.02-mL increments of titrant are added near the end point. Explain how to determine which color indicates the correct end point.

Titration Curves of Polyprotic Acids

13. Indicate the general shape of the curve produced when each of the following dibasic acids is titrated with $0.1M$ sodium hydroxide. In each case, what is the pH when 1 mole of sodium hydroxide has been added for each mole of the acid?

(a) Maleic acid ($pK_1 = 1.92$, $pK_2 = 6.23$)
(b) Oxalic acid ($pK_1 = 1.27$, $pK_2 = 4.27$)
(c) Adipic acid ($pK_1 = 4.41$, $pK_2 = 5.28$)

14. Sketch the approximate curve for the titration of nitrilotriacetic acid, $N(CH_2COOH)_3$, with $0.1M$ sodium hydroxide. Calculate the pH at the sharpest end point and select a suitable indicator from Figure 8–9. Acid-ionization constants for nitrilotriacetic acid are $pK_1 = 2.0$, $pK_2 = 2.6$, $pK_3 = 9.8$.
15. Sketch the titration curve for disodium hydrogen phosphate, Na_2HPO_4, titrated with $0.1M$ hydrochloric acid (see the ionization constants for phosphoric acid in Appendix 2). Suggest a method for titrating a sample containing NaH_2PO_4 and Na_2HPO_4.
16. Calculate the pH at the end point in the titration of $0.2M$ potassium acid maleate (KHM) with $0.2M$ sodium hydroxide. Choose a suitable indicator for detecting the end point. The ionization constants for maleic acid (H_2M) are $K_1 = 1.2 \times 10^{-2}$ and 9×10^{-7}.
17. For each of the following titrations with standard base, calculate the pH at the equivalence point(s) listed.

(a) H_3PO_4, titration of first hydrogen and titration of second hydrogen. $pK_1 = 2.1$, $pK_2 = 7.2$, $pK_3 = 12.3$.
(b) Oxalic acid, titration of first hydrogen. $pK_1 = 1.1$, $pK_2 = 4.3$.
(c) Nitrilotriacetic acid, titration of sum of first and second hydrogens. $pK_1 = 2.0$, $pK_2 = 2.6$, $pK_3 = 9.8$.
(d) EDTA, titration of sum of first and second hydrogens, titration of third hydrogen. $pK_1 = 2.1$, $pK_2 = 2.7$, $pK_3 = 6.2$, $pK_4 = 10.3$.

18. For each equivalence point pH in the previous problem, select a suitable indicator from the following list and calculate the mole ratio of basic to acidic forms of the indicator that corresponds to the desired pH.
     Indicators and their pK_a values: methyl orange, $pK_a = 3.46$; bromophenol blue, $pK_a = 4.10$; chlorophenol blue, $pK_a = 4.43$; chlorophenol red, $pK_a = 6.25$; phenol red, $pK_a = 8.00$; thymol blue, $pK_a = 9.20$.

19. The curve for a titration of malonic acid with sodium hydroxide is shown in Figure 8–7. Describe what would happen if a mixture of disodium malonate and sodium hydroxide were titrated with hydrochloric acid.

Numerical Calculations

20. Following Problem 19, the titration of a disodium malonate-sodium hydroxide mixture required 17.12 mL of $0.0977M$ hydrochloric acid to reach the first potentiometric end point and an additional 12.30 mL to reach the second potentiometric end point. Calculate the milligrams of sodium hydroxide (form wt = 40.00) and of disodium malonate (form wt = 148.03) in the sample titrated.

21. A 308.5-mg sample contains malonic acid (form wt = 104.06), sodium hydrogen malonate (form wt = 126.04), and water. Titration to the first potentiometric end point required 18.06 mL of $0.1000M$ sodium hydroxide; an additional 25.14 mL of sodium hydroxide was required to reach the second potentiometric end point. Calculate the percentage of malonic acid and sodium hydrogen malonate in the sample.

22. A sample containing disodium malonate plus sodium chloride was passed through a cation-exchange column in the hydrogen form. Titrating the effluent acid with standard base gave a titration curve similar to Figure 8–7, but 28.60 mL of $0.1000M$ sodium hydroxide was required to reach the first potentiometric end point, while only an additional 14.16 mL was required to reach the second end point. Calculate the milligrams of disodium malonate (form wt = 148.03) and of sodium chloride (form wt = 58.44) in the sample.

23. A pure, unknown organic base is titrated with standard perchloric acid. From the following data, calculate the equivalent weight of the base: sample weight = 0.6650 g; 24.20 mL of $0.1000M$ perchloric acid used.

24. Exactly 427 mg of 100% pure unknown organic acid is titrated with $0.1000M$ sodium hydroxide using a pH meter. The titration curve shows that the acid is monoprotic and that the end point occurs at 35.00 mL of titrant. The pH at the point at which 17.50 mL of titrant have been added is 4.10 (the ionic strength μ is 0.10). (a) Calculate the ionization constant of this acid. (b) Calculate the formula (equivalent) weight of this acid. (c) If the acid contains only C, H, and O, give a logical empirical formula. (d) Assuming the acid group is —COOH, write a logical structure for this acid.

25. A 1.000-g impure sample containing sodium carbonate and sodium bicarbonate is dissolved in water and titrated with $0.10000M$ hydrochloric acid. The buret reading at the phenolphthalein end point is 17.5 mL; at the methyl red end point, it is 40.1 mL. Calculate the percentages of sodium carbonate and sodium bicarbonate in the sample.

26. An organic chemist wishes to know the formula weight of a newly synthesized alcohol. A 52.0-mg sample requires 8.48 mL of $0.1000M$ sodium hydroxide for titration by the acetic anhydride method. The acetic anhydride blank requires 12.58 mL to titrate the acid produced. Calculate the formula weight of the alcohol, assuming that there is only one alcoholic —OH group in the molecule.

27. An antacid preparation claims that one 3.0-g tablet contains enough antacid to neutralize a "full stomach" of stomach acid. Assuming that stomach acid is $0.1M$ hydrochloric acid and that the average stomach contains 0.75 L of acid, calculate the equivalent weight of the antacid preparation.

28. The level of CO_2 in blood (as H_2CO_3) can be determined by accurately meauring the pH and then titrating the amount of bicarbonate. Calculate the molar concentration of CO_2 for a sample that has a pH = 6.220 and a HCO_3^- concentration = $0.0020M$. Assume an

H_2CO_3–HCO_3^- equilibrium and K_1 for H_2CO_3 of 4.30×10^{-7}. (Ignore K_2 for ionization of H_2CO_3.)

Methods of Analysis

29. A certain series of samples may contain sodium hydroxide, sodium carbonate, sodium bicarbonate, or a mixture of two of these compounds. From the data given, decide which compound or compounds each of the following samples contains

HCl Buret Reading, mL

Sample	Phenolph.	Methyl Red
1	0.00	20.05
2	14.67	22.80
3	16.02	32.05
4	15.16	39.08

30. Ammonium chloride is too weak an acid ($pK_a = 9.25$) to titrate accurately with standard base. Outline two procedures, each involving an acid base titration for determining the amount of ammonium chloride in solution.
31. Oxalic acid ($pK_1 = 1.1$, $pK_2 = 4.3$) and sulfuric acid often occur together in analytical samples. Outline an acid-base titration scheme for determining the amount of each acid in an aqueous solution.
32. The amount of carbon dioxide in a gas also containing O_2 and N_2 is determined by bubbling the gas through a standard solution of sodium hydroxide.

   (a) If a large excess of sodium hydroxide is used, in what chemical form will the CO_2 be after absorption?
   (b) If the volume and molarity of sodium hydroxide and the volume of gas are known, outline a titration method for determining the concentration of CO_2 in the gas (in g/L).

33. Air discharged from an industrial process contains SO_3 and SO_2. A measured sample of the air is passed through an "impinger," which absorbs the gases and converts them to H_2SO_4 and H_2SO_3, respectively. Outline a procedure for determining the amounts of SO_3 and SO_2 in the air sample, based on a differentiating acid-base titration (see Appendix 2 for ionization constants).
34. Boric acid is too weakly acidic to be titrated (see Appendix 2). However, an excess of sorbitol (a neutral sugarlike compound) forms a complex with boric acid in which one hydrogen is sufficiently acidic to be titrated. Suggest a method for the quantitative analysis of a mixture of acetic acid and boric acid.
35. Outline a method for determining the amounts of boric acid, H_3BO_3, and sodium borate, NaH_2BO_3, in a mixture.
36. Richard was unable to obtain a decent end point when he titrated a very weak acid ($pK_a = 7.8$) with standard sodium hydroxide.

   (a) Susan suggested that this difficulty could be overcome by simple adding an excess of standard NaOH and backtitrating with standard HCl. Explain whether this idea would or would not work.
   (b) Robin suggested using an instrument that would automatically plot the first derivative of the titration curve. Sketch such a plot in the vicinity of the equivalence point. Would this approach work?

Kjeldahl Method

37. In the Kjeldahl method, why must ammonia be distilled into an acid? Explain how the boric acid modification works.

38. A 1.000-g food sample is analyzed for nitrogen by the Kjeldahl method. After digestion of the sample, the ammonia is distilled and collected in a receiver containing exactly 50.00 mL of 0.10000M hydrochloric acid. The *unreacted* hydrochloric acid requires 24.60 mL of 0.1200M sodium hydroxide for back-titration. Calculate the percentage of nitrogen (N) in the sample.

39. A 0.5000-g sample of impure ammonium sulfate is analyzed by a Kjeldahl method in which the digestion step is omitted. After concentrated sodium hydroxide is added, the ammonia is distilled into a receiver containing 50.00 mL of 0.2000M hydrochloric acid. The unreacted hydrochloric acid is back-titrated with 19.70 mL of 0.2000M sodium hydroxide. Calculate the percentage purity as percent $(NH_4)_2SO_4$.

40. A 0.2500-g sample of the same impure ammonium sulfate as in Problem 39 is dissolved in water and passed through a cation-exchange column, converting the ammonium sulfate to sulfuric acid. The sulfuric acid is titrated with 35.00 mL of 0.1000M sodium hydroxide. Calculate the percentage purity of the ammonium sulfate and compare it with your previous answer. If there is a significant difference, comment on which method is more likely to be correct and what might cause the results of the incorrect method to be higher or lower.

41. An enzyme, urease, will selectively hydrolyze urea according to the following equation:

$$NH_2CONH_2 + H_2O \xrightarrow{\text{urease}} 2NH_3 + CO_2$$

How might you determine (a) the total combined nitrogen and (b) the urea nitrogen in a sample of urine?

9

Acid-Base Titrations in Nonaqueous Solvents

There is no reason why titrations and other chemical reactions useful to the analytical chemist must always be carried out in water solvent. Many organic solvents are available that can be used in place of water. A compound can be dissolved in a suitable nonaqueous solvent and titrated with a standard solution of a strong acid or base that is also dissolved in nonaqueous solvent. The end point of such a titration can be detected with a visual indicator or with a pH meter. The accuracy of acid-base titrations in nonaqueous solvent is as good as that of titrations carried out in water, and sometimes even better.

There are two main reasons why acid-base titrations are often performed in nonaqueous solvents. One reason is solubility. Many acids and bases are organic compounds that are barely soluble in water but readily soluble in a suitable organic solvent. Another reason is that many compounds too weakly basic or acidic to be titrated in water can be titrated very accurately in a suitable nonaqueous solvent. For example, a base weaker than about $K_b = 10^{-7}$ ($pK_b = 7$) cannot be titrated accurately in water. In glacial acetic acid, a base having a pK_b (in water) of 11 can be titrated with excellent accuracy (± 0.1–0.2%). Acids weaker than about $pK_a = 7$ cannot be titrated in water, but they can be determined accurately by titration with a strong base in a solvent such as pyridine, t-butyl alcohol, or acetone.

A brief treatment of the principles and scope of acid-base titrations in nonaqueous solutions is given in this chapter. More detailed information is available in specialized books on the subject [1, 2].

The most important reason that such titrations are necessary is that an acid (or base) whose K_a (or K_b) is smaller than 1×10^{-7} cannot be titrated *accurately* (*quantitatively*) at the usual $0.1 M$ level in water. The end point break of the titration curve of a weak acid or a weak base (Figure 8–4) is not sharp enough when the K_a is smaller than 1×10^{-7} (a pK_a of 7.0). What limits such titrations is the reaction of the

[1] W. Huber, *Titrations in Nonaqueous Solvents*, New York: Academic Press, 1967.

[2] J. S. Fritz, *Acid-Base Titrations in Nonaqueous Solvents*, Boston: Allyn and Bacon, 1973.

salt formed at the end point with the water solvent. At the end point of a weak acid (HA) titration, this reaction is:

$$A^- + H_2O \rightleftharpoons HA + OH^- \tag{9–1}$$

When the K_a of HA is less than 1×10^{-7}, the basicity of A^- is high enough to convert a significant portion of A^- to HA. This reduces the percentage neutralization below 99.9%, the limiting value for quantitative neutralization. The value of 1×10^{-7} is a limiting value only for titrations conducted at the usual $0.1M$ level; for titrations at the $0.01M$ level, the limiting value for K_a is 1×10^{-6}.

The limiting value of K_a can be calculated for any given concentration level by substituting into the equilibrium expression for Equation 9–1 as follows:

$$\frac{K_w}{K_a} = \frac{[HA][OH^-]}{[A^-]}$$

For $0.1M$ concentration and 99.9% reaction, $[HA] = [OH] = 1 \times 10^{-4}M$, and $[A^-] = 1 \times 10^{-1}M$. Substituting gives:

$$\frac{1 \times 10^{-14}}{K_a} = \frac{[1 \times 10^{-4}][1 \times 10^{-4}]}{[1 \times 10^{-1}]}$$

So the limiting value of K_a is $1 \times 10^{-14}/1 \times 10^{-7} = 1 \times 10^{-7}$.

9–1. SOLVENTS

Solvents may be divided into three general types:

Amphiprotic. This type undergoes self-ionization, or *autoprotolysis*, and has both acidic and basic properties. Autoprotolysis may be represented as follows:

$2SH \rightleftharpoons SH_2^+ + S^-$	*General Case*
$2H_2O \rightleftharpoons H_3O^+ + OH^-$	*In Water*
$2CH_3OH \rightleftharpoons CH_3OH_2^+ + OCH_3^-$	*In Methanol*
$2CH_3CO_2H \rightleftharpoons CH_3CO_2H_2^+ + CH_3CO_2^-$	*In Acetic Acid*

In each case, the products of autoprotolysis are the solvated proton (sometimes written simply as H^+ or H^+_{SH}, for convenience) and the solvent anion, which is sometimes called the *lyate* ion. The autoprotolysis constant, K_S, is defined as follows for the general case:

$$K_s = [SH_2^+][S^-]$$

Nonionizable, with Basic Properties. Examples of this type include ethers and pyridine. Ethers can react with an acid through the weakly basic oxygen and with pyridine through the basic nitrogen. Other than being weakly solvated, bases apparently do not react with this type of solvent.

Aprotic, or Inert. This category includes solvents such as toluene, petroleum ether, and carbon tetrachloride. Aprotic solvents do not interact with either acids or bases, except perhaps for weak solvation effects.

According to Brönsted's definition, an acid is any substance that can give up protons, and a base is any substance that can combine with protons. When an acid, HA, is dissolved in an amphiprotic solvent, SH, the resulting ionization is actually an acid-base reaction and increases the concentration of solvated protons, SH_2^+.

$acid_1$ $base_2$		$acid_2$	$base_1$	
HA + SH	$\rightleftharpoons$	SH_2^+	+ A^-	*General Case*
HA + H_2O	$\rightleftharpoons$	H_3O^+	+ A^-	*In Water*
HA + CH_3OH	$\rightleftharpoons$	$CH_3OH_2^+$	+ A^-	*In Methyl Alcohol*
HA + CH_3CO_2H	$\rightleftharpoons$	$CH_3CO_2H_2^+$	+ A^-	*In Glacial Acetic Acid*

The extent of ionization depends on several things. One is the inherent acidity of HA. A strong acid, such as hydrochloric acid, is completely ionized in water, whereas a weak acid, such as acetic acid, is only slightly ionized. Another is the basicity of the solvent. A basic solvent will promote ionization of an acid by virtue of an acid-base reaction with the dissolved acid. All the solvents listed above have some basic properties, although glacial acetic acid is a much weaker base than water, for example. Finally, the dielectric constant of the solvent, which is a measure of the electrical insulating ability of the solvent, has an effect on ionization. Water has an unusually high dielectric constant; and in water, ions are relatively free from attraction and repulsion effects (some effects do exist, as will be recalled from the difference between ionic concentration and activity in water). In most organic solvents, however, ions have a tendency to be present as *ion pairs* (cation–anion pairs). Thus, in glacial acetic acid solution even strong acids exist mostly as ion pairs, with very few *free* $CH_3COOH_2^+$ or A^- being present.

Each species in an ion pair bears a full positive charge and may have the same color as the free ion; however, an ion pair, like a molecule in solution, conducts essentially no current.

When a base is dissolved in an amphiprotic solvent, SH, the solvent acts as an acid and the resulting ionization increases the solvent anion concentration, S^-.

$base_1$ $acid_2$		$acid_1$	$base_2$	
B + SH	$\rightleftharpoons$	BH^+	+ S^-	*General Case*
B + H_2O	$\rightleftharpoons$	BH^+	+ OH^-	*In Water*
B + CH_3OH	$\rightleftharpoons$	BH^+	+ CH_3O^-	*In Methyl Alcohol*
B + CH_3CO_2H	$\rightleftharpoons$	BH^+	+ $CH_3CO_2^-$	*In Glacial Acetic Acid*

The reaction of an acid and base in solution may well take place by combining a solvated proton (with an ionization of the acid) from solvent anion (from ionization of the base).

$$HA + SH \rightleftharpoons SH_2^+ + A^-$$

$$B + SH \rightleftharpoons BH^+ + S^-$$

$$SH_2^+ + S^- \rightleftharpoons 2SH$$

The sum of these equations, however, is the simple Brönsted reaction

$$\underset{acid_1}{HA} + \underset{base_2}{B} \rightleftharpoons \underset{acid_2}{BH^+} + \underset{base_1}{A^-}$$

In nonionizable solvents, acids also form a solvated proton. Because of the low dielectric constants of such solvents, the solvated proton exists primarily as an ion pair with the acid anion,

$$HA + S \rightleftharpoons SH^+A^-$$

A base, B (which may be weakly solvated) reacts with the ion pair as follows:

$$SH^+A^- + B \rightleftharpoons BH^+A^- + S$$

This reaction occurs because B is a stronger base than the solvent S.

In aprotic solvents, an acid exists either as a molecular compound (HA) or as an undissociated ion pair (H^+A^-). These may be weakly solvated by the solvent. The acid can give up its proton to a base added to the aprotic solvent.

$$HA \text{ (or } H^+A^-) + B \rightleftharpoons BH^+A^-$$

Water. Water is an amphiprotic solvent with a very high dielectric constant (78.5). The autoprotolysis constant, K_s, for water has the value 10^{-14} at room temperature.

$$K_s = [H_3O^+][OH^-] = 10^{-14}$$

In water, the pH scale of 0 to 14 is determined by the 10^{-14} value of K_s.

Water is both a weak acid and a weak base. An acid reacts with the solvent to increase the H_3O^+ concentration, and a base reacts to increase the OH^- concentration. Strong acids dissociate completely in water.

$$\underset{acid_1}{HClO_4} + \underset{base_2}{H_2O} \rightarrow \underset{acid_2}{H_3O^+} + \underset{base_1}{ClO_4^-}$$

$$\underset{}{HCl} + H_2O \rightarrow H_2O^+ + Cl^-$$

$$HNO_3 + H_2O \rightarrow H_3O^+ + NO_3^-$$

Although these three "strong" acids actually have different intrinsic acid strengths, they all form the same acid in water (H_3O^+) and thus appear to be of equal acidic strength. The reaction of the acids with water reduces their acidic strength to form the weaker acid, H_3O^+. This is called the *leveling effect*.

Strong bases also react completely with water.

$$\underset{acid_1}{(H_2O)} + \underset{base_2}{NaOH} \rightarrow \underset{acid_2}{Na^+} + \underset{base_1}{OH^-}$$

$$(H_2O) + R_4NOH \rightarrow R_4N^+ + OH^-$$

The ions are solvated by water, but water is usually included in the chemical formula for an ion only in the case of a solvated proton. These examples show that OH^- is the strongest base that can exist in water; stronger bases are leveled to the basic strength of hydroxyl ion in aqueous solution.

Weak acids and bases are partially ionized in water. The strength of the acid or base is given by its ionization constant, K_a or K_b.

Acetic Acid.   Glacial acetic acid is another amphiprotic solvent with an autoprotolysis constant only slightly different from that of water.

$$2HAc \rightleftharpoons H^+{}_{HAc} + Ac^-$$

$$K_s = [H^+{}_{HAc}][Ac^-] = 10^{-14.45}$$

Acetic acid differs from water in that it is much more acidic and has a considerably lower electric dielectric constant (DK = 6.1).

In acetic acid, strong acids ionize completely (or almost so) but the low dielectric constant causes the positively and negatively charged ions to remain associated primarily as ion pairs. The strongest acid in acetic acid is perchloric acid, which has a dissociation constant of only $10^{-4.87}$.

$$K_{HClO_4} = \frac{[H^+{}_{HAc}][ClO_4{}^-]}{[HClO_4]} = 10^{-4.87}$$

In this and in the following instances, the equilibrium constant expressions are *overall* dissociation constants. Thus, $[HClO_4]$ represents the analytical concentration of undissociated perchloric acid and includes the concentration of the undissociated ion pair, $H^+{}_{HAc}ClO_4{}^-$.

Acetic acid is less basic than water and does not level strong acids. Since titrations of bases are best performed with as strong an acid as possible, the advantage of perchloric acid over the other acids listed is apparent.

Acetic acid solvent is acidic enough to cause bases of medium strength to react more or less completely with the solvent.

$$B + HAc \rightleftharpoons BH^+Ac^-$$

The ion pair is only partially dissociated because of the low dielectric constant of acetic acid.

$$BH^+Ac^- \rightleftharpoons BH^+ + Ac^-$$

The overall dissociation constant expression is

$$K_B = \frac{[BH^+][Ac^-]}{[B]}$$

where [B] represents the sum of the free base and the undissociated ion pair.

For bases that are strong enough to react completely with the solvent, K_B tends to be about the same, even though the bases may be of different strength in a less acidic solvent, such as water. Acetic acid thus acts as a leveling solvent for aliphatic amines and simple aromatic amines. Aromatic amines with electron-withdrawing substituents such as —NO_2 or —Cl are more weakly acidic and are not leveled by acetic acid.

In acetic acid, the product of the titration of a base B with a strong acid such as $HClO_4$ is a salt, $BHClO_4$. Because of the lower dielectric constant of acetic acid, this salt is only slightly dissociated into free ions. This low salt ionization shifts the titration equilibrium to the right.

$$HClO_4 + B \rightleftharpoons BH^+ + ClO_4^- \rightleftharpoons BHClO_4$$

9–2. TITRANTS

Perchloric Acid. Perchloric acid is the preferred titrant for titrations carried out in acetic acid and other nonbasic solvents because it is the strongest of the common mineral acids. Because of the leveling effect of water and other solvents having significant basic properties, other strong acids serve as well as perchloric acid. In acetic acid, however, perchloric acid gives a longer potentiometric break than hydrochloric acid and a much longer break than nitric acid.

Perchloric acid titrant is made up in various solvents, depending on the titration to be carried out. Perchloric acid in acetic acid is commonly used to titrate weak bases in acetic acid, nitromethane, chloroform, and many other solvents. The titrant is prepared simply by dissolving the required amount of 70–72% perchloric acid (which is approximately $HClO_4 \cdot 2H_2O$) in acetic acid. Usually, a 0.01–$0.5M$ titrant is used. If a very weak base is to be titrated, the water introduced with the perchloric acid is removed by adding a calculated amount of acetic anhydride to combine with the water. The acid-catalyzed reaction of acetic anhydride with water is fairly rapid. When titrating a primary or secondary amine that might react with acetic anhydride, it is important to avoid any excess acetic anhydride in the titrant. Properly prepared titrants are stable for long periods of time.

Sometimes it is desirable to exclude acetic acid from a titration system because of its leveling effect on mixtures of certain bases. In such cases, perchloric acid in 1,4-dioxane is a good titrant. The brown color that sometimes develops in these solutions causes no titration errors, but it can be avoided by using reagent-grade dioxane or by first purifying the dioxane by shaking it with cation-exchange resin. As with acetic acid, titrants are prepared simply by adding the calculated amount of 70–72% perchloric acid to dioxane. The small amount of water introduced has a negligible effect on the titration of most bases. Solutions of perchloric acid in dioxane are stable and may be used to titrate bases in almost any solvent.

Potassium acid phthalate (KHP), equivalent weight 204.2, is a well-established primary standard acid for basic titrants in aqueous solution. In glacial acetic acid, KHP serves as a primary standard *base* for standardizing perchloric acid titrants.

KHP is sparingly soluble in acetic acid; heat must be used to dissolve it completely.

Alkali Metal Bases.   Solutions of various sodium or potassium alcoholates in an appropriate alcohol can be used to titrate weak acids in a nonaqueous solvent. A solution of sodium or potassium methoxide in methanol or benzene-methanol is perhaps the best of this class of titrants. The methoxide titrant is prepared by carefully reacting freshly cut sodium or potassium metal with a small volume of methanol and then diluting with methanol or benzene and methanol.

$$2CH_3OH + 2Na \rightarrow 2CH_3ONa + H_2$$

The use of sodium or potassium methoxide titrants has declined in recent years. Methanol is almost as acidic as water, and its presence limits the sharpness with which a weak acid can be titrated in a nonaqueous solvent. Replacement of part of the methanol with benzene (which serves as an inert diluent) is helpful, but benzene is now known to be a toxic substance that should be avoided whenever possible.

Quaternary Ammonium Hydroxides.   Tetrabutylammonium hydroxide in 2-propanol is probably the most widely used titrant for acids in nonaqueous solution. This and other quarternary ammonium hydroxides have at least two major advantages over other titrants. In almost every case, the tetraalkylammonium salt of the titrated acid is soluble in the solvents commonly used. Sodium or potassium salts of titrated acids frequently form gelatinous precipitates. The other advantage of tetraalkylammonium hydroxides is that excellent potentiometric curves are obtained using ordinary glass and calomel electrodes. (The "alkali error" limits the use of the glass electrode in conjunction with alkali-metal titrants, particularly in basic solvents.) Harlow, Noble, and Wyld [3] prepared quaternary ammonium hydroxide titrants in 2-propanol by passing a solution of the quaternary ammonium iodide through a large anion-exchange column in the hydroxyl form. (See Chapter 24 for information on anion exchange resins and the principles of ion exchange.) The iodide anion in solution is exchanged for the hydroxide anion of the ion-exchange resin.

$$R_4N^+I^- + \text{Anex—OH}^- \rightarrow R_4N^+OH^- + \text{Anex—I}^-$$
(in 2-PrOH)   (in 2-PrOH)

They found it necessary to use a very slow flow rate to obtain good conversion of the iodide to the hydroxide. A modification of their method uses a macroreticular anion-exchange resin, which gives faster exchange rates in nonaqueous solution than the older gel-type resin.

In preparing any strongly basic titrant, care must be taken to ensure that it does not contain carbonate as an impurity. Carbonate is a moderately weak base and reduces the sharpness of the end point of a titration. In the ion-exchange method for preparing tetrabutylammonium hydroxide, carbonate-free sodium

[3] G. A. Harlow, C. M. Noble, and G. E. A. Wyld, *Anal. Chem.*, *28*, 787 (1956).

hydroxide (see Chapter 8) should be used in converting the ion-exchange resin to the hydroxide form.

One drawback of quaternary ammonium hydroxides is their instability. For example, solutions of tetrabutylammonium hydroxide slowly decompose to tributylamine, which is a weaker base than the quaternary ammonium hydroxide:

$$Bu_4N^+OH^- \rightarrow Bu_3N + BuOH$$

However, a carefully prepared solution of $Bu_4N^+OH^-$ in 2-propanol is stable for 2–4 weeks at room temperature and for longer periods when kept in a refrigerator.

Usually, basic titrants are standardized against benzoic acid, which is an excellent primary standard. The standardizing titration is best done under conditions that are similar to those that will be used for titrating acid samples. Any acidic impurities in the solvent must be measured and taken into account in calculating the molarity of the titrant.

9–3. TITRATION OF BASES

When organic amines are titrated as bases in nonaqueous solution, the relative basicities tend to be the same as in water. Thus o-chloroaniline (pK_b in water, 11.2) is a weaker base than aniline (pK_b in water, 9.4) when titrated in acetic acid. p-Nitroaniline (pK_b in water, 12.1) is yet a weaker base and gives a shorter potentiometric break when titrated in acetic acid (see Figure 9–1). Ionization constants in water are available in the literature for a great many organic bases; these constants are useful in predicting their titration behavior in a nonaqueous solvent such as acetic acid. However, the *leveling effect* of acetic acid toward stronger bases must be kept in mind. Bases stronger than about pK_b 9.2 in water all give similar titration curves in acetic acid because they react with the solvent to form the acetate ion, which is weaker than the parent amine.

$$RNH_2 + HAc \rightarrow RNH_3^+ + Ac^-$$

This leveling effect of stronger bases can be avoided by titrating in a nonacidic solvent, such as acetonitrile (CH_3CN) or acetone (CH_3COCH_3).

Hundreds of organic aliphatic amines can be titrated in nonaqueous solution. These include aliphatic amines (RNH_2), amino acids ($RCH(NH_2)COOH$), aromatic amines ($ArNH_2$), and cyclic nitrogen bases such as pyridine (C_5H_5N).

Salts of weak acids, A^-, can be titrated with a strong acid.

$$H^+ClO_4^- + A^- \rightarrow HA + ClO_4^-$$

In aqueous solution, only the salts of very weak acids (weaker that $pK_a \approx 6$) can be titrated in this way. In glacial acetic acid, the alkali metal and amine salts of carboxylic acids, and even of some "strong" mineral acids, may be successfully titrated. For example, amine nitrates are titrated to nitric acid (which is fairly weak acid in acetic acid solution) as a neutralization product, and sulfates are titrated to

Figure 9–1. Titration of amines in acetic acid with $0.1\,M$ perchloric acid measured with glass-calomel electrodes: (*a*) aniline, pK_b in $H_2O = 9.4$; (*b*) *p*-bromoaniline, pK_b in $H_2O = 10.1$; (*c*) *o*-chloroaniline, pK_b in $H_2O = 11.2$; (*d*) *p*-nitroaniline, pK_b in $H_2O = 12.1$; (*e*) quinoxaline, pK_b in $H_2O = 13.2$.

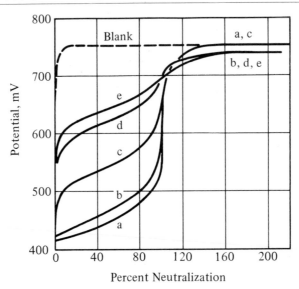

bisulfates:

$$H^+ClO_4^- + RNH_3^+NO_3^- \rightarrow RNH_3^+ClO_4^- + H^+NO_3^-$$

$$H^+ClO_4^- + (RNH_3^+)_2SO_4^{2-} \rightarrow RNH_3^+ClO_4^- + RNH_3^+HSO_4^-$$

An ingenious modification introduced by Pifer and Wollish [4] permits the titration of amine halide salts, which are too weakly basic to be titrated directly in acetic acid. This titration is made possible by first adding mercuric acetate to convert the halide (X^-) to undissociated HgX_2.

$$2RNH_3^+X^- + HgAc_2 \rightarrow 2RNH_3^+Ac^- + HgX_2$$

$$H^+ClO_4^- + RNH_3^+Ac^- \rightarrow RNH_3^+ClO_4^- + HAc$$

Mercuric acetate is undissociated in acetic acid and does not interfere in the titration if a modest excess is used. The method is applicable to amine salts of hydrochloric, hydrobromic, and hydriodic acid. Some dioxane mixed with the acetic acid enhances the sharpness of the potentiometric end point, particularly if the titrant is as dilute as $0.01\,M$. For this reason, perchloric acid in dioxane is the recommended titrant.

Direct titration of amine salts is especially important in the pharmaceutical industry because it permits an assay of the total amine content of a drug whether it is

[4] C. W. Pifer and E. G. Wollish, *Anal. Chem.*, **24**, 300 (1952).

Figure 9–2. Potentiometric titration curve of a mixture of butylamine and pyridine titrated in acetonitrile with perchloric acid in dioxane. [From J. S. Fritz, *Anal. Chem., 25*, 407 (1953). By permission.]

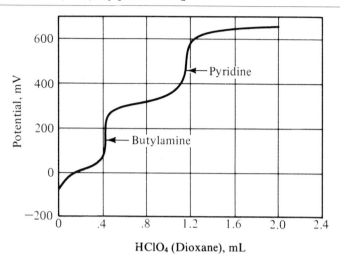

present as the free amine or as the salt. A small excess of acid (HX) will not interfere in the amine titration, because it is converted to acetic acid by the mercuric acetate.

$$2H^+X^- + HgAc_2 \rightarrow HgX_2 + 2HAc$$

Quaternary ammonium salts may also be titrated according to the methods just described.

Amines in mixtures can often be titrated separately if their basicities differ enough. Thus, a mixture of butylamine and pyridine in acetonitrile solution shows two potentiometric breaks when titrated with perchloric acid in dioxane (see Figure 9–2). Acetonitrile is used as the solvent because it is not acidic and exerts no leveling effect on the two amines. When this same mixture is titrated in acetic acid solution, only a single end point is observed, corresponding to the sum of the amines. Butylamine reacts with acetic acid to form the acetate ion, which has about the same basicity as pyridine:

$$C_4H_9NH_2 + HAc \rightarrow C_4H_9NH_3^+ + Ac^-$$

9–4. TITRATION OF ACIDS

Many organic compounds are acidic enough to be titrated in nonaqueous solvents, provided that the proper conditions are selected. The types of compounds that can be titrated include sulfonic acids ($ArSO_3H$), carboxylic acids (RCOOH), phenols (ArOH), enols ($—COCH_2CO—$), imides ($—CONHCO—$), certain nitro compounds, and various sulfur-containing compounds. The solvent should readily

Figure 9–3. Differentiating titration of picric acid, 2,4-dinitrophenol, *o*-nitrophenol, and phenol with tetrabutylammonium hydroxide in benzene-isopropanol. [From J. S. Fritz and L. W. Marple, *Anal. Chem., 34*, 921 (1962). By permission.]

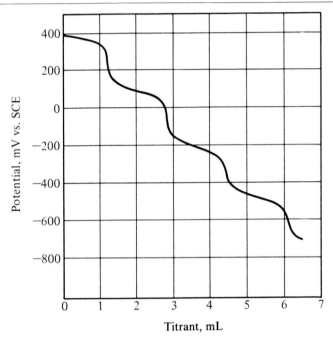

Titrant, mL

dissolve the acidic compounds to be titrated but should have no acidic properties itself. For the successful titration of mixtures, the solvent should not be a strong enough base to level the strengths of the acids in the mixture.

The titrant should be a strong base dissolved in a nonacidic solvent and should be stable on standing. Sodium methoxide (CH_3ONa) in benzene-methyl alcohol [5], tetrabutylammonium hydroxide $((C_4H_9)_4NOH)$ in benzene-methyl alcohol, or isopropyl alcohol [6] are commonly used as the titrant.

Although alcohols are suitable solvents for titrating acids of moderate strength, most alcoholic solvents are too acidic to use in titrating weak acids. An exception is *tert*-butyl alcohol $((CH_3)_3COH)$, which is an excellent solvent for titrating all types of organic acids [7]. Pyridine [6], acetone [8], and dimethylformamide are also useful solvents for a wide variety of acids.

Acids are best titrated with tetrabutylammonium hydroxide $((C_4H_9)_4NOH)$ in isopropyl alcohol or in benzene-methyl alcohol. Titrations may be followed potentiometrically using an ordinary glass-calomel electrode combination. Figure 9–3 shows the titration curve obtained for a mixture of four acids of different

[5] J. S. Fritz and N. M. Lisicki, *Anal. Chem., 23*, 589 (1951).
[6] R. H. Cundiff and P. C. Markunas, *Anal. Chem., 28*, 792 (1956).
[7] J. S. Fritz and L. W. Marple, *Anal. Chem., 34*, 921 (1962).
[8] J. S. Fritz and S. S. Yamamura, *Anal. Chem., 29*, 1079 (1957).

Figure 9–4. Indicator transition ranges of some common indicators in tertiary butyl alcohol solvent. [From L. W. Marple and J. S. Fritz, *Anal. Chem.*, *35*, 1305 (1963). By permission.]

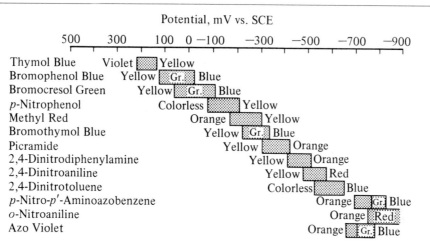

strengths. Increasing basicity is indicated by more negative potentials (expressed in millivolts). The strongest acid titrated is picric acid (represented by the first potentiometric break), while the weakest is phenol (represented by the fourth potentiometric break). The total potentiometric range covered in this titration is approximately 1100 millivolts (equivalent to a pH range of more than 18 units).

A variety of visual indicators is available for titrations performed on acids in *tert*-butyl alcohol, pyridine, and other solvents. In Figure 9–4, the transition ranges (expressed in millivolts) of several indicators in *tert*-butyl alcohol are given. The indicator selected for a particular titration should be one whose transition range encompasses the equivalence-point potential obtained in a potentiometric titration of the acid.

QUESTIONS AND PROBLEMS

Solvents and Titrants

1. Define the following types of solvents.

   (a) Amphiprotic
   (b) Nonionizable

   (c) Aprotic (inert)

2. Describe how the following titrants may be prepared for nonaqueous titrations. For each, list a suitable primary standard.

   (a) Perchloric acid
   (b) Sodium methoxide

   (c) Tetrabutylammonium hydroxide

3. List two weakly basic impurities that could be in a tetrabutylammonium hydroxide titrant. Suggest an experimental method for determining whether this titrant contains any weakly basic impurities.

4. Perchloric acid is used as titrant in glacial acetic acid solvent. Explain why perchloric acid, a strong acid in water, is not strongly ionized in acetic acid.

5. Sodium hydroxide is a strong base in water and is commonly used as a titrant for acid-base titrations. Suggest a sodium-containing base for acid-base titrations in acetic acid.

6. What does the large potential range covered by acid-base titrations in t-butyl alcohol suggest about the autoprotolysis constant of this solvent? Explain briefly.

7. Dichloroacetic acid ($CHCl_2COOH$) is a liquid at room temperature and is a stronger acid than acetic acid. List any advantages that dichloroacetic acid might have over acetic acid as a solvent for titration of bases. List any likely disadvantages.

8. Explain why the autoprotolysis constant (see Chapter 7) is an important factor in choosing a solvent for an acid-base titration. Suggest a simple experimental way to estimate the autoprotolysis constant of an amphiprotic solvent.

9. Ammonium acetate ($NH_4{}^+Ac^-$) cannot be titrated as an acid or a base in aqueous solution, but it can be titrated as either an acid or a base in a nonaqueous solvent. Give the chemical reaction and conditions for titrating ammonium acetate as an acid in a nonaqueous solvent. Do the same for titrating it as a base in a nonaqueous solvent.

10. State whether each of the following acids can be titrated accurately (99.9%) in water with standard base at the 0.1M level. If not, indicate a suitable nonaqueous solvent and titrant.

(a) Phenol, $K_a = 1.4 \times 10^{-10}$
(b) p-Nitrophenol, $K_a = 5.2 \times 10^{-8}$
(c) Maleic acid (1st and 2nd hydrogens), $K_1 = 1.00 \times 10^{-2}, K_2 = 5.50 \times 10^{-7}$
(d) Carbonic acid (1st H only), $K_1 = 4.3 \times 10^{-7}$
(e) p-Chlorophenol, $K_a = 4.19 \times 10^{-10}$

11. State whether each of the following bases can be titrated accurately (99.9%) in water with standard acid at the 0.1M level. If not, indicate a suitable nonaqueous solvent and titrant.

(a) Aniline, $K_b = 4.20 \times 10^{-10}$
(b) Ethylenediamine (1st and 2nd N atoms), $K_1 = 1.28 \times 10^{-4}, K_2 = 2.0 \times 10^{-7}$
(c) Hydrazine, $K_b = 1.00 \times 10^{-6}$
(d) Pyridine, $K_b = 1.49 \times 10^{-9}$

Titration Methods

12. Explain how an amine hydrochloride ($RNH_3{}^+Cl^-$) can be titrated as a base in nonaqueous solution. Suggest a scheme for determining separately the amount of each component in the following mixtures:

(a) Amine and amine hydrochloride
(b) Amine hydrochloride and hydrochloric acid

13. When a mixture of two bases of different basicity is to be titrated in acetonitrile to give two potentiometric breaks (see Figure 9–2), explain why the perchloric acid titrant should be made up in dioxane rather than in acetic acid.

14. From Figure 9–4, select a suitable indicator for each of the four acids in the mixture titrated in Figure 9–3.

10

Precipitate-Formation
Titrations

The formation of a precipitate can be used as the basis of a titration, provided that there is a suitable way of determining when a stoichiometric amount of titrant has been added. It is also necessary for the system to reach equilibrium rapidly after each addition of titrant. (Unfortunately, this is not always possible in precipitation reactions.) Although many such titration procedures have been developed, those for the titration of halide ions with silver(I) and of sulfate with barium(II) are the most important.

10-1. POTENTIOMETRIC TITRATIONS WITH SILVER(I)

The titration of chloride, bromide, iodide, and other anions with silver(I) can be followed potentiometrically with a silver indicator electrode. The potential E, in volts, of an immersed silver electrode is a function of the concentration of silver ions in the solution (see Chapter 16):

$$E = 0.800 + 0.059 \log a_{Ag^+} \qquad (10-1)$$

where a_{Ag^+} is the activity of the silver. In a potentiometric titration, we measure the *difference* in potential between the silver indicator electrode and a reference electrode whose constant potential is unaffected by the composition of the solution titrated.

Curves for the titration of chloride, bromide, and iodide with silver nitrate are shown in Figure 10-1. Data for these titration curves are obtained experimentally by measuring the potential of a silver electrode as various volumes of silver nitrate are added. Then pAg is calculated from a_{Ag^+} using Equation 10-1, and the various data points are joined to form a smooth curve as shown in Figure 10-1. Note that

Figure 10–1. Curves for the titration of 40 mL of a 0.01 M sodium halide solution with 0.01 M silver nitrate.

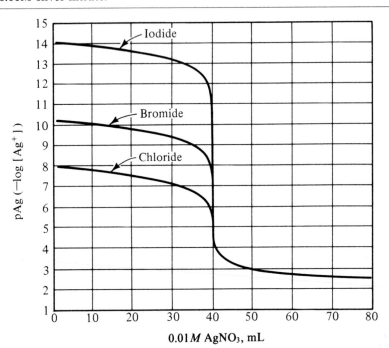

0.01 M AgNO₃, mL

log a_{Ag^+} (or log[Ag^+]) is a function of the potential of the silver electrode. The term pAg is analogous to the term pH for acid-base titrations.

Exact Definition	Approximate Definition
$pH = -\log a_{H^+}$	$pH = -\log[H^+]$
$pAg = -\log a_{Ag^+}$	$pAg = -\log[Ag^+]$

Convenience dictates the use of the approximate definitions—especially in dilute solutions, where the difference between activity and concentration is small.

Curves like those shown in Figure 10–1 can be calculated by using the solubility product of the silver halide.

Example: A 40-mL aliquot of 0.01 M sodium chloride is titrated with 0.010 M silver nitrate. Using 1×10^{-10} as the solubility product for silver chloride, calculate what the pAg is (a) when 20 mL of silver nitrate has been added and (b) when 40 mL of silver nitrate has been added.

(a) The amount of Cl⁻ left in solution is 20 mL × 0.010 M = 0.20 mmole. The concentration of Cl⁻ is

$$\frac{0.20 \text{ mmole}}{60 \text{ mL}} = 0.0033 M$$

The $[Ag^+]$ is found by substituting $[Cl^-]$ into the solubility-product expression:

$$K_{sp} = [Ag^+][Cl^-]$$

$$10^{-10.0} = [Ag^+](0.0033)$$

$$[Ag^+] = 3 \times 10^{-8}$$

$$\log[Ag^+] = 0.47 - 8 = -7.5_3$$

$$pAg = 7.5$$

(b) All of the chloride has been precipitated except for the low-concentration amount that remains in solution because of the solubility of the silver chloride precipitate. Neither Ag^+ nor Cl^- is in excess:

$$K_{sp} = [Ag^+][Cl^-] = [Ag^+]^2$$

$$[Ag^+] = \sqrt{10^{-10.0}} = 10^{-5.0}$$

$$pAg = 5.0$$

Titration of mixtures of halide ions with silver nitrate will give a separate potentiometric break for each halide. Iodide is titrated first because its silver salt is the least soluble; then bromide is titrated, and finally chloride. Unfortunately, this titration is not very accurate because the more soluble halide will coprecipitate by isomorphous inclusion (Section 4–3). For example, chloride is coprecipitated during the titration of bromide with silver nitrate, giving high results for bromide and low results for chloride. Mixtures of halide ions can be separated and determined by ion chromatography, as described in Chapter 24.

10-2. MOHR METHOD FOR HALIDES

The Mohr method for chloride was published more than a hundred years ago and is still used [1]. Chloride is titrated with a standard silver nitrate solution; a soluble chromate salt is added as the indicator. When precipitation of the chloride is complete, the first excess of silver(I) reacts with chromate to form a red precipitate of silver chromate:

$$Ag^+ + Cl^- \rightarrow AgCl(s) \qquad \textit{Titration Reaction}$$
$$\text{(AgNO}_3)$$

$$2Ag^+ + CrO_4^{2-} \rightarrow Ag_2CrO_4(s) \qquad \textit{End Point Reaction}$$
$$\text{(red)}$$

Titrant added near the equivalence point causes local excesses of silver(I) that result in flashes of red color, but the correct end point is the first permanent darkening of the yellow chromate color. The end point is not as sharp as might be desired. Some excess silver nitrate must be added to form enough silver chromate to be seen over

[1] Mohr, *Annalen der Chemie und Pharmacie, 97*, 335 (1856).

the heavy white precipitate and yellow chromate solution. This makes it necessary to determine an indicator blank and subtract this from the volume of silver nitrate used in titrating the sample.

The concentration of chromate indicator is important. If too much chromate is added, the end point occurs before the equivalence point; if not enough, the end point comes late.

Example: From $K_{sp,AgCl}$, it is calculated that the pAg should be 5.0 at the titration end point (Figure 9-1). From the K_{sp,Ag_2CrO_4} expression, calculate what concentration of chromate should be present to ensure that the first precipitation of silver chromate occurs at pAg = 5.0. The K_{sp,Ag_2CrO_4} is 1.1×10^{-12}.

$$Ag_2CrO_4(s) \rightleftharpoons 2Ag^+ + CrO_4^{2-}$$

$$K_{sp} = [Ag^+]^2[CrO_4^{2-}]$$

$$[Ag^+] = 10^{-5.0}$$

Substituting into the solubility-product expression,

$$1.1 \times 10^{-12} = (10^{-5.0})^2[CrO_4^{2-}]$$

$$[CrO_4^{2-}] = 1.1 \times 10^{-2}M = 0.011M$$

Acidity is important in the Mohr titration. In acidic solutions, part of the indicator is present as $HCrO_4^-$ instead of as CrO_4^{2-}; hence, more indicator is needed to form a silver chromate precipitate. About pH 8 is ideal for the titration; solid calcium carbonate will buffer the solution at about this pH. With more alkaline solutions, there is danger of precipitating some silver carbonate or silver hydroxide.

10-3. VOLHARD METHOD

The Volhard method is a procedure for titrating silver(I) with standard potassium thiocyanate. Indirectly, it is a method for determining the amount of halide or of any anion precipitated quantitatively by silver nitrate. Its most important use is for the quantitative determination of halide anions.

Equations for the Volhard titration of silver are given below:

$$SCN^- + Ag^+ \rightarrow AgSCN(s) \qquad \textit{Titration Reaction}$$
$$\text{(KSCN)} \qquad\qquad \text{(white)}$$

$$SCN^- + Fe^{3+} \rightarrow Fe(SCN)^{2+} \qquad \textit{End Point Reaction}$$

The titration with thiocyanate is carried out in acidic solution. When the silver(I) has been precipitated as white silver thiocyanate, the first excess of titrant and the iron(III) indicator react and form a soluble red complex. (The color change at the end point is not extremely sharp, but it can be detected with a little practice.)

In the Volhard method for determining chloride and other anions, a measured volume of standard silver nitrate solution is added to the sample solution. The volume of silver nitrate is in excess of the amount needed to react with the halide:

$$\underset{(AgNO_3)}{Ag^+} + X^- \rightarrow AgX(s) + excess\ Ag^+ \qquad \textit{Titration Reaction}$$

The excess silver(I) is then back-titrated with standard thiocyanate:

$$SCN^- + excess\ Ag^+ \rightarrow AgSCN(s) \qquad \textit{Back-titration Reaction}$$

$$SCN^- + Fe^{3+} \rightarrow Fe(SCN)^{2+} \qquad \textit{End Point Reaction}$$

If the silver halide precipitate is *less* soluble than silver thiocyanate, the excess silver can be titrated directly with potassium thiocyanate. Such is the case if X^- is Br^- or I^-. However, silver chloride is more soluble than silver thiocyanate and must be removed by filtration so that it will not be converted to silver thiocyanate during the titration:

$$SCN^- + AgCl(s) \rightarrow AgSCN(s) + Cl^-$$

Instead of filtering out the silver chloride precipitate, it is faster to add nitrobenzene and shake vigorously. Nitrobenzene is a liquid that is immiscible with water and heavier than water. Shaking with nitrobenzene coagulates and coats the silver chloride precipitate, so that it cannot react with the aqueous layer of solution during the titration. The excess silver nitrate stays in the aqueous layer and is titrated with standard thiocyanate.

10–4. ADSORPTION INDICATOR METHOD FOR HALIDES

In the adsorption indicator method for halide ions, the end point reaction occurs *on the surface* of the silver halide precipitate. The stoichiometric titration reaction is simply a precipitation of the silver halide; the end point reaction is between silver(I) and a colored indicator anion, such as dichlorofluorescein:

$$Ag^+ + X^- \rightarrow AgX(s) \qquad \textit{Titration Reaction}$$

$$Ag^+ + AgX(s) + \underset{(yellow)}{Indic^-} \rightarrow \underset{(red)}{AgX:Ag^+|Indic^-(s)} \qquad \textit{End Point Reaction}$$

To understand better the mechanism of an adsorption indicator end point, the student should recall the rules of adsorption coprecipitation discussed in Chapter 4. During the titration (before the end point), the halide ion is in excess and is adsorbed on the surface of the precipitate as the primary adsorbed ion. The indicator anion is repelled from the negatively charged precipitate:

$$\underset{(AgNO_3)}{Ag^+} + \underset{(NaX)}{2X^-} \rightarrow AgX:X^-|Na^+(s)$$

When the equivalence point of the titration is reached and the first excess of titrant is added, an abrupt change occurs on the surface of the precipitate. Now Ag^+ is the primary adsorbed ion, and the surface of the precipitate has a positive charge. When this happens, the indicator anion is adsorbed on the precipitate surface as the counter ion:

$$AgX:Ag^+(s) + Indic^- \rightarrow AgX:Ag^+ \!\mid Indic^-(s)$$
$$\quad\quad\quad \text{(yellow)} \quad\quad\quad\quad \text{(red)}$$

A color change accompanies the adsorption of the colored indicator by the precipitate. The reason for this is not completely understood; many colored organic anions are adsorbed *without* any color change. The probable explanation is that a suitable indicator must form a colored complex with silver(I). This complex is too unstable to form in solution at low concentrations, but the equilibrium is shifted favorably by the precipitation of the complex on a precipitate surface:

$$Ag^+ + Indic^- \rightleftharpoons AgIndic \xrightarrow{\text{AgX(s)}} AgX:AgIndic(s)$$
$$\quad\quad\quad\quad \text{(in solution)}$$

The indicator anion $(Indic^-)$ must not displace the primary adsorbed ion (X^-) during the titration but must be adsorbed as the counter ion at the end point. Fluorescein can be used as an indicator for titrating any of the halides because it will not displace halide from $AgX:X^-(s)$. If the pH is lowered to 4.0 with acetic acid, dichlorofluorescein can be used as indicator for the titration of chloride and other halide ions.

Too high an acidity interferes in the adsorption indicator titration by reducing the concentration of indicator anion. A slight acidity is often beneficial, as discussed before for dichlorofluorescein.

The adsorption indicator method is one of the best for determining halides. The end point is sharp, and the results are generally accurate.

10–5. ADSORPTION INDICATOR METHOD FOR SULFATE

The determination of sulfate is very important, yet the gravimetric method is slow and tedious. A titration method with an adsorption indicator end point [2] is quick and is reasonably accurate when preceded by ion-exchange removal of interfering cations.

The titration is carried out at about pH 3.5 in a mixture of approximately 50:50 water and methyl alcohol. From this solvent mixture, barium sulfate precipitates as a fluffy, highly adsorptive precipitate that is very different in appearance from the fine, crystalline precipitate obtained from aqueous solutions. A proper physical form of barium sulfate is needed for the adsorption indicator to function; in water alone, no end point is obtained.

[2] J. S. Fritz and M. Q. Freeland, *Anal. Chem.*, 26, 1593 (1954). See also J. S. Fritz and S. S. Yamamura, *Anal. Chem.*, 27, 1461 (1955).

Alizarin red S is used as indicator. The indicator is yellow in solution but forms a pink complex on the surface of the precipitate when the first excess of barium(II) is added. The color change is sharp and distinct. The mechanism of the end point is analogous to that in the adsorption indicator method for halide:

$$Ba^{2+} + SO_4^{2-} \rightarrow BaSO_4{:}SO_4^{2-}{\vdots}2Na^+(s) \qquad\qquad \textit{Titration Reaction}$$
$$\text{(BaCl}_2\text{)} \quad \text{(Na}_2\text{SO}_4\text{)}$$

$$Ba^{2+} + BaSO_4(s) + 2Indic^- \rightarrow BaSO_4{:}Ba^{2+}{\vdots}2Indic^-(s) \qquad \textit{End Point Reaction}$$

Coprecipitation errors are great in the gravimetric sulfate determination, but they are even more serious in titration methods for sulfate. Foreign cations coprecipitate as sulfates, causing underconsumption of barium(II) titrant:

$$BaSO_4{:}SO_4^{2-}{\vdots}2M^+(s)$$

Since sulfate is calculated from the amount of titrant used, the results for the sulfate determination are low. The error caused by the coprecipitation of sodium, ammonium, and (especially) potassium ions is serious (see Table 10-1). A few metal cations also interfere by forming a colored complex with the indicator.

Foreign anions coprecipitate as the barium salts and cause high results for sulfate. Chloride, bromide, or perchlorate causes only a small error, but nitrate must be avoided, as it causes the results to be quite high.

Ion exchange is used to remove cations that would interfere with the end point or cause error by coprecipitation. A column containing cation exchange resin in the hydrogen form is used. The cations in the sulfate solution are exchanged for the hydrogen ions of the resin:

$$2M^+SO_4^{2-} + 2Resin\!-\!H^+ \rightarrow 2Resin\!-\!M^+ + 2H^+SO_4^{2-}$$

The sulfuric acid is partially neutralized with magnesium acetate,

$$MgAc_2 + 2H^+SO_4^{2-} \rightarrow Mg^{2+}SO_4^{2-} + 2HAc$$

Table 10-1.   Effect of Foreign Salts on the Titration of Sulfate

	Difference from Theory, %	
Salt Added	Direct Titration	Titration After Ion Exchange
Al(ClO$_4$)$_3$	interferes	-0.4
Cu(ClO$_4$)$_2$	-0.4	-0.3
Fe(ClO$_4$)$_3$	interferes	-0.2
NaClO$_4$	-2.8	$+0.3$
NH$_4$ClO$_4$	-2.2	±0.0
KCl	-6.3	±0.0
Zn(ClO$_4$)$_2$	-1.2	±0.0

and the sulfate is titrated with standard barium(II). Magnesium acetate is used to neutralize the excess acidity, because Mg^{2+} is one of the *least* coprecipitated cations in the sulfate titration.

QUESTIONS AND PROBLEMS

1. Look up the solubility of silver fluoride. Would you expect that fluoride could be determined by titration with standard silver nitrate?

2. Write the end point reaction for each of the following: (a) Mohr chloride determination. (b) Volhard silver determination. (c) Adsorption-indicator bromide determination. (d) Adsorption-indicator sulfate determination.

3. Silver(I) forms insoluble precipitates with mercaptans (type formula, RSH). The precipitate dissociates as follows:

$$RSAg(s) \rightleftharpoons RS^- + Ag^+$$

Suggest an experimental method of determining the solubility product. How might the basic properties of the mercaptide ion (RS^-) complicate the determination?

4. Suggest a scheme for analyzing a sample containing water, hydrochloric acid, sulfuric acid, and perchloric acid. Give only titrimetric methods. Water may be calculated by difference.

5. Write equations for the determination of oxalate by the Volhard method. The solubility product for silver oxalate is 4.0×10^{-11}. In this case, will it be necessary to shake with nitrobenzene before the final titration?

6. The solubility product for silver hydroxide, AgOH(s), is 1.5×10^{-8}. If a $1 \times 10^{-4}M$ solution of silver nitrate is gradually made more alkaline with sodium hydroxide, at what pH will silver hydroxide begin to precipitate? What bearing does this have on the pH at which a titration of chloride by the Mohr method is carried out?

7. The solubility product for silver bromide is 4×10^{-13}. Do the following calculations for the titration of $0.01M$ bromide with $0.01M$ silver(I): (a) Calculate pAg when 99% of the bromide has been titrated. (b) Calculate pAg at the equivalence point (100% titration). (c) Calculate the change in silver electrode potential (in volts) between 99% and 100% titration.

8. Silver(I) can be determined by titration with a standard thiocyanate solution using a silver metal indicating electrode. If $0.01M$ Ag^+ is titrated with $0.01M$ SCN^-, calculate pAg at each of the following points: 0%, 50%, 90%, 99%, 100%, 101%, 110%, and 200% titration. (Use 1.10×10^{-12} as the solubility product of AgSCN). Also plot the titration curve.

9. A sample is analyzed for chloride by the Volhard method. From the following data, calculate the percentage of chloride present.

> Weight of sample: 314.0 mg
> Silver nitrate added: 40.00 mL of 0.1234M
> Thiocyanate back-titration: 13.20 mL of 0.0930M

10. A *pure* organic compound has the formula $C_4H_8SO_x$. A sample of a compound was decomposed, then the sulfur was converted to sulfate and titrated in a modified version of the adsorption indicator procedure. From the following data, determine the correct

formula for the compound.

>Weight of sample: 12.64 mg
>Ba(ClO_4)$_2$ needed for titration: 10.60 mL of 0.0100M

11. Specify the indicator and write equations for the adsorption indicator end point in the titration of thiocyanate with silver nitrate.
12. The indicator methyl violet, which has the general formula $R_4C^+ClO_4^-$, may be used to titrate Ag(I) with sodium chloride. Write the chemical equations describing the adsorption-indicator end point.

11

Complexes and Complex-Formation Titrations

The formation of complexes in solution is used in at least three major areas of analytical chemistry. Metal ions can be determined by forming highly colored complexes that can be determined spectrophotometrically (Chapter 17). A second area is the determination of metal ions by titration with a standard solution of a complexing reagent, such as ethylenediaminetetraacetic acid (EDTA). Such titrations are probably the best general method available for determining selected metal ions in solution with high accuracy. Complexing reagents are often added as "masking agents" to prevent the complexed metal ions from precipitating or otherwise interfering with an analytical procedure. Finally, many separation procedures, such as solvent extraction (Chapter 20) or ion-exchange chromatography (Chapter 24), are based on selective complex formation.

In this chapter, the nature of complexes and their stability will be considered first. Then the principles and methods of titration of metal ions with EDTA will be considered.

11-1. FORMATION OF COMPLEXES

General Principles. Most, if not all, metal ions can form coordination complexes or compounds with molecules or anions. Such molecules and anions are referred to as *ligands*. To act as a ligand, a species must donate an unshared electron pair to the metal ion to form a metal-ligand bond. Thus, molecules such as water and ammonia can act as ligands, but methane (CH_4) and carbon tetrachloride (CCl_4) cannot.

There are various theories, or points of view, as to the nature of the metal-ligand bond. The *valence-bond theory* views the metal-ligand bond as a polar covalent bond in which the electron density of the bond is closer to the ligand than to the metal ion. This theory predicts that metal-ligand bonds will be formed by using ligands with an unshared pair of electrons. It also helps in predicting the number of ligands that will bond to certain metal ions.

The electrostatic *crystal-field theory* views the metal-ligand bond as ionic and predicts that stronger metal-ligand bonds will be formed between ions of higher charges, or between a metal ion and a ligand with a large dipole moment. The *ligand-field theory*, a molecular orbital approach to bonding, accommodates both covalent and ionic bonding of the ligands to the metal ions. It is a more advanced theory than the other two and will not be used in our approach to complexation reactions.

Ligands. A number of common anionic ligands and molecular ligands are listed in Table 11-1. Note that anionic ligands may consist of just one atom, such as the halide ions, or of two or more atoms. Of course, both anionic and molecular ligands must contain at least one donor atom to share an electron pair with the central metal ion. The donor atom may be a halide ion, an oxygen atom, a sulfur atom, a nitrogen atom, a phosphorus atom, or any atom that can share an electron pair.

Complexes with Unidentate Ligands. Ligands that possess only one donor atom, or that donate just one of their available pairs of electrons to a metal ion, are called *unidentate* ligands. Water is the most common unidentate ligand, since all metal ions exist in aqueous solution as *aquo* complex ions. The general formula of an aquo complex ion is $M(H_2O)_n^{z+}$, where n is the number of water molecules complexing the metal ion. (Common values for n are four and six, as in $Be(H_2O)_4^{2+}$ and $Cu(H_2O)_6^{2+}$.)

Complexation reactions of an aquo metal ion and a unidentate ligand, L, involve replacing the same number of water molecules as the number of ligands bonding to the metal ion. This occurs in a *stepwise fashion*, so that the formation of ML_2 really involves:

$$M(H_2O)_n^{z+} + L = [M(H_2O)_{n-1}L]^{z+} + H_2O$$

$$[M(H_2O)_{n-1}L]^{z+} + L = [M(H_2O)_{n-2}L_2]^{z+} + H_2O, \text{ etc.}$$

Table 11-1. Common Ligands

Anionic Ligands	Molecular (Neutral) Ligands
F^-, Cl^-, Br^-, I^- (halides)	H_2O
	NH_3
SCN^-	RNH_2 (aliphatic amines)
CN^-	C_5H_5N (pyridine = py)
OH^-	$H_2N-C_2H_4-NH_2$ (ethylenediamine = en)
$RCOO^-$	
S^{2-}	
RS^- (anion of mercaptan, RSH)	$C_{12}H_8N_2$ (1,10-phenanthroline = ferroin or phen)
$C_2O_4^{2-}$ (oxalate)	$(C_8H_{17})_3P-O$ (trioctylphosphine oxide = TOPO)
	$(C_8H_{17})_3P-S$ (trioctylphosphine sulfide = TOPS)

Usually, such reactions are simplified by omitting the water and writing ML, ML_2, etc.

A metal ion frequently forms a series of complexes with the same ligand. For example, the cupric ion forms complexes stepwise that contain as much as 5 moles of ammonia per mole of copper(II). The solution contains a mixture of these cupric-amino complexes. The relative amounts of the various complexes depend on the concentration of ammonia present (see Figure 11–1).

An equilibrium-constant expression for each step in the complex formation can be written. Stepwise formation constants for many metal complexes with either inorganic or organic ligands have been measured. As an example, the expressions and numerical values of the stepwise constants for cadmium(II) iodide complexes are given below:

$$K_1 = \frac{[CdI^+]}{[Cd^{2+}][I^-]} = 10^{2.4}$$

$$K_2 = \frac{[CdI_2]}{[CdI^+][I^-]} = 10^{1.0}$$

$$K_3 = \frac{[CdI_3^-]}{[CdI_2][I^-]} = 10^{1.6}$$

$$K_4 = \frac{[CdI_4^{2-}]}{[CdI_3^-][I^-]} = 10^{1.2}$$

Often the *product* of stepwise formation constants is reported by using the symbol β with a subscript to denote the highest stepwise constant in the product. For the stepwise constants given above: $\beta_1 = K_1, \beta_2 = K_1K_2, \beta_3 = K_1K_2K_3, \beta_4 = K_1K_2K_3K_4$. Some of the useful calculations that can be made with formation constants will be illustrated later in this chapter.

The most complete source is *Stability Constants* by J. Bjerrum, G. Schwarzenbach, and L. G. Sillen, London: The Chemical Society, 1957. A less extensive, but

Figure 11–1. The distribution of various copper(II) amine complexes. (From A. Ringbom, *Complexation in Analytical Chemistry*, New York: Wiley-Interscience, 1963, p 29. By permission.)

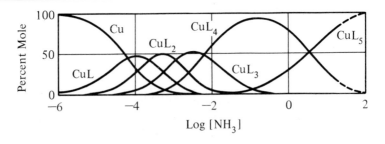

more convenient, listing of constants is given by A. Ringbom in *Complexation in Analytical Chemistry*, New York: Wiley-Interscience, 1963.

Chelates with Multidentate Ligands. Ligands that possess two or more donor atoms may share more than one pair of electrons with a single metal ion by coordinating to two or more positions around the central metal ion. These ligands are, in general, known as *multidentate* ligands. Specifically, they are known as *bidentate, tridentate, tetradentate,* or *hexadentate* if one such ligand coordinates to two, three, four, or six positions around the metal ion, respectively.

Multidentate ligands complex metal ions to form complex ions known more properly as *chelates;* instead of a linear structure, the chelates have a *chelate ring* structure. For example, zinc(II) ion and the organic ligand ethylenediamine (see Table 11–1) form $Zn(en)_2^{2+}$ with the following chelate ring structure:

$$Zn^{2+} + 2H_2NCH_2CH_2NH_2 \rightarrow$$
(ethylenediamine)

Note that both of the chelate rings of $Zn(en)_2^{2+}$ are five-membered rings. In general, for a stable chelate to form, the ligand must be of such a structure that a 5- or 6-membered ring will be formed with the chelated metal atom. Also note that the overall charge of the chelate of zinc(II) is still two, the same as the charge on the unchelated zinc(II) ion. In general, if all of the donor atoms are *basic* nitrogens, or other atoms that cannot lose a proton during chelation, there is no charge neutralization of the central metal ion.

A more common situation is when the chelating ligand has at least one acidic group or donor atom, as well as one or more basic donor atoms such as nitrogen. Here the acidic group loses a proton during the chelation and becomes an anionic donor, thus resulting in *charge neutralization.* A good example of such a chelating ligand is 8-hydroxyquinoline (Table 11–1). The −1 anion of this ligand reacts as follows with zinc(II) ion:

$$Zn^{2+} + 2$$
(8-hydroxyquinolate)

Note that the zinc(II) oxinate chelate has no overall charge; the charge on the zinc(II) ion has been completely neutralized to form a *neutral* chelate.

Some examples of common acidic groups and groups containing a basic nitrogen (or similar atom that does not lose a proton) are as follows (the anionic form of the acidic group is given in parentheses):

Acidic Groups	Basic Groups
—COOH (—COO$^-$)	—NH$_2$
enolic —OH (—O$^-$)	$>$NH
—SH (—S$^-$)	$\geqslant$N and aromatic N
phenolic —OH (—O$^-$)	$>$C=NOH
	$>$C=O

Chelation reactions may also exhibit stepwise reactions involving more than one ligand molecule. Where the ligand has one acidic group and basic group, the stepwise reactions usually terminate in the formation of a *neutral* chelate. Thus, zinc(II) reacts stepwise with oxine, but the neutral chelate is the usual product.

$$Zn^{2+} + oxine \rightleftharpoons Zn(oxine)^+$$

$$Zn(oxine)^+ + oxine \rightleftharpoons Zn(oxine)_2$$

In general, we find that a ligand with one basic group and one acidic group will form 2:1 (2 ligand:1 metal ion) chelates with $+2$ cations and 3:1 chelates with $+3$ metal ions. When there is more than one acidic group in a ligand, *negatively charged* chelates are usually formed. For example, zinc(II) and the oxalate ion form $Zn(C_2O_4)^{2-}$. EDTA also forms negatively charged chelates with metal ions, as will be discussed in the next section.

Coordination Number and Structure.   The coordination number of a metal ion may be defined as the number of metal-ligand bonds it forms. Strictly, the coordination number includes the number of metal-water bonds, as well as the number of bonds to other ligands. Often, the number of water molecules is omitted to simplify things. For example, copper(II) ion can coordinate to six ammonia molecules, but the most predominate form is the tetraamminediaquocopper(II) ion. Many authors write this latter ion as $Cu(NH_3)_4^{2+}$ instead of $Cu(NH_3)_4(H_2O)_2^{2+}$, for simplicity.

In considering coordination number, there is the question: How can a metal ion bond to more donor atoms, or groups, than its positive charge (or valence)? Near the beginning of the twentieth century Alfred Werner proposed that, in addition to their charge or primary valence, metal ions exhibit a secondary valence or coordination sphere of a definite number of ligands. Werner's ideas were ultimately fashioned into the valence bond theory previously mentioned. According to this theory, the metal ion accepts electron pairs from the ligands via its unfilled orbitals. Each ion forms complex ions with the most stable structure possible for a given ligand. A metal ion may in fact react with two different ligands to form two complex ions with different structures in order to form the most stable structure in each case. To illustrate these ideas, we will discuss a few examples of how coordination number and structure are related.

Coordination Number Two.   Metal ions with a coordination number of two exhibit a *linear* structure in which the two ligands are on opposite sides of the metal ion—for example, $[H_3N: \rightarrow Ag \leftarrow :NH_3]^+$. Univalent metal ions of Periodic Group IB often exhibit this structure. Thus silver(I), copper(I), and gold(I) form complexes such as $Ag(CN)_2^-$, $Cu(NH_3)_2^+$, and $AuCl_2^-$.

Coordination Number Four. Metal ions with a coordination number of four form complexes with a *tetrahedral* structure, or occasionally with a *square planar* structure. Except for $Cu(NH_3)_4^{2+}$ and certain nickel(II) complexes, most metal ions exhibit the tetrahedral structure. Thus, nickel(II) is the only common metal ion that exhibits more than one ion of each structure. The tetrahedral and square planar nickel(II) structures are:

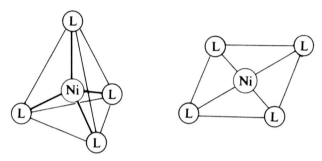

The divalent metal ions of Groups IB, IIB, and VII exhibit a CN of four as their most common coordination number. Thus, ions such as nickel(II), cobalt(II), copper(II), zinc(II), and even iron(III), coordinate to four halide ligands, four SCN⁻ ligands, etc. However, nickel(II), cobalt(II), copper(II), and cadmium(II) exhibit a CN of six with water and ammonia.

Coordination Number Six. Metal ions with a coordination number of six have an *octahedral* structure, as shown:

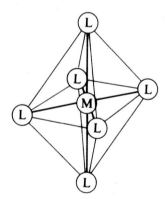

Four of the ligands are in the same plane as the metal ion; this part of the structure is the same as the square planar structure for CN four. The other two ligands are above and below the plane. In $Cu(NH_3)_4(H_2O)_2^{2+}$, for example, the four ammonia ligands are in the same plane as the copper(II) ion, and the water molecules are above and below this plane. Thus, some authors write $Cu(NH_3)_4^{2+}$ and indicate a square planar structure, which is consistent with the structure of this ion written as $Cu(NH_3)_4(H_2O)_2^{2+}$.

Many metal ions have a coordination number of six toward ligands such as water, ammonia, and ethylenediamine. Some of these are iron(III), cobalt(III), chromium(III), copper(II), nickel(II), cobalt(II), and calcium(II) ions. EDTA is a ligand that has the ability to coordinate to all six octahedral positions around a metal ion, but frequently it only coordinates to five positions, leaving the sixth to be occupied by water. Thus, we find the formula $Co-EDTA^-$ for cobalt(III)-EDTA and the formula $[Ni-EDTA(H_2O)]^{2-}$ for nickel(II)-EDTA. EDTA reactions will be discussed in detail in the next section.

11-2. THEORY OF COMPLEXOMETRIC TITRATIONS

Many metal ions can be determined by titrating them with a reagent that complexes them in solution. The solution to be titrated is buffered at a suitable pH, an indicator is added, and the metal ion is titrated with a standard solution of the complexing agent. Usually, a sharp color change marks the end point of the titration. Titrations of this type are convenient and accurate; in many instances, they have replaced time-consuming gravimetric procedures. Except for the alkali metals, most metal cations can be determined by titration with a suitable complexing agent.

Choice of Titrant. For titration of a metal ion with a complexing ligand, the formation constant of the complex must be large so that the titration reaction will be stoichiometric and quantitative. In the case of unidentate ligands that form several complexes with a metal ion, the overall constant (product of the stepwise constants) is frequently large, but the stepwise constants are too small. The result is a gradual change in metal-ion concentration as more ligand is added. For a titration reaction, however, there must be a sharp change in metal-ion concentration at some stoichiometric point in the titration.

A few multidentate ligands form strong 1:1 complexes with various metal ions. The complexation occurs in a single step, so titration of a metal produces a sharp change in metal-ion concentration at the equivalence point. Multidentate ligands useful in the titration of metal ions include EDTA (ethylenediaminetetraacetic acid) and related compounds, and polyamines such as trien (triethylenetetramine):

$$NHCH_2CH_2NH_2$$

HOOCCH$_2$ CH$_2$COOH CH_2

$\qquad$ N—CH$_2$CH$_2$—N CH_2

HOOCCH$_2$ CH$_2$COOH $NHCH_2CH_2NH_2$

$\qquad\qquad$ EDTA   trien

Trien is a quadridentate ligand that forms a coordinate bond to a metal through each of its four nitrogen atoms. It is useful for titrating metal ions such as copper(II), mercury(II), and nickel(II) in alkaline solution. In acidic solutions, trien loses most of its chelating ability because of protonation of the nitrogen atoms.

The ligand EDTA, which forms stable chelates with a large number of metal ions, is by far their most important titrant. The structural chemistry of metal-EDTA complexes has been reviewed [1]. Most EDTA complexes in solution are intermediate between penta- and hexadentate. In a hexadentate complex, a bond to the metal is formed from each of the four carboxyl groups and each of the two nitrogens in EDTA. Magnesium(II) forms a heptadentate complex anion with EDTA, H_2OMgY^{2-}, in which the seventh bond is to a water molecule.

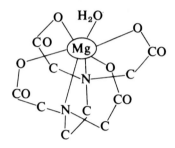

Regardless of how many coordination positions EDTA occupies around a metal ion, the important thing is that EDTA always reacts with metal ions in a 1:1 molar ratio. All such EDTA chelates are soluble in water; most of them are colorless, with the exception of the chelates of a few transition metal ions.

The acid form of EDTA is frequently represented as H_4Y, H representing the acidic hydrogen atoms and Y the rest of the molecule. When a strong base such as sodium hydroxide is added to H_4Y, the neutralization occurs in steps, and H_3Y^-, H_2Y^{2-}, HY^{3-}, and Y^{4-} are formed. The free acid (H_4Y) and the monosodium salt (NaH_3Y) are not soluble enough in water to use as titrants, but the disodium salt of EDTA (Na_2H_2Y) is soluble and can be used. Hydrogen ions are liberated during the titration of metal ions with disodium EDTA; for example:

$$Mg^{2+} + H_2Y^{2-} \rightarrow MgY^{2-} + 2H^+$$

$$Al^{3+} + H_2Y^{2-} \rightarrow AlY^- + 2H^+$$

$$Th^{4+} + H_2Y^{2-} \rightarrow ThY + 2H^+$$

Because of this liberation of hydrogen ions, the solution is buffered and so prevents a large change in pH during titration.

Titration Curves: Effect of pH. The strength or stability of complexes with EDTA is different for different metal ions. The formation constant (often called stability constant) is a measure of the strength of a complex. The reaction for the formation of a metal-EDTA complex is

$$M^{2+} + Y^{4-} \rightleftharpoons MY^{2-}$$

Written without charges, this reaction is

$$M + Y \rightleftharpoons MY$$

[1] R. H. Nuttall and D. M. Stalker, *Talanta*, 24, 355 (1977).

Table 11–2. Formation Constants of Some Metal–EDTA Complexes

Metal Ion	$\log K_{MY}$	Metal Ion	$\log K_{MY}$	Metal Ion	$\log K_{MY}$
Fe^{3+}	25.1	Ni^{2+}	18.6	Ce^{3+}	16.0
Th^{4+}	23.2	Pb^{2+}	18.0	La^{3+}	15.4
Cr^{3+}	23	Cd^{2+}	16.5	Mn^{2+}	14.0
Bi^{3+}	22.8	Zn^{2+}	16.5	Ca^{2+}	10.7
VO^{2+}	18.8	Co^{2+}	16.3	Mg^{2+}	8.7
Cu^{2+}	18.8	Al^{3+}	16.1	Sr^{2+}	8.6
				Ba^{2+}	7.8

The formation-constant expression is

$$K = \frac{[MY]}{[M][Y]} \tag{11–1}$$

Note that in the representation of formation constants Y^{4-} (written simply Y), rather than H_2Y^{2-}, is considered the reactive EDTA species. The formation constants for some metal-EDTA complexes are given in Table 11–2.

Titration curves are useful in understanding EDTA titrations. In some cases, they can be obtained experimentally by measuring the potential of an ion-selective electrode (Chapter 16) at various points during the titration of a metal ion with EDTA. But curves calculated from available equilibrium constants are helpful in predicting the feasibility of a particular EDTA titration performed under a given set of conditions. Usually, the titration curve is a plot of pM (negative logarithm of the metal-ion concentration) versus milliliters of EDTA or percent titration. Figure 11–2 shows titration curves for metal ions having different formation constants for the metal-EDTA complex. Note that the larger the formation constant, the greater the break at the equivalence point.

The titration equilibrium will be modified by the effect of pH on the availability of Y. Qualitatively, increasing acidity weakens the MY complex by protonating Y and thus decreasing the proportion of uncomplexed EDTA present as Y^{4-}:

$$M + Y \rightleftharpoons MY$$
$$\text{H} \, \Big\|$$
$$HY, H_2Y, \text{ etc.}$$

Quantitatively, the effect of hydrogen ions on the equilibrium may be calculated by using α_Y, the fraction of all forms of uncomplexed EDTA present as Y^{4-}:

$$\alpha_Y = \frac{[Y^{4-}]}{[H_4Y] + [H_3Y^-] + [H_2Y^{2-}] + [HY^{3-}] + [Y^{4-}]} = \frac{[Y]}{[Y']} \tag{11–2}$$

Figure 11-2. Theoretical curves for the titration of M with EDTA as a function of the conditional formation constant, $K_{MY'}$ (see Equation 11-3). It is assumed that no ligand other than Y complexes M.

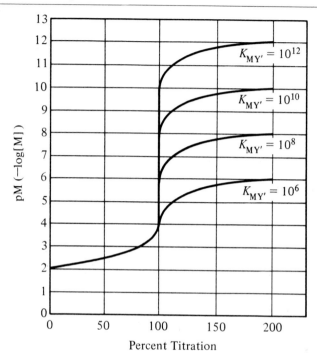

The method of calculating α_Y from the ionization constants for H_4Y is explained in Chapter 7, pp. 128-129.

Example: Calculate α_Y for EDTA at pH 5.0. The ionization constants for H_4Y are as follows: $pK_1 = 2.07$, $pK_2 = 2.75$, $pK_3 = 6.24$, $pK_4 = 10.34$.

Substituting these ionization constants and $[H^+] = 10^{-5.0}$ into Equation 7-20 on pages 128-129,

$$\frac{1}{\alpha_Y} = \frac{10^{-20.0}}{10^{-21.4}} - \frac{10^{-15.0}}{10^{-19.3}} + \frac{10^{-10.0}}{10^{-16.6}} + \frac{10^{-5.0}}{10^{-10.3}} + 1$$

$$\frac{1}{\alpha_Y} = 10^{1.4} + 10^{4.3} + 10^{6.6} + 10^{5.3} + 1$$

If α_Y is calculated to one decimal place, only the third exponential term is significant. Therefore, $\alpha_Y = 10^{-6.6}$.

For any given complexing agent (such as EDTA) α_Y can be calculated and plotted as a function of pH. This is done in Figure 11-3. By means of α_Y, a

Figure 11–3. Fraction α_Y of all forms of uncomplexed EDTA present as Y^{4-}, versus pH.

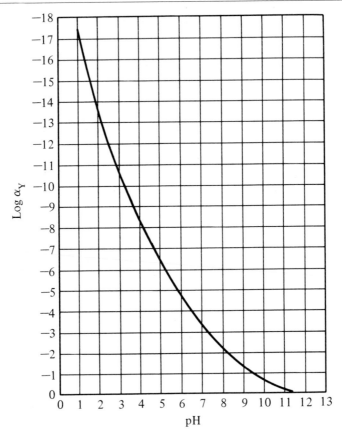

conditional constant may be calculated for a metal-EDTA complex at any pH:

$$K_{MY} = \frac{[MY]}{[M][Y]}$$

$$[Y] = \alpha_Y[Y']$$

$$K_{MY} = \frac{[MY]}{[M][Y']\alpha_Y}$$

$$K_{MY}\alpha_Y = K_{MY'} = \frac{[MY]}{[M][Y']} \qquad (11\text{–}3)$$

In Equation 11–3, $K_{MY'}$ is the conditional constant at a specified pH. It is actually the value of $K_{MY'}$ that determines the course of the titration curves in Figure 11–2; the larger the conditional constant, the more favorable the titration should be. This will be illustrated by calculating the pM at various points in a titration of M with EDTA at a fixed pH.

Example: A 0.0100M solution of nickel(II) is buffered at pH 5.0 and titrated with 0.010M EDTA. Calculate the pNi at 50%, 100%, and 200% titration.

50% Titration: Here we have NiY and free Ni^{2+} that has not yet been titrated with EDTA. The pNi is calculated from the concentration of untitrated Ni^{2+} because the dissociation of NiY to Ni^{2+} is negligible by comparison.

$$[Ni^{2+}] = 50\% \text{ of } 0.0100M = 0.0050M, \text{ neglecting dilution}$$

$$[Ni^{2+}] = 0.0050 \times 2/3 = 3.3 \times 10^{-3}M \text{ including dilution of } 50\%$$

$$pNi = 3 - \log 3.3 = 2.48$$

100% Titration: Essentially all of the nickel has been converted to NiY; the very low concentration of Ni^{2+} remaining can be calculated using the conditional formation constant.

At pH 5.0, $\alpha_{Ni} = 10^{-6.6}$. The conditional constant is calculated by using Equation 11-3:

$$K_{NiY'} = K_{NiY}\alpha_{Ni} = 10^{18.6} \times 10^{-6.6} = 10^{12.0}$$

$$[NiY] = 0.010 \times \tfrac{1}{2} = 0.0050M$$

taking into account a 100% volume increase. Since the NiY dissociates slightly to give $Ni^{2+} + Y'$, it follows that $[Ni] = [Y']$. Substituting into the $K_{NiY'}$ expression,

$$10^{12.0} = \frac{[NiY]}{[Ni][Y']} = \frac{0.0050}{[Ni]^2}$$

$$[Ni] = \sqrt{50 \times 10^{-16}} = 7.1 \times 10^{-8}$$

$$pNi = 8 - \log 7.1 = 7.15$$

200% Titration: The first 100% of EDTA reacts to form NiY and the second 100% remains as Y'. Therefore, the ratio of [NiY] to [Y'] is 100 to 100, or 1 to 1. Substituting into the $K_{NiY'}$ expression,

$$10^{12.0} = \frac{[NiY]}{[Ni][Y']} = \frac{100}{[Ni]100}$$

$$[Ni] = 10^{-12.0}, pNi = 12.0$$

A plot of the complete titration curve for this example above is given in Figure 11-4. Note that at 200% titration, $pNi = \log K_{NiY'} = 12.0$. For EDTA titration of any metal at 200% titration,

$$pM = \log K_{MY'} \tag{11-4}$$

This is very convenient for estimating the shape of any given titration curve.

Example: A 0.01M solution of calcium(II) is buffered at pH 5.0 and titrated with 0.01M EDTA. Sketch approximately the expected titration curve, using pCa at 0% and 200% titration.

At the start of the titration, $[Ca^{2+}] = 0.01M$ and $pCa = 2.0$. From Table 11–2, $K_{CaY} = 10^{10.7}$. At pH 5.0, $\alpha_Y = 10^{-6.6}$ (Figure 11–3). Thus, $K_{CaY'} = 10^{10.7} \times 10^{-6.6} = 10^{4.1}$ (Equation 11–3). From the discussion above, we know that at 200% titration, $pCa = 4.1$. Since the pCa at 0% and 200% titration is known, the titration curve can be sketched as shown in Figure 11–4.

Titration Curves; Effect of Complexing Buffers. We have seen that acidity has a major effect on the titration curve of a metal ion with EDTA. The titration curve may also be affected by complexing the free metal ion with a complexing buffer such as ammonia, or with hydroxide ions.

$$M^{2+} + 4NH_3 \rightarrow M(NH_3)_4^{2+}$$

$$M^{2+} + OH^- \rightarrow M(OH)^+$$

The formation of such complexes reduces the concentration of free metal ions but need not interfere with the EDTA titration. For example, zinc(II), cadmium(II), and other metal ions frequently are titrated with EDTA in basic solutions containing ammonia. An ammonium ion–ammonia buffer maintains the solution at a nearly fixed pH, and the ammonia prevents precipitation of metal hydroxides by complexation.

The qualitative effect of ammonia, hydroxyl, or other weak complexing ligands on the metal ion–EDTA equilibrium is to shift the equilibrium to the left, as

Figure 11–4. Curves for titration of nickel(II) and calcium(II) with EDTA at pH 5.0.

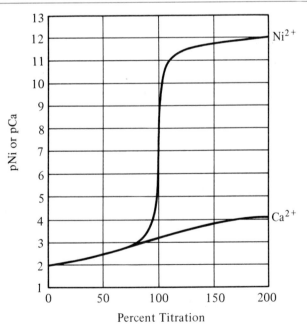

seen from the following equation:

$$M + Y \rightleftharpoons MY$$

$$L \Big\Updownarrow$$

$$ML$$

Quantitatively, the effect of L on the equilibrium is determined by calculating α_M, which is the fraction of all forms of M and ML in solution present as the free metal ion:

$$\alpha_M = \frac{[M]}{[M']} \tag{11-5}$$

Here, $[M]$ is the concentration of free metal cation M^{z+}, and $[M']$ is the total concentration of metal in solution not complexed by EDTA.

Sometimes the substance that does the complexing is written in parentheses after the subscript on the alpha. Thus, $\alpha_{M(L)}$ indicates the ratio of $[M]/[M']$ resulting from complexing metal M with ligand L and $\alpha_{Y(H)}$ is the ratio of $[Y]/[Y']$ resulting from the action of H on Y.

For a given concentration of free, unbound ligand (L), the value of α_M is calculated from the stepwise-formation constants of ML_n with the following formula:

$$\frac{1}{\alpha_M} = 1 + \beta_1[L] + \beta_2[L]^2 + \cdots + \beta_n[L]^n \tag{11-6}$$

where β_1, β_2, and β_n are the products of stepwise-formation constants for the metal-ligand complexes.

Example: Calculate α_{Cd} for cadmium(II) in a solution containing $0.10M$ NH_4^+ and $0.10M$ unbound NH_3. Logarithms of the formation constants for the cadmium-ammonia complexes are:

$$\beta_1 = 2.60, \quad \beta_2 = 4.65, \quad \beta_3 = 6.04, \quad \beta_4 = 6.92, \quad \beta_5 = 6.60$$

Substituting these formation constants and $[L] = 0.10$ into Equation 11-6,

$$\frac{1}{\alpha_{Cd}} = 1 + (10^{2.6})(10^{-1}) + (10^{4.65})(10^{-2}) + (10^{6.04})(10^{-3}) + (10^{6.92})(10^{-4})$$

$$+ (10^{6.6})(10^{-5})$$

$$\frac{1}{\alpha_{Cd}} = 1 + 40 + 450 + 1100 + 830 + 40 = 2461$$

$$\alpha_{Cd} = \frac{1}{2461} = 10^{-3.4}$$

Once the value of α_M is known, it can be used to calculate the pM during the early stages of a titration with EDTA.

Example: Calculate pCd when a $0.01M$ cadmium(II) solution is buffered with $0.10M$ NH_4^+ and $0.10M$ free NH_3, and 50% of the cadmium(II) has been titrated with EDTA. Neglecting any dilution from the EDTA titrant,

$$[Cd'] = \tfrac{1}{2} \times 0.010 = 0.0050M$$

(The CdY is so stable that any Cd' from its dissociation is negligible.) From the previous example,

$$\alpha_{Cd} = 10^{-3.39}$$

Since

$$\alpha_{Cd} = [Cd]/[Cd']$$

$$[Cd] = \alpha_{Cd}[Cd'] = 10^{-3.39} \times 5.0 \times 10^{-3} = 5.0 \times 10^{-6.39}$$

$$pCd = 6.39 - \log 5 = 5.69$$

If the Cd^{2+} were not complexed with ammonia, the [Cd] at 50% titration would be $0.0050M$ and $pCd = 2.30$.

Acidity can also reduce the concentration of L in a complexing buffer by protonation to form HL (or a series of protonated ligand species). This may be taken into account by using conditional constants (β_1, β_2, etc.) at the titration pH to calculate α_M.

In summary, the course of the EDTA titration of a metal ion may be affected by acidity, which reduces the effective concentration of Y^{4-}, and by a complexing buffer, which reduces the concentration of M^{z+}. These effects on a titration curve are shown in Figure 11–5. The latter effect is given by α_M and increases pM before the equivalence point. The acidity effect is given by α_Y and reduces pM after the equivalence point.

Selection of Indicator. The indicator for the titration of a metal ion with a chelating agent such as EDTA is usually a highly colored dye that forms a complex of a different color with the metal ion being titrated. The titration of a metal ion with EDTA thus consists of two consecutive reactions: (1) titration of the free metal ion with EDTA, and (2) breaking up the metal-indicator complex by forming a stronger metal-EDTA complex.

$$MInd + Y' \rightarrow MY + Ind'$$

The second reaction may occur at widely varying pM values, depending on the conditional-formation constant of the metal indicator at the pH of the titration.

Like acid-base indications (which change color over a pH range), metal indicators change color over a range of about 2 pM units. To obtain a sharp and accurate end point, the indicator chosen should have a pM transition range that includes the pM value at the equivalence point of the EDTA titration. It is also necessary that the *rate* at which the metal-indicator reacts with EDTA be rapid, so that the end point will not be overrun.

The pM transition range of a metalochromic indicator can be calculated from the conditional-formation constant of the metal-indicator complex.

Figure 11-5. The effect of α_M and α_Y on a complexometric titration curve, shown by the solid line. The dashed line depicts the titration curve if M were not complexed by L or Y were not complexed by H.

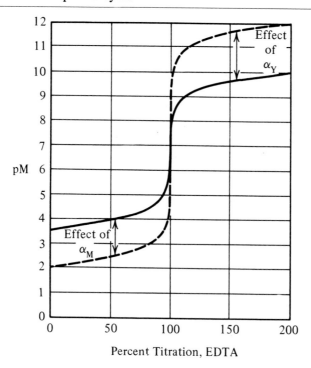

Percent Titration, EDTA

Example: Calculate the approximate transition range of an indicator used in the titration of nickel at pH 5.0 if the indicator's conditional-formation constant at this pH is $10^{8.0}$. Decide whether this indicator would be suitable for titrating nickel(II) with EDTA at pH 5.0 when the pNi is 7.15 at the equivalence point.

The nickel-indicator formation-constant expression is

$$K = \frac{[\text{NiInd}]}{[\text{Ni}][\text{Ind}']} = 10^{8.0} \tag{11-7}$$

Near the end point, the NiInd complex is broken up by the reaction of the nickel with EDTA. It is generally assumed that the human eye can detect about 1 part of one color in 10 parts of another (see p. 149); therefore, the first discernible color change will occur at a $[\text{NiInd}]:[\text{Ind}']$ ratio of about 10. Substituting into Equation 11-7,

$$10^{8.0} = \frac{10}{[\text{Ni}]}$$

$$\text{pNi} = 7.0$$

The *last* discernible color change will occur when this ratio is reversed (a

[NiInd]:[Ind'] ratio of about 0.1). Substituting this ratio, we get

$$10^{8.0} = \frac{0.1}{[Ni]}$$

$$pNi = 9.0$$

The approximate pNi transition range at pH 5.0 is 7.0–9.0. The pNi at the equivalence point in the EDTA titration was calculated to be 7.15. Thus, this indicator may be used, provided the *first* discernible color change is taken as the end point.

11–3. DETERMINATION OF TOTAL HARDNESS IN WATER

Water Hardness. Hard water contains dissolved calcium and magnesium salts, mostly as the bicarbonates. Water heaters, faucets, and other objects that are in long contact with hard water often "lime up" with a deposit of precipitated metal salts. Calcium and magnesium salts of soaps are insoluble and form a scum called "bathtub ring." In a great many industries, the hardness of the water used must be kept under strict control.

Titration with standard EDTA is used throughout most of the world to determine the amount of hardness in water. Usually, *total hardness* is determined, which is the *sum* of calcium and magnesium. The color change of a visual indicator is used to mark the end point of the titration. An "inhibitor" is often added to complex iron or other metal salts that might interfere with the action of the indicator.

Although both calcium and magnesium are titrated with the standard EDTA, total hardness is calculated as if it was all calcium carbonate, $CaCO_3$. The hardness, as mg/L of $CaCO_3$, is calculated as follows.

$$Hardness = \frac{(mL \ of \ EDTA)(M \ of \ EDTA)(100.09)(1000 \ mL/L)}{mL \ Sample}$$

Titration Methods. Total hardness (the sum of calcium(II) and magnesium in the water) is usually determined by titration with standard $0.01 M$ EDTA using a visual indicator. The conditional formation constants of calcium(II) and magnesium(II) with EDTA are too small in acidic or neutral solutions, so the solution is buffered to maintain a pH of 10.0 throughout the titration. In the most popular procedure, Calmagite (or a closely related compound, Eriochrome Black T) is the indicator.

Indicator solutions of Calmagite are more stable than solutions of Eriochrome Black T, which decomposes slowly on standing. Calmagite and Eriochrome Black T may be used interchangeably.

The free indicator is blue, but the magnesium complex formed in the solution is red. When magnesium is titrated with EDTA, the stoichiometric reaction is

$$Mg^{2+} + H_2Y^{2-} \ \rightarrow \ MgY^{2-} + 2H^+$$

The titrant, represented here as H_2Y^{2-}, is a solution of the disodium or diammonium salt of EDTA. When the free magnesium ions are all titrated, a drop or two of EDTA causes the end point reaction to occur. This results in a color change from red to blue:

$$MgIndic^- + H_2Y^{2-} \rightarrow MgY^{2-} + HIndic^{2-} + H^+$$

$$\text{(red)} \qquad\qquad \text{(colorless)} \qquad \text{(blue)}$$

Calmagite is not a satisfactory indicator for the EDTA titration of calcium(II) in the complete absence of magnesium(II). The reason is that the calcium-Calmagite complex is too weak—the end point is gradual and appears too soon. (This is shown in Figure 11–6a.) The magnesium-Calmagite complex is stronger, and the end point for the titration of magnesium(II) with EDTA comes later, at a higher pM value (Figure 11–6b).

The titration curves in Figure 11–6 are for the titration of approximately $0.001M$ calcium(II) and $0.001M$ magnesium(II) with $0.01M$ EDTA at pH 10.0. At pH 10.0, $\alpha_Y = 10^{-0.5}$ (Figure 11–3). $K_{CaY} = 10^{10.7}$ and $K_{MgY} = 10^{8.7}$ (Table 11–1).

Figure 11–6. Theoretical curves for the titration of (a) calcium(II) and (b) magnesium(II) with EDTA.

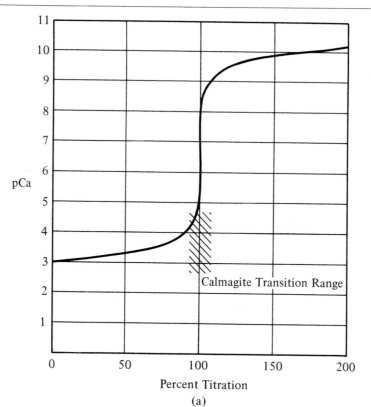

pCa

Calmagite Transition Range

0   50   100   150   200

Percent Titration

(a)

Figure 11–6. (*continued*)

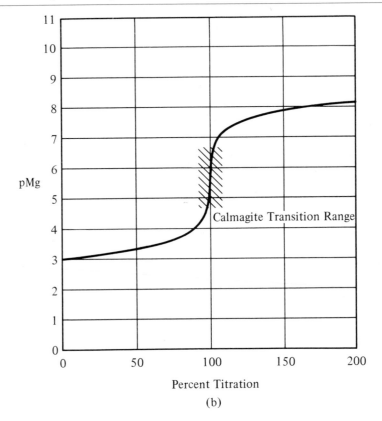

Percent Titration

(b)

Thus, the conditional constants at pH 10.0 are

$$K_{\text{CaY}'} = 10^{10.2} = \frac{[\text{CaY}]}{[\text{Ca}][\text{Y}']} \qquad K_{\text{MgY}'} = 10^{8.2} = \frac{[\text{MgY}]}{[\text{Mg}][\text{Y}']}$$

At the equivalence point in the titration,

$$[\text{CaY}] = 0.001M \qquad [\text{MgY}] = 0.001M$$

$$[\text{Ca}] = [\text{Y}'] \qquad\qquad [\text{Mg}] = [\text{Y}']$$

By substituting into the conditional-constant expression, the pM at the equivalence point can be calculated.

$$10^{10.2} = \frac{0.001}{[\text{Ca}]^2} \qquad 10^{8.2} = \frac{0.001}{[\text{Mg}]^2}$$

$$[\text{Ca}] = \sqrt{10^{-13.2}} \qquad [\text{Mg}] = \sqrt{10^{-11.2}}$$

$$\text{pCa} = 6.6 \qquad\qquad \text{pMg} = 5.6$$

At the middle of the magnesium(II)-Calmagite transition range, the reported value of pMg is 5.7, which is in excellent agreement with pMg = 5.6 at the equivalence point in the EDTA titration. However, at the middle of the calcium(II)-Calmagite transition, the pCa is 3.7, which is in poor agreement with the equivalence-point pCa of 6.6. These calculations confirm that magnesium must be present if Calmagite indicator is to be used.

If calcium(II) and magnesium(II) are present in the same sample, the free calcium and magnesium ions are titrated together. The end-point reaction is between EDTA and the magnesium(II)-Calmagite complex (Calmagite reacts with the magnesium in preference to calcium), and the end point is sharp. If calcium alone is to be determined with Calmagite as indicator, a small amount of magnesium(II) is added; this forms a magnesium(II)-Calmagite indicator system that will give a good end point. The amount of EDTA equivalent to the magnesium(II) is subtracted from the total EDTA used in the titration. Alternatively, magnesium(II)-EDTA is added before the titration, in which case no subtraction is necessary.

For real water samples, some modification in the titration procedures is necessary to avoid interference from impurities that are likely to be present in water samples [2]. Water frequently contains small amounts of iron and other dissolved metals, which react irreversibly with Calmagite and prevent an end point in the titration of calcium and magnesium. If the iron concentration is quite low, the ammonia buffer will form hydroxo complexes with the iron and prevent interference. Sodium cyanide will sometimes prevent interference from slightly larger amounts of iron and from copper(II) by forming cyanide complexes of the metal ions. Triethanolamine is effective with dissolved aluminum but somewhat less effective than cyanide with iron.

A water-hardness procedure that uses Arsenazo I as the indicator avoids most of the problems associated with Calmagite [3]. With Arsenazo, it is not necessary to add cyanide to prevent interference from small amounts of iron or copper salts. The Arsenazo color change at the end point is much faster than Calmagite, which has rather slow kinetics. With either calcium(II) or magnesium(II), the Arsenazo end point comes near the theoretical pM; therefore, it is not necessary to add magnesium(II) when titrating a solution containing only calcium(II).

Example: Calculate the ratio of Ca-Arsenazo (violet) to free Arsenazo (orange-yellow) at the stoichiometric point in the titration of $0.001\,M$ calcium(II) with $0.01\,M$ EDTA at pH 10.0. The formation constant for Ca-Arsenazo at pH 10.0 is $10^{5.68}$.

Earlier, the pCa at the stoichiometric point in this same titration was calculated to be 6.6. Substituting $[Ca] = 10^{-6.6}$ into the conditional-constant expression for Ca-Arsenazo,

$$10^{5.68} = \frac{[CaArs]}{10^{-6.6}[Ars]}$$

$$\frac{[CaArs]}{[Ars]} = 10^{5.68} \times 10^{-6.6} = 10^{-0.9} \approx \tfrac{1}{8}$$

[2] H. Diehl, C. A. Goetz, and C. Hach, *J. Am. Waterworks Assoc., 42*, 40 (1950).

[3] J. S. Fritz, J. P. Sickafoose, and M. A. Schmitt, *Anal. Chem., 41*, 1954 (1969).

Thus, the correct end point is when most of the violet Ca-Arsenazo has been converted to orange-yellow free Arsenazo.

Solutions of EDTA are frequently standardized with calcium carbonate as primary standard, especially if they are to be used primarily to titrate calcium and magnesium. Weighed portions of calcium carbonate are dissolved carefully in acid:

$$CaCO_3(s) + 2H^+ \rightarrow Ca^{2+} + CO_2(g) + H_2O$$
$$\text{(2HCl)}$$

Then the solution is buffered to pH 10.0 and titrated with $0.01M$ EDTA using Arsenazo indicator, or with Calmagite indicator after adding some magnesium(II). The exact molarity of the EDTA is calculated from the volume of titrant used and from the weight of calcium carbonate taken.

Besides determining total hardness, it is possible to determine only calcium hardness. In this procedure, the solution is made strongly alkaline with sodium hydroxide so that magnesium hydroxide may precipitate. Calcium hydroxide is soluble, so calcium(II) can be titrated in the presence of the precipitated magnesium hydroxide. An indicator known as Calcein is used for the calcium-hardness determination [4].

11-4.   OTHER EDTA TITRATION METHODS

Although calcium and magnesium are titrated with EDTA in alkaline solution, many other metal ions are best titrated in neutral or acidic solutions. One reason is that many metal ions form insoluble hydroxides in alkaline solutions. If a complexing agent is added to prevent precipitation, the metal ion may be complexed so strongly that the conditional constant for the titration reaction is not large enough. When acidic solutions are used, it is frequently possible to titrate metal ions that form very stable EDTA complexes in the presence of ions—such as magnesium(II) and calcium(II)—that do not complex (or else complex very weakly) with EDTA at acidic pH values.

Although many indicators have been proposed and used for complexometric titrations, most EDTA titrations can be carried out with one of the following: Calmagite, Arsenazo I, NAS, or xylenol orange.

Calmagite.   The structure of this indicator is as follows:

[4] H. Diehl and J. L. Ellingboe, *Anal. Chem.*, 28, 882 (1956).

Calmagite forms colored complexes with most metal cations. The encircled part of the molecule is responsible for the complexing. Both of the oxygen atoms in the encircled part lose their protons and become bonded to the metal.

Like most metal-ion indicators, Calmagite has acid-base properties, which are summarized as follows:

$$\underset{\text{(red)}}{H_2Indic^-} \underset{}{\overset{\text{pH 8.1}}{\rightleftharpoons}} \underset{\text{(blue)}}{HIndic^{2-}} \underset{}{\overset{\text{pH 12.4}}{\rightleftharpoons}} \underset{\text{(orange)}}{Indic^{3-}}$$

Since the metal complexes are red, Calmagite is a useful indicator only in the blue range (pH 8.1–12.4).

Probably the chief difficulty with Calmagite is that many metal ions form such strong complexes that they "block" the indicator and prevent an end point. Examples of ions that block Calmagite are copper(II), nickel(II), iron(III), and aluminum(III). Interference from the first three can be prevented by complexing them with cyanide, although cyanide is effective in complexing only small amounts of iron(III).

Arsenazo I. This indicator has the following structure:

As was mentioned earlier, Arsenazo I is a satisfactory indicator for the EDTA titration of calcium(II), magnesium(II), or total hardness at pH 10.0. Because traces of copper(II) and iron(III) do not block this indicator, it is not necessary to use any cyanide. Arsenazo I is also an excellent indicator for titrating rare earths and thorium(IV) with EDTA [5].

NAS. The structural formula of this indicator is as follows:

NAS is red-violet in very acidic solutions and red-orange in solutions of pH 3.5 and above. It forms pale-yellow complexes with copper(II), zinc(II), and lead(II) and yellow or pale orange complexes with several other metal ions. In forming complexes, the metal probably bonds to the oxygen of the —OH group and to the ring nitrogen. NAS is a useful indicator in the pH range of about 3–9.

NAS is recommended for titrating copper(II), cobalt(II), nickel(II), cadmium(II), zinc(II), aluminum(III), and several other metal cations with EDTA

[5] J. S. Fritz, R. T. Oliver, and D. J. Pietrzyk, *Anal. Chem.*, 30, 1111 (1958).

Table 11–3. Titration of Metal Ions with EDTA and NAS Indicator (with Pyridine–Pyridinium or Acetate–Acetic Acid Buffer)

Metal Ion	pH	Other Conditions
Al^{3+}	6.4	Back-titration with copper
Cd^{2+}	6–8	——
Co^{2+}	6–8	——
Cu^{2+}	4–8	——
Fe^{3+}	5.5	Back-titration with zinc
Ni^{2+}	6	Slow titration near end point
Pb^{2+}	5.5–6.5	——
Rare earths^{3+}	6	——
Ti^{4+}	4–5	Back-titration with copper; 3 drops H_2O_2 (30%) added before EDTA
VO^{2+}	6	Back-titration with zinc; ascorbic acid added before EDTA
Zn^{2+}	6–8	——

[6]. Some of the elements titrated using this indicator and some of the special conditions required are given in Table 11–3.

In most titrations with NAS indicator, it is necessary or advantageous to have a small amount of copper(II) present to react with the indicator. The copper(II)-indicator complex is not broken until the titration of the metal ion is complete. This is analogous to titrating calcium(II) with Calmagite indicator, where magnesium is added to form a suitable indicator system. However, with NAS it is best to wait until the titration is nearly complete before adding the copper(II); even without any copper(II), the approximate end point can be seen by a color change from orange to red.

The rate at which various metal ions react with EDTA varies. For example, aluminum(III) reacts too slowly for a direct titration to be convenient. Aluminum(III) is determined by adding an excess of standard EDTA and heating to ensure complete reaction, then cooling, buffering, and back-titrating the solution with a standard copper(II) solution and NAS indicator. Several other elements are determined by a similar back-titration, but most of them do not require heating.

Xylenol Orange. This indicator is made by reacting an acid-base indicator (o-cresolsulfonphthalein) with formaldehyde and iminodiacetic acid, so that one or two chelating groups,

$$\left(-CH_2N\begin{array}{l}CH_2COOH\\CH_2COOH\end{array}\right)$$

are added to the molecule. This makes the indicator react with metal ions; the structure of the indicator is such that the free indicator is yellow (below pH 6.4) and the metal complexes are violet.

[6] J. S. Fritz, J. E. Abbink, and M. A. Payne, *Anal. Chem., 33,* 1381 (1961). J. S. Fritz, W. J. Lane, and A. S. Bystroff, *Anal. Chem., 29,* 821 (1957).

Although xylenol orange may be used to titrate several metal ions [7], it is most useful as an indicator in titrating metal ions that form very strong EDTA complexes about pH 1.5–3.0. This includes the direct titration of bismuth(III) and thorium(IV) and the determination of zirconium(IV) and iron(III) by back-titration with thorium(IV) or bismuth(III).

QUESTIONS AND PROBLEMS

1. Define each of the following terms: (a) Ligand, (b) Bidentate ligand, (c) Hexadentate ligand, (d) Coordination number, (e) Chelate, (f) Aquo complex ion.
2. Give the coordination number of the metal ion in each of the following complex ions.

   (a) AlF_6^{3-} (d) $Cu(NH_3)_4(H_2O)^{2+}$
   (b) $Co(SCN)_4^{2-}$ (e) $[Cr(H_2O)_5Cl]Cl_2$
   (c) $Cu(CN)_3^{2-}$ (f) $[MgY_2(H_2O)]^{2-}$

3. Is it correct to write a single formula for a metal complex such as $Zn(NH_3)_4^{2+}$ or FeF_6^{3-}? Why or why not?
4. Which form of EDTA is used in preparing a titrating solution? Why is a solution containing a metal ion buffered before titration with EDTA?
5. Explain why increased acidity (lower pH) results in a smaller pM change at the end point when a metal ion is titrated with EDTA.
6. Suggest a possible reason why the coordination number of many metals is four toward halide ions, whereas the coordination number of the same metals is six toward water and ammonia.
7. It is desired to form -1 or -2 complex ions of the following metal ions before extracting them from water. Suggest a suitable ligand for each metal ion: (a) Fe^{3+}, (b) Co^{2+}, (c) Co^{3+}, (d) Cu^{2+}.
8. If we titrate a $0.01M$ solution of a metal ion with EDTA and assume a $0.005M$ solution of MY at the equivalence point, calculate the minimum value of the conditional formation constant, $K_{MY'}$, for a quantitative (99.9% complete) titration.
9. Calculate the minimum value of the formation constant in Problem 8 if the solutions are a bit more concentrated and the concentration of MY at the equivalence point is $0.01M$.
10. Using the value of $K_{MY'}$ calculated in Problem 9 for an MY concentration of $0.01M$ at the stoichiometric point, estimate to the nearest 0.5 pH unit the most acidic pH that can be used for a quantitative titration of each of the following ions: (a) Ca^{2+}, (b) Mg^{2+}, (c) Mn^{2+}, (d) Fe^{3+}.
11. It can be shown that a $0.01M$ solution of a metal ion will not interfere in the EDTA titration if its conditional formation constant is 10^2 or less. Using Figure 11–3, estimate the most basic pH at which each of the following ions will cause no interference: (a) Ca^{2+}, (b) Mg^{2+}, (c) Mn^{2+}, (d) Al^{3+}.
12. Choose a pH range in which each of the following EDTA titrations might be performed. Review Problems 10 and 11 for the requisite conditions for a quantitative titration.

    (a) $0.01M$ Mn^{2+} in the presence of $0.01M$ Mg^{2+}
    (b) $0.01M$ Fe^{3+} in the presence of $0.01M$ Mn^{2+}

[7] J. Körbl and R. Pribil, *Chemist-Analyst*, 45, 102 (1956).

(c) $0.01M$ Zn^{2+} in the presence of $0.01M$ Sr^{2+}
(d) $0.01M$ Cu^{2+} in the presence of $0.01M$ Ca^{2+}

13. (a) Estimate to the nearest 0.5 pH unit the most acidic pH at which a $0.01M$ solution of aluminum(III) can be titrated quantitatively with EDTA. Assume an AlY^- concentration of $0.01M$ at the equivalence point and a conditional formation constant of 10^8.
    (b) Estimate the most basic pH at which a $0.01M$ solution of aluminum(III) can be titrated with EDTA without forming a precipitate of $Al(OH)_3$ at the start of the titration.

14. Using data from Table 11–2 and Figure 11–3, calculate the pM at the equivalence point for each of the following EDTA titrations. Assume $[MY] = 0.005M$ at the equivalence point.

(a) Ca^{2+} at pH 10.0   (c) Mn^{2+} at pH 8.0
(b) Ca^{2+} at pH 8.0   (d) Zn^{2+} at pH 5.0

Effect of pH and Complexing on EDTA Titrations

15. Calculate the pNi $(-\log Ni^{2+})$ of a solution of $0.001M$ nickel(II) to which sufficient oxalate has been added to form nickel-oxalate complexes and have a $0.01M$ concentration of excess oxalate. Assume that the pH is basic enough that all excess oxalate is present as the $2-$ anion. The formation constants for nickel-oxalate complexes are

$$\beta_1 = 10^{4.1}, \qquad \beta_2 = 10^{7.2}, \qquad \beta_3 = 10^{8.5}.$$

16. Calculate the pCd $(-\log Cd^{2+})$ of a $0.01M$ solution of cadmium(II) with a free iodide-ion concentration of $0.1M$. See p. 186 for the formation constants of cadmium iodide complexes.

17. Using the data from Problem 16 and from Figure 11–3, calculate the conditional constant for the titration of cadmium(II) with EDTA at pH 6.0 in the presence of $0.1M$ free iodide. Predict whether it should be possible to analyze a sample of Cs_2CdI_4 for cadmium(II) by displacing the iodide.

18. Using data from Table 11–2 and Figure 11–3 and the values of $\alpha_{M(OH)}$ given in the following table, calculate and graph $K_{AlY'}$ as a function of pH. From the curve obtained, suggest the optimum pH for the titration of aluminum(III) with EDTA.

pH	$\alpha_{Al(OH)}$
5	$10^{-0.4}$
6	$10^{-1.3}$
7	$10^{-5.3}$
8	$10^{-9.3}$
9	$10^{-13.3}$

19. The copper in a copper(II) glycine complex is to be titrated with EDTA.

(a) Calculate pCu in a solution containing $0.01M$ copper glycine and $0.01M$ excess glycine at pH 10.0. At this pH, the conditional formation constants for copper glycine are: $\beta_1 = 10^{8.1}, \beta_2 = 10^{15.1}$.
(b) Calculate pCu at 200% titration with EDTA in a solution buffered at pH 10.0. $K_{CuY'} = 10^{18.3}$ at pH 10.0.
(c) From the results in parts a and b, sketch the approximate curve for the titration of copper(II) in glycine solution, buffered at pH 10.0, with EDTA. Comment on the probable success of this titration.

20. (a) Calculate pCd in a solution containing $0.01M$ cadmium acetylacetonate complexes and $0.02M$ excess acetylacetonate at pH 7.0. At this pH, the conditional formation constants for cadmium acetylacetonate are $\beta_1 = 10^{1.6}$, $\beta_2 = 10^{7.6}$.

(b) Calculate pCd at 200% titration with EDTA in a solution buffered at pH 7.0. $K_{CdY'} = 10^{13.1}$ at pH 7.0.

(c) From the results in parts (a) and (b), sketch the approximate curve for titration of cadmium in acetylacetonate solution, buffered at pH 7.0, with EDTA. Comment on the probable success of this titration.

21. A complexing titrant known as EGTA complexes calcium(II) much more strongly than magnesium(II). The EGTA formation constants have the following values: $K_{CaY} = 10^{11.0}$, $K_{MgY} = 10^{5.2}$. At pH 8.0, $\alpha_Y = 10^{-2.5}$ for EGTA.

(a) Calculate pCa at 0% titration and 200% titration when $0.01M$ Ca^{2+} is titrated with EGTA at p 8.0.

(b) Calculate pMg at 0% titration and 200% titration when $0.01M$ Mg^{2+} is titrated with EGTA at pH 8.0.

(c) Sketch the approximate titration curves for Ca^{2+} and Mg^{2+} and decide whether it is possible to titrate calcium without interference from magnesium under these conditions.

Indicators

22. Calculate the conditional formation constant of a suitable indicator for each of the titrations in Problem 14 assuming that the ratio of MIndic:Indic is 1:5 at the observed end point.

23. Indicators used in complex-formation titrations also undergo an acid-base color change. Often the basic color of an indicator is almost the same as the metal-indicator complex color. What limitation does this place on the pH at which an indicator is used in a titration?

24. Xylenol orange indicator is yellow in acidic solution and changes to red at pH 6.4 and above. Lead(II) forms a red-orange complex with conditional formation constants as follows: $K = 10^{4.2}$ at pH 3.0, $10^{4.8}$ at pH 4.0, $10^{7.0}$ at pH 5.0, and $10^{8.2}$ at pH 6.0. From data in Table 11–2 and Figure 11–3, calculate pPb at the equivalence point (assuming $[PbY] = 1.0 \times 10^{-3}M$) for pH 3.0, 4.0, 5.0, and 6.0. At which pH does the xylenol orange end point most closely correspond to the pPb at the equivalence point in the titration?

25. Magnesium(II) forms a 1:1 complex with the indicator Arsenazo I at pH 10.0. The formation constant of this complex was estimated by using a spectrophotometer to follow the titration of very dilute Arsenazo with magnesium(II). This titration showed that, at the equivalence point, 50% of the Arsenazo was present as the magnesium complex and 50% as uncomplexed Arsenazo. If $[Mg\text{-}Arsenazo] + [Arsenazo] = 1 \times 10^{-5}$ at the equivalence point, calculate the value of the formation constant.

26. Using the formation constant from Problem 23, calculate the ratio of Mg-Arsenazo to Arsenazo at the correct end point for the titration of magnesium with EDTA at pH 10. The concentration of magnesium-EDTA at the end point is $0.001M$ (see p. 202). The conditional constant for Mg-EDTA at pH 10 is $10^{8.1}$.

Water Hardness

27. Explain why magnesium(II) must be added when EDTA is standardized using $CaCO_3$ as a primary standard and Calmagite as indicator.

28. In a total water-hardness determination using Calmagite indicator, explain how interference from dissolved iron and aluminum salts can be avoided.

29. List some advantages of Arsenazo I Indicator over Calmagite for a water-hardness titration.
30. Titrating a 50.0 mL water sample for total hardness requires 4.08 mL of $0.0100M$ EDTA. Calculate the hardness of the water as parts per million calcium carbonate.
31. A 50.00-mL aliquot of a hard-water sample is titrated with 15.00 mL of $0.0100M$ EDTA using Arsenazo indicator. A second 50.00-mL aliquot is made strongly alkaline with sodium hydroxide. A precipitate results. Calcein indicator is added to the solution. The sample then requires 10.00 mL of $0.0120M$ EDTA for titration. Calculate for this sample (a) the molarity of the calcium present, (b) the molarity of the magnesium present, and (c) the total hardness (as $CaCO_3$) of the water.
32. A 20.00-mL volume of EDTA is required to titrate 25.00 mL of standard $0.0112M$ calcium carbonate. A 25.00-mL hard-water sample requires 9.78 mL of this EDTA for a total hardness determination. Calculate the total hardness as ppm of $CaCO_3$ in the sample. (Recall that 1 ppm is 1 mg/L.)

Methods of Analysis

33. Why is iron(III) titrated with EDTA in acidic, rather than basic, solution? Name a suitable indicator for this titration.
34. Explain why some metal ions are best determined by adding excess EDTA and then back-titrating. List some metal ions that fall into this category. Explain how to select a suitable metal ion for the back-titration.
35. Suggest an EDTA method for determining the first substance in the presence of the second substance(s). Include the appropriate indicator, pH, and any other necessary reagents. Other end point detection methods may also be used.

    (a) Cu^{2+} in Ca^{2+}
    (b) $Ca(NO_3)_2$ in $Mg(NO_3)_2$
    (c) Fe^{3+} in Ca^{2+}
    (d) Ca^{2+} in small amounts of Fe^{3+} and Cu^{2+}
    (e) Bi^{3+} in Cd^{2+}
    (f) Mg^{2+} in Ni^{2+}
    (g) Al^{3+} in K^+

36. Metal ions such as Al^{3+}, Fe^{3+}, and Ni^{2+} react slowly with EDTA, so a direct titration is slow with several false, fading end points. Explain how such metal ions can be determined more rapidly and accurately by EDTA titration.

12

Theory of Oxidation-
Reduction Reactions
and Titrations

An oxidation-reduction reaction is one in which one or more of the reacting chemicals goes to a more positive oxidation state by losing electrons and one or more reactants is converted to a less positive oxidation state by gaining electrons. For example, when zinc metal is dissolved in aqueous hydrochloric acid, the zinc is oxidized from Zn^0 to Zn^{+2}, and the acid hydrogen is reduced from H^+ to $H_2(g)$:

$$Zn(s) + 2HCl \rightarrow Zn^{+2} + H_2(g) + 2Cl^-$$

Oxidation-reduction reactions are closely tied in with chemical analysis in a number of ways. Many inorganic ions and some organic substances can be determined quantitatively by oxidation-reduction titration (Chapter 13). Electrochemical methods for precipitation and quantitative determination of various chemical substances (Chapter 15) are really oxidations and reductions brought about by the use of electricity. The potentials of electrodes used to measure the concentration of various ions in solution (Chapter 16) are usually associated with oxidation-reduction processes. Oxidation-reduction reactions have a number of additional uses in analytical chemistry, such as in dissolving samples, adjustment of oxidation state before an analytical determination, and some color-forming reactions for spectrophotometry (Chapter 17).

12–1. OXIDATION-REDUCTION (REDOX)
REACTIONS

When an oxidation-reduction reaction occurs, there is a net change in the oxidation numbers (see Appendix 4) of one or more elements in the reacting substances. Such a *redox* reaction must involve both oxidation and reduction. If oxidation occurs where electrons are lost, there must be an accompanying reduction reaction to absorb the electrons released during oxidation.

An oxidizing agent is one that gains electrons and goes to a more negative oxidation state:

$$A_{ox} + ne^- + A_{red} \qquad \textit{General Case}$$

$$Fe^{3+} + e^- = Fe^{2+} \qquad \textit{Specific Example}$$

A reducing agent loses electrons and goes to a more positive oxidation state:

$$B_{red} = B_{ox} + ne^- \qquad \textit{General Case}$$

$$Zn^0(s) = Zn^{2+} + 2e^- \qquad \textit{Specific Example}$$

Each is called a *half-reaction.* The oxidized and reduced forms involved in a half-reaction are sometimes called a redox *couple.* It should be emphasized that half-reactions do *not* represent equilibria, because free electrons do not exist in solution (at least not for long). A redox reaction occurs only when suitable oxidizing and reducing agents are brought together. Then a reaction takes place that is the sum of the two half-reactions:

$$A_{ox} + ne^- = A_{red}$$
$$\underline{+ \; B_{red} = B_{ox} + ne^-}$$
$$A_{ox} + B_{red} = A_{red} + B_{ox}$$

Since free electrons cannot exist in solution, the electron gain must equal the electron loss. In the preceding example, zinc metal releases electrons two at a time on being oxidized to Zn^{2+}. To balance the complete reaction, it is necessary to multiply the ferric-ferrous half-reaction by 2:

$$2Fe^{3+} + 2e^- = 2Fe^{2+}$$
$$\underline{+ \; Zn^0(s) = Zn^{2+} + 2e^-}$$
$$2Fe^{3+} + Zn^0(s) = 2Fe^{2+} + Zn^{2+}$$

Because there is usually a *transfer* of electrons, oxidation-reduction processes are closely associated with electricity. This may be illustrated by the following experiment. If a solution of orange cerium(IV) ion in H_2SO_4 is added to a solution containing an equal amount of light green iron(II) ion, both of the colors disappear. The reaction is

$$\underset{\text{(orange)}}{Ce^{+4}} + \underset{\text{(light green)}}{Fe^{+2}} \;\; \rightarrow \;\; \underset{\text{(colorless)}}{Ce^{+3}} + \underset{\text{(colorless in } H_2SO_4)}{Fe^{+3}}$$

Now suppose that acidic solutions of cerium(IV) ion and iron(II) ion are placed in *separate* beakers, connected by a salt bridge. Next an assembly of two electrodes connected by wires to a galvanometer is obtained, and the electrodes are simultaneously placed in each of the beakers. As soon as the electrodes contact the two solutions, the galvanometer indicates that an electric current is flowing through circuit of the beakers, salt bridge, and electrodes. The orange color in the Ce^{+4} beaker gradually decreases to colorless, and the light-green color in the Fe^{+2} beaker also gradually turns colorless.

The process is as follows. The iron(II) ion is oxidized at the surface of the platinum electrode in the left-hand beaker:

$$Fe^{2+} \rightarrow Fe^{3+} + e^-$$

The electrons released are taken up through the wire to the right-hand beaker. In the right-hand beaker, the *simultaneous* reduction of cerium(IV) ion is occurring:

$$Ce^{4+} + e^- \rightarrow Ce^{3+}$$

Electrons for this reduction are extracted from the wire. The net result is an electron flow through the wire from left to right, as indicated by the galvanometer. For the current to be carried through the solutions, positive ions (cations) must pass through the salt bridge into the cerium solution, or negative ions (anions) must pass through the salt bridge into the iron solution. This completes the electric circuit.

In practice, *both* anions and cations carry current through the solution. The relative amounts of anion and cation crossing the boundary from one compartment to the other depend on the relative *mobilities* of the anion and cation. For example, the small, fast-moving hydrogen ion carries a considerably larger portion of the current than the larger, slower chloride ion.

If chemical redox reactions take place as described in the previous paragraph, then it follows that an electric current from some external source, such as a battery, can be used to cause chemical reactions. In the experiment described, the electrons that reduced cerium(IV) could just as well have come from a battery as from the oxidizing of iron(II) to iron(III) in the other beaker. Chemical reactions caused by an external electrical source are indeed useful. Some of the analytical applications of this type of oxidation-reduction will be discussed in Chapter 15.

The experiment described above is an example of an electrochemical cell. A *galvanic*, or *voltaic*, cell generates a spontaneous voltage by a chemical redox reaction and will therefore produce a current if the electrodes are connected by a conducting wire. An *electrolytic cell* is a cell in which electricity from an external source, such as a battery, causes a chemical reaction to take place.

The fact that a chemical redox reaction can produce an electric current means that chemical energy can be converted to electric energy, at least in some cases. The force (or voltage) of the electric current produced is proportional to the energy of the chemical redox reaction. Electrical voltages can be measured very precisely and so provide a way of measuring the energy of a chemical oxidation-reduction reaction.

12–2. STANDARD REDUCTION POTENTIALS

We have seen that oxidation-reduction reactions can be carried out in such a way that an electric current will be produced. The voltage associated with this current is a measure of the difference in potential of the two electrodes making up the cell. In this section, we will consider the potentials of various electrodes under certain "standard" conditions and see how these potentials can be used to predict the relative *oxidizing* ability of oxidizing agents and the relative *reducing* ability of reducing agents.

For our present purposes, an *electrode* will be defined as one of the following:

1. A conducting metal in contact with a solution of its ions. An example is lead metal dipping into a solution containing a soluble lead salt:

$$Pb^0(s)/Pb^{2+}$$

The half-reaction associated with this electrode, written as a reduction, is

$$Pb^{2+} + 2e^- = Pb^0(s)$$

Written as an oxidation, the half-reaction is

$$Pb^0(s) = Pb^{2+} + 2e^-$$

2. An inert metal, such as platinum, in contact with an oxidation-reduction couple in solution. An example is a bright platinum wire in contact with a solution containing ferric and ferrous ions:

$$Pt^0(s)/Fe^{3+}, Fe^{2+}$$

The half-reaction for this electrode, written as a reduction, is

$$Fe^{3+} + e^- = Fe^{2+}$$

Written as an oxidation, the half-reaction is

$$Fe^{2+} = Fe^{3+} + e^-$$

These electrodes may be used to measure the *potential, E,* or tendency of a half-reaction to lose or gain electrons. Unfortunately, it is impossible to measure E for a half-reaction directly because electrons cannot exist in water. Instead such E's are measured indirectly by measuring the voltage of an electric cell consisting of two half-reactions; such a voltage is called a *cell voltage* or *cell electromotive force* (emf) and may be symbolized as E_{cell}. Specifically, such a cell is composed of an arbitrary reference half-reaction and a reference electrode, in addition to the half-reaction and electrode whose potential is to be measured. The reference half-reaction universally used is the reduction of the hydrogen ion:

$$2H^+ + 2e^- = H_2(g)$$

and the reference electrode is the hydrogen reference electrode.

Because neither hydrogen nor hydrogen ion is metallic, the hydrogen electrode must be fabricated as the second type of electrode above. The inert metal is a platinum sheet coated with amorphous platinum metal (platinum black) which will absorb hydrogen gas. Hydrogen gas is bubbled onto the electrode at 1-atmosphere pressure through a solution containing hydrogen ions at unit activity.

When the hydrogen electrode is used under STP ($a_{H^+} = 1$; pressure$_{H_2} = 1$ atm) the *standard electrode potential,* or E^0 of this half-reaction is defined arbitrarily to be exactly zero volts. (Such a number is an *exact* number, and we may write as many zeros after the decimal point as desired without violating significant figure rules: $E^0 = 0.0$ V, 0.00 V, 0.000 V, etc.)

Standard electrode potentials of redox couples such as the Fe^{3+}/Fe^{2+} couple may then be measured relative to the E^0 of the hydrogen electrode with a potentiometer. The setup is shown schematically in Figure 12–1. The spontaneous cell reaction would be

$$2Fe^{3+} + H_2(g) \xrightarrow[\text{pH 0.0}]{\text{STP}} 2Fe^{2+} + 2H^+ \tag{12–1}$$

If the activities of all species are unity, the cell emf or cell voltage will be 0.771 V. Such a cell emf may correctly be called the *standard electrode potential* of the Fe^{3+}/Fe^{2+} couple, according to the recommendation of the International Union of Pure and Applied Chemistry (IUPAC). In Appendix 3, we list this information as follows:

$$Fe^{3+} + e^- = Fe^{2+} \qquad E^0 = +0.771 \text{ V}$$

It is against internationally accepted practice to call the cell emf of -0.771 V for the reverse reaction of Equation 12–1 the standard electrode potential for the oxidation of Fe^{2+} to Fe^{3+}. Therefore, we will use the following rules to operate in accord with IUPAC practice.

1. Write all half-reactions as reductions (as in Appendix 3) with the oxidized form (oxidizing agent) on the left and the reduced form on the right; e.g.,

$$Fe^{3+} + e^- = Fe^{2+} \qquad E^0 = 0.771 \text{ V}$$

$$2H^+ + 2e^- = H_2(g) \qquad E^0 = 0.000 \text{ V}$$

$$V^{3+} + e^- = V^{2+} \qquad E^0 = -0.255 \text{ V}$$

Figure 12–1. Oxidation-reduction experiment.

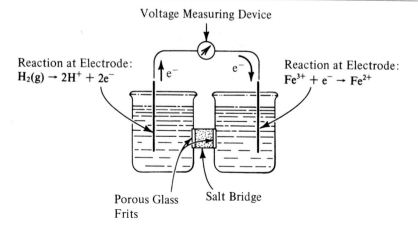

Voltage Measuring Device

Reaction at Electrode:
$H_2(g) \rightarrow 2H^+ + 2e^-$

Reaction at Electrode:
$Fe^{3+} + e^- \rightarrow Fe^{2+}$

Porous Glass Frits

Salt Bridge

For purposes of comparing electrode reactions, all half-reactions are written the same way—that is, as reductions. In a complete reaction, of course, oxidation as well as reduction must occur. A complete reaction is obtained by subtracting one reduction half-reaction from another, as will be discussed in Section 12–3.

2. In a table of standard reduction potentials, a plus sign before the E^0 value indicates that the oxidized form is a better oxidizing agent than hydrogen ion; a minus before the E^0 means that the oxidized form is a poorer oxidizing agent than the hydrogen ion.

> Some authors reverse the sign of the E^0 of the reducing agent and call this a standard oxidation potential. Although we will not do this, this is acceptable according to IUPAC as long as such a potential is *not* called a standard *electrode* potential.

3. A standard electrode potential, E^0, can be used as such only when each substance involved in the half-reaction is in its *standard state*. For example, an elemental solid is in its standard state. The standard state of a gas is at 1-atm pressure. An ion in solution is in its standard state when its activity is unity ($a_i = 1$). However, we shall consider an ion to be *nearly* in its standard state when its concentration is $1M$. Recall that the activity of an ion (a_i) is equal to the product of its activity coefficient (f_i) times its molar concentration (C_i):

$$a_i = f_i C_i$$

> The difference between a_i and C_i is negligible in very dilute solutions but becomes greater as the concentration increases. However, if both forms of the redox couple are soluble, the activity coefficients of the oxidized and reduced forms often nearly cancel each other when the Nernst equation (Section 12–4) is used. In this case, very little error results from using concentrations instead of activities.

4. For a complete (cell) reaction, symbolize the cell voltage, or cell emf, as E_{cell}; where all species have unit activity, symbolize it as E^0_{cell}, the standard cell voltage (emf). E^0_{cell} is related to the standard free energy change, ΔG^0, as follows:

$$\Delta G^0 = nFE^0_{cell} \qquad (12\text{--}2)$$

where n is the number of electrons transferred and F is the Faraday constant. See the units for the Faraday constant in Table 12–1.

Measurement of Standard Reduction Potentials.   Standard reduction potentials have been carefully measured by a variety of techniques. One is to compare the difference in potential of an oxidation-reduction couple of interest and that of a suitable reference electrode of known potential. Another is to calculate the energy of the reduction (or oxidation) half-reaction from thermodynamic data, and then convert this into electrical energy (volts). Standard reduction potentials are in a sense idealized; that is, they are calculated by taking into account such things as complex-formation constants, hydrolysis constants, and activity coefficients.

Some chemists prefer to use a table of *formal potentials*. These are potentials obtained experimentally in solutions that contain a 1-formal concentration of various oxidized and reduced substances. A 1-formal solution (1F) contains one

Table 12–1. Basic Electrochemical Parameters and Their Units

Parameter	Symbol	Description and/or Units
Joule	J	Quantity of work done when a force of one newton $(kg \cdot m/s^2)$ acts through a distance of one meter.
Coulomb	C	Quantity of electric charge passing a point in 1 sec when current is 1 ampere (A).
Ampere	A	Quantity of electric current; $A = C/s$
Volt	V	Quantity of electrical potential difference; $V = J/C$
Watt	W	Quantity of electric power; $W = J/s$
Equivalent	eq	$eq = n \times$ moles, where n is the number of electrons in an oxidation or reduction; $n = eq/mole$
Faraday constant	F	$F = C/mole\ e^- = C/eq$ $(F = 96,487\ C/mole\ e^-$ or eq)
Ohm	Ω	Quantity of electrical resistance; $\Omega = V/A$
Gas constant	R	$R = J/K \cdot mole$ ($R = 8.314\ J/K \cdot mole$)

gram-formula weight per liter of the substances listed. No provision is made for activity coefficients, and no attempt is made to specify the exact chemical species in solution. Formal potentials of a number of redox half reactions are listed in the *Handbook of Analytical Chemistry* and other standard references.

Some standard electrode potentials are tabulated in Appendix 3. Those with which the acid medium is listed are formal potentials. The more positive the potential of the half-reaction, the greater the oxidizing power of the *oxidized* form of the redox couple. The more negative the potential, the greater the reducing power of the *reduced* form of the redox couple.

Strong and Weak Oxidizing Agents. Before leaving the discussion of E^0 values, we should emphasize the significance of the magnitude of an E^0. In general, the more positive E^0, the greater the oxidizing power of the oxidized form (oxidizing agent) of the redox couple. As E^0 becomes more negative, the reduced form (reducing agent) will have greater reducing power.

Strong oxidizing agents are those oxidized forms that have an E^0 greater than 0.90 V. The three classical quantitative oxidizing agents are: potassium permanganate (1.51 V), cerium (IV) in sulfuric acid (1.44 V), and potassium dichromate (1.33 V).

Weak oxidizing agents of course have E^0 values less than 0.9 V. Some classical quantitative weak oxidizing agents are the iron(III) ion (0.771 V), iodine (0.535 V), and $Fe(CN)_6^{-3}$ (0.36 V).

In general, a strong oxidizing agent is used to oxidize an inorganic species with a positive E^0 or to oxidize an inorganic species to its highest oxidation state. Weak oxidizing agents will generally be used to oxidize inorganic species to lower oxidation states or to oxidize most organic molecules. Such oxidizing agents have the advantage of fewer interferences because they will not react with as many inorganic species and they will not cause side reactions in oxidizing organic molecules.

12–3. COMBINING HALF-REACTIONS TO FORM A COMPLETE REACTION

No oxidation-reduction reaction will occur unless both an oxidizing and a reducing agent are combined. A half-reaction such as

$$Fe^{3+} + e^- = Fe^{2+}$$

does *not* indicate that an equilibrium exists in which there is a certain concentration of free electrons in solution. It means only that the half-reaction will occur, in either the forward or reverse direction, provided that some second half-reaction takes place *simultaneously*. The equation for a complete oxidation-reduction reaction may be written by combining two half-reactions in such a way that the electrons are cancelled. This is done as follows.

The chemical equation for a complete oxidation-reduction reaction can be obtained by algebraically subtracting the two appropriate redox half-reactions. By also subtracting the E^0 values of the half-reactions, the E^0_{cell} of the complete reaction is obtained. A positive sign for this E^0 means that the reaction will take place as written if all reactants and products are present at unit activity. The larger the numerical value of E^0, the more likely the reaction will be rapid and complete. A negative sign for the E^0 of the complete oxidation-reduction reaction means that the reaction does *not* occur as written.

The mechanics of combining two half-reactions to form a complete reaction are as follows:

Step 1. Write the reduction half-reaction with the proposed oxidizing agent first. Below it, write the reduction half-reaction containing the proposed reducing agent. Also write down the standard reduction potentials for the two half-reactions.

$$A_{ox} + e^- = A_{red} \qquad E^0 = 0.75 \text{ V}$$
$$B_{ox} + 2e^- = B_{red} \qquad E^0 = 0.15 \text{ V}$$

Step 2. If necessary, multiply one of the half-reactions by a number that will make the electron charges equal. Do *not* multiply the E^0 values because E^0 is an experimental value that does not depend on how the reduction half-reaction is written.

$$2A_{ox} + 2e^- = 2A_{red} \qquad E^0 = 0.75 \text{ V}$$
$$B_{ox} + 2e^- = B_{red} \qquad E^0 = 0.15 \text{ V}$$

Step 3. Enclose both the second half-reaction and the E^0 for the second half-reaction in brackets and place a minus sign in front of each. (Note that the E^0 part should be enclosed in brackets.) Using the rules of algebra, subtract the two half-reactions to form a complete reaction; also subtract the two E^0 values to obtain E^0_{cell} for the complete reaction:

$$
\begin{array}{ll}
2A_{ox} + 2e^- = 2A_{red} & E^0 = 0.75 \text{ V} \\
-[B_{ox} + 2e^- = B_{red}] & -[E^0 = 0.15 \text{ V}] \\
\hline
2A_{ox} + B_{red} = 2A_{red} + B_{ox} & E^0_{cell} = 0.60 \text{ V}
\end{array}
$$

In this case, the E^0_{cell} is positive, indicating that the reaction does proceed spontaneously in the direction written. A negative sign for the E^0 of a complete redox reaction indicates that the reaction does *not* go spontaneously in the direction written; the spontaneous reaction will go in the *opposite* direction.

Example: Calculate the E^0 for the reaction of cerium(IV) with iron(II) and decide whether the reaction is spontaneous.

$$
\begin{array}{ll}
Ce^{4+} + e^- = Ce^{3+} & E^0 = 1.44 \text{ V} \\
-[Fe^{3+} + e^- = Fe^{2+}] & -[E^0 = 0.77 \text{ V}] \\
\hline
Ce^{4+} + Fe^{2+} = Ce^{3+} + Fe^{3+} & E^0_{cell} = 0.67 \text{ V}
\end{array}
$$

The reaction is spontaneous.

Example: Calculate the E^0 for the reaction of chromium(III) and vanadium(II) and decide whether the reaction is spontaneous.

$$
\begin{array}{ll}
Cr^{3+} + e^- = Cr^{2+} & E^0 = -0.41 \text{ V} \\
-[V^{3+} + e^- = V^{2+}] & -[E^0 = -0.255 \text{ V}] \\
\hline
Cr^{3+} + V^{2+} = Cr^{2+} + V^{3+} & E^0_{cell} = -0.15_5 \text{ V}
\end{array}
$$

The reaction does not take place spontaneously.

From the discussion and examples, the following rule becomes apparent: An oxidant (A_{ox}) will react with the reduced form of a second substance (B_{red}) provided B_{red} is *below* A_{ox} in a table of standard reduction potentials. Put another way, the reduction potential of $A_{ox} - A_{red}$ must be more positive than that of $B_{ox} - B_{red}$ for the reaction to occur spontaneously.

Effect of H^+ or OH^-. The E^0 for the complete reaction is the predicted cell voltage that should be obtained when $1.0M$ solutions of the reactants are first mixed. If either of the redox half-reactions involve H^+ or OH^-, these should also be $1.0M$.

Example: Will iodine oxidize hydroquinone to quinone when the solution contains $1.0M$ H^+?

$$
\begin{array}{ll}
\text{Quinone} + 2H^+ + 2e^- = \text{Hydroquinone} & E^0 = 0.699 \text{ V} \\
-[I_2 + 2e^- = 2I^-] & -[E^0 = 0.535] \\
\hline
\text{Quinone} + 2H^+ + 2I^- = \text{Hydroquinone} + I_2 & E^0_{cell} = 0.164 \text{ V} \\
I_2 + \text{Hydroquinone} = \text{Quinone} + 2H^+ + 2I^- & E^0_{cell} = -0.164 \text{ V}
\end{array}
$$

Therefore, I_2 will *not* oxidize hydroquinone in $0.1M$ acid. However, reducing the H^+ concentration will make a difference, as we shall see in the next section.

Oxidation of Species with Multiple Oxidation States. A number of elements can exist in aqueous solution in more than one oxidation state. Some examples are iron(III) and (II); chromium(II), (III), and (VI); and vanadium(II), (III), (IV), and (V). It is possible to predict to what oxidation state an element will be oxidized by calculating E^0_{cell} for each possible step in the oxidation sequence. The highest oxidation state reached will occur in the complete reaction having the smallest positive E^0_{cell}. A specific example follows.

Example: Predict the highest oxidation state to which metallic chromium will be oxidized when it is dissolved in acid solution ($a_{H^+} = 1$).

1. $Cr^0 \rightarrow Cr^{2+}$:

$$
\begin{array}{ll}
2H^+ + 2e^- = H_2(g) & E^0 = 0.000 \text{ V} \\
-[Cr^{2+} + 2e^- = Cr(s)] & -[E^0 = -0.56 \text{ V}] \\
\hline
2H^+ + Cr(s) = H_2(g) + Cr^{2+} & E^0_{cell} = +0.56 \text{ V (spontaneous)}
\end{array}
$$

2. $Cr^{2+} \rightarrow Cr^{3+}$:

$$
\begin{array}{ll}
2H^+ + 2e^- = H_2(g) & E^0 = 0.000 \text{ V} \\
-[2Cr^{3+} + 2e^- = 2Cr^{2+}] & -[E^0 = -0.41 \text{ V}] \\
\hline
2H^+ + 2Cr^{2+} = H_2(g) + 2Cr^{3+} & E^0_{cell} = +0.41 \text{ V (spontaneous)}
\end{array}
$$

3. $Cr^{3+} \rightarrow Cr^{6+}$:

$$
\begin{array}{ll}
6H^+ + 6e^- = 3H_2(g) & E^0 = 0.000 \text{ V} \\
-[Cr_2O_7^{2-} + 14H^+ + 6e^- = 2Cr^{3+} + 7H_2O] & -[E^0 = 1.33 \text{ V}] \\
\hline
2Cr^{3+} + 7H_2O = 3H_2(g) + Cr_2O_7^{2-} & E^0_{cell} = -1.33 \text{ V} \\
& \text{(not spontaneous)}
\end{array}
$$

Therefore, metallic chromium will dissolve spontaneously in acid to produce first chromium(II) and then chromium(III), at which point the oxidation stops. (This is the type of approach needed for working Problem 8 at the end of this chapter.)

Limitations of Redox Theory. Suppose that a positive E^0 value has been calculated for a given oxidation-reduction reaction, indicating that the reaction should occur spontaneously. If the E^0_{cell} is at least several tenths of a volt, experimentation will generally show that the reaction does indeed take place as predicted. But sometimes the *rate* of a predicted reaction is so slow that it is almost as if the reaction does not occur at all. Often a higher temperature or the addition of some chemical catalyst will speed up a slow oxidation-reduction reaction. Some oxidation-reduction reactions are highly irreversible and will go readily in only one direction. For example, oxalic acid ($H_2C_2O_4$) can be oxidized to carbon dioxide and water, but it is not possible to combine carbon dioxide and water to produce oxalic acid. Sometimes the chemical nature of the oxidizing (or reducing) chemical seems to determine whether a given reaction will proceed.

The actual pathway, or mechanism, by which an oxidation-reduction reaction takes place is sometimes rather complicated. It is necessary to have some understanding of the mechanism in order to make intelligent predictions as to which redox reactions are likely to occur. Mechanisms of several redox reactions are discussed in Section 12-8.

12–4. VARIATION OF ELECTRODE POTENTIAL WITH CONCENTRATION

The standard potential of an electrode is the potential existing when both the oxidized and reduced forms of the oxidation-reduction couple are in their standard states. The potential of an electrode is different from its standard potential if the redox substances are not in their standard states. It is useful to know how the electrode potential changes as the concentrations of the soluble species change. The change of potential with concentration can be determined experimentally by measuring the electromotive force of a cell consisting of the electrode in question (the *indicator electrode*) connected by a salt bridge to a reference electrode of known potential. The electrode potential at different concentrations of redox species can be calculated theoretically with the aid of the *Nernst equation*.

For the half reaction, $A_{ox} + ne^- = A_{red}$, the Nernst equation is

$$E = E^0 - \frac{RT}{nF} \ln \frac{a_{A_{red}}}{a_{A_{ox}}} \tag{12-3}$$

where E is the potential in volts, E^0 is the standard electrode potential, R is the gas constant (8.314 Joules per degree-mole), T is the absolute temperature (273 + degrees C), n is the number of electrons involved (eq/mole), F is the Faraday constant 96,487 (J/V), ln is the natural logarithm (base e), $a_{A_{red}}$ is the activity of the reduced form of A, and $a_{A_{ox}}$ is the activity of the oxidized form of A.

Note that by using the units in Table 12–1, the RT/nF ratio in Equation 12–3 reduces to the units of volts:

$$\frac{(J/°K - mole)(°K)}{(eq/mole)(C/eq)} = \frac{J}{C} = V$$

If values for all the constants are substituted into Equation 12–3 at room temperature (273 + 25 = 298°K), the natural logarithm (ln) is converted base 10 logarithm (log), and concentrations are used instead of activities, the Nernst equation becomes

$$E = E^0 - \frac{0.05915 \, V}{n} \log \frac{[A_{red}]}{[A_{ox}]} \tag{12-4}$$

This form of the Nernst equation is very convenient and is widely used, even though the use of concentrations instead of activities make it slightly less accurate than Equation 12–3. Note that the constant at 25°C, 0.0591, still has the units of volts.

In a half-reaction in which A_{ox} combines with H^+ or OH^-, the concentration of that ion appears in the denominator of the Nernst equation. When H^+ or OH^- is a product in the half-reaction, it appears in the numerator.

Example: Write the Nernst equation expression for each of the following half-reactions at 25°C:

      (a)   Quinone + $2H^+$ + $2e^-$ = hydroquinone $E^0 = 0.699 \, V$

(b) $OCl^- + H_2O + 2e^- = Cl^- + 20H^-$ $E^0 = 0.89$ V

Substituting into Equation 12–4,

(a) $E = 0.699 \text{ V} - \dfrac{0.0591 \text{ V}}{2} \log \dfrac{[\text{hydroquinone}]}{[\text{Quinone}][H^+]^2}$

(b) $E = 0.89 \text{ V} - \dfrac{0.0591 \text{ V}}{2} \log \dfrac{[OCl^-]}{[Cl^-][OH^-]^2}$

The potential of an electrode varies with the ratio of the reduced to the oxidized species in solution and can be calculated by inserting appropriate values or percentages into the Nernst equation.

Example: Calculate the potential (versus standard hydrogen electrode) of a platinum wire dipping into a solution containing 0.0010M quinone, 0.00050M hydroquinone, and 1.0M H^+.

$$E = 0.699 - \frac{0.0591 \text{ V}}{2} \log \frac{(0.00050)}{(0.0010)(1.0)^2} = 0.708 \text{ V}$$

Example: Calculate the potential of the platinum electrode in the previous example if the quinone and hydroquinone concentrations are kept the same but the hydrogen ion concentration is reduced to $1.0 \times 10^{-5}M$ (pH = 5.0).

$$E = 0.699 - \frac{0.0591 \text{ V}}{2} \log \frac{(0.00050)}{(0.0010)(1.0 \times 10^{-5})^2} = 0.412 \text{ V}$$

Note the large change in potential resulting from the acidity change.

In principle, it would be useful to set up an electrode, measure its potential, and use the Nernst equation to calculate the activity of the electrode ion(s) in solution. This is a practical way to measure the activity of silver(I) ions in solution. A silver wire (or any small piece of silver metal to which an electrical connection has been made) dipping into a solution containing Ag^+ assumes a potential that is proportional to the activity of the silver irons in solution:

$$E = E^0 - 0.05915 \text{ V} \log \frac{1}{a_{Ag^+}} = 0.799 - 0.05915 \text{ V} \log \frac{1}{a_{Ag^+}} \qquad (12\text{–}5)$$

The potential of this electrode is obtained by measuring the difference in potential between this electrode and a reference electrode of accurately known potential. The activity of the silver(I) in solution is then calculated from the Nernst equation.

Example: The measured voltage of a cell consisting of a Ag^0/Ag^+ electrode and a saturated calomel reference electrode ($E = +0.246$ V) is $+0.374$ V. Calculate the activity of Ag^+ in the first electrode.

The Ag^0/Ag^+ is 0.374 V more positive than the reference electrode, which in turn is 0.246 V more positive than the standard hydrogen electrode. Therefore, the silver electrode is $0.374 + 0.246 = +0.620$ V compared with the hydrogen electrode. Substituting into the Nernst equation,

$$+0.620 \text{ V} = 0.799 \text{ V} - 0.05915 \text{ V} \log \frac{1}{a_{Ag^+}}$$

$$\log a_{Ag^+} = -3.03 \qquad a_{Ag^+} = 9.4 \times 10^{-4} M$$

In a similar fashion, the activity of Hg^{2+} in solution can be obtained by measuring the potential of a small pool of mercury in contact with mercury(II) ions. The Nernst equation expression is

$$E = 0.854 \text{ V} - \frac{0.05915 \text{ V}}{2} \log \frac{1}{a_{Hg^{2+}}} \qquad (12\text{–}6)$$

Except for Ag^+ and Hg^{2+}, it is impractical to use the potential of a metal-metal ion electrode to measure the activity of metal ions in solution. In many cases, a very long time is needed to establish an equilibrium potential between the metal and its ions. The physical form of a metal seems to have a major effect on the potential that is finally obtained. Small amounts of oxides on the surface often affect the potential.

The *ratio* of two ions or chemical compounds in solution can usually be measured from the electrode potential provided that the oxidation-reduction half-reaction is reversible. For example, the ratio of $Fe^{2+}:Fe^{3+}$ or $Ce^{3+}:Ce^{4+}$ in solution can be calculated from the potential of a platinum electrode in contact with a solution containing either pair of ions. Likewise, the ratio of hydroquinone:quinone can be obtained from the potential of a platinum electrode immersed in solution because the quinone-hydroquinone couple is reversible.

Quantitative calculation of the ratio of reduced to oxidized forms of ions or compounds is not possible for either of the following situations: (1) When the redox couple is highly irreversible, as in the earlier example of oxalic acid–carbon dioxide. (2) When the redox half-reaction involves more than one or two electrons. When a larger number of electrons is involved, the oxidation or reduction tends to occur as a *series* of 1-electron or 2-electron steps. For example, the actual mechanism for reduction of dichromate to Cr^{3+} is quite complicated, as will be shown in Section 12–8. Application of the Nernst equation to the overall half-reaction,

$$Cr_2O_7^{2+} + 14H^+ + 6e^- = 2Cr^{3+} + 6H_2O$$

gives highly erroneous results.

12–5. COMPLETENESS OF A REDOX REACTION

For a reversible oxidation such as

$$A_{ox} + B_{red} \rightleftharpoons A_{red} + B_{ox}$$

the reaction continues until an equilibrium is reached in which the potential of one oxidation-reduction pair $(E_A{}^0)$ is equal to that of the other pair $(E_B{}^0)$. The equilibrium constant expression is

$$K = \frac{[A_{red}][B_{ox}]}{[A_{ox}][B_{red}]}$$

The value of K can be calculated from the standard electromotive force of the complete reaction ($E_A^0 - E_B^0$):

$$\log K = \frac{n(E_A^0 - E_B^0, V)}{0.059 \text{ V}} \tag{12-7}$$

where n is the number of electrons transferred during the stoichiometric reaction.

This equation is derived as follows. The complete reaction is a combination of the following half-reactions (written as reductions):

$$A_{ox} + ne^- = A_{red}$$

$$B_{ox} + ne^- = B_{red}$$

At equilibrium, the potential of the $A_{ox} - A_{red}$ couple, E_A, is equal to the potential of the $B_{ox} - B_{red}$ couple, E_B. Therefore,

$$E_A^0 + \frac{0.059}{n} \log \frac{A_{ox}}{A_{red}} = E_B^0 + \frac{0.059}{n} \log \frac{B_{ox}}{B_{red}}$$

$$E_A^0 - E_B^0 = \frac{0.059}{n} \left(\log \frac{B_{ox}}{B_{red}} - \log \frac{A_{ox}}{A_{red}} \right)$$

$$\frac{n(E_A^0 - E_B^0)}{0.059} = \log \frac{B_{ox}}{B_{red}} + \log \frac{A_{red}}{A_{ox}} = \log K$$

The equilibrium constant can be used to calculate the completeness of an oxidation-reduction reaction. It can also be used to calculate the minimum difference in E^0 values needed for a quantitative reaction.

Example: Calculate the minimum difference in standard potentials ($E_A^0 - E_B^0$) needed for a quantitative reaction in which both reactants undergo a 1-electron change.

Assume that, in a quantitative reaction, no more than one part in a thousand of either A_{ox} or B_{red} must remain when the titration is complete:

$$A_{ox} + B_{red} \rightleftharpoons A_{red} + B_{ox}$$

$$K = \frac{[A_{red}][B_{ox}]}{[A_{ox}][B_{red}]}$$

$$\text{minimum } K = \frac{(1000)(1000)}{(1) \quad (1)} = 10^6$$

$$\log K = \frac{n(E_A^0 - E_B^0)}{0.059}$$

Substituting $n = 1$ and $\log K = 6$ into this equation,

$$E_A^0 - E_B^0 = 6(0.059) = 0.354 \text{ V}$$

Thus, for this case, a difference of 0.354 V is required for a quantitative reaction. A somewhat smaller difference in $E_A{}^0$ and $E_B{}^0$ is required when one or both reactants undergo an electron change greater than 1 (see Table 12–2).

When one half-reaction undergoes a 1-electron change and the other a 2-electron change, a higher equilibrium constant is needed to ensure a quantitative reduction ($K = 10^9$), but a smaller difference in E_A and E_B is required. The minimum differences in E^0 values for different values of n are tabulated in Table 12–2. These tabulations assume that equivalent concentrations of reactants are present initially and that all reactants and products are soluble.

The completeness of an oxidation-reduction reaction under various experimental conditions can be estimated by calculating the numerical value of the equilibrium constant (Equation 12–7) and then calculating the equilibrium concentration of the reaction products. The method and reasoning used can perhaps best be illustrated by examples.

Example: Nickel(II) is to be reduced to the metal by reaction with chromium(II):

$$Ni^{2+} + 2Cr^{2+} \rightleftharpoons Ni^0(s) + 2Cr^{3+}$$

Calculate the concentration of Ni(II) remaining after addition of excess Cr(II). The conditions are such that the concentration of Cr(II) at equilibrium is $0.0080M$ and the concentration of Cr(III) is $0.0020M$.

From Appendix 3, the E^0 for the Ni^{2+}–Ni^0 half-reaction = -0.25 V, and E^0 for the Cr^{3+}–Cr^{2+} half-reaction = -0.41 V.

$$\log K = \frac{2[(-0.25) - (-0.41)]}{0.0591} = 5.41 \qquad K = 2.57 \times 10^5$$

Substituting into the equilibrium constant expression,

$$2.57 \times 10^5 = \frac{[Cr^{3+}]^2}{[Cr^{2+}]^2[Ni^{2+}]} = \frac{(0.0020)^2}{(0.0080)^2[Ni^{2+}]}$$

$$[Ni^{2+}] = 2.4_3 \times 10^{-7}M$$

Table 12–2. Minimum difference in $E_A{}^0$ and $E_B{}^0$ for a quantitative reaction (99.9% complete) when all species are soluble

n for A couple	n for B couple	$E_A{}^0 - E_B{}^0$
1	1	+0.354 V
1	2	+0.2655 V
2	2	+0.177 V
2	3	+0.1475 V

Note: These values apply only when *equivalent* amounts of A and B are mixed, as in a titration. If H^+ or OH^- appear in the equation, each must be $1M$.

Example: Solutions containing 0.01 mole of VO_2^+ and 0.01 mole of Fe^{2+} are mixed and quickly diluted to 1 liter with dilute acid so that the final concentration of H^+ is $2.0M$. Calculate the equilibrium concentrations of both reactants and decide whether the reaction is quantitative (99.9%) under these conditions.

Using E^0 values from Appendix 2,

$$\log K = \frac{1(1.00 - 0.77)}{0.0591} = 3.89 \qquad K = 7.73 \times 10^3$$

At equilibrium, $[Fe^{3+}] = [VO^{2+}] \cong 0.010M$, $[H^+] = 2.0M$, $[VO_2^+] = [Fe^{2+}] = xM$. Substituting into the equilibrium constant expression,

$$7.73 \times 10^3 = \frac{[Fe^{3+}][VO^{2+}]}{[VO_2^+][Fe^{2+}][H^+]^2} = \frac{(0.010)^2}{x^2(2.0)^2}$$

$$x = 5.7 \times 10^{-5}M$$

Thus, the reaction is not quite quantitative. $(x = 1 \times 10^{-5})$ would be needed for a quantitative reaction.

12-6. POTENTIOMETRIC TITRATIONS

The end point in an oxidation-reduction titration can be detected by the color change of a visual indicator or by plotting data taken by using a potentiometer. In a *potentiometric titration*, as the latter method is called, the difference in potential between an indicator electrode and a reference electrode is plotted against the volume of titrant added. This is analogous to using a pH meter to determine the titration curve for an acid-base titration.

A potentiometer is designed to measure the difference in potential between two electrodes *without* drawing appreciable electric current from the chemical system. This condition is important if the equilibrium potentials of the electrodes are to be maintained during the measurement. More information on potentiometers and potentiometric titrations is given in Chapter 16.

An electrode of bright platinum metal often serves as the indicator electrode for potentiometric oxidation-reduction titrations. The potential of the platinum electrode depends on the ratio of oxidizing and reducing agents in solution; therefore, it changes as the titration proceeds. A calomel electrode is used as reference; its potential remains constant throughout the titration. The calomel electrode is kept isolated from the solution being titrated; contact is made through a capillary salt bridge in the tip of the electrode.

A standard hydrogen electrode (SHE) could, of course, be used as the reference electrode. However, the calomel electrode is less bulky and much more convenient to use. The saturated calomel electrode (SCE) has a potential of $+0.246$ V when compared with the standard hydrogen electrode. The voltage measured represents the difference in potential between the indicator and reference electrodes:

$$E_{meas} = E_{ind} - E_{ref}$$

The potential of the indicator electrode can, therefore, be calculated from the measured voltage by substituting the potential of the calomel reference electrode into the previous equation:

$$E_{ind} = E_{meas} + 0.246$$

Example: A potentiometric titration is being performed with platinum and calomel electrodes. At one point, the voltage measured is 1.142 V. Calculate the potential of the platinum electrode vs. SHE at this point in the titration.

$$E_{Pt} = 1.142 + 0.246 = 1.388 \text{ V}$$

Potentiometric titration curves for typical titrations are shown in Figure 12–2. Potentiometric titration is a good means of following the course of an oxidation-reduction titration and detecting the end point of a titration. Titration curves are also useful in selecting the proper indicator for any given oxidation-reduction titration.

Potentiometric titration curves can be calculated with the Nernst equation from a table of standard electrode potentials. For example, consider the titration of iron(II) with cerium(IV) represented in Figure 12–2. The E^0 values for the two half-

Figure 12–2. Curves for potentiometric titrations using cerium(IV) to determine iron(II) (dashed line), and iron(II) plus tris(1,10-phenanthroline) iron(II) (solid line) in sulfuric acid solutions. SHE = standard hydrogen electrode; SCE = saturated calomel electrode.

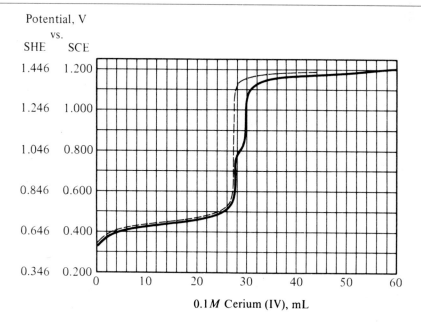

0.1M Cerium (IV), mL

reactions are

$Ce^{4+} + e^- = Ce^{3+}$	$E_A^0 = 1.44$ (in H_2SO_4)	*Half-reaction A*
$Fe^{3+} + e^- = Fe^{2+}$	$E_B^0 = 0.771$	*Half-reaction B*
$Ce^{4+} + Fe^{2+} \rightleftharpoons Ce^{3+} + Fe^{3+}$	$E^0 = 0.66_9$ V	*Complete*

At any point during the titration, cerium(IV) reacts with Fe^{2+} according to the complete reaction until equilibrium is reached. At equilibrium, the E for half-reaction A equals the E for half-reaction B.

The Nernst equation expressions for E_A and E_B are

$$E_A = E_A^0 - 0.0591 \log \frac{[Ce^{3+}]}{[Ce^{4+}]}$$

$$E_B = E_B^0 - 0.0591 \log \frac{[Fe^{2+}]}{[Fe^{3+}]}$$

Because $E_A = E_B$, the potential of a platinum indicator electrode dipping into the solution can be calculated from the Nernst equation expression for either half-reaction, provided that the ratio of ions in the Nernst equation can be deduced.

Before the end point, the ratio of Ce^{4+} to Ce^{3+} is very small because practically all of the Ce^{4+} is reduced by the excess Fe^{2+} present. However, this ratio cannot be conveniently calculated, and the indicator-electrode potential is not calculated from the Nernst equation for E_A. Before the end point, the ratio of Fe^{3+} to Fe^{2+} is known, assuming that an amount of Fe^{2+} equivalent to the added Ce^{4+} reacts to form Fe^{3+}. Thus, the Nernst equation for E_B can be used to calculate the potential of the platinum indicator electrode before the end point.

Example: Calculate the potential of the indicator electrode when 50% of the Fe^{2+} has been titrated with Ce^{4+}. At this point, the $Fe^{3+}:Fe^{2+}$ ratio is 50:50. From the Nernst equation for E_B,

$$E = 0.771 - 0.059 \log \frac{50}{50}$$

$$= 0.771 \text{ V}$$

In *any* oxidation-reduction titration at 50% titration, $E = E_B^0$, where E_B^0 is the standard redox potential for the substance being titrated.

Example: Calculate the potential of the indicator electrode when 80% of the Fe^{2+} has been titrated.

At this point, the ratio $Fe^{3+}:Fe^{2+} = 80:20$

$$E = 0.771 - 0.059 \log \frac{20}{80} = 0.807 \text{ V}$$

If no H^+ or OH^- is involved in the reaction, the potential at the equivalence point can be calculated by using the equation

$$E_{eq} = \frac{n_A E_A^0 + n_B E_B^0}{n_A + n_B}$$

where E_A^0 refers to the redox couple of the oxidizing agent, n_A is the electron change in this half-reaction, E_B^0 refers to the redox couple of the reducing agent, and n_B is the electron change in this half-reaction.

Example: Calculate the equivalence-point potential for the titration of Fe^{2+} with Ce^{4+}.

$$n_A = 1, \qquad n_B = 1$$

$$E_{equiv} = \frac{1.44 + 0.77}{2} = 1.10_5 \text{ V}$$

After the end point, virtually all of the Fe^{2+} has been oxidized to Fe^{3+}. The $Fe^{3+}:Fe^{2+}$ ratio is very high, but it cannot be conveniently calculated. However, the $Ce^{4+}:Ce^{3+}$ ratio can be easily estimated and the potential calculated from the Nernst equation for E_A.

Example: Calculate the potential of the indicator electrode at 140% titration. At this point, a 40% excess of Ce^{4+} has been added, and the $Ce^{4+}:Ce^{3+}$ ratio is 40:100. From the Nernst equation for E_A,

$$E = 1.44 - 0.059 \log \frac{100}{40}$$

$$= 1.44 - 0.02 = 1.42 \text{ V}$$

Example: Calculate the potential at 200% titration. At this point, a 100% excess of Ce^{4+} has been added and the $Ce^{4+}:Ce^{3+}$ ratio is 100:100.

$$E = 1.44 - 0.059 \log 1$$

$$= 1.44 \text{ V}$$

12–7. OXIDATION-REDUCTION INDICATORS

An oxidation-reduction indicator is a highly colored substance that changes color when oxidized or reduced. Each redox indicator changes color over a particular potential range, just as an acid-base indicator has a given pH transition range for its color change. The indicator chosen for an oxidation-reduction titration should have a transition potential that corresponds as closely as possible to the potential at the equivalence point in the titration. It is also important that the indicator be oxidized or reduced *quickly* and *reversibly*. If the indicator reacts slowly, the end point may be overshot. If the indicator is not reversible, temporary local excesses of titrant will oxidize the indicator gradually, and no sharp color change will occur.

Table 12–3.   Oxidation-Reduction Indicators

Indicator	Reduced Color	Oxidized Color	E^0, V
Tris(5-nitro-1,10-phenanthroline) iron(II) sulfate (nitro ferroin)	red	pale blue	1.25
Tris(1,10-phenanthroline) iron(II) sulfate (ferroin)	red	pale blue	1.06
Tris(2,2′-bipyridine) iron(II) sulfate	red	pale blue	0.97
Tris(4,7-dimethyl-1,10-phenanthroline) iron(II) sulfate	red	pale blue	0.88
Diphenylaminesulfonic acid	colorless or green	purple	0.84
Diphenylamine	colorless	violet	0.76
Methylene blue	blue	colorless	0.53
1,10-Phenanthroline vanadium(II) ion[a]	blue	pale green	0.15

a. W. P. Schaefer, *Anal. Chem.*, **35**, 1746 (1963).

Only a few really good oxidation-reduction indicators are available. Tris(1,10-phenanthroline) iron(II) sulfate, commonly called "ferroin," is probably the best. The blood-red ferrous complex ferroin is oxidized to a pale-blue ferric complex ferriin. The standard potential for the ferriin-ferroin couple is + 1.06 V, which makes ferroin ideal as an indicator for titrations with cerium(IV). In Figure 12–2, a very large amount of ferroin indicator was added to show the potential at which the indicator is oxidized. In an actual titration, only a trace of indicator is added, and the color change at the end point (caused by oxidation of the indicator) is quite sharp.

For titrations with standard dichromate, an indicator is needed that changes color at a lower potential than ferroin. Diphenylaminesulfonate is widely used, especially for titrating ferrous iron with dichromate. Both diphenylaminesulfonate and diphenylamine are oxidized irreversibly to an intermediate green form. At the end point, this green form is oxidized reversibly to a purple color. The E^0 for diphenylaminesulfonate is + 0.86 V.

The standard potentials for various oxidation-reduction indicators are given in Table 12–3.

12–8.   RATES AND MECHANISMS OF OXIDATION-REDUCTION REACTIONS

Rates of Oxidation-Reduction Reactions.   Many uncatalyzed oxidation-reduction reactions are too slow for analytical purposes. Unfortunately, there is no way to predict with absolute certainty whether a particular reaction will be slow. It is sometimes assumed that a large positive difference in E^0 values means that the reaction will be fast. This is misleading; there is no relationship between E^0 differences and rates because the two are different aspects of a reaction.

This is best understood in terms of the collision theory of reaction rates. Collison theory says that for reactants to form products, the reactants must collide to form a transitory complex, called the activated complex. This can occur only if the colliding reactants have a certain minimum amount of energy, called the activation energy. Only after the activated complex is formed do the reactants decompose to products, releasing energy to the solution. A reaction with a large activation energy will be slow and have a small rate constant.

Figure 12–3 illustrates the application of collison theory to the iron(III)-tin(II) reaction in perchloric acid. The difference in E^0 values is $+0.62$ V, but a large activation energy is needed to form the activated complex, so the reaction is slow. Once the iron(II) and tin(IV) products are formed, a large amount of energy is released. The difference in energy between the reactants and products is proportional to the $+0.62$ V difference in E^0 values, but this is not related to the large activation energy required for the reaction. The rate constant k for such a reaction will be quite small.

If the rate constant for a reaction is known, it is possible to calculate the time needed for a quantitative (99.9%) reaction using the kinetic rate laws (see

Figure 12–3.   Energy changes during the iron(III)-tin reaction. Energy difference A is the activation energy of the reaction in perchloric acid (it is much smaller in hydrochloric acid). Energy difference B is proportional to $(E^0_{Fe} - E^0_{Sn})$.

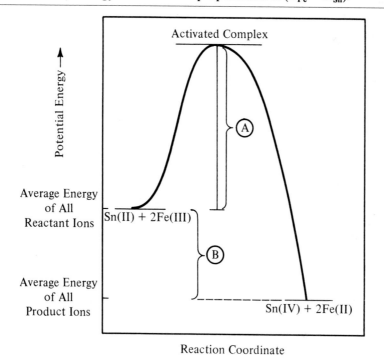

Table 12–4. Approximate Times for 99.9% Reactions, after Mixing, and Catalysts

Reactants (0.1N after mixing)	Reaction Time	Catalyst
$Ce^{4+} + Fe^{2+}$ (0.5M H_2SO_4)	$\sim 10^{-2}$ seconds	——
$Ce^{4+} + Fe(CN)_6^{4-}$ (H_2SO_4)	$\sim 10^{-2}$ seconds	——
$Cu^{2+} + I^-$ (pH 3–4)	~ 1 second	——
$IO_4^- + I^-$ ($IO_3^- + I_2$)	~ 1 second	——
$Fe^{2+} + MnO_4^-$ (H_2SO_4)	~ 1 second	——
$Ce^{4+} + VO^{2+}$ (1M H_2SO_4)	~ 10 seconds	——
$Ce^{4+} + Sn^{2+}$ (H_2SO_4)	~ 100 seconds	——
$H_3AsO_3 + I_2$ (pH < 5)	> 100 seconds	OH^-
$IO_3^- + I^-$ (pH > 7)	Very slow (less than 0.1% reaction in 30 minutes)	H^+
$Fe^{3+} + Sn^{2+}$ ($HClO_4$)	Very slow	OH^-, halide ions
$MnO_4^- + C_2O_4^{2-}$	Very slow	Mn^{3+}
$Ce^{4+} + As(III)$ (H_2SO_4)	Very slow (less than 0.1% reaction in 100 seconds)	OsO_4, I^-
$Ce^{4+} + Ferroin$ (0.5M H_2SO_4)[a]	~ 10 seconds	——

a. Calculated for $10^{-5}M$ reactants (for 90% reaction) after mixing.

Chapter 14). Table 12–4 lists several reactions and the approximate time for a 99.9% reaction after mixing.

Thus, the time it takes for a quantitative oxidation-reduction reaction to take place can vary tremendously. Given a choice, the analytical chemist usually chooses a reaction that is known to be fairly fast. A slow reaction can often be speeded up by gently heating the solution or by adding an excess of one of the reactants and back-titrating the excess when the reaction is complete. Usually, oxidation-reduction reactions are fairly fast if both reactants undergo a 1-electron or a 2-electron change and if the difference in reduction potentials is considerable. However, prediction of the rates of redox reactions is difficult because the rate is often influenced by the *mechanism* of the reaction.

Introduction to Oxidation-Reduction Reaction Mechanisms. A general knowledge of the mechanism or pathway of a redox reaction is often useful [1]. Examination of the mechanism may suggest a suitable catalyst or a change in conditions to increase the rate of the reaction.

The reaction mechanism is often more complicated than the simple stoichiometric equation would indicate. For example, the equation

$$Cr_2O_7^{2-} + 6Fe^{2+} + 14H^+ \rightarrow 2Cr^{3+} + 7H_2O$$

suggests that a simultaneously seven-body collision (six Fe^{2+} and one $Cr_2O_7^{2-}$) occurs in which six electrons are transferred from iron(II) to chromium in dichromate. However, this does not happen. For one thing, the statistical probability of

[1] G. H. Schenk, *J. Chem. Ed. 41*, 32 (1964).

a simultaneous seven-body collision is almost nil. What actually does occur is a series of one-electron transfer steps, the sum of these intermediate reactions being the preceding stoichiometric reaction.

Studies of the mechanism of the reduction of dichromate in acidic solution with a suitable reducing agent such as iron(II) indicate that the hydrogen chromate anion is the reactive chromium(VI) species.

$$Cr_2O_7^{2-} + H_2O \rightleftharpoons 2HCrO_4^-$$

Then the hydrogen chromate ion undergoes a series of three 1-electron reductions, each involving hydrogen ions:

$$HCrO_4^- + 2H^+ + e^- \rightleftharpoons H_3CrO_4$$

$$H_3CrO_4 + 3H^+ + e^- \rightarrow CrO^{2+} + 3H_2O$$

$$CrO^{2+} + 2H^+ + e^- \rightarrow Cr^{3+} + H_2O$$

The first reaction in which chromium is reduced from 6^+ to 5^+ is reversible, but the other two reductions are believed to be irreversible. For this reason, the Nernst equation cannot be applied quantitatively.

A mechanism for redox reactions between two metal ions involves an anion acting as an *electron bridge*. In one such mechanism, the anion may form an intermediate in which both metal-ion reactants are complexed by the same anion. The electron *appears* to be transmitted from the reducing agent through the electronic system of the anion, or electron bridge, to the oxidizing agent. In another metal-ion reaction mechanism, the reaction is catalyzed by a third metal ion that is oxidized and reduced over and over again. Finally, we have nonmetal-ion reaction mechanisms; these are quite different from metal-ion reaction mechanisms and will be illustrated with the iodine-thiosulfate reaction mechanism.

The Mechanism of a Metal-Ion Reaction Catalyzed by Electron Bridging. The slow reaction of iron(III) and tin(II), which forms iron(II) and tin(IV), is of great analytical interest. Research has shown that this reaction very probably takes place in two steps. First tin(II) is oxidized to tin(III), and then tin(III) is oxidized to tin(IV):

$$Fe^{3+} + Sn^{2+} \xrightarrow{\text{(slower)}} Fe^{2+} + Sn^{3+}$$

$$Fe^{3+} + Sn^{3+} \xrightarrow{\text{(faster)}} Fe^{2+} + Sn^{4+}$$

The net reaction is

$$2Fe^{3+} + Sn^{2+} \rightleftharpoons 2Fe^{2+} + Sn^{4+}$$

In perchloric acid solution, the reaction of iron(III) and tin(II) is so slow it is hardly perceptible. A likely reason is that the perchlorate ion, an extremely poor ligand, is not effective in transferring electrons between the two metal ions. In hydrochloric acid, the reaction is extremely fast. It can be shown that the rate of the reaction is proportional to the concentrations of iron(III) and tin(II) and to the *fourth power* of the chloride concentration:

$$\text{Rate} = k[Fe^{3+}][Sn^{2+}][Cl^-]^4$$

The mechanism of the slower reaction step is one in which Fe(III) (as Fe^{3+} or as $FeCl^{2+}$) comes together with Sn(II) (as $SnCl_3^-$ or as $SnCl_4^{2-}$) to form an intermediate of the following type:

$$\left[Fe:Cl:\overset{\overset{\textstyle Cl}{\cdot\cdot}}{\underset{\underset{\textstyle Cl}{\cdot\cdot}}{Sn}}:Cl \right]^+$$

A chloride atom serves as an *electron bridge* for an electron to pass from tin(II) to iron(III). The intermediate then decomposes into iron(II) and tin(III). The latter stabilizes by forming a chloride complex:

$$Fe_3^+ + Sn^{2+} + 4Cl^- \rightleftharpoons \left[Fe:Cl:\overset{\overset{\textstyle Cl}{\cdot\cdot}}{\underset{\underset{\textstyle Cl}{}}{Sn}}:Cl \right]^+ \rightleftharpoons Fe^{2+} + Sn^{3+} \text{ (probably } SnCl_4^-)$$

The tin(III) is then oxidized by another iron(III) to tin(IV) by a similar mechanism.

Mechanisms of Metal-Ion Reactions Catalyzed by Another Metal Ion. Many metal ions act as catalysts by being continually oxidized and reduced in a cycle. There are two requirements for this type of catalyst: It must react rapidly with both the oxidizing and reducing agents, and it must react spontaneously with both agents. It need not react quantitatively because it is being continually regenerated in the catalyst cycle.

A good example is the silver(I) catalysis of the peroxydisulfate oxidation of cerium(III) discussed in Section 13–3. In the first step of the cycle, the silver(I) is oxidized to silver(II) by peroxydisulfate:

$$Ag^+ \xrightarrow[\text{fast}]{S_2O_8^{2-}} Ag^{2+}$$

(Note that the E^0 for this reaction is only $+0.03$ V, indicating that it is spontaneous but not quantitative.) In the second step, the silver(II) is reduced to silver(I) by cerium(III):

$$Ag^{2+} \xrightarrow[\text{fast}]{Ce^{3+}} Ag^+$$

The E^0 for this reaction is $+0.54$ V, indicating that it is both spontaneous and quantitative.

Mechanism of a Nonmetal-Ion Reaction. Unlike reactions between metal ions, reactions between nonmetal anions and molecules are frequently irreversible, resembling organic nucleophilic displacement reactions. For example, iodine and thiosulfate react irreversibly and produce tetrathionate and iodide ions. (This is also a good example of a reaction that occurs rapidly without a catalyst.) According to

Awtrey and Connick [2], the mechanism is as follows:

$$I{-}I + :SSO_3{}^{2-} \overset{faster}{\rightleftharpoons} I:SSO_3{}^- + I^-$$

$$O_3SS^{2-}: + I:SSO_3{}^- \overset{slower}{\longrightarrow} O_3SS:SSO_3{}^{2-} + I^-$$

It should be noted that the two electrons transfer in the second step, when a second thiosulfate ion attacks the $I:SSO_3{}^-$ intermediate. The two electrons bonding iodide to sulfur in the intermediate are transferred to iodide as it is displaced from the intermediate by the second thiosulfate:

$$O_3\underset{\frown}{SS}:^{2-}\;\overset{\curvearrowright}{(I:)}\;SSO_3{}^-$$

The student of organic chemistry may recognize this as a bimolecular nucleophilic displacement reaction (S_N2).

Once the reaction is complete, no equilibrium concentration of thiosulfate ions can be detected because of the second, irreversible step. It would be difficult to convert tetrathionate back to thiosulfate by the pathway described earlier. Even though the second step is slower than the first, the overall reaction rate is still very fast in comparison to the slow reactions listed in Table 12–4.

The mechanism of the reactions of iodine with certain other sulfur compounds is similar to that of the iodine-thiosulfate reaction. As an example, consider an organic mercaptan RSH, which can ionize to RS:$^-$ and H^+. The RS:$^-$ anion reacts rapidly with iodine to form RSSR, an organic disulfide. The mechanism of this reaction also occurs in two steps.

$$I{-}I + RS:^- \rightleftharpoons RS:I + I^-$$

$$RS:^- + RS:I \rightarrow RS:SR + I^-$$

Any organic mercaptan can be expected to behave in the same way. Because of the similarity in reaction mechanism and in potential, mercaptans interfere with an iodide-thiosulfate titration, and vice versa.

Sulfate does not interfere with iodine titrations because it contains sulfur in its highest oxidation state. Sulfur in the sulfite ion has an available electron pair but is oxidized to sulfate, not to a dimer, by iodine. Thiocyanate undergoes a complicated oxidation to sulfate and iodine cyanide rather than to a dimer.

QUESTIONS AND PROBLEMS

Questions on Concepts

1. Define briefly, and give an example of, each of the following: (a) half-reaction, (b) oxidation-reduction couple, (c) standard electrode potential, (d) indicator electrode, (e) reference electrode.

[2] A. D. Awtrey and R. E. Connick, *J. Am. Chem. Soc.*, 73, 1341 (1951).

2. Explain how both cations and anions can complete a circuit by carrying a current through a solution. What is a transport number?

3. An oxidation-reduction couple is somewhat analogous to a conjugate acid-base pair (see the following equations). In what ways are they *not* analogous?

$$\text{Oxidized} + e^- = \text{reduced}$$

$$\text{Base} + H^+ \rightleftharpoons \text{acid}$$

4. List and discuss the limitations of the Nernst equation in describing potential changes in a potentiometric titration.

5. Review the leveling effect in nonaqueous titrations (Chapter 9). (a) Indicate why this is not usually observed in oxidation-reductions. (b) Oxidants such as Ag^{2+} and F_2 are not stable in aqueous solution, and both produce the OH radical. Explain whether this might be considered a form of leveling.

Predicting Which Reactions Will Occur

6. Consult the table of standard reduction potentials and decide which of the following reactions should occur spontaneously.

(a) $Cl_2 + 2Br^- \rightarrow Br_2 + 2Cl^-$
(b) $2Fe^{3+} + SO_3^{2-} + H_2O \rightarrow SO_4^{2-} + 2Fe^{2+} + 2H^+$
(c) $2VO^{2+} + 2H^+ + H_2O_2 \rightarrow 2V^{3+} + O_2 + 2H_2O$
(d) $VO_2^+ + 2H^+ + H_2O_2 \rightarrow VO^{2+} + O_2 + 2H_2O$
(e) $I_2 + \text{Ascorbic acid} \rightarrow 2I^- + \text{Dehydroascorbic acid} + 2H^+$

7. Consult the standard reduction potentials in Appendix 3 and predict which of the following substances will react with iodine (I_2) in an oxidation-reduction reaction. Explain your reasoning.

(a) Ag^0 in a solution containing NO_3^- or ClO_4^-.
(b) Ag^+ in a solution containing I^-.
(c) HQ (hydroquinone) in a solution containing $1.0M$ H^+.
(d) HQ in a solution buffered at pH 5.0.

8. Vanadium can exist in aqueous solution as vanadium(II), (III), (IV), or (V). Use the standard electrode potentials in Appendix 3 to predict the highest oxidation state to which vanadium(II) will be oxidized by each of the following reagents.

(a) Sn^{4+} (c) Fe^{3+}
(b) Ce^{4+} (in $1N$ H_2SO_4)

9. The reaction $Fe^{3+} + Co^{2+} \rightarrow Fe^{2+} + Co^{3+}$ is highly unfavorable ($E^0 = -1.05$ V). However, the oxidation-reduction reaction of their 1,10-phenanthroline complexes is quantitative and has been used to titrate cobalt(II):

$$Fe(phen)_3^{3+} + Co(phen)_3^{2+} \rightarrow Fe(phen)_3^{2+} + Co(phen)_3^{3+}$$

From this information, what do you deduce regarding the relative strengths of the $Co(phen)_3^{2+}$ and $Co(phen)_3^{3+}$ complexes?

10. Vanadium can exist in aqueous solution in the $+2, +3, +4$, and $+5$ oxidation states. From the standard potentials given in Appendix 3 (plus the following one), predict which oxidation state will predominate when vanadium metal is dissolved in acid solution ($a_{H^+} = 1$)

$$V^{2+} + 2e^- = V^0 \qquad E^0 = -1.5 \text{ V}$$

Electrode Potentials; The Nernst Equation

11. In solutions of total ionic strength 0.2, the single-ion activity coefficients of iron species are $f_{Fe^{3+}} = 0.18$ and $f_{Fe^{2+}} = 0.40$. Calculate the potential of a platinum electrode dipping into a solution containing $0.01M$ Fe^{3+}, $0.01M$ Fe^{2+}, and acid so that total $\mu = 0.2$ (a) using ion activities and (b) using ion concentrations.

12. Calculate the potential of each of the following electrodes.

  (a) $Ag^0(s)/Ag^+$ $(1.0 \times 10^{-4}M)$
  (b) $Hg^0(s)/Hg^{2+}$ $(1.0 \times 10^{-4}M)$
  (c) $Pt^0(s)/Fe^{2+}$ $(0.001M)$, Fe^{3+} $(0.0025M)$
  (d) $Pt^0(s)/HQ$ $(0.001M)$, Q $(0.002M)$, H^+ $(1.0 \times 10^{-5}M)$
  (e) $Pt^0(s)/Ascorbic$ acid $(0.001M)$, Dehydroascorbic acid $(0.001M)$, H^+ $(1.0 \times 10^{-5}M)$

13. (a) Calculate the potential (versus SHE) of a platinum wire dipping into a solution of $0.0010M$ Q (Quinone), $0.0005M$ HQ (hydroquinone), and $1.0M$ H^+.

$$Q + 2H^+ + 2e^- = HQ \qquad E^0 = +0.699 \text{ V}$$

  (b) Calculate the potential for part (a) if $[H^+]$ is now $1.0 \times 10^{-5}M$.

14. A mercury metal–Hg^{++} electrode is reversible and can be used to measure the concentration of Hg^{++} in solution.

  (a) The potential of a mercury metal electrode in contact with a solution containing an unknown concentration of Hg^{++} was measured to be 0.777 V. Calculate the unknown concentration of Hg^{++}.
  (b) A calibration curve for Hg^{++} was prepared by measuring the potential of the mercury electrode in contact with several standards and plotting potential versus the log of the concentration of Hg^{++}. What is the expected slope of the resulting straight-line plot?

$$Hg^{++} + 2e^- = Hg^0 \qquad E^0 = +0.854 \text{ V}$$

15. An organic chemist is studying the effect of chemical structure on the reduction potential of a series of quinones. All of these undergo a reversible reduction of the type:

$$Q + 2H^+ + 2e^- = HQ$$

(HQ stands for hydroquinone.) Tell how to measure the standard reduction potential of each of the quinones. Explain why it is necessary to measure the pH and to keep it constant.

Calculation of Concentration and Solubility Product

16. The measured potential at a silver wire dipping into a saturated solution of a slightly soluble silver mercaptide (AgSR) is $+0.440$ V versus SHE. Calculate the $[Ag^+]$ of the solution and the K_{sp} of the silver mercaptide.

17. A silver indicating electrode and a standard hydrogen reference-electrode are dipped into a saturated solution of silver bromide. The measured potential (E_{meas}) is $+0.434$ V. From the standard potential of the $Ag^0(s)/Ag^+$ electrode given in the following equation, calculate (a) the $[Ag^+]$ to the correct number of significant figures and (b) the solubility-product constant for silver bromide.

$$Ag^+ + e^- = Ag^0(s) \qquad E^0 = +0.800 \text{ V}$$

18. A silver indicating electrode and a standard hydrogen reference electrode are dipped into a saturated aqueous solution of silver oxalate. The potential measured on the

potentiometer (E_{meas}) is $+0.589$ V. From the standard potential of the Ag^0/Ag^+ electrode given in Problem 14, calculate (a) the $[Ag^+]$ to the correct number of significant figures, and (b) the solubility-product constant for silver oxalate, $Ag_2C_2O_4$.

19. An electrode that is capable of measuring dissolved oxygen gives a potential of $+1.057$ V (versus SHE) for an aqueous sample buffered at pH 2.00. Calculate the molarity of dissolved oxygen, assuming that molarity = activity in this case.

$$O_2 + 4H^+ + 4e^- = 2H_2O \qquad E^0 = +1.229 \text{ V}$$

Completeness of a Reaction; Equilibrium Constants

20. Decide whether the following reactions will be quantitative when equivalent amounts are mixed together and equilibrium has been attained.

   (a) $Ag^{2+} + Ce^{3+} \rightarrow Ag^+ + Ce^{4+}$ (in $1M$ $HClO_4$)
   (b) Same as (a), but in $1M$ HNO_3
   (c) $I_2 + 2V^{3+} + 2H_2O \rightarrow 2I^- + 2VO^{2+} + 4H^+$ (when $H^+ = 1.0M$)
   (d) $2Fe^{3+} + Sn^{2+} \rightarrow 2Fe^{2+} + Sn^{4+}$

21. Suppose a newly invented buret has a relative accuracy of 0.1 ppt (1 part in 10,000), so that a "quantitative" reaction must now be defined as one in which no more than 0.1 ppt of A_{ox} and B_{red} remain when the following titration reaction is complete:

$$A_{ox} + B_{red} \rightleftharpoons A_{red} + B_{ox}$$

Assuming that both reactants undergo a one-electron change, calculate the minimum difference in standard potentials ($E_A^0 - E_B^0$) for a "quantitative" reaction.

22. It is desired to conduct an analysis by adding a twofold excess of reagent A to a flask containing the desired constituent C and an equal amount of interference B. What must be the difference in standard potentials ($E_A^0 - E_B^0$) so that no more than 1 ppt of B will react with A? Assume that all reactants undergo 1-electron change and that C reacts quantitatively with A.

23. Calculate the equilibrium constant for the following reaction, carried out in $1M$ acid:

$$VO_2^+ + Fe^{2+} + 2H^+ \rightleftharpoons VO^{2+} + Fe^{3+} + H_2O$$

Is the oxidation of Fe^{2+} to Fe^{3+} quantitative ($>99.9\%$)? How might conditions be adjusted to make the oxidation more complete?

24. Calculate the equilibrium constant for the following reaction, carried out in $1M$ perchloric acid:

$$2Fe^{3+} + 2I^- \rightleftharpoons 2Fe^{2+} + I_2$$

25. Calculate the equilibrium constant for the following reaction:

$$S_2O_8^{2-} + 2H_2O \rightarrow 2SO_4^{2-} + H_2O_2 + 2H^+$$

Choose the appropriate half-reaction for oxidation of water from these:

$$O_2 + 4H^+ + 2e^- = 2H_2O \qquad E^0 = 1.29 \text{ V}$$
$$H_2O_2 + 2H^+ + 2e^- = 2H_2O \qquad E^0 = 1.770 \text{ V}$$
$$H_2O_2 + 2e^- = 2OH^- \qquad E^0 = 0.88 \text{ V}$$

26. The equilibrium constant for the following reaction is less than 1×10^6, indicating that it is not quantitative at $1M$ H^+.

$$Fe(CN)_6^{3-} + Ti^{3+} + H_2O \rightarrow Fe(CN)_6^{4-} + TiO^{2+} + 2H^+$$

(a) Calculate the numerical value of K for the reaction.

(b) Calculate the value of the $[H^+]$ that will permit 99.9% conversion $(10^3/L)$ of $Fe(CN)_6^{3-}$ to $Fe(CN)_6^{4-}$ and 99.9% conversion $(10^3/L)$ of Ti^{3+} to TiO^{2+}.

$$E^0 \text{ Ferriin-Ferroin} = 0.36$$

27. A solution containing $0.01M$ Pb^{2+} is to be reduced to lead metal by the addition of a Cr^{2+} solution. What must be the equilibrium ratio of Cr^{2+} to Cr^{3+} in order for the lead(II) concentration to be reduced to $1.0 \times 10^{-6}M$?

28. (a) Calculate the percentage of VO^{2+} that will be reduced to V^{3+} by adding excess Ag^0 to a dilute solution of VO^{2+} in $1M$ HCl. (See Appendix 3 for electrode potentials.)

(b) Calculate the percentage of VO^{2+} that will be reduced in part (a) if the solution contains $1M$ HBr instead of $1M$ HCl.

29. The amino acids cystine and cysteine constitute a reversible oxidation system:

$$\text{Cystine} + 2H^+ + 2e^- = 2 \text{ Cysteine}$$

At pH 7.0, the reduction potential is -0.340 V. Calculate the standard reduction potential at $1.00M$ acid.

30. Ascorbic acid (vitamin C) is often determined by titration with standard iodine (I_2) solution. Calculate what the minimum pH should be so that the oxidation-reduction reaction will be 99.9% complete when equivalent amounts of I_2 and ascorbic acid have been mixed. (Consult Appendix 3 for the appropriate standard electrode potentials.)

31. The equilibrium constant for the following reaction is 1.1×10^4 in $1M$ HCl.

$$\text{Bi(s)} + 4Cl^- + \text{As(III)} \rightarrow BiCl_4^- + \text{As(s)}$$

(a) If excess Bi(s) is added to a solution of $0.01M$ As(III) in $1M$ Cl^-, calculate whether the conversion to As(s) will be quantitative $(>99.9\%)$.

(b) Do the same calculation for a solution of $0.01M$ As(III) in $0.10M$ Cl^-.

Titration Curves

32. A sample containing iron(III) is buffered at pH 1.00 and titrated with a standard solution of ascorbic acid.

(a) Calculate the potential of a platinum indicator electrode (versus SHE) at the following points in the titration: 10%, 50%, 90%, 110%, 150%, and 200% titration.

(b) Also calculate E_{cell} at each of the titration points listed in part (a) when a platinum indicator electrode is used in conjunction with a saturated calomel reference electrode (SCE).

(c) Plot the titration curve, E_{cell} versus percentage titration.

33. In a potentiometric titration with chromium(II), the measured emf using a platinum indicator electrode and a saturated calomel reference-electrode was -0.590 V. If the calomel electrode potential is $+0.246$ V versus the standard hydrogen electrode, what is the potential of the platinum electrode?

34. At what point in a potentiometric redox titration should the potential of the indicator electrode be equal to the E^0 of the redox couple titrated? At what point should the potential be equal to the E^0 of the titrant redox couple? (Assume that $[H^+] = 1$.)

35. A solution containing titanium(III) and iron(II) in $1N$ sulfuric acid is titrated potentiometrically with cerium(IV). Calculate the potential of the platinum indicator electrode versus SHE when (a) 50% of the titanium(III) has been titrated, (b) 50% of the

iron(II) has been titrated, and (c) a 100% excess of cerium(IV) has been added. Sketch the approximate curve for this titration.

36. In titrations with cerium(IV) and ferroin indicator, the disappearance of the red color is taken as the end point. Since the red color of ferroin is several times more intense then the blue oxidized color, the visible disappearance of red occurs when the oxidized form has a concentration about 10 times that of the reduced form. At what potential does the visible disappearance of the red color take place? The E^0 for ferriin-ferroin is 1.06 V.

37. Using the table of standard electrode potentials (Appendix 3), calculate the equivalence-point potential for the titration of U^{4+} with $VO_2{}^+$ in $1M$ acid. Using Table 12–3, select a suitable indicator for this titration.

38. A reducing agent with $E^0 = 0.00$ V is titrated to an equivalence point of $E = 1.35$ V using an oxidizing agent whose E^0 is unknown. The electron change for both half-reactions is unknown, but it is known that the oxidizing agent gains three times as many electrons as the reducing agent loses. Calculate the numerical value of E^0 for the oxidizing agent.

39. The *absorbance* of a colored solution (see Chapter 17) is linearly proportional to the concentration of the colored chemical in solution. The intensely red-colored ion, tris(bipyridine)iron II, is titrated with a dilute cerium(IV) solution, forming a pale blue tris(bipyridine)iron III ion. Assuming that the absorbance is due only to the red tris(bipyridine)iron II, sketch the expected titration curve. How is the titration end point determined from this titration curve?

40. Following the previous problem, uncomplexed iron(II) is titrated with $MnO_4{}^-$, which is an intense violet color, to form iron(III) and Mn^{++}. Assuming that the absorbance of Fe(II), Fe(III), and Mn(II) is negligible, sketch the titration curve and indicate how the end point is determined.

Rates of Oxidation-Reduction Reactions

41. Calculate the difference in E^0 for the two half-reactions involved in the reaction

$$2Ce^{4+} + As(III) \rightarrow 2Ce^{3+} + As(V)$$

Is the activation energy for this reaction (in sulfuric acid) relatively large or small? Draw an energy diagram similar to Figure 12–3 for this reaction.

42. From the generalizations regarding reaction rates and the reaction times listed in Table 12–4, suggest possible catalysts for the following slow reactions.

(a) As(III) + $IO_4{}^-$ (basic solution)   (c) $Cr_2O_7{}^{2-} + I^-$ (acid solution)
(b) $Fe^{3+} + Sn^{2+}$ ($HClO_4$ solution)   (d) Ce^{4+} + As(III)(H_2SO_4)

43. Suggest a reagent and reaction conditions for the redox determination of $IO_4{}^-$ in the presence of $IO_3{}^-$. The reagent chosen should react very slowly (or not at all) with $IO_3{}^-$ under the reaction conditions. Consult Table 12–4 for ideas.

44. Applying the concepts of fast and slow reactions and differences in E^0 values, suggest titrants for determining iron(II) in the presence of (a) cerium(III); (b) silver(I); (c) arsenic(III); (d) oxalic acid ($E^0 = -0.49$ V for $2CO_2 + 2H^+ + 2e^- = H_2C_2O_4$).

45. Explain why the oxidation of tin(II) by iron(III) is faster in dilute than in concentrated $HClO_4$.

46. Predict whether iodine will react rapidly or slowly with ethyl xanthate, $C_2H_5OCSS^-$ Write the products for any reaction that occurs.

47. To measure the rate of a slow reaction, it is necessary to "quench" (stop) the reaction by quickly pipetting some quenching reagent into the reaction mixture. The quenching reagent must react rapidly with one of the reactants to prevent the slow reaction from

proceeding further. To measure the rate of the slow cerium(IV)-arsenic(III) reaction in sulfuric acid, Yates and Thomas (1956) decided to add a reducing agent to react *rapidly and quantitatively* with cerium(IV). Suggest a suitable reducing agent and indicate how it fulfills these requirements.

48. The uncatalyzed cerium(IV)-thallium(I) reaction is extremely slow in sulfuric acid. Suggest a way to titrate iron(II) in the presence of thallium(I). Alternatively, write a report on a specific method for this titration. (See *Anal. Chem.*, *40*, 162 (1968) for details.)

13

Oxidation-Reduction Titrations

13–1. INTRODUCTION

General Principles. A number of inorganic and organic substances can be determined by titration with a standard solution of an oxidizing reagent, or sometimes with a reducing titrant. For example, the determination of iron in solution by first reducing it to iron(II) and then titrating it with standard potassium permanganate is one of the oldest of all titration methods. The presence of the first permanent violet color of excess permanganate can serve as a convenient indication of the end point in this titration. The development of modern analytical methods such as atomic spectroscopy (Chapter 19) and EDTA titrations (Chapter 11) has somewhat reduced our dependence on oxidation-reduction titrations for inorganic analysis. However, the methods discussed in this chapter continue to be useful, especially for determinations in which a high degree of accuracy is required.

The requirements for a quantitative determination by oxidation-reduction are as follows:

1. The substance to be titrated must be in a definite oxidation state. Sometimes this necessitates some kind of a preliminary chemical treatment. For example, iron(III) salts in solution are reduced to iron(II) by zinc metal (which has a thick surface coating of zinc amalgam).

$$2\,Fe^{3+} + Zn^0 \;\rightarrow\; 2\,Fe^{2+} + Zn^{2+}$$

2. The oxidation-reduction reaction that occurs during the titration must be quantitative. Redox equilibrium was discussed in Chapter 12.
3. For a direct titration, the oxidation reaction must also be quite fast. Unlike acid-base reactions, which always occur very rapidly, some oxidation reactions are fairly slow and others are very slow. If an oxidation-reduction reaction is too slow for a direct titration to be practicable, an indirect titration may be feasible. Here, a measured excess of the standard oxidizing solution is added to the sample, and

sufficient time is allowed for the reaction to be complete. Then the excess of the oxidizing chemical is back-titrated by a standard reducing solution that reacts rapidly with the oxidant. The determinations of organic compounds by bromination and glycols by periodate oxidation (see Section 13–3) are examples.

4. There must be a good way to detect the end point. The use of a visible indicator or a potentiometric end point is commonly used (see Chapter 12).

Calculations. Up to now, we have recommended the use of molarity, millimoles, and formula weight in calculating the results of titrations. These may be used for any type of titration including oxidation-reduction, providing the combining ratio of the reactants is taken into account. However, calculations with the use of normality are so widely used (especially in connection with oxidation-reduction titrations) that the student should become familiar with them.

Calculations with normality are based on the use of *equivalent weight* instead of formula weight and on *equivalents* (or milliequivalents) instead of moles (or millimoles). The equivalent weight of a substance involved in a redox reaction is the formula weight of the substance (weighed or calculated), divided by the number of electrons transferred in the redox reaction:

$$\text{Equivalent weight} = \frac{\text{formula weight}}{e^- \text{ transferred}}$$

As examples, the equivalent weights are indicated for the following reactions:

Reaction	Equivalent Weight of Reactant
$Fe^{2+} \rightarrow Fe^{3+} + e^-$	Form wt $Fe \div 1$
$KMnO_4 + 5e^- \rightarrow Mn^{2+}$	Form wt $KMnO_4 \div 5$
$Na_2S_2O_3 \cdot 5\,H_2O \rightarrow \frac{1}{2}S_4O_6^{2-} + e^-$	Form wt $Na_2S_2O_3 \cdot 5\,H_2O \div 1$
$H_2S \rightarrow SO_4^{2-} + 8e^-$	Form wt $H_2S \div 8$
$Cr_2O_7^{2-} + 6e^- \rightarrow 2Cr^{3+}$	Form wt $Cr_2O_7^{2-} \div 6$

Other definitions and relationships necessary for calculations with normality are:

$$\text{Equivalents} = \frac{\text{grams}}{\text{equiv wt}}$$

$$eq = \frac{g}{eq\ wt}$$

$$\text{Normality} = \frac{\text{equivalents}}{\text{liters}}$$

$$N = \frac{eq}{L}$$

$$\text{Milliequivalents} = \frac{\text{milligrams}}{\text{equiv wt}}$$

$$meq = \frac{mg}{eq\ wt}$$

$$\text{Normality} = \frac{\text{milliequivalents}}{\text{milliliters}}$$

$$N = \frac{meq}{mL}$$

The result of a titration of a substance B with a titrant A is calculated as follows:

$$(mL_A)(N_A)(eq\ wt_B) = mg_B$$

$$\frac{(mg_B)(100)}{mg_{sample}} = \%\ B$$

Or, combining these two equations,

$$\frac{(mL_A)(N_A)(eq\ wt_B)(100)}{mg_{sample}} = \%\ B$$

Notice that calculations made with normality are very similar to those made with molarity, but the reacting ratio considered with molarity is taken care of mechanically with normality and equivalent weight. However, normality must not be used carelessly because the number of electrons transferred (and therefore its equivalent weight) may be different in different reactions. For example, molybdenum may be titrated either from $3+$ to $6+$ or from $5+$ to $6+$, depending on the method used to reduce the molybdenum prior to titration. Permanganate is usually used as an oxidizing agent in acid solution and is reduced to Mn^{2+}. Here, the equivalent weight is the formula weight divided by 5. In neutral solution, permanganate is reduced to MnO_2, in a solution containing fluoride it is reduced to a manganese(III) fluoride complex, and in strongly basic solution it is reduced to the manganate ion, MnO_4^{2-}.

13–2. ADJUSTMENT OF OXIDATION STATE

Prior Reduction of Substance to be Titrated.   The reduced forms of several elements may be titrated to a higher oxidation state by means of a standard solution of an oxidizing agent such as potassium permanganate, potassium dichromate, cerium(IV), or iodine. Unless the substance is entirely present in a single reduced state, a reduction step is necessary before titration. The substance to be determined must be reduced quantitatively to a definite ion or valence state, but there must be some way of removing the excess reducing agent so that it will not interfere in the titration. One way to do this is with a metal reductor. The most common metal reductor is the Jones reductor, a column filled with granular zinc coated with zinc amalgam. The zinc surface is amalgamated to prevent acid from reacting with the zinc metal, forming hydrogen gas:

$$2H^+ + Zn(s) \rightarrow H_2(g) + Zn^{2+}$$

Strong acids and zinc readily react and form hydrogen. However, mercury and metal amalgams have such a high hydrogen overpotential that the hydrogen is not discharged. (See Chapter 15.)

The sample solution is passed slowly through the column, after which the column is rinsed with several portions of dilute acid, which are also collected for analysis. Ferric iron is reduced quantitatively to ferrous iron:

$$2Fe^{3+} + Zn(Hg) \rightarrow 2Fe^{2+} + Zn^{2+}$$

Titanium(IV) in sulfuric acid solution is quantitatively reduced to violet-colored titanium(III):

$$2Ti(IV) + Zn(Hg) \rightarrow 2Ti(III) + Zn^{2+}$$

The reduction of some other metal ions is indicated in Table 13–1.

A column packed with spongy, granular silver metal is a useful means of reducing some metal ions prior to titration. For example, ferric iron is reduced in hydrochloric acid solution to ferrous iron:

$$Fe^{3+} + Ag(s) + HCl \rightarrow Fe^{2+} + AgCl(s) + H^+$$

In this reduction, the precipitation of silver ions as silver chloride shifts the equilibrium farther to the right. Without chloride, a silver metal reductor cannot reduce iron(III) because the E^0 for the reaction is negative.

$$Ag^+ + e^- = Ag(s) \qquad E^0 = +0.8000 \text{ V}$$

$$Fe^{3+} + e^- = Fe^{2+} \qquad E^0 = +0.77 \text{ V}$$

$$AgCl(s) + e^- = Ag(s) + Cl^- \qquad E^0 = +0.222 \text{ V}$$

The reactions of other metal ions with a silver reductor are summarized in Table 13–1.

Other reduction methods are frequently used. A lead reductor reduces uranium(VI) to uranium(IV) and is preferred to the Jones reductor, which gives a mixture of uranium(III) and uranium(IV). Aluminum metal (in a flask, not in a reductor column) reduces titanium(IV) to titanium(III). The excess aluminum metal is removed by being allowed to dissolve in the acid present. Stannous chloride is used for reducing iron(III) to iron(II) (see Chapter 30, Experiment 21).

Prior Oxidation of Substance to be Titrated. Potassium peroxydisulfate, $K_2S_2O_8$ (often called potassium persulfate), can be used to oxidize any of several elements to a higher oxidation state before they are titrated with a standard reducing agent.

For example, chromium(III) is oxidized to chromium(VI), dichromate. The excess peroxydisulfate is decomposed by boiling the solution for a few minutes after

Table 13–1. Action of Zinc Amalgam and Silver Reductors on Various Metal Ions

Metal Ion	Reduction Product	
	Zn(Hg)	Ag(HCl)
Cr(III)	Cr(II)	Not Reduced
Cu(II)	Cu⁰	Cu(I)
Fe(III)	Fe(II)	Fe(II)
Mo(VI)	Mo(III)	Mo(V)
Ti(IV)	Ti(III)	Not Reduced
U(VI)	U(III) + U(IV)	U(IV)
V(V)	V(II)	V(IV)

the oxidation is complete:

$$S_2O_8{}^{2-} + 2H_2O \rightarrow 2SO_4{}^{2-} + O_2 + 4H^+$$

Peroxydisulfate oxidations are carried out in hot, acidic solution. A small amount of silver(I) must be added as a catalyst in most of these oxidations. Peroxydisulfate oxidizes silver(I) to silver(II), or even to some silver(III). The silver(II) oxidizes chromium(III) to chromium(VI), then the silver(I) formed is reoxidized by peroxydisulfate and the cycle is repeated.

$$Ag(I) \xrightarrow{S_2O_8{}^{2-}} Ag(II)$$

$$3Ag(II) + Cr(III) \rightarrow Cr(VI) + 3Ag(I)$$

Peroxydisulfate also quantitatively oxidizes cerium(III) to cerium(IV) and vanadium(IV) to vanadium(V). Manganese(II) is oxidized to permanganate, but part of the permanganate is always reduced as the solution is boiled to decompose the excess peroxydisulfate.

Hot, concentrated perchloric acid is a powerful oxidizing agent and has been used to oxidize chromium to dichromate and cerium to cerium(IV) prior to titration. Dilute perchloric acid at room temperature has no oxidizing properties; hence, the excess perchloric acid need not be removed. Chlorine, formed as a decomposition product of hot, concentrated perchloric acid, must be boiled off after dilution. *Caution!* Hot, concentrated perchloric acid can be explosive in the presence of organic matter.

After oxidation with persulfate, the cerium(IV), chromium(VI), or vanadium(V) may be determined by titration with a standard solution of iron(II):

$$Ce^{4+} + Fe^{2+} \rightarrow Ce^{3+} + Fe^{3+}$$

$$Cr_2O_7{}^{2-} + 6Fe^{2+} + 14H^+ \rightarrow 2Cr^{3+} + 6Fe^{3+} + 7H_2O$$

$$VO_2{}^+ + Fe^{2+} + 2H^+ \rightarrow VO^{2+} + Fe^{3+} + H_2O$$

13-3. TITRATIONS WITH STRONG OXIDIZING AGENTS

A standard solution of cerium(IV), permanganate, dichromate, or occasionally vanadium(V) is used to titrate iron(II), titanium(III), arsenic(III), and other reducing substances.

Potassium Permanganate. In acidic solution, a reducing substance will reduce deep magenta permanganate to colorless manganese(II), as is indicated by the following redox half-reaction:

$$MnO_4{}^- + 8H^+ + 5e^- = Mn^{2+} + 4H_2O \qquad E^0 = 1.51 \text{ V}$$

The actual mechanism of the reduction is much more complicated than is indicated by this simple chemical equation, so the Nernst equation cannot be used for quantitative potential calculations. Nevertheless, permanganate is a powerful oxidizing agent and will oxidize many substances quantitatively.

Since potassium permanganate is not a primary standard, a stock solution of approximately known strength is first prepared. The aqueous solution is allowed to stand overnight and then is filtered or is heated to boiling and then cooled and filtered. A porous glass or porcelain filter is used because filter paper readily reduces permanganate. The purpose of filtration is to remove dust or lower oxides of manganese, which catalyze the decomposition of permanganate. A properly prepared solution of permanganate retains its normality for a long period of time; a poorly prepared solution decomposes on standing.

Sodium oxalate, $Na_2C_2O_4$, or pure iron metal is generally the primary standard for permanganate. Sodium oxalate is dissolved in acid (forming oxalic acid, $H_2C_2O_4$) and is titrated with permanganate according to the following equation:

$$2MnO_4^- + 5H_2C_2O_4 + 6H^+ \rightarrow 2Mn^{2+} + 10CO_2 + 8H_2O$$

Each carbon atom in the oxalate is oxidized from $3+$ to $4+$; therefore, the equivalent weight of sodium oxalate is the formula weight divided by 2.

Iron metal is weighed accurately, dissolved in acid, reduced to iron(II), and titrated with permanganate. (A more complete description of this procedure is given in the section dealing with the determination of iron.) In all of the standardizing titrations, the end point is detected by the first permanent appearance of the permanganate magenta color.

Cerium (IV) Titrants. Cerium(IV) solutions are usually prepared in sulfuric acid or in perchloric acid solution. Cerium(IV) is a stronger oxidizing agent in perchloric acid ($E^0 = 1.70$ V) than in sulfuric acid ($E^0 = 1.44$ V). This is probably caused by complex formation between the cerium and sulfuric acid. Some authorities refer to cerium(IV) solutions as *cerate* solutions because of anionic complexes such as $Ce(SO_4)_3^{2-}$ and $Ce(NO_3)_6^{2-}$. Others think that the name *ceric* is more apt because most of the cerium(IV) is present as cationic species. Under most conditions, cerium(IV) in solution is probably present as a mixture of several ionic species.

Usually ceric hydroxide, $Ce(OH)_4$, or ammonium hexanitratocerate(IV), $(NH_4)_2Ce(NO_3)_6$, is used in preparing cerium(IV) solutions. To prevent the formation of insoluble cerium salts by hydrolysis, the salt is thoroughly mixed with concentrated acid and then diluted with water, which at first is added in small increments.

Cerium(IV) solutions can be standardized by titration of arsenic(III), which is obtained from primary standard arsenious oxide, As_2O_3. Arsenious oxide is sparingly soluble in acids, but it dissolves readily in a few milliliters of water to which one or two pellets of sodium hydroxide have been added:

$$As_2O_3 + 2NaOH + H_2O \rightarrow 2Na^+ + 2H_2AsO_3^-$$

The solution is acidified with sulfuric acid and forms soluble arsenious acid

(H_3AsO_3), which can be titrated. However, the titration with cerium(IV) succeeds only in the presence of a catalyst such as osmic acid:

$$2Ce^{4+} + H_3AsO_3 + H_2O \xrightarrow{\text{osmic acid}} 2Ce^{3+} + H_3AsO_4 + 2H^+$$

Ferroin is used as indicator in this titration. In this titration, each arsenic atom is oxidized from $3+$ to $5+$; however, since there are two arsenic atoms in the substance weighed, As_2O_3, the equivalent weight of the primary standard is the molecular weight of As_2O_3 divided by 4.

Pure iron metal may also be used as a primary standard for cerium(IV). The iron is dissolved, reduced to iron(II), and titrated with cerium(IV) with ferroin as indicator.

Potassium Dichromate. Dichromate is reduced to chromium(III) by a reducing agent through a series of reactions summarized by the half-reaction:

$$Cr_2O_7{}^{2-} + 14H^+ + 6e^- = 2Cr^{3+} + 7H_2O \qquad E^0 = 1.36 \text{ V}$$

The Nernst equation is not applicable to quantitative calculations involving potential in this case. Although the E^0 is given as 1.36 V, the practical formal potential in $1M$ hydrochloric acid is only 1.09 V. Standard solutions can be prepared by accurately weighing out primary standard $K_2Cr_2O_7$ and diluting to a known volume.

Dichromate is a somewhat weaker oxidizing agent than permanganate or cerium(IV). However, potassium dichromate is widely used to titrate ferrous iron and certain other reducing agents.

Determination of Iron. The determination of iron by oxidation-reduction titration is an important analytical method. Furthermore, the method illustrates several techniques used in other redox determinations.

Iron ore or steel samples are usually dissolved in hydrochloric acid. Usually, part (or all) of the iron is oxidized to iron(III) during the dissolution step. The iron can be reduced to iron(II) and then titrated to iron(III) in the presence of dilute hydrochloric acid, provided that dichromate or cerium(IV) is used as titrant.

Two methods of reducing iron from iron(III) to iron(II) in hydrochloric acid are widely used. In one procedure, iron is reduced by the dropwise addition of stannous chloride. Complete reduction is marked by the disappearance of the yellow color of iron(III):

$$2Fe^{3+} + Sn^{2+} \rightarrow 2Fe^{2+} + Sn^{4+}$$

The excess tin(II) is removed by rapidly adding a large excess of mercuric chloride. The mercury(II) is reduced to mercury(I), which forms a white precipitate of mercurous chloride:

$$Sn^{2+} + 2HgCl_2(\text{excess}) \rightarrow Sn^{4+} + 2HgCl(s) + 2Cl^-$$

If not enough mercuric chloride is added, or if it is not added rapidly enough, some of the mercury(II) is reduced to mercury metal (gray precipitate), which interferes in the subsequent titration of iron(II):

$$Sn^{2+} + HgCl_2 \rightarrow Sn^{4+} + Hg(s) + 2Cl^-$$

This reaction can be avoided by cooling the solution below 5°C before adding $HgCl_2$.

The other way of reducing iron(III) to iron(II) in hydrochloric acid solution is with a silver reductor. After reduction, the iron(II) is titrated with either a cerium(IV) solution or a dichromate solution:

$$Ce^{4+} + Fe^{2+} \rightarrow Fe^{3+} + Ce^{3+}$$

$$Cr_2O_7{}^{2-} + 6Fe^{2+} + 14H^+ \rightarrow 6Fe^{3+} + 2Cr^{3+} + 7H_2O$$

In the titration with cerium(IV), the ferroin indicator gives a sharp color change from red to pale blue at the end point. In the titration with dichromate, it changes at too high a potential for it to be used as indicator. The only other readily available indicator, diphenylaminesulfonate, changes color at too low a potential! (See Figure 13–1.) This dilemma is resolved by using diphenylaminesulfonate indicator and carrying out the titration in the presence of phosphoric acid. Phosphoric acid complexes iron(III) but not iron(II); thus, the potential of the ferric-ferrous couple is

Figure 13–1. Effect of phosphoric acid on the dichromate titration of iron(II) in $1M$ sulfuric acid.

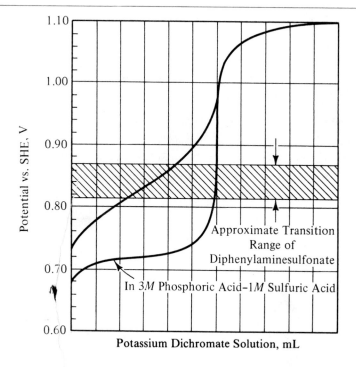

Potential vs. SHE. V

Approximate Transition Range of Diphenylaminesulfonate

In 3M Phosphoric Acid–1M Sulfuric Acid

Potassium Dichromate Solution, mL

lowered:

$$E = 0.77 + 0.059 \log \frac{[\text{Fe}^{3+}]}{[\text{Fe}^{2+}]} \quad \nwarrow \quad \begin{pmatrix} \text{concentration lowered} \\ \text{by complexation} \end{pmatrix}$$

The lower titration curve in Figure 13–1 shows that the diphenylaminesulfonate indicator changes color at the proper point in the presence of phosphoric acid.

Titration of iron(II) with permanganate has long been a standard procedure:

$$\text{MnO}_4^- + 5\text{Fe}^{2+} + 8\text{H}^+ \rightarrow 5\text{Fe}^{3+} + \text{Mn}^{2+} + 4\text{H}_2\text{O}$$

The end point in this titration is simply the magenta color produced by the first drop of excess permanganate. Titration in the presence of hydrochloric acid results in the following reaction, which consumes permanganate:

$$2\text{MnO}_4^- + 10\text{HCl} + 6\text{H}^+ \rightarrow 5\text{Cl}_2 + 2\text{Mn}^{2+} + 8\text{H}_2\text{O}$$

This unwanted side reaction is avoided by adding Zimmerman–Reinhardt reagent (MnSO_4, H_2SO_4, H_3PO_4) before titration with permanganate.

Alternatively, after a metal or ore sample has been dissolved in hydrochloric acid, sulfuric or perchloric acid is added and the HCl volatilized by evaporating to fumes of the higher-boiling acid. After being cooled, the solution is diluted and passed through a Jones reductor, which reduces the iron(III) to iron(II). The redox reaction in the column is

$$2\text{Fe}^{3+} + \text{Zn}(\text{Hg})_x \rightarrow 2\text{Fe}^{2+} + \text{Zn}^{2+}$$

The zinc ions that go into the solution have no effect on the titration of iron(II) with permanganate. The titration can also be carried out with cerium(IV) or with dichromate, as previously described.

Other Determinations.   Many elements that have more than one oxidation state in solution can be determined by oxidation-reduction titration. After oxidation to their highest oxidation state, cerium(IV), chromium(VI), and vanadium(V) can be titrated with a standard iron(II) solution (see Section 13–2). The reduced forms of several elements that can be determined by a direct titration are listed in Table 13–2.

Table 13–2.   Summary of Some Oxidation-Reduction Methods

Substance	Reduced Form	Oxidized Form	Titrant
As	As(III)	As(V)	Ce(IV) ($+$ osmic acid)
Cu	Cu(I)	Cu(II)	Ce(IV)
Fe	Fe(II)	Fe(III)	Ce(IV), MnO_4^-, or $\text{Cr}_2\text{O}_7^{2-}$
H_2O_2	H_2O_2	O_2	Ce(IV) or MnO_4^-
Mo	Mo(V)	Mo(VI)	Ce(IV) or MnO_4^-
Ti	Ti(III)	Ti(IV)	Fe(III)
U	U(IV)	U(VI)	MnO_4^- or Ce(IV)
V	V(IV)	V(V)	Ce(IV)

13–4. DIRECT TITRATIONS WITH IODINE

Elemental iodine, I_2, is a solid that is soluble in a number of organic solvents but is insoluble in water. However, iodine is extremely soluble in an aqueous solution containing sodium or potassium iodide. The iodine reacts with the iodide ion and forms a soluble anionic complex, which is deep red-brown in color:

$$I_2 + I^- \rightarrow I_3^-$$

A standard solution can be prepared by dissolving a carefully weighed portion of elemental iodine in an aqueous iodide solution. For convenience in writing and balancing chemical equations, we shall refer to iodine dissolved in iodide solution simply as I_2.

The standard reduction potential indicates that iodine is a fairly weak oxidizing agent.

$$I_2 + 2e^- = 2I^- \qquad E^0 = +0.535 \text{ V}$$

Nevertheless, several substances are conveniently determined by titration with a standard iodine solution. Actually, the mild oxidizing action of iodine is often an advantage for titrations where a more powerful titrant might cause nonstoichiometric oxidations to occur.

The titration reactions for several substances that can be determined by direct titration with iodine are summarized as follows:

$$I_2 + As(III) \rightarrow As(V) + 2I^-$$
$$I_2 + Sb(III) \text{ tartrate} \rightarrow Sb(V) \text{ tartrate} + 2I^-$$
$$I_2 + Sn(II) \rightarrow Sn(IV) + 2I^-$$
$$I_2 + H_2S \rightarrow S + 2I^-$$
$$I_2 + SO_3^{2-} + H_2O \rightarrow SO_4^{2-} + 2H^+ + 2I^-$$
$$I_2 + 2S_2O_3^{2-} \rightarrow S_4O_6^{2-} + 2I^-$$
$$I_2 + 2Fe(CN)_6^{4-} \rightarrow 2Fe(CN)_6^{3-} + 2I^-$$

The indicator for these titrations is a starch suspension. The first trace of excess iodine marks the end point of the titration by forming an intense blue complex with the starch.

Direct iodine titrations are usually carried out in neutral or acidic solution. At pH values higher than 11, the stoichiometry of iodine titrations is erratic and uncertain because iodine oxidizes to hypoiodite (IO^-), which is unstable and decomposes to iodate (IO_3^-) plus iodide. The pH is fairly critical in some direct iodine titrations. For example, oxidizing arsenic(III) (arsenite), to arsenic(V) (arsenate) involves hydrogen ions:

$$I_2 + H_3AsO_3 + H_2O \rightarrow HAsO_4^{2-} + 4H^+ + 2I^-$$

In slightly alkaline solution, the titration of arsenite with iodine proceeds smoothly,

as indicated by the equation. However, in strongly acidic solution, the reaction goes
in the reverse direction:

$$H_3AsO_4 + 2I^- + 2H^+ \rightarrow H_3AsO_3 + I_2 + H_2O$$

This reaction, followed by back-titration of the iodine with thiosulfate, is used to
determine arsenate.

The effect of acidity on the iodine-arsenite reaction can be predicted from
oxidation-reduction theory; the theoretical reduction potential of the arsenic(V)-
arsenic(III) couple as a function of pH is shown in Figure 13–2. Since $n = 2$ for the
arsenic(III)-iodine reaction, the minimum difference in potentials required for a
quantitative reaction is $+0.18$ V (Table 12–2, p. 225). This means that the potential
for the arsenic(V)-arsenic(III) couple needs to be $+0.36$ V—0.18 V less than the
iodine-iodide E^0 of $+0.54$ V. The pH at which the arsenic(V)-arsenic(III) couple is
$+0.36$ V is about 3 (Figure 13–2). Thus, a pH of 3 is theoretically the lowest pH at
which 99.9% of arsenic(III) can be oxidized to arsenic(V). However, the rate at
which equilibrium is attained is so slow that a pH of 5 or greater is used for rapid
reaction.

Oxidation of Vitamin C.   Iodine can also be used to titrate vitamin C (ascorbic
acid) directly:

$$I_2 + C_4H_6O_4(OH)C{=}C(OH) \rightarrow 2I^- + 2H^+ + C_4H_6O_4C({=}O){-}C{=}O$$

Figure 13–2.   Potential pH diagram for arsenic(V)–arsenic(III).

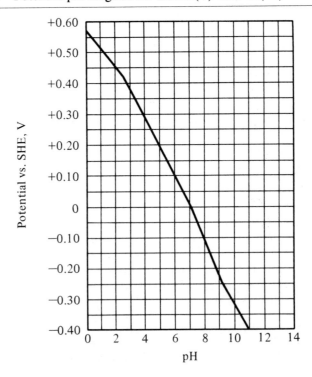

Note that the iodine oxidizes the (OH)C=C(OH) functional group to an alpha-diketone group in the dehydroascorbic acid product. The vitamin C content of tablets and solids such as Tang can be found by using an iodine titrant and a starch end point [1].

Karl Fischer Method for Water.   A quantitative determination of water is frequently needed, particularly in organic materials. The Karl Fischer method, being almost specific for water, is very famous; an entire book has been devoted largely to its analytical applications [2].

The Karl Fischer method is a titration of water with an anhydrous methyl alcohol solution containing iodine, sulfur dioxide, and excess pyridine. The titration is based on a reaction between iodine and sulfur dioxide that will only occur if water is present:

$$I_2 \cdot Pyr + SO_2 \cdot Pyr + H_2O \xrightarrow{Pyr} SO_3 \cdot Pyr + 2PyrH^+I^-$$

where *Pyr* represents pyridine. The product, $SO_3 \cdot Pyr$, reacts further with the methyl alcohol to form the methane-sulfate anion:

$$SO_3 \cdot Pyr + CH_3OH \rightarrow PyrH^+CH_3SO_4^-$$

Note from these equations that each mole of water requires one mole of I_2. The sample is titrated with the Karl Fischer reagent until a permanent iodine color is observed. Because of other reaction products, the color change is usually from a yellow to a brownish color that may be difficult to detect visually. A much sharper end point can be obtained using the electrometric "dead stop" end point described on p. 302.

The Karl Fischer reagent is not very stable and must be standardized frequently. Sometimes it is prepared just before use by mixing solutions of iodine in methyl alcohol with sulfur dioxide in pyridine. A proprietary commercial reagent available as a single solution is more stable. In any water determination, all glassware and solvents must be scrupulously dried and protected from atmospheric moisture.

13-5.   INDIRECT TITRATIONS INVOLVING IODINE

General Scheme.   An important general method for determining oxidizing chemicals involves iodine indirectly. An excess of sodium or potassium iodide solution is added to the sample. The oxidizing chemical is reduced by the iodide, resulting in the formation of an equivalent amount of iodine, which is titrated with standard

[1] J. W. Stevens, *Ind. Eng. Chem., Anal. Ed. 10*, 269 (1938); D. N. Bailey, *J. Chem. Educ. 51*, 488 (1974).
[2] J. Mitchell and D. M. Smith, *Aquametry*, New York: Interscience, 1948.

sodium thiosulfate. The general scheme is summarized as follows:

$$A_{ox} + I^- \text{ (excess)} \rightarrow A_{red} + I_2$$

$$2S_2O_3{}^{2-} + I_2 \rightarrow S_4O_6{}^{2-} + 2I^-$$

The iodine formed in the first reaction is equivalent to the amount of oxidizing agent, A_{ox}, in the sample. In this reaction, the iodide (added as potassium iodide or sodium iodide) is in large excess and is not a standard solution. The iodine formed is then titrated with standard sodium thiosulfate, and iodide and tetrathionate ions are formed as the products. Starch is the indicator in this titration. The disappearance of the blue starch-iodine complex marks the end point. It is important not to add the starch indicator until most of the iodine has been titrated. If the starch is added too soon, iodine is adsorbed onto the starch, making the end point very slow and hard to detect.

Reactions for a few of the specific substances that can be determined by indirect iodine titration follow:

$$Cr_2O_7{}^{2-} + 6I^- + 14H^+ \rightarrow 2Cr^{3+} + 3I_2 + 7H_2O$$

$$Cl_2 + 2I^- \rightarrow 2Cl^- + I_2$$

$$Br_2 + 2I^- \rightarrow 2Br^- + I_2$$

$$ClO^- + 2I^- + 2H^+ \rightarrow Cl^- + I_2 + H_2O$$

$$IO_3{}^- + 5I^- + 6H^+ \rightarrow 3I_2 + 3H_2O$$

$$IO_4{}^- + 7I^- + 8H^+ \rightarrow 4I_2 + 4H_2O$$

$$2Cu^{2+} + 4I^- \rightarrow 2CuI(s) + I_2$$

In each case, the iodine formed is titrated with standard thiosulfate, starch being the indicator:

$$I_2 + 2S_2O_3{}^{2-} \rightarrow 2I^- + S_4O_6{}^{2-}$$

Chlorine and bromine are much stronger oxidizing agents than iodine and do not oxidize thiosulfate stoichiometrically. The indirect iodine method, however, works very well for determining either chlorine or bromine:

$$Br_2 + 2I^- \rightarrow 2Br^- + I_2$$

$$I_2 + 2S_2O_3{}^{2-} \rightarrow 2I^- + S_4O_6{}^{2-}$$

Bromine is widely used to determine organic unsaturated compounds and phenols. Carbon-carbon double bonds add bromine quantitatively:

Phenols react with bromine at the *ortho* and *para* positions:

In both of these determinations, standard bromide is added in excess, and the excess is determined by the indirect iodine method.

Periodate is a very selective oxidizing agent for organic compounds. Alpha glycols (compounds with hydroxy groups on adjacent carbon atoms) are oxidized quantitatively without interference from alcohols or other types of glycol. Ethylene glycol, which is used as an antifreeze, is oxidized to two molecules of formaldehyde.

$$\begin{matrix} CH_2OH \\ | \\ CH_2OH \end{matrix} + HIO_4 \;\rightarrow\; 2HC\!\!\overset{O}{\diagup}\!\!H + HIO_3 + H_2O$$

Because organic oxidations require several minutes for quantitative reaction, an excess of periodate is added, and the excess is determined in acidic solution by the indirect iodine method:

$$HIO_4 + 7I^- + 7H^+ \;\rightarrow\; 4I_2 + 4H_2O$$

The reduction product of periodate in the glycol oxidation is iodate. This complicates matters because in acidic solution iodate also reacts with iodide to form iodine:

$$HIO_3 + 5I^- + 6H^+ \;\rightarrow\; 3I_2 + 3H_2O$$

At pH 7.5, *only periodate* is reduced by the iodide or arsenic(III) (to iodate). This fact is used in the laboratory procedure discussion in Chapter 30, Experiment 18.

The total iodine produced is titrated by thiosulfate. The amount of glycol is calculated from the *difference* between the milliequivalents of periodate taken and the milliequivalents of thiosulfate needed for the final titration.

An indirect iodine procedure is widely used in determining copper quantitatively. Copper(II) reacts with excess iodide to form a precipitate of copper(I) iodide plus free iodine. The iodine is titrated with thiosulfate:

$$2Cu^{2+} + 4I^- \;\rightarrow\; 2CuI(s) + I_2$$

$$I_2 + 2S_2O_3^{2-} \;\rightarrow\; 2I^- + S_4O_6^{2-}$$

Iron(III) oxidizes iodide slowly and thus interferes with the procedure. However, small or moderate concentrations of iron(III) cause no difficulty if complexed with fluoride:

$$Fe^{3+} + F^- \;\rightarrow\; FeF^{2+} \qquad \text{(plus other iron(III)-fluoride complexes)}$$

The "oxygen error" is frequently a problem in analyses done by the indirect iodine method. In acid solution, oxygen from the air oxidizes iodide to iodine:

$$O_2 + 4I^- + 4H^+ \rightarrow 2I_2 + 2H_2O$$

This reaction causes high results in determinations made by the indirect iodine method. The magnitude of the error increases with increasing acidity.

The oxygen error is avoided by working in an inert atmosphere. Adding solid carbon dioxide or sodium bicarbonate to an acid solution produces a protective blanket of carbon dioxide. It is also helpful to avoid excess acidity.

QUESTIONS AND PROBLEMS

Oxidation States and Equation Balancing

1. For each of the following compounds or ions, give the oxidation state of the element indicated.

   (a) S in $K_2S_4O_6$
   (b) I in H_5IO_6
   (c) Mn in MnO_4^{2-}
   (d) Co in $CoCl_4^{2-}$
   (e) W in $WO_2F_4^{2-}$
   (f) C in $H_2C_2O_4$

2. Balance each of the following equations.

   (a) $HIO_4 + H^+ + I^- \rightarrow I_2 + H_2O$
   (b) $Mn(OH)_2 + O_2 + H_2O \rightarrow Mn(OH)_3$
   (c) $Sb_2S_3 + HNO_3 \rightarrow SbO_4^{3-} + SO_4^{2-} + NO_2 + H_2O + H^+$

3. Balance each of the following equations. Indicate the equivalent weight of the underlined substances in terms of formula or atomic weight.

   (a) $\underline{KMnO_4} + Sn^{2+} + H^+ \rightarrow Sn^{4+} + Mn^{2+} + H_2O + K^+$
   (b) $\underline{H_3AsO_3} + I_2 + H_2O \rightarrow HAsO_4^{2-} + I^- + H^+$
   (c) $\underline{K_2Cr_2O_7} + Fe(CN)_6^{4-} + H^+ \rightarrow Cr^3 + Fe(CN)_6^{3-} + H_2O + K^+$
   (d) $\underline{Mn^{2+}} + S_2O_8^{2-} + H_2O \rightarrow MnO_4^- + SO_4^{2-} + H^+$
   (e) $\underline{KBrO_3} + \underline{NaBr} + H^+ \rightarrow Br_2 + H_2O$

4. Write balanced equations for the reactions involved in (a) standardizing cerium(IV) against pure As_2O_3 as primary standard and (b) standardizing $KMnO_4$ against primary standard iron metal.

5. Balance each of the following equations involving organic compounds.

   (a) $HIO_4 + CH_2(OH)CH_2OH \rightarrow HIO_3 + H_2CO + H_2O$
   (b) $HIO_4 + CH_2(OH)CH(OH)CH_2OH \rightarrow HIO_3 + H_2CO + HCO_2H + H_2O$
   (c) $Ce^{4+} + CH_3CHO + H_2O \rightarrow Ce^{3+} + CH_3CO_2H + H^+$
   (d) $Ce^{4+} + CH_3CH(OH)CH_3 \rightarrow Ce^{3+} + CH_3COCH_3$

Quantitative Calculations: Permanganate, Cerium, and Dichromate

6. A solution in the laboratory bears the label "0.0102N potassium permanganate." Does this label unequivocally define the concentration of the solution? Explain.

7. A solution of potassium permanganate is standardized with pure iron metal as primary

standard. The iron is dissolved in acid, reduced to Fe^{2+}, and titrated with permanganate:

$$MnO_4^- + 5Fe^{2+} + 8H^+ \rightarrow 5Fe^{3+} + Mn^{2+} + 4H_2O$$

If 33.00 mL of permanganate is needed to titrate a 0.5585-g sample of iron, what is the normality of the permanganate?

8. The iron in a 1.0000-g rock sample is dissolved, reduced to Fe^{2+}, and titrated with dichromate. If the titration requires 12.40 mL of $0.0500N$ dichromate, what is the percentage of Fe_2O_3 in the sample?

9. What is the weight of an iron ore sample containing 55.85% Fe if 30.0 mL of $0.1000N$ $K_2Cr_2O_7$ is required for titration?

10. A 1.0000-g sample of limonite iron ore ($2Fe_2O_3 \cdot 3H_2O$) is dissolved, reduced to Fe^{2+}, and titrated with 20.00 mL of $0.20000N$ cerium(IV). Calculate the percentage of $2Fe_2O_3 \cdot 3H_2O$ (form wt, 373.38) in the sample.

11. A cerium(IV) solution is standardized with As_2O_3 as primary standard according to the following reactions:

$$As_2O_3 + 6NaOH \rightarrow 6Na^+ + 2AsO_3^{3-} + 3H_2O$$
$$AsO_3^{3-} + 3H^+ \rightarrow H_3AsO_3$$
$$2Ce^{4+} + H_3AsO_3 + H_2O \rightarrow H_3AsO_4 + 2Ce^{3+} + 2H^+$$

A 0.1980-g sample of As_2O_3, when dissolved, required 20.00 mL of cerium(IV) for titration. Calculate the normality of the cerium(IV) solution.

12. Chemical oxygen demand (COD) is the oxygen equivalent of the chemically oxidizable organic matter in water. It is an important parameter for studies on rivers and industrial wastes and the control of waste treatment plants. COD is determined by heating the water sample with standard dichromate and sulfuric acid and then titrating the remaining dichromate with a standard iron(II) solution. A 20.00-mL water sample required 36.40 mL of $0.0250M$ iron(II) for titration of the chromium (VI) remaining after the oxidation. A distilled water blank with exactly the same amount of dichromate as was initially added to the sample required 45.17 mL of $0.0250M$ iron(II) for titration. Calculate the COD in mg/L of the sample. (*Note:* The total milliequivalents of oxygen demand is the same as the milliequivalents of dichromate used in the oxidation. The equivalent weight of oxygen is 8.000.)

13. The lead in a 0.2000-g sample is precipitated as $PbCrO_4$. The precipitate is filtered, washed, and dissolved in acid, giving a solution containing dichromate ions and lead ions. The dichromate requires 15.50 mL of $0.1N$ ferrous sulfate for titration. Calculate the percentage of lead in the sample.

Quantitative Calculations: Iodine Methods

14. A 1.2500-g sample of pure As_2O_3 is weighed into a 250-mL volumetric flask, dissolved, and diluted to volume. A 25.00-mL aliquot of this solution requires exactly 26.00 mL of iodine solution for titration to the starch end point. Calculate the percentage of As in a 1.0000-g ore sample that requires 15.00 mL of this same iodine solution for titration.

15. In the classic Winkler method for measuring dissolved oxygen in water, sodium hydroxide is added to the sample containing Mn^{2+} to form a precipitate of $Mn(OH)_2$, which will be partially oxidized to $Mn(OH)_3$ by the dissolved oxygen. Sodium iodide is also added so that when sulfuric acid is later added to dissolve the manganese hydroxide precipitates, the manganese(III) will oxidize iodide to iodine:

$$2Mn(III) + 2I^- \rightarrow 2Mn(II) + I_2$$

If a 25-mL water sample, treated as described above, requires 11.70 mL of $0.1008M$ sodium thiosulfate for the titration, calculate the oxygen concentration of the sample in mg/mL.

16. Potassium iodate is reduced with excess iodide, and the resulting iodine is titrated with standard sodium thiosulfate. A 25-mL aliquot of potassium iodate solution requires 28.60 mL of $0.1000N$ sodium thiosulfate for titration. Calculate the concentration of the solution in milligrams of KIO_3 per milliliter of solution.

17. A 1.0000-g sample containing As_2O_3, As_2O_5, and an inert salt is dissolved and titrated in neutral solution to the starch end point with 20.00 mL of $0.2000N$ iodine. The resulting solution is made strongly acid, and excess KI is added, releasing iodine. The iodine is titrated with 40.00 mL of $0.1500N$ sodium thiosulfate. Calculate the percentage of As_2O_3 and the percentage of As_2O_5 (form wt = 229.84) in the sample.

18. A 0.3100-g sample of liquid ethylene glycol (form wt = 62) is treated with 50.00 mL of $0.1200M$ HIO_4. Back-titration of unreacted HIO_4 to HIO_3 (Chapter 30, Experiment 20) indicates that 2.00 meq of HIO_4 remain. Calculate the percent purity of the ethylene glycol.

19. A 0.4700-g sample of impure phenol (form wt = 94.1) is treated with 40.00 mL of $0.3000N$ bromine (Br_2). Iodometric titration of the unreacted bromine indicates that 3.00 meq of bromine ($Br^0 \rightarrow Br^-$) remain. Calculate the percent purity of the phenol.

20. A hair-waving preparation contains ethyl thioglycolate, which is quantitatively oxidized by iodine as follows.

$$2HSCH_2CO_2C_2H_5 + I_2 \rightarrow [SCH_2CO_2C_2H_5]_2 + 2H^+ + 2I^-$$

A 2.560-g sample, when diluted, requires 18.15 mL of $0.0518M$ I_2 for titration. Calculate the weight percentage of the ethyl thioglycolate (form wt = 120.0) in the preparation.

Prior Oxidation and Reduction

21. Catalysis of peroxydisulfate oxidations by a low concentration of Ag^+ involves oxidation to Ag^{2+}, yet the E^0 for the reduction of $S_2O_8^{2-}$ to SO_4^{2-} is 2.01 V, while that for the reduction of Ag^{2+} to Ag^+ is almost the same, 1.98 V. Explain.

22. In the determination of chromium: (a) Write a balanced chemical equation for the oxidation of Cr^{3+} to $Cr_2O_7^{2-}$ by $S_2O_8^{2-}$. (b) What catalyst is necessary for this oxidation? (c) How is excess $S_2O_8^{2-}$ decomposed after the chromium oxidation is complete? (d) Write a balanced chemical equation for titration of the $Cr_2O_7^{2-}$.

23. In reducing iron(III) to iron(II) in a Jones reductor column containing zinc with zinc amalgam on the particle surfaces, a student finds that a great many gas bubbles form, which make it impossible to pass the sample solution through the column. What should be done to prevent this excessive "gassing"?

24. A student, disgusted with gas formation in a Jones reductor, decides to reduce iron(III) to iron(II) by pouring some amalgamated zinc into each sample flask. The student proved that the procedure was correct by observing a green solution of iron(II) that formed when a fairly concentrated solution of iron(III) was reduced in this manner. When reduction was complete, the student then attempted to titrate the iron(II) with permanganate. What is wrong with this procedure?

Determination of Iron

25. Suppose that no diphenylaminesulfonate indicator were available for the titration of iron(II) with potassium dichromate. Suggest a way to detect the end point under these conditions.

26. Explain why phosphoric acid must be added in titrating Fe^{2+} with diphenylaminesulfonate as indicator.

27. What difficulty would be encountered if the stannous chloride reduction method for iron(III) were combined with a permanganate titration of the iron(II)?

28. It can easily be calculated from Beer's law that a $10^{-5}M$ solution of $KMnO_4$ ($\epsilon = 10^3$) will barely give a detectable scale reading ($A = 0.001$) in a 1-cm spectrophotometer cell. Yet a flask containing such a solution will be perceptibly violet to the eye at the end point of an iron(II) titration. Explain.

Direct and Indirect Iodine Methods

29. Explain why starch indicator must be added just before the end point in an indirect iodine titration with thiosulfate, but can be added at the beginning of a direct titration with iodine.

30. What is the "oxygen error" in an indirect iodine titration? How can this error be minimized or avoided?

31. Air can be pumped through a small chemical scrubber at a known flow rate (m^3/hr) and the SO_2 in the air quantitatively absorbed into the aqueous scrubber solution as sulfite (SO_3^{2-}). Outline an analytical procedure for determining the concentration of SO_2 in atmospheric air.

32. An organic chemist wishes to determine bromine in acetic acid solution. Explain how the bromine content could be determined by an indirect iodine method after diluting the solution with water. Explain what error could occur if the acetic acid were not buffered to reduce the acidity of the aqueous solution.

Methods of Analysis

33. A sample solution contains both titanium(IV) and iron(III) in dilute sulfuric acid. Outline a quantitative method for determining both elements.

34. The manganese in a sample containing fluoride was oxidized to a purple solution. Initially, this was believed to be permanganate, but a glance at a reference book indicated the possibility of a similarly colored manganese(III) fluoride complex. Outline a chemical method to tell which of these possibilities is correct.

35. A sample solution contains both vanadium(V) and vanadium(IV). Outline a quantitative method for determining how much of each is present.

36. Outline a procedure for determining the amount of chromium in chrome-tanned leather. Organic matter, such as leather, can be decomposed by heating with conc. nitric and perchloric acids.

37. In the presence of manganese(III) catalyst, the following reaction is fast and stoichiometric. However, the color of the catalyst prevents the use of a visual indicator. Suggest at least two methods for detecting the end point, using cerium(IV) as the titrant. (Both thallium(I) and thallium(III) are colorless.)

$$2Ce^{4+} + Tl^+ \rightarrow 2Ce^{3+} + Tl^{3+}$$

38. Argentic oxide dissolves in aqueous nitric acid to give a solution containing silver(II). The silver(II) is a powerful oxidant (see Appendix 3) that decomposes in acidic solution; however, complete decomposition may require one-half hour (or longer) at room temperature. Outline an oxidation-reduction method for determining cerium(III) with argentic oxide.

39. The sum of *meta*-cresol and *para*-cresol in a mixture can be determined by an acid-base

titration in a nonaqueous solvent. Suggest a way to determine quantitatively the amount of each of the two isomeric cresols present in the mixture.

meta-cresol   *para*-cresol

40. A battery contains both AgO and Ag_2O. Outline a method for determining the AgO content. Both oxides are readily soluble in dilute acid, but the Ag^{2+} from the AgO is unstable and is reduced by water to Ag^+ within a few minutes.
41. Benzoyl peroxide, $C_6H_5CO_3H$, is used in polymerization reactions such as the making of synthetic rubber. Outline a method for quantitative data of benzoyl peroxide in solution, keeping in mind that it can be readily reduced to benzoic acid, $C_6H_5CO_2H$.
42. Barium(II) is precipitated quantitatively and selectively as $Ba(IO_3)_2$. The precipitate will dissolve in acidic aqueous solution. Outline a quantitative titrimetric method for barium(II) based on first precipitating barium iodate.
43. Chlorinated water is said to contain hypochlorous acid, HOCl, rather than elemental clorine. What is the oxidation state of Cl in HOCl? Outline a method for determining the concentration of HOCl in a heavily chlorinated water.

14

The Use of
Reaction Rates

For the most part, quantitative analyses utilize reactions that are quite fast. For example, a sample of any acid will react with any strongly basic titrant just as fast as these two reactants can be mixed together. Similarly, many, but not all, oxidation-reduction and complex formation titrations are rapid.

In contrast, a great many reactions of organic compounds occur more slowly. These may require anywhere from one second or less to several hours for quantitative reaction. Some inorganic reactions are slow and require comparable times for reaction. In polarography (Chapter 15), the rate of a reaction at an electrode surface is *diffusion-controlled*; that is, the reaction rate depends on the rate at which reactive ions diffuse from the bulk solution to the electrode surface. Such diffusion may be very slow.

It is important, then, to become acquainted with reaction rate theory. Fairly simple rate theory has been worked out that applies to slow chemical reactions and to certain other analytical processes that do not occur instantaneously.

With the help of appropriate constants, called *rate constants*, the analyst can calculate the time required for a slow reaction to react quantitatively. He or she can also decide whether a substance can be determined in the presence of a second, slower-reacting, substance. Examples of analytical methods based on measurement or use of known reaction rates will also be discussed to illustrate what analysts have already done.

14–1. KINETIC THEORY: RATE CONSTANTS

In determining some sample constituents in the presence of an interference, the analyst normally depends on a separation or on masking the interference with some reagent that displays a favorable *difference in equilibrium constant* between the two substances. Analytical chemists can also utilize favorable *kinetic differences*, or differences in reaction rates, to accomplish the same thing [1].

The Order of a Reaction.   To comprehend how reaction rates may be used to analyze for a sample constituent, it is necessary to understand the effect of concentration on reaction rates. In the language of kinetics, it is the *order* of a reaction that controls the rate of a reaction. This is *not* the same as the stoichiometry of the reaction. The order of a reaction refers to the experimentally determined *rate law* of a reaction, whereas the stoichiometry refers to the numerical relation of molecules in the overall reaction, regardless of rate. The *individual order* for a reaction species refers to the exponent of the concentration term for that species in the rate law. The *reaction order*, or *overall order*, refers to the sum of all the individual orders that appear in the rate law.

To illustrate this, consider the general rate law below for the reaction of A and B:

$$\text{rate} = k[A]^a[B]^b$$

If experimental measurements demonstrate that $a = 0$ and $b = 1$, then the individual order of A is zero and that for B is one, making this an overall first-order reaction. If, however, the measurements show that $a = 1$ and $b = 1$, then the individual order for A is one and that for B is one, making this a second-order reaction.

It should again be stressed that the stoichiometry of a reaction is not necessarily the same as the order. For instance, the reduction of iron(III) by tin(II),

$$2Fe(III) + Sn(II) \rightleftharpoons 2Fe(II) + Sn(IV)$$

involves *three* reactant molecules stoichiometrically. However, the reaction is *second*-order kinetically because the rate (in chloride media) is proportional only to the first power of the iron(III) and tin(II) concentrations. The reason is that the slow, rate-determining step in this reaction involves only one iron(III) and one tin(II) (see p. 233.

First-Order Reactions.   Consider the case of substance A reacting to give product P_A:

$$A \rightarrow P_A$$

When the rate of reaction is directly dependent on [A], and not on $[A]^2$, etc., the rate law or rate expression can be written as

$$\text{rate} = k[A] \tag{14-1}$$

Mathematically, the rate can be expressed by the disappearance of A with respect to

[1] G. A. Rechnitz, *Anal. Chem. 36*, 453R (1964).

Figure 14–1. Rate of a first-order reaction.

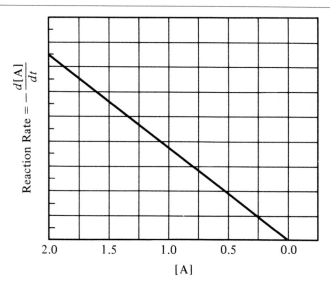

time and can thus be written in differential form as

$$\frac{-d[A]}{dt} = k[A] \tag{14-2}$$

This rate expression is said to exhibit a first-order dependence on A, since the [A] term is raised to the first power. The symbol k represents the specific reaction-rate constant, and has the dimensions of time^{-1} (sec^{-1} or min^{-1}).

Equation 14–2 states that the reaction rate at any instant depends only on the value of k and on the concentration of A. Since the concentration of A becomes smaller as the reaction proceeds, the reaction rate also decreases with time. This is shown in Figure 14–1.

Instead of determining the reaction rate at various times, it is often more useful to measure how much A has reacted or how much of A remains after a given reaction time. This can be calculated with the help of the *integrated* form of the differential rate equation. Integrating Equation 14–2 between the limits $t = 0$ to $t = t$, and $A = A_0$ (concentration of A at zero time) and $A = A$ (instantaneous concentration) gives:

$$\ln \frac{[A_0]}{[A]} = kt \tag{14-3}$$

where ln = the natural logarithm ($= 2.303 \log_{10}$).

Equation 14–3 can be rearranged to

$$\ln [A] = -kt + \ln [A_0] \tag{14-4}$$

Thus, a plot of ln [A] versus t yields a straight line with the slope $= -k$ and the intercept $(t = 0) = \ln [A_0]$, as shown in Figure 14–1.

By substituting $2.303 \log = \ln$ into Equation 14–3, a similar equation can be obtained:

$$2.303 \log \frac{[A_0]}{[A]} = kt \qquad (14\text{–}5)$$

A number of chemical reactions exhibit a first-order reaction rate. The reaction of many metal ions with an excess of an anion such as thiocyanate (SCN^-) exhibits such first-order rates; for example;

$$Cr^{3+} + SCN^-(xs) \rightarrow CrSCN^{2+}$$

In organic chemistry, the rate of hydrolysis of organic halides (in water or water-alcohol solvent) is another well-known example:

$$R_3C\text{–}X + H_2O \rightarrow R_3C\text{–}OH + H^+X^-$$
$$\text{(halide)} \qquad\qquad \text{(alcohol)} \qquad \text{(strong acid)}$$

In nuclear chemistry, simple radioactive decay of radioactive isotopes depends only on the amount of the isotope and is, therefore, first-order. Diffusion of a gas into a vacuum and diffusion of an electroactive species to an electrode surface are other examples of first-order rate processes.

Second-Order Reactions.   Consider the case of a substance A reacting with a reagent R to give one or more products, P.

$$A + R \rightarrow P$$

If the individual orders of A and R are both one, then the overall order of the reaction is second-order, and the rate law will have the form

$$\text{rate} = k_A[A][R] \qquad (14\text{–}6)$$

The symbol k_A is again the specific reaction-rate constant with the dimensions of time^{-1} molarity^{-1} (for example, $\sec^{-1} M^{-1}$).

The rate of the reaction also can be expressed mathematically as a decrease in either [A] or [R]:

$$\frac{-d[A]}{dt} = \frac{-d[R]}{dt} = k_A[A][R] \qquad (14\text{–}7)$$

This is known as the *differential form* of the second-order rate law.

The *integrated form* of this differential equation depends on whether the concentration of A at zero time, $[A_0]$, equals the concentration of R at zero time, $[R_0]$. If $[A_0] = [R_0]$, the equation is

$$k_A t = \frac{[A_0] - [A]}{[A_0][A]}$$

If $[A_0] \neq [R_0]$, the equation is

$$k_A t = \frac{1}{[R_0] - [A_0]} 2.303 \log \frac{[A_0][R]}{[A][R_0]} \qquad (14\text{–}8)$$

The integrated second-order rate law (Equation 14–8) is more complex than that of the first-order case. However, a second-order reaction such as Equation 14–6 can be made to follow first-order kinetics by using a large excess of reagent R. The concentration of R will then not change significantly after reaction with A. The rate will then vary only with the changing concentration of A; Equation 14–7 simplifies to

$$\frac{-d[A]}{dt} = k[A][R] \tag{14–9}$$

$$= k'[A]$$

where $k' = k[R]$. Analogous to Equation 14–4, the integrated first-order rate law for this case is

$$\ln[A] = -k't + \ln[A_0] \tag{14–10}$$

14–2. SOME APPLICATIONS OF RATE CONSTANTS

Reaction Times. A rather lengthy time may be needed for a slow reaction to be used for a chemical analysis. In particular, for a slow second-order reaction, one should increase the concentration of the reagent to change the kinetics to first order, and then evaluate the first-order rate constant, k', from experimental data. This approach is best illustrated with a numerical example.

Example: The hydrolysis of an ester in base to an alcohol and acid salt is a slow second-order reaction. The rate law is

$$\text{rate} = k[OH^-][\text{ester}] \tag{14–11}$$

If a large excess of base is used, the reaction follows first-order kinetics:

$$\text{rate} = k'[\text{ester}]$$

Analysis gave the following concentrations of ester remaining after the given reaction period; calculate the value of k' from the data.

$$
\begin{array}{ll}
0.0 \text{ min:} & 0.100M \\
2.5 \text{ min:} & 0.061M \\
5.0 \text{ min:} & 0.035M \\
7.5 \text{ min:} & 0.022M
\end{array}
$$

Substitute the data into Equation 14–10 and solve for k':

$$\ln(0.061M) = -k'(2.5 \text{ min}) + \ln(0.100M)$$

$$-2.7969 = -k'(2.5) - 2.3026$$

$$k' = 0.4943/2.5 = 0.1977 \text{ min}^{-1}$$

Similar calculations give $k' = 0.2100 \text{ min}^{-1}$ for 5.0 min and $k' = 0.2018 \text{ min}^{-1}$ for 7.5 min, with an average k' of 0.2031 min^{-1}.

Time for Quantitative Reaction.   For first-order reactions, a general equation can easily be derived to relate the rate constant to the time required for quantitative reaction. The definition of a quantitative reaction time is arbitrary, but it is often defined as 99.9% (or 99.0%).

In terms of Equation 14–3, $[A_0] = 100.00\%$, and $[A] = 100.00 - 99.90\% = 0.10\%$. Substituting into this equation gives

$$\ln\frac{[100.00\%]}{[0.10\%]} = kt_{99.9\%}$$

$$\ln 1000 = 6.907 = kt_{99.9\%}$$

$$t_{99.9\%} = \frac{6.907}{k}$$

By using this equation, the estimated time for quantitative reaction can easily be calculated for any time of reaction.

Example: The hydrolysis of *tert*-butyl chloride in 60% water–40% ethyl alcohol at 25°C follows first-order kinetics and has a specific rate constant of 1.29×10^{-3} sec^{-1}. Calculate the time for quantitative (99.9%) hydrolysis to *tert*-butyl alcohol. Substituting the value for k into the previous equation gives

$$t_{99.9\%} = \frac{6.907}{1.29 \times 10^{-3} \text{ sec}^{-1}} = 5.354 \times 10^3 \text{ sec} = 1.48_7 \text{ h}$$

This is a fairly slow reaction for quantitative purposes, but it could be used if no faster method were avaliable.

Relative Reaction Rates.   Much kinetic information in the chemical literature is in the form of reaction rates *relative* to that of a standard compound, whose rate constant is arbitrarily chosen as unity (1.0). These relative rate constants are useful in that they can be used to calculate the amount of one component that has reacted relative to a second in a two-component mixture. If we assume the faster-reacting component reacts completely, then we can calculate the error from the amount of the slower-reacting component that has reacted in the same time.

For example, if the hydrolysis rate of methyl acetate ester is defined as unity (1.0), the relative hydrolysis rate or rate constant, of methyl chloroacetate ester (C) is 761, and that of ethyl acetate ester (A) is 0.60 [2].

We can now calculate how much of slow-reacting A has reacted when fast-reacting C has reacted quantitatively (99.9%). We assume excess base to achieve first-order kinetics for both esters and divide Equation 14–3 for C by Equation 14–3 for A:

$$\frac{\ln([C_0]/[C])}{\ln([A_0]/[A])} = \frac{k_c't}{k_a't} \tag{14–12}$$

[2] L. P. Hammett, *Physical Organic Chemistry*, New York: McGraw-Hill, 1940, p. 24.

We then substitute $999/1 \ (= 99.9\%/0.1\%)$ for $[C_0]/[C]$ and $1268/1$ for k_c'/k_a':

$$\frac{\ln(999)}{\ln([A_0]/[A])} = \frac{6.9067}{\ln([A_0]/[A])} = \frac{1268}{1}$$

Rearranging gives

$$\ln([A_0]/[A]) = \frac{6.9067}{1268} = 0.005447$$

$$[A_0]/[A] = 1.0054/1$$

$$[A]/[A_0] = 0.9945/1$$

The fraction of $[A_0]$ reacted in this time is

$$\text{Fraction reacted} = [A_0] - [A] = 1 - 0.9945 = 0.0055$$

$$\text{Percent reacted} = 0.55\%$$

The error in measuring C from the reaction of A is thus only 0.55%, provided that the reaction time is just sufficient for 99.9% reaction of A and the initial concentrations are about equal: $[C_0] = [A_0]$.

This time can be determined on a sample of C alone as outlined in the example on the hydrolysis of *tert*-butyl chloride (p. 266). If the time is too short or too long for convenient measurement, then the $[OH^-]$ or temperature must be changed appropriately. For example, raising the temperature will increase the rate constant and decrease the time. Lowering the $[OH^-]$ can increase the time, and so on.

Differential Kinetic Analysis. The success of the approach used for relative reaction rates depends on how large k_c' is in comparison to k_a'. If the rate constants differ only by a factor of 10, it will be impossible to obtain a quantitative reaction for C without a significant amount of A also reacting. However, it is possible to *differentiate* C and A by allowing both to react and analyzing for the amount of each reacted at different time intervals. An approach called *differential kinetic* is used to determine the amount of each present as long as both exhibit the same kinetic order, preferably first-order kinetics.

The reactions of C and A to form a common product, P, can be summarized as

$$C \ \rightarrow \ P; \quad \text{rate} = k_c[C]$$

$$A \ \rightarrow \ P; \quad \text{rate} = k_a[A]$$

The integrated forms of these rate equations are

$$\ln([C_0]/[C]) = k_c t \quad \text{or} \quad \ln[C] = -k_c t + \ln[C_0] \qquad \textbf{(14–13)}$$

$$\ln([A_0]/[A]) = k_a t \quad \text{or} \quad \ln[A] = -k_a t + \ln[A_0] \qquad \textbf{(14–14)}$$

If $\ln[C]$ is plotted against t for a sample containing only C, a linear plot will result with an intercept at $t = 0$ of $[C] = [C_0]$. A sample containing only A will also give a linear plot of $\ln[A]$ against t, but the slope will be lower because A reacts at a slower rate than C. (See Figure 14–2.)

Figure 14–2.   First order reaction rate plots.

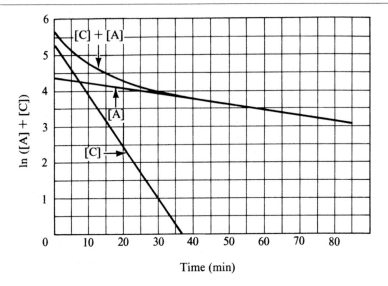

Time (min)

In a sample mixture containing both A and C, it will be possible to measure only the *sum* of [A] + [C]. This is done by measuring the product, P, at various reaction times and subtracting it from the total [A_0] + [C_0]. (The latter sum can be obtained by measuring P under different reaction conditions where both A and C have *completely reacted* to form P.) A plot of log ([A] + [C]) versus t will follow the curved line in Figure 14–2.

During the early part of the reaction, the plot in Figure 14–2 is not linear because both A and C are reacting significantly. However, after the concentration of C has dropped to essentially zero (a time of the order of the 35 min in Figure 14–2), the slope of the plot becomes that of the slower-reacting A. Graphical extrapolation to $t = 0$ gives the initial concentration of A, namely, [A_0]. Subtraction of this from the sum of the initial concentrations (determined previously) gives [C_0].

14–3.   KINETICS OF ENZYME-CATALYZED REACTIONS

The field of biochemical analysis using enzyme-catalyzed reactions provides many excellent uses for kinetic principles discussed above. By way of definition, enzymes are protein catalysts for biochemical reactions. The mechanism for enzyme reactions was formulated by Michaelis and Menten [3]. For the simplest case, it involves the reaction of S (a substrate) with E (an enzyme) to give ES (an addition complex),

[3] M. Dixon and E. C. Webb, *Enzymes*, New York: Academic Press, 1960, p. 75.

which then decomposes to P (product) and the original enzyme:

$$E + S \rightleftharpoons ES \rightarrow P + E$$

Numerous analytical methods based on the kinetics of this reaction have been devised [4, 5]. Before discussing these methods, the kinetic rate law for this reaction will be discussed.

The rate constants for each of the three steps in the reaction of an enzyme and substrate are as follows:

$$E + S \underset{k_2}{\overset{k_1}{\rightleftharpoons}} ES \overset{k_3}{\rightarrow} P + E$$

The rate law for the formation of ES has been found to be

$$\text{rate}_f = k_1[E_0 - ES][S] \tag{14–15}$$

where E_0 is the enzyme concentration at time zero, $[E_0 - ES]$ is the concentration of the enzyme at any other time, and $[S]$ is the concentration of the substrate at any time.

The rate law for the decomposition of ES into E and S has been found to be

$$\text{rate}_d = k_2[ES] \tag{14–16}$$

where $[ES]$ is the concentration of the enzyme-substrate complex. The rate law for the formation of products has been found to be

$$\text{rate}_p = k_3[ES] \tag{14–17}$$

In general, these reactions reach an equilibrium condition, at which point the concentrations of ES and the other reactants attain some constant value. When this occurs, the rate of formation of ES must equal the total rate of the disappearance of ES:

$$\text{rate}_f = \text{rate}_d + \text{rate}_p \tag{14–18}$$

Substituting Equations 14–15, 14–16, and 14–17 into Equation 14–18,

$$k_1([E_0] - [ES])[S] = k_2[ES] + k_3ES] \tag{14–19}$$

Equation 14–19 can be rearranged to give

$$[ES] = \frac{k_1[E_0][S]}{(k_2 + k_3)/k_1 + k_1[S]} \tag{14–20}$$

The term containing the rate constants in Equation 14–20 is the same as the well-known [6] Michaelis constant, K_m:

$$K_m = \frac{k_2 + k_3}{k_1} \tag{14–21}$$

[4] G. Guilbaut, "Kinetic Methods of Analysis," in *Fluorescence: Theory, Instrumentation, and Practice*, New York: Dekker, 1967, pp. 297–358.

[5] H. B. Mark and G. A. Rechnitz, *Kinetics in Analytical Chemistry*, New York: Wiley-Interscience, 1968, pp. 22–60.

[6] G. E. Briggs and J. B. S. Haldane, *Biochem. J.*, *19*, 338 (1925).

The step of greatest analytical interest is the formation of products from ES (Equation 14–17). Substituting Equations 14–20 and 14–21 into the rate law (Equation 14–17) for that step,

$$\text{rate}_p = k_3[E_0]\frac{[S]}{K_m + [S]} \tag{14–22}$$

Since the substrate is usually what is measured, the $k_3[E_0]$ product will control the rate. This product is usually termed V_{max}, the maximum obtainable. Equation 14–22 can then be written as follows [4]:

$$\text{rate}_p = V_{max}\frac{[S]}{K_m + [S]} \tag{14–23}$$

The relation between the product formation rate (the overall reaction rate) and substrate concentration is shown in Figure 14–3.

Determination of Substrates.   For best accuracy, [S] should be measured in the part of the concentration range where the reaction rate is directly proportional to [S]. As shown in Figure 14–3, this is the region where [S] is smaller than K_m. This is because the reaction rate increases linearly as [S] increases—that is, the reaction rate is directly proportional to the substrate concentration. In this region, K_m is much larger than [S], and Equation 14–23 simplifies to

$$\text{rate}_p = V_{max}\frac{[S]}{K_m} \tag{14–24}$$

This rate law is essentially the same as the first-order rate law discussed earlier, except that the "rate constant" is now V_{max}/K_m.

Figure 14–3.   A plot of the reaction rate (rate of product formation) of an enzyme-substrate reaction against the concentration of substrate. The concentration of the enzyme is held constant. $V_{max} = k_3[E_0]$; this is the maximum rate obtainable with the particular concentration of enzyme used. K_m is the Michaelis constant.

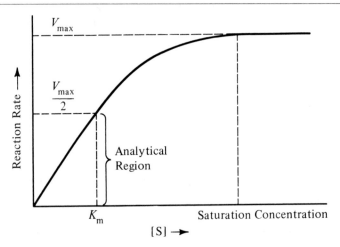

An unsuitable region for measuring [S] is the region where [S] is equal to or greater than K_m. (Note that when $[S] = K_m$, $rate_p = V_{max}/2$.) In this region, the substrate concentration is so large that there is only a small, nonlinear increase in the reaction rate. As the substrate concentration reaches its so-called saturation concentration, the reaction rate is constant. At this point, essentially all of the enzyme is tied up in the addition complex, so it cannot increase the rate any further. This is therefore a poor region in which to measure [S] by measuring the reaction rate.

Analytical Methods for Substrates. The concentration of a substrate can be found using one of two approaches. In one approach, large amounts of enzyme are added to a small amount of substrate, and the substrate is allowed to react completely. The amount of substrate can be found by measuring the amount of product formed. This approach is expensive because so much of the relatively expensive enzyme is required. The second approach requires very little enzyme and is based entirely on kinetics. In this approach, the *initial* rate of the reaction is measured by following the rates of either product production or disappearance of substrate. This method is faster because the reaction need not be complete. However, the temperature and other conditions must be carefully controlled. To illustrate how substrates may be determined, several examples will be discussed.

Determination of Urea. The hydrolysis of urea is catalyzed by the enzyme urease to the following reaction:

$$NH_2CONH_2 + 2H_2O + H^+ \xrightarrow{\text{urease}} 2NH_4^+ + HCO_3^-$$

Note that H^+ is consumed in the reaction and that the pH will increase unless a buffer is present or acid is added. Thus, the rate of the increase in pH or of the consumption of H^+ indicates the reaction rate. Unfortunately, the rate changes somewhat if the pH changes too much. An ingenious instrumental method [7] has been devised to measure urea using this reaction and yet keep the pH constant. This is a pH stat, an automated instrument that contains a pH meter, a provision for adding a measured amount of acid, and a recorder. After urease is added to the urea sample, acid is continuously added to the sample to maintain a pH of about 6.2 for a preset time period of 2.5 minutes. A linear plot of the amount of acid added versus known urea concentration is used to find the concentration of urea in a sample once the amount of acid added to the sample has been measured.

A much simpler approach to the same problem is to use a cation-sensitive glass electrode (Chapter 16) to measure the ammonium ion produced [8]. The only changes in the cation concentration during the hydrolysis are the increase in the NH_4^+ concentration and the decrease in the H^+ concentration. Since the electrode does not respond to H^+, it can measure the increase in NH_4^+ accurately.

Another approach is to use the urease electrode discussed in Chapter 17. The electrode itself contains the urease enzyme, and there is no need to add it to the solution.

Determination of Glucose. Glucose is determined so often in blood and urine

[7] H. V. Malmstade and E. H. Piemeier, *Anal. Chem. 37*, 34 (1965).
[8] S. A. Katz and G. A. Rechnitz, *Z. Anal. Chem., 196*, 248 (1963).

samples that its analysis is extremely important in clinical laboratories. Glucose may be determined very quickly and accurately using enzymes. Two such methods, both kinetic, will be discussed here.

By using an enzyme, glucose may be determined by means of a so-called *coupled reaction* [9]. In the first step of the reaction, the enzyme glucose oxidase is used to catalyze the oxidation of glucose by oxygen:

$$\text{glucose} + H_2O + O_2 \xrightarrow{\text{glucose oxidase}} \text{gluconic acid} + H_2O_2 \qquad \textbf{(14-25)}$$

Since neither gluconic acid nor hydrogen peroxide can be measured *directly* by any quick means, this reaction is *coupled* to a second reaction:

$$H_2O_2 + \text{reduced form of dye} \xrightarrow{\text{peroxidase}} H_2O + \text{oxidized dye}$$
$$\text{(colorless)} \qquad\qquad\qquad\qquad\qquad \text{(colored)}$$

The enzyme peroxidase is added to catalyze the normally slow oxidation of the organic dye by hydrogen peroxide. When the dye is oxidized, it forms an intensely colored species that can be measured spectrophotometrically or colorimetrically. The rate of increase in absorbance is thus proportional to the glucose concentration. Measurements taken in the first few minutes or less of the reaction are sufficient to give the glucose concentration.

Another kinetic method for glucose involves measuring the rate of oxygen consumption as glucose is oxidized in the presence of glucose oxidase enzyme (Equation 14–25).

The rate of decrease in glucose concentration is the same as the rate of decrease of dissolved oxygen, provided that the solution is not in rapid equilibrium with the atmosphere. A commercial *glucose analyzer instrument* has been devised to measure glucose by taking advantage of this rapid decrease.

This instrument uses a polarographic method (see Chapter 15) to measure oxygen, which is reduced at a polarographic electrode as follows:

$$\tfrac{1}{2}O_2 + 2H^+ + 2e^- = H_2O$$

The instrument is first *calibrated* in two steps by using two different glucose standards and glucose oxidase. In Step 1, the enzyme and a standard glucose solution containing 150-mg glucose are added to the container around the electrode. After a set time, usually 10 sec, the instrumental readout is adjusted so that it corresponds to a set scale reading for 150-mg glucose. This corresponds to the decrease in oxygen concentration around the electrode after 10 sec and is essentially a rate measurement. Then the solution around the polarographic electrode is drained, and a fresh enzyme solution and a second standard glucose solution containing 300-mg glucose are added to the sample container. After the same set time of 10 sec, the instrumental readout is again adjusted so that it corresponds to a fixed scale reading for 300-mg glucose. The total time for calibration is about a minute, after which the instrument is ready to analyze up to 60 samples per hour for glucose.

[9] G. G. Guilbault, *Anal. Chem.*, *38*, 527R (1966).

Because the instrument measures oxygen and does not respond to the other organic compounds in body fluids that interfere with colorimetric methods for glucose, it gives essentially a true measure of glucose. (In general, colorimetric methods for glucose have always given slightly high results because of the reactions of other oxidizable compounds in body fluids.) The only known interference in the instrumental method results from intact red blood cells. Because the common anticoagulants do not interfere, a whole-blood sample may be coagulated and the serum used for the analysis instead.

14–4. DETERMINATION OF CATALYSTS

A more sensitive spectrophotometric trace-analysis method for determining a catalyst involves measuring the rate of a reaction catalyzed by the substance. The sensitivity is increased because the spectrophotometer measures not just a few colored molecules but the cumulative effect produced by a few catalyst molecules participating in a reaction over and over again.

Suppose molecules A and B react slowly, but their reaction is catalyzed by a molecule C. A general rate law defining the rate is

$$\text{rate} = k_c[C]^c[A]^a[B]^b$$

where k_c is a specific rate constant, $[C]$ is the concentration of the catalyst, $[A]$ and $[B]$ are concentrations of the reactants, and a, b, and c are the respective reaction orders of each substance.

If A or B absorbs some form of light, one can directly measure the reaction rate (and, indirectly, $[C]$) by spectrophotometrically measuring the rate of disappearance of either reactant. A calibration curve can be made by plotting the reaction rate against $[C]$. Or, instead of plotting the reaction rate, the absorbance of A or B after a given time or the time required for the absorbance of A or B to become zero may be plotted against $[C]$.

The more effective the catalyst (the larger the k_c), the greater the rate of the catalyzed reaction and the lower the measurable concentration of the catalyst. The ultimate catalysts are enzymes, which are much more effective than inorganic substances [10].

Catalytic Determination of Iodide Ion. Protein-bound iodine in blood or trace levels ($10^{-6}M$) of iodide ion are best determined by measuring the catalytic effect of iodide ion on the slow cerium(IV)-arsenic(III) reaction [11]:

$$2\text{Ce(IV)} + \text{A(III)} \xrightleftharpoons{1\,M\,H_2SO_4} 2\text{Ce(III)} + \text{As(V)}$$

The reaction is first-order in cerium(IV), arsenic(III), and iodide. Normally, the

[10] W. J. Blaedel and G. P. Hicks, *Advances in Analytical Chemistry and Instrumentation*, Vol. 3, New York: Wiley, 1964, pp. 105–142.

[11] A. L. Chaney, *Ind. Eng. Chem., Anal. Ed.*, 12, 179 (1940).

arsenic(III) concentration is adjusted so as to be larger than the cerium(IV) concentration, since the rate of the reaction is followed by measuring the fading of the yellow color of cerium(IV) at 420 nm.

In determining protein-bound iodine, the blood sample is first ashed, so that the organic iodine is converted to inorganic iodide. Then the sample is taken up in sulfuric acid and an aliquot added to a solution of cerium(IV) and arsenic(III). In one procedure [11], a calibration curve is developed by measuring the cerium(IV) remaining exactly 15 min after the iodide sample has been added. In another procedure [12], the time required for the quantitative reduction of cerium(IV) is measured and used to find the concentration of iodide ion. A further refinement in this method avoids errors arising from impurities that also catalyze the reaction. A known amount of iodide ion, besides the original amounts of cerium(IV) and arsenic(III), is added to the same solution. The time required for the quantitative reduction of cerium(IV) is again determined spectrophotometrically, and an equation is used to calculate the concentration of iodide in the original sample.

Oxidizing and reducing agents that react rapidly with cerium(IV) or arsenic(III) will interfere, but few such substances are found in the blood. High concentrations of sodium chloride appear to enhance the catalysis. Mercury(II), however, will interfere with the determination of iodide in aqueous samples because it strongly complexes iodide ions and thus inhibits the catalysis.

SUGGESTIONS FOR LABORATORY EXPERIMENTS

W. H. Cone and R. A. Hermens, *J. Chem. Ed., 40,* 421 (1963). *A study of the effect of silver(I) on the rate constant for oxidizing benzoic acid with peroxydisulfate.*

J. F. Davies and A. F. Trotman-Dickenson, *J. Chem. Ed., 43,* 483 (1966). *The rate of dissolution of tin in solutions of iodine.*

C. E. Hedrick, *J. Chem. Ed., 42,* 479 (1965). *Formation rate of the chromium-EDTA complex.*

P. C. Moews, Jr., and R. H. Petrucci, *J. Chem. Ed., 41,* 549 (1964). *Kinetics of the oxidation of iodide with peroxydisulfate.*

R. D. Whitaker, *J. Chem. Ed., 40,* 264 (1963). *Rate of decomposition of tris-(1,10-phenanthroline)nickel(II) by acid.*

PROBLEMS

Rate Law Problems

1. A first-order reaction has a rate constant of 10^0 sec^{-1}. Calculate the time needed for a (a) 99.9% reaction and (b) 99% reaction.

2. Substance A and substance B undergo first-order reactions with rate constants of

[12] E. B. Sandell and I. M. Kolthoff, *Mikrochim. Acta, 1,* 9 (1937).

10^1 sec^{-1} and 10^{-1} sec^{-1}, respectively. Calculate the percentage of B remaining when A has reacted 99.9%.

3. In the second example, the hydrolysis of *tert*-butyl chloride with a $k = 1.29 \times 10^{-3}$ sec^{-1} was shown to be a slow reaction (1.487 h for 99.9% reaction). Suppose that it was important to take no more than one hour for the hydrolysis. Calculate the time needed for each of the following reaction percentages and state which would require less than one hour and which more than one hour.

(a) 99.5% reaction　　　　(b) 99.0% reaction　　　　(c) 98.0% reaction

4. In $1M$ SCN$^-$, the first-order rate constant for the reaction of Co(II) and SCN$^-$ is 1×10^4 sec^{-1} at 25°C. The rate law in which the excess SCN$^- = 1M$ has the form, rate $= k$ [Co(II)]. Calculate the time in seconds and minutes for the following reaction percentages.

(a) 50.0% reaction　　　　(b) 99.0% reaction　　　　(c) 99.9% reaction

5. When [SCN$^-$] $= 0.200M$ and [Co(II)] $= 0.100M$, the reaction between Co(II) and SCN$^-$ is second order and $k = 1 \times 10M^{-1}$ sec^{-1}. Calculate the time for 99.0% reaction and compare it to that in the previous problem.

Analysis Problems

6. When certain second-order reactions are used for the colorimetric determination of metal ions, the rate law changes to an apparent first-order (in metal ion) rate law. At 25°C and 0.01M KSCN, the first-order rate constants for the formation of FeSCN^{2+} and CrSCN^{2+} are 1.27 sec^{-1} and 1.1×10^{-7} sec^{-1}, respectively. Calculate the time for a 99% reaction of (a) Fe^{3+} with 0.01M KSCN and (b) Cr^{3+} with 0.01M KSCN. Are these times suitable for colorimetric analysis? If not, suggest improvements.

7. It is important that an indicator change color fast enough at an end point to avoid overshooting it. The rate law for the reaction of Ce(IV) titrant and the indicator ferroin is

$$\text{rate} = k[\text{Ce(IV)}][\text{ferroin}]$$

and

$$k = 1.4 \times 10^5 \ M^{-1} \text{ sec}^{-1} \text{ at } 25°C \qquad (0.5M \ H_2SO_4)$$

Calculate the time for a 99.0% reaction of $10^{-4}M$ Ce and $10^{-4}M$ ferroin at the end point. Is this time short enough to avoid overshooting? (Recall the rate of titration at the *end point* compared to the initial titration rate.)

8. Methyl chloroacetate ester hydrolyzes much faster (relative $k = 761$) than methyl acetate ester (relative $k = 1.0$) and ethyl acetate ester (relative $k = 0.60$). Calculate the percentage reacted of the slower-reacting ester under the conditions given, assuming equal molarities.

(a) 99.9% reaction methyl chloroacetate: % of ethyl acetate reacted is?
(b) 99.9% reaction methyl chloroacetate: % of methyl acetate reacted is?
(c) 99.0% reaction methyl chloroacetate: % of ethyl acetate reacted is?
(d) 99.0% reaction methyl chloroacetate: % of methyl acetate reacted is?

9. A mixture of A and C form a common product, P. Since their rate constants differ by less than 10, their mixture is analyzed by differential kinetic analysis. A 50.0-mL aliquot is titrated for the total, consuming 40.00 mL of 0.100M titrant. Graphical extrapolation of

the differential plot gives an initial concentration of A of 0.0300M. What is the concentration of C in the mixture?

10. Decide whether each mixture can be analyzed by allowing only the first substance to react 99.9% or whether differential kinetic analysis must be used. Assume that all metals are reacting with SCN^- at the same molarities.

   (a) Co(II), $k = 1 \times 10^4 M^{-1}\ sec^{-1}$ and Ni(II), $k = 5 \times 10^3 M^{-1}\ sec^{-1}$.
   (b) Fe(III), $k = 1.27 \times 10^2 M^{-1}\ sec^{-1}$ and V(II), $k = 4 \times 10^1 M^{-1}\ sec^{-1}$.

15

Electroanalytical Chemistry

Electrochemistry, which deals with the relationship between electrical energy and chemical reactions, is useful to the analytical chemist in several ways. Some metal ions in solution can be reduced electrically and deposited onto a weighed electrode (electrodeposition) for purposes of separation and quantitative determination. The *potential* at an electrode, measured while drawing almost no current, is proportional to the activity (or concentration) of certain ions in solution. This type of measurement (potentiometry) is used in following the course of titrations and detecting the end point (potentiometric titrations). It is also useful in measuring the activity of certain ions in solution, as will be discussed in Chapter 16 on ion-selective electrodes.

When an electrical potential is applied so that an electrochemical reaction occurs, the current produced can be measured and related to the concentration of the reactive species in a sample solution. *Coulometry* refers to methods in which the total amount of electricity is measured for a quantitative electrochemical reaction. Coulometric methods are discussed in Section 15–3. *Voltammetry* refers to methods in which the magnitude of current, measured at a particular applied potential, is related to the concentration of a chemical substance in solution. Voltammetry is covered in Section 15–4. A number of electrochemical devices, such as detectors used in liquid chromatography (Chapter 23), are based on the use of coulometry or voltammetry.

15–1. PRINCIPLES OF ELECTROLYSIS

Galvanic and Electrolytic Cells. It is necessary to distinguish clearly between the two fundamental types of electrochemical cell. The first type is known as a *galvanic*, or *voltaic*, cell.

Figure 15–1. A galvanic cell.

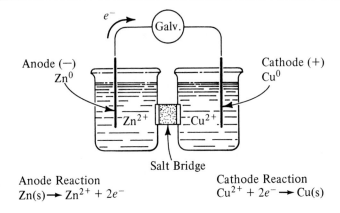

Anode Reaction
$$Zn(s) \longrightarrow Zn^{2+} + 2e^-$$

Cathode Reaction
$$Cu^{2+} + 2e^- \longrightarrow Cu(s)$$

Galvanic reactions within a galvanic cell produce an electric current. The conventional shorthand way of writing such a cell (see Figure 15–1) is to represent the electrode with the more negative standard reduction potential at the left:

$$Zn(s)/Zn^{2+} // Cu^{2+}/Cu(s)$$

The single line represents a phase boundary, such as between a solid metal and a liquid solution. The double line represents a liquid junction or a salt bridge.

The cell reaction and the cell voltage of a galvanic cell can be calculated as follows:

1. Write the reduction half-reaction for the more positive oxidation-reduction couple. Write the reduction potential, which is either the standard reduction potential or the reduction potential calculated with the Nernst equation.
2. From this, subtract the reduction half-reaction for the less positive half-reaction and the corresponding reduction potential.

As an illustration, the complete reaction and the voltage of the following cell is calculated:

$$Zn(s)/Zn^{2+}(0.1M) // Cu^{2+}(1M)/Cu(s)$$

$$Cu^{2+} + 2e^- = Cu(s) \qquad\qquad\qquad\qquad E = E^0 = 0.345$$

$$-[Zn^{2+} + 2e^- = Zn(s)] \qquad \left| -\left| E = E^0 + \frac{0.059}{2}\log[Zn^{2+}] = -0.792 \right. \right|$$

$$\overline{Cu^{2+} + Zn(s) = Cu(s) + Zn^{2+}} \qquad E = 0.345 - (-0.792) = 1.137\ V$$

Thus far, nothing has been said about *liquid-junction potential* in electrochemical cells. This is a potential that develops at the interface of two solutions of different composition; frequently, it is several hundredths of a volt. Liquid-junction potentials arise from an unequal distribution of cations and anions across the liquid junction, which in turn is caused by difference in the diffusion rates of various ions.

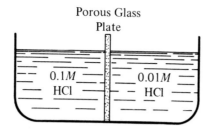

Porous Glass
Plate

A simple case of liquid-junction potential is two hydrochloric acid solutions of different concentrations separated by a porous glass plate, which prevents the solutions from mixing but permits ions to diffuse through the plate. Both hydrogen ions and chloride ions will diffuse from the more concentrated to the more dilute solution. However, hydrogen ions are much more mobile than chloride ions and diffuse at a faster rate [in dilute HCl, $t_{H^+} = 0.83$, $t_{Cl^-} = 0.17$]. As a result, the solution on the right-hand side of the liquid junction builds up a net positive charge (from an excess of hydrogen ions), and the left-hand solution has a net negative charge. Because of the resulting electrostatic effects, the rate of diffusion of H^+ is slowed, and that of Cl^- is increased. A steady state is established in which the potential difference between the two solutions has a constant value; the charge difference is just enough to counteract the difference between the mobilities of the hydrogen and chloride ions.

Liquid-junction potentials are largely (but not entirely) eliminated by inter-posing a salt bridge between the solutions of a cell. A salt bridge is most effective in reducing liquid-junction effects when the salt solution is concentrated and when the mobilities of the salt's cation and anion are nearly equal. Potassium chloride fulfills the latter requirement and is widely used in salt bridges.

The other type of cell is an *electrolytic* cell. In an electrolytic cell, the chemical reactions are made to occur by applying an external voltage. This voltage *opposes* the current flow (if any) in a galvanic cell. Thus, the individual electrode reactions and the net cell reaction are the *opposite* of the spontaneous galvanic cell reaction. The minimum voltage required for electrolysis (neglecting overvoltage, which will be discussed later) is slightly greater than the opposing voltage of the galvanic cell.

As an example, consider the following cell:

$$Cu(s)/Cu^{2+}//H^+(1M), O_2/Pt(s)$$

$$Cu^{2+} + 2e^- = Cu(s) \qquad E^0 = 0.345 \qquad \tfrac{1}{2}O_2 + 2H^+ + 2e^- = H_2O \qquad E^0 = 1.229$$

$\xleftarrow{\text{reaction}}$ $\xrightarrow{\text{reaction}}$

The *galvanic cell* electrode reaction is in the direction indicated by the arrow because electrons are taken up by the electrode and flow from the minus $(-)$ electrode to the plus $(+)$ electrode. The external wire offers almost no resistance to this flow.

The galvanic cell reaction is in the direction indicated by the arrow because of the inflow of electrons from the minus $(-)$ electrode.

In the electrolytic cell, the copper electrode is still the negative electrode, and the platinum is still the positive. However, the applied voltage forces electrons in the opposite direction (from the $-$ electrode through the solution to the $+$ electrode).

This makes the electrode reactions the reverse of the galvanic cell reactions (see Figure 15–2). The net electrolytic cell reaction is the sum of the reactions at the two electrodes:

$$Cu^{2+} + H_2O \rightarrow Cu(s) + \tfrac{1}{2}O_2 + 2H^+$$

The voltage of either a galvanic or an electrolytic cell is the *difference* in potential between the electrodes. To obtain electrode potential from a table of standard potentials, always use the reduction potential and the reduction half-reaction. In this case, assuming the activity coefficients are unity,

$$E_{cell} = 1.229 - 0.345 = +0.884 \text{ V}$$

The minimum voltage needed to operate an electrolytic cell is essentially that of the galvanic cell, but an external electromotive force must be applied to oppose the spontaneous cell emf. In practice, an electrolytic cell requires a greater applied emf than the minimum because of overvoltage and cell resistance.

The cell emf prior to the actual electrolysis of a Cu^{2+} solution is indeterminate because of the indeterminate concentration of dissolved oxygen. However, as electrolysis proceeds, the stirred solution becomes saturated with oxygen, and a finite back emf (galvanic cell emf) develops. If the cell reaction is reversible, the magnitude of this back emf can be estimated from the Nernst equation.

In both galvanic and electrolytic cells, the *cathode* is the electrode at which reduction occurs and the *anode* is the electrode at which oxidation occurs (see Figure 15–2). Note that the anode is negative with respect to the other electrode in a galvanic cell, but is positive in an electrolytic cell. In turn, the cathode is positive in a galvanic cell and negative in an electrolytic cell.

Figure 15–2. An electrolytic cell.

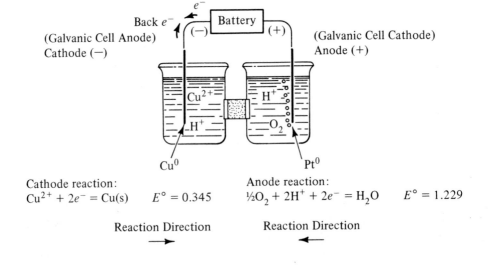

(Galvanic Cell Anode) Cathode (−) Back e^- Battery (Galvanic Cell Cathode) Anode (+)

Cu^0 Pt^0

Cathode reaction: $Cu^{2+} + 2e^- = Cu(s)$ $E° = 0.345$

Anode reaction: $\tfrac{1}{2}O_2 + 2H^+ + 2e^- = H_2O$ $E° = 1.229$

Reaction Direction   →

Reaction Direction   ←

*Cat*ions (used to describe positively charged ions) are so named because they are attracted to the *cat*hode ($-$) in an electrolytic cell; *an*ions (used to describe negatively charged ions) because they are attracted to the *an*ode ($+$).

Voltage Required for Electrolysis. By measuring the voltage of a galvanic cell potentiometrically, the concentration of chemical species in the cell can be calculated with the aid of the Nernst equation. In a potentiometric titration, a galvanic cell is formed, and the change in concentration of certain ions is measured at various times during the titration. Since virtually no current is drawn from the cell during a potentiometric measurement, the relationship between cell composition and cell voltage is fairly simple.

The situation is more complicated in an electrolytic cell used in electrodeposition. In this case, a considerable amount of electrochemical work is done in depositing the metal from solution. The voltage needed to operate the cell changes as electrolysis proceeds and the cell's composition changes. Furthermore, a voltage is required over and above that needed to overcome the voltage of the opposing galvanic cell. This additional voltage is called *overvoltage*. The overvoltage of a cell changes during electrolysis and is difficult to reproduce exactly.

The voltage that must be applied to operate an electrolytic cell at a given current is given by the equation

$$E_{appl} = E_{back} + iR$$

where E_{appl} is the externally applied voltage, E_{back} is the back emf, i is the current in amperes, and R is the resistance of the total circuit in ohms. Most of the resistance occurs between the two electrodes.

The back emf of a cell is given by the equation

$$E_{back} = E_{rev} + \text{overvoltage}$$

where E_{rev} is the reversible back emf of the opposing cell, as calculated from the Nernst equation.

Overvoltage and Overpotential. An electrode is said to be *polarized* if its potential is different from its reversible potential as calculated from the Nernst equation. The amount of polarization is the overpotential, η, of an electrode:

$$\eta = E - E_{rev}$$

Overpotential is of two types: (a) concentration polarization or concentration overpotential and (b) activation overpotential.

We speak of the potential of a single electrode, but we express the difference in potential of electrodes in a cell as the cell voltage, or emf. In like manner, the term *overvoltage* applies to a complete cell, while the term *overpotential* refers to the situation at a single electrode.

Concentration polarization occurs when the rate of reduction (or oxidation) at an electrode is so fast that the concentration of reducible (or oxidizable) substances is different at the electrode's surface from that in the bulk of the solution.

Example: At an electrode where Cu^0 is being deposited, the Cu^{2+} at the electrode surface is $1.0 \times 10^{-4}M$, but the Cu^{2+} in the bulk of the solution is $1.0 \times 10^{-2}M$. Calculate the concentration overpotential.

The electrode potential should be

$$E_{rev} = 0.345 \text{ V} + \frac{0.059 \text{ V}}{2} \log 1.0 \times 10^{-2}$$

$$= +0.286 \text{ V}$$

Because of concentration overpotential, the actual E is

$$E = 0.345 + \frac{0.059}{2} \log 1.0 \times 10^{-4}$$

$$= 0.227 \text{ V}$$

The concentration overpotential η is

$$\eta = E - E_{rev} = 0.227 - 0.286 = -0.059 \text{ V}$$

The concentration overpotential can be reduced or eliminated by efficient stirring and by operating at a small current density. (Ordinarily, current density is measured in amperes per square centimeter of electrode surface.)

It is necessary to raise the potential at an electrode over the reversible potential by a certain amount to cause the passage of a measurable current at the electrode. This shift is called the *activation overpotential*. The activation overpotential ranges from a very small value for plating almost any metal on bright platinum to a value of several tenths of a volt for discharging certain gases at an electrode. For any given electrode, the activation overpotential is affected by the condition of the electrode surface, the amount of current flowing (η is less at small currents), and other factors. Table 15–1 lists values for the overpotential of hydrogen and oxygen under various conditions.

Overpotential may be observed at both electrodes. It causes a more positive potential for oxidation at the anode and a more negative potential for reduction at the cathode. Thus,

$$\eta_a \text{ is } + \text{ in sign}$$

$$\eta_c \text{ is } - \text{ in sign}$$

Table 15–1. Overpotential of Hydrogen and Oxygen

Current Density, Amperes/cm²	Hydrogen		Oxygen	
	Smooth Pt	Platinized Pt	Smooth Pt	Platinized Pt
0.001	0.024	0.015	0.72	0.40
0.01	0.068	0.030	0.85	0.52
0.10	0.29	0.041	1.28	0.64
1.00	0.68	0.048	1.49	0.77

The voltage necessary for electrolysis is given by the equation

$$E = (E_a + \eta_a) - (E_c + \eta_c) + iR$$

where E_a and E_c are the reversible potentials at the anode and cathode calculated from standard reduction potentials by using the Nernst equation. The iR term is fairly small and represents the voltage drop caused by the cell resistance. This is frequently called the iR drop.

Electrode Reactions. The electrode reactions occurring in an electrolytic cell are always those that require the smallest applied voltage. Thus, if more than one reaction is possible at an electrode, the oxidation reaction that requires the lowest positive potential (including overpotential) will occur at the anode. Likewise, the reduction reaction that requires the smallest negative potential (relative to the anode) will take place at the cathode. Overpotential often plays an important role in determining which reactions will occur.

To illustrate these points, consider the electrolysis of a solution of Zn^{2+} in $1M$ hydrochloric acid with bright platinum electrodes. What are the electrode reactions? Possible oxidation reactions (at the anode) are

$$2Cl^- \rightarrow Cl_2 + 2e^-$$

$$H_2O \rightarrow \tfrac{1}{2}O_2 + 2H^+ + 2e^-$$

If either of these oxidation reactions is to take place, the potential of the anode must be made more positive than that of the reduction half-reaction.

An electrode with given amounts of A_{ox} and A_{red} will be poised at its reversible potential if equilibrium has been attained. If a more positive potential is impressed ("e^- sink"), the equilibrium will shift to the left:

$$\overleftarrow{\text{reaction direction}}$$
$$A_{ox} + ne^- = A_{red}$$
$$\downarrow$$

$$e^- \text{ sink}$$

If a more negative potential is impressed ("e^- source"), the equilibrium shifts to the right:

$$\overrightarrow{\text{reaction direction}}$$
$$A_{ox} + ne^- = A_{red}$$
$$\downarrow$$

$$e^- \text{ source}$$

The oxidation of water to oxygen requires a less positive anode potential than does that of chloride to chlorine:

$$Cl_2 + 2e^- = 2Cl^- \qquad\qquad E^0 = 1.36$$

$$\tfrac{1}{2}O_2 + 2H^+ + 2e^- = H_2O \qquad E^0 = 1.23\ (H^+ = 1M)$$

Thus, the evolution of oxygen is favored thermodynamically. However, the over-potential for the discharge of oxygen at a platinum electrode is at least 0.4 or 0.5 V, whereas that for the chloride-chlorine oxidation is small. Thus, the following anode reaction occurs:

$$2Cl^- \rightarrow Cl_2 + 2e^-$$

The possible reduction reactions at the cathode are

$$2H^+ + 2e^- = H_2(g) \qquad E^0 = 0.00 \ (H^+ = 1M)$$

$$Zn^{2+} + 2e^- = Zn(s) \qquad E^0 = -0.67$$

The reduction of hydrogen ions to hydrogen gas takes place at a less negative cathode potential than does the plating with zinc metal. Even assuming a hydrogen overpotential on platinum of 0.3 or 0.4 V, the cathode reaction will be

$$2H^+ + 2e^- = H_2(g)$$

Zinc from alkaline solution can be deposited on a platinum cathode because the hydrogen-ion concentration is so low that the discharge of hydrogen takes place at a more negative potential:

$$H^+ + e^- = \tfrac{1}{2}H_2(g) \qquad E^0 = 0.00 \ V \ (H^+ = 1M)$$

At pH 10 ($H^+ = 10^{-10}$), the potential is

$$E = 0.00 + 0.059 \log 10^{-10}$$

$$= -0.59 \ V$$

Because of overvoltage, a potential still more negative than -0.59 V can be applied in alkaline solution without an appreciable discharge of hydrogen.

In electrolytic separations, the essential *cathode* reaction is the reduction of the selected metal-ion to the free metal,

$$M^{2+} + ze^- = M^0(s)$$

Other common cathode reactions are the reduction of a metal ion to an ionic form with a lower oxidation state,

$$Fe^{3+} + e^- \rightarrow Fe^{2+}$$

$$Cu^{2+} + e^- + 3Cl^- \rightarrow CuCl_3^{2-}$$

and the reduction of hydrogen ions to hydrogen gas

$$H^+ + e^- \rightarrow \tfrac{1}{2}H_2(g)$$

The discharge of hydrogen gas at the cathode represents a major limitation on electrodeposition in analytical separations. In acid solution, only the few metals that reduce at a more positive potential than hydrogen can be deposited on a platinum cathode. Of these, the separation of copper by electrodeposition is most useful.

In electrodeposition, the *anode* reaction is usually the discharge of O_2:

$$\tfrac{1}{2}O_2 + 2H^+ + 2e^- = H_2O \qquad E = 1.23 \ (H^+ = 1M)$$

The overvoltage for the discharge of oxygen at a platinum electrode is high; thus, the anode potential will have to be more positive than 1.23 V by several tenths of a volt.

In solutions of more alkaline pH, the potential needed to discharge O_2 at the anode is less positive. For example, at pH 7.0,

$$E = 1.229 + \frac{0.059}{2} \log [H^+]^2$$

$$= 1.229 + 0.059 \log 10^{-7.0}$$

$$= 1.229 - 0.41_3 = 0.81_6 \, V$$

This value must be more positive than indicated because of the overpotential required.

15–2. ELECTRODEPOSITION

The apparatus used for electrodeposition is shown in Figure 15–3. Direct current is supplied by a battery. The voltage applied to the cell is measured with voltmeter V and the current flowing with ammeter A. The voltage can be adjusted with the variable resistor R. The solution is stirred with a mechanical or magnetic stirrer.

The electrodes are cylinders of platinum gauze. The metal is deposited on the weighed cathode (the negative electrode). When deposition is complete, the cathode is dried and weighed again. The amount of metal deposited is calculated from the weight gain of the cathode.

Current-Time Requirements. One Faraday of electricity, 96,487 coulombs, is required to plate out one gram-equivalent weight of a metal. The number of Faradays is equal to the number of coulombs divided by F, where F is the Faraday constant, 96,487 coul/eq.

Figure 15–3. Setup for electrodeposition.

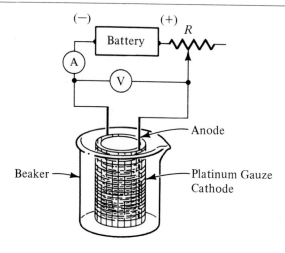

The coulomb is a *quantity* of electricity; electric current (usually measured in amperes) is the *rate of flow* of electricity:

$$amperes = \frac{coulombs}{seconds}, \qquad i = \frac{coul}{t}$$

$$equivalents\ deposited = \frac{coulombs}{96,487} = \frac{it}{96,487}$$

$$equivalents = \frac{grams}{equiv\ wt} = \frac{(grams)(n)}{form\ wt}$$

where n is the number of electrons involved in the reduction. Equating the right-hand sides of the latter two equations,

$$\frac{(grams)(n)}{form\ wt} = \frac{it}{96,487}$$

$$t = \frac{96,487\ (grams)(n)}{i(form\ wt)}$$

From this equation, the time required to deposit a given weight of metal can be estimated.

Typical Determinations.   Only the few metals with a positive standard reduction potential can be deposited from acid solution without a discharge of hydrogen gas. Furthermore, the deposits of some of these metals *adhere loosely* to the cathode and thus are difficult to weigh. For this reason, silver is usually deposited from ammoniacal or alkaline cyanide solution. Copper is almost the only metal commonly separated by cathodic deposition from acidic solution. It is usually deposited from a solution of dilute nitric or sulfuric acid.

In the last stages of copper deposition, the cathode often attains a potential at which hydrogen gas is discharged as a side reaction. If nitrate ion is present, nitrous acid and other products are formed rather than the disturbing hydrogen-gas evolution.

Nickel and cobalt are sometimes determined by deposition from an ammoniacal solution. Although this is an accurate way of determining these elements (by weighing the metal deposited), it is not especially valuable as a separation method. Zinc can be determined by its deposition from solutions of ammonia or strong sodium hydroxide.

Lead is unique in that it can be determined by deposition as PbO_2 on an *anode*. The deposit contains slightly more than the theoretical amount of oxygen and thus requires an empirical factor for converting the weight of the dioxide deposit to that of lead. The anodic deposition of lead dioxide is very good as a separation method.

Separation by Electrodeposition.   In electrodeposition, the cathode potential must be such that deposition is quantitative. This means that, aside from overpotential, the *cathode potential* needed for quantitative deposition must be more negative than that required in the earlier stages of electrolysis. For example, the minimum cathode potential to *start* the electrodeposition of $0.1M$ Cu^{2+} is given by the Nernst equation as

$$E = 0.35 + \frac{0.059}{2} \log 10^{-1}$$

$$= 0.35 - 0.03 = +0.32 \text{ V}$$

A quantitative deposition should not leave more than one part per thousand of copper undeposited. This means that the final Cu^{2+} concentration should not exceed $10^{-4}M$. The reversible cathode potential at this concentration is

$$E = 0.35 + \frac{0.059}{2} \log 10^{-4}$$

$$= +0.23 \text{ V}$$

Because of overvoltage and the iR drop across the cell, the cathode potential needs to be somewhat more negative than $+0.23$ V.

Copper(II) can be separated from any other metal, provided that the metal does not *start* to be deposited at the cathode potential at which the deposition of copper is complete. This principle applies in separating other metals from each other by electrodeposition. However, it is very difficult to control the cathode potential throughout an electrodeposition. As the composition of the solution changes, so does the resistance, the current flow, and the overvoltage. Because of these difficulties, ordinary electrodeposition is a useful separation method only when there are large differences in potential required for depositing metals.

Other separations can be achieved by electrodeposition with controlled cathode potential. A further reference electrode is inserted into the working electrolytic cell and used to measure the actual potential of the cathode. This is much more accurate than estimating the cathode potential from the difference in potential between the two cell electrodes. Because of the changes in overvoltage and resistance of the solution, frequent adjustments are required to maintain a constant cathode potential during the electrodeposition. Thus, unless the instrument is automated, electrodeposition by controlled cathode potential is a tedious process.

15–3. COULOMETRIC METHODS OF ANALYSIS

Instead of weighing the substance plated on the electrode, the analyst may base a determination on the quantity of electricity (the number of coulombs) needed for a quantitative electrochemical reaction. The quantity of the reactant is calculated from Faraday's law of electrolysis. This, of course, requires a known current efficiency; that is, that there be a known relationship between the number of coulombs of electricity and the number of equivalents of substances oxidized or reduced. Because the number of coulombs of electricity is the quantity measured, methods of this type are known as coulometric methods of analysis.

Coulometric procedures are more versatile than electrodeposition because they include both electrochemical reactions in which a gas is formed and those in which both the reactant and the product are soluble species. With these procedures, it is possible to accurately measure quantities too small to be weighed or titrated by ordinary methods.

Direct Methods.   Direct coulometric methods are those in which the substance to be determined is oxidized or reduced directly at one of the electrodes. The coulometric determination of copper at a constant applied voltage is an example. The electrode reactions are the same as when copper is determined by electrodeposition. The current decreases continuously as the electrolysis proceeds.

As the plating proceeds, the concentration of copper(II) in solution decreases. The rate of reaction at the cathode depends, among other things, on the rate at which copper(II) becomes available for reduction at the electrode surface. This is proportional to the concentration of copper(II) in the bulk of the solution, and also upon the efficiency with which the solution is stirred. The rate of reaction decreases as the concentration decreases; therefore, the current decreases with time.

The electroreduction of copper(II) is complete when the current has become negligibly small. The number of coulombs of electricity used in the reaction can be calculated from the area under the current-time curve (Figure 15–4).

Coulometric determinations at controlled potential are often time-consuming. Coulometric methods in which a *constant current* is employed are accurate and easier to carry out. A known constant current, an electric stopwatch for measuring the time elapsed, and a means of locating the equivalence point provide the data from which are calculated the number of coulombs used and, hence, the amount of the substance being determined. Direct coulometric analysis of thin films or plated deposits is possible by electrolytic stripping (the reverse of electroplating). The equivalence point in the process is marked by an inflection in the voltage-time curve.

Direct coulometric methods are applicable only in some cases. Their use is limited by the fact that side reactions often occur before the desired electrochemical reaction is complete; thus, a constant current efficiency is no longer attainable. Consider, for example, the determination of iron(II) by direct coulometric oxidation. As the concentration of unoxidized iron(II) decreases, the current normally also decreases. In a constant-current procedure, this cannot occur; therefore,

Figure 15–4.   A coulometric current-time curve.

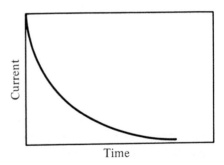

the potential increases and a side reaction (oxidation of water to oxygen) takes place before all of the iron(II) has been oxidized.

Indirect Methods (Coulometric Titrations). Indirect coulometric processes constitute by far the largest number of analytical applications of coulometry. In these, a redox "titrant" is generated at a constant rate by the anodic or cathodic electrolysis of a reagent present at high concentration in the solution. As it is being generated, the titrant reacts stoichiometrically with the substance to be determined. This technique overcomes the difficulties associated with direct coulometry at constant current because the high concentration of the chemical from which the titrant is generated stabilizes, or poises, the potential of the working electrode and so prevents side reactions and loss of 100% current efficiency.

For example, iron(II) can be readily determined by constant-current coulometry. First, a large excess of cerium(III) is added to an acidic solution of the sample. As soon as cerium(IV) is formed from cerium(III) oxidizing at the working electrode, it reacts with iron(II):

$$Ce^{4+} + Fe^{2+} \rightarrow Ce^{3+} + Fe^{3+}$$

Because there is a large supply of cerium(III) available for oxidation at the anode, the potential is stabilized below that at which cerium(IV) oxidizes water to oxygen, and also below that at which O_2 is produced electrolytically. The end point is detected by potentiometric or visual means, as will be explained later.

Some direct electrooxidation of iron(II) may occur in the early stages of the electrolysis before the oxidation of cerium(III) becomes predominant, but this causes no error. The coulometric measurement gives the equivalents of iron(II) present, no matter what the reaction.

Indirect coulometric methods are often referred to as "coulometric titrations." The method just described, for example, may be considered a titration of iron(II) with cerium(IV). The difference between this and a conventional titration is that *no buret* and no standard solution of cerium are required in this titration. The number of coulombs of electricity required for the reaction, rather than the volume of titrant, is the quantity measured.

Of course, the equivalence point in a constant-current coulometric titration must be determined, but the methods employed for a conventional titration generally are applicable. Thus, either a visual indicator or a potentiometer may be used. An amperometric end point (see p. 301) is advantageous for coulometric titrations of very small amounts of a substance.

A block diagram of a simple constant-current coulometric apparatus is shown in Figure 15–5. Note that the second (counter) electrode is isolated from the sample solution to avoid reoxidation or reduction of the products at the generating electrode or in the sample solution. The constant-current source can be a 45-V B-battery with a large resistance in series, or it can be a conventional DC power supply fed from an ordinary AC line through a constant-voltage transformer. The approximate current flowing through the cell is indicated by a milliammeter (not shown), but the exact current is calculated by measuring the voltage drop across a precision 100-ohm resistor using a potentiometer. If the voltage drop E and the

Figure 15–5. Coulometric apparatus.

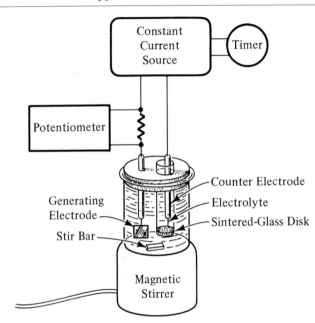

resistance R are known, the current i may be calculated:

$$i = E/R$$

Time is measured with a precision electric stopwatch. It is important that the electrolysis current and the timer be turned on and off in perfect synchronization. To prevent coasting after the current is turned off, the timer is equipped with a magnetic brake.

A typical commercial apparatus has a relative accuracy and precision of 0.1%. It provides various cell currents, from 4.82 to 193 milliamperes, equivalent to an electrochemical reaction of 0.05 to 2 μeq/sec. A one-minute reaction at the lowest current will titrate only 0.003 meq of a substance to be determined; a ten-minute reaction at the highest current will titrate 1.2 meq of substance.

Applications and Advantages of Coulometric Methods. The titration of iron(II) with electrically generated cerium(IV) has been mentioned. Other reducing agents can be determined in much the same manner. Bromine produced by oxidizing bromide ion at a platinum electrode has been used for the coulometric titration of numerous substances:

$$2Br^- \rightarrow Br_2 + 2e^-$$

Arsenic(III), antimony(III), uranium(IV), hydrazine, and others can be determined by quantitative oxidation with bromine. For some organic compounds, such as phenols, the reaction is one of substitutive bromination. The titration of oxine

(8-hydroxyquinoline) is an example:

$$2Br_2 + \text{(8-hydroxyquinoline)} \rightarrow \text{(5,7-dibromo-8-hydroxyquinoline)} + 2HBr$$

Chlorine and iodine have also been produced electrochemically for titrating various substances.

Because the titrating species is used up almost as soon as it is generated, it is possible to perform coulometric titrations with substances that are too unstable to use in conventional titrations. For example, silver(II) produced by the anodic oxidation of silver(I) has been used to titrate cerium(III), vanadium(IV), and other metals. Oxidizing agents have been titrated with copper(I) (chlorocuprous ion) and with titanium(III), both of which have limited stability.

Metal ions may be titrated coulometrically by the cathodic reduction of a mercury(II)-EDTA complex:

$$Hg\text{-EDTA} + 2e^- \rightarrow Hg^0 + \text{EDTA}$$

The liberated EDTA is then available to titrate metal ions in the sample.

Acid-base titrations may be carried out by coulometric procedures. The titration of small quantities of acid with electrically generated hydroxide ion is particularly useful. The coulometric method has the advantage that very small amounts of titrant can be accurately prepared and measured; furthermore, the titrant is free from carbonate impurities. An excess of sodium or potassium bromide is added to the sample solution. The cathode reaction produces hydroxide ions that react with the acid:

$$2H_2O + 2e^- \rightarrow H_2 + 2OH^-$$

The anode should be isolated from the solution (see Figure 15–5) to avoid possible acid-base or other reactions of the anodic oxidation products.

Small amounts ($\sim 1-10$ mg) of weak acids may be titrated coulometrically in nonaqueous solvents with electrically generated tetrabutylammonium hydroxide [1].

Coulometric titrations are capable of extremely high precision and accuracy. For several types of chemical reactions, the Faraday of electricity, $96,486.6 \pm 0.5$ coulombs per chemical equivalent [2], ranks as the best primary standard yet discovered. Primary standard potassium dichromate has been titrated coulometrically with electrically generated iron(II) and found to be 99.975% pure with a standard deviation of 2 parts per 100,000 [3]. Cerium(IV) salts have been titrated

[1] J. S. Fritz and F. E. Gainer, *Talanta*, *15*, 939 (1968).
[2] W. F. Koch, W. C. Hoyle, and H. Diehl, *Talanta*, *22*, 717 (1975); W. F. Koch and H. Diehl, *Talanta*, *23*, 509 (1976).
[3] G. Marinenko and J. K. Taylor, *J. Res. Natl. Bur. Stds*, *67A*, 453 (1963).

with a standard deviation of 5 parts per 100,000 [4]. In these titrations, the end point is determined amperometrically (see Section 15–4).

In summary, several advantages of coulometric methods (especially coulometric titrations) can be cited.

1. The method is especially useful for determining very small quantities of substances because small currents and time can be measured very accurately. Although conventional titrations can be performed on a micro scale, it is frequently more convenient and more accurate to do a coulometric titration.
2. No standard solutions are needed if the current efficiency is known; the coulomb becomes the primary standard. At constant current, the number of coulombs can easily be calculated from the measured time of reaction.
3. In a coulometric titration, it becomes feasible to use rather unstable reagents because they are consumed almost as soon as they are generated.
4. The method is easily adapted to automation. This is an advantage in handling radioactive materials because the analysis can be performed in a shielded, remote place.

15–4. VOLTAMMETRY

Voltammetry is a type of electroanalytical in which an electrical current is measured as a function of time as a function of applied electrical potential. The magnitude of current is related to the concentration of an oxidizable or reducible species in the sample solution.

In contrast to electrodeposition, the working electrode in voltammetry is of very small area. A reference electrode of large area is connected to the sample solution via a salt bridge. Usually the solution is not stirred, so electroactive material in solution (usually reducible, but sometimes oxidizable) can reach the working electrode only by diffusion. As a consequence, currents of only a few *micro*amperes are produced, and the amount of material actually reduced (or oxidized) during the course of analysis is only a small fraction of the total electroactive material present.

Although platinum or other noble-metal electrodes are often used in voltammetry, the dropping-mercury electrode (to be described) is the most common. Mercury has a very high overpotential for discharge of hydrogen gas and thus can be used at much more negative applied potentials than other working electrodes. Voltammetry at a dropping-mercury electrode is referred to as *polarography*.

Polarography.   In polarography, the analysis depends on plotting a curve, called a polarogram. The polarographic cell current produced during electrolysis is measured at various applied voltages, and a current-voltage curve is plotted manually or recorded automatically. A typical curve (polarogram) obtained with the usual dropping-mercury working electrode (to be described) is shown in Figure 15–6.

[4] J. Knoeck and H. Diehl, *Talanta, 16*, 181 (1969).

Figure 15–6.   Polarogram of a reducible species.

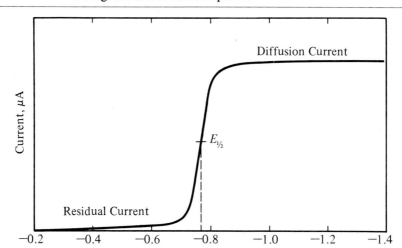

(For simplicity, the current oscillations produced with this electrode are not shown.) Note that as the voltage increases, the current first increases, then levels off. This increase is called a *polarographic wave*. A separate wave is observed for each substance reduced (or oxidized) at the working electrode. Substances that are reduced in steps may exhibit more than one wave.

Ordinarily, the dropping mercury working electrode is operated at negative applied potentials (compared to the reference electrode) so that reduction occurs and *cathodic* waves are observed. At less negative or positive applied potentials, oxidation can occur at the working electrode, giving rise to an *anodic* wave. However, anodic waves are not often seen at the dropping mercury electrode because the mercury metal will oxidize at too positive an applied potential. On the plateau region of either wave, the current is proportional to the concentration of the electroactive material in solution. By making a plot of current versus concentration, the concentration of the electroactive material in a sample can be determined quantitatively.

The polarographic method is subject to fairly high relative error ($\pm 1\%$ or more), but it is useful for analyzing small amounts of substances, usually about 10^{-5}–$10^{-2}M$ sample solutions. The method can be used to determine inorganic ions reducible to the solid metal, inorganic substances reducible to soluble species at a lower oxidation state, and numerous organic compounds. Mixtures can often be analyzed because any two given substances are usually reduced at different applied voltages.

Polarographic Apparatus.   The basic circuit used in polarography is shown in Figure 15–7. The voltage applied to the polarographic cell is varied by adjusting resistance R and adjusting the slide wire, which acts as a voltage divider. The current flowing through the cell is measured in microamperes with galvanometer G.

Figure 15–7.   Basic circuit for polarography.

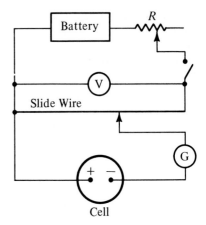

A typical polarographic cell is shown in Figure 15–8. The reference electrode (usually calomel) contains a large amount of the oxidized and reduced forms of its redox couple so that the potential of this electrode will not change during the electrolysis. It might be said that this electrode is buffered against potential changes, much as a solution is buffered against changes in pH.

The dropping mercury electrode shown in Figure 15–8 is the usual working, or analytical, electrode. This is simply a fine capillary tube filled with mercury metal. The tube is open at the bottom so that small droplets of mercury continuously fall from the tube. The drop rate is controlled by adjusting the height of a mercury reservoir (not shown), which exerts a given head of pressure on the capillary. As each drop grows, its surface area increases simultaneously. When it falls off, there is a sharp decrease in surface area until the next drop begins to form. The increase in surface area increases the rate of the electrochemical reaction and the current; the

Figure 15–8.   A polarographic cell

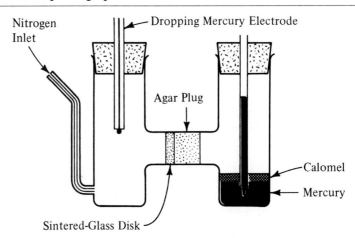

decrease in surface area decreases both of these. Therefore, a polarogram made with a dropping mercury electrode exhibits a series of oscillations, as will be shown later.

The analytical (working) electrode is always very small. One reason for this is to prevent the electrolysis from becoming significant enough to decrease the concentration of the electroactive material appreciably. Thus, a polarogram is recorded at nearly constant concentration. Another reason is that a very small electrode is more easily polarized than a large electrode and can thus be operated at any desired potential within a range of about two volts.

Explanation of Polarographic Waves. While a polarogram is being made, the applied voltage is gradually increased so that the potential of the analytical electrode becomes progressively more negative. At potentials that are too positive for an electrochemical reaction to occur at the analytical electrode, no current should be detected. In practice, a small "residual" current is observed (see Figure 15–6).] When the applied potential becomes large enough that the applied emf is greater than the back emf of the cell, an electrochemical reaction (usually reduction) occurs at the analytical electrode and current begins to flow through the cell. At this point, Ohm's law is followed and the current is a function of applied potential.

$$i = \frac{E}{R} \quad \text{i.e.,} \quad i \propto E$$

When the electrolysis rate is small (small i), the depletion of electroactive ions near the electrode is small and i responds dramatically to changes in E. This is reflected by the steep rise in the curve in Figure 15–6. The increase in i with applied E does not continue forever because the electrolysis rate (as indicated by i) becomes large enough that the supply of ions in the vicinity of the electrode becomes exhausted. Further increases in E cause no increase in i, and there is a plateau region in the curve (Figure 15–6). In this plateau region, the applied E is much greater than that needed for exhaustive electrolysis near the electrode surface. Therefore, i becomes limited by the rate of *mass transport*—that is, by the rate at which additional ions are transported from the bulk of the solution to the vicinity of the electrode surface.

Mass transport of ions from the bulk solution to the vicinity of the electrode can occur by thermal convection, mechanical convection (stirring), electrochemical migration, and diffusion. In polarography, mechanical agitation is avoided, and the polarographic cell is immersed in a constant-temperature bath to prevent temperature changes. The *supporting electrolyte* contains a high concentration of nonreducible ions (at least 100 times that of the reducible ions), so the fraction of current carried through the cell by the reducible species (the *migration current*) is a negligible factor. Therefore, the only way the reducible ions can reach the vicinity of the analytical electrode is by diffusion. The limiting current produced by diffusion and subsequent reaction at the analytical electrode is called the *diffusion current*.

The Ilkovic Equation. The diffusion current is in the plateau region of a polarogram and, thus, is independent of potential. The magnitude of the diffusion current, i_d, depends on several factors summarized by the Ilkovic equation:

$$i_d = 706nCD^{1/2}m^{2/3}t^{1/6}$$

In this equation, i_d is the maximum value of the diffusion current during the life of the drop, in microamperes; 706 is a combination of several constants; n is the number of electrons involved in the reaction; C is the concentration of electroactive substance, in millimoles per liter; D is the diffusion coefficient of the ion, in square centimeters per second; m is the mass of mercury, in milligrams per second; and t is the time between drops, in seconds.

The Ilkovic equation states the simple fact that i_d is proportional to the *concentration* of an electroactive ion in the bulk solution, provided that the mercury drop time and the temperature of the solution are held constant. (Changes in temperature will affect the value of the diffusion coefficient, D.)

For a quantitative polarographic analysis, an applied voltage is selected from the plateau region of the ion's polarogram. The currents of several standards of different concentration are measured at this voltage, and a graph of i_d versus concentration is made. The i_d values of samples of unknown concentration are measured under the same conditions, and the concentrations are read directly from the graph.

Scope of Polarography.   Several aspects of polarography can perhaps best be illustrated by looking at actual chromatograms. Figure 15–9 shows the polarogram of $0.1M$ potassium nitrate in an aqueous solution saturated with nitrogen to remove dissolved oxygen. A dropping mercury electrode is used with a saturated calomel electrode (SCE) as the reference electrode. Potassium nitrate is a supporting electrolyte, so this polarogram constitutes essentially a blank that defines the usable applied potential range for aqueous solutions. At a slightly positive applied potential (around $+0.1$ V versus SCE), there is an anodic wave resulting from the oxidation of mercury metal. This oxidation imposes a limit at the positive end of the applied potential range. The large cathodic wave starting around -1.8 V results from the reduction of potassium ions from the supporting electrolyte. Thus, the usable polarographic range with a dropping mercury electrode is approximately $+0.1$ to -1.8 V versus SCE. With other supporting electrolytes, this may be extended slightly at the negative end. If a platinum electrode is used instead of a dropping mercury electrode, more positive potentials can be employed, but the negative end of the potential range is severely limited because of the ease with which hydrogen is discharged at a platinum electrode.

> Note that the potential of the dropping mercury electrode is usually given in volts with respect to the saturated calomel electrode (SCE). The potential versus the standard hydrogen electrode (SHE) can be easily calculated. For example, if the applied potential is -0.500 V versus SCE, then it is $-0.500 + 0.246$, or -0.254 V versus SHE.

Figure 15–10 shows the polarogram for a $0.1M$ potassium nitrate aqueous solution saturated with air. Here two cathodic waves are observed. The first wave represents a two-electron reduction of oxygen to hydrogen peroxide:

$$2H_2O + O_2 + 2e^- \rightarrow H_2O_2 + 2OH^-$$

The second wave (at a more negative applied potential) is for the further reduction of

Figure 15–9. Polarogram of 0.1M potassium nitrate saturated with N$_2$, using a dropping mercury electrode. (Courtesy of Dennis Johnson, Iowa State University.)

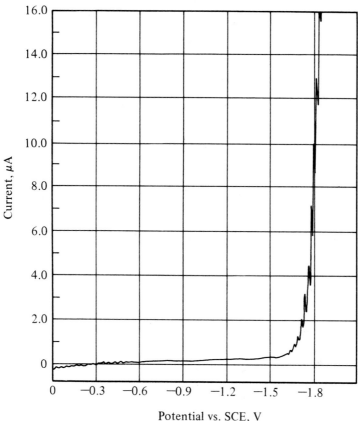

hydrogen peroxide:

$$H_2O_2 + 2e^- \rightarrow 2OH^-$$

The overall reduction is the sum of the two steps.

$$2H_2O + O_2 + 4e^- \rightarrow 4OH^-$$

Since the reduction of oxygen covers a large part of the available potential range, it is *almost always* necessary to remove dissolved oxygen from water before making a polarographic determination.

Figure 15–11 shows a well-defined polarographic wave for the reduction of cadmium (II) in 0.1M potassium nitrate. The dissolved oxygen was first removed by purging the solution with nitrogen for several minutes. Cadmium(II) could be determined quantitatively by measuring the peak diffusion current at an applied potential on the plateau region and then reading the concentration from a calibration curve run under the same conditions.

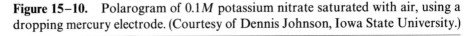

Figure 15–10.   Polarogram of $0.1M$ potassium nitrate saturated with air, using a dropping mercury electrode. (Courtesy of Dennis Johnson, Iowa State University.)

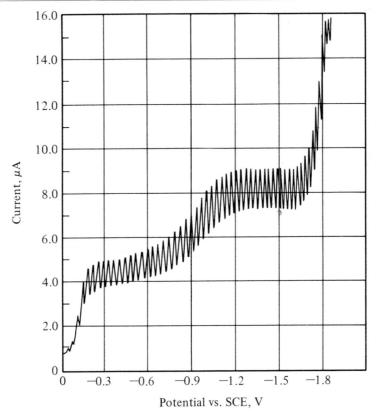

Half-Wave Potentials.   In polarography, an electroactive substance is characterized by its half-wave potential. This is the potential at a point on the polarogram where the current i is equal to one-half of the diffusion current i_d (see Figure 15–6). Note that the residual current is measured (or extrapolated) and subtracted, so it is not included in either i or i_d.

For a reversible system, the polarographic half-wave potential is related to the amalgam standard reduction potential (E^0); the half-wave potential, E_{half}, differs from E^0 somewhat, depending on the relative magnitude of the diffusion coefficients for the oxidized and reduced species. A few half-wave potentials are given in Table 15–2. Note that the supporting electrolyte is often buffered at a certain pH and sometimes also contains a complexing agent. It is possible to alter the half-wave potential (and, thus, the selectivity of a polarographic method) by the proper choice of complexing agent and acidity in the supporting electrolyte.

In Table 15–2, a *maximum suppressor* is mentioned. The currents of many polarographic waves have a disturbing tendency to go through a maximum before returning to the plateau region. To prevent this, a trace of gelatin or some surface-active agent is added as a maximum suppressor.

Figure 15–11. Polarogram of $1.0 \times 10^{-3}M$ cadmium(II) plus $0.1M$ potassium nitrate saturated with N_2. (Courtesy of Dennis Johnson, Iowa State University.)

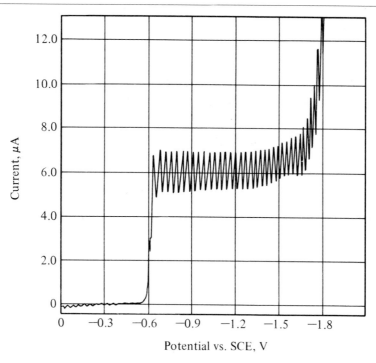

Potential vs. SCE, V

Table 15–2. Some Polarographic Half-Wave Potentials

Reaction	$E_{1/2}$ vs. SCE	Supporting Electrolyte
$Cu^{2+} \rightarrow Cu$	$+0.04$	$0.1M$ KCl
$Sn^{4+} \rightarrow Sn^{2+}$	-0.25	$4M$ NH$_4$Cl-$1M$ HCl
	-0.52	
$Pb^{2+} \rightarrow Pb$	-0.40	$0.1M$ KCl
$Pb^{2+} \rightarrow Pb$	-0.50	$0.5M$ sodium tartrate (pH 9)
$Cd^{2+} \rightarrow Cd$	-0.60	$0.1M$ KCl
$Pb^{2+} \rightarrow Pb$	-0.76	$1M$ NaOH
$Zn^{2+} \rightarrow Zn$	-1.00	$0.1M$ KCl
$Ni^{2+} \rightarrow Ni$	-1.1	$0.01M$ KCl
$Zn^{2+} \rightarrow Zn$	-1.15	$0.5M$ sodium tartrate (pH 9)
$Mn^{2+} \rightarrow Mn$	-1.51	$1M$ KCl
$Zn^{2+} \rightarrow Zn$	-1.53	$1M$ NaOH

Note: In most cases, the supporting electrolyte also contains 0.01% gelatin as a maximum suppressor.

In general, a substance can be determined polarographically if its half-wave potential is within the range specified (some substances undergo reduction in steps and give more than one polarographic wave). Mixtures of two or more substances can often be analyzed from the heights of the successive waves, provided there is a reasonable difference (say 0.2 V) between their half-wave potentials. The curve for a mixture of lead(II) and cadmium(II) (Figure 15–12) is an example. Organic compounds containing any of several functional groups can be determined by the polarographic method, despite the fact that the polarographic reduction of most organic compounds is irreversible.

Perhaps it should be repeated that polarography is used only to determine small amounts of substances in solution, and that a relative error of 1% or more is encountered. However, larger amounts of many substances can be determined by amperometric titration with accuracy comparable to ordinary titrations.

Amperometric Titrations. The polarographic method can be used to locate the equivalence point in a titration involving a reducible substance. A constant potential

Figure 15–12. Polarogram of lead(II) and cadmium(II).

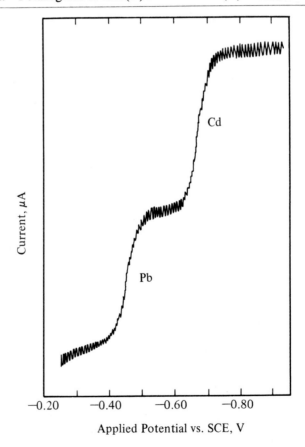

Applied Potential vs. SCE, V

Figure 15–13. Amperometric titration of bismuth(III) and EDTA.

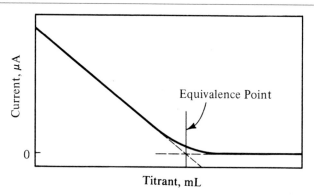

difference is applied between the indicator and reference electrodes, and the current passing through the cell is measured and plotted against the volume of titrant added. The applied potential is chosen so that the current will be on the plateau (diffusion current) region of the polarogram of either the substance titrated or the titrant. Under these conditions, the current is proportional to the concentration of the substances reduced at the applied potential.

The amperometric titration of bismuth with EDTA is a good example of an amperometric titration procedure. Bismuth(III) is titrated at pH 1–2 with a standard EDTA solution. The potential of the indicator electrode (-0.16 to -0.20 V versus SCE) is such that bismuth(III) is reduced, but the bismuth-EDTA complex and the titrant are not. The titration curve has the shape shown in Figure 15–13. The current decreases linearly as titrant is added and the concentration of uncomplexed bismuth(III) is decreased. After the equivalence point, virtually all of the bismuth(III) has been complexed by EDTA, and the current does not change.

Two other types of curves are possible. In Figure 15–14 the amperometric titration curve is shown for a situation in which only the titrant is reduced at the

Figure 15–14. Amperometric titration of magnesium(II) with oxine: $E = -1.6$ V versus SCE.

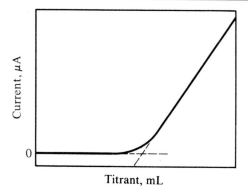

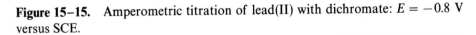

Figure 15–15. Amperometric titration of lead(II) with dichromate: $E = -0.8$ V versus SCE.

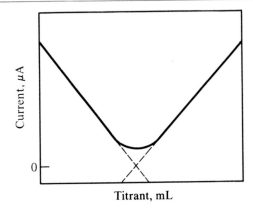

applied potential. Figure 15–15 shows an example of an amperometric titration in which both the substance titrated and the titrant are reduced under the conditions of the titration. Note that amperometric titration curves usually show some curvature in the vicinity of the equivalence point. The equivalence point is obtained by extrapolating the straight-line portions of the curve. Such extrapolation is useful because it permits a quantitative titration in cases where the equilibrium constant for the titration reaction is too small for an accurate titration with either a visual or a potentiometric end-point indicator.

The equipment for an amperometric titration can be somewhat simpler than for a polarographic analysis. The applied potential need only be controlled to about ± 0.1 V, and no thermostat is required because temperature changes are usually slight during a titration. Either a conventional dropping mercury electrode or a motor-driven rotating platinum microelectrode [5] serves as the indicator electrode. Unfortunately, it is still necessary to purge oxygen from the solution with nitrogen gas; furthermore, in an amperometric titration, the purging must be repeated (for one or two minutes) after adding each increment of titrant.

Amperometric Titrations Using an Indicating Electrode Pair. An interesting modification of the amperometric titration method employs two small platinum electrodes dipping into the titration cell. A small, constant voltage is impressed across these electrodes, and any current that flows is measured with a galvanometer. At the end point, the current either goes to a minimum or else increases suddenly from nearly zero. This is often referred to as a *biamperometric* or *dead-stop* end point [6].

Titrating iodine (in a solution containing iodide) with arsenic(III) illustrates this method very well. A small, constant potential of perhaps 0.100 V is impressed

[5] H. A. Laitinen and I. M. Kolthoff, *J. Phys. Chem.*, *45*, 1061 (1941).
[6] C. W. Foulk and A. T. Bawden, *J. Am. Chem. Soc.*, *48*, 2045 (1926).

electrically across the two small platinum electrodes immersed in the solution. The normal potential of each electrode would be that calculated from the Nernst equation for the iodine-iodide couple. The impressed potential will cause one electrode to become slightly more negative than the reversible potential, and iodine will be reduced to iodide. The other electrode becomes more positive than the reversible potential, and iodide is oxidized to iodine. Thus, the galvanometer will indicate a current flow between the two electrodes.

> It should be pointed out that the arsenic(V)-arsenic(III) system is irreversible; hence, arsenic will not be oxidized or reduced at the platinum electrodes and will not affect the potential except through the reaction of iodine with arsenic(III).

As the titration proceeds, the current will remain almost constant through most of the titration. A little before the stoichiometric point, however, the concentration of free iodine will decrease enough that iodine cannot reach the negative electrode fast enough to maintain the previous reduction rate. The iodine concentration at the electrode surface is now diffusion controlled. In this region, the current decreases almost linearly with additional increments of titrant, as in a conventional amperometric titration. The end point of the titration comes when the iodine concentration has been reduced essentially to zero. On the titration curve, this is indicated by the current dropping to a minimum that does not change with the addition of more titrant (Figure 15–16).

> At the platinum anode, plenty of iodide remains for oxidation to iodine. However, it must be remembered that oxidation at one electrode can take place only if there is an accompanying reduction at the other. Thus, the current in this case is controlled by the availability of iodine at the microcathode.

Titrating a substance that is not part of a reversible redox system with a reversible titrant gives a curve that is the reverse of the example in Figure 15–16; at the end point, the current rises abruptly from near zero. If both the titrant and the

Figure 15–16. Curve for the titration of iodine with arsenic(III), using two indicating electrodes. The magnitude of the plateau current depends on the voltage impressed across the electrode pair.

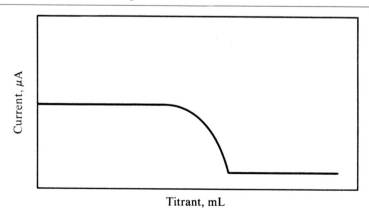

Titrant, mL

substance titrated are part of reversible systems, a "V"-shaped curve results. One advantage of the biamperometric end point is that graphical plotting is not required and that the biamperometric end point is often sharper than that in an ordinary amperometric titration.

QUESTIONS AND PROBLEMS

General Principles

1. Define *cathode* and *anode* in terms of the type of electrode reaction occurring at each. Which is the negative electrode in a galvanic cell? Which is the negative electrode in an electrolytic cell?
2. Define each of the following: (a) back emf, (b) polarization, (c) overpotential, (d) overvoltage.
3. Explain what is meant by concentration polarization. How does the efficiency of stirring affect concentration polarization?
4. A schematic diagram for a galvanic cell is shown below:

$$Hg, Hg_2Cl_2(s)/Cl^-(0.10M)//Fe^{3+}(1.0mM), Fe^{2+}(1.0mM)/Pt$$

for which

$$E^0_{Hg_2Cl_2(s),Hg} = 0.268 \text{ V versus SHE}$$

$$E^0_{Fe^{3+},Fe^{2+}} = 0.771 \text{ V versus SHE}$$

  (a) Calculate the cell potential assigning the polarity of the Pt electrode (assume activity coefficients are unity).
  (b) Write the balanced chemical equation for the spontaneous cell reaction.

Electrode Position

5. A current of 0.050 A was applied to a cell for 45.0 sec to deposit some silver metal from a solution of Ag^+. At 100% current efficiency, calculate the number of grams of silver deposited. At wt Ag = 107.87, F = 96,487 coulombs/equivalent.
6. Explain why zinc(II) in acidic solution can be electrolytically reduced to the metal at a mercury cathode but not at a platinum cathode. Explain why zinc metal can be deposited quantitatively on a platinum cathode from *alkaline* solution.
7. Electrodeposition may be used to concentrate trace impurities prior to their analytical measurement. Outline a possible method for separating and determining traces of copper and silver in nickel metal.

Coulometry

8. What advantages do coulometric methods of analysis have over electrogravimetric methods? What disadvantage?
9. Explain why cerium(III) is added in the constant-current coulometric determination of iron(II).
10. Suggest a constant-current method for the coulometric titration of iron(III). How might the end point in this titration be detected?

11. A coulometric titration of lactic acid, $CH_3CH(OH)COOH$, is performed with hydroxide ion coulometrically generated at a constant current of 19.3 mA. If a sample requires 154.1 sec for titration, what is the number of milligrams of lactic acid in the sample? Assume 100% current efficiency.

12. Outline an oxidation-reduction method suitable for the coulometric titration of iron(III). The method should be for iron(III) without prior reduction to iron(II).

13. Suppose you use the primary standard acid KHP (MW = 204.2 g/mole) to check a coulometric titration with electrically-generated OH^-.

$$KHP \rightarrow K^+ + HP^-$$

$$\underline{\text{Generator Electrode}}$$

$$2H_2O + 2e \rightarrow 2OH^- + 2H_2$$

$$OH^- + HP^- \rightarrow H_2O + P^{-2} \qquad \text{(titration reaction)}$$

At a constant current of 0.100 A, it takes 20.0 min, 2 sec to titrate a 0.2550-g KHP sample to a phenolpthalein end point. Calculate the mass of KHP as determined by the coulometric titration.

$$F = 96,487 \text{ coulombs/equivalent}$$

Voltammetry and Polarography

14. Explain what is meant by each of the following terms:
(a) cathodic wave, (b) half-wave potential, (c) residual current, (b) diffusion current, (e) supporting electrolyte.

15. Explain why a dropping mercury electrode is so widely used. When must a platinum electrode be used instead?

16. In a polarographic determination, explain why measuring the sample and the calibration standards at different temperatures will cause an error.

17. In a polarographic determination, a metal ion is reduced at the dropping mercury electrode. Write the electrochemical reaction that occurs simultaneously at the calomel reference electrode. What effect, if any, does this reaction have on the potential of the reference electrode?

18. Explain why dissolved oxygen must be removed from aqueous solution before making a polarographic determination. How is the oxygen removed?

19. Suggest a polarographic method for determining both oxygen and hydrogen peroxide in aqueous solution.

20. An organic compound in aqueous solution has a well-defined cathodic wave at $E_{1/2} = -0.23$ V versus SCE. If the diffusion current, concentration of the organic compound, and its approximate diffusion coefficients are all known, tell how to determine the number of electrons, n, involved in the cathodic wave.

21. Explain why the current oscillates with each mercury drop in polarography.

22. In voltammetry, an electroactive chemical in solution can reach the working electrode by diffusion, convection, and electromigration. Explain how the two latter transport mechanisms can be avoided.

23. Explain what effect rapid stirring will have on the current obtained in voltammetry.

24. Stripping voltammetry utilizes a preconcentration of a species to be determined by electrodeposition. The deposited material is subsequently stripped (removed) from the electrode surface and determined by the current peak produced during the stripping process. Find a published paper that uses stripping voltammetry. Give the journal reference and a brief description of the method.

25. Find a published paper on chromatography that uses an electrochemical detector. Give the journal reference and briefly describe the detector.

Amperometric Titrations

26. Compare and contrast polarography and amperometric titration as quantitative analytical methods.
27. Compare the amperometric method for end-point detection with the biamperometric (dead-stop) end point. What advantages and disadvantages does the latter method have?
28. In a titration with a biamperometric end point, explain how an increase in the potential applied to the platinum electrode pair will affect the magnitude of the current measured.

16

Potentiometric Determinations with Ion-Selective Electrodes

In Chapter 12 (Section 12–2) an electrical potential was defined as the measured tendency of the oxidized and reduced forms of an element, molecule, or ion to gain or lose electrons, as represented by a redox half-reaction:

$$A_{ox} + ne^- = A_{red} \qquad (16–1)$$

The potential of such a system developed at an electrode, which we will call the *indicating electrode*, is given at 25°C by the following form of the Nernst equation:

$$E = E^0 - \frac{0.05915 \text{ V}}{n} \log \frac{a_{A_{red}}}{a_{A_{ox}}} \qquad (16–2)$$

In this equation, E is the potential compared to the SHE, E^0 is the standard reduction potential, and $a_{A_{red}}$ and $a_{A_{ox}}$ are the activities of the reduced and oxidized forms of the electrode chemicals, respectively. By connecting the indicating electrode to an electrical circuit with a reference electrode such as the saturated calomel electrode (SCE) and measuring the resulting voltage with a *potentiometer* (or a modern voltmeter), the ratio of $a_{A_{red}}$ to $a_{A_{ox}}$ can be calculated. This can be done periodically to follow the course of an oxidation-reduction titration (See Chapter 12, Section 12–6).

Often the potential of an indicating electrode will depend only on the activity of a single ionic or molecular species in solution. For example, the potential of a silver metal electrode in contact with silver ions in solution varies with the activity of these ions:

$$E = 0.800 - 0.05915 \log \frac{1}{a_{Ag^+}}$$

In such a case, the voltage at an *indicating* electrode (in an electrical circuit with a reference electrode) can be used to measure the activity of the indicating electrode ions (Ag^+ in this case) in solution. Such a measurement is called a *direct potentiometric measurement*.

Just as a silver metal electrode can be used to measure the activity of silver(I) ions in solution, other electrodes have been devised that will measure only the activity of certain ions in solutions. These are called ion-selective electrodes and are now very widely used in analytical chemistry. The glass electrode is used for measuring hydrogen-ion activity in solution (pH) and is unaffected by other ions over most of the pH range in water. Ion-selective electrodes are available that will measure the activity of fluoride, chloride, nitrate, sodium, and potassium ions in solution, each with reasonably good selectivity for the desired ion.

Direct potentiometric measurement, including direct measurement of pH, is one of the truly fascinating methods of chemical analysis. No titration is required, and not even the addition of energy such as light energy in spectrophotometry (Chapter 17) is necessary. The electric cell generates its own electrical energy, but no significant amount of electricity need be passed to make a potentiometric measurement. In fact, the current passed must be almost infinitesimal, or else the concentration of the ion determined would change during the measurement. How then is it possible for an ion to be measured by just inserting the above two electrodes into a solution? The answer lies in the nature of a potentiometric measurement, which can be understood by reading Section 16–1 thoroughly.

Before turning to that section, consider some of the unique features of modern potentiometric measurement. In contrast to the measurement of *amounts* in gravimetry and titrimetry, potentiometry measures the *activity* (which is related to *concentration*) of a selected ion. Direct potentiometric measurements are also unique in that they yield the activity of only *uncomplexed* metal ions or anions, not the *complexed* forms. For example, in an equilibrium mixture of Ag^+, $Ag(NH_3)^+$, and $Ag(NH_3)_2^+$, potentiometric measurement with a silver metal indicating electrode gives the activity of only the Ag^+ ion. A precipitation titration, in contrast, would shift the equilibrium and precipitate the sum of all three silver(I) species.

Current research on potentiometric measurements with ion-selective electrodes is reviewed every two years in the April "Fundamental Reviews" issue of *Analytical Chemistry* [1]. Writings by Freiser [2], Durst [3], and Bates [4] are also important.

16–1.   THE POTENTIOMETER, pH METER, AND MICROPROCESSOR pH/mV METER

The three components required for a potentiometric measurement are the reference electrode, the indicating electrode, and some type of potentiometer or electronic voltmeter. The indicating electrode *senses* the activity or concentration of an ion;

[1] See, for example, M. A. Arnold and M. E. Meyerhoff, *Anal. Chem. 56*, 20R (April 1984).

[2] H. Freiser, ed., *Ion-Selective Electrodes in Analytical Chemistry*, New York: Plenum Press, Vol. 1, 1978; Vol. 2, 1980.

[3] R. A. Durst, Ion-Selective Electrodes in Science, Medicine, and Technology, *Am. Scientist 59*, 353 (1971).

[4] R. M. Bates, *Determinations of pH: Theory and Practice*, 2nd ed, New York: John Wiley, 1964.

such an electrode can take many forms and will be discussed separately in the next section. The reference electrode is less complicated and so can be discussed briefly before the various types of meters.

The Saturated Calomel Reference Electrode (SCE).   A reference electrode has a fixed potential that is not affected by the test solution. Most commonly, a *saturated calomel reference electrode* (SCE) is used. A drawing of the SCE is shown in Figure 16–1. Actually, this is an electrode and a salt bridge encased in a single unit. The lead is attached to the inner tube containing mercury metal and calomel (Hg_2Cl_2) paste. This is in contact with a potassium chloride solution. The potassium chloride solution also serves as a bridge to the test solution via an asbestos fiber or a porous ceramic cylinder. The potential of the calomel electrode depends on the concentration of the potassium chloride solution used. The saturated calomel electrode has a potential of $+0.246$ V versus the standard hydrogen electrode. The potential of a calomel electrode containing $1.0M$ potassium chloride is $+0.281$ V; one containing $0.1M$ potassium chloride has a potential of $+0.333$ V. For measurements when chloride ions might leak into the sample solution and affect the indicator-electrode potential, a double-junction reference electrode is used in which a second electrolyte is interspersed between the potassium chloride salt bridge and the sample solution.

The Potentiometer.   If an indicator electrode and a reference electrode are connected through a *voltmeter* and inserted into a test solution, a galvanic cell is set up. The emf (voltage) of the cell is determined by the concentration of the ions to be measured. Unfortunately, a conventional voltmeter draws enough current from the cell to cause a decrease in the concentration of the ion to be measured, so the emf in this type cell cannot be measured with a voltmeter. To maintain the equilibrium

Figure 16–1.   A calomel electrode.

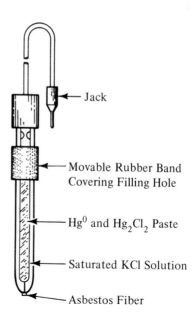

- Jack

- Movable Rubber Band Covering Filling Hole

- Hg^0 and Hg_2Cl_2 Paste

- Saturated KCl Solution

- Asbestos Fiber

potentials at the electrodes, the emf must be measured without drawing appreciable current from the cell. The instrument classically used for this is a *potentiometer*. Titrations in which the course of the titration is followed by means of potentiometric measurements are called *potentiometric titrations*.

A diagram of a simple potentiometer is shown in Figure 16–2. In a potentiometer, a variable voltage opposes the voltage of the cell being measured. A sensitive galvanometer, G, acts as a null-point detector, to indicate when the variable voltage has been made exactly equal to the opposing cell-voltage.

The variable voltage in a potentiometer is provided by a linear voltage divider, a high-resistance, linear slide wire. The resistance of any given length of the slide wire is proportional to the resistance of the entire length. The *potential difference* between any two points on the slide wire is also proportional to the ratio of the distance between the points and the total length of the wire, because the current that flows through any given length of wire is the same as that flowing through the entire wire. (This is because a finite current flows from the negative battery terminal through the slide wire and back to the positive battery terminal. No appreciable current flows through the lower loop of the circuit shown in Figure 16–2, because the emf of the cell is opposed by the equal emf from the battery.) Thus, a linear voltage scale can be placed alongside the slide wire. To make this scale read accurately in volts, a *standard cell* (such as a Weston cell) is switched into the circuit in place of the cell to be measured. The slide-wire scale is set at the exact voltage of the Weston cell (1.0186 V) and resistance R is adjusted so that no current passes through the galvanometer G. By switching the unknown cell into the circuit in place of the Weston cell and adjusting the slide wire until the null point of the galvanometer is again attained, the emf of the unknown cell can be read from the position of the slide wire.

In principle, the potentiometric measurement should always be 99.99+% accurate. In practice, however, the standard cell and the unknown cell will be out of balance until the null point is reached. Small or even significant amounts of current may flow before the final null point is attained. Thus, the true activity or concentration may not always be measured. For this reason, electronic voltmeters are now used instead of potentiometers.

The pH/mV Meter. The modern pH/mV meter measures pH or mV (millivolts)

Figure 16–2. Diagram of a simple potentiometer.

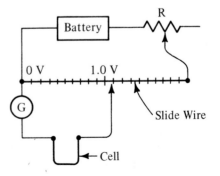

and is an electronic voltmeter with an operational amplifier designed to have a high input impedance. Although some current does flow through the voltmeter, its resistance is so high that the amount of current that flows is too small to affect significantly the concentration of the ion measured. The voltage readout can be then calibrated to read in pH units, mV (millivolts), or activity or concentration of a specific ion.

Conventional Direct-Reading pH Meters. The conventional pH meter is simply a special electronic voltmeter that is calibrated to read directly from pH 0 to 14. (The differences between it and the microprocessor pH meter will be discussed later in this section.) The pH meter uses the *glass electrode* (Section 16–2) as an indicating electrode and the SCE reference electrode. The glass electrode has an unusually high electrical resistance (1–100 megohms), requiring that the voltmeter be designed to handle such a resistance.

All pH meters have a pH 0–14 readout calibrating for measuring hydrogen-ion activity. However, several steps are required for standardizing the pH meter to obtain accurate pH readings:

1. The pH meter and electrodes must be allowed to warm up and stabilize before use.
2. The temperature of the buffer(s) and test solution must be entered into the pH meter.
3. The pH meter must be standardized by using two standard buffers, one on either side of the pH of the test solution.

Standardization is done by immersing the electrodes in a standard buffer solution of *accurately known pH* and adjusting the pH meter readout so that it reads the same pH as that of the buffer. Periodic standardization is required to compensate for the asymmetry potential (see p. 314) of the glass electrode. For best accuracy, two standardizing buffers should be used, one on either side of the pH of the solution to be measured. Ammonium acetate (pH 7.00), potassium acid phthalate (pH 4.01), and sodium tetraborate (pH 9.18) buffers are widely used for standardizing purposes.

Error can arise in potentiometric measurements when the activity of foreign ions is high enough to affect the potential of the indicator electrode. In pH measurements, the *alkali error* occurs above approximately pH 10, where the hydrogen ion concentration is extremely low and the glass electrode begins to respond to alkali-metal ions. The magnitude of this error increases as the pH becomes higher; the error may be as much as 1 pH unit for highly alkaline solutions containing sodium ions. Alkali-metal ions of larger ionic radius cause less error than sodium ions. Special glass electrodes have been developed for measuring pH in alkaline solutions; these reduce, but usually do not eliminate, the alkali error. Readings at the lower end of the pH scale (near pH 0) tend to be somewhat high and unreliable. This is known as the *acid error*.

A potential known as *liquid-junction potential* develops at the interface of any two solutions of different composition. The salt bridge used in the calomel electrode minimizes the liquid-junction potential because K^+ and Cl^- carry almost identical fractions of the current, but there is still some uncertainty in the measurement of cell emf because of this effect. This limits the accuracy of measurements with an ordinary pH meter to ± 0.01 to 0.02 pH units.

The Microprocessor pH/mV Meter.   Many modern pH meters are now interfaced with a microprocessor to simplify operation and prevent operator errors. When using the conventional pH meter, the operator allows the electrodes and meter to warm up to stabilize, inserts the electrodes into the buffer for standardization after adjusting the temperature control on the meter, and finally measures the pH of the test solution. All of these steps require judgment and may involve errors, especially on the part of an inexperienced operator.

When using the microprocessor pH/mV meter, the operator still turns the instrument on and inserts the electrodes into the buffer (and later the test solution), but the microprocessor performs certain functions involving arbitrary operator judgment (when is the instrument ready, etc.). When the instrument is turned on, the microprocessor continuously monitors electrode and instrument stability and notifies the operator, often by illuminating a pushbutton, when readings may be taken. Many microprocessors have a preprogrammed table of buffer-temperature pH values that avoid errors from temperature differences. The microprocessor also can sense if the operator has used a wrong buffer, such as pH 4.2, for a desired buffer pH, such as 7.2. A warning indicator lights up on the display of the pH meter to signal this. Finally, the microprocessor pH/mV meter differs from the conventional digital pH meters in that it uses a built-in program to calculate the pH from the measured voltage rather than the pH being obtained from analog circuit manipulation. The user sees no difference, of course, but this allows the introduction of pH buffers in any order when the meter is being standardized.

THREE ION-SELECTIVE INDICATING ELECTRODES

In the next three sections, we will discuss three indicating electrodes used for measurement *of ions*: the glass electrode, the liquid-membrane electrode, and the solid-state electrode. Briefly, these electrodes measure the following types of ions:

Glass electrodes.   In practice, these measure only $+1$ cations (H^+, Na^+, etc.).

Liquid membrane electrodes.   These measure mainly $+2$ cations and a few $+1$ anions.

Solid-state electrodes.   These measure mainly -1 anions and a few $+2$ cations.

16-2. THE GLASS ELECTRODE: MEASUREMENT OF $+1$ CATIONS

Ions Measured.   Commercial glass electrodes respond strongly only to $+1$ ions, including H^+; electrodes containing some Al_2O_3 also respond weakly to $+2$ ions, but their greater response to $+1$ ions makes them useless for $+2$ ions. However,

Vlasov and Bychkov [5] have shown that glass electrodes based on $Fe_2(Ge_{28}Sb_{12}Se_{60})_{98}$ can measure the Fe^{3+} and Cu^{2+} ions. In addition, Nomura and Nakaguwa [6] have shown that glass electrodes made from alkali-free magnesium phosphate glass containing Ag_2O can measure ligands that complex the silver(I) ion, including ammonia and certain anions.

The commercial glass electrodes measure +1 cations only, using glasses with favorable physical properties and a specific composition such as 65% SiO_2, 28% Li_2O, 4% La_2O_3, and 3% Cs_2O. Some +1 ions that can be measured potentiometrically are H^+, Group IA ions such as Li^+, Na^+, and K^+, and ions such as Ag^+ and NH_4^+. The present commercial glass electrodes respond weakly or not at all to +2 and +3 cations because the -1 reactive sites (see below) of the glass are *fixed*. Thus two such sites cannot move together to attract a +2 ion. Because the H^+ ion has the strongest attraction to the -1 reactive sites of the glass electrode, the measurement of any other +1 cation will be subject to error in acid solution. Thus, such samples are often buffered at a neutral or basic pH to eliminate this error.

Construction.   The essential elements of a glass electrode are schematically shown in Figure 16–3. To provide electrical contact, an internal reference electrode of silver wire coated with silver chloride is immersed in an electrolyte solution. For a hydrogen ion–sensitive glass electrode, the electrolyte is dilute hydrochloric acid.

Figure 16–3.   A glass electrode.

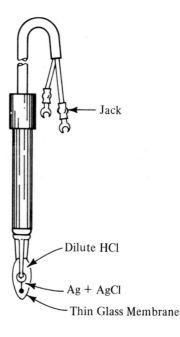

Jack

Dilute HCl

Ag + AgCl

Thin Glass Membrane

[5] Y. G. Vlasov and E. A. Bychkov, *Hung. Sci. Instrum.* *53*, 35 (1982).
[6] T. Nomura and G. Nakaguwa, *Bull. Chem. Soc. Jpn.*, *56*, 943 (1983).

All this is surrounded by a bulb containing a membrane made of a special glass.

glass membrane
↓
Ag, AgCl(s), 0.1M HCl // test solution
(internal reference electrode)

The potential of the internal reference electrode is constant, but a potential difference develops *across the glass membrane* that is a function of the difference between the hydrogen ion activity of the hydrochloric acid solution inside the electrode and that of the test solution.

The potential of an electrode selective for an ion is given by the following equation:

$$E = K + \frac{RT}{z_iF} \ln \frac{a_i}{a_i'} \qquad (16\text{--}3)$$

where E is the potential of the glass electrode, K is a constant, z_i is the charge on the ion i, F is the Faraday (96,487 coul/eq), a_i is the activity of the ion in the test solution, and a_i' is the activity of the ion in the internal filling solution. If the internal filling solution is a salt of constant concentration, Equation 16–3 can be written:

$$E = \text{constant} + \frac{2.3RT \log}{z_iF} a_i \qquad (16\text{--}4)$$

The constant in Equation 16–4 is made up of K from Equation 16–3 and the logarithmic function of a_i'.

At 25°C, Equation 16–4 may be written as follows for an electrode sensitive to a univalent ion such as H^+, Na^+, or K^+:

$$E = \text{constant} + 0.059 \log a_i \qquad (16\text{--}5)$$

For a hydrogen ion–sensitive glass electrode, this equation may be written in terms of pH:

$$E = \text{constant} - 0.059 \text{ pH} \qquad (16\text{--}6)$$

One of the components of the constant, an *asymmetry potential*, is a small potential across the glass membrane that is present even when the solutions on either side of the membrane are identical.

Theory of the Glass Electrode.   Consider a section of the bulb of a glass electrode (Figure 16–4). In contrast to a metallic electrode, the glass membrane of this bulb has *two surfaces*, both of which are hydrated. For measurements to be made at the outer surface, electrical and chemical processes at the inner surface have to be kept constant by filling the bulb with a hydrochloric acid solution of constant concentration.

Glass Membrane −1 Reactive Sites.   The glass used for electrodes always contains SiO_2 (60–80%); to measure pH, Li_2O, La_2O_3, and Cs_2O are added. Hydration of the glass in a pH glass electrode forms *fixed −1 silicate reactive sites* ($-SiO^-$). Lithium ions and a few hydrogen ions also form; both are relatively free

Figure 16–4. Schematic drawing of a glass membrane.

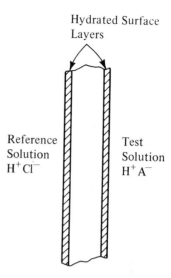

Hydrated Surface Layers

Reference Solution H⁺Cl⁻

Test Solution H⁺A⁻

to migrate within the hydrated layer. When a new glass electrode is immersed in an aqueous solution, most of the Li^+ in the outermost hydrated layer is exchanged for H^+ from the solution:

$$-SiO^-Li^+ + H^+ \rightarrow -SiO^-H^+ + Li^+$$

(glass)   (solution)   (glass)   (solution)

The outer surface of the pH electrode may be thought of as an ion exchanger with a much stronger affinity for H^+ than for Li^+ or other cations.

After a new glass electrode has been soaked in an aqueous solution to hydrate the outer surface and convert the outer ion-exchange layer largely to the hydrogen form, it is ready for use. A potential quickly builds up on the outer surface; this potential is a function of the hydrogen-ion concentration of the test solution. A plausible explanation is that some hydrogen ions from the test solution enter the hydrated glass layer, where the activity is less than in the aqueous solution. Anions are repelled from the hydrated glass by the negatively charged silicate sites, which are fixed and not as free to move about as the cations in the hydrated glass. Additional hydrogen ions are prevented from entering the glass by the positive charge of the microscopic layer of excess hydrogen ions that builds up. This is the chief source of the potential at the glass-solution interface.

*Macro*scopic electroneutrality must be maintained, however. This means that the positive cation charge and the negative anion charge must be equal both for the bulk of solution and for the glass.

Of course, there are two glass-liquid interfaces, one with the test solution and the other with the interior electrolyte. One surface is negatively charged *with respect to the other surface.* Since the test solution is usually more dilute in H^+ than in the rather concentrated inside solution, the outer surface is negatively charged with respect to the inside. The potential changes only as the hydrogen concentration of the test solution changes. This change is given by Equation 16–5 or 16–6.

To make a potentiometric measurement with a glass electrode in conjunction with a calomel or other reference electrode, *some* current must be drawn, even though a very minute amount. If hydrogen ions do not pass through the membrane of a glass electrode, how can any current be obtained? The answer seems to be that current is carried through the glass by monovalent cations in the glass (mostly lithium). No single lithium ion moves through the glass; instead, each ionic charge carrier moves only a few atomic diameters before passing on its energy to the next carrier. A mechanical analog is a row of bowling balls, each touching the next. On hitting the row at one end with another ball, the energy is transferred to the ball on the other end, which rolls away from the group.

Foreign ions in solution may affect the potential of a glass electrode. This is true whether the electrode is sensitive to H^+, Na^+, K^+, Ag^+, or other ions, although it turns out that the hydrogen ion–selective glass electrode is much less affected by foreign cations than the others. The following equation gives the potential for a glass electrode in a test solution containing the test ion A and a foreign ion of the same charge B, assuming an interior filling solution of constant concentration:

$$E = \text{constant} + \frac{2.3RT}{z_A F} \log \left[a_A + \left(\frac{u_B}{u_A} K_{A,B} \right) a_B \right] \qquad (16\text{--}7)$$

where u_A and u_B represent the mobilities of the ions *in the membrane* and $K_{A,B}$ is the ion-exchange equilibrium constant for the following exchange:

$$B_{(aq)} + A_{(membrane)} \rightleftharpoons B_{(membrane)} + A_{(aq)}$$

A value of 0.1 for the term $((u_B/u_A)K_{A,B})$ means that B has 0.1 times the effect in determining the electrode potential as the same molar concentration of A.

There are usually opposing effects between mobility and ion-exchange affinity in establishing the selectivity of an electrode. For example, the tenfold selectivity of K^+ over Na^+ for a potassium-selective glass electrode is the result of nearly a hundredfold preference in $K_{A,B}$ for K^+ opposed by a tenfold lower mobility of K^+ than Na^+.

Often the mobility and the ion-exchange equilibrium are combined into a single term called the *selectivity coefficient*, which we shall write simply as $k_{A,B}$. Then Equation 16–7 can be written as follows:

$$E = \text{constant} + \frac{2.3RT}{z_A F} \log \left[a_A + k_{A,B}(a_B)^{z_A/z_B} \right] \qquad (16\text{--}8)$$

Equation 16–8 is a fundamental equation applicable to *all* types of ion-selective electrodes. At 25°C, Equation 16–8 becomes

$$E = \text{constant} + \frac{0.059}{z_A} \log \left[a_A + k_{A,B}(a_B)^{z_A/z_B} \right] \qquad (16\text{--}9)$$

If there is negligible interference from a foreign ion B (because the selectivity coefficient is very small or the activity of B is very low), a plot of E versus a_A gives a linear plot with a slope of 0.059 V per tenfold change in a_A. When the ion A measured by an ion-selective electrode is an *anion*, Equation 16–9 applies, but the sign in front of the 0.059 term is changed to a negative sign.

Potentiometric Measurement of Sodium Ion.   Glass electrodes for measurement of sodium ion are made of about 71% SiO_2, 18% Al_2O_3, and 11% Na_2O in contrast to potassium glass electrodes, which contain much less Al_2O_3 and much more Na_2O [7]. The value of the selectivity coefficient, $k_{Na,K}$, for the sodium electrode measurement of sodium in the presence of potassium varies with pH. At neutral pH, $k_{Na,K} = 3.3 \times 10^{-3}$, and at high pH, $k_{Na,K} = 1 \times 10^{-3}$. Thus, potassium ions cause little error in the measurement of sodium in body fluids because potassium is present in lower concentration. Although the concentration of Ca^{2+} is significant in body fluids, and although Al_2O_3-glasses (aluminosilicates) respond to Ca^{2+} ions, their response is much weaker to Ca^{2+}. For a glass of 70% SiO_2–10% Al_2O_3–6% CaO–14% Li_2O, $k_{Ca,Na} = 15$, indicating that the response to Na^+ is much higher than the response to Ca^{2+}.

Because the measurement of sodium in body fluids is so widespread, complete potentiometric systems have been made commercially. One such system consists of a sodium glass electrode, reference electrode, and microprocessor mV meter (potentiometer). It uses a microprocessor to correct for the presence of chloride ion in body fluids containing NaCl and provides direct readout of sodium activity or concentration.

16–3. LIQUID MEMBRANE ELECTRODES: MEASUREMENT OF +2 CATIONS AND ANIONS

Ions Measured.   Commercial liquid membrane electrodes are designed to respond strongly to +2 cations or to anions. However, researchers [8] have shown that certain organic "crown ether" polymers can be used in liquid membrane electrodes to measure potassium ion selectively in the presence of sodium ion ($k_{K/Na} = 6 \times 10^{-4}$) in urine samples.

Commercial liquid membrane electrodes are, like glass electrodes, designed to measure a single specific ion. Thus liquid membrane electrodes are made to measure Ca^{2+}, Mg^{2+}, Cu^{2+}, Pb^{2+}, Cl^-, NO_3^-, and ClO_4^-.

The key difference [9] between liquid membrane electrodes and glass membrane electrodes is that the *liquid membrane reactive sites are mobile and able to travel to the ion*, whereas such sites are *fixed* in the glass membrane. This enables two -1 reactive sites in the liquid membrane to travel to a $+2$ calcium ion at the outer surface of the electrode and react to form a neutral species, something a fixed -1 site in glass cannot do. Such reactive sites in liquid membrane electrodes in general have a stronger attraction for H^+ than for other cations, so measurements must be made in weakly acidic to weakly basic solution.

[7] G. Eisenman, *Adv. Anal. Chem. Instrum. 4*, 213 (1965).
[8] J. P. Willis, C. C. Young, L. Olson-Mank, and L. Radle, *Clin. Chem. 29*, 1193 (1983).
[9] G. A. Rechnitz, *J. Chem. Educ. 60*, 282 (1983).

Construction and Theory.   The essential elements in a liquid ion-exchange membrane electrode are shown schematically in Figure 16–5. Schematically, it is similar to a glass electrode in that it contains an internal reference electrode and an internal reference solution of fixed composition. Instead of glass, the membrane is a thin, porous organic polymer saturated with a liquid ion-exchanger dissolved in a water-immiscible organic solvent.

A liquid-membrane electrode without an internal reference electrode can also be made. This is accomplished simply by coating a platinum wire with a liquid ion exchanger in a polyvinylchloride polymer. The ion exchanger serves as the sensor for the ion to be measured. Freiser and associates (reference 2, Vol. 2, Chap. 2) give an excellent review of coated-wire ion-selective electrodes and their applications. Moody and Thomas (reference 2, Vol. 1, Chap. 4) also describe ion-selective electrodes that have a polyvinylchloride film containing a sensor.

There is a potential at each aqueous membrane-interface, but because the internal aqueous solution does not change, the potential varies only with the composition of the test solution, which affects the potential at the outer interface. The potential of a membrane selective for an ion A (in a solution also containing a foreign ion of the same charge B) is given by the equation

$$E = \text{constant} + \frac{2.3RT}{z_A F} \log \left[a_A + \left(\frac{u_B D_B}{u_A D_A} \right) a_B \right] \qquad (16\text{–}10)$$

Figure 16–5.   Liquid-membrane ion-selective electrode. (Courtesy of Orion Research, Inc.)

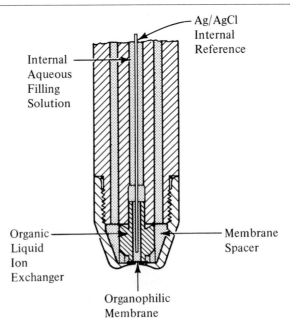

Internal Aqueous Filling Solution

Ag/AgCl Internal Reference

Organic Liquid Ion Exchanger

Membrane Spacer

Organophilic Membrane

where u_A and u_B are the mobilities of the two ions in the membrane and D_A and D_B are the distribution ratios of A and B, respectively, between the aqueous and membrane phases. (Distribution ratio is defined in Chapter 20.) For practical purposes, it is usually more convenient to use Equation 16–6 or Equation 16–7, remembering that the first plus sign is changed to minus when an anion-selective electrode is used.

The selectivity of a membrane electrode for a given ion is determined primarily by the liquid ion-exchanger used. Selectivity increases as the preference of the ion-exchanger for that ion over other ions increases. The organic solvent in which the liquid ion-exchanger is dissolved also affects D_A and D_B and hence the selectivity.

In Table 16–1, information on several liquid-membrane electrodes is summarized. The type of liquid ion-exchanger is indicated, and a partial list of selectivity coefficients (see Equation 16–8) is given.

The tolerance of an electrode for a given foreign ion may be calculated from the selectivity coefficient with the aid of Equation 16–9.

Example: Calculate the error caused by nitrite, $a_{NO_2^-} = 0.010$, in measuring nitrate, $a_{NO_3^-} = 0.001$, using a nitrate ion–selective electrode. Use the selectivity coefficient given for NO_2^- in Table 16–1.

Substituting into Equation 16–9,

$$E = \text{constant} - 0.059 \log [0.001 + (0.06)(0.010)]$$

$$= \text{constant} - (0.059)(-2.80)$$

$$= \text{constant} + 0.165 \text{ V}$$

The error can be ascertained by calculating the apparent activity of nitrate if the electrode potential was caused *only* by nitrate. Substituting into Equation 16–9:

$$\text{constant} + 0.165 = \text{constant} - 0.059 \log a_{NO_3^-}$$

$$\log a_{NO_3^-} = -2.80; \ a_{NO_3^-} = 0.0016$$

$$\text{Error} = \left(\frac{0.0016 - 0.0010}{0.0010} \right) \times 100 = +60\%$$

Example: Estimate whether $0.20M$ sodium(I) will cause a significant interference in measuring calcium(II), $a_{Ca^{2+}} = 0.0010$, other than the effect caused by an increase in total ionic strength.

From Table 16–1, $k_{Ca,Na} = 0.0016$. Substituting into Equation 16–9,

$$E = \text{constant} + \frac{0.059}{2} \log [0.0010 + 0.0016(0.20)^2]$$

$$E = \text{constant} + \frac{0.059}{2} \log [0.0010 + 0.00006]$$

The last term is equivalent to $a_{Ca^{2+}} = 1.06 \times 10^{-3}$, instead of the correct value of 1.00×10^{-3}. The error is thus

$$\frac{0.06 \times 10^{-3}}{1.00 \times 10^{-3}} \times 100 = 6\%$$

Table 16–1. Liquid-Membrane Electrodes. (R represents an organic group of several carbon atoms.)

Ion Measured	Exchange Site	Selectivity Coefficients
K^+	Valinomycin[a]	Na^+, 0.0001
Ca^{2+}	$(RO)_2POO^-$	Na^+, 0.0016
		Mg^{2+}, Ba^{2+}, 0.01
		Sr^{2+}, 0.02
		Zn^{2+}, 3.2
		H^+, 10^7
Ca^{2+} and Mg^{2+}	$(RO)_2POO^-$	Na^+, 0.01
		Sr^{2+}, 0.54
		Ba^{2+}, 0.94
Cu^{2+}	$RSCH_2COO^-$	Na^+, K^+, 0.0005
		Mg^{2+}, 0.001
		Ca^{2+}, 0.002
		Ni^{2+}, 0.01
		Zn^{2+}, 0.03
NO_3^-	$\left[\text{Ni}(\text{phen-R}) \right]^{2+}_3$	F^-, 0.0009
		SO_4^{2-}, 0.0006
		PO_4^{3-}, 0.0003
		Cl^-, Ac^-, 0.006
		HCO_3^-, CN^-, 0.02
		NO_2^-, 0.06
		Br^-, 0.9
ClO_4^-	$\left[\text{Fe}(\text{phen-R}) \right]^{2+}_3$	Cl^-, SO_4^{2-}, 0.0002
		Br^-, 0.0006
		NO_3^-, 0.0015
		I^-, 0.012
		OH^-, 1.0

Source: Data from R. A. Durst, ed., *Ion-Selective Electrodes*, National Bureau of Standards Special Publication 314, Washington, 1969, pp 70–71.
a. M. S. Frant and J. W. Ross, *Science*, *167*, 987 (1970).

Potentiometric Measurement of Calcium Ion. Liquid membrane electrodes for the measurement of calcium ion are usually made from phosphate ion exchange resins. Such resins are good choices for the measurement of calcium in body fluids because they have a greater affinity for calcium than magnesium, which is also present in such fluids.

The probable mechanism by which these electrodes operate can be illustrated by using the calcium-ion electrode as an example. The long-chain phosphoric acid molecules are oriented at each membrane interface so that the organic groups are sticking into the organic liquid phase while the more polar —POOH ends of the molecules extend into the aqueous part of the liquid junction. Soaking in an aqueous calcium(II) solution undoubtedly converts the compound to a calcium salt or a calcium chelate,

$$\left[(RO)_2P{\overset{O}{\underset{O}{\diagup}}}\right]_2 Ca^{2+} \quad or \quad \left[(RO)_2P{\overset{O}{\underset{O}{\diagup}}}\right]_2 Ca$$

If the test solution contains a low concentration of Ca^{2+}, there should be some tendency for calcium ions to migrate from the membrane surface to the test solution. Migration of the organic anion, $(RO)_2POO^-$, would be greatly hindered by its large size and by the strong solvation of the R groups by the organic solvent. Thus, the aqueous side of the liquid junction becomes positively charged with respect to the organic side. A similar junction potential (but of different magnitude) would be established at the junction between the membrane and the inside reference solution.

A big question is, why does this electrode respond so much more to calcium ions than to magnesium and sodium ions? The answer lies in the greater chemical affinity of the organic phosphate for calcium ions; hence, the ion-exchange equilibrium between cations on the membrane surface and in the aqueous solution depends mostly on the activities of calcium ions in the two phases.

These membrane electrodes measure the *activity* of the ion in question. The concentration of an ion such as Ca^{2+} can be calculated if the activity coefficient is known:

$$[Ca^{2+}] = a_{Ca+2}/f_{Ca+2}$$

In doing this calculation, it must be emphasized that the activity coefficient of an ion

Figure 16–6. Calibration plots for calcium electrode using activity and concentration. (Courtesy of Orion Research, Inc.)

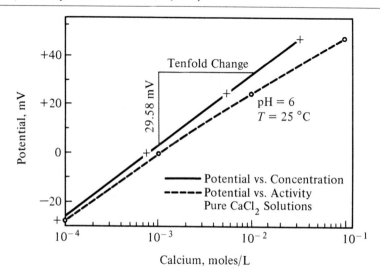

such as Ca^{2+} depends on the *total ionic strength* of the solution, not just on the ionic strength of the calcium salt. Thus, foreign ions do affect the response of selective membrane electrodes, but this effect can be corrected for by estimating the solution's ionic strength and using the appropriate activity coefficient to calculate the ion concentration being measured (Figure 16–6).

16–4. SOLID-STATE MEMBRANE ELECTRODES: MEASUREMENT OF ANIONS AND SOME CATIONS

Ions Measured. Commercial solid-state membrane electrodes are designed to respond strongly to certain anions or to a few $+2$ cations. However, research has shown that a solid-state electrode based on ammonium molybdophosphate, $(NH_4)_3PMo_{12}O_{40} \cdot nH_2O$, can be used to measure the NH_4^+ ion [10], and one based on fused $Li_2CO_3-V_2O_5$ can be used to measure the Li^+ ion [11].

Commercial solid-state membrane electrodes are mainly designed to measure a single ion such as a halide ion or a $+2$ cation like Cu^{2+}. Like the glass membrane electrodes, the solid-state membrane electrodes have *fixed* reactive sites that cannot travel to the ions measured. Thus, to measure -1 anions, a $+1$ reactive site like the Ag^+ ion must be imbedded in the membrane. To measure a $+2$ cation, a -2 anion like S^{2-} must be arrayed in the membrane as a reactive site. Table 16–2 lists the major ions measured with commercial electrodes.

Construction and Theory. The essential elements in a solid-state membrane electrode are shown schematically in Figure 16–7. The *sensing* element is a conducting solid, either a single crystal or a pellet pressed from crystalline material.

Table 16–2.   Solid Membrane Electrodes.

Ion Measured	Membrane	Major Interferences
F^-	LaF_3	OH^-
S^{2-}, Ag^+	Ag_2S	Hg^{2+}
Cl^-	$AgCl-Ag_2S$	Br^-, I^-, S^{2-}, CN^-, NH_3
Br^-	$AgBr-Ag_2S$	I^-, S^{2-}, CN^-, NH_3
I^-	$AgI-Ag_2S$	S^{2-}, CN^-
SCN^-	$AgSCN-Ag_2S$	Br^-, I^-, S^{2-}, CN^-, NH_3
Cd^{2+}	$CdS-Ag_2S$	Ag^+, Hg^{2+}, Cu^{2+}
Cu^{2+}	$CuS-Ag_2S$	Ag^+, Hg^{2+}
Pb^{2+}	$PbS-Ag_2S$	Ag^+, Hg^{2+}, Cu^{2+}

[10] Y. G. Vlasov, M. S. Miloshova, P. P. Antonov, E. A. Bychkov, and A. U. Efa, *Elektrokhimiya 19*, 1049 (Russ-1983); *Chem. Abstr. 99*, 148310r (1983).

[11] L. I. Manakova, N. V. Bausova, V. L. Volkov, *Zh. Anal. Khim. 37*, 539 (Russ-1982); *Chem. Abstr. 96*, 209940h (1982).

Figure 16–7.   Solid-state ion-selective electrode. (Courtesy of Orion Research, Inc.)

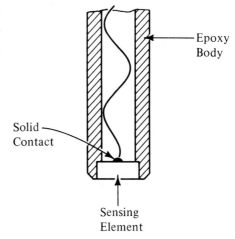

Epoxy Body

Solid Contact

Sensing Element

Some electrodes have an internal reference electrode and reference solution, but other types apparently have only a simple electrical contact to the inner surface of the membrane.

The most common type of solid-state membrane electrode is made from a solid silver sulfide membrane. This can be manufactured by pressing polycrystalline silver sulfide powder in a pellet press like that used to prepare samples for infrared spectroscopy. In this membrane, silver ions are the mobile species. The electrode is excellent for measuring silver-ion activity because of the low solubility and inert nature of silver sulfide. Sulfide ions may also be measured because the sulfide ion activity of the sample solution affects the silver ion activity according to the solubility-product principle:

$$K_{sp} = (a_{Ag^+})^2 a_{S^{2-}}$$

$$a_{Ag^+} = \left(\frac{K_{sp}}{a_{S^{2-}}}\right)^{1/2}$$

The silver sulfide membrane electrode measures silver and sulfide levels down to approximately $10^{-8} M$ in simple salt solutions. However, in solutions containing complexes the *free* silver ion activity can be measured as low as $10^{-20} M$. Similarly, S^{2-} in H_2S solutions has been measured as low as $10^{-19} M$.

An electrode with a membrane consisting of a finely divided silver halide (AgX) in a silver sulfide matrix responds selectively to halide ions (X^-). The silver halide salt must be more soluble than silver sulfide but insoluble enough that its equilibrium solubility gives a lower halide ion activity than that of the sample solution. Several electrodes of this type are described in Table 16–2.

Several silver-sulfide–metal-sulfide electrodes have been developed that respond to the appropriate metal ion (see Table 16–2). The electrodes contain enough silver sulfide to provide silver ion conducting pathways through the membrane and, thus, function as silver ion detectors. However, the silver ion activity is determined by the metal ion activity of the sample solution. For example, the lead ion electrode contains PbS, which is more soluble than the Ag_2S in the membrane.

Therefore, $a_{Pb^{2+}}$ determines $a_{S^{2-}}$:

$$K_{sp,PbS} = a_{Pb^{2+}} a_{S^{2-}}$$

$$a_{S^{2-}} = \frac{K_{sp,PbS}}{a_{Pb^{2+}}}$$

The activity of Ag^+ is, in turn, determined by $a_{S^{2-}}$:

$$K_{sp,Ag_2S} = (a_{Ag^+})^2 a_{S^{2-}}$$

$$a_{Ag^+} = \left(\frac{K_{sp,Ag_2S}}{a_{S^{2-}}}\right)^{1/2}$$

Combining these expressions, and substituting into the Nernst equation for a silver ion conducting system, gives a simple equation for the electrode potential:

$$E = \text{constant} + \frac{2.3RT}{2F} \log a_{Pb^{2+}} \qquad (16\text{--}11)$$

Potentiometric Measurement of Chloride Ion: Cystic Fibrosis. Solid-state membrane electrodes for the measurement of chloride are made from the silver sulfide membrane discussed above. This measurement illustrates how this electrode functions. Since AgCl is more soluble than Ag_2S, the activity of Ag^+ in the membrane will be controlled by the activity of Cl^- in the sample solution, via the solubility product of AgCl:

$$K_{sp,AgCl} = a_{Ag^+} a_{Cl^-}$$

$$a_{Ag^+} = \frac{K_{sp,AgCl}}{a_{Cl^-}}$$

The potential of the Ag_2S matrix depends directly on a_{Ag^+} (as in the simple Ag_2S electrode), but indirectly on a_{Cl^-}.

Diagnosis of Cystic Fibrosis. The potentiometric analysis of body perspiration for chloride ("sweat chloride") is an important clinical measurement. The diagnosis of cystic fibrosis in young children depends in part on the measurement of sweat chloride.

First, sweating is induced by rubbing the wrist with an alkaloid called pilocarpine, $C_{11}H_{16}O_2N_2$. Then a mild, safe electrical current is applied to stimulate rapid sweating. The first layer of sweat is often removed by wiping with a pad to reduce the possibility of high results from evaporation that often affects the first layer more than subsequent layers. Then a special flat-headed combination electrode containing the chloride-indicating electrode and a reference electrode is placed directly on the skin. The chloride level in meq/L is read directly from the potentiometer.

The presence of high chloride levels in sweat is of great significance in helping the physician diagnose cystic fibrosis, a lung disease characterized by such levels. The analytical results are interpreted with reference to a mean normal chloride level for infants of 20 meq/L ($2.0 \times 10^{-2} M$). The following approximate concentration regions are relevant:

   0–40 meq Cl/L: Safe region

   40–60 meq Cl/L: Cystic fibrosis is not indicated; more examination needed

   > 60 meq Cl/L: Cystic fibrosis likely; confirmed by examination

Potentiometric Measurement of Fluoride Ion.   The fluoride-indicating electrode is made from a *single crystal* of lanthanum fluoride, rather than from a pressed pellet. The fluoride ion, being the lattice ion with the smallest ionic radius and smallest charge, is the ion involved in electrical conduction. Conduction occurs by a lattice defect mechanism whereby a mobile ion next to a vacancy defect moves into the vacancy. The size, shape, and charge requirements of the vacancy are such that the mobile lattice ion is the only one that can move into it. Thus, chloride, bromide, and iodide ions would be too big to fit into the vacancy. The hydroxide ion, being the size of the oxide ion, is about the same size as the fluoride ion and is the only anionic interference.

   The fluoride-ion electrode is highly specific for fluoride ion down to an activity of about 10^{-6}. Below about $10^{-6}M$, the response begins to level off because of the solubility of LaF_3 in the sample solution. The pH limits the use of the electrode; below about pH 3.5, fluoride forms HF, which is not measured by the electrode. At $[OH^-]$ much greater than $[F^-]$, the OH^- also limits the measurement of fluoride.

TWO MOLECULE-SELECTIVE INDICATING ELECTRODES

In the next two sections, we will discuss two indicating electrodes used mainly for the measurement of molecules: the gas-sensing electrode and the enzyme electrode. Briefly, these electrodes measure the following types of molecules:

   Gas electrodes. These measure *dissolved* gases like $SO_2(aq)$, $CO_2(aq)$, etc.

   Enzyme electrodes. These measure biochemically important molecules that undergo reactions catalyzed by enzymes.

16–5.   GAS-SENSING ELECTRODES

Gas-sensing electrodes measure the concentration of certain gases dissolved in aqueous solution, or the concentration of an ion in solution that can be converted to a dissolved gas by a simple chemical reaction. For example, sulfur dioxide (SO_2) in solution can be measured by a gas-sensing electrode, or a bisulfite salt can be measured after acidifying the solution to form sulfur dioxide.

$$HSO_3^- + H^+ \rightleftharpoons H_2SO_3 \rightleftharpoons H_2O + SO_2(g)$$

   A diagram of a typical gas-sensing electrode is shown in Figure 16–8. The permeable membrane is the key to the electrode's gas selectivity. The membrane is

Figure 16–8. Schematic diagram of gas-sensing electrode. (Courtesy of Orion Research, Inc.)

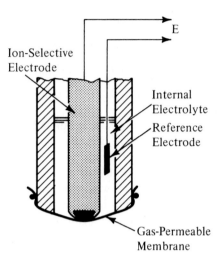

made of a hydrophobic porous plastic that prevents water from entering the pores or passing through the membrane. The gas in the sample solution diffuses through the membrane and comes to equilibrium with a liquid film inside the electrode, where it chemically reacts with some substance to form ions. These are detected by an ion-selective electrode inside the gas-sensing electrode. The difference in potential between the ion-selective electrode and an internal reference electrode is measured in millivolts by using a potentiometric instrument similar to a pH meter.

The operation of the sulfur dioxide electrode illustrates well the principles of such an electrode. When the electrode is immersed in a sample solution (as little as 1.0 mL), SO_2 from the sample diffuses through the membrane so that the SO_2 concentration in the liquid film is the same as in the sample solution. The SO_2 equilibrates with water in the liquid film to form sulfurous acid, H_2SO_3, which partially ionizes, forming H^+.

$$SO_2(g) + H_2O \rightleftharpoons H_2SO_3 \rightleftharpoons H^+ + HSO_3^-$$

The HSO_3^- concentration is kept constant by $Na^+HSO_3^-$ in the internal (liquid-film) electrolyte. The ion-selective electrode responds to $[H^+]$, which is proportional to the $[SO_2]$ produced by the ionizing H_2SO_3.

The proportionality of the electrode voltage (E) to SO_2 concentration can be seen from the following equilibrium in the liquid-film electrolyte:

$$K_a = \frac{[H^+][HSO_3^-]}{[SO_2]} \tag{16–12}$$

$$[H^+] = \frac{K_a[SO_2]}{[HSO_3^-]} = K'[SO_2] \tag{16–13}$$

since HSO_3^- is held constant.

The potential, E, for the ion-selective electrode is

$$E = \text{constant} + 0.059 \log [H^+] \tag{16–14}$$

Substituting Equation 16–13 into Equation 16–14,

$$E = \text{constant} + 0.059 \log [SO_2] + 0.059 \log K' \qquad (16–15)$$

From Equation 16–15, a plot of E versus $\log [SO_2]$ is linear with a slope of 0.059 V (59 mV). In practice, a linear curve is obtained from about $10^{-2}M$ to $10^{-6}M$ SO_2.

Other selective gas-sensing electrodes are commercially available for measuring in a similar manner such compounds as carbon dioxide (or carbonate and bicarbonate salts), ammonia (or ammonium salts), hydrogen sulfide (or sulfides), and nitrogen oxide (or nitrite salts).

16–6. POTENTIOMETRIC METHODS INVOLVING ENZYMES

Enzymes are remarkable in their ability to catalyze only specific, selected chemical reactions. By combining the selectivity of an enzyme-catalyzed reaction with potentiometric measurement of a reaction product using an ion-selective electrode, many rapid and highly selective analytical determinations are possible. These have been reviewed by R. K. Kobos (reference 2, Vol. 2, Chap. 1). Three different modes of potentiometric determinations employing enzymes will now be described.

Methods Using a Soluble Enzyme. The scheme here can be represented as follows:

$$\text{Substrate} \xrightarrow{\text{Enzyme}} \text{Products}$$

The substance to be determined (the substrate) reacts in solution in the presence of an enzyme. The concentration of one of the products is measured with an ion-selective electrode.

The determination of urea is a specific example of this kind of determination:

$$\text{Urea}((NH_2)_2C{=}O) + H_2O \xrightarrow{\text{Urease}} 2NH_3 + CO_2$$

In various procedures, NH_3 or CO_2 has been measured by a gas-sensing electrode, and the change in pH has been measured by a glass electrode. Other examples include the determination of L-aminoacids using L-aminoacid oxidase as the enzyme and the determination of glucose using glucose oxidase.

Determination of Enzymes. In clinical analyses, it is often necessary to determine the amounts of certain enzymes in samples. Pioneering work by Katz and Rechnitz [12, 13] and by Guilbault and co-workers [14] demonstrated that enzymes can be measured by adding a suitable substrate and reading off the concentration of a

[12] S. A. Katz and G. A. Rechnitz, Z. Anal. Chem., 196, 248 (1963).
[13] S. A. Katz, Anal. Chem., 36, 2500 (1964).
[14] G. G. Guilbault, R. K. Smith, and J. G. Montalvo, Anal. Chem., 41, 600 (1969).

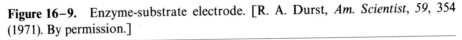

Figure 16–9. Enzyme-substrate electrode. [R. A. Durst, *Am. Scientist, 59*, 354 (1971). By permission.]

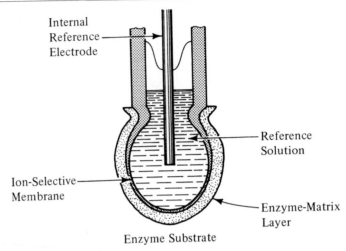

reaction product with an ion-selective electrode. This is essentially the reverse of the scheme for measuring a substrate by an enzyme-catalyzed reaction.

Enzyme Electrodes. The enzyme electrode is an ingenious device for measuring the concentration of an enzyme substrate or (upon adaption) the enzyme itself. For example, the urease electrode developed by Guilbault and Montalvo [15] provides a rapid method for determining urea in solution. The electrode is a cation-sensitive glass electrode surrounded by a gel impregnated with the enzyme urease (Figure 16–9). The gel is supported by a piece of nylon net cut from a nylon stocking. When inserted in a sample solution containing urea, the urea diffuses into the gel and undergoes the following urease-catalyzed reaction:

$$NH_2CONH_2 + 2H_2O + H^+ \xrightarrow[\text{(pH 7 buffer)}]{\text{urease}} 2NH_4^+ + HCO_3^-$$

After allowing 30–60 sec for the system to reach a steady state, the potential of the electrode is measured. The glass electrode responds to the NH_4^+ produced; but at fixed urease concentration and other fixed conditions, the potential is a linear function of log [urea] in the sample solution.

Another enzyme electrode enables glucose to be measured in blood serum or plasma by catalyzing the oxidation of glucose with glucose oxidase. Penicillin can be determined in drug formulations or fermentation broths with an electrode containing penicillinase. An electrode for measuring the concentration of L-amino acids in solution uses a gel containing L-amino acid oxidase (L-AOO) to surround the cation-sensitive glass electrode. Ammonium ions are produced via the following

[15] G. G. Guilbault and J. G. Montalvo, Jr., *J. Am. Chem. Soc., 91*, 2164 (1969).

reaction:

$$2RCH(\overset{+}{N}H_3)COO^- + O_2 \xrightarrow{\text{L-AOO}} 2RCOCOO^- + 2NH_4^+$$

(Another enzyme, catalase, is also added to destroy the hydrogen peroxide produced by the oxidation.)

Penicillin can be determined by a pH electrode that is coated with the enzyme penicillinase [16]. Many other examples are given in reference 2, Vol. 2, pp. 32–35.

QUESTIONS AND PROBLEMS

Principles

1. Define each of the following terms: (a) alkali error, (b) liquid-junction potential, (c) asymmetry potential, (d) selectivity coefficient.
2. Explain how a potentiometer works. Why can't an ordinary potentiometer be used for pH measurement with a glass electrode?
3. Write the reactions that occur to a minute extent at each electrode when the pH of a pH 6 solution is measured potentiometrically using a glass and a calomel electrode. Indicate by a diagram the direction of current flow in the external wire connecting the electrodes, and how the current is carried through the solution.
4. Explain how buffers are used to standardize a pH meter using a glass indicator electrode and a calomel reference electrode. Why is this necessary?
5. Explain why a glass electrode behaves erratically when it becomes dehydrated.
6. Explain how a Ag_2S solid-membrane electrode can respond either to a_{Ag^+} or to $a_{S^{2-}}$. Show that the sign in the equation for the potential of this electrode is plus when silver ion is measured and minus when sulfide ion is measured.
7. List the five types of indicating electrodes and state how the ion-sensing or molecule-sensing tip of each functions.
8. Modern pH glass electrodes have a very low error from response to sodium because they contain lithium oxide, not sodium oxide. Would you predict that this electrode would have the same low error from response to lithium ion? If not, why is this not a serious problem?
9. Why does a pH measurement give only the approximate $[H^+]$?
10. Explain how a sweat chloride measurement is made to assist in diagnosing cystic fibrosis.

pH and pM Measurements

11. Calculate the activity of the hydrogen ion corresponding to each of the following pH values, using correct significant figures.

(a) pH = 1.96   (c) pH = −0.04   (e) pH = 4.996
(b) pH = 0.939   (d) pH = 0.00   (f) pH = 6.9

12. Assuming that no additional ions are present other than the H^+ and corresponding anion

[16] G. J. Papariello, A. K. Mukherji, and C. M. Shearer, *Anal. Chem.*, *45*, 790 (1973).

in each part of Problem 11, state whether the $[H^+]$ will be larger than, smaller than, or the same as the activity of the H^+ in each part.

13. Calculate the activity of each ion below from the pM reading using an ion-selective indicating electrode.

(a) pNa = 1.23

(b) pK = 10.96

(c) pF = 0.20

(d) pCa = −0.301

Calculations of Ion Activities

14. An equation for the potential of a calcium-sensitive membrane electrode containing a calcium solution at a fixed concentration is

$$E = \text{constant} + \frac{0.059}{2} \log a_{Ca^{2+}} = \text{constant} - \frac{0.059}{2} pCa$$

What will be the slope if E (in millivolts) is plotted against pCa? Compare this with the slope of a plot of E versus pH for a glass electrode.

15. Referring to the example on p. 319, calculate the error caused by $0.0010M$ nitrite when nitrate ion, $a = 0.0010$, is measured with a nitrate ion–selective electrode.

16. Changes in the total ionic strength of a sample can affect the output of an ion-selective electrode. Calculate the potential change of a fluoride ion–selective electrode in a solution containing $2.0 \times 10^{-5}M$ sodium fluoride when the sodium nitrate concentration of the solution is increased from $0.020M$ to $0.20M$. Assume that the contribution of sodium fluoride to total ionic strength is negligible and that the selectivity coefficient, k_{F^-,NO_3^-}, is almost zero. See Table 1–2, p. 11 for the necessary activity coefficients.

17. Calculate the percentage error caused by Mg(II), $a_{Mg^{2+}} = 0.01$, when Ca(II), $a_{Ca^{2+}} = 0.001$, is measured with a liquid-membrane electrode (see Table 16–1).

18. With the aid of Table 1–2, calculate the Ca^{2+} activity (a) in a solution containing $6.66 \times 10^{-4}M$ $CaCl_2$, and (b) in the same solution to which enough sodium chloride has been added to make it $0.05M$ in NaCl ($f_{Ca^{2+}} = 0.485$).

Methods

19. State which type of indicating electrode can be used to measure the first ion in each mixture. Then state whether the concentration of the second ion will interfere and whether a buffer will be needed.

(a) Na^+ in $10^{-7}M$ H^+

(b) K^+ in $10^{-8}M$ OH^-

(c) Na^+ in $10^{-9}M$ OH^-

(d) $10^{-5}M$ F^- in $10^{-4}M$ OH^-

20. State whether the first ion in each mixture can be measured by using a solid-state indicating electrode. If not, explain why not.

(a) $10^{-4}M$ I^- in $10^{-4}M$ Cl^-

(b) $10^{-4}M$ Cl^- in $10^{-4}M$ Br^-

(c) $10^{-4}M$ Br^- in $10^{-10}M$ I^-

21. State which type of indicating electrode can be used to determine the first compound in the presence of the second, and state which ion is measured.

(a) $10^{-3}M$ $CaSO_4$ in $10^{-4}M$ $MgSO_4$

(b) $10^{-1}M$ NaCl in $10^{-1}M$ KBr

(c) $10^{-6}M$ NaF in $10^{-4}M$ KOH

17

Spectrophotometric
Methods of Analysis

The concentration of a colored substance in solution can be estimated by comparing visually the intensity of its color with that of several standard solutions of known concentrations. Such methods, using the eye as a detector, are known as *colorimetric methods*. In various forms, they have played a major role historically in analytical chemistry. Now an instrument called the *spectrophotometer* is used instead of the eye, and *spectrophotometric methods* have replaced the visual methods. Occasionally, however, the term "colorimetric analysis" is used to describe methods based on the absorption of radiation by naturally colored solutions or solutions in which a color is formed.

A spectrophotometer is an instrument in which radiant energy (such as visible light) of a very narrow wavelength range is selected from a source and passed through the sample solution, which is contained in a glass or quartz "cell." Some of the radiant energy is absorbed by the chemicals in the sample, and the rest passes on through. The *ratio* of the radiant power of the transmitted beam (P) to that of the incident beam (P_0) is measured by means of a photoelectric detector (such as a phototube, as described in Section 17–3). Spectrophotometers range from simple instruments that use only visible light to more sophisticated UV-visible instruments that contain a microcomputer.

The quantitative basis of spectrophotometry is that the amount of radiation absorbed (the absorbance) at an appropriate wavelength is proportional to the concentration of the light-absorbing chemical in the sample. Since absorption occurs in less than one second, it can be measured very rapidly; thus, spectrophotometry is a very fast and convenient method of quantitative analysis.

Literally thousands of spectrophotometric methods for analyzing elements and most organic compounds are described in the chemical literature. Usually, these methods are used to determine small amounts, or even traces, of substances. They are usually not used to determine macro quantities of substances, but rather for amounts below 2% of solids and some liquids. For example, the iron in steel or iron ore is best determined by a titration method rather than by a colorimetric method.

The latter approach would be somewhat inaccurate and inconvenient because of the dilutions required to reduce the iron concentration to the level needed for the method.

There are two major approaches to spectrophotometric analysis. One is to measure the radiant energy absorbed by the ion or molecule itself. For example, highly colored species such as the permanganate ion, the dichromate ion, or organic dyes obviously absorb light and can be measured by spectrophotometric or colorimetric analysis. Colorless species, such as most organic molecules or colorless cations, do not absorb light, but they may absorb ultraviolet or infrared radiation and so can be measured only by spectrophotometric analysis.

The other major approach is used with species that do not absorb *significant* amounts of light (or other desired radiant energy). A suitable chemical reagent is then added to these species to convert them to a new species that absorbs light intensely. For example, the iron(II) ion, $Fe(H_2O)_6^{2+}$, is a very light green. At low concentrations, it does not absorb a significant amount of light and is virtually colorless in dilute acid solution. However, an organic compound, 1,10-phenanthroline, reacts with it to form an intense red complex ion that is suitable for colorimetric measurement. Other colorless inorganic ions can be similarly determined after reaction with a suitable color-forming reagent.

The spectrophotometric methods that will be described in this chapter are based on the principles of *molecular* spectroscopy. For example, infrared spectroscopy is widely used as a tool for qualitative identification of *molecular* structure and less widely used for quantitative measurements. Molecular fluorescence and phosphorescence are types of molecular spectroscopy in which emitted light is measured after a molecule has absorbed shorter-wavelength light or UV radiation (Chapter 18).

Atomic spectroscopy involves the spectroscopy of gaseous atoms and so will be discussed separately from this chapter (see Chapter 19). To obtain the gaseous atoms, a high-temperature flame, plasma, or electric spark is used to atomized aqueous ions to the gas state. The absorption of light by such atoms can be measured, or the atoms can be excited by light absorption to emit light. Quantitative analytical methods based on atomic spectroscopy are very important for chemical analysis, as will be seen in Chapter 19.

17-1.   ABSORPTION OF RADIANT ENERGY

Wave-Particle Nature of Radiant Energy.   Light and other forms of radiant energy appear to have a dual nature. The wave theory is required to explain diffraction, refraction, and other optical effects, but the particle theory is also required to explain the absorption and emission of radiant energy.

The wave theory of radiation pictures radiant energy as a periodic, or oscillating, force field having electric and magnetic components that maintain each other and interact with matter. Figure 17-1a shows the oscillation of the electric component, which is vibrating in the plane of the paper. (The magnetic component is

Figure 17–1. Wave motion of light.

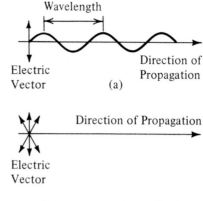

not shown, but its vibration would be perpendicular to the plane of the paper.) The combination of vibration and propagation through space gives radiant energy a wave motion. Figure 17–1b shows the vibration directions in an unpolarized beam of radiation consisting of several waves.

Wave motion may be described in terms of wavelength or in terms of frequency, which is the number of waves passing a fixed point per unit time:

$$\lambda v = c \tag{17–1}$$

where λ is the wavelength in centimeters, v is the frequency in $\sec^{-1}$, and c is the speed of light in vacuo, 2.998×10^{10} cm/sec.

According to the particle theory, a beam of radiation consists of a stream of discrete particles called *photons*, each possessing the energy hv, where h is Planck's constant (see below). The photons have essentially no rest mass but are viewed as particles of radiant energy. Photons are absorbed by many chemical species in solution. The energy of the absorbing species is increased temporarily; the excess is lost, usually in the form of heat. Absorption of radiant energy of course cannot increase the mass of the absorbing species.

The Electromagnetic Spectrum.   Radiant energy is characterized by its wavelength. The following units for wavelength are in common use:

$$nm = nanometer = 10^{-9} \text{ meter } = 10^{-7} \text{ centimeter}$$

$$\text{Å} = angstrom = 10^{-10} \text{ meter} = 10^{-8} \text{ centimeter}$$

$$\mu m = micrometer = 10^{-6} \text{ meter } = 10^{-4} \text{ centimeter}$$

The nanometer and the micrometer replace the older units, the millimicron ($m\mu$), and the micron (μ), respectively.

The energy of radiant energy is directly proportional to the frequency, as shown in the equation

$$E = h\nu \tag{17-2}$$

where E is the energy of a photon in joules (J), ν is the frequency in hertz (Hz) (cycles per second), and h is Planck's constant, 6.62×10^{-34} J-s.

Combination of Equations 17–1 and 17–2 gives

$$E_{(J)} = \frac{hc}{\lambda_{(nm)}} = \frac{6.62 \times 10^{-34} \text{ J-s} \times 2.998 \times 10^{-17} \text{ nm-s}^{-1}}{\lambda_{(nm)}}$$

$$= \frac{1.885 \times 10^{-16} \text{ J-nm}}{\lambda_{(nm)}} \tag{17-3}$$

Sometimes the unit of radiant energy is given in electron volts. In this case, Equation 17–3 becomes

$$E_{(eV)} = \frac{1.240 \times 10^3 \text{ eV-nm}}{\lambda_{(nm)}} \tag{17-4}$$

In either case, note that the wavelength and energy of radiant energy are inversely proportional:

$$\xrightarrow{\begin{array}{c} \lambda \text{ increasing} \\ \hline E \text{ decreasing} \end{array}}$$

We can now discuss the various regions of the electromagnetic spectrum in terms of wavelength and energy (see Figure 17–2). At the top are microwaves and radio waves, both with very long wavelengths and very small energies. These are followed by the infrared region, which consists of the far infrared and near infrared subregions. The 2.5–25 μm (2500–25,000 nm) region of the infrared is the region that is most frequently used in analysis.

After the infrared comes the visible region (light), the radiation perceived by the human eye. The borderline wavelengths for light are dependent on how small the eye's response must become before it is considered insignificant. The borderline wavelengths chosen herein are 380–750 nm (Figure 17–2). Note that light occupies only a very small region of the electromagnetic spectrum. On the other side of the visible region is ultraviolet radiation; it covers approximately 10–380 nm, although analysis is usually done in the 200–380 nm subregion. At the bottom of the spectrum are X-rays and gamma rays, both with very short wavelengths and high energies.

The reasons that radiation outside the 380–750 nm region is not visible have to do with the nature of the eye. At the short-wavelength borderline, the eye's response is limited by absorption of protein in the lens. At the long-wavelength borderline, its perception is limited by the declining response of the visual pigments and (possibly) by the composition of the eye (water in the aqueous humor, etc.). The response of the eye in the visible region also varies with wavelength (Table 17–1). The color of a chemical species in solution depends on the relative response of the eye to the various colors not absorbed by the solution and (to some extent) on the makeup of the white light striking the solution. Some authors use a chromaticity diagram to relate the color of a solution and the colors absorbed.

Figure 17–2. The electromagnetic spectrum.

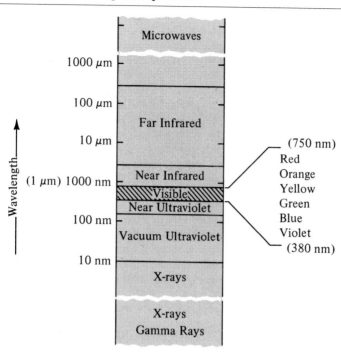

The colors of the various forms of copper(II) ion provide an instructive application of Table 17–1. A dilute solution of $Cu(H_2O)_6^{2+}$ absorbs red, orange, yellow, yellow-green, and some green light; it transmits some green, blue, and violet. The combined effect of the blue and green overpower the weak effect of violet on the eye, and the solution appears light green-blue. Similarly, a concentrated solution of $Cu(NH_3)_4^{2+}$ transmits only violet and some blue light, appearing violet-blue.

The Absorption Process. In the following discussion, radiant energy will be referred to as "light," even though ultraviolet radiation may be involved. It is

Table 17–1. Wavelength Regions for Each Color

Wavelength Region	Color	Relative Eye Response
380–450 nm	Violet	0.0022
450–495 nm	Blue	0.10
495–550 nm	Green	0.83
550–570 nm	Yellow-Green	1.00
570–590 nm	Yellow	0.87
590–620 nm	Orange	0.57
620–750 nm	Red	0.10
———	Purple[a]	———

a. Purple is seen by the eye when equal numbers of photons of blue and red light strike the eye.

important to remember that the energy of any type of radiation is inversely proportional to its wavelength.

Light is absorbed by a chemical species only when its wavelength corresponds to the energy needed to cause some change in the electronic configuration of the species. This concept is illustrated most simply by the absorption of light by gaseous metal atoms. (Gaseous metal atoms can be obtained by spraying a metal-salt solution into a high-temperature flame. The flame evaporates the solvent and, in most cases, decomposes the salt into gaseous metal atoms.)

In the flame, the lowest energy state of a gaseous atom such as sodium is known as the ground state and is symbolized by a symbol such as Na_0. Absorption of a photon by such an atom causes the outermost electron to "jump" to a higher-energy orbital, producing an *excited state*. The first excited state is designated by a subscript one, the second (next higher energy) excited state by a subscript two, and so on.

Two examples of such electron jumps are shown in Figure 17–3. A ground state sodium atom may absorb a 589-nm photon to produce sodium in its first excited state. Another ground state sodium atom may absorb a 330-nm photon and become sodium in its second excited state. As can be seen, the energy for the second electron jump is $E_2 - E_0$, making it a higher-energy process than the $E_1 - E_0$ energy for the first electron jump.

Absorption by Molecules in Solution.   Absorption of light by molecules or ions causes three types of energy changes: *electronic* (change in the energy of the electrons of a molecule), *vibrational* (change in the average internuclear distance of two or more atoms in the molecule), and *rotational* (change in the energy of a molecule as it rotates around a center of gravity). Absorption of light by molecules in solution is more complicated than absorption by gaseous atoms because energy *sublevels* are involved. Each electronic state of a molecule is subdivided into a number of *vibrational sublevels*; each vibrational sublevel is in turn divided into a number of *rotational sublevels*. The difference in energy between the various vibrational sublevels and rotational sublevels is much less than the difference in energy of electronic states.

Figure 17–3.   Absorption of light by the gaseous sodium atom.

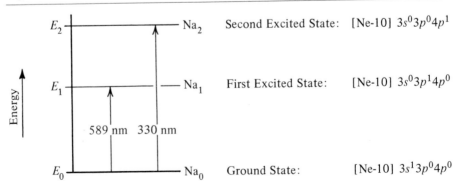

Figure 17–4 is an energy-level diagram for a typical chemical substance showing several energy levels. Also shown are three different *transitions*, in which the substance absorbs a ground state species absorbs a photon and undergoes a transition to an excited state.

The transition on the left in Figure 17–4 (T_{uv}) is a relatively high-energy transition resulting from the absorption of ultraviolet radiation. In this transition, one electron of the molecule will be promoted to a higher energy level; the molecule is then said to be in an excited state. Because of the different vibrational energy levels within the molecular excited state, ultraviolet radiation of a range of wavelengths (energies) will be absorbed. Very soon (10^{-13}–10^{-11} sec) after the transition, the excited molecule undergoes *relaxation* to the lowest vibrational level in the excited state, as indicated by the wavy line in the figure.

The transition in the center of Figure 17–4 (T_{vis}), like T_{uv}, involves a transition to an excited state, but with a smaller change in energy level because the visible light absorbed has a lower energy than ultraviolet. Finally, the transition on the right (T_{ir}) is a relatively low-energy transition resulting from the absorption of infrared radiation, which has a lower energy (and higher wavelength) than visible or ultraviolet radiation. In this transition, the molecule is promoted to higher vibrational and rotational levels.

Vibrational and rotational changes in a molecule result from the absorption of low-energy (infrared) radiation. Electronic transitions require more energy and thus occur in the visible and ultraviolet (UV) spectral regions. Since each type of

Figure 17–4. Typical transitions involving light absorption. Transitions T_{uv} and T_{vis} are electronic transitions caused by absorption of ultraviolet and visible light. T_{ir} is a vibrational transition caused by infrared radiation. The wavy lines indicate nonradiative transitions (vibrational relaxation).

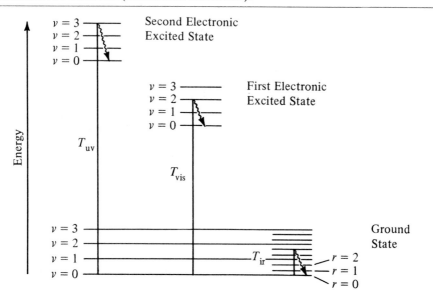

electronic transition, in reality, involves many transitions from the ground state to many vibrational (and rotational) sublevels of a given excited state, visible and/or ultraviolet radiation of a rather broad wavelength range is absorbed by molecules. Such molecular absorption, illustrated in Figure 17–5b, is in sharp contrast to the very narrow absorption peaks of gaseous atoms shown in Figure 17–5a.

Molecular Absorption and Color.   The color of a molecule in solution depends on the wavelength of light it absorbs. Thus, when a sample solution of an organic compound or an inorganic ion is exposed to white light (polychromatic light, or light of all colors), certain wavelengths are absorbed, and the remaining wavelengths are transmitted to the eye. The color perceived by the eye is determined only by the wavelengths transmitted. (In simple terms, the color seen is the complementary color of the color(s) absorbed.)

　　If more than one color is transmitted, the response of the eye (Table 17–1) determines which color(s) will be perceived. In general, yellow and/or green will always be seen because of their high relative response. If yellow, green, and yellow-green are absorbed, then other colors may be perceived. For example, the carotene solution in Figure 17–5b transmits both orange and red light only and will appear orange because the response of the eye is much greater to orange than to red. It will not appear yellow, green, or yellow-green because these are all absorbed.

Absorption Spectra.   In spectrophotometric analysis, the sample solution is ideally irradiated with light of a single wavelength (actually light of a wavelength *range* of a

Figure 17–5.   The absorption of light as a function of wavelength. (a) *Idealized* atomic absorption peaks of gaseous sodium atoms resulting from two different electronic transitions. (b) Molecular absorption bands of β-carotene ($C_{40}H_{56}$), the colored pigment in carrots.

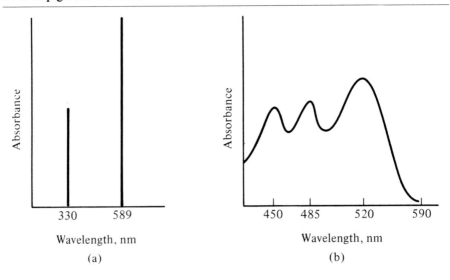

Figure 17–6. Absorption spectra of two compounds: (a) tris-(1,10-phenan-throline)iron(II) sulfate; (b) potassium dichromate in 0.1M sulfuric acid.

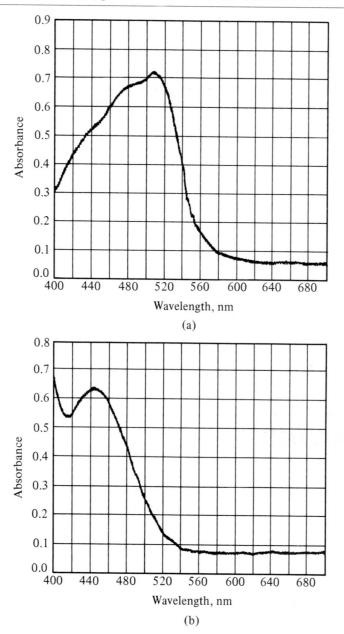

Wavelength, nm

(a)

Wavelength, nm

(b)

few nanometers is used), and the amount of absorption at each wavelength is measured as the wavelength is varied. By plotting the absorbance, or percent transmittance, against the wavelength, an *absorption spectrum* is obtained. An instrument called a *spectrophotometer* is used to make the measurements.

The absorption spectra for two compounds that absorb light are shown in Figure 17–6 (ultraviolet and infrared spectra will be considered later). An important use of absorption spectra is in selecting a wavelength that is suitable for quantitative analysis. The wavelength chosen is generally in a region where the substance to be determined absorbs strongly and other substances absorb negligibly.

17–2.   BEER'S LAW

From the discussion of absorption in the previous section, it can be seen intuitively that the amount of light absorbed by a species in solution will depend on the number of ions or molecules of the species in the pathway (light path) of the photon beam. It follows that more light will be absorbed as the *concentration* of the absorbing species increases. Similarly, the longer the light path followed by the photon beam through the solution, the more photons absorbed. The third factor that governs the amount of light absorbed is the *probability* of a photon being absorbed and causing an electronic transition in a chemical species. Different chemical species have different probabilities; the species with the highest probability will absorb more light than another species at the same concentration.

These facts are the basis of the fundamental law of spectrophotometry, which is known as the Lambert–Beer law or simply as *Beer's law*. This law states that the amount of light or ultraviolet or infrared radiant energy absorbed or transmitted by a solution is an exponential function of the concentration of absorbing substance present and the sample path length. The following discussion will try to explain why this is so.

Suppose that a beam of light of radiant power P is passed through a solution containing N absorbing ions or molecules. The amount of light absorbed will be directly proportional to the number of absorbing species in the light path. If we divide the solution into a number of small, equal sections, the change in radiant power (ΔP) will depend on the number of absorbing species in this section (ΔN). The radiant power of the beam that enters succeeding sections will be diminished by the absorption by the preceding sections of solution. The amount of light absorbed by each section depends on the number of absorbing species in that section and is proportional to the radiant power of light *entering that section*:

$$\Delta P = -kP(\Delta N) \tag{17–5}$$

$$-\frac{\Delta P}{\Delta N} = kP \tag{17–6}$$

where k is a proportionality constant. The minus sign indicates a decrease in radiant power of the beam.

If the sections are infinitely small, Equation 17–5 may be written in differential form:

$$-\frac{dP}{dN} = kP \qquad (17\text{–}7)$$

If we rearrange and integrate between the limits P_0 and P (the initial and final radiant power of the light beam) and between zero and N for the number of absorbing species in the light path, the following results are obtained:

$$\int_{P_0}^{P} \frac{dP}{P} = -k \int_{0}^{N} dN \qquad (17\text{–}8)$$

$$\ln \frac{P}{P_0} = -kN \qquad (17\text{–}9)$$

N depends on (1) the concentration of absorbing species in solution (c) and (2) the thickness of absorbing solution traversed by the light beam (b).

$$N = k'bc \qquad (17\text{–}10)$$

If Equation 17–8 is converted to base 10 logarithms and combined with Equation 17–9, we get the following expression for Beer's law:

$$\log \frac{P}{P_0} = -abc \qquad (17\text{–}11a)$$

or

$$\log \frac{P_0}{P} = abc \qquad (17\text{–}11b)$$

where a is the proportionality constant, b is the length of light path in centimeters through the solution, and c is the concentration of absorbing species in the solution. The absorptivity, a, is characteristic of a particular absorbing species and will change as the wavelength changes. In other words, Beer's law applies only to *monochromatic* radiant energy.

The term $\log(P_0/P)$ is commonly called the *absorbance* and given the symbol A. Thus, Equation 17–11b becomes

$$A = abc \qquad (17\text{–}12)$$

Absorptivity and Molar Absorptivity. The numerical value of the absorptivity, a, depends on the units used for expressing the concentration of the absorbing solution. Concentration units such as parts per million (ppm), which is milligrams per liter, grams per 100 mL, etc. are often used. However, a different symbol, ϵ (epsilon), is used in place of a when the concentration is expressed as *molarity*:

$$A = \epsilon bc \qquad (17\text{–}13)$$

where ϵ is the molar absorptivity, b is the light path in centimeters, and c is the molar concentration of the absorbing species.

Like a, the molar absorptivity is characteristic of a particular chemical species at a given wavelength, usually the wavelength of maximum absorbance. The molar absorptivity always has the units of L-mole^{-1} cm^{-1}. Chemists report numerical values of *molar* absorptivity as they do melting point, refractive index, etc. Molar absorptivities of various chemical species vary from 10^{-2} to an upper limit of 10^5 L-mole^{-1} cm^{-1} for a single absorbing group in a molecule or ion. For example, $Mn(H_2O)_6^{2+}$ and $Co(H_2O)_6^{2+}$ have molar absorptivities of 0.02 (532 nm) and 10 (530 nm), respectively, whereas the permanganate ion (MnO_4^-) has a much larger molar absorptivity, 2×10^3 (525 nm). Metal-organic complexes usually have molar absorptivities ranging from 1×10^3 to 5×10^4 and higher.

Transmittance and Percent Transmittance.   Although Equation 17–13 (based on absorbance) is the most useful form of Beer's law, it is also helpful to use the transmittance of a solution to formulate Beer's law. Transmittance is the ratio of the radiant power of the transmitted beam to that of the incident beam:

$$T = \frac{P}{P_0} \qquad\qquad (17\text{--}14)$$

Combining Equations 17–9, 17–13, and 17–14 gives the following form of Beer's law:

$$\log\frac{P_0}{P} = -\log T = \epsilon bc \qquad\qquad (17\text{--}15)$$

Note that the logarithm of the transmittance must be calculated to obtain a linear plot involving concentration.

17–3. SPECTROPHOTOMETERS

Components.   Although various types of instruments are used to measure the absorption of radiant energy, they are all basically spectrophotometers in optical terms. The essential optical components are (1) a source of radiant energy, (2) a monochromator for wavelength selection, (3) cells for holding the sample solution and blank, (4) a detector for measuring the transmitted radiation, and (5) an electronic readout for the detector. We will now discuss each of these components.

   Ultraviolet-Visible Radiation Sources.   Radiation sources can be classified as *line sources*, which give atomic spectral lines, and *continuum sources*, which emit radiation whose intensity varies smoothly over an extended range of wavelengths. The latter is used in molecular spectroscopy (spectrophotometry in solution). A monochromator is used to select radiation of a narrow wavelength range from the continuum source. This range, usually measured in nanometers, is called the *spectral bandpass* or *bandwidth*. For convenience, the light selected is expressed in terms of the mean wavelength. For example, if the monochromator dial reads 520 nm and the bandpass is 10 nm, the beam essentially consists of radiation from 515 to 525 nm. The two main requirements for a source are that it emit continuous radiation over the desired wavelength range and that this band be intense enough for accurate measurement. More than one source is needed for the 200–750 nm region (ultraviolet-visible) because no single source meets the above requirements over the entire region.

Some common sources are listed in Table 17–2. The tungsten filament lamp is the only common source for the visible region. It also covers part of the ultraviolet region, but it is generally not used below 330 or 340 nm. At wavelengths shorter than this in the ultraviolet, the deuterium lamp is used. This lamp is preferred to the hydrogen discharge lamp, which was formerly used in the ultraviolet. The tungsten halogen lamp with an outer quartz envelope emits more intensely in the ultraviolet than the tungsten filament lamp and can be used down to 220 nm on Varian Cary spectrophotometers.

Monochromator. The function of a monochromator is to select a beam of monochromatic (one-wavelength) radiation that can be varied over a wide range. The essential components of a monochromator are as follows: (1) entrance slit, (2) collimating device (a lens or mirror that causes light to travel as parallel rays), (3) dispersion device (to select light of different wavelengths), (4) focusing lens or mirror, and (5) exit slit.

Dispersion devices include diffraction gratings, prisms, and various optical filters. A diffraction grating is a surface ruled with a large number of parallel grooves that are approximately one light-wavelength wide. Light striking the grating is diffracted so that different wavelengths come off at different angles. Rotating the grating allows radiation of the desired wavelength to be selected; this is done by turning a wavelength dial on the instrument. Standard diffraction gratings may be of either the transmission or reflection type; both allow a significant amount of stray light to reach the detector at high absorbance values. In some newer instruments, a *holographic grating* is used to reduce the degree of stray light as well as transmit more radiant energy to the detector.

Table 17–2. Ultraviolet and Visible Radiation Sources

Source	Wavelength Range	Intensity
Tungsten Filament Lamp	320–2,500 nm	Weak below 400 nm Strong above 750 nm
Tungsten Halogen Lamp (quartz envelope)	250–2,500 nm (220–2,500 nm at high-intensity setting)	——
Hydrogen Discharge Lamp	180–375 nm	Weak everywhere, but best in 200–325 nm region
Deuterium Discharge	180–400+ nm	Moderate

Any of the gratings above offer the best general dispersion of wavelengths. A monochromator with a grating gives a narrow radiation bandwidth of 1–20 nm, depending on the quality of the spectrophotometer. In addition, the bandwidth is *constant* over the entire wavelength range.

The *prism* disperses radiation by means of refraction. Radiation of different wavelengths is bent at different angles on entering and emerging from the prism. The effective bandwidth of a prism varies, being small in the ultraviolet but extremely large in the red and near-infrared spectral regions.

Optical filters are used instead of monochromators in most colorimeters. These are usually of the "narrow (band)pass" type and transmit light of about a 50-nm bandwidth. A special type, called an interference filter, is also available. It rejects unwanted wavelengths by destructive interference and transmits a narrow bandwidth of 10–20 nm.

Cells. The cells used to contain the sample (or blank) in spectrophotometry commonly have a 1.00-cm by 1.00-cm cross section and are several centimeters in height. This is so the light path, b, is always 1.00 cm. Some inexpensive spectrophotometers use circular test tube cells with diameters slightly larger than 1.00 cm, but these tend to reflect more light than the better cells. Cells used in the visible region are made of optical quality borosilicate glass. These can be used down to about 320 nm, at which point the glass begins to absorb most of the radiant energy. For lower wavelengths, it is necessary to use more expensive cells made of quartz or some other form of silica. These cells can also be used above 320 nm.

Studies have shown that removing a cell and then replacing it in the compartment of the spectrophotometer adds to the variance of spectrophotometric readings. To avoid this, at least one spectrophotometer (Hach DR/3000) is designed with a fixed "flow-through" cell. The solution to be measured requires 30–60 sec to flow through the cell. The reading can be taken anytime after the first few seconds, which are needed to wash out the cell and fill it with the new sample.

Detectors. The simplest type of detector for the visible and ultraviolet spectral regions is the photodiode. There are two classes of phototubes: the vacuum photodiode or phototube and the semiconductor photodiode. Various phototubes and photodiodes cover different spectral regions, as shown by Table 17–3.

The *vacuum photodiode*, or *phototube*, contains a cathode that acts as the photoresponsive element (Figure 17–7). The blue-sensitive phototube (Table 17–3)

Table 17–3. Ultraviolet and Visible Radiation Detectors

Detector	Useful Range
Human Eye	380–750 nm
Blue-Sensitive Phototube (Vacuum Photodiode)	330–625 nm
Red-Sensitive Phototube (Vacuum Photodiode)	600–975 nm
Wide-Range Phototube (Vacuum Photodiode)	400–800 nm
Silicon Photodiode (Semiconductor Photodiode)	350–1170 nm
Silicon Photodiode, UV-Enhanced	200–1170 nm
Photomultiplier Tube	200 or 300–1100 nm

Figure 17-7. Schematic diagram of a phototube (photoemissive tube) showing the emission of an electron from the cathode to the anode.

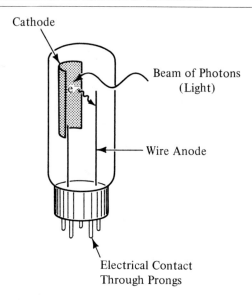

is standard equipment on routine-use spectrophotometers such as the Spectronic-20 spectrophotometer, except when measurements must be made above 625 nm; then the red-sensitive phototube must be used.

A typical phototube consists of a half-cylinder cathode and a wire anode in a sealed evacuated glass tube (see Figure 17-7). Because the cathode emits electrons when photons strike it, the phototube is termed a photoemissive tube. The response of the phototube to different wavelengths depends on the composition of the cathode coating.

A typical cathode consists of nickel plated with silver and silver oxide. The surface is covered with a layer of cesium metal, which partly interacts with the silver oxide to produce some cerium oxide.

The phototube measures light as follows: When the instrument is turned on, a high voltage is impressed across the phototube circuit; the cathode may be heated at the same time, By means of a resistance, the anode is maintained at a positive voltage relative to the cathode. A beam of photons passes through the sample and strikes the inner surface of the cathode. Some of these photons have enough energy to ionize electrons away from the cesium nuclei and thus eject them from the cathode. (Not all of the photons have enough energy to eject electrons, but under operating conditions, the number of such photons is proportional to the number of photons striking the cathode.) The electrons migrate through the vacuum to the positive wire anode and produce a current. The response time of the phototube is less than one microsecond (10^{-6} second). This is still much longer than the 10^{-18} second required for the absorption process.

A *photomultiplier* detector contains a photoemissive cathode, a multiplying chain of dynodes, and an anode. The incident radiation impinges on the cathode, causing photoelectrons to be ejected. These are focused and accelerated toward the first dynode, a curved electrode coated with BeO, GaP, or CsSb. The first dynode emits additional photoelectrons, which are focused on the second dynode. This process is repeated over several dynodes to produce a current avalanche that finally impinges on the anode. It will be seen that a photomultiplier is a much more sensitive detector than a simple phototube.

A semiconductor phototube contains a solid-state p-n junction as the photoresponsive element. Typically it consists of a small (1.5 × 1 cm) silicon chip on a circuit board. A photon striking the p-n junction generates an electron-hole pair, producing current. This class of diode is used in the Spectronic-21 series of spectrophotometers.

> Because a given semiconductor has a characteristic *band-gap energy*, the semiconductor used in this photodiode will limit its long-wavelength response. For silicon, the energy is 1.06 eV, limiting its long-wavelength response to 1170 nm (cf. Equation 17–4). Semiconductors suitable for longer wavelengths are PbS (0.34 eV), PbTe (0.30 eV), and PbSe (0.27 eV). CdS (2.42 eV) and AgCl (3.2 eV) are limited to shorter wavelengths than silicon.

Thus far, we have considered spectrophotometers in which a single detector is used for whatever wavelength is selected for the measurement. More recently, instruments have become available that use an *array* of detectors. The detector array consists of more than 200 photodiodes, each of which responds to one or two different wavelengths. The advantage of this arrangement is that an entire spectrum can be measured and displayed in less than 1 sec.

Instrument Readout.   The minute electric current produced in the detector is amplified and either fed to a recorder or displayed on the spectrophotometer via a digital or scale readout. Digital display is now used except on instruments of the lowest cost. The readout can be in either transmittance or absorbance.

The measurement procedure on most instruments is as follows:

1. The reading is adjusted to $0\% \ T$ (infinite absorbance) by blocking radiation from the source. A small amount of current will still flow through the detector (the "dark current").

2. The $100\% \ T$ (zero absorbance) reference point is established by inserting a cell containing pure solvent (or an appropriate blank) into the light path and then manipulating the slit width (or other instrumental parameter) to obtain a reading of $100\% \ T$. This adjusts P_0 to an arbitrary, but convenient, $100\% \ T$ reading.

3. The absorbance of the sample is measured by placing the sample solution in the light path without changing the slit width or wavelength. In effect, this gives a reading of P, from which the absorbance (or transmittance) is measured by comparison of P with P_0.

Types of Spectrophotometers.   Spectrophotometric instruments vary greatly in price, performance, and sophistication. Basically, spectrophotometers can be divided into three classes: (1) those that operate mainly in the visible spectral region,

(2) those that operate in both the ultraviolet and visible regions, and (3) those that have the capability of recording or otherwise displaying an entire spectrum. Each of these will be discussed briefly.

Visible Instruments. These instruments tend to have lower-cost glass optical components and operate from 325 nm (upper end of the ultraviolet) to 900–1000 nm (near-infrared). Older instruments such as the Spectronic-20 spectrophotometer select wavelengths mechanically through a wavelength knob, whereas the modern digital-readout instruments (LKB Novaspec, etc.) offer electronic wavelength selection via a *keyboard*. Older instruments rely on blue- and red-sensitive phototubes; modern instruments may use a silicon photodiode as a detector to avoid switching phototubes.

Ultraviolet-Visible Instruments. An ultraviolet-visible spectrophotometer is especially designed for measurements in the ultraviolet, as well as the visible, region. Such instruments may be equipped to measure absorption anywhere in the 200–1000 nm region.

By definition, ultraviolet spectrophotometry involves the absorption of radiant energy in the 200 (or 180) to 380 nm region. Some ultraviolet measurements may be made as low as 320 nm by using a visible spectrophotometer such as the Spectronic-20 . However, for measurements below 320 nm, the spectrophotometer must be equipped with a source of radiation other than the tungsten filament lamp. The most common ultraviolet source is the deuterium discharge lamp (Table 17–2). On some spectrophotometers, the tungsten halogen lamp can be used as low as 250 nm, or sometimes 220 nm.

Most ultraviolet spectrophotometers utilize gratings in the monochromators; the radiation striking the sample is thus restricted to a much smaller bandwidth (1–8 nm) than in visible spectrophotometers. Since glass absorbs ultraviolet radiation, the sample cells and the optical system of the spectrophotometer must be made of quartz. Generally, a photomultiplier is used as the detector, although some models use a silicon photodiode. The operating principles for this type of spectrophotometer are essentially the same as described above for the visible spectrophotometers, with the exception of the ultraviolet source.

Spectrophotometers may be either single-beam or double-beam instruments. In the latter, the monochromatic light beam is split by the optical system so that one part passes through the sample cell while the other passes through the reference cell. In simple single-beam instruments, the sample cell and then the reference cell have to be placed in the beam to measure the sample absorbance.

"Recording" Spectrophotometers. Often it is desirable to measure the spectra of absorbing substances. A spectrum (see Figure 17–6) is useful for qualitative identification of a substance because each substance has a unique absorption spectrum. The spectrum is also useful for selecting a suitable wavelength for a quantitative analysis.

Although a spectrum can be plotted manually by measuring absorbance at a number of wavelengths and connecting the points with a smooth line, it is much easier to use an instrument that is designed for this particular task. Historically, this has been done by a small motor in the instrument to scan the usable wavelength region at a fixed rate and to record (on chart paper) the absorbance (or transmittance) as a function of wavelength.

At least one instrument (Hewlett-Packard 8450) has a diode-array detection system and can display an entire spectrum on an oscillometer in about 1 sec. "White light" (light of many wavelengths) is passed through the sample. The transmitted light is separated into various wavelengths by a focusing holographic grating and measured simultaneously by an array of photodiode detectors. An instrument of this type has many unique uses. Spectral changes can be used to follow the course of very fast chemical reactions. In liquid chromatography, the spectrum of a single peak can be monitored several times between the time a peak enters the detector and the time it has all passed through the detector. An unchanged spectrum means that the peak contains only a single compound, while a changing spectrum indicates that the peak contains two or more unresolved compounds.

Instruments with Microprocessors or Microcomputers. In using a conventional spectrophotometer, the operator inserts one or more cells into the instrument, chooses the light source, selects the wavelength using a control on the mono-chromator, and measures the absorbance. The absorbance is then used to calculate the concentration of the measured sample component or to prepare a Beer's law plot. Manipulation of the data requires more time than the actual measurements.

Modern spectrophotometers with a built-in microprocessor or microcomputer can do rapid computations (data processing), store information for later use, and control many of the spectrophotometric operations. In such instruments, the operator still inserts one or more cells into the instrument but uses the *keyboard* to punch in the necessary operating instructions such as the wavelength to be used. Alternatively, the keyboard may be used to activate a programmed format for a particular analytical method. Calibration plots are made from standards by linear regression (see Chapter 3) or may be graphically displayed. The major functions performed by a microprocessor in spectrophotometers are listed in Table 17–4.

Table 17–4. Microprocessor Functions in Microprocessor-Spectrophotometer

Data Processing	Self-Diagnosis	Operating Status
1. Adjust readout to known concentration	1. Wavelength accuracy check (when turned on)	1. Automatic lamp selection
2. Calculation and printout of unknown concentration	2. Recalibration of wavelength (if needed)	2. Status light for lamps
3. Calculation of mean and standard deviation	3. Constant monitoring of $0\% \ T$	3. Wavelength selection
4. Calculation of molar absorptivity, Beer's law plot	4. Correction of $0\% \ T$ setting (if needed)	4. Keyboard for operator instructions to instrument
	5. Corrects for wavelength change in baseline	5. Setting of $100\% \ T$

17–4. SPECTROPHOTOMETRIC METHODS

Steps in an Analysis

Selection of a Color-Forming Reagent. Usually, several spectrophotometric methods can be used for the selective determination of any given substance. Where a colored substance is to be determined, the light absorbed by the substance is measured directly. However, most substances are colorless or weakly colored, so spectrophotometric methods usually involve adding a reagent to form an intensely colored reaction product. The following two examples will illustrate the methods used in inorganic and organic-biochemical analysis.

If a colorless metal ion, M^{z+}, is to be determined, a reagent (R) that produces a colored product such as a complex ion must be used:

$$M^{z+} + nR \rightarrow \underset{\text{(colored complex ion)}}{MR_n^{z+}} \qquad (17\text{--}16)$$

If a colorless organic molecule such as glucose is to be determined, a reagent such as o-toluidine (an aromatic amine) must be used to form a colored product. In this case, the colored product is a green Schiff base resulting from the reaction between the aldehyde group of glucose and the amino group of o-toluidine:

$$\underset{\text{(glucose)}}{C_5H_{11}O_5\underset{|}{\overset{}{-}}\underset{H}{\overset{}{C}}=O} + \underset{\text{(o-toluidine)}}{C_7H_8-NH_2} \rightarrow \underset{\text{(green product)}}{C_5H_{11}O_5\underset{|}{\overset{}{-}}\underset{H}{\overset{}{C}}=NC_7H_8} + H_2O \quad (17\text{--}17)$$

In choosing the reagent, several points should be considered:

1. The reagent should react selectively with the substance to be determined.
2. Conditions must be chosen to obtain optimum color formation.
3. The colored product chosen for measurement should have a molar absorptivity that is large enough that the substance can be determined in the concentration range of the actual samples.

Each of these points will be discussed below.

1. The reagent chosen should not cause interferences by forming a color with other substances that are likely to be in the sample. For example, a reagent used to determine calcium(II) in a hard-water sample should react only with calcium and not with magnesium(II), which is also likely to be present. In Equation 17–17, the o-toluidine reagent should react only with glucose and not with fructose or other sugars that might be in a biological sample.
2. Among the parameters that are often crucial for accurate analysis are pH, solvent composition, order of adding reagents, time required for color development, and stability of color. In aqueous spectrophotometric procedures, the pH usually must be controlled within certain limits to obtain optimum, reproducible color formation. In developing a new method or checking a standard method, it is usually best to measure the absorbance of solutions of the colored substance buffered at

several different pH's. A plot of absorbance versus pH will often define a plateau region of constant absorbance, which will then be the optimum pH range.

The order of adding reagents is sometimes critical. For example, it may be better to add a color-forming reagent to the sample before a buffer because prior addition of the buffer might raise the pH to a point at which the metal ion being determined hydrolyzes and reacts slowly and incompletely with the reagent.

The time required for color development and the stability of the color may require that the absorbance be measured within a certain time period. While a rapid color-forming reaction producing a stable color is needed, some reactions are kinetically slow, and a wait of some minutes may be required for the color to reach its full intensity. For example, determining fluoride by measuring the bleaching of a highly colored zirconium(IV) alizarin complex (to produce a colorless zirconium(IV) fluoride complex) requires many minutes to attain equilibrium absorbance. Other zirconium dye complexes are bleached more quickly and are therefore preferred for this determination.

3. The minimum value of the molar absorptivity depends on the lowest absorbance reading that can be made with reasonable accuracy and on the concentration level of the substance being determined. Spectrophotometric measurement of absorbance (or transmittance) is very inaccurate at both very low and very high readings. For this reason, the concentration of absorbing substance should always be adjusted until the absorbance is in the range 0.10–1.00 (or to 1.50 for some precision spectrophotometers). The corresponding transmittance range is 0.80–0.10 (or 0.03 for $A = 1.50$). Thus, the minimum absorbance that should be measured is 0.03 to 0.10.

Next, the approximate level of concentration of the substance being determined should be estimated. From this information, the minimum value of the molar absorptivity can be calculated from Beer's law. For example, for 1.0-cm cells and $10^{-5}M$ concentration levels, the minimum molar absorptivity is calculated as follows:

$$\epsilon_{min} = \frac{A_{min}}{bc} = \frac{0.10}{(1.0 \text{ cm})(1 \times 10^{-5}M)} = 1 \times 10^4 \text{ L-mole}^{-1} \text{ cm}^{-1}$$

If the actual molar absorptivity is slightly less than 1×10^4, then cells with a longer path length, such as 2.0 cm, may be used.

Selection of Analytical Wavelength. In the absence of interferences, the wavelength chosen for a quantitative determination is the wavelength of maximum absorbance. Unfortunately, this wavelength is not always usable. It is fairly common for the color-forming reagent to absorb somewhat at the wavelength of maximum absorbance of the complex being measured (see Figure 17–8). Since absorbances are additive, it may be possible to use λ_{max}, the wavelength of maximum absorbance (558 nm in this instance), and subtract the absorbance caused by the excess reagent. However, it is difficult to know the concentration of excess reagent precisely; therefore, if the reagent absorbs very strongly at λ_{max}, the correction will be too large, and the probable error will be too great. A better procedure is to use a wavelength at which the complex absorbs rather strongly but at which the

Figure 17-8. Spectra of lanthanum-Arsenazo complex and Arsenazo reagent.

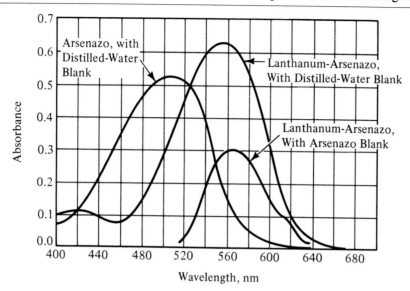

absorbance caused by the excess reagent is zero, or at least small. A small correction for reagent absorbance can usually be made without adversely affecting the accuracy of the determination.

Often a spectrum is determined by using a reagent blank instead of the usual pure-solvent blank (see Figure 17-8). The absorbance is negative (or off the scale) at lower wavelengths, when the absorbance versus solvent blank of the reagent is greater than the complex. The maximum appears at 565 nm, where the *difference* between the absorbance versus solvent blank of the complex and the reagent is at a maximum.

It is well to avoid an analytical wavelength that is on the side of a steep absorbance "hill," where the absorbance changes sharply with wavelength. In such a case, a small lack of reproducibility in wavelength setting is likely to cause a fairly serious error in measured absorbance.

Preparation of a Calibration Plot (Working Curve). Although the concentration of an unknown can be estimated by comparison with one standard solution, it is most accurately found by means of a calibration plot or working curve. To prepare such a plot, the absorbances of several standard solutions are measured after the color has been developed.

> In practice, one standard solution of the substance to be measured is prepared. Then varying volumes of the standard solution are pipetted into as many volumetric flasks. The color-forming reagent is added, and conditions are adjusted for optimum color formation. Then each solution is diluted to the volume of the volumetric flask, and the absorbance of each solution is measured at the chosen wavelength.

A plot of absorbance versus the molarity of the standard substance is then made on graph paper. If Beer's law is obeyed, the plot will be (within experimental error) a straight line with a slope of ϵ/b, the molar absorptivity divided by the internal cell length. If standard 1.0-cm cells are used, the slope of the line is just the molar absorptivity. If it is not possible to draw a straight line through all the points, a straight line is drawn as close as possible to all points that do not deviate greatly from the line. This tends to average out and reduce the small errors inherent in preparing and measuring the various standard solutions. Any points that deviate greatly from the line are usually ignored and rechecked if necessary.

The accuracy of the graphical method depends on how precisely absorbance and concentration can be located on the graphical scales and on how well the straight line can be drawn through the data points. For a more accurate measurement of concentration, the slope and intercept of the plot can be determined by a least squares calculation of the equation of the best straight line through the data points. Such a computation can be made easily with a programmable electronic calculator.

Regardless of whether a calibration plot or least squares calculation is used to determine concentration, the absorbance of the standards should fall in the range of 0.1–1.0 (1.5 for precision instruments) because measurements outside this range are much less accurate on standard spectrophotometers.

Sample Measurement and Calculations. The concentration of the sample should also be adjusted until the final solution of the sample has an absorbance between 0.1 and 1.0 (or 1.5 for precision instruments). Often this will require one or more dilutions. After the final dilution, the color-forming reagent, buffer, etc., are added under the same conditions used for the standards, and the solution is diluted to volume. The absorbance is measured at the same wavelength used for the standards, and the concentration is read from the calibration curve. The concentration or percentage of the substance in the original sample is then calculated, using a *dilution factor* to correct for any dilution used.

Example: The percentage of iron in an aluminum alloy is to be determined spectrophotometrically. A 1.0000-g (1000-mg) sample of the alloy is dissolved in acid and diluted to exactly 250 mL. A 10-mL aliquot of this solution is taken for analysis. The iron(III) ion in the solution is reduced to iron(II) ion, which is then complexed with 1,10-phenanthroline reagent to form a red complex ion. When the color formation is complete, the solution is diluted to exactly 100 mL. A portion of the diluted solution is measured at the analytical wavelength and found to contain 1.14 mg of iron per liter. Calculate the percentage of iron in the original sample.

First calculate the amount of iron in the final solution of 100 mL:

$$1.14 \text{ mg/L} \times 0.100 \text{ L} = 0.114 \text{ mg}$$

Next use the dilution factor of 250/10 to calculate the iron in the original sample dissolved in the 250 mL:

$$0.114 \times 250/10 = 2.85 \text{ mg}$$

The percentage of iron in the original sample is:

$$\frac{2.85}{1000 \text{ mg}}(100) = 0.285\% \text{ iron}$$

Steps in an Analysis: An Actual Case.   An actual case history will be used to illustrate the principles discussed above. In this instance, a spectrophotometric method was needed for measuring calcium(II) in biological fluids where as little as 20 μg of calcium per sample volume will be present. The fluids also contain magnesium(II).

   Selection of the Color-Forming Reagent.   To determine colorless calcium(II), a reagent must be added to produce a colored complex ion with calcium. Of the several color-forming reagents described in the literature, Arsenazo III appeared to be the most promising [1]. The predominate form of this reagent under most analytical conditions is the H_3L^{5-} species; its reaction with calcium ion splits out a proton:

$$Ca^{2+} + H_3L^{5-} \rightarrow CaH_2L^{4-} + H^+$$

Therefore, the formation of the colored product and the molar absorptivity of the product depend on the pH. At the optimum pH of 9.1, the molar absorptivity of the calcium–Arsenazo-III complex ion is 4.40×10^4 at 650 nm.

   The first point to be evaluated was whether the reagent would react selectively with calcium(II)—in this case, whether magnesium(II) would interfere. It was found that magnesium(II) did interfere at the optimum pH of 9.1; therefore, optimum color formation could not be achieved. However, further study showed that magnesium(II) did not interfere at a pH of 5.6–5.8. The molar absorptivity of the colored calcium–Arsenazo-III complex ion at this pH was 7.0×10^3 at 590 nm against an Arsenazo-III reagent blank.

   Next to be considered was whether the above molar absorptivity was large enough to accurately determine the 20-μg amount of calcium. Since a minimum molar absorptivity of 1×10^4 is required for $10^{-5}M$ levels (Example, p. 350), we know that at this pH, the reagent will not quite measure $1 \times 10^{-5}M$ levels of calcium. By assuming a minimum absorbance, A_{min}, of 0.10 and a cell light path of 1.00 cm, we can calculate by Beer's law the lowest molarity of calcium measurable:

$$c = \frac{A_{min}}{b\epsilon} = \frac{0.10}{(1.00)(7.0 \times 10^3)} = 1.4 \times 10^{-5}M$$

Since the sample size is 20 μg (2.0×10^{-2} mg) of calcium, the above calculation does not give the lowest weight of calcium measurable. We can convert molarity of calcium to mg Ca^{2+}/mL by rewriting molarity as mmole/mL and multiplying by the atomic weight of calcium in mg/mmole:

$$c = \frac{1.4 \times 10^{-5} \text{ mmole } Ca^{2+}}{mL} \times \frac{40 \text{ mg } Ca^{2+}}{\text{mmole}} = \frac{5.6 \times 10^{-4} \text{ mg } Ca^{2+}}{mL}$$

Suppose the biological sample was 1 mL; the concentration of calcium would be 2.0×10^{-2} mg calcium per milliliter—well above the lowest measurable weight of calcium. In fact, the sample would have to be diluted to give an absorbance between 0.10 and 1.0 absorbance unit. The final volume of a sample solution that

[1] V. Michayova and N. Kouleva, *Talanta 21*, 523 (1974).

would have an absorbance near 0.100 is calculated as follows:

$$\frac{2.0 \times 10^{-2} \text{ mg Ca}^{2+}}{5.6 \times 10^{-4} \text{ mg Ca}^{2+}/\text{mL}} = 36 \text{ mL} \qquad (A = 0.10)$$

Selection of the Analytical Wavelength. Because of interference from magnesium, the calcium-Arsenazo III complex must be measured at 590–600 nm at a pH of 5.8. Unfortunately, the free Arsenazo-III reagent absorbs somewhat in this region, although its maximum absorbance is at a lower wavelength. To compensate for this, the complex is measured at 590 nm against an Arsenazo-III reagent blank. The measurement simply involves replacing the pure solvent used to set zero absorbance with a solution containing the same concentration of Arsenazo III and buffer as the sample solution. Under these conditions, the Ca in a sample with a 100:1 Mg:Ca ratio can be determined without interference from Mg.

Preparation of Calibration Curve. As calculated above, the concentration of calcium standard that will have an absorbance of 0.100 is $1.4 \times 10^{-5}M$. The upper limit is about $A = 1.00$, or $c = 14 \times 10^{-5}M$. Standards containing 2.00, 6.00, 10.00, and $14.0 \times 10^{-5}M$ calcium cover this range fairly well. These can be prepared by pipetting given volumes of a standard calcium(II) solution, adding the Arsenazo-III reagent and buffer, and diluting to volume in a volumetric flask. For example, the volume of $0.00104M$ calcium(II) solution needed to prepare a $2.00 \times 10^{-5}M$ solution diluted to volume in a 100-mL volumetric flask is

$$\frac{\text{mL} \times 0.00104}{100} = 2.00 \times 10^{-5}M$$

$$\text{mL} = 1.92 \text{ of Ca solution}$$

The absorbances of various calcium standards were measured in a 1.00-cm cell at 590 nm.

Ca Concentration, M	A
2.00×10^{-5}	0.142
6.00×10^{-5}	0.416
10.0×10^{-5}	0.698
14.0×10^{-5}	0.985

The calibration plot for these data is shown in Figure 17–9. From a least squares computation (Appendix 5), the molar absorptivity was found to be 7.03×10^3.

Measurement of Sample. The necessary reagents are added to a sample to form the colored calcium complex, and the sample is diluted to exactly 25 mL in a volumetric flask. A portion of this solution measured at 590 nm against a reagent blank gave an absorbance of 0.512. From the calibration curve or from the molar absorptivity of 7.03×10^3, a calcium concentration of $7.28 \times 10^{-5}M$ is calculated. The sample therefore contains

$$7.28 \times 10^{-5} \times 40.0 = 2.91 \times 10^{-3} \text{ mg/mL of Ca}$$

$$2.91 \times 10^{-3} \text{ mg/mL} \times 25.0 \text{ mL} = 7.28 \times 10^{-2} \text{ mg} = 72.8 \text{ } \mu\text{g of Ca}$$

Figure 17–9. Calibration curve for Calcium–Arsenazo-III complex at 590 nm versus a reagent blank.

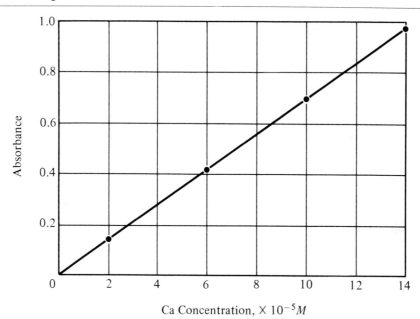

Ca Concentration, $\times 10^{-5} M$

Deviations from Beer's Law. When Beer's law is obeyed, a linear calibration plot will be obtained. If there are deviations from Beer's law, the calibration plot will curve upward (positive deviation) or downward (negative deviation) at higher concentrations. The reasons for such deviations are both instrumental and chemical. A very common instrumental cause of deviation is the use of polychromatic radiation. Because Beer's law is defined strictly for monochromatic radiation and because most of the time spectrophotometric measurements fall just short of being monochromatic, deviations may occur in the calibration plots.

Chemical deviations from Beer's law are usually more serious than instrumental deviations. Chemical deviations generally arise because of a change in the chemical makeup of the absorbing species. At any given wavelength, even two slightly different absorbing species of the same metal ion will have different molar absorptivities. (For example, $Fe(SCN)_2^+$ has a larger molar absorptivity than $Fe(SCN)^{2+}$.)

In practice, these deviations occur when the absorbing species in a sample solution exists in two (or more) chemical forms, the relative proportion of the two forms changing over the concentration range of the calibration plot. For example, at low concentrations, $Fe(SCN)_2^+$ may be predominate; but as the amount of iron(III) increases, less SCN^- is available for complexation, and $Fe(SCN)^{2+}$ may predominate. Such deviations may be caused by intermolecular interactions, formation of complex ions with varying numbers of ligands, concentration-dependent dissociations or associations, or reactions with solvent or hydrogen ions.

Light absorption by dichromate and chromate ions is a good example of both the effect of hydrogen ions and the deviation caused by concentration-dependent dissociation. The pertinent equilibrium reaction is

$$Cr_2O_7{}^{2-} + H_2O \rightleftharpoons 2H^+ + 2CrO_4{}^{2-}$$

If the pH is not held constant in the sample and in standards for the calibration plot, an error will result because the relative proportions of dichromate and chromate ions will vary from standard to standard. Unfortunately, the acidity can never be increased sufficiently to force all of the chromium(VI) into the dichromate form. Because of this, there will be a higher proportion of dichromate ion present at the lowest concentrations used for the plot than at the highest concentrations (see equation above). This will yield a curved calibration plot instead of a straight line; that is, the measured molar absorptivity will be different at different concentrations.

Accurate spectrophotometric determination of chromium(VI) is done by measuring the chromate ion, not the dichromate ion. It is possible to adjust conditions chemically so that at equilibrium only the chromate ion is present over the entire range of the calibration plot. It is left to the student to decide how this may be done.

Approaches to Spectrophotometric Determinations.   Several basic approaches are available for determining a species that does not absorb light appreciably. Three general chemical methods for forming a highly absorbing species are listed below, and a detailed example of each is then given. A fourth approach (two-component analysis) is used when a sample solution contains two color-absorbing species.

1. *Complexation.*   Metal ions are often reacted with one of the following types of color-forming reagents:
(a) A simple ligand that forms a colored complex ion, or a series of complex ions such as $Fe(SCN)^{2+}$, $Fe(SCN)_2{}^+, \ldots, Fe(SCN)_6{}^{3-}$.
(b) An organic ligand that forms a colored *chelate* (Section 11–1), such as 2,2'-bipyridine $(C_{10}H_8N_2)$, which forms a red chelate with iron(II) having the formula $(C_{10}H_8N_2)_3Fe^{2+}$. Ligands that form chelates are generally preferred because they usually form only one stable colored species. In contrast, a simple ligand often forms a series of complexes, as shown for the thiocyanate-iron(III) complexes above.

Except for group IA ions (Na^+, K^+, etc.), most metal ions will form colored complexes that can be determined spectrophotometrically.

2. *Oxidation.*   A number of metal ions or organic molecules in low oxidation states can be oxidized to higher oxidation states that can be measured spectro-photometrically. The lower oxidation states are often faintly colored, but oxidation converts them to a stable, intensely colored product. For example, chromium(III) is oxidized to chromium(VI), which is measured as the $CrO_4{}^{2-}$ ion; manganese(II) is oxidized to manganese(VII), which is measured as the $MnO_4{}^-$ ion.

3. *Indirect Spectrophotometry.*   Colorless inorganic ions and organic molecules

can often be determined by reacting them with an excess of a colored reagent. The reagent's color loss measures the concentration of the colorless ion or molecule.

4. *Two-Component Analysis.* Mixtures of two colored species are often difficult to analyze because their absorption spectra overlap significantly. Both can often be determined by measuring the total absorbance of the mixture at two different wavelengths and then setting up and solving two equations in two unknowns.

An Example of Complexation: Determination of Iron. Usually, traces of iron are determined spectrophotometrically as a complex of iron(II). Although iron(III) forms many colored complexes, it is frequently difficult to control the number of ligands per iron atom, with the result that deviations from Beer's law can occur.

The classic reagent used to determine iron (after reduction to iron(II)) is the organic reagent 1,10-phenanthroline (Phen). This reagent has the empirical formula $C_{12}H_8N_2$ and the structural formula

The complex contains three molecules of 1,10-phenanthroline to one of iron:

$$Fe^{2+} + 3Phen \rightarrow Fe(Phen)_3^{2+}$$

Each nitrogen atom in the phenanthroline forms a coordinate covalent bond with the iron(II), for a total of six such bonds. The complex is named tris(1,10-phenanthroline)iron(II) ion, according to complex-ion nomenclature.

Several other metal ions also form complexes with 1,10-phenanthroline, but none is as intensely colored as the ferrous complex. The principal interferences are silver(I), cobalt(II), copper(II), and nickel(II). Yamamura and Sikes [2] have shown that a mixture of citric acid and ethylenediaminetetraacetic acid (see Chapter 11) can be used to mask silver(I), copper(II), nickel(II), and large amounts of other common metal ions. The method can be used for determining very small amounts of iron because of the large molar absorptivity (1.11×10^4) of the iron complex. The detection limit ($A = 0.01$) is about $1 \times 10^{-6} M$, and the usual range for analysis is 0.4 to 8 ppm of iron in a 1-cm cell.

In the procedure, iron is first reduced to Fe^{2+} with an excess of a reducing agent, such as hydroxylammonium chloride or ascorbic acid. Then, 1,10-phenanthroline solution is added in excess of the amount required to react with Fe^{2+}, and the pH of the solution is adjusted by means of an acetate buffer. (The last step is important because full color development will occur only in the proper pH range.) Finally, the solution is diluted to volume, and the absorbance is measured at 512 nm with a spectrophotometer.

More recently, the reagent FerroZine has been introduced for the spectrophotometric determination of iron. The procedure is similar to that using 1,10-phenanthroline, but the molar absorptivity of the iron(II)-FerroZine complex is

[2] S. S. Yamamura and J. H. Sikes, *Anal. Chem.,* 38, 793 (1966).

much greater ($\epsilon = 2.79 \times 10^4$[3]), and lower concentrations of iron in solution can be measured.

An Example of Oxidation: Determination of Manganese. Manganese is present in many alloys and steels; but when it is dissolved in acid, even in nitric acid, it forms the nearly colorless manganese(II) aquo ion:

$$Mn(s) + 2H^+ + 6H_2O \rightarrow Mn(H_2O)_6{}^{2+} + H_2(g)$$

The aquo ion is fairly stable and does not readily form intensely colored complex ions with chloride (or thiocyanate), as do cobalt(II) or copper(II) ions. Although manganese(II) ion does form colored chelates with certain organic ligands, it can be determined more simply by oxidation. Oxidation is useful because manganese has a number of higher oxidation states, manganese(VII) (as the permanganate ion) being intensely colored and the most stable. The manganese(II) ion is oxidized to the permanganate ion by oxidation with periodate in a hot, acidic solution:

$$2Mn^{2+} + 5IO_4{}^- + 3H_2O \rightarrow 2MnO_4{}^- + 5IO_3{}^- + 6H^+$$

The method is almost specific for manganese and can be used to determine quite small amounts of it. If excess periodate is present, the permanganate color is stable but dilutions must be carried out with distilled water that is entirely free from organic matter, which might reduce the permanganate. The acidity of the solution does not affect the color.

Manganese in steel is determined colorimetrically by oxidizing it to permanganate. The steel is first dissolved in nitric acid. After the dissolution, any carbon remaining is oxidized with potassium persulfate, $K_2S_2O_8$. If the persulfate oxidizes part of the manganese to manganese dioxide or to permanganate, a bit of bisulfite will return all of the manganese to the divalent state. Then manganese is oxidized by being boiled with periodate (see the preceding equation). The large quantity of ferric iron present gives the solution a strong yellow color; this can be prevented by adding phosphoric acid, which forms a colorless complex with iron(III). The absorbance of permanganate is measured, and the amount of manganese in the sample is calculated from a calibration plot.

Indirect Spectrophotometry: Determination of Olefins. The typical olefin possesses a single, unconjugated, double bond (—C=C—) and is not colored. Such olefins form few complexes, and those that are formed are generally colorless or weakly colored. The oxidation-reduction approach is also useless, since the oxidation products are generally colorless as well. Thus, an indirect spectrophotometric determination appears to be the best choice. An indirect determination of a colorless species is made by measuring how much it destroys the color of a colored species via a chemical reaction. A stoichiometric reaction is preferred, but even a nonstoichiometric reaction may be utilized if it is reproducible. The calibration plot, or working curve, obtained is shown in Figure 17–10.

For an indirect determination of olefinic unsaturation (—C=C— group) in colorless organic compounds, an excess of bromine in 90% acetic acid–10% water is added to the sample. The bromine reacts with the carbon-carbon double bond to

[3] L. L. Stookey, *Anal. Chem., 42,* 779 (1970).

Figure 17–10. Calibration plot for the direct determination of a colorless species.

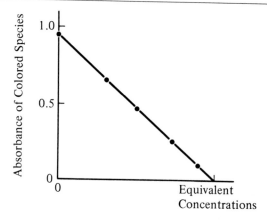

Concentration of Colorless Species

form a colorless organic product. The major reaction is

$$Br_2 + -\overset{|}{C}=\overset{|}{C}- \rightarrow -\overset{|}{\underset{Br}{C}}-\overset{|}{\underset{Br}{C}}-$$

but some of the olefin reacts as follows:

$$Br_2 + -\overset{|}{C}=\overset{|}{C}- + H_2O \rightarrow -\overset{|}{\underset{Br}{C}}-\overset{|}{\underset{OH}{C}}- + HBr$$

It should make no difference how much of the olefin reacts according to the first reaction and how much by the second because the combining ratio of bromine to olefin is the same in both reactions. Measurement of the olefin is based on the decrease in absorbance of the bromine at 410 nm. Unfortunately, a plot of bromine absorbance versus concentration of olefin was not linear but showed marked deviations from Beer's law. The difficulty was finally traced to HBr, which is formed in the second reaction given above but not in the first. The bromide ion reacts with bromine to set up the following equilibrium:

$$Br_2 + Br^- \rightleftharpoons Br_3^-$$

The tribromide ion is also colored but has a much higher molar absorptivity at 410 nm than does bromine. The relative concentrations of Br_2 and Br_3^- vary with the concentration of olefin because higher concentrations of olefin result in more bromide, shifting the equilibrium to the right.

Once the reason for the deviations from Beer's law was recognized, the difficulty was fairly easy to correct. In the method finally developed, HBr is added to the bromine solution in sufficient excess to convert most of the bromine to tribromide. Thus the small amount of HBr formed by reaction with the olefin does not affect the bromine-tribromide equilibrium. A linear calibration plot similar to Figure 17–10 is obtained, and the method is widely applicable for the analysis of organic material for small amounts of simple olefinic unsaturation [4].

Two-Component Analysis: Measurement of Forms of an Indicator. It is easy to analyze mixtures of two colored substances if there is a wavelength at which just one substance absorbs and a second wavelength at which only the second absorbs. But what if the spectra of the two substances are such that there is no wavelength at which one absorbs appreciably while the other does not absorb at all? A two-component mixture can still be analyzed if there are two wavelengths at which there is a considerable *difference* in absorbance of the two substances. A good illustration is measuring the acidic and basic forms of an indicator contained in a particular solution. Here, the first step is to obtain the spectrum of a solution containing a known concentration of the acidic form only and also to obtain the spectrum of a known concentration of the basic form. Full conversion of the indicator to the acidic form may be accomplished by adding a considerable excess of acid to the solution; likewise, the spectrum of the full basic form may be measured on a second solution of the indicator to which excess base has been added. The spectra obtained are shown in Figure 17–11.

From these spectra and the concentration of dye known to be present in each case, ϵ_A and ϵ_B (molar absorptivity of the acidic and basic forms, respectively) are calculated at two wavelengths—at 560 nm, where B absorbs strongly and A absorbs but slightly, and at about 430 nm, where the acidic form A absorbs more strongly than the basic form, B. With this information, the absorbances of the sample containing a *mixture* of A and B are measured at 430 nm and at 560 nm. The amount of each form present may be calculated by solving the following simultaneous equations for c_A and c_B:

$$A_{(560)} = \epsilon_{A_{(560)}} b c_A + \epsilon_{B_{(560)}} b c_B$$

$$A_{(430)} = \epsilon_{A_{(430)}} b c_A + \epsilon_{B_{(430)}} b c_B$$

Automated Analysis. Most quantitative spectrophotometric methods require several manipulative operations, each of which must be performed precisely. In laboratories where large numbers of analyses must be carried out, almost all spectrophotometric methods are automated to ease the load on the chemists and technicians. A good example of an automated analysis is the spectrophotometric determination of glucose in body fluids, as done in clinical laboratories. One of the several methods for glucose is the oxidation of glucose to gluconic acid by ferricyanide:

$$C_5H_6(OH)_5CHO + 2Fe(CN)_6^{3-} + H_2O \rightarrow C_5H_6(OH)_5CO_2H + 2Fe(CN)_6^{4-} + 2H^+$$

The aldehyde group (—CHO) of glucose is rapidly oxidized at 95°C by a $10^{-3} M$ solution of the intensely yellow ferricyanide ion. The unreacted ferricyanide is measured at 420 nm, and the destruction of the yellow color is a measure of the

[4] J. S. Fritz and G. E. Wood, *Anal. Chem.*, 40, 134 (1968).

Figure 17-11. Spectral curves for alizarin: (a) acidic form; (b) basic form.

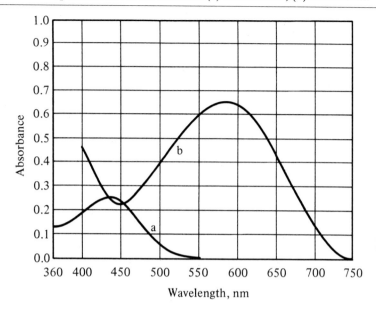

glucose. Although the ferricyanide is also yellow, its molar absorptivity at 420 nm is only about 1.0, which is too small to cause a measurable effect in determining the ferricyanide.

An instrument used frequently for automated analysis is the Technicon® AutoAnalyzer®. It is capable of making 60 determinations of glucose in blood samples per hour. Samples are withdrawn from small beakers on a turntable. The samples are mixed with a diluent and pumped down the same tubing by a proportioning pump. A large air bubble separates each sample from the next, thus preventing it from mixing with another sample. In the constant temperature dialyzer, the glucose in each sample diffuses away from most of the blood through a membrane and mixes with the ferricyanide reagent. Still separated by air bubbles, the glucose-ferricyanide mixtures are pumped into a 95°C heating bath, where they react rapidly. Each mixture passes into the colorimeter, where the unreacted ferricyanide is measured and recorded on the strip-chart recorder.

Other Spectrophotometric Determinations. Literally thousands of spectrophotometric procedures are described in scientific journals and books. They are for the determination of inorganic, organic, and biochemical substances, and they include methods with ultraviolet, infrared, and visible radiation. Table 17-5 lists a few widely used color-forming reagents and the elements determined with them.

Ultraviolet Measurement of Inorganic Species. Most of the determinations listed in Table 17-5 involve reaction to form a complex or other product that absorbs strongly in the visible spectral region. However, many inorganic ions and molecules

Table 17–5. Typical Colorimetric Methods

Element Detected	Reagent	Color Formed (λ used, nm)
Al	8-Hydroxyquinoline	Yellow (395)
Bi	Thiourea	Yellow
Ca	Calcein[a]	Y-green fluoresc. (520)
Cl(Cl$_2$)	o-Tolidine	Yellow
Co	Ammonium thiocyanate	Blue (620)
Cr	Diphenylcarbazide	Red-violet (540)
Cu	FerroZine	Brown (470)
F(F$^-$)	Cerous alizarin complexone[b]	Wine red
Fe	1,10-phenanthroline	Red (512)
Fe	FerroZine	—— (562)
Mg	o,o'-Dihydroxyazobenzene[a]	Orange (485)
Mn	Periodate	Purple (520)
Mn	Thiothenoyltrifluoroacetone	—— (450)
Mo	Thiolactic acid[c]	Yellow-brown
P(PO$_4^{3-}$)	Molybdate, hydrazine	Blue (830)
Pb	Dithizone	Pink
S(SO$_3^{2-}$)	Iodine (reduction)	I$_3^-$ decreased
Ti	Hydrogen peroxide	Yellow
U	Arsenazo I or III	Violet-blue (640)
Zn	Dithizone	Pink

a. H. Diehl, Calcein, Calmagite and o,o-Dihydroxyazobenzene..., G. Frederick Smith Chemical Co., Columbus, Ohio, 1964.

b. S. S. Yamamura, M. A. Wade, and J. H. Sikes, *Anal. Chem.*, *34*, 1308 (1962).

c. J. S. Fritz and D. R. Beuerman, *Talanta*, *19*, 366 (1971).

either absorb appreciably in the ultraviolet region or can be converted to species that do.

Ions and molecules in which a nonmetal atom is doubly bonded to oxygen absorb in the ultraviolet region; some examples are NO_3^-, NO_2^-, NO_2, SO_4^{2-}, SO_3^{2-}, and O_3. In addition to this group, there are a number of ions that contain various double-bond configurations that cause ultraviolet radiation to be absorbed. These include CO_3^{2-}, SCN^-, MnO_4^-, and CrO_4^{2-}. Several of the 11 substances listed absorb in the "practical" ultraviolet region of approximately 190–380 nm, while others absorb only in the UV region below 190 nm, where measurement with ordinary UV-visible spectrophotometers is impossible.

In contrast to the above species, inorganic species without double bonds usually do not absorb in the accessible 190–380 nm region. Some examples are NH_3, NH_4^+, HF, F^-, H_2O, H_3BO_3, and $HOCl$. (The outer electrons of the fluoride ion are capable of absorbing radiation but require photons of higher energy than 180 nm.)

Let us consider the determination of an ion with a double bond in the presence of a second ion that also has a double bond. Since it is possible that both could absorb in the accessible 190–380 nm region, it would be necessary to obtain the ultraviolet spectrum of each species alone. Then the two spectra would be examined to find a region in which only the one ion absorbs. If such a region exists, then that ion could be determined by ultraviolet spectrophotometry.

As an example of the above situation, consider the determination of traces of nitrate ion in carbonate solutions. The structures of the nitrate ion and carbonate ion are respectively:

We would initially predict that both ions could absorb in the ultraviolet because both possess double bonds. However, examination of the ultraviolet spectrum of each ion showed that carbonate did not absorb significantly in the 200–380 nm region, but nitrate exhibited absorption bands at 203 nm ($\epsilon = 1 \times 10^4$) and at 300 nm ($\epsilon = 7.5$), with a trough or valley at 260 nm. Traces of nitrate ion could thus be measured at 203 nm, although 220 nm was actually chosen to avoid the interference of the perchlorate ion (ClO_4^-), which was also present. Ions such as NH_4^+, Na^+, and Cl^- did not interfere at 220 nm.

The ultraviolet determination of other inorganic ions is also possible, although certain common ions such as iron(III) absorb so strongly in the ultraviolet that their presence makes ultraviolet measurements less accurate than measurements in the visible region. For example, the permanganate ion absorbs in the ultraviolet but is measured more accurately in the visible region (Section 17–4). Many metal cations can be converted to an anionic chloride complex and the metal determined quantitatively by ultraviolet spectrophotometry (see Section 18–1).

QUESTIONS AND PROBLEMS

Absorbance and Transmittance Calculations

1. Calculate the absorbance for each situation, observing significant figure rules.

    (a) A colored solution has a transmittance of 0.72.
    (b) A certain type of sunglasses has a transmittance of 0.17.

2. Calculate the absorbance of each solution, observing significant figure rules.

    (a) A colored solution has an 88% transmittance.
    (b) A faintly colored solution has a 99% transmittance. After calculating the absorbance, explain why all the significant figures are zero.

3. Calculate the absorbance of each solution after it is transferred to a 2.0-cm cell.

    (a) A solution of copper(II) sulfate with an absorbance of 0.20 in a 1.0-cm cell.
    (b) A solution of an organic dye with a %T of 28% in a 5.0-cm cell.

4. An organic dye has an absorbance of 0.40 in a 1.0-cm cell. The dye solution is diluted to

one-half (0.50) the original concentration and placed in a 3.0-cm cell. Calculate its absorbance and transmittance at the same wavelength.

5. By mistake, the transmittance scale reading of a distilled water blank is set at 95% instead of 100%. With this setting, a colored solution has a transmittance scale reading of 35.2% T. Calculate the correct transmittance of the colored solution.

6. A certain solution has an absorbance that is numerically the same as its transmittance to two significant figures. Find this value. (*Hint:* One approach is trial-and-error.)

7. For convenience, several glass or quartz cells are generally used in spectrophotometric analyses; these cells should be "matched" with regard to their optical properties. Suppose one of the cells has a light path that is slightly longer than the usual 1.00 cm. How would this affect the results of an analysis if the defect was not recognized? Explain how this cell could be calibrated so that it could be used without error.

8. The difference between any two absorbance readings, $A_1 - A_2$, is equal to the log of the inverse ratio of their corresponding transmittances, $\log(T_2/T_1)$. Verify that this is true for:

   (a) $A_1 = 1.00$ ($T_1 = 0.10$) and $A_2 = 0.10$ ($T_2 = 0.80$).
   (b) $A_1 = 1.00$ and $A_2 = 0.20$ (calculate T_1 and T_2).

Beer's Law Calculations

9. Calculate the log of each molar absorptivity; use the correct number of significant figures.

   (a) $\epsilon = 2.0 \times 10^4$ (c) $\epsilon = 11$
   (b) $\epsilon = 0.02$ (d) $\epsilon = 1.02$

10. Calculate the molar absorptivity from each log value, using the correct number of significant figures.

    (a) $\mathrm{Log}\,\epsilon = 4.0$ (c) $\mathrm{Log}\,\epsilon = 0.06$
    (b) $\mathrm{Log}\,\epsilon = 4$ (d) $\mathrm{Log}\,\epsilon = -0.02$

11. Calculate the molar absorptivity of each species, assuming a 1.00-cm cell.

    (a) At 540 nm, a $2.0 \times 10^{-4}M$ solution of $KMnO_4$ has an $A = 0.40$.
    (b) At 400 nm, a $2.00M$ solution of $Mn(H_2O)_6{}^{2+}$ has an $A = 0.070$.
    (c) A $2.0 \times 10^{-4}M$ solution of acetylsalicylic acid has an A of 0.28 at 280 nm and an A of 0.22 at 235 nm.

12. After calculating the molarity of the copper(II) ion in each of the following solutions, calculate the molar absorptivity using the correct number of significant figures. Formula weights are: $CuSO_4 \cdot 5H_2O$, 250; $CuSO_4$, 160; and H_2O, 18.

    (a) A solution of 0.400 g of $CuSO_4 \cdot 5H_2O$ in 100 mL of water has an absorbance of 0.576 at 790 nm in a 3.00-cm cell.
    (b) A solution containing 600 mg/dL of $CuSO_4 \cdot 5H_2O$ has a $\%T$ of 33% at 820 nm in a 2.00-cm cell. (Recall that 1 dL = 0.1 L.)

13. After calculating the molarity of each solution, calculate the molar absorptivity at 258 nm.

    (a) A solution of 5.19 mmoles of phenylalanine in 1.000 L has a $\%T$ of 9.5% in a 1.00-cm cell.
    (b) A solution of 2.475 mg/dL of phenylalanine (form wt = 165) has a $\%T$ of 81% in a 3.00-cm cell.

Calculation of Unknown Concentration from One Standard

14. Calculate the concentration of MnO_4^- in each unknown, assuming that the same cell (unknown diameter) is used for both standard and unknown. A $1.00 \times 10^{-4}M$ permanganate standard has $A = 0.20$ (525 nm) and the unknown permanganate has $A = 0.70$ (525 nm).

15. At 760 nm, a $5.3 \times 10^{-2}M$ standard copper(II) solution has an A of 0.577. An unknown copper(II) solution has an A of 0.233 (760 nm) in the same cell.

   (a) Calculate the molarity of the copper(II) in the unknown.
   (b) Calculate the mg of Cu^{2+}/dL in the unknown using the atomic weight of copper. (Recall that 1 dL = 0.1 L.)

16. A 10.0-mL aliquot of an Fe^{3+} unknown is diluted to 50.0 mL. A 5.00-mL portion of the 50.0 mL is then reduced to Fe^{2+} and treated with 1,10-phenanthroline; the resulting solution has an $A = 0.233$ (512 nm). A $2.2 \times 10^{-5}M$ standard solution of Fe^{3+} reduced and treated with 1,10-phenanthroline without dilution has $A = 0.577$ (512 nm).

   (a) Calculate the molarity of Fe^{3+} in the 50.0 mL solution.
   (b) Calculate the molarity of the Fe^{3+} in the original unknown.
   (c) Calculate the ppm of Fe^{3+} in the 10.0-mL aliquot.

17. A standard solution of permanganate has an absorbance $= 0.20$ (525 nm) in a 1.00-cm cell and a concentration of $1.00 \times 10^{-4}M$. Calculate the concentration of permanganate in each of the two unknowns below.

   (a) A permanganate unknown with $A = 0.70$ (525 nm) in a 1.00-cm cell.
   (b) A permanganate unknown with $A = 0.25$ (525 nm) in a 2.00-cm cell.

Instrumental Measurement of Radiant Energy

18. List four sources of ultraviolet radiation and the wavelength range emitted by each. Also give one advantage and one disadvantage each source has for use in spectroscopy.

19. List three sources of 800-nm radiation. What type of spectrophotometer should be used to measure this radiation?

20. List three detectors of ultraviolet-visible radiation and the wavelength range of response of each. Which detector does not need to operate in a vacuum or envelope protected from air?

21. Suggest a specific detector that will respond to the entire wavelength region for each of the following regions.

   (a) 300–600 nm   (b) 700–1000 nm   (c) 500–700 nm

22. Explain why the tungsten halogen lamp is preferred to the conventional tungsten filament lamp. Why is it important that the envelope for the tungsten halogen lamp be quartz and not glass if measurements are to be done over most of the ultraviolet region?

23. Compare the vacuum photodiode and the silicon semiconductor photodiode on the following points.

   (a) The process by which photons interact with the detector.
   (b) The range of the response, which is larger, and what limits the response.
   (c) The advantages of each over the other.

24. Compare the conventional spectrophotometer with the microprocessor spectrophoto-

meter. What advantage would the latter have when a standard method with fixed wavelength is to be used many times?

25. Compare the microprocessor spectrophotometer (no memory) with the microcomputer spectrophotometer (large memory). Which spectrophotometer would be used for multicomponent analysis? Which would be more expensive?

Basic Analysis Problems

26. The following data were obtained for a colored zinc complex at 465 nm with a 1-cm cell:

Standard Number	Zn, ppm	Absorbance
1	2.0	0.105
2	4.0	0.205
3	6.0	0.310
4	8.0	0.415
5	10.0	0.515

(a) Prepare a calibration plot (use graph paper).
(b) Calculate the absorptivity in $L\text{-mg}^{-1}$ cm^{-1} and the molar absorptivity in $L\text{-mol}^{-1}$ cm^{-1}.
(c) Calculate the concentration of an unknown that has an absorbance of 0.200.

27. The following data were obtained for $KMnO_4$ at 540 nm with a 1-cm cell:

$KMnO_4$, Molarity	Absorbance
0.00005	0.101
0.00010	0.202
0.00020	0.405
0.00030	0.606
0.00040	0.809

(a) Prepare a calibration plot (use graph paper).
(b) Calculate the molar absorptivity at this wavelength.
(c) Using the plot from (a), find the concentration of an unknown that has a transmittance of 50.0%.

28. A 0.2000-g sample of a metal alloy was dissolved in acid and diluted to exactly 200 mL in a volumetric flask. Periodate was added to a 25-mL aliquot, and the solution was boiled so that manganese was oxidized to permanganate. After being cooled, this solution was diluted to exactly 100 mL. A portion of the final solution was measured colorimetrically and found to contain 1.85-ppm manganese. Calculate the percent of manganese in the metal alloy.

29. In each case, a 0.200-g sample of a manganese alloy is dissolved in acid and diluted to exactly 200 mL in a volumetric flask. The specified aliquot is withdrawn and oxidized to permanganate. After correcting for the dilutions specified, calculate the percentage of manganese in the alloy.

(a) A 25.00-mL aliquot is taken from a 200-mL solution of the alloy, and measurement gives 1.85 ppm of manganese in the 25.00 mL.
(b) A 25.00-mL aliquot from the 200-mL solution of the alloy is diluted to 100.0 mL before measurement. Measurement gives 1.80 ppm of Mn in the 100.0 mL.

30. At 540 nm, the absorbance of the dye pararosaniline as a function of pH is as follows: $A = 0.00$ at pH 2.0, 0.84 at pH 4.0, 0.88 at pH 4.6, 0.90 at pH 5.2, 0.90 at pH 6.2, 0.78 at pH 7.2, and 0.46 at pH 8.0. Prepare a plot of absorbance versus pH and indicate the best pH range for a spectrophotometric determination of pararosaniline.

31. At 475 nm, the molar absorptivity of the acidic form of an acid-base indicator is 120. At the same wavelength, the molar absorptivity of the basic form is 1050. A $1.00 \times 10^{-3}M$ solution of this indicator has an absorbance of 0.864 in a 1-cm cell at 475 nm when the pH is such that both the acidic and basic forms of the indicator are present. Calculate the concentration of each form and the ratio of the basic to acidic forms of the indicator.

32. From the following data, calculate the correct molar absorptivity for a 1:1 metal-dye complex at 530 nm. Absorbance of $1.0 \times 10^{-2}M$ dye at 530 nm $= 0.055$; absorbance of $1.0 \times 10^{-4}M$ metal plus $1.0 \times 10^{-3}M$ dye at 530 nm $= 0.725$. (Both measurements were made in 1-cm cells.)

33. Often a metal ion is titrated with a reagent that reacts with the metal ion to form a colored product. If only the metal complex absorbs at the wavelength used, a plot of absorbance versus milliliters of reagent will give two straight lines. The stoichiometric point is where the two straight lines intersect. Exactly 10-mL aliquots of a $3.63 \times 10^{-4}M$ metal-ion solution was titrated with a $4.0 \times 10^{-3}M$ reagent (R). After addition of the reagent, each aliquot was diluted to exactly 50 mL, and the absorbance was measured. The following data were obtained:

mL Reagent	Absorbance
0.00	0.000
0.20	0.220
0.40	0.440
0.60	0.660
0.90	0.882
1.40	0.999
1.60	1.000
1.80	1.000

(a) Plot absorbance versus milliliters of reagent and find the volume of reagent at the stoichiometric point. Calculate the ratio of reagent to metal at this point.

(b) Calculate the molar absorptivity of the complex at the wavelength used.

(c) The absorbance at the stoichiometric point is less than the theoretical absorbance owing to dissociation of the complex. Calculate the concentrations of metal complex, uncomplexed metal ion, and free reagent at the stoichiometric point. From these data, calculate the value of the formation constant of the complex.

34. A common error in spectrophotometric determinations is incomplete reaction of a metal ion with a color-forming reagent in the very dilute solutions that are commonly used. A 1:1 metal-reagent complex is formed by adding two moles of reagent for each mole of metal ion and diluting to a total metal concentration (M + MR) of $1.0 \times 10^{-5}M$. Calculate the percentage of the total metal that exists as the complex, MR, when the formation constant for MR is: (a) 10^4, (b) 10^6, (c) 10^8.

35. Exactly 10 mL of $5.0 \times 10^{-4}M$ reagent (R) is added to 2.00 mL of $5.0 \times 10^{-4}M$ metal ion (M) and diluted to 50.00 mL in a volumetric flask. A complex, MR_2, is completely formed and has an absorbance of 0.875 at 468 nm in a 1-cm cell. A reagent blank containing 10.00 mL, after dilution to 50.00 mL, has absorbance of 0.215 at 468 nm. Calculate the correct molar absorptivity at 468 nm of: (a) The MR_2 complex, (b) The reagent, R.

36. Even distilled water of good quality often contains impurities of several metal ions in the

low ppb concentration range. A reagent that is highly sensitive for zinc(II) is added to distilled water and gives an absorbance of 0.130 when measured at an appropriate wavelength in a 1.0-cm cell. Outline a scheme for determining the concentration of zinc in the distilled water given the restraint that the same distilled water cannot be further purified and must be used for any zinc standards that are prepared.

37. Compounds in solution that have a visible yellow or orange-yellow color often have a more intense absorptium maximum in the ultraviolet spectral region. From an examination of its visible spectrum in Figure 17–5b, decide whether this is probably the case for potassium dichromate. If a suitable instrument is available, ask your instructor whether it may be possible to measure the UV spectrum of a potassium dichromate solution.

Analysis of Mixtures

38. Suppose that the alizarin solutions shown in Figure 17–11 are each $1.0 \times 10^{-4}M$ and are measured in 1-cm cells. The basic form has $A = 0.650$ at 580 nm and $A = 0.250$ at 430 nm. The acidic form has $A = 0$ at 580 nm and $A = 0.250$ at 430 nm. At a certain pH, an alizarin solution of unknown concentration has $A = 0.505$ at 580 nm and $A = 0.413$ at 430 nm. Calculate the concentrations of both the acidic and basic forms at this pH.

39. Suggest a method for measuring the correct molar absorptivites at 400 nm for Br_2 and for Br_3^- in 90% acetic acid–10% water. Using this information, explain how you could estimate the relative concentrations of Br_2 and Br_3^- in a mixture.

40. A mixture of *ortho* and *para* nitroanilines is analyzed by using ultraviolet spectrophotometry to make absorbance measurements at two different wavelengths. From the following data, calculate the molar concentrations of the *ortho* and *para* isomers. At 285 nm: A(mixture $= 1.040$, $\epsilon(ortho) = 5260$, $\epsilon(para) = 1400$. At 347 nm: A (mixture) $= 0.916$, $\epsilon(ortho) = 1280$, $\epsilon(para) = 9200$.

41. It is proposed to determine chromium in a manganese-free steel by dissolving the steel in nitric acid and oxidizing the resulting chromium(III) to $Cr_2O_7^{2-}$.

(a) At what wavelength might $Cr_2O_7^{2-}$ be measured? (See Figure 17–6.)

(b) What possible interference might be encountered?

(c) How might this interference be eliminated?

18

Spectroscopy of Organic Compounds

In this chapter, we will discuss the theory of spectral transitions and the analysis of organic compounds by ultraviolet spectrophotometry. This will provide a good foundation for the next topic, fluorescence methods of analysis. Finally, both quantitative and qualitative infrared methods will be discussed briefly.

18–1. NATURE AND PROBABILITY OF ELECTRONIC TRANSITIONS

In Chapter 17, the theory of visible and ultraviolet absorption spectra was discussed briefly. The energy of the absorbed radiation (which is inversely proportional to wavelength) matches that needed to cause an electronic transition in the absorbing chemical species. Gaseous atoms absorb energy in a very narrow wavelength range, producing a set of very narrow absorption peaks called a "line spectrum." Chemical species in solution have numerous vibrational and rotational energy levels, both in the ground state and in electronic excited states. The absorption spectra of substances in solution thus have broader peaks because transitions can occur to various vibrational and rotational levels within a given excited state.

In Figure 17–4, the general nature of a *molecular* energy diagram for the ground state and electronic excited states was depicted. However, electronic transitions were not discussed at the level needed for understanding ultraviolet absorption.

Orbital Theory and Electronic Transitions. Now we will examine electronic transitions in somewhat greater detail. Because some of the following explanations invoke atomic and molecular orbital theory, a brief review of orbital theory seems advisable [1].

According to the Bohr–Sommerfeld model, electrons rotate about the nucleus of an atom in orbits whose radii are determined by a principal quantum number n (an integer: 1, 2, 3, etc.). Increasing values of n represent higher energy levels because more energy is required to place negatively charged electrons in orbits farther away from the positively charged nucleus.

The various atoms contain several electrons at each value of n; two electrons at $n = 1$, eight electrons at $n = 2$, etc. These electrons occupy different portions of space, called *orbitals*, within the atom. The s orbitals are spherically symmetrical around the nucleus of the atom. The three p orbitals, each shaped roughly like a figure 8 spun around its long axis, are oriented along the x, y, and z axes of a three-dimensional representation of the atom. The five d orbitals are directed between, or along, the x, y, and z axes in various manners.

Finally, it should be recalled that electrons may spin in either a clockwise or a counterclockwise direction. The Pauli exclusion principle states that a maximum of two electrons may occupy an orbital and then only if the spins of the two electrons are in opposite directions. (This is called *pairing* of spins.)

According to present theory, chemical bonds are formed by the overlap of atomic orbitals, thus forming *molecular orbitals* in which the electrons are, on the average, closer to the atomic nuclei than they were in the atomic orbitals and therefore have lower energy. Since each atomic orbital can contain a maximum of two electrons, the combination of two atomic orbitals to form a chemical bond must result not only in the formation of a molecular *bonding orbital* (containing two electrons), but also in the formation of another molecular orbital to accommodate the other two electrons that may have been present in the original atomic orbitals. This latter molecular orbital is called an *antibonding orbital*; its electrons are, on the average, farther from the atomic nuclei than before and are therefore in a higher energy level. Thus, when a chemical bond forms, the outer orbitals can be divided into three types: (1) antibonding orbitals (at higher energy), (2) bonding orbitals (at lower energy), and (3) orbitals not involved in bond formation (nonbonding orbitals, at a different energy level from the other two types).

Two common types of chemical bonds should be mentioned. A σ(sigma) bond is cylindrically symmetrical about a line joining the centers of the two bonded atoms; it results from the overlap of two s orbitals, of an s and a p_z orbital, or of two p_z orbitals (see Figure 18–1a). A π(pi) bond lies outside the line joining the two bonded atoms and arises from overlap of p_x or p_y atomic orbitals (Figure 18–1b).

The double bond in ethylene, $CH_2=CH_2$, illustrates these two types of bonds. Carbon has two $2s$ electrons, one $2p_z$ electron, and one $2p_x$ electron. The s and p_z electrons are involved in forming the carbon-hydrogen bond and the sigma bond between the two carbon atoms; the other carbon-carbon bond is a pi bond formed from the overlap of the p_x electrons. Even though they are vacant in this case, there are pi antibonding orbitals in ethylene.

---

[1] For a more detailed but elementary treatment of orbital theory, we recommend the book by M. Orchin and H. H. Jaffe, *The Importance of Antibonding Orbitals*, Boston: Houghton Mifflin, 1967.

Figure 18–1. (a) Combination of an s and a p_z orbital to form a sigma bond. (b) Combination of two p_x orbitals to form a pi bond.

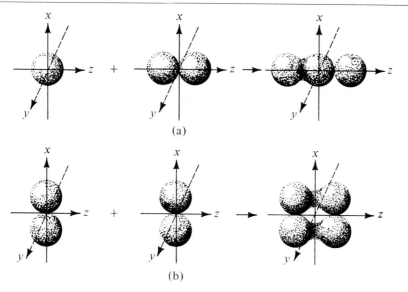

Commonly used nomenclature designates molecular orbitals involved in sigma bonds as σ orbitals and the antibonding orbitals created by sigma bond formation as σ^*(sigma star) orbitals. Pi bonding orbitals are designated simply as π orbitals, and the corresponding antibonding orbitals as π^*(pi star). Uninvolved outer electrons are often called n, or nonbonding, electrons.

The foregoing discussion has, of course, omitted a great many details because the application of orbital theory to specific cases can become extremely complicated. Many compounds have d orbitals of various atoms involved in bonding. In many instances, the use of *hybrid* orbitals must be invoked to explain chemical structure. Although the present molecular orbital theory is widely accepted, a few doubts are creeping in. A description of the "electron repulsion" theory of chemical bonding makes interesting reading [2].

Nature and Probability of Electronic Transitions. When a chemical molecule absorbs light in the visible or ultraviolet spectral region, an electron in the molecule is promoted from its most stable orbital (in the ground state) to a higher-energy orbital (in the excited state). Molecular orbital theory will usually indicate the nature of the electronic transition and its energy relative to other transitions. It may also indicate the probability of the electronic transition. Since the molar absorptivity ϵ is related to the probability, application of this theory may indicate the value of ϵ within an order or two of magnitude. For example, a high-probability electronic

[2] W. F. Luder, *J. Chem. Ed.*, *44*, 206 (1967).

transition will have an ϵ in the 10^3–10^5 range. A low-probability electronic transition will have an ϵ below 10^2. Information about the nature, energy, and probability of an electronic transition is helpful in correlating chemical structure with absorption spectra.

The correlation of structure with spectra may be illustrated by considering the absorption of ultraviolet radiant energy by an organic carbonyl group, $>$C$=$O, in an aldehyde $\left(\begin{smallmatrix} O \\ \| \\ R-C-H \end{smallmatrix}\right)$ or ketone $\left(\begin{smallmatrix} O \\ \| \\ R-C-R \end{smallmatrix}\right)$. First, consider the electronic structure of carbon with four electrons in its outer shell. Bonding to two R or H groups and a single sigma bond to oxygen (in the z axis) accounts for three electrons. The remaining electron is a p_x electron that overlaps with a p_x electron of oxygen to form a second bond (pi bond) between carbon and oxygen. Oxygen has six electrons in its outer shell. One p_z electron forms a sigma bond with carbon (in the z axis), and one p_x electron is used to form the pi bond with carbon. The four remaining electrons form lone pairs that are not involved in bonding. According to orbital theory, one of these electron pairs is in an sp hybrid orbital. The other pair occupies a p_y orbital, which, being perpendicular to the C—O sigma bond (z axis), cannot combine with any of the carbon orbitals. Since its energy is not changed by bonding (see Figure 18–2), it is termed a *nonbonding* orbital and given the symbol n.

In considering possible electronic transitions caused by absorption of radiant energy, we will neglect the lone-pair electrons in the sp hybrid orbital of oxygen and the electrons involved in sigma bonding because promotion of electrons out of these two orbitals requires a great deal of energy. It could also result in breaking the sigma bond that holds the molecule together. An energy level diagram of the remaining outer shell electrons is given in Figure 18–2.

The electronic transition requiring the lowest energy is the promotion of a nonbonding (n) electron to the antibonding π^* orbital. However, this is a "forbidden," or low-probability, transition because the nonbonding orbital is in a plane (y axis) perpendicular to the plane (z axis) of the π^* orbital. (The π^* orbital is in the same plane as the π orbital.) Since the overlap of these orbitals is so small, the

Figure 18–2. Molecular-orbital energy-level diagram for the carbonyl group. Electrons are indicated by arrows.

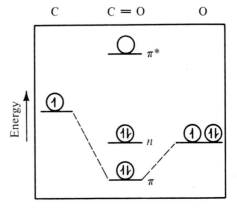

probability of the transition (and hence the resulting molar absorptivity) is quite low. For instance, an n to π^* transition takes place in formaldehyde, producing an absorption peak at about 270 nm with a molar absorptivity of only about 17.

Another possible transition is a π to π^* transition. Since both these orbitals are in the same plane, their overlap is large, and the probability of this transition occurring is high. However, this transition requires more energy than the n to π^* transition and thus should occur at a shorter wavelength (see Figure 18–2). With special instrumentation that permits measurement in the vacuum ultraviolet region, a very strong absorption peak ($\epsilon \approx 10^4$ to 10^5) is observed for formaldehyde at about 185 nm.

Whenever two double bonds are conjugated, one of the bonding orbitals is raised in energy and the other lowered relative to the energy of a double bond by itself. The same applies to antibonding orbitals. Because of this, the shift in absorption maxima caused by conjugation may be predicted qualitatively from molecular-orbital considerations.

Example: Acrylaldehyde,

$$
\begin{array}{c}
H \\
\diagdown \\
C{=}C{-}C{=}O \\
\diagup | | \\
H H H
\end{array}
$$

is similar to formaldehyde except that the carbon-carbon double bond is conjugated with the carbonyl group. Draw a molecular orbital energy-level diagram for acrylaldehyde showing the π and n orbitals. Compare the $n \rightarrow \pi^*$ and $\pi \rightarrow \pi^*$ transitions with those in formaldehyde and decide whether these transitions in acrylaldehyde occur at shorter or longer wavelengths than those of formaldehyde.

In acrylaldehyde, there is one pi bonding orbital and one pi antibonding orbital for each double bond. As in formaldehyde, there are also nonbonding electrons. The energy-level diagrams in Figure 18–3 show that both $n \rightarrow \pi^*$ and $\pi \rightarrow \pi^*$ transitions can occur at lower energies and hence at longer wavelengths in acrylaldehyde than in formaldehyde.

The Nature of Inorganic Electronic Transitions. It is also useful to know something about the transitions that occur when inorganic substances absorb radiant energy. First of all, the *probability* of a transition is of interest because the probability determines what the molar absorptivity will be. In general, a high-probability electronic transition will have a molar absorptivity in the 10^3–10^5 range, 10^5 being the upper limit predicted by spectroscopic theory. A medium-probability transition will have a molar absorptivity from 10^2 to 10^3, and a low-probability transition from 10^{-1} to 10^2.

For the lanthanides, absorption of radiant energy results in transitions among the partially filled $4f$-electron levels. In the actinides (thorium, uranium, neptunium, plutonium, etc.), transitions occur in the $5f$-electron levels. The spectral bands are sharp and are not affected by changes in chemical environment, such as the solvent. These $f \rightarrow f$ transitions are mostly of low probability, so solutions of $+3$ cations

Figure 18-3.   Energy-level diagrams for formaldehyde and acrylaldehyde.

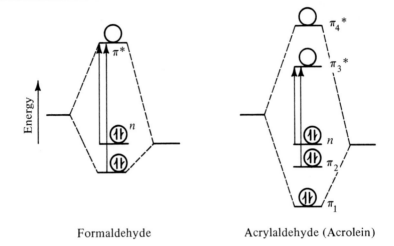

Formaldehyde   Acrylaldehyde (Acrolein)

are weakly colored. Ions with f^0, f^1, f^{13}, and f^{14} configurations are colorless; f^1 ions exhibit a high probability of $f \to d$ transitions in the ultraviolet.

Colors of transition metal ions and their complexes are largely the result of $d \to d$ transitions, and these are also mostly transitions of low probability. For octahedral ions, ϵ is of the order of 10^1; for example, ϵ is 5.0 (476 nm) for $Mn(H_2O)_6{}^{2+}$, 13.0 (428 nm) for $Cr(H_2O)_6{}^{3+}$, and 2.15 (640 nm) for $CrCl_2(H_2O)_4{}^+$.

This generalization about the average ϵ permits one to calculate the detection limit (absorbance = 0.01) for any transition metal ion except d^0, d^5, and d^{10} metal ions.

Example: It is desired to develop a trace analysis method for cobalt(II). Is the light absorption of the red $Co(H_2O)_6{}^{2+}$ ion intense enough at 530 nm for traces of cobalt(II) to be detected?

Assuming that ϵ at this wavelength is about $10 \text{ cm}^{-1}M^{-1}$ and that the cell is 1 cm long, the detection limit ($A = 0.01$) is calculated as

$$c = \frac{A}{b} = \frac{0.01}{1 \text{ cm} \times 10 \text{ cm}^{-1}M^{-1}} = 1 \times 10^{-3}M$$

This indicates that concentrations of $Co(H_2O)_6{}^{2+}$ lower than $10^{-3}M$ will not even deflect the spectrophotometer needle. Since trace analysis encompasses the 10^{-7}–$10^{-3}M$ range, traces of cobalt(II) cannot be measured as $Co(H_2O)_6{}^{2+}$.

A different type of electronic transition, the *charge transfer* transition, occurs with high probability in complex ions containing ligands more easily reduced or oxidized than water [3]. In this transition, light absorption *promotes* or *transfers* an electron from an occupied orbital of one atom or ion to an unfilled orbital of

[3] H. Gray, *J. Chem. Ed.*, *41*, 2 (1964); G. H. Schenk, *Rec. Chem. Progress*, *28*, 135 (1967).

another atom or ion. In a complex ion, the electron may be transferred from the central metal ion to the ligand or (more commonly) from the ligand to the central metal ion. Such transitions are mainly $p \rightarrow d$ transitions; a p electron of the ligand is transferred to an unfilled d orbital of the central metal ion.

The red iron(III)-thiocyanate complex is a well-known example of charge transfer. Absorption of light causes an electron to be transferred from the thiocyanate to an orbital state largely associated with the iron(III) ion. Thus the product is predominately iron(III) and the thiocyanate radical, SCN. This is a high-probability transition, the molar absorptivity for $FeSCN^{2+}$ being 5×10^3 at 453 nm. The violet color of the permanganate ion is another example of a charge transfer transition; the molar absorptivity of MnO_4^- is 2×10^3 at 525 nm.

Most metal ions in solution, when complexed with a suitable ligand, can undergo charge transfer absorption. Since charge transfer transitions are of high probability, such complexes are widely used in making spectrophotometric measurements.

The determination of metal ions as their chloro complexes in $6M$ hydrochloric acid is a very useful application of charge transfer transitions. It was observed that 11 chloro complexes had molar absorptivities greater than 10^4 in the 215–400 nm range [4]. A multicomponent analysis of an alloy of iron, copper, lead, and tin was devised by using $6M$ hydrochloric acid as the solvent [5].

Many complex ions have a charge transfer band as well as d-d absorption bands. Thus, $CrCl_2^+$ has a charge transfer band at 210 nm ($\epsilon = 1.4 \times 10^4$) as well as the d-d band at 640 nm. Obviously, a charge transfer band is more suitable for trace analysis, as the following example will indicate.

Example: A trace analysis method is being developed for cobalt(II) using the blue $Co(NCS)_4^{2-}$ complex ion. Because the thiocyanate ion is easily oxidized, this gives rise to a charge transfer band at 620 nm. Is the light absorption of $Co(NCS)_4^{2-}$ intense enough for trace analysis?

Assuming that ϵ at this wavelength is about $10^3 \text{ cm}^{-1}M^{-1}$ and that the cell is 1 cm long, the detection limit ($A = 0.01$) is calculated as

$$c = \frac{A}{b} = \frac{0.01}{1 \text{ cm} \times 10^3 \text{ cm}^{-1}M^{-1}} = 1 \times 10^{-5}M$$

Since trace analysis encompasses the 10^{-7}–$10^{-3}M$ range, traces of cobalt(II) can be measured as $Co(NCS)_4^{2-}$, although not below $10^{-5}M$. (Compare this with the previous example involving $Co(H_2O)_6^{2+}$.)

Fate of Absorbed Energy. The time required for light absorption is fantastically short. According to classical physics, a photon traveling at the speed of light ($3 \times 10^{18} \text{ Å sec}^{-1}$) would be near a molecular electronic system 3 Å in diameter for 10^{-18} sec. It then requires about 10^{-15} sec for the electron to undergo a transition to an antibonding orbital, and even longer for the electron to return to its original

[4] L. Goodkin, M. D. Seymour, and J. S. Fritz, *Talanta, 22,* 245 (1975).
[5] G. E. James, *American Laboratory, 14,* 79 (Nov. 1982).

orbital. The absorption step is thus by far the most rapid of the various events involved in the excitation process.

It can be shown that only a very small fraction of the molecules in solution are excited by the absorption of radiant energy under the conditions employed in a spectrophotometric analysis.

A rough estimate of the concentration of excited molecules can be made by assuming that each quantum of light striking the solution excites one molecule. If a tungsten lamp having a flux of 1.1×10^{17} quanta per sec at 400 nm is used [5], the number of moles excited in 10^{-15} sec is

$$10^{-15} \text{ sec} \times 1.1 \times 10^{17} \frac{\text{quanta}}{\text{sec}} \times 1 \frac{\text{molecule}}{\text{quantum}} \times \frac{1 \text{ mole}}{6 \times 10^{23} \text{ molecules}}$$

Assuming a cell of 2-mL volume, the molarity of the excited molecules is $10^{-19} M$.

Quantitative spectrophotometric methods are based on the assumption that the number of molecules that absorb light, even though a minute fraction of the total, is proportional to the total number of molecules present. This is really another qualitative statement of Beer's law.

18–2. DETERMINATION OF ORGANIC COMPOUNDS BY ULTRAVIOLET SPECTROPHOTOMETRY

In the previous section the electronic transitions of aldehydes and ketones were discussed in some detail. A chemical group that has the ability to absorb radiant energy is called a *chromophore*. For example, in aldehydes of the type RCHO, the carbonyl group is the chromophore. In conjugated aldehydes, RCH=CHCHO, the combined carbon-carbon double bond and the carbonyl group is the chromophore. Organic compounds are often determined quantitatively by measuring the absorbance of their chromophore groups.

Among the well-known chromophores are the nitro group (weak $n \rightarrow \pi^*$ at 280 nm), amide (weak $n \rightarrow \pi^*$ at 214 nm), and carboxyl (weak $n \rightarrow \pi^*$ at 204 nm). Conjugated carbon-carbon double bonds absorb more strongly and will now be discussed.

Olefins.   The simplest olefin, ethylene, absorbs ultraviolet radiation at 160 nm, which is below the range of most ultraviolet spectrophotometers. Simple substituted olefins with only one double bond likewise absorb below 200 nm. Olefins with conjugated double bonds absorb above 200 nm and thus can be measured by most common ultraviolet spectrophotometers. The simplest conjugated olefin is butadiene:

$$H_2C=CH-HC=CH_2$$

It absorbs strongly at 217 nm ($\epsilon = 2 \times 10^4$).

As the number of conjugated double bonds increases in a molecule, two changes occur in the spectra:

1. The molecule absorbs over a *wider* region of the ultraviolet; its spectrum has several peaks, rather than just one as for butadiene.
2. The long-wavelength peak shifts to longer wavelengths, eventually into the visible region.

The net result is that molecules with three or more conjugated double bonds absorb over the entire 200 to 300+ nm region. Thus such species are relatively easily detected after having been separated via liquid chromatography (Chapter 23) because the chromatographic detector usually operates at 245 nm. This is especially true for conjugated molecules like aromatic hydrocarbons, to be discussed below.

Measurement of Aromatic Hydrocarbons. Aromatic hydrocarbons are very similar to conjugated olefins in their properties and hence absorb ultraviolet radiation intensely. Most aromatic hydrocarbons have several ultraviolet bands, although frequently only one band is found in the 220–380 nm region. A good example is benzene, C_6H_6, whose absorption band is split into several sharp peaks centered around 250 nm (Figure 18–4). Benzene is colorless because its absorption

Figure 18–4. Ultraviolet absorption spectrum of benzene in cyclohexane solution.

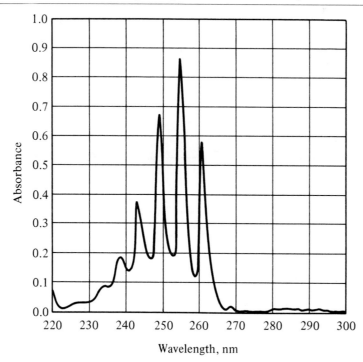

does not extend into the visible region, but some higher aromatics are colored because an absorption "tail" extends into the visible region.

The ultraviolet spectra of aromatic hydrocarbons [6] are sometimes useful in the qualitative characterization of samples, such as air or water pollution samples. In this connection, it is often useful to refer to a plot of the log of the molar absorptivity, instead of absorbance, against wavelength [6] (see Figure 18–5). The molar absorptivity of an aromatic at a given wavelength can usually be estimated to one significant figure from such plots.

It is clear from Figure 18–5 that as the number of rings is increased in an aromatic hydrocarbon, there is a shift of the longest wavelength absorption bands towards the visible region. There is also an increase in the molar absorptivities at all absorption bands. Benzene, with one ring, absorbs at about 250 nm; naphthalene, with two rings, absorbs at about 314 nm; and anthracene, with three rings, absorbs at 357–380 nm.

Figure 18–5. Ultraviolet absorption spectra of benzene, naphthalene, and anthracene in ethyl alcohol. [Adapted from Mayneord and Roe, *Proc. Roy. Soc.*, A152. 299 (1935). By permission of the senior author and The Royal Society.]

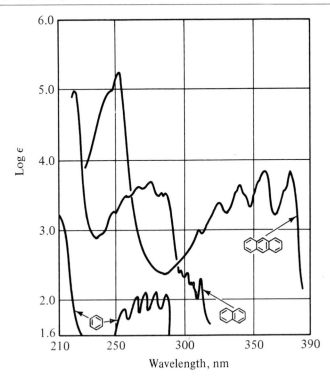

[6] A. E. Gillam and E. S. Stern, *An Introduction to Electronic Absorption Spectroscopy*, London: E. Arnold, 1957; and E. Clar, *The Aromatic Sextet*, New York: Wiley, 1972, pp. 17–31.

This shift is not as large when the aromatic hydrocarbon is angular rather than linear. For example, phenanthrene has three rings but is angular rather than linear like anthracene; the longest wavelength absorption band of phenanthrene is at 338–346 nm, in contrast to the 357–380 nm band of anthracene.

To decide whether an unknown molecule contains a benzene, naphthalene, or anthracene chromophore, one would record the complete ultraviolet absorption spectrum and compare it with the known spectra of the above molecules. Usually, the *wavelengths* of the absorption bands as well as the relative *intensities (in absorbance units)* of the bands are compared. For example, if a particular unknown had an anthracene ring, we would expect two bands: one at 250 nm and the other having peaks at 340, 357, and 380 nm. We would also expect from the ratio of the respective molar absorptivities that the ratio of the absorbance at 250 nm to that at 380 nm would be greater than 20:1.

The presence of substituent groups, such as $-CH_3$, $-C(CH_3)_3$, $-Cl$, or $-OH$, will not appreciably change the fundamental ultraviolet absorption spectrum of the parent compound. For instance, toluene (methylbenzene) has a peak at 270 nm, among others, and a cutoff wavelength similar to that of benzene; the general shape of its spectrum also is similar to that of benzene. These facts lead to the conclusion that the toluene molecule contains a benzene nucleus.

To simplify the quantitative determination of one aromatic hydrocarbon in the presence of another, the ultraviolet cutoff wavelength and the position of the peak with the longest wavelength should be established. Although it is possible to determine lower concentrations of an aromatic compound by making measurements at the absorbance peak with the largest molar absorptivity, other aromatics in the sample generally also absorb at that wavelength. Suppose that anthracene is to be determined in the presence of large amounts of benzene and naphthalene. The preferred wavelength would be 357 or 380 nm because naphthalene and benzene absorb at lower wavelengths. Even though the molar absorptivity of anthracene at 357 or 380 nm is only about one tenth of what it is a 250 nm, a simple analysis is possible only at one of the former wavelengths.

18–3. FLUORESCENCE ANALYSIS

Nature of Fluorescence.   We have seen how an electron in a molecule can be promoted to a higher-energy orbital through the absorption of visible or ultraviolet radiation. The resulting excited state is transient; in many cases, the excess energy is lost as heat when the molecule returns to its ground state. However, the excited states of some substances return to the ground state with *emission* of radiant energy. *Luminescence* is a general term for such behavior. *Fluorescence* is a type of luminescence in which the emission from a photoexcited state occurs within a nanosecond to a microsecond after excitation. *Phosphorescence* is a type in which there is a delay from 10^{-4} to 10 sec before emission occurs. Fluorescence in particular is a most valuable quantitative analytical technique and will be discussed further in this section.

In luminescence spectrometry, the process by which a molecule reaches an excited state by photon absorption is referred to as photoexcitation, or excitation. In this process, radiation is absorbed, resulting in an $n \to \pi^*$ or a $\pi \to \pi^*$ transition. The spin of the excited electron in such a transition is also important. For the usual organic molecule in the ground state, all electron spins are paired so that the net spin is zero. Such a state is called a *singlet* state, symbolized as S_0. One important type of excited state of organic molecules retains the same spin pairing and thus is also a singlet. A given molecule may have several excited singlet states, symbolized as S_1, S_2, etc. In another important type of excited state, the spin of the excited electron is not paired with the spin of the electron with which it was paired in the ground state; hence, the net spin is not zero. This is called a *triplet* state (Figure 18–6), symbolized as T. The energy of an excited triplet state is less than that of an excited singlet state of a given molecule.

> The triplet "state" is actually a level of three states, each involving a different spin energy. The three states are observable only when resolved in a magnetic field.

The rules for transitions occurring during the excitation of *organic molecules* may be summarized as follows:

$$S_0 \to S_1 \qquad \textit{Allowed Transition}$$
$$S_0 \to T_1 \qquad \textit{Forbidden Transition}$$

The $S_0 \to T_1$ transition is forbidden because of the low probability of a spin flip of the excited electron; typically, this transition is only about 10^{-6} times as probable as a singlet-singlet transition.

The time required for singlet-singlet transitions is only about 10^{-15} sec. Since the lifetime of a singlet excited state is about 10^{-9}–10^{-6} sec, we summarize

Figure 18–6.   Energy-level diagram for the ground state and for the excited singlet and triplet states resulting from an $n \to \pi^*$ transition. Each arrow represents an electron; the direction of the arrow indicates the spin direction.

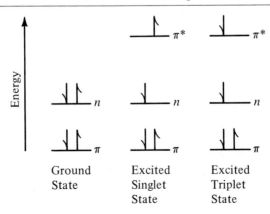

the fluorescence excitation and emission steps as follows:

$$S_0 + photon \xrightarrow{10^{-15} \, sec} S_1 \tag{18–1}$$

$$S_1 \xrightarrow{10^{-9} \, to \, 10^{-6} \, sec} S_0 + photon \tag{18–2}$$

During the lifetime of the S_1 state, it undergoes a very rapid *vibrational relaxation* (10^{-12} sec) to the lowest vibrational level of the excited singlet. After relaxation, the excited singlet can return to the ground state by emitting a photon (Equation 18–2), which, by conservation of energy, must have a lower energy than the photon used in excitation (Equation 18–1). For this reason, the wavelengths of the fluorescence band are always longer than the wavelength(s) of the exciting photon(s).

Another possible step is for the S_1 state in its lowest vibrational level to cross over to the unstable T_1 state. (This process is called *intersystem crossing.*) The excited triplet state then undergoes relaxation to *its* lowest vibrational level. The transition from the unstable excited triplet state to the ground state has a low probability, so the triplet remains on this level for a relatively long time (10^{-4}–10 sec) before emitting a photon and returning to the S_0 state (a process called *phosphorescence*). The triplet lifetime is so long that there is good opportunity for loss of excitation energy (*quenching*) by collision with oxygen and with solvent molecules. For this reason, phosphorescence is rarely observed at room temperature in solutions of low viscosity; it must be observed in the solid state [7].

Unlike phosphorescence, fluorescence in solution can be an efficient process and is the basis for many useful analytical methods. The efficiency of fluorescence is expressed in terms of the *quantum efficiency*, Φ. This is the fraction of excited molecules that return to the ground state by fluorescence emission of a photon. The theoretical range for quantum efficiency is from 0 to 1, but in practice it ranges from about 0.01 to 0.9.

Species That Fluoresce. A large number of organic molecules and a small number of inorganic ions fluoresce. In general, those species that absorb strongly in the ultraviolet region (usually at wavelengths longer than 250 nm) *can* fluoresce. Whether they actually do depends on the rate of fluorescence compared to the rate of nonradiative processes. When the latter are much faster than fluorescence, no fluorescence is observed.

In general, species that absorb at wavelengths shorter than 250 nm are subject to *photodecomposition* because of the high energy of the absorbed radiation. The absorbed energy breaks one or more of the bonds, keeping the species from reaching the excited state. This is particularly common for organic molecules.

> The energy of a carbon-carbon double bond is 146 kcal per mole. Ultraviolet radiation of 180 nm absorbed by such a bond provides 157 kcal/mole to the olefin. Clearly, it is possible for the olefin to undergo direct photodecomposition rather than reach the first excited state. Whether it does so depends on the rate of photodecomposition and the rate of other processes.

[7] G. H. Schenk, *Absorption of Light and Ultraviolet Radiation: Fluorescence and Phosphorescence Emission*, Boston: Allyn & Bacon, 1973.

Of the few inorganic ions that fluoresce, the most important are uranium(VI) [as the uranyl (UO_2^{2+}) ion], the cerium(III) ion, and certain chelates. These will be left for your outside reading. In general, organic molecules with aromatic rings are the most likely to fluoresce. Common classes of fluorescent organic molecules include aromatic hydrocarbons, alkyl-substituted aromatic hydrocarbons, aromatic amines, aromatic amino acids, some halo-substituted aromatic hydrocarbons, phenols and their eithers, heterocyclic molecules, and a few aromatic acids. The presence of $-CHO$, $-C=O$, $-CO_2H$, $-NO_2$, $-Br$, and $-I$ groups on an aromatic ring tends to eliminate fluorescence. However, appropriately placed groups in an aromatic compound can react with a metal ion to form a chelate ring; such formations are fluorescent.

The Relation Between Concentration and Fluorescence Intensity. Fluorescence depends on the absorption of ultraviolet (or visible) radiation, the emitted radiation always being at a longer wavelength. The intensity (number of photons) of radiation emitted will depend on the amount (number of photons) of exciting radiation absorbed, among other factors. The light absorbed may be calculated by a rearrangement of Beer's law (Chapter 17):

$$\text{Fraction of light transmitted} = \frac{P}{P_0} = e^{-\epsilon bc}$$

$$\text{Fraction of light absorbed} = \left(1 - \frac{P}{P_0}\right) = 1 - e^{\epsilon bc}$$

Rearranging the above equation,

$$P_0 - P = P_0(1 - e^{-\epsilon bc})$$

The term $P_0 - P$ is the *rate* (in quanta per unit time) at which light is absorbed.

Note that the radiant power terms P_0 and P are also expressed as rates (quanta per unit time). When the fraction of light transmitted, (P/P_0) or of light absorbed $(1 - P/P_0)$ are used, the units of quanta per time cancel out.

The intensity [8] of fluorescence emitted (F) equals the rate of light absorption $(P_0 - P)$ multiplied by the quantum efficiency of fluorescence (Φ).

$$F = (P_0 - P)\Phi = \Phi P_0(1 - e^{-\epsilon bc})$$

The last term in this equation may be expressed by a series that, at low concentrations, reduces to $(2.3\epsilon bc)$. A proportionality constant, k, must also be added because fluorescence is emitted in all directions but only that of a limited aperture is measured. Combining these factors, we obtain the equation

$$F \approx k\Phi P_0(2.3\epsilon bc)$$

[8] Strictly speaking, the intensity of fluoroescence emitted (F) should be termed the *radiant power* of fluorescence emitted, since it is the rate at which fluorescence emission energy is absorbed in the detector of the fluorometer. See H. K. Hughes *et al.*, *Anal. Chem.*, 24, 1349 (1952). However, most workers and standard works on fluorescence use *intensity* and F. See A. L. Conrad, *Treatise on Analytical Chemistry*, Part I, Volume 5, New York: Wiley-Interscience, 1964, pp. 3057–3078.

Since all the terms on the right side are kept constant under analytical conditions except c, the instrumental readout of F is a function of the concentration of a compound:

$$F = k'c$$

A plot of F versus c should give a straight-line calibration plot analogous to a Beer's law plot.

Some important limitations of these simple equations must be pointed out. The intensity of fluorescence approaches a limit equal to kP_0 (or k'). This limit is absolute because the power of emitted radiant energy cannot exceed the power of exciting radiant energy, P_0. Another limitation is the *inner-filter effect* occurs at the high-concentration end of the calibration plot, causing a decrease in the slope of the curve.

> The inner-filter effect is usually caused by a disproportionate absorption of the exciting radiation by a concentrated solution in the front part of the cell. Fluorescence occurs as in other parts of the cell, but the fluorescence is emitted away from the slit exit to the detector. Thus, a smaller proportion of the fluorescence is measured, even though the same total amount is emitted.

Because of the inner-filter effect, a nearly linear fluorescence versus concentration curve is obtained only at relatively low concentrations. Even at low concentrations, standards should bracket the unknown in concentration to assure accurate results.

The sensitivity of fluorometry is its great advantage over spectrophotometry. Because the instrumental reading F is a direct function of P_0, the intensity of the exciting radiation can be increased over that used in spectrophotometry to obtain higher readings for a given concentration. (Recall that in spectrophotometry, the *ratio* of P_0/P is measured.) Thus, fluorometric analysis can be used for $10^{-8} M$ or ppb measurements, compared to $10^{-5} M$ or ppm measurements in the average spectrophotometric method. However, scrupulously clean glassware and extremely pure solvents and reagents are necessary.

Excitation and Fluorescence Spectra. Different chemical molecules have characteristic excitation and fluorescence spectra. The excitation spectrum of a molecule is quite similar to the ultraviolet absorption spectrum. Since fluorescence is the reemission of absorbed light, it is not surprising that a fluorescence spectrum is usually similar to its excitation spectrum; often, they are mirror images of each other with the fluorescence spectrum shifted to higher wavelengths. The excitation and fluorescence spectra of anthracene, shown in Figure 18–7, are good examples of this. Although only part of the excitation spectrum is shown, it is quite similar to the ultraviolet spectrum of anthracene in Figure 18–5. In both spectra, the band is split into four peaks (maxima) at 320, 340, 357, and 377 nm.

To obtain the excitation and fluorescence spectra, an instrument called a spectrophotofluorometer must be used. This is similar to the spectrophotometers discussed in Chapter 17, in that prisms or gratings are used to select the appropriate wavelengths of excitation and fluorescence radiation. The spectrophotofluorometer

Figure 18–7.   Part of the excitation spectrum (solid line) and all of the fluorescence spectrum (dashed line) of anthracene. The left axis is the intensity of the exciting radiation absorbed, and the right axis is the intensity, F, of the fluorescence. Although the fluorescence band overlaps the excitation band slightly, all four maxima or peaks of the fluorescence band are at longer wavelengths than the excitation band.

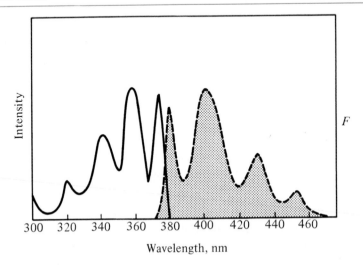

may also be used for quantitative analysis, although a less expensive instrument (the filter fluorometer) is often used instead. In the filter fluorometer, filters (see the discussion of optical filters in Chapter 17) are used instead of gratings to select excitation radiation (the primary filter) and fluorescence radiation (the secondary filter).

The fluorometer is a more versatile instrument than the spectrophotometer because it has two instrumental variables instead of one. Suppose molecule A is to be determined in the presence of molecule B. Using the fluorometer, two possible techniques may be investigated:

1. If A absorbs ultraviolet radiation in a spectral region where B does not, then the *grating* or *primary filter* selecting the excitation wavelength can be adjusted so that only A is excited. Then only A will emit fluorescence, which is then measured by the phototube of the fluorometer.
2. If A and B both absorb light in the same spectral region but A emits fluorescence in a different region than B, then the *grating or secondary filter* can be adjusted to respond only to fluorescence from A. Even though both A and B are emitting light, the phototube only "sees" light from A.

These techniques will be illustrated in the organic analysis section below.

Organic Analysis and Air Pollution.   One the most useful applications of fluorescence has been the determination of aromatic hydrocarbons. A good example is the analysis of a mixture of two similar aromatic hydrocarbons like the following:

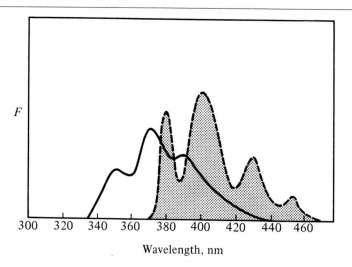

Anthracene   Phenanthrene

It has been shown [9] that anthracene can be determined by the first of the two techniques just described. The ultraviolet absorbance of phenanthrene cuts off below 360 nm. It can be seen from Figure 18–7 that anthracene has an excitation band above 360 nm, so it is possible to excite only anthracene. The actual wavelength used is 365 nm [9]. The fluorescence emission at 400 nm is measured in the fluorometer to determine anthracene. It is also possible to determine phenanthrene by the second technique above. Phenanthrene absorbs ultraviolet light intensely in the 265-nm region, but so does anthracene (see Figure 18–5). It can be seen from Figure 18–8 that only phenanthrene emits fluorescent light at 350 nm. Light of 265-nm wavelength is used to excite phenanthrene (and anthracene), and

Figure 18–8.   The fluorescence spectra of phenanthrene (solid line) and anthracene (dashed line), both $C_{14}H_{10}$ aromatic hydrocarbons. Phenanthrene may be determined in the presence of anthracene by exciting both compounds at 265 nm and measuring the fluorescence of phenanthrene alone at 350 nm. Anthracene may be determined by exciting the anthracene alone at 365 nm and measuring its fluorescence at any maxima above 380, 400, or 430 nm, since phenanthrene does not fluoresce under these conditions.

F

300   320   340   360   380   400   420   440   460

Wavelength, nm

[9] G. A. Thommes and E. Leininger, *Talanta*, **7**, 181 (1961).

the fluorescence emission at 350 nm is measured with the fluorometer to determine the phenanthrene.

Since aromatic hydrocarbons are important air pollutants, the fluorometer has been used by the Public Health Service [10] for determining these compounds. Fluorometry is more useful than ultraviolet spectrophotometry in pollution analysis, not only because it is more versatile, but also because it can measure the very low concentrations of hydrocarbons found in samples of particles collected from the air. This analysis is all the more important because some aromatic hydrocarbons are carcinogenic—that is, they are known to cause cancer in test animals. One of the best known carcinogenic aromatic hydrocarbons is benzo[a]pyrene, or 3,4-benzpyrene:

A very specific determination of benzo[a]pyrene in the aromatic hydrocarbon fraction of polluted air samples has been developed that uses sulfuric acid as a solvent [11]. In this acid, this compound forms a cation with a strong absorption band at 520 nm in the visible region. A few other aromatic hydrocarbons have weak absorption bands at 520 nm, but none of these emit fluorescent light at 545 nm as does benzo[a]pyrene. To determine benzo[a]pyrene, the sample is excited at 520 nm and the fluorescence emission at 545 nm is measured with the fluorometer. Benzo[a]pyrene may be estimated without separation in the presence of over 40 compounds in artificial mixtures. This analysis is also unusual in that visible, not ultraviolet, radiation, is used to excite the molecule. For further examples of organic analysis and a more theoretical treatment of fluorescence, the student is referred to two monographs [12].

Fluorometric Screening for Lead Poisoning.   Lead poisoning is always a serious threat to children living where a lead-based paint has been used. Although flame methods (Chapter 19) measure lead *directly*, they also require a relatively large sample, since interferences have to be removed. (When testing the blood of children, it is important to use as small a sample as possible.) An unusual indirect method [13] measures lead using only one drop of blood.

The drop of blood is transferred to a glass slide, which is inserted into a specially designed fluorometer using a horizontal, rather than vertical, sample compartment and appropriate optics to accurately measure the fluorescence. The

[10] E. Sawicki, W. Elbert, T. W. Stanley, T. R. Hauser, and F. T. Fox, *Anal. Chem.*, *32*, 811 (1960).

[11] E. Sawicki, W. Elbert, T. W. Stanley, T. R. Hauser, and F. T. Fox, *Int. J. Air Poll.*, *2*, 273 (1960).

[12] A. L. Conrad, *Treatise on Analytical Chemistry*, Part I, Volume 5, New York: Wiley-Interscience, 1964, pp 3057–78. D. M. Hercules, *Fluorescence and Phosphorescence Analysis*, New York: Wiley-Interscience, 1966.

[13] *Chem. & Engr. News 53*, 18 (Feb. 3, 1975); W. E. Blumberg, J. Eisinger, A. A. Lamola, and D. M. Zuckerman, *J. Lab. Clin. Med. 89*, 712 (1977).

exciting radiation is permanently fixed at a wavelength near 424 nm in the visible region. The method is indirect because the action at the lead(II) on blood protoporphyrin produces zinc(II)-protoporphyrin, which is excited at 424 nm and emits red fluorescence centered at 625 nm. The fluorescence is so intense that the toxic level of lead (70 micrograms of lead(II) per 100 mL) can easily be detected.

The zinc(II)-protoporphyrin is produced in the blood because the action of an iron-insertion enzyme, Enz-Fe, is blocked by lead(II) ion. In normal blood, this enzyme catalyzes the reaction of iron(II) and protoporphyrin:

$$\text{Fe(II) + protoporphyrin} \xrightarrow{\text{Enz-Fe}} \text{Fe(II)-protoporphyrin}$$

When lead(II) ion is present, it reacts with the Enz-Fe so that iron(II) and the protoporphyrin cannot react at a rapid rate. Because zinc(II) ion is present in the blood and can react much faster than ion(II) in the absence of the enzyme catalyst, it reacts first:

$$\text{Zn(II) + protoporphyrin} \rightarrow \text{Zn(II)-protoporphyrin}$$

Since zinc(II) is a d^{10} metal ion, the zinc(II)-protoporphyrin chelate will fluoresce. On the other hand, since iron(II) has a partially filled d sublevel, the iron(II) chelate will not.

18–4. INFRARED METHODS AND CHEMICAL STRUCTURE

Infrared spectrophotometry generally covers the 1–15 μm wavelength region because this is the spectral region available on most commercial instruments. All organic molecules and some inorganic ions absorb in this region; often the spectra are complicated, with many sharp absorption bands or peaks of varying intensity. Infrared spectrophotometry is most useful for identifying organic compounds and deducing the structure of newly synthesized compounds. A brief introduction to the absorption of infrared radiation will be presented first, followed by a discussion of the instruments and applications of infrared spectrophotometry.

Absorption of Infrared Radiant Energy. From molecular orbital theory, we have shown that the absorption of the visible and ultraviolet radiant energy is accompanied by electronic transitions. The absorption of infrared energy results in changes in *vibrational* motion in a molecule. As a mechanical analog, view a molecule as a set of balls (representing atoms) connected by springs (representing chemical bonds). In the ground state, the atom balls vibrate somewhat on the bond springs. Absorption of infrared energy at particular wavelengths (i.e., frequencies) increases this motion. The frequency of the incoming radiation must match that of the atom balls vibrating on their bond springs. This is called a condition of *resonance*, analogous to that in music where a tone of a particular pitch will start an undamped piano string or a wine glass vibrating at the same frequency.

Although the vibrational changes in a chemical molecule are caused by the absorption of only certain frequencies of infrared energy, there are also numerous rotational energy changes. Because of these, spectra consist of absorption *bands* rather than lines. For example, an absorption band might be the result of a number of vibrational transitions similar to the one (T_{ir}) shown in Figure 17–4. A molecule might undergo transitions from the lowest vibrational sublevel to the r_0, r_1, r_2, etc., rotational sublevels of the $v = 1$ vibrational sublevel.

The molecular vibrations observed are of two types:

1. *Stretching.*   This is a rhythmical movement of the atoms back and forth along the bond axis.
2. *Bending.*   This may be of two types. One is a change in bond angles between two atoms, each bonded to a third atom. A second type involves the movement of a group of atoms with respect to the rest of the molecule. The terms "scissoring," "rocking," "wagging," and "twisting" are often used to describe the various bending motions.

The general regions of some of the common stretching and bending frequencies are shown in Figure 18–9 by bar graphs overlayed on a blank infrared spectral chart. The stretching band frequencies observed for the various atom combinations are in fairly good agreement with those predicted using Hooke's law for harmonic oscillators:

$$\bar{v} = \frac{1}{2\pi c} \sqrt{\frac{k(m_1 + m_2)}{m_1 m_2}}$$

where $\bar{v}$ is the wavenumber in cm^{-1}, k is the force constant in dynes per cm, c is the speed of light in cm/sec, and m_1 and m_2 are the masses of the individual two adjacent atoms. Further verification is provided by substituting a hydrogen atom (mass = 1)

Figure 18–9.   Correlation of group vibrations with general regions of infrared absorption. (From Robert T. Conley, *Infrared Spectroscopy*, 2nd ed., Boston: Allyn and Bacon, 1972. By permission.)

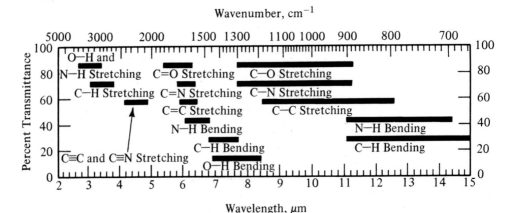

with deuterium (mass = 2) in a certain place in an organic molecule. As predicted from this equation, the frequency of the stretching vibration shifts by a factor of approximately $\sqrt{2}$.

In ultraviolet and visible spectra, absorbance or transmittance is plotted against wavelength. In infrared spectra, many workers prefer to use wavenumbers (v) instead of wavelength. Wavenumber is the reciprocal of wavelength expressed in cm, and therefore has the unit of cm^{-1}. The relationship between wavenumber and wavelength in micrometers (μm) is

$$\bar{v} = \frac{1}{\mu m} \times 10^4$$

In this text, we shall study infrared spectra primarily in terms of wavelength rather than frequency because with most infrared instruments the wavelength scale is linear (while the frequency scale is not) and is easier to read.

> However, some instruments do have a linear presentation of frequency and a nonlinear wavelength scale. This results in sharper peaks in the higher-wavelength region (lower-frequency region), but bands in the important carbonyl and low-wavelength region are not as sharp.

A graph for interconversion of wavelength and wavenumber is given in Figure 18–10.

Infrared Instrumental Components. Infrared spectrophotometers have the same basic components as ultraviolet-visible spectrophotometers, but the radiation sources, the material used in the optical system and sample cells, and the detectors are different. We will discuss these components separately.

Infrared Radiation Sources. The only important infrared sources are *continuum sources*, which emit radiation whose intensity varies smoothly over an extended range of wavelengths. In such "hot-body" sources, the emission of infrared radiation depends on the temperature of the source. The tungsten lamp is actually a source of infrared, as well as visible radiation, but its useful output is limited to about 2500 nm (2.5 μm). This source would be useful for the near-infrared region (750–2500 nm), but other sources are used for most infrared spectrophotometers.

The two most important infrared sources are the *Nernst glower* and the *globar*. The Nernst glower is a short rod of zirconium and yttrium oxides that is heated electrically to 1500–2000°K. Its most intense emission is at about 1.5 μm, but it falls

Figure 18–10. Graph for interconversion of wavelength and wavenumber.

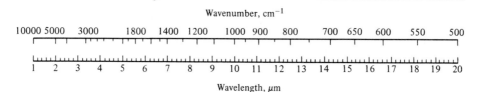

off rapidly at higher wavelengths. It is more intense than the globar between 1 and 10 μm. The glober, a rod of sintered silicon carbide, is also heated electrically to 1200–1400°K. At operating temperature, its maximum output is at 1.9 μm. Its output is about as intense as that of the Nernst glower from 10 to 15 μm and more intense above 15 μm. Unfortunately, the intensity of both sources is small in comparison to ultraviolet-visible sources, so shorter path lengths are required. For both sources, the wavelength of maximum energy decreases with increasing temperature, even though the total energy output (number of photons emitted) continues to increase. Although this effect is less pronounced with the globar than with the Nernst glower, increasing the temperature of either source too much will shift its output to wavelengths that are too long to be useful.

Monochromators. The basic elements of an infrared monochromator system are the same as those for ultraviolet-visible spectrophotometers, the key element being the dispersing device. At present, the only dispersing device used for wavelength selection in the infrared is the diffraction grating.

Infrared Cells and Sample Handling. Both glass and quartz absorb a great deal of infrared radiation. Hence, the monochromator optics and the cells must be made of an inorganic crystalline solid such as salt (sodium chloride), which does not absorb in the infrared. Salt can be used over a broad range, up to 17 μm. Cells of lithium fluoride or calcium fluoride provide better resolution at lower wavelengths; cells of potassium bromide or cesium iodide are more useful above 10 μm.

Since infrared cells are made of salt, samples and solvents must be dried carefully so that the cell walls will not be etched by water impurities. Because of the relatively weak intensity of infrared sources, the path length is typically 0.1 nm, or 1 mm for trace analysis, rather than the 1 cm of ultraviolet-visible cells. For samples of widely varying concentrations, a variable-thickness cell can be used.

Liquid samples may be examined without dilution (neat) or dissolved in a solvent. Solid samples can also be dissolved in an appropriate solvent. However, since the infrared absorption of the solvent can interfere with the analysis, a solvent must be chosen that has as few absorption bands as possible. There are only a few of these; chloroform, carbon tetrachloride, and carbon disulfide are most often used.

Insoluble samples or ones that, for some reason, cannot be analyzed in solution can be determined in two different ways. In one, a "mull" is prepared by dispersing a finely ground sample in a refined mineral oil (Nujol). In the other, the sample is ground with potassium bromide and pressed in a special press at 8–20 tons per square inch. The resulting pellet is a transparent half-inch disc that can be placed directly in the radiation path.

Infrared Detectors. Infrared radiation may be measured by detecting the temperature change of a material in the infrared beam; this type of detector is a *thermal detector.* Another detector, currently less useful, is the *photon detector,* of which the semiconductor photodiode discussed in Chapter 17 is a good example. Examples of the thermal detector are the thermocouple, the bolometer, the Golay cell, and the pyroelectric detector. The thermocouple contains two wires welded together at two junction points. One junction is kept at constant temperature, and the other is heated by incident infrared radiation, developing an electrical potential relative to the other junction. The bolometer changes its electrical resistance as absorbed infrared causes a temperature increase. The Golay cell operates as a gas thermometer; gas in this detector increases in pressure when warmed by absorbed infrared, and this increase can be converted into an electrical signal.

Because the radiant power of infrared radiation is so weak, the response of most thermal detectors is quite low. A preamplifier is usually necessary to obtain a good signal-to-noise ratio in the amplifier. Another problem is the heat radiated from objects in the room; to minimize this source of error, the detector may be housed in a vacuum and shielded from direct exposure to heat.

Infrared Spectrophotometers. Infrared spectrophotometers generally employ a *double-beam* optical arrangement, the sample cell and solvent (reference) cell, each being exposed to equivalent beams from the same infrared source. To do this, a rotating half-circle mirror is used to direct an equivalent beam alternately through the two cells many times a second. Thus, any condition that affects the sample beam P will affect the reference beam P_0 to an equal extent, and the condition will be cancelled out in the readout of $\log (P_0/P)$.

Most infrared spectrophotometers are equipped with a recorder to display a complete infrared spectrum. Some special instruments operating at a fixed wavelength are used to determine one specific compound. Such instruments usually possess only a scale or digital readout and cannot be used to obtain a complete spectrum.

An infrared spectrophotometer should be placed in a room where the temperature and humidity can be controlled. Water vapor absorbs strongly in the infrared, and excess humidity can etch the faces of the crystalline-salt optical surfaces. Temperature changes in the room are undesirable if accurate measurements are desired.

Quantitative Infrared Analysis. A compound in a mixture can be measured quantitatively if there is a wavelength in the infrared spectrum where only the compound of interest absorbs strongly. Because of the complexity of infrared spectra, other sample components often absorb slightly at the chosen analytical wavelength. In addition, the baseline for measuring the absorbance peak height is almost never at zero absorbance, or $100\%T$ (see Figure 18–11), and the *apparent* baseline will change as the wavelength is varied. For quantitative purposes, a working baseline can be established by drawing a line connecting two points of low absorbance on either side of the analytical wavelength (Figure 18–11). The areas in which the points are located should be reproducible from spectra to spectra and ideally should be flat rather than curved. The resulting baseline may be either horizontal or slanted.

Quantitative infrared analysis is also limited by relatively weak absorption bands compared to the ultraviolet-visible region. Molar absorptivities for all but the various carbonyl compounds are 100 or less. Thus, compounds without a carbonyl group cannot be measured below $10^{-2}M$ (in a 1-mm cell) or $10^{-1}M$ (in a 0.1-mm cell), assuming a minimum absorbance of 0.10. Aldehydes and ketones ($\epsilon = 500$), amides and acids ($\epsilon = 1300$), and esters ($\epsilon = 800$) can of course be measured at lower concentrations.

Figure 18–11. Drawing a correct baseline (line *A-B*) for quantitative measurements involving infrared absorption band *P*. The absorbance is calculated from point *P* to the baseline.

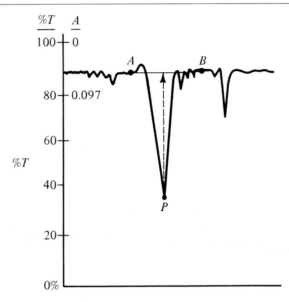

The measurement of hydroxyl compounds is also subject to interference from intermolecular hydrogen bonding (see below). To avoid this, a hydrogen-bonding solvent like pyridine is used [14] instead of carbon tetrachloride. This shifts the O—H stretching frequency to lower energies (3280 cm^{-1}, or 3.05 μm). The molar absorptivity per one OH group on a molecule is about 100 ± 10 in pyridine solvent. Assuming a minimum absorbance of 0.10 in a 1.0-mm infrared cell, the lowest measurable concentration of an alcohol is

$$c_{OH} = \frac{0.10}{(0.10 \text{ cm})(100 \text{ L-mole}^{-1} \text{ cm}^{-1})} = 1.0 \times 10^{-2} M$$

Correlation of Infrared Spectra with Molecular Structure. Although quantitative spectrophotometric determinations may be carried out with infrared spectroscopy, the complexity of infrared spectra presents many practical difficulties not encountered with visible and ultraviolet radiation. However, a more important use of infrared is in the *qualitative* identification of chemical substances, especially for deducing the structural features of organic and some inorganic compounds. Unknown compounds can be identified by matching the unknown spectrum with that of a known compound. However, structural features may be inferred even without a direct matching by noting that certain groups of atoms give rise to bands at wavelengths characteristic of that group. Although the infrared spectrum is a

[14] P. Kabaskalian, E. R. Townley, and M. D. Yudis, *Anal. Chem. 31*, 375 (1959).

Figure 18-12. Infrared spectrum of 1-octene.

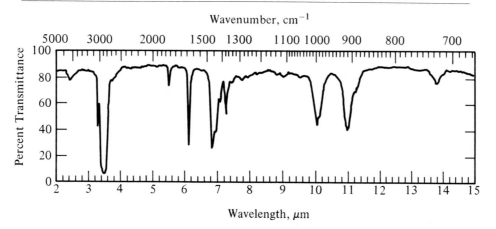

characteristic of the entire molecule, the absorption band for a given group of atoms occurs at or near the same wavelength regardless of the rest of the molecule. It is thus possible to deduce structural features simply by comparing the observed bands with charts of group absorption wavelengths.

This discussion of infrared spectra and chemical structure is designed merely to introduce the student to a somewhat complex, but intriguing, aspect of analytical chemistry. Excellent books by Silverstein and Bassler [15], Conley [16], and others deal more extensively with analytical uses of infrared. It should also be pointed out that infrared spectroscopy is only one of several tools available for deducing chemical structure. Nuclear magnetic resonance (NMR) is extremely useful, and mass spectroscopy is in increasing use for structural work. Of course, chemical information regarding composition or simple structural features is useful, as are ultraviolet and sometimes visible spectra. Thus, for qualitative structure work, infrared should be employed in conjunction with any available information from these other techniques.

The absorption bands for some of the common organic functional groups are shown in Table 18-1. In simple and favorable cases, the main groupings in a chemical compound may be identified from the infrared spectrum simply by noting whether the bands listed in the table for each functional group are present or missing.

The infrared spectrum for a simple olefinic hydrocarbon, 1-octene, is illustrated in Figure 18-12. The bands at 3.4 μm, 3.5 μm, and around 6.8 μm are caused by the methyl and methylene groups. The presence of a methyl group is further confirmed by the absorption peak at 7.25 μm. The sharp peaks at 3.25 μm and 6.1 μm are from olefinic C—H stretching and C=C stretching, respectively. Since this is a vinyl-type olefin, there are also strong bands at 10.1 μm and at 11.0 μm.

[15] R. M. Silverstein and G. C. Bassler, *Spectrometric Identification of Organic Compounds*, 2nd ed., New York: Wiley, 1967.

[16] R. T. Conley, *Infrared Spectroscopy*, Boston: Allyn & Bacon, 1966.

Table 18–1.   Some Infrared Absorption Bands of Organic Groups. [Abbreviation: (s) = strong band, (m) = medium-intensity band, (w) = weak band.]

SATURATED HYDROCARBONS

	C—H *Stretch*	C—H *Bend*
Methyl (—CH$_3$)	3.4, 3.5(s)	6.8–6.9(m), 7.2–7.3(s)
Methylene (—CH$_2$—)	3.5(s)	6.8–6.9(m)

UNSATURATED HYDROCARBONS

	C—H *Stretch*	C=C *Stretch*	*Out-of-Plane Bending*
Olefin (—CH=CH—)	3.25(m)	6.0–6.2(m-w)	
Vinyl (—CH=CH$_2$)	3.25(m)	6.0–6.2(m-w)	10.1(s), 11.0(s)
Acetylene (—C≡CH)	3.0–3.1(s) (sharp)		

AROMATICS (MONONUCLEAR)

=C—H *Stretch*	*Overtone & Combination*	C=C *Stretch*	*In-Plane Bending*	*Out-of-Plane Bending*
3.3(m)	5.0–6.0(w) (May be several bands, pattern affected by substitution)	6.1–6.9(m) (2 Bands)	7.7–10 (1 or 2 Bands)	11.0–15.0 (1 or 2 Bands)

ALCOHOLS AND PHENOLS (—OH)

(—OH)	O—H *Stretch*	O—H *Bend*	C—O *Stretch*
	2.7–3.15	7.0–7.5(w)	8.3–9.9(s)

	O—H *Stretch*		C—O *Stretch*
Free —OH	2.7–2.8(s) (Sharp)	Saturated, Primary	9.2–9.5(s)
Intramolec. Bonded	2.8–2.9(m) (Sharp)	Saturated, Secondary	8.3–9.2(s)
Intermolec. Bonded	2.8–3.15(s) (Broad)	Saturated, Tertiary	8.3–8.9(s)

AMINES (—N⟨)

	N—H *Stretch*	N—H *Bend*	C—N *Stretch*
Pri. aliphatic	2.9(m) (doublet)	6.2(m)	8.1–9.8(m), 11.8–12.2 (broad)
Pri. aromatic	3.0(m) (doublet)	6.2(m)	7.5–8.0(s)
Sec. aliphatic	3.0(w)		8.1–8.5(m), 13.1–14.0(m)
Sec. aromatic	3.0(w)	6.5(w)	7.5–7.9(s)
Tert. aliphatic			8.0–9.0(m)
Tert. aromatic			7.3–7.7(s)

(continued)

Table 18–1. *(continued)*

CARBONYL COMPOUNDS

	C=O Stretch	C—O Stretch
Ester $\left(\begin{matrix} O \\ \parallel \\ -C-OR \end{matrix}\right)$	5.7–5.8(s)	8.0–8.35(s)

	C=O Stretch	C—O Stretch
Ketones $\left(\begin{matrix} O \\ \parallel \\ R-C-R \end{matrix}\right)$		
Dialkyl	5.8–5.9(s)	8.1–9.0(m)
Aryl alkyl	5.8–5.9(s)	7.5–8.3(m)

	C=O Stretch	C—H Stretch
Aldehydes $\left(\begin{matrix} O \\ \parallel \\ R-C-H \end{matrix}\right)$	5.8–5.9(s)	3.5–3.7(m) (Often a doublet), 7.0–7.25(m)

	C=O Stretch	C—H Stretch
Carboxylic acids $\left(\begin{matrix} O \\ \parallel \\ -C-OH \end{matrix}\right)_2$	5.9–6.0(s)	3.0–3.5(s), 7.5–8.0(s), 10.9(s) (broad)

ETHERS (—OR)

	C—O—C Stretch
Aliphatic	8.7–9.2(s)
Aryl alkyl	7.8–8.3(s), 9.2–9.8(m)

NITRILES (—C≡N)

C≡N Stretch
4.4–4.5(w–m)

NITRO (—NO₂)

Asym. Stretch	Sym. Stretch	C—N Stretch
6.2–6.5(s)	7.2–7.6(s)	10.9

The weak peak at 5.5 μm is probably an overtone of the strong 11.0-μm band because the wavelengths differ by a multiple of two.

Unfortunately, the task is not always this easy. In some compounds, atom groupings not shown in Table 18–1 may be present. These may contribute bands that are not listed in the table or may cause some of the groups listed to shift in wavelength. While some bands (such as the carbonyl band around 5.7–6.1 μm) are remarkably stable and dependable, others move around in a most disconcerting manner. In part, this is because of the different interactions between different groups of atoms.

One such interaction is the so-called "coupled" oscillation. For example, in alcohols, the strong C—O stretching band in the 8.2–9.9 μm region is actually a coupled vibration of a C—C—O stretch rather than an isolated C—O stretch. In aromatic compounds, the ring C—H bending is also a coupled vibration; its wavelength depends on the number of adjacent hydrogen atoms in the ring.

Hydrogen bonding may have an important effect on both the wavelength and the shape (sharpness or broadness) of a band. This is especially noticeable in the case of a hydroxyl group, which readily enters into hydrogen bonding with carbonyl, nitro, or a few other groups. The hydrogen bonding may be *intra*-molecular, as in the case of salicylaldehyde, or it may be *inter*molecular, as in carboxylic acid dimers.

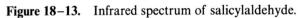

Salicylaldehyde   A carboxylic acid dimer

In either case, the OH stretching band is found at a higher wavelength than that of a "free" OH group, and the hydrogen bonded OH band is broadened distinctively. The stretching wavelength of the other group involved in hydrogen bonding (C=O in the examples above) is also increased, but not as much as the hydroxyl.

The spectrum of salicylaldehyde (Figure 18–13) illustrates some of these difficulties. The OH band around 3.2 μm is broadened by intramolecular hydrogen bonding. The aldehyde carbonyl band at 6.0 μm is also broadened and is shifted to a slightly higher wavelength than usual by hydrogen bonding. The doublet at 3.55 and 3.65 μm is characteristic of the aldehyde. The aromatic absorption peak at 3.3 μm is nearly obscured by the broad OH band, but peaks at 6.3 μm, 6.7 μm, 8.7 μm, 9.7 μm and the broad bands in the 13–15 μm region confirm the presence of an aromatic ring.

Figure 18–13.   Infrared spectrum of salicylaldehyde.

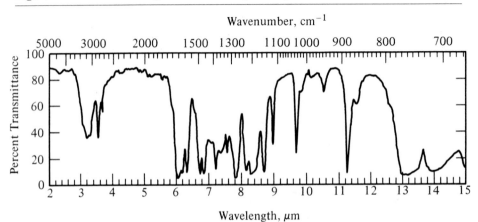

QUESTIONS AND PROBLEMS

Electronic Absorption

1. What type of transition is caused by the absorption of ultraviolet radiant energy? by visible? by infrared?
2. Distinguish between atomic orbitals and molecular orbitals.
3. What are antibonding orbitals, and how do they differ from bonding orbitals?
4. Show by appropriate diagrams the difference between sigma and pi bonds.
5. Silver salts form useful pi complexes with organic olefinic $(-\overset{|}{C}=\overset{|}{C}-)$ compounds. Explain how such a complex might be formed.
6. Explain why molar absorptivity is a quantitative estimate of the probability of an electronic transition. What is a "forbidden" transition?
7. Draw molecular-orbital energy diagrams for ethylene ($CH_2{=}CH_2$) and butadiene ($CH_2{=}CH{-}CH{=}CH_2$) showing the pi orbitals of ethylene and the two bonding and two antibonding pi orbitals of butadiene. Compare the transition energy of ethylene with the lowest transition energy of butadiene. Knowing that ethylene absorbs ultraviolet light at 170 nm, predict the position of the absorption band of butadiene relative to that of ethylene.
8. Iron(III) in acidic aqueous solution forms the $FeSCN^{2+}$ and $Fe(SCN)_2{}^+$ complexes with the thiocyanate ion. The $FeSCN^{2+}$ ion has λ_{max} at 453 nm with an ϵ of 5000. The $Fe(SCN)_2{}^+$ ion has λ_{max} at 485 nm with an ϵ of 9800.

    (a) Explain why the molar absorptivity of the $Fe(SCN)_2{}^+$ ion is almost twice that of the other ion.
    (b) Explain how the existence of two absorbing species could affect the development of a spectrophotometric method for iron(III) using SCN^- as the color-forming reagent.

9. Although aqueous solutions of HCl do not absorb in the ultraviolet region, the chloro complexes of a number of metal ions do absorb strongly. Some metal ions form a series of chloro complexes with varying λ_{max} and ϵ values. For example, mercury(II) ion can form $HgCl^+$, $HgCl_2$, $HgCl_3{}^-$, and $HgCl_4{}^{2-}$ [log of formation constants: $\beta_1 = 6.8$, $\beta_2 = 6.6$, $\beta_3 = 0.57$, and $\beta_4 = 1.46$ (Chapter 10)]. What might be done to form only one chloro complex so as to determine mercury(II) spectrophotometrically?

Ultraviolet Absorption

10. The ultraviolet spectrum of benzene in an inert hydrocarbon solvent shows several absorbance bands (Figure 18–4). In the vapor state, the spectrum of benzene is more complicated, with more "fine structure" such as splitting of the bands into many peaks. In a polar organic solvent such as ethyl alcohol, the spectrum of benzene is even less complicated than that in Figure 18–4. Explain these differences.
11. Acetic acid, CH_3CO_2H, exhibits only a single absorption band, $\lambda_{max} = 208$ nm, $\epsilon = 17$. However, propenoic acid, $CH_2{=}CHCO_2H$, exhibits an absorption band at 200 nm with $\epsilon = 1 \times 10^5$.

    (a) What type of transition is occurring for each acid?
    (b) Why does the acetic acid band have such a low molar absorptivity compared to that of propenoic acid?

12. Acetone, $CH_3C{=}OCH_3$, has a $\pi \rightarrow \pi^*$ transition at 188 nm and an $n \rightarrow \pi^*$ transition

at 279 nm. Calculate the energy required in electron volts for each transition (see Chapter 17).

13. Flow-injection analysis (FIA) is an attractive alternative to conventional spectro-photometry. In FIA, an absorbing sample can be introduced as a "plug" into a flowing solvent stream in an analysis tube. The absorbance profile of the sample plug is measured as it flows through a spectrophotometric detector cell (such a detector is so sensitive that it can be set to move to full-scale deflection at 0.003 absorbance unit). Assuming a detection limit of 1/100th of full scale ($A = 0.00005$), calculate the detection limit for a molecule with a molar absorptivity of 13.0 in a 1.0-cm cell. Compare that to the detection limit in a spectrophotometer at $A = 0.010$.

14. From Figure 18–5, estimate the log of the molar absorptivity, and calculate a numerical value for the molar absorptivity for each molecule. Then calculate the detection limit in a 1.0-cm cell at $A = 0.010$.

(a) Benzene, 260 nm
(b) Naphthalene, 314 nm and 275 nm
(c) Anthracene, 380 nm

15. By consulting Figures 18–4 and 18–5, devise a spectrophotometric method for determining naphthalene in the presence of an unknown amount of benzene and a mono-substituted benzene. Specify the wavelength at which naphthalene may be measured without interference, and calculate the detection limit for naphthalene in a 1.0-cm cell at $A = 0.010$.

16. Estimate to two significant figures, the location of the absorption band of each organic aromatic hydrocarbon:

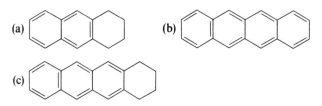

17. The rate of a reaction such as a hydrolysis of an ester of an aromatic acid can often be followed by measuring the change in ultraviolet absorbance with time. The *decrease* in absorbance of the ester reactant or the *increase* in absorbance of the acid product can be measured:

$$\text{Ester} + H_2O \xrightarrow[\text{NaOH}]{\text{heat}} \text{Acid anion} + \text{alcohol}$$

Typically, the hydrolysis of methyl salicylate (oil of wintergreen) can be followed by measuring the increase in absorbance at 298 nm from the salicylic acid anion.

(a) Explain how to calculate the actual molar concentration of the anion as a function of time.
(b) Consult Chapter 14 and specify what quantity to plot and how to calculate the rate constant (such as a first-order rate) for this hydrolysis under fixed conditions of temperature, solvent, and concentration of NaOH.

Fluorescence Analysis

18. Compare the two techniques for fluorescence analysis. Which technique avoids the possibility of error from measurement of a small amount of fluorescence from interfering

molecule B? Which technique avoids the possibility of error from absorption of a significant portion of exciting radiation by a large concentration of interfering molecule B?

19. For each mixture, decide whether (1) molecule A can be determined by selectively exciting only A, (2) molecule A can be determined by measuring only the fluorescence of A, or (3) molecules A and B must be separated before determining A.

   (a) A absorbs from 200 to 300 nm and emits from 300 to 380 nm.
       B absorbs from 250 to 300 nm and emits from 300 to 390 nm.
   (b) A absorbs from 250 to 380 nm and emits from 380 to 460 nm.
       B absorbs from 250 to 390 nm and emits from 390 to 425 nm.
   (c) A absorbs from 200 to 300 nm and emits from 300 to 380 nm.
       B absorbs from 200 to 310 nm and emits from 300 to 390 nm.

20. In the fluorometric screening for lead(II) ion, the presence of lead(II) is indicated by the fluorescence of zinc(II)-protoporphyrin. Suppose a patient has an extremely low level of blood iron(II) ion. Will the fluorescence of that patient's blood be normal, lower than normal, or higher than normal? What does this indicate about the accuracy of the fluorescence measurement for lead(II) ion?

21. Carcinogenic hydrocarbons in air are collected by passing air through an absorption tube and then washing out the absorbed matter with 100 mL of an organic solvent. If a $10^{-8}M$ solution can be determined by a fluorometric method, estimate the minimum weight, in micrograms, that must be collected to determine a hydrocarbon of formula weight 300.

22. In pentane, benzo[a]pyrene is excited at wavelengths from 295 to 381 nm and emits fluorescence at 403, 427, and 454 nm. Propose a fluorometric method for analyzing benzo[a]pyrene in the presence of anthracene (see Figure 18–7). Can anthracene be determined by fluorescence in the presence of benzo[a]pyrene?

23. Phenanthrene has an ultraviolet absorption band with a maximum at 331 nm ($\log \epsilon = 2.5$). Naphthalene emits fluorescence in the 310–370 nm region. Suggest two methods for the fluorometric determination of phenanthrene in the presence of naphthalene. (See the ultraviolet absorption spectra of naphthalene in Figure 18–5.)

24. Suggest two fluorometric methods for determining anthracene in the presence of naphthalene. (See the previous problem and the absorption spectra in Figure 18–5.)

Infrared Instrumentation

25. Name three sources of infrared radiation and indicate why increasing the temperature of such sources too much will not increase the output of photons in the important 2–15 μm region.

26. From an instrumental viewpoint, explain why water cannot be used as a solvent in infrared spectrophotometry.

27. Explain why the detectors used in ultraviolet-visible spectrophotometry cannot usually be used in infrared spectrophotometry.

28. How would you obtain the infrared spectrum of a polymer that is insoluble in carbon tetrachloride, carbon disulfide, and other common infrared solvents?

Infrared Analysis

29. Stretching band frequencies can be determined by writing a ratio of two equations for Hooke's law. (a) Calculate the C—D frequency in cm^{-1} for CH_3—CO—CH_2—CD_3, knowing that the corresponding C—H frequency occurs at 2980 cm^{-1}. (Assume that the force constant k_{C-H} is the same as k_{C-D}.) (b) Calculate the C=O^{18} frequency in cm^{-1} for

$(C_6H_5)_2C{=}O^{18}$, knowing that the corresponding $C{=}O^{16}$ frequency occurs at 1667 cm^{-1}. (Assume that the force constants for each group are the same.) Check your calculation by referring to the experimental value found in *Anal. Chem. 36*, 1980 (1964).

30. A $1M$ solution of ethanol in carbon tetrachloride exhibits an intense broad band at about 3.0 μm. As the ethanol solution is diluted to $0.2M$, a sharp band appears at 2.8 μm in addition to the band at 3.0 μm. At $0.028M$, ethanol solution exhibits only a sharp band at 2.8 μm. Explain why two bands are observed and why only one band is observed at $1M$ and $0.028M$ concentrations. What significance does this have for preparing a calibration curve for quantitative analysis?

31. The infrared spectrum of 2,4-pentanedione has a broad absorption band at 3.3 μm and bands at 5.8 μm and 6.1 μm. (a) Explain why the first band is observed at all, since both oxygens are doubly bonded to carbon in this compound. Then explain why the bond is broad. (b) Explain why two bands are observed instead of one in the carbonyl region.

32. From the infrared spectrum (Figure 18–14), identify the structural groups present in Compound 4. (The weak band at 3.9μ is $-SH$).

Figure 18–14.   Infrared Spectrum of Compound 4.

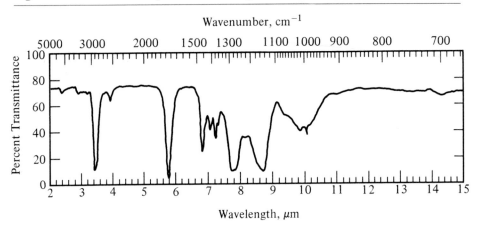

Figure 18–15.   Infrared Spectrum of Compound 5.

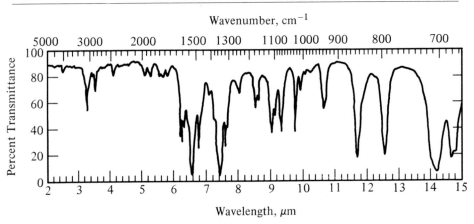

33. The infrared spectrum for Compound 5 is shown in Figure 18–15. If the molecular weight is 120 ± 5, identify the compound.
34. The empirical formula of Compound 6 is $C_5H_{10}O$. From this and the infrared spectrum in Figure 18–16, write the probable structural formula for this compound.
35. From the infrared spectrum in Figure 18–17, identify the groups present in Compound 7.

Figure 18–16.   Infrared Spectrum of Compound 6.

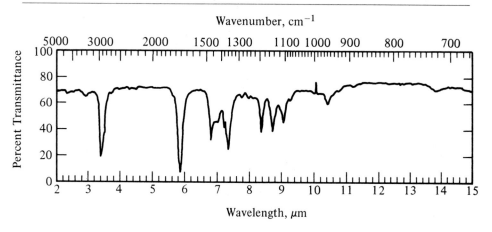

Figure 18–17.   Infrared Spectrum of Compound 7.

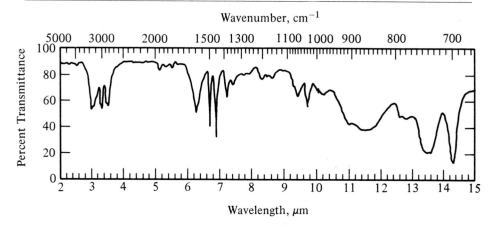

19

Analytical Atomic Spectrometry

In Chapters 17 and 18, we discussed quantitative analytical methods based on molecular spectra, produced when groups of chemically combined atoms in solution absorb radiant energy at characteristic wavelengths. This is often called *absorption spectrophotometry*. In Chapter 18, fluorescence methods were also covered in which a molecular species in solution absorbs radiant energy at characteristic wavelengths and reemits radiant energy at higher wavelengths. The analytical methods are based on measurement of the emitted radiation at wavelengths that are characteristic of the analyte molecule(s).

In this chapter, we will discuss analytical methods using *atomic spectra*. To produce an atomic spectrum, a compound must first absorb enough energy to vaporize it to a molecular gas and dissociate the molecules into free atoms. Atomic absorption spectra are produced when the free atoms absorb radiant energy at characteristic wavelengths. Atomic emission spectra are produced when the free atoms are excited by the thermal energy or a flame, arc, spark, or plasma and emit radiant energy at similar wavelengths.

Three types of emission spectra are observed. Continuous spectra may be emitted by glowing solid bodies; excited molecules emit band spectra; excited atoms emit line spectra. Line spectra consist of sharply defined (and often widely spaced) lines. Band spectra actually consist of groups of lines that become more closely spaced as they approach the head of the band. Atomic line spectra are usually employed for analytical purposes, although molecular band spectra are occasionally used.

The atomic emission spectra produced at the very high temperatures in an electrical arc or spark are the basis of a useful analytical method for determining mostly metallic elements in solid samples. This is a rather old method that requires rather bulky and expensive instrumentation, but it is still widely used for both qualitative and quantitative analyses in many industrial laboratories. However, this form of atomic emission will not be covered here.

In Section 19–1, methods are discussed in which atomic emissions are produced by nebulizing (spraying very fine particles of) a sample solution into a flame or plasma. Methods in which an inductively coupled plasma is used as the excitation source for the atomic spectra are capable of simultaneous quantitative analysis of a large number of elements in aqueous samples.

In any type of atomic emission, the excitation conditions are controlled and kept as constant as possible over a series of analyses. The intensity of a suitable spectral line (the analytical line) of an element is measured. The concentration of this element can then be read from a plot of the analytical line intensity against concentration. (The ratio between the intensity of the analytical line and that of an internal standard can also be used.)

Another major analytical method, called *atomic absorption spectrometry*, is described in Section 19–2. Here, the sample is sprayed into a flame and converted into gaseous atoms largely in the ground state. A beam of ultraviolet or visible radiation is then directed into the flame. The amount of the light absorbed at a particular wavelength is a function of the concentration of a particular element in the sample.

19–1. EMISSION SPECTROMETRY WITH FLAME EXCITATION

Flame excitation will also produce characteristic emission spectra for the various metallic elements. Measuring a selected spectral line (the analytical line) by means of a spectrometer is a very useful quantitative analytical method, properly called "atomic emission spectrometry" or "flame atomic emission spectrometry." For brevity, we shall frequently refer to it as "flame emission."

> In the older literature, this method is often called "flame photometry." This name comes from rather crude, early instruments in which the desired spectral line was isolated from many of the other spectral lines by an optical filter and its intensity measured by a photometer. The recommended method is to employ a spectrometer, which combines a monochromator with a device for measuring the radiant power of the analytical line.

Flame emission spectrometry is an attractive analytical method for determining small amounts of metal salts in solution. All of the metallic elements can be determined by flame emission; better instruments with fewer interferences make the method less sensitive to variations in sample composition.

Analysis with the flame spectrometer is widely used in geological and agricultural laboratories; it is a standard tool in clinical laboratories, where it is used to rapidly determine sodium ions in blood. It is so sensitive to sodium that in one procedure, an 80-μL serum sample is diluted 200 times with a solvent before the final measurement. The entire analysis requires only 30 seconds.

In flame emission, a solution of the sample is nebulized as a fine mist and sprayed into a flame, producing emission spectra of the susceptible elements in the sample. Light from the flame enters a monochromator and is resolved into the different spectral lines. The radiant power of characteristic lines is measured with a phototube connected to a current-measuring device. The concentration of an element is then calculated from the radiant power of a given spectral line, with reference to a working curve prepared from standard solutions.

Principles.   When a sample is sprayed into a flame, the following events occur in rapid succession.

1. The solvent is evaporated or burned, leaving airborne particles of the solid compounds that were dissolved in the sample.

2. The solid compounds are vaporized and are partially converted into gaseous atoms. For example,

$$MgCl_2(s) \rightarrow MgCl_2(g) \rightarrow Mg(g) + 2Cl(g)$$

3. A small fraction of the gaseous atoms are excited by the thermal energy. These excited atoms are not stable for long and return to the ground state by emitting ultraviolet or visible radiation:

$$Mg(g) \xrightarrow{\text{heat}} Mg^*(g) \rightarrow Mg(g) + hv$$

The radiant power of emitted light depends on the number of excited atoms in the flame. Since light is propagated in all directions, there is a geometrical consideration—only a certain fraction of the light emitted can be focused on the detector. If a constant set of conditions is maintained, the radiant power measured by the instrument will be directly proportional to the concentration of the desired element in the sample.

4. Some of the metal atoms may become ionized:

$$Mg(g) \rightarrow Mg^+(g) + e^-$$

Considerable ionization may occur with metals such as the alkali metals, which have a low ionization potential. Ionization is also favored by a high temperature flame.

Extensive ionization will reduce the emission at the desired wavelength. At the same time, some of the ions may become excited and emit light of other wavelengths. The ionization of an element may be suppressed by adding a large excess of a salt of an easily ionized element (see example, p. 418).

5. Unwanted side reactions may occur that reduce the number of simple metal atoms in the flame and, hence, the number of excited atoms. One such reaction is the formation of metal monoxides. If the oxides are very stable, like those of rare earths, zirconium, etc., the emission from excited metal atoms may be greatly reduced. Fortunately, metal monoxide formation is lessened by the use of a fuel-rich flame (see p. 416).

In Chapter 17, it was pointed out that, in atomic spectrometry, the frequency of emitted radiation depends on the *difference* in energies of the excited and ground states of the atom:

$$E_2 - E_1 = hv = \frac{hc}{\lambda}$$

where E_2 is the energy of the higher excited state, E_1 is the energy of the lower excited state or ground state, h is Planck's constant, v is the frequency of emitted ultraviolet or visible light, c is the speed of light, and λ is the wavelength of the emitted light.

When the energies are in electron volts (eV) and the wavelength λ in Ångstrom units, this equation becomes

$$E_2 - E_1 = \frac{12,400}{\lambda}$$

Example: The excitation potential for the $4p$ level in calcium is 2.95 eV. Estimate the wavelength of the spectral line resulting from a transition from the $4p$ back to the $4s$ ground state.

Since the energy difference between the ground state E_1 and the excitation state E_2 is 2.95 eV, we obtain the following:

$$2.95 = \frac{12,400}{\lambda}$$

$$\lambda = 4200 \text{ Å}$$

In most atoms, a number of transitions are possible, each giving rise to a spectral line. However, the most prominent line is that resulting from an electron returning from the lowest excited state to the ground state. In sodium, for example, this is the $3p$ (excited) to $3s$ (ground state) transition, at a wavelength of 5893 Å. The next most intense line in sodium is the $4p$ to $3s$ transition (3300 Å). Transitions between excited states, such as the $3d$ to $3p$ (8190 Å) in sodium, produce much less intense lines.

Instrumentation.   The essential equipment for flame atomic-emission spectrometry is as follows:

1. *Gas Pressure Regulator.*   Most flames used in flame emission are produced by
    burning a mixture of a fuel gas and air or oxygen. For the proper flame charac-
teristics, the gases should be mixed in the correct proportion and fed at a suitable flow rate. Control is accomplished with the aid of valves, pressure gauges, and flowmeters.

2. *Nebulizer and Burner.*   Usually, the nebulizer (often called atomizer) and burner
    are a single unit. The purpose of the nebulizer is to suck up the sample and spray
it into the flame at a constant and reproducible rate. The two basic types of burners used in flame spectrometry are the "total consumption" burner and the premix burner. In the total consumption type, the fuel gas, oxidizing gas, and sample are all passed through separate channels to a single opening from which the flame emerges. The flame produced is turbulent and relatively small in cross-section.

The premix burner produces a quieter flame that is less turbulent and has lower background emission than the other burner. Here, the sample and the two fuel gases are mixed before entering the flame (Figure 19–1). This burner is so designed

Figure 19–1.   A nebulizer-burner. (Courtesy of Varian Techtron.)

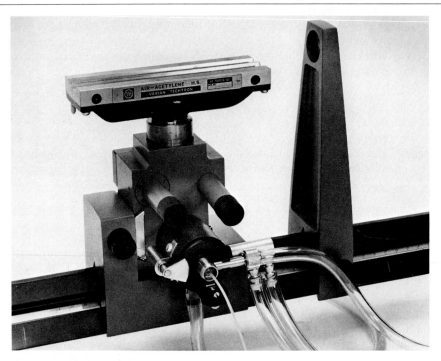

that only the fine particles of the nebulized sample enter the flame; the larger droplets are caught by baffle plates and rejected.

3. *Optical System and Detector.*   In the simpler and cheaper instruments, a filter or interference filter is used to isolate light in the desired spectral band. This light impinges on a phototube and is converted to an electric current, which is measured directly or with amplification. Usually, instruments of this type are limited to determining sodium and potassium; sometimes calcium may also be determined. More sophisticated instruments contain a prism or grating monochromator, a narrow entrance slit for better line resolution, and sensitive detecting circuits. Spectral interferences from foreign elements are much less prevalent when quality flame-emission instruments, rather than makeshifts or cheaper, filter-type instruments, are used.

Analytical Procedure

Preparation of Sample and Standards.   Inorganic samples are dissolved in a suitable acid or other solvent and diluted. Sensitivity is improved if the final solution contains at least 80% of some organic solvent such as methyl alcohol. Biological samples are usually decomposed before analysis by ashing them or by treating them with oxidizing acids. Organic samples are dissolved or diluted by a suitable organic solvent.

Standards are prepared containing known amounts of a compound containing the element(s) to be measured. If possible, the compound in the standard is the same as that present in the samples to be analyzed. Other salts are added to the standards to approximate as closely as possible the composition of the unknown samples.

Choice of Experimental Variables.   Although conditions for various analyses are available in the chemical literature and in the operating procedures accompanying various commercial instruments, the student should know at least some of the important variables in those conditions. In general, lower-cost instruments tend to be less versatile and have more fixed variables than more expensive instruments. A spectral line should be selected that provides good sensitivity and is as free as possible from background emission or interference from the spectral lines of other elements. The region of maximum emission should be located in the flame. Fuel ratios and flow rates should be adjusted to excite the desired spectral lines with steady intensity and minimum background.

The chemical composition of a sample frequently has a significant effect on the intensity of the spectral lines of any given element. It is often possible to isolate the desired element by a selective extraction from aqueous solution into an organic solvent (see Chapter 20). Then the nonaqueous solution can be sprayed directly into a flame and the element determined. (A significant increase in sensitivity usually results when a sample is dissolved in an organic solvent rather than in water.) This technique avoids many interferences and often makes separate working curves for each type of sample unnecessary.

Preparation of Analytical Calibration Curves.   Quantitative flame emission methods require the preparation of an analytical calibration (working) curve. For spectrophotometers with a flame attachment, the percent-transmittance scale is used. When the correct wavelength setting has been ascertained and the flame properly adjusted, a blank solution is sprayed into the flame.

> A *blank* has the same solvent composition as the sample. The same reagents may also be added to it, or it may be taken through the same separation procedure as the real sample.

With the shutter open, the dark-current rheostat is adjusted to a transmittance scale reading of zero. Then the most concentrated standard to be used is introduced into the flame, and the percent-transmittance scale is set at 100. The scale readings of several standards with progressively lower concentrations of the analytical element are read, and an analytical curve of emission intensity (transmittance-scale readings) against concentration is prepared. Ideally, a straight line is obtained, but frequently the line curves somewhat.

> Note the differences between operations here and in absorption spectrophotometry. Using a pure solvent blank, P_0 is set at 100 in absorption spectrophotometry, while in emission work the scale is set at 0. In absorption spectrophotometry, the scale reading is set as 0 in total darkness; in emission spectrometry, the scale reading is set to an arbitrary value such as 100 using the most concentrated standard.

Sample solutions are introduced into the flame under exactly the same conditions as those used in preparing the analytical curve. To ensure that constant conditions are being maintained, standards are checked before, after, and often during, the analysis of each series of samples.

In the absence of interferences, the standards used contain only known concentrations of the element to be determined. If interfering effects are likely, the standards should contain the same amounts of all of the major constituents as the unknown samples to be analyzed.

In quantitative spectrographic analysis, the use of an internal standard is almost mandatory because of the difficulty in maintaining constant excitation conditions. In flame emission, it is easier to maintain and reproduce excitation conditions; however, an internal standard is often used here also. Because a smaller number of spectral lines are excited by a flame than by electrical means, it is sometimes difficult to find a suitable homologous pair of lines for flame emission spectrometry. Nevertheless, many flame emission procedures do use an internal standard. As in spectrography, the logarithm of the intensity ratio between the analytical line and the internal standard line is plotted against the logarithm of the concentration in preparing the analytical calibration curve.

Interferences in Flame Emission.   Several effects complicate flame emission analysis. One of these is background from the flame. This is small in most spectral regions. There is a strong OH band near 3050–3200 Å, but this does not interfere with many lines of metallic elements. Relatively high concentrations of most elements exhibit a continuous radiation in addition to the background from the flame. In quantitative analysis, it is usually necessary to measure the background from a blank and subtract it from the measured intensities of the various lines.

Emitted radiant energy may be absorbed by other atoms of the same element, causing these atoms to become excited. This effect is termed "self-absorption"; it will, of course, decrease the intensity of the spectral lines from the element involved. Self-absorption can be minimized by using dilute solutions.

Other elements may also affect the intensity of the emission spectrum of an element being measured. Usually, this results in reduced line intensities, but in some cases the amount of light emitted is enhanced.

Certain anions reduce the intensities of metallic spectral lines significantly. An important example is the effect of phosphate on a calcium determination. It has been speculated that a stable, nonvolatile compound of calcium and phosphate is formed that prevents the excitation of calcium spectral lines.

Applications.   With proper excitation conditions, any of the metallic elements can be determined by flame atomic emission. The smallest concentration that can be determined varies widely from element to element, ranging from less than 0.001 $\mu g/mL$ to about 100 $\mu g/mL$. The sensitivity adjustment on most instruments is flexible enough that calibration curves may be extended to much higher concentrations of an element. However, the relative accuracy of the flame method is about ± 2–5%, which is satisfactory for small quantities of an element but inferior to chemical methods for larger amounts.

Determining the individual alkali metals (lithium, sodium, potassium, etc.) is

Table 19–1. Analysis of Rare Earths by Flame Atomic Emission Spectrometry.

Element	Analytical Wavelength, Å	Detection Limit, μg/mL
Cerium	——	10
Dysprosium	4211.7	0.1
Erbium	4008.0	0.3
Europium	4594.0	0.003
Gadolinium	3684.1	2
Holmium	4163.0	0.1
Lanthanum	3927.6	1
Lutetium	3312.1	0.2
Neodynium	4634.2	1
Praseodynium	4951.4	2
Scandium	3911.8	0.07
Samarium	4296.7	0.6
Terbium	4326.5	1
Thulium	4094.2	0.2
Yttrium	4077.4	0.3
Ytterbium	3988.0	0.05

probably the most widely used application of flame emission. These elements are easily determined by flame methods, whereas their analysis by chemical methods is mostly slow and unselective. These and other metals may be determined in biological and medical samples, soil extracts, water, glasses, and other types of samples.

Perhaps the crowning achievement of flame atomic emission has been the analysis of complex mixtures of rare earths and scandium [1]. The chemical properties of the rare earths are so similar that individual rare earths in complex mixtures can, in general, be determined only after a rather lengthy chromatographic separation. (Indeed, it is only within the last few years that efficient analytical separation of rare earths has been possible.) With the use of a fuel-rich oxyacetylene flame, however, mixtures may be analyzed for individual rare earths without separation. In Table 19–1, the analytical lines and the detection limits [2] for the entire rare-earth series are given.

Plasma Excitation of Emission Spectra. The analytical method involving a plasma as the excitation source is called "inductively coupled plasma atomic emission spectrometry," or more commonly "ICP–AES." The principles and technology of this technique have been reviewed [3–5].

[1] A. P. D'Silva, R. N. Kniseley, V. A. Fassel, R. H. Curry, and R. B. Myers, *Anal. Chem.*, *36*, 532 (1964).
[2] R. N. Kniseley, V. A. Fassel, and Constance Butler (Unpublished work).
[3] V. A. Fassel and R. N. Kniseley, *Anal. Chem.*, *46*, 1110A–1120A, 1155A–1164A (1974).
[4] V. A. Fassel, *Anal. Chem.*, *51*, 1290A (1979).
[5] V. A. Fassel, *J. Assoc. Off. Anal. Chem.*, *67*, 212 (1984).

Figure 19–2 depicts the sample nebulizer and plasma torch, and Figure 19–3 shows a more detailed schematic of a typical torch used in analytical work. Windings of an induction coil outside the torch cylinder are connected to a RF generator; these create a magnetic field. The plasma is created from argon gas inside a quartz tube. Because of the very high temperatures, the quartz tube is cooled somewhat by a tangential flow of argon coolant. The plasma itself (at the top of the figure) has a very brilliant white cone and a flamelike tail. The area where the sample is excited may reach a temperature of 7000°K—about a factor of two higher than that attainable in a gas flame.

The sensitivity of ICP is excellent, detection limits generally being lower than those for flame emission. These vary according to the element determined; typically, they are in the lower ppb range. In flame emission, conditions have to be changed to accommodate elements requiring different excitation conditions. However, with the ICP, one set of experimental parameters provides essentially optimal conditions for

Figure 19–2. Plasma torch and nebulizer. (Courtesy of R. N. Kniseley.)

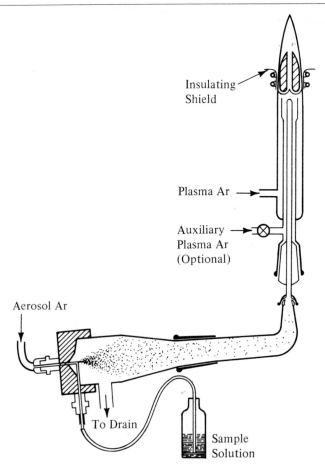

Insulating Shield

Plasma Ar →

Auxiliary → Plasma Ar (Optional)

Aerosol Ar

To Drain

Sample Solution

Figure 19–3. Schematic representation of a typical ICP torch used in analysis. (Reprinted with permission from *Analytical Chemistry, 51,* 1290A, 1979. Copyright 1979 American Chemical Society.)

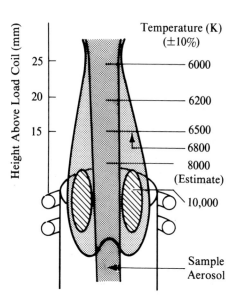

Height Above Load Coil (mm)

Temperature (K) (±10%)

- 6000
- 6200
- 6500
- 6800
- 8000 (Estimate)
- 10,000

Sample Aerosol

all of the metals and metalloids. Because the free atoms are released in a noble gas environment, interfering reactions such as metal monoxide formation are minimized. The depressing effect of phosphate on calcium or aluminum determination that is noted in flame emission also does not occur in plasma emission [6].

When the plasma torch is mated to a classical polychromator, many chemical elements can be determined simultaneously. In these instruments, a diffraction grating disperses the emission spectra excited by the plasma. Precisely located exit slits around a Rowland circle isolate the spectral lines of interest. Each exit channel has its own photomultiplier detector and electronic integrator. As many as 50 channels are provided in commercial instruments, thus permitting the simultaneous integration of many signals and determination of many elements. Such instruments are expensive but are now widely used in analytical laboratories because of their speed, sensitivity, and freedom from most interferences from other elements.

19–2. ATOMIC ABSORPTION SPECTROMETRY

Atomic absorption (often referred to as "AA") is an analytical method based on the absorption of ultraviolet or visible light by gaseous atoms. The sample is converted (at least partly) into atomic vapor by spraying the solution into a flame. A hallow-cathode lamp containing the element to be determined is used as the light source. The atoms of this element in the flame absorb at precisely the wavelength emitted by the light source. The wavelength spread is extremely narrow, for both the emission line of the light source and the absorption line of the same element in the flame. For this reason, interference from the spectral lines of other elements is almost nil.

[6] R. W. Wendt and V. A. Fassel, *Anal. Chem., 37,* 920 (1965).

Atomic absorption is essentially the opposite of flame emission spectrometry. Although the sample is sprayed into a flame in both methods, flame emission methods depend on the *emission* of light by excited atoms in the flame, whereas atomic absorption is based on the *absorption* of light by neutral, unexcited atoms in the flame. Perhaps a better name for atomic absorption would be "absorption flame spectrometry."

Several advantages of atomic absorption over flame emission spectrometry may be cited.

1. There is less interference from salts of foreign metals than in ordinary flame atomic emission.
2. Atomic absorption is less dependent on experimental conditions, and there is no need (as in flame emission) to use an internal standard to correct for minor differences in the flame or other conditions.
3. For some elements, atomic absorption has better sensitivity and precision.

A disadvantage of atomic absorption is the necessity of acquiring and maintaining an expensive light source for each element to be determined.

At this writing, 68 different elements (including 15 rare earths) have been determined by atomic absorption. Except for boron, silicon, arsenic, selenium, and tellurium, which are sometimes classed as nonmetals, only the metallic elements can be determined directly. Like absorption spectrophotometry and flame emission spectrometry, atomic absorption is used primarily for measuring small to trace amounts of elements. The sensitivity for most elements is about 1 ppm (1 μg per mL of solution) or less. The accuracy of this technique is $\pm 2\%$, which is quite satisfactory for small amounts but often inadequate for analyzing samples for major constituents.

Recommended literature on atomic absorption includes a book [7], a review article on instrumentation [8], and a compilation of laboratory procedures [9].

Principles.   When a sample solution is aspirated into the flame, the solvent is evaporated or burned, and the sample compounds are thermally decomposed and converted into a gas of the individual atoms present. Most of these are in the ground state, although a few of the atoms become excited and emit light. The neutral atoms will absorb light only from the hollow-cathode source that emits the characteristic wavelength of the single element being determined.

Absorption of light by gaseous atoms is analogous to the absorption of radiant energy by ions and molecules in absorption spectrophotometry. The chief difference is that the absorption spectra (also the emission spectra) of gaseous atoms consist of sharp lines, in contrast to the broad absorption peaks characteristic of ions and molecules in solution. This is because there are fewer possible energy-level

[7] J. W. Robinson, *Atomic Absorption Spectroscopy*, New York: Marcel Dekker, 1966.
[8] H. L. Kahn, "Instrumentation for Atomic Absorption," *J. Chem. Ed.*, Jan. and Feb. 1966.
[9] *Analytical Methods for Atomic Absorption Spectroscopy*, Norwalk, Conn.: Perkin Elmer Co., 1966.

Figure 19–4. Lines emitted by the light source are much narrower than the absorption line to be measured.

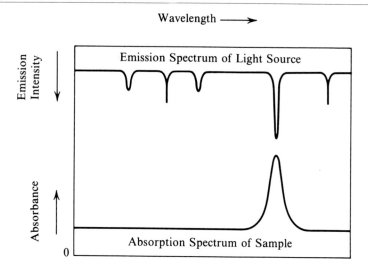

transitions in atoms than in molecules (which have a more complicated electronic structure) and also because the solute and solvent do not interact, as in solutions.

The lines from a hollow-cathode lamp (see below) are extremely narrow, approximately 0.01 Å in width. This is one of the advantages of a hollow-cathode source; it is impossible to obtain such a narrow-wavelength band from a continuous light source by any existing monochromator. Although the absorption line of an element to be measured is also narrow, it is still broad in comparison with the emission line used. This arrangement is advantageous in that the absorption line can be measured at its peak (see Figure 19–4).

Only one of several lines emitted by a hollow cathode is used for analysis. For example, the 2852 Å line of magnesium is used both as the light source and as the analytical line for measuring the magnesium content of the sample. The line in question may be isolated by a comparatively simple monochromator, since no other line is very close to it.

The hollow-cathode light source is directed at the flame into which the sample solution is sprayed. Some of the light is absorbed by an atomized sample component; the rest passes through. The selected spectral line is isolated from the emerging beam by a monochromator and the ratio of its intensity to that of the source is measured by a photocell or photomultiplier.

The light must be modulated to avoid interference from light emitted from the flame. This will be explained in the instrumentation section.

The ratio between the radiant power of the incident beam (P_0) and that of the transmitted beam (P) is measured.

The amount of light absorbed depends on the number of atoms in the light path. Provided the flame is hot enough to convert a chemical compound to free atoms, the light absorbed is almost independent of the flame temperature and the absorption wavelength. In some instances, an excessively hot flame may promote so many free atoms to the excited state that the number of unexcited atoms, and hence the light absorption, is reduced. If flame conditions and the rate of aspiration of sample into the flame are kept approximately constant, the absorbance ($\log P_0/P$) will be directly proportional to the concentration of the given metal in the sample.

Instrumentation

Hollow-Cathode Lamp. A diagram of a hollow-cathode source is shown in Figure 19–5. The tube is evacuated but contains some neon or argon gas. The cathode is made of the element to be determined. A high voltage (approximately 400 V, 2–20 mA) charges atoms of helium or argon, which bombard the cathode. The bombardment "sputters" metal atoms into the tube atmosphere, where they become excited by collision and emit light with wavelengths characteristic of the cathode metal.

Modulation. When an element is atomized in a flame, a certain low percentage of the atoms will be excited and emit light. Since an element emits at exactly the same wavelengths that it absorbs, the light of the analytical wavelength emitted in the flame will be added to the beam emerging from the flame. If this is not compensated for, we will measure absorbance minus emission, instead of absorbance. This lowers the sensitivity somewhat; but, more important, it introduces the errors encountered in flame emission spectrometry, caused by minor changes in flame temperature and other experimental conditions.

Figure 19–5. Diagram of a hollow-cathode lamp.

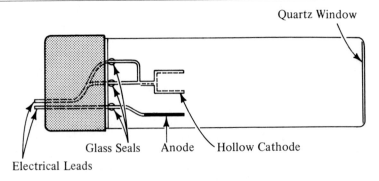

Quartz Window

Glass Seals   Anode   Hollow Cathode

Electrical Leads

To illustrate, suppose that for each 1000 atoms in the flame there is 1 excited atom. A change in conditions that increases the number of excited atoms to 2 per 1000 doubles the emission signal but only decreases the absorption (from unexcited atoms) by 1 ppt.

To avoid these difficulties, the power supply to the hollow-cathode lamp is modulated, and the radiant power of the source beam is measured with an AC amplifier tuned to the same frequency. Unmodulated light from the flame does not have any effect on the output signal of the AC amplifier. An alternative method of modulation is to use DC power for the hollow cathode but to mechanically "chop" the light beam. The detector is synchronized and tuned to the intermittent beam and is unaffected by any steady emitted light from the flame.

Burner-Nebulizer. With minor modifications, the burner-nebulizer described in Section 19–1 for flame emission photometry can be used for atomic absorption as well. A burner with a long slot increases the sensitivity of atomic absorption by providing a longer light-path through the gaseous atoms in the flame.

Monochromator and Detector. The purpose of the monochromator is to select the chosen analytical line from nearby lines. This task is simplified by the narrow lines of the source and the almost-zero background. For a general-purpose instrument, a resolution of approximately 1.0 Å is desirable.

Photomultipliers have been extensively used; their higher sensitivity is desirable because the analytical lines of most elements lie in the ultraviolet or blue spectral region. As was pointed out earlier, an AC amplifier (tuned or untuned), rather than a DC amplifier, is used to amplify the signal from the detector.

Quantitative Procedure

Sample Preparation. All samples should be diluted with an appropriate solvent to bring the absorbance into the 0.1–0.9 range. Aqueous solutions are aspirated into the flame directly. Plant and animal tissues are decomposed by oxidizing acids, or dry ashed and the ash dissolved in hydrochloric acid and diluted with water. Metal alloys, ores, etc., are dissolved in acid and diluted. Oils and other nonaqueous liquids are diluted with an organic solvent, usually methyl isobutyl ketone, before analysis.

Separation methods such as solvent extraction or ion exchange are useful in separating minor and trace elements from matrix elements. After this is done, simple calibrating standards for minor constituents can often be used without the addition of major constituents.

(Omitting the major constituents in the standards, however, may change the viscosity and, thus, the rate of atomization.) Also, trace elements may be concentrated by separation techniques to make their detection possible. Solvent extraction employing pyrrolidine dithiocarbamate is often used to concentrate traces of "H_2S group" metals.

Calibration Curve. Preparing a calibration, or working, curve for atomic absorption is similar to preparing a calibration plot in absorption spectrophotometry. In the simplest cases, standard metal-salt solutions may be used in preparing standards. The salt chosen should be in approximately the same form as the metal in the analytical samples. When a metal to be measured is a trace or minor constituent and separations are not contemplated, it is usually best to add the appropriate concentrations of the major sample constituents to the calibrating solutions. Since it is difficult to duplicate exactly flame and aspiration conditions from one day to the next, it is well to run a set of standards with each group of samples.

Experimental Variables.

1. *Type of Flame.* For a majority of metallic elements, an air-acetylene flame is satisfactory and commonly used. However, several metals are refractory and difficult to convert into gaseous atoms. Among these are boron, rare earths, thorium, titanium, uranium, and zirconium. This group of metals forms unusually strong metal to oxygen bonds, and it is believed that the monoxides are particularly difficult to decompose thermally. Fassel, Kniseley, and co-workers have solved this problem by the rather simple expedient of using a fuel-rich oxyacetylene flame [10] and a special burner [11]. Another satisfactory method is to use the fuel-rich nitrous oxide–acetylene flame first proposed by Willis [12]. The latter gives a flame almost as hot as oxygen-acetylene, but with greater safety and convenience.

2. *Height in Flame.* The part of the flame at which the hollow-cathode source is focused may be an important variable. Some metals absorb much more strongly in one part of the flame than in another. A suggested procedure is to adjust the position of the source beam with respect to the flame so that the maximum absorbance is obtained for a solution of the element to be measured.

3. *Solvent.* The absorbance of a given concentration of metal in a sample is always greater when the sample is in an organic solvent rather than in aqueous solution. (The absorbance also tends to vary considerably from one organic solvent to another.) Apparently, a greater proportion of the metal salts are converted into absorbing neutral atoms in organic solvents. This increased sensitivity demonstrates another advantage of first separating the desired element or group of elements by solvent extraction—after extraction, the organic layer can be aspirated directly into the flame.

4. *Aspiration Rate.* Within certain limits, an increase in the rate of aspiration will increase the number of atoms in the light path and, hence, the sensitivity. However, too rapid a rate will swamp the flame and reduce the signal. Therefore, an intermediate rate is best. This should be kept as constant as possible to ensure reproducible results. To do this, it is important to keep the burner tip clean.

Scope and Limitations. Prior to the development of techniques employing a fuel-rich oxygen-acetylene flame or a nitrous oxide–acetylene flame, some 26 elements were all but impossible to detect by atomic absorption. Now, however, virtually all of the metallic elements may be determined.

[10] V. A. Fassel, R. B. Myers, and R. N. Kniseley, *Spectrochim. Acta, 19,* 1187 (1963).

[11] J. A. Fiorino, R. N. Kniseley, and V. A. Fassel, *Spectrochim. Acta, 23B,* 413 (1968).

[12] J. B. Willis, *Nature, 207,* 715 (1965); M. D. Amos and J. B. Willis, *Spectrochim. Acta, 22,* 1325 (1966); J. A. Bowman and J. B. Willis, *Anal. Chem., 39,* 1210 (1967).

The precision and accuracy attainable by atomic absorption vary with the element measured and the conditions and equipment used. However, an average deviation of approximately 1.5 pph and accuracy within $\pm$ 2.0 pph are fairly typical.

Virtually no cases are recorded in which a foreign metal interferes by absorbing at the same wavelength as the element determined. This is because of the extremely narrow emission band and the narrow absorption lines. However, other effects can cause interference if not compensated for. Magnesium and aluminum interfere seriously with each other by forming an intermetallic compound. An interference may occur that reduces the number of absorbing atoms through excitation. An example of this and a technique for its elimination are discussed in the following section with regard to determining strontium in rocks.

Certain anions may interfere by forming a compound that is difficult to dissociate into atoms. For instance, in determining lead, the iodide or carbonate ion reduces the absorbance considerably from when other common anions are present. Phosphate ion may interfere in the determination of calcium and certain other metals. This problem can be solved by complexing the metal ion with a reagent such as EDTA to break the metal-anion bond before analysis.

A high concentration of foreign salts, unless compensated for in preparing the calibration standards, may cause error by altering the aspiration rate and the atomization efficiency of the flame.

Typical Determinations

Calcium in Blood Serum, Animal Tissue, and Blood [13]. This determination is made with a fuel-rich air-acetylene flame using a calcium hollow-cathode lamp operating at 4227 Å. An excess of lanthanum salt is added to all standards and samples to combine with any phosphate and prevent it from interfering with the production of calcium atoms in the flame. For analyzing blood serum, a 0.25-mL sample is diluted with lanthanum solution and distilled water to 5.0 mL. Standards containing 0, 2, 5, 8, and 10 ppm of calcium are prepared, each containing the same concentration of lanthanum solution as the sample. Tissue and bone samples are ashed in a muffle furnace overnight; the ash is dissolved in hydrochloric acid and diluted with lanthanum solution and distilled water until the calcium concentration is in the 1–10 ppm range.

Strontium and Barium in Rocks [14]. Strontium and barium are present at 100–1000 ppm in igneous rocks. The samples are dissolved in hydrofluoric and perchloric acids, or by fusion in a molten salt (flux), before analysis by atomic absorption. Strontium and barium are incompletely atomized in the air-acetylene flame, but the nitrous oxide–acetylene flame is hotter (2950°C) and increases the sensitivity appreciably. However, a good many neutral barium and strontium atoms are ionized at this temperature. To suppress the ionization, a relatively concentrated solution of a salt of a more easily ionized metal (such as potassium) is added to both

[13] Modification of the method of J. B. Willis, *Spectrochim. Acta, 16,* 259 (1960), *Anal. Chem., 33,* 556 (1961), and of E. Neubrun, *Nature, 192,* 1182 (1961).

[14] J. A. Bowman and J. B. Willis, *Anal. Chem., 39,* 1210 (1967).

Figure 19–6. Effect of potassium concentration on the calibration curve for strontium. [From *Anal. Chem.*, *39*, 1210 (1967). By permission.]

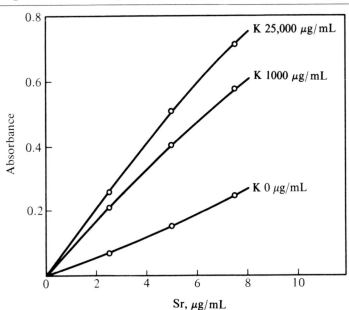

sample and standard solutions. The effect of this added salt is illustrated in Figure 19–6.

This is a beautiful example of the law of mass action. We wish to suppress the reaction

$$Ba(g) \rightleftharpoons Ba^+(g) + e^- \qquad \textit{Ionization Potential} = 5.2 \text{ eV}$$

that results in the ionizing of approximately 80% of the barium atoms in a nitrous oxide–acetylene flame. This is done by adding a large excess of a potassium salt that ionizes even more readily than barium:

$$K(g) \rightleftharpoons K^+(g) + e^- \qquad \textit{Ionization Potential} = 4.3 \text{ eV}$$

The large supply of electrons from ionization of potassium atoms reverses the Ba–Ba^+ equilibrium, so only a negligible number of barium atoms ionize.

Vanadium in Petroleum [15]. Determining traces of vanadium and nickel in fuel oil and certain other products is a problem of major importance in the petroleum industry. The analytical problem is difficult because the metals are present in low concentrations, often in the form of porphyrin complexes that are resistant to chemical attack.

[15] J. A. Bowman and J. B. Willis, *Anal. Chem.*, *39*, 1210 (1967).

By using atomic absorption, the vanadium content of petroleum may be measured fairly easily. The sample is diluted 1:10 with methyl isobutyl ketone (MIBK) and sprayed into a nitrous oxide–acetylene flame. In this determination, standards present something of a problem. Adding vanadium organic salts as standards is unreliable, probably because the vanadium is in a different chemical form in the available salts than it is in the oil. Therefore, the sample solutions are compared with an oil whose vanadium content is accurately known from other measurements.

Atomic Absorption Using a Graphite Furnace.   A flame, or even an ICP, is actually a turbulent excitation source, and many conditions must be controlled to obtain reproducible results. Heated graphite furnaces and carbon-rod analyzers offer an attractive alternative for atomic absorption analyses. The sensitivity of these nonflame methods is very high (10^{-8}–10^{-11} g), sample volumes of only 5–100 μL can be used, and solid samples can also be handled. Offsetting disadvantages are greater effects from matrix elements in some cases and lower precision than with flame methods.

In nonflame AA, a few microliters of a liquid sample or a few milligrams of a solid sample are placed in a graphite tube or in a small hole drilled in a carbon rod. The tube is then heated electrically so that the temperature is increased at an extremely fast rate to a limiting value and the sample is converted to gaseous atoms. These are swept out within a few seconds by an inert gas that passes continuously through the tube. However, while the gaseous atoms are still in the tube, the amount of monochromatic light from a hollow-cathode lamp that is absorbed by the atoms is measured. The amount of a desired sample element is calculated from this absorbance in essentially the same manner as in flame atomic absorption.

The graphite furnace technique involves three carefully programmed heating steps. In the first, or dry, cycle, the system is heated for 20–30 sec at around 120°C to evaporate the solvent from liquid samples or any highly volatile components. A dry residue of the sample remains. The second heating step is the ash or char cycle. This step often converts the sample components to a different chemical state. In the third stage, the furnace is raised rapidly to the desired temperature for atomization. This cycle lasts only 2–8 sec, depending on the instrument used, but it is sufficient to produce an atom cloud for the atomic absorption of the monochromatic light.

Many elements can be determined by nonflame atomic absorption. Such elements as calcium, cadmium, magnesium, sodium, silver, and zinc can be detected at less than 0.1 μg/mL. Mercury can also be determined with excellent sensitivity with a conventional graphite-furnace instrument or with a special mercury analyzer. In the latter, sample solutions are treated with an acidic solution of tin(II) to reduce the mercury salts to the free metal.

$$Hg^{2+} + Sn(II) \; \rightarrow \; Hg^0 + Sn(IV)$$

The finely divided particles of mercury metal are volatilized by passing air through the liquid sample. The mercury is measured from the absorbance of mercury atoms in a long optical cell in the instrument.

QUESTIONS AND PROBLEMS

Principles and Comparisons

1. Two of the transitions that occur in a flame-excited lithium atom are the $3p \rightarrow 2s$ transition (3.85 eV → 0.0 eV) and the $4d \rightarrow 2p$ transition (4.60 eV → 1.90 eV). Calculate the wavelengths associated with these transitions, in Ångstrom units.
2. The ionization potential for potassium is 4.32 eV. Calculate whether atomic emission lines below about 3000 Å are likely to occur. Explain your answer.
3. For sodium, the spectral line corresponding to a $3p \rightarrow 3s$ transition has a wavelength of 5893 Å, whereas the line corresponding to a $4p \rightarrow 3s$ transition has a wavelength of 3300 Å. Calculate the energies of these transitions in electron volts.
4. Explain how ionization in a flame might affect the determination of a given element by (a) atomic emission or (b) atomic absorption.
5. In tabular form, compare the relative advantages of quantitative analysis by atomic emission using (a) flame excitation and (b) plasma excitation.
6. Compare the advantages and disadvantages of (a) atomic absorption and (b) atomic emission spectrometry for quantitative determinations.

Flame Emission and Plasma Emission

7. Describe the preparation of an analytical calibration (working) curve for flame atomic emission spectrometry.
8. Traces of sodium are to be determined in a CaO sample. The analytical calibration curve is prepared using standards containing CaO to compensate for the CaO in the sample. The following analytical data are obtained:

mg Na/L (ppm)	Emission Intensity at 5893 Å
74.3	100
55.7	87
37.0	69
18.5	46
7.4	22
0	3

1.00 g CaO sample dissolved in 100 mL	28

Calculate the percentage of sodium in the sample.
9. For a number of years, the determination of rare earth elements by flame spectrometry was thought impossible because the rare earths formed metal oxides instead of gaseous metal atoms. Explain how to partly overcome these difficulties by adjusting the flame conditions. Also explain whether it is possible to determine rare earths using ICP. Why or why not?
10. The higher temperature of a plasma compared to a flame often results in extensive ion formation. Explain how this might be minimized.
11. Explain briefly why the sensitivity of determining the various elements is usually better when plasma excitation is used rather than flame excitation in atomic emission.
12. Explain how simultaneous determination of several elements can be achieved in plasma atomic emission spectrometry.

Atomic Absorption

13. Describe the preparation of an analytical calibration (working) curve for flame atomic absorption spectrometry.
14. Briefly describe the setup for flameless AA analysis. Prepare a one-paragraph report from the recent literature on an example of this method.
15. Extensive production of gaseous metal ions, rather than gaseous atoms, will reduce the strength of the signal obtainable in atomic absorption spectrometry. Explain how production of these metal ions can be reduced by adding another element to the sample. What properties must the added element have?
16. Explain why the precise location of the hollow-cathode beam in the flame is necessary for reproducible results in atomic absorption.
17. Briefly explain how a graphite furnace is used in atomic absorption spectrometry. Explain the three essential heating steps in a graphite-furnace analytical instrument.

Methods

18. Consult recent chemical journals and find one example of each of the following techniques. For each, give the complete journal reference and list the essential conditions used in determining one element. (a) Plasma emission spectrography, (b) flame atomic emission, (c) flame atomic absorption, (d) graphite-furnace AA.
19. The rate of import duty for imported lead crystal glassware varies with the content of lead (as lead monoxide). From the methods described in this chapter, suggest a *rapid*, accurate method for determining lead in such glassware.
20. Minute traces of copper affect the taste of butter markedly. From the methods described in this chapter, suggest a *rapid* analytical method for determining copper in butter. What treatment might be necessary to obtain a sample for analysis?

20

Liquid-Liquid Extraction

It is difficult to overemphasize the importance of separations in analytical chemistry. In determining the composition of a chemical sample, it is frequently necessary to separate some or all of the components before attempting their quantitative measurement. To study the structure and physical properties of a substance, it must be obtained free from impurities.

The material in earlier chapters included some separation techniques; precipitation was discussed in Chapter 4 and electrodeposition in Chapter 16. In this chapter, we will discuss liquid-liquid solvent extraction, which is a useful separation method in itself. Solvent extraction is also important in understanding chromatography because chromatographic methods also involve the distribution of solutes between two phases. The chapters immediately following this one deal with the principles of various types of chromatography: gas chromatography, liquid chromatography, and ion-exchange chromatography.

20-1. GENERAL PRINCIPLES

Many organic liquids are not miscible with water. When such a liquid is added to water, two layers are formed. Whether the organic liquid is the upper or lower layer depends on whether the density of the liquid is less than or greater than that of water.

Suppose an aqueous solution containing two dissolved solutes, A and B, is shaken vigorously with an immiscible organic liquid and the mixture is allowed to stand until the two solvent layers settle out. If the organic liquid has a much greater affinity for one of the dissolved solutes than does water, most (or all) of that solute will pass from the aqueous phase to the organic liquid phase. We say that this solute is *extracted*. If the other solute prefers the aqueous phase to the organic, it will not be extracted. If extraction is carried out in a separatory funnel, the lower liquid can be carefully drained off and the two solutes will be physically separated. If the extraction of A is not complete, one or more additional extractions with fresh solvent can be performed in the same way. A quantitative separation of A and B requires an "all or nothing" situation, namely, that *all* of the A is in the extract (or combined extracts) and that all of B remain in the aqueous layer.

An organic liquid used for solvent extractions must be a good solvent for the solute(s) to be extracted. After being shaken with an aqueous solution, the droplets of organic liquid should coalesce quickly and settle out as a separate layer. To do this, the specific gravity (sp. gr.)—the density of the liquid divided by the density of water—should be substantially greater or substantially less than 1.

The vapors of organic solvents are toxic to varying degrees. For this reason, solvent extractions should be performed in a hood and the use of the more toxic solvents should be avoided. Chloroform ($CHCl_3$, sp. gr.:1.49, density:1.50) is a heavy solvent that has been widely used for extracting organic compounds and metal coordination complexes from aqueous solution. However, chloroform has now been designated a cancer suspect agent, and its use has declined. Dichloromethane (CH_2Cl_2, dens.:1.325), which is also called methylene chloride, is now used extensively even though it is quite volatile (b.p.:40°C). Benzene (C_6H_6, dens.:0.874) is another cancer suspect agent, so toluene ($C_6H_5CH_3$, dens.:0.867) is used in its stead as a lighter-than-water solvent. Methyl isobutyl ketone ($CH_3COCH_2CH(CH_3)_2$ (dens.:0.800, abbreviation MIBK) is another excellent solvent for many types of extraction. Tributylphosphate ($C_4H_9O)_3PO$ (dens.:0.979, abbreviation TBP) is used extensively for the extraction of many metal association complexes, even though its density is close to that of water. Sometimes TBP is mixed with toluene or kerosene to make the solvent layers separate more readily.

When a solute dissolved in a solvent such as water is shaken with a second, immiscible solvent, there is a competition between the two solvents for the solute. What determines then whether the solute ends up primarily in one solvent phase or the other? Why do some solutes prefer an organic phase while others stay in the aqueous phase?

Completely satisfying answers to these questions are not available. A general answer is that a solute will go into the phase in which it can form the more stable chemical species. A qualitative empirical rule that often works is that "like dissolves like." Thus, in an equilibration between water and a solvent such as toluene, molecular organic solutes are apt to be mostly in the toluene phase. Ionized organic solutes and inorganic salts usually remain in the aqueous solution. For example, the extraction of phenol depends on two sequential equilibria:

$$C_6H_5OH \;\rightleftharpoons\; C_6H_5OH \;\xrightleftharpoons{NaOH}\; C_6H_5O^-Na^+$$
$$\text{(toluene)} \qquad \text{(aqueous)} \qquad\qquad \text{(aqueous)}$$

In neutral or acidic aqueous solution, phenol is largely in the molecular form and the distribution strongly favors the toluene layer. However, a strongly basic aqueous solution converts the phenol to the phenolate anion and results in extraction from the toluene back into the aqueous later.

20–2. COMPLETENESS OF EXTRACTION

Suppose we start with an aqueous solution of a solute, A, and shake it with an immiscible organic solvent. When equilibrium has been reached, the relative concentrations of solute in the organic and water phases is described by the *concentration distribution ratio*, D_c:

$$D_c = \frac{\text{concn. of solute in organic phase}}{\text{concn. of solute in water phase}} = \frac{[A]_o}{[A]_w} \qquad (20-1)$$

The concentration distribution ratio is closely related to the distribution coefficient, K_d, distribution constant, K_D, and partition coefficient, P, which are frequently seen in chemical publications. However, these terms are usually concerned with the distribution of only a single species of A, while the distribution ratio includes the total concentration of *all* forms of A (such as A, HA^+, A^-, and dimers of A).

Since concentration is mmoles of solute per volume (mL) of solvent, we can write Equation 20–1 as follows:

$$D_c = \frac{(\text{mmoles A})_o/V_o}{(\text{mmoles A})_w/V_w} = \frac{(\text{mmoles A})_o V_w}{(\text{mmoles A})_w V_o} \qquad (20-2)$$

Another useful constant for any solute is the mass distribution ratio, D_m, which is defined as the ratio of the *amounts* of solute in the two phases:

$$D_m = \frac{(\text{mmoles A})_o}{(\text{mmoles A})_w} \qquad (20-3)$$

On combining Equations 20–2 and 20–3, we get

$$D_c = D_m \frac{V_w}{V_o} \qquad (20-4)$$

$$D_m = D_c \frac{V_o}{V_w} \qquad (20-5)$$

In solvent extraction, conditions are often chosen so that $V_o = V_w$; then D_m and D_c are equivalent. In this event, D_c can be substituted for D_m in the following equations. (Otherwise, D_m must be evaluated from D_c, V_o, and V_w.)

By using Equation 20–3, the fraction of solute *not extracted*, f, can be defined as

$$f = \frac{(\text{mmoles A})_w}{(\text{mmoles A})_o + (\text{mmoles A})_w} = \frac{1}{D_m + 1} \qquad (20-6)$$

Frequently, a single extraction of a solute will not be quantitative. In such cases, the solvent layers from the first extraction are separated and the aqueous layer is again extracted with a *fresh* portion of the organic solvent. This process may be repeated several times. After n extractions with fresh solvent, the fraction remaining in the aqueous layer is

$$f = \frac{1}{(D_m + 1)^n} \qquad (20\text{--}7)$$

The percentage extracted by the total of n extractions is obtained by subtracting the fraction remaining from 1.0 and multiplying by 100.

$$\%E = 100\left[1 - \frac{1}{(D_m + 1)^n}\right] \qquad (20\text{--}8)$$

For most purposes, an extraction is considered quantitative if it is at least 99.9% complete. If at least 90% of the desired solute is extracted each time ($D = 9$), an essentially quantitative extraction is achieved after two or three extractions with fresh solvent. When the distribution ratio is less favorable, complete extraction may still be possible if enough extractions are carried out. To do this conveniently requires a continuous extractor.

20–3. EXTRACTION OF ORGANIC COMPOUNDS

Very small amounts of organic compounds frequently need to be determined in aqueous samples such as drinking water or industrial waste water. By solvent extraction of a relatively large volume of water with a small volume of an organic solvent, the extracted organic solutes are *concentrated*. Further concentration is often possible by partial evaporation of the extractive solvent.

The determination of chloroform and other volatile chlorinated hydrocarbons in drinking water is a good example. These substances in drinking water are generally in the low parts-per-billion (ppb) concentration range, which is far too low for direct analysis. Extraction of the water sample with a small volume of pentane transfers a reproducible fraction of each chlorinated hydrocarbon from the aqueous to the pentane phase. The individual compounds are then separated and measured by injecting a portion of the pentane extract into a gas chromatograph (see Chapter 22).

The extraction of organic compounds that have acidic or basic properties often varies greatly with the pH of the aqueous solution. Consider an organic acid, HA. If the aqueous solution is acidic, the acid will be primarily in the molecular form, HA, which is apt to be well extracted by an organic solvent. However, as the pH of the aqueous solution increases, more and more of the acid ionizes to form A^-. Usually, the anion A^- will be very poorly extracted (often 0%). The net result is that extraction of the organic acid will decrease with rising pH and may go to zero.

We can use this principle to separate organic compounds into acidic, basic, and neutral fractions. First, the water sample is acidified with an inorganic acid to convert organic acids entirely to the molecular form (HA) and organic bases to the ionic form (BH^+). On equilibration of the aqueous phase with a water-immiscible organic liquid, the neutral organic compounds plus molecular acids are extracted. Then the organic liquid is "back-extracted" with an alkaline solution of distilled water. The organic acids are converted to anions (A^-) and are extracted from the organic to the aqueous phase. However, neutral organic compounds remain in the organic phase.

The basic organic compounds remain in the original aqueous sample. By making this basic (with an inorganic base), the ionic bases (BH^+) are converted to neutral bases (B), which can now be extracted into an organic solvent.

This method for "fractionating" organic compounds seldom works perfectly because the various extraction steps are usually not entirely quantitative. But it does work well enough to have been used extensively to simplify the analysis of complicated samples.

20–4. EXTRACTION OF METAL-ORGANIC COMPLEXES

Complex Formation and Extraction. A considerable number of organic compounds form complexes with metal cations. Many of these complexes are more organic than inorganic in nature and are more soluble in organic solvents than in water; hence, they are extractable. For example, the iron(III)-cupferron chelate or the aluminum(III)-oxine complex is extracted by chloroform.

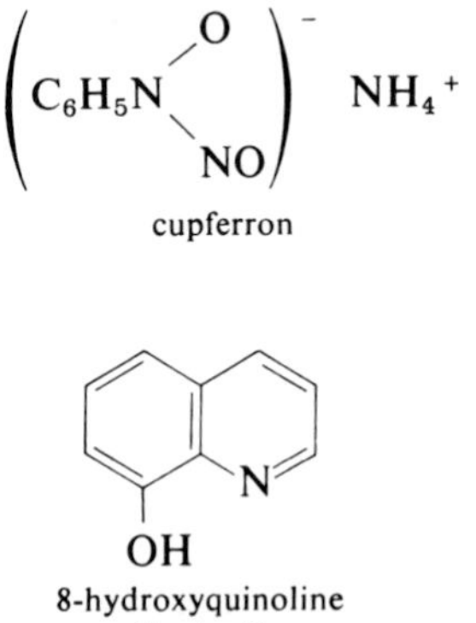

cupferron

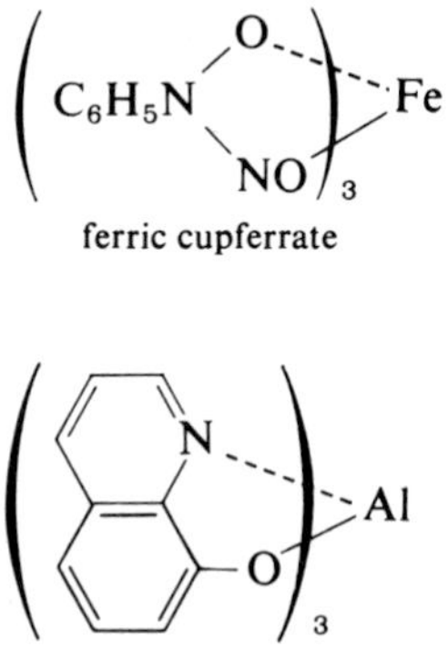

ferric cupferrate

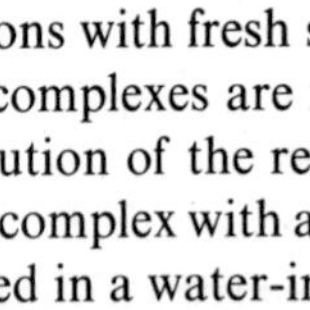

8-hydroxyquinoline
("oxine")

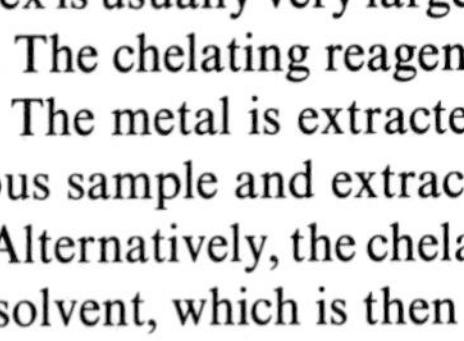

aluminum oxinate

The distribution ratio for this type of complex is usually very large, so only one or two extractions with fresh solvent are needed. The chelating reagents are solids, and the metal complexes are insoluble in water. The metal is extracted by adding an aqueous solution of the reagent to the aqueous sample and extracting the precipitated metal complex with an organic solvent. Alternatively, the chelating reagent may be dissolved in a water-immiscible organic solvent, which is then shaken with the sample solution. Shaking produces a temporary emulsion of the two liquid phases. Metal ions in the sample solution react with the chelating reagent at the interfaces between phases and then are extracted into the organic phase.

Finely divided solvent droplets are necessary for efficient extraction, but the phases must nonetheless coalesce and settle out quickly after shaking. Solvent systems that produce long-lasting emulsions should be avoided.

Equilibria.   A number of equilibria are involved in forming a metal complex with a chelating reagent and extraction of that complex.

1. Ionization of the reagent to give an anion:

$$HR \rightleftharpoons H^+ + R^-$$

2. Formation of the chelate:

$$M^{2+} + 2R^- \rightleftharpoons MR_2$$

3. Competing reactions for the metal ion. These include hydrolysis (reaction with water) and reaction with any additional complexing ligand (L) that may have been added:

$$M^{2+} + H_2O \rightleftharpoons M(OH)^+ + H^+$$

$$M^{2+} + L^- \rightleftharpoons ML^+, \quad \text{etc.}$$

4. Distribution of the reagent between the aqueous and organic phases:

$$D_R = \frac{[HR]_o}{[HR]_w}$$

5. Distribution of the metal chelate between the two phases:

$$D_{MR_2} = \frac{[MR_2]_o}{[MR_2]_w}$$

The pH affects the proportion of the reagent that exists as the anion (equilibrium 1). A higher concentration of R^- shifts equilibrium 2 to the right and thereby promotes the formation of the metal chelate. Distribution of the reagent and metal-reagent complex between the two phases (equilibria 4 and 5) depends on the chemical nature of the reagent and of the extractive solvents used.

Formula of a Complex.   Solvent extraction experiments are a good way to obtain information about the formula of a metal-organic complex. In particular, we can often determine the ratio of reagent to metal ion and the number of hydrogen ions given off in forming the complex. Consider the reaction

$$M^{2+}(w) + xH_2L(o) \rightleftharpoons \left. \begin{array}{l} ML_x(o) \\ \text{or } MHL_x(o) \\ \text{or } M(OH)L_x(o) \end{array} \right\} + yH^+(w) \qquad (20\text{–}9)$$

where L is the chelating ligand and (w) and (o) indicates that the reactant or product is in the water or organic phase. Although it may be an oversimplification, we will assume that this reaction occurs at the aqueous–organic solvent interface as the mixture is shaken. We will also assume that the uncomplexed metal ions and hydrogen ions are entirely in the aqueous (water) solution and that the distribution ratios for H_2L and ML_x are also so great that these species are essentially all in the organic phase. Then the equilibrium constant for this reaction is

$$K = \frac{[ML_x]_o[H^+]_w{}^y}{[M^{2+}]_w[H_2L]_o{}^x} \qquad (20\text{--}10)$$

If we assume that ML_x is the only form of M in the organic layer, the concentration distribution ratio D_c is given by the equation

$$D_c = \frac{[ML_x]_o}{[M^{2+}]_w} \qquad (20\text{--}11)$$

Substituting this into Equation 20–10,

$$K = \frac{D_c[H^+]_w{}^y}{[H_2L]_o{}^x} \qquad (20\text{--}12)$$

and solving for D_c,

$$D_c = \frac{[H_2L]^x\, K}{[H^+]^y} \qquad (20\text{--}13)$$

Taking the log of both sides of this equation,

$$\log D_c = x\log[H_2L]_o - y\log[H^+]_w + \log K \qquad (20\text{--}14)$$

If $[H^+]_w$ is held constant while the concentration of H_2L in the organic phase is varied, a plot of $\log D_c$ versus $\log[H_2L]_o$ will give a straight line with a slope of x.

Figure 20–1.   The distribution of copper(II) extracted from an acetate solution (pH 4) as a function of the concentration of DHDO in the organic phase.

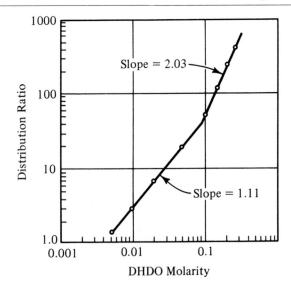

Slope = 2.03

Slope = 1.11

DHDO Molarity

Table 20-1.   Some Chelating Reagents Used in Solvent Extraction.

Name or Abbreviation	Formula	Typical Extractions/Conditions
Oxine		Al, Bi, Cu, Ni, Sn(IV), U(VI), Zn, etc./ dil acid
TTA		Most metal ions/pH 3 or higher
PBHA		Mo(VI), Sn(IV), Ti(IV), V(V), W(VI), Zr(IV)/ acidic solns
IOTG	$\overset{\text{SH}}{\underset{}{\mid}}\ \overset{\text{O}}{\underset{}{\parallel}}$ $CH_2—C—OC_8H_{17}$	Ag, Au(III), Hg, Bi, Cu/0.1–2M HNO$_3$[a]
DHDO	$\overset{\text{OH}}{\underset{}{\mid}}\ \overset{\text{NOH}}{\underset{}{\parallel}}$ $C_4H_9—CH—CH—C—CH—C_4H_9$ $\overset{}{\underset{}{\mid}}\qquad\overset{}{\underset{}{\mid}}$ $C_2H_5\qquad C_2H_5$	Cu(II), Mo(VI)/Cu, pH 4–7; Mo, pH 0–2[b]
TOPS	$(C_8H_{17})_3\ P \rightarrow S$	Ag, Hg, Au(III)/acidic solns[c]
Sodium diethyldithio- carbamate	$\overset{\text{S}}{\underset{}{\parallel}}$ $(C_2H_5)_2\ N—C—S^-\ Na^+$	Metal ions that form insoluble sulfides/dil acid

a. J. S. Fritz, R. K. Gillette, and H. E. Mishmash, *Anal. Chem.*, *38*, 1869 (1966).
b. J. S. Fritz, D. R. Beuerman, and J. J. Richard, *Talanta*, *18*, 1095 (1971); *Anal. Chem.*, *44*, 692 (1972).
c. D. E. Elliott and C. V. Banks, *Anal. Chim. Acta.*, *33*, 237 (1965).

Then, if another series of extractions is done with $[H_2L]_o$ constant while $[H^+]_w$ is varied, a plot of $\log D_c$ versus $\log [H^+]_w$ will give a straight line with a slope of $-y$.

Example: A new oxime reagent, H_2L, was found to extract copper(II) selectively, giving a green complex at low oxime concentrations and a brown complex at higher concentrations [1].

A plot of $\log D_c$ versus $\log H_2L$ gave slopes (x) of 1 and 2 for the green and brown complexes, respectively (Figure 20–1). Similarly, a plot of $\log D_c$ versus $\log [H^+]$ gave a slope (y) of -2 for each complex. Substituting these values into Equation 20–9,

$$\text{Green complex:}\qquad Cu^{2+} + H_2L \ \rightarrow\ CuL + 2H^+$$

$$\text{Brown complex:}\qquad Cu^{2+} + 2H_2L \ \rightarrow\ Cu(HL)_2 + 2H^+$$

[1] J. S. Fritz, D. R. Beuerman, and J. J. Richard, *Talanta*, *18*, 1095 (1971).

Separations.   Metal ions can be separated from each other by solvent extraction provided that one forms an extractable complex while the other does not. Sometimes a chelating reagent can be selected that only reacts with one of the metal ions present. In other cases, the acidity may be increased so that only the metal ion that forms the more stable complex will react with the chelating reagent. For example, titanium(IV) forms a complex with 8-hydroxyquinoline at more acidic pH values than aluminum(III); the titanium complex can be extracted between pH 0 and 1.4 and separated from aluminum(III). Aluminum can then be extracted as the 8-hydroxyquinoline complex between pH 2.8 and 9.5.

Many other examples of organic reagents used to extract metal ions may be found in the chemical literature. Table 20–1 lists some typical reagents used for separating metal ions. Solvent extraction is also used along with color-forming reagents for avoiding interferences and (sometimes) for concentrating the colored species by shaking a rather large volume of aqueous solution with a small volume of organic solvent. Sometimes the metal is back-extracted into aqueous solution for the spectrophotometric analysis; but in other cases, the color is formed and the absorbance measured directly in the extracting solvent.

20–5.   EXTRACTION OF ION-ASSOCIATION COMPLEXES

Under the proper conditions, some metal ions form extractable ion pairs. The metal may be incorporated in a complex cation or in a complex anion. An example is the extraction of iron(III) from aqueous $6M$ hydrochloric acid by ether or methyl isobutyl ketone (MIBK). Iron(III) is extracted as a chloro complex or as a mixture of chloro complexes, $FeCl_4^-$ predominating. This is extracted as the ion pair with the hydronium ion:

$$H_3O^+ FeCl_4^-$$

In this type of extraction, two major requirements appear to be necessary: (1) Conditions must favor formation of an ion pair of large size. (In the preceding example, a high concentration of hydrochloric acid is needed to form the complex $FeCl_4^-$ ion.) (2) The organic solvent must solvate the ion pair strongly.

In the latter connection, it is interesting to compare the effectiveness of various solvents in extracting iron(III) from hydrochloric acid solutions. Inert solvents such as benzene, chloroform, and carbon tetrachloride do not extract iron(III) at all. Among oxygen-containing organic solvents, extraction is in the order: alcohols < ethers < esters < ketones < phosphate esters.

Diluting a ketone solvent (MIBK) with an organic liquid such as toluene (which does not solvate the complex ion pair) reduces the extractability of the metal. The distribution ratio decreases as the concentration of MIBK in toluene is decreased. This is similar to the concentration effects noted with chelating reagents in Section 20–4.

Figure 20–2. Plot of distribution ratio versus molarity of pure solvent for the extraction of iron(III) from $4M$ hydrochloric acid with methyl ketones. $[Fe^{3+}] \approx 10^{-5}M$.

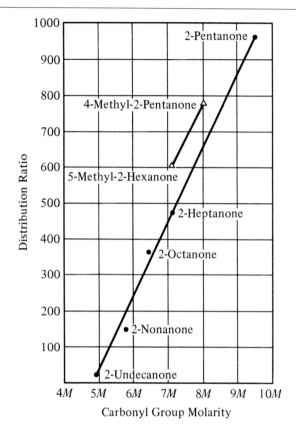

The molecular weight of a solvating organic liquid is an interesting variable in solvent extraction. In one series of experiments [2], iron(III) was extracted from aqueous solutions of fixed hydrochloric acid concentration with pure methyl

$$\underset{\substack{\| \\ }}{\overset{O}{}}$$

ketone solvents, CH_3C—R. When R was a small group (ethyl), the extraction was influenced by the rather large solubility of the ketone in the aqueous phase. But as the carbon chain length of R was increased from C_3 (propyl) to C_8 (octyl), the distribution ratio for iron(III) became progressively smaller. A plot of distribution ratio versus molarity of carbonyl (C=O) in the pure ketone is linear (Figure 20–2). This shows that the carbonyl concentration in a ketone solvent is the governing factor in solvating the metal complex and causing it to be extracted.

[2] M. D. Seymour, Unpublished work.

Figure 20–3. Effect of $(C_4H_9)_4N^+$ on the extraction of tin(IV) from aqueous hydrochloric acid by MIBK.

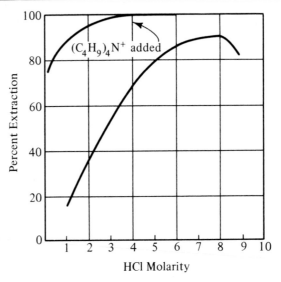

Some extractions are made possible by the addition of a strongly solvating reagent other than the organic solvent itself. For example, uranium(VI) salts such as uranyl chloride may be extracted with the aid of a solvating reagent such as trioctyl phosphine oxide (TOPO) or dioctyl sulfoxide (DOSO). With the latter reagent, the extracted species is the ion pair

$$
\begin{bmatrix}
\text{Oct}—\text{S}—\text{Oct} \\
| \\
\text{O} \\
| \\
\text{O}—\text{U}—\text{O} \\
| \\
\text{O} \\
| \\
\text{Oct}—\text{S}—\text{Oct}
\end{bmatrix}^{2+}
$$

with two chloride or nitrate ions. Many cations may be extracted into an organic solvent containing TOPO [3].

Another technique to increase the extraction of a metal is to add a large organic ion to form an ion pair with the complex metal ion. For example, the cation of a quaternary ammonium salt such as tetrabutylammonium chloride, $(C_4H_9)_4N^+$, forms an ion pair with $SnCl_6^{2-}$ and permits extraction at lower hydrochloric acid concentration than when H_3O^+ is the cation in the ion pair (Figure 20–3).

With higher molecular weight quaternary ammonium or tertiary amine salts, it is also possible to extract iron(III), tin(IV), and several other metals from aqueous

[3] J. C. White and W. J. Ross, U.S. At. Energy Comm., NAS-NS, 3102 (1961).

Table 20–2. Some Examples of Solvent Extraction Using Ion-Association Systems

Acid	Acid Concentration	Solvent	Elements Extracted
HCl	6–8M	Diisopropyl ether	As(III), Fe(III), Ga(III), Ge(IV), Sb(V)
HCl	6–8M	MIBK	Mo(VI), Sn(IV)
HNO$_3$	5–6M	Tributyl phosphate	Au(III), Th(IV), U(VI)
H$_2$SO$_4$	0.5M	Toluene + trioctylamine	U(VI)
HF[a]	5–6M	MIBK	Ta(V), Nb(V)

a. Another strong acid such as HCl or HNO$_3$ must also be present.

hydrochloric acid into solvents (such as toluene and aliphatic hydrocarbons) that normally would not extract metal ion association complexes at all. The term "liquid ion exchangers" has been given to these organic amines and quaternary ammonium compounds because their function is so similar to that of anion exchange resins, which also contain an amine group. Several reviews on solvent extraction using liquid ion exchangers have been published [4].

Table 20–2 lists some examples of metals that can be extracted as ion-association complexes. Extraction is complete in one or two extractions with fresh solvent; the metals extracted are separated from most other metal ions. MIBK is a more powerful extracting agent than an ether and will also quantitatively extract the metals extracted from hydrochloric acid by diisopropyl ether. In general, the metals extracted may be returned to aqueous solution by extracting the organic layer with water. An exception is uranium(VI), which is back-extracted with stronger sulfuric acid than was used for the original extraction.

QUESTIONS AND PROBLEMS

Principles

1. Explain how an organic base such as aniline, $C_6H_5NH_2$, can be separated from a neutral compound such as toluene, $C_6H_5CH_3$.
2. An ion-association complex such as $H_3O^+FeCl_4^-$ is well extracted by methyl isobutyl ketone but is not extracted at all by toluene. Explain what would happen if this complex were extracted by a *mixture* of methyl isobutyl ketone and toluene.
3. Gallium(III) forms an extractable ion-association complex with halides. It is reported that gallium(III) is 55% extracted into diethyl ether from 4M aqueous HBr. List as many ways as you can to obtain an improved (hopefully quantitative) extraction of gallium(III) from aqueous HBr into an organic solvent.
4. Liquid mercaptans (RSH) can extract silver(I) by forming silver mercaptides (RSAg) that dissolve in the organic liquid. Determine which of the following compounds, hexyl mercaptan ($C_6H_{13}SH$) or lauryl mercaptan ($C_{12}H_{25}SH$), is more effective in extracting silver(I) from aqueous solution. (Neither mercaptan is appreciably soluble in water.)

[4] H. Green, *Talanta, 11*, 1561 (1964).

5. Students were extracting the copper(II) from a dilute aqueous solution with a chelating reagent dissolved in methyl isobutyl ketone (MIBK). George reported that over 90% of the copper was extracted, but Julie reported that only about 10% of the copper was extracted. On questioning, it was found that George used a $1.0M$ solution of the reagent (in MIBK) while Julie used a $0.1M$ solution. Could this reasonably account for the large difference in their result? What was a probable ratio of the reagent to copper in the extracted complex?

Completeness of Extraction

6. Assuming a distribution ratio of 20, make calculations to show which is more effective: (a) extracting 10 mL of an aqueous solution with 20 mL of organic solvent or (b) extracting 10 mL of aqueous solution with 10 mL of organic solvent, followed by another extraction with 10 mL of fresh organic solvent.
7. Drinking water is often contaminated with a trace of chloroform. Repeated experiments show that when 100 mL of water is shaken with 1.0 mL of pentane, 53% of the chloroform is extracted into the pentane. Calculate the percentage of chloroform extracted when 10.0 mL of drinking water is shaken with 1.0 mL of pentane.
8. Under a particular set of conditions, an organic compound is 72.1% extracted from water into an equal volume of toluene. Calculate the value of D_m. Also calculate the number of extractions with fresh solvent needed to extract 99.8% of the organic compound.
9. Compound A has $D_m = 20.0$, while compound B has $D_m = 0.02$. Calculate the percentage of each that is extracted after two extractions with fresh solvent. Is the separation of A and B quantitative?
10. Benzoic acid, when extracted from water into a certain organic solvent, has $D_m = 40$ at pH 1.0 where essentially all of the benzoic acid is present as the molecular acid, HA. If the ionization constant in water is 1.58×10^{-4}, calculate the values of D_m at pH 2, 3, 4, 5 and 6. Assume that the anion, A^-, is not extracted at all.

Formula of a Complex

11. The uranyl ion (UO_2^{2+}) forms a complex that is extracted from an aqueous solution containing nitrate ions into a solution of organic ligand L in toluene. The extracted complex has the formula $UO_2L_x(NO_3)_y$. Outline experiments that would indicate the values of x and y.
12. The Co^{2+} ion is not extractable, but an intensely blue cobalt thiocyanate complex, $Co(SCN)_x^{-x+2}$, is well extracted. Explain how to experimentally determine the value of x in this complex.
13. The ion pair $H_3O^+FeCl_4^-$ is believed to be solvated by a definite number (x) of molecules of a liquid organic amide. Thus, the solvated complex is $H_3O^+FeCl_4^- \cdot x$ amide. Tell how to experimentally measure the value of x. Assume that the liquid amide is not miscible with water but is miscible with an inert organic solvent such as toluene.
14. A reagent, HL, reacts with copper(II) as follows:

$$Cu^{2+} + 2HL \rightleftharpoons CuL_2 + 2H^+$$

At pH 3.00, $D_c = [CuL_2]/[Cu^{2+}] = 2.17$. Calculate the value of D_c at pH = 3.50.
15. Human urine contains many inorganic and organic salts that are not extracted by an organic solvent. A few organic constituents are extracted. Explain how solvent extraction can help in sample cleanup when urine is to be analyzed for a drug metabolite that is extractable and happens to be a base.

16. Consult Tables 20–1 and 20–2 and devise a scheme for separation of each of the following samples using solvent extraction.

(a) Cu^{2+}, Hg^{2+}, Fe^{3+}
(b) Fe^{3+}, Mo(VI), U(VI)
(c) Separation of UO_2^{++} and Al^{3+} after dissolving aluminum-clad fuel rods in nitric acid

21

Principles of Chromatography

21–1. BASIC PRINCIPLES

Introduction. The separation methods we have encountered thus far—precipitation, electrodeposition, and solvent extraction—are "single-stage" separation processes. This means that there is really only a single equilibration of sample solutes between two phases. A major limitation of a single-stage process is that it must be "all or nothing" to achieve a quantitative separation. Thus, if two constituents are to be extracted, one must be extracted completely in two or three extractions with fresh solvent, while the other must not be extracted at all in the same number of extractions. Frequently, conditions cannot be found for this; in many instances, complete extraction of one sample constituent results in the partial extraction of other sample constituents.

Now we are going to study chromatography, which is a "multistage" separation process. The device used most often is a tube (called a *column*) containing a solid granular material; often a liquid phase is held tightly by physical or chemical forces. A second phase, called the eluent, flows continuously through the column. When a small amount of sample is added to the top of the column, the various sample chemicals divide themselves (partition) in varying degrees between the two phases. Since one phase (the eluent) is moving, the sample chemicals undergo many equilibrations between the two phases as the chemical moves down the column. An "all or nothing" situation is not required; only a slight difference in distribution ratios is needed for solutes to move at different rates through a column and thereby be separated from one another. The separation power of multistage methods is much greater than that of single-stage methods. This is perhaps the main reason why chromatography in its various forms is so widely used for analytical separations and analyses.

The term "chromatography" was coined by Tswett in 1906 from two Greek words meaning "color" and "to write." Tswett devised a method for separating chlorophylls and other plant pigments using a tube (a column) filled with a dry, solid adsorbent such as calcium carbonate. The first step in the method involved extracting the pigments from the plant material by using an organic solvent. When an extract of plant material was added to the column and the column was then washed (eluted) with an appropriate organic solvent, the constituents of the extract moved down the column at different rates and resolved themselves into colored rings (called bands). When the resolution of the bands was complete, the moist column material was pushed out of the tube as an intact cylinder and the bands were cut apart (fractionated) with a knife. An English translation of one of Tswett's original papers has been published [1].

Terms. The term *chromatography* refers to any separation method in which the components are distributed between a *stationary phase* and a moving (*mobile*) phase. The stationary phase is a porous solid used alone or coated with a liquid stationary phase. (In the latter case, it is called the solid support.) Separations occur because sample components have differing affinities for the stationary and mobile phases and therefore move at different rates along a column (or other adsorbent such as a piece of filter paper). The mobile phase is called the *eluent* or sometimes the carrier. The process by which the eluent causes a compound to move along the column is called *elution*. The separated substances need not be colored, although some suitable method for their detection or quantitative measurement must be used. In modern column-chromatographic practice, the column material is usually not extruded and the bands cut apart; instead, elution is continued until the sample constituents come off the column one at a time. The fractions are then collected and analyzed, or the constituents are measured by some kind of a sensor (detector) located in the exit tube from the column. In either case, a plot like the one shown in Figure 21–1 is plotted or recorded automatically. A curve of this type is called a chromatogram. Each peak in the plot represents a separate sample constituent. The area under each peak is a measure of the relative amount of that constituent.

The separation capability of modern chromatography is tremendous. In favorable cases, a complete separation of a sample containing 10 or 20 constituents can be accomplished in a few minutes. Organic compounds of closely related structure can be separated, as can inorganic substances having nearly identical chemical properties. Even nitrogen-14 and -15 isotopes in ammonia have been separated by ion-exchange chromatography. For these reasons, chromatography is indispensible in modern chemical analysis, and chromatographic methods are among the most widely used of all analytical procedures.

Types of Chromatographic Methods. Chromatographic separations are brought about by differences in the movement of the sample constituents resulting from differences in partitioning between two phases. Several combinations are possible and have been realized in actual practice. Listing the mobile phase first, and then the stationary phase, these are: liquid-liquid, liquid-solid, gas-liquid, and gas-solid. Gas

[1] M. Tswett, "Adsorption Analysis and Chromatographic Methods. Application to the Chemistry of Chlorophylls," *Ber. Deut. Botan. Ges., 24*, 384 (1906). English translation by H. H. Strain and J. Sherma, *J. Chem. Ed., 44*, 238 (1967).

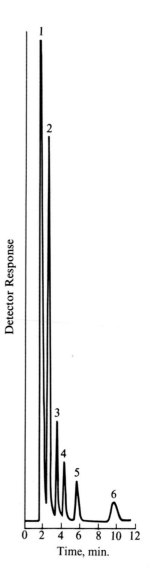

Figure 21–1. Chromatographic elution curve. The horizontal scale may also be plotted as volume of eluent.

chromatography (GC) includes gas-liquid (GLC) and gas-solid (GSC) systems and will be discussed in the next chapter. Liquid chromatography (LC) includes systems in which sample compounds partition between two liquid phases (LLC) or between a liquid and a solid phase (LSC) and will be discussed in Chapter 23. Separations in which the solid phase is an ion exchanger (as opposed to solids that retain a solute by sorptive forces) are categorized under ion-exchange chromatography; these will be covered in Chapter 24.

Instead of using a column, chromatographic separations can be carried out by using a sheet of some material that holds the stationary phase. This is called *paper chromatography* (PC) when a flat bed of filter paper is used and *thin-layer chroma-*

tography (TLC) when a sheet of plastic or glass, coated with a solid absorbent, is used. These techniques will be covered in Chapter 22.

Elution Chromatography. Virtually all analytical chromatography is actually "elution chromatography." In column elution chromatography, substances are separated by differences in their partition (distribution) between a stationary phase on the column packing and a mobile phase flowing through the column. In liquid-liquid partition chromatography, for example, the sample constituents may partition between an aqueous mobile phase and a stationary organic phase held in the pores of a sorptive solid resin. Or the phases may be reversed so that the stationary liquid is aqueous and the mobile phase is organic.

A scheme for a simple, classical chromatographic separation is given in Figure 21–2. A glass column is packed with a granular solid such as silica gel, which holds water as a stationary (nonmoving) phase. In the sorption step, the sample is dissolved in a suitable organic solvent and added to the column. The sample constituents A and B are adsorbed by the silica gel at the top of the column, forming a band of mixed solutes.

> The sample constituents A and B may be presumed to be organic compounds because partition chromatography is usually (but not always) concerned with separating mixtures of organic substances.

At this point, little, if any, separation of A and B will have occurred. Now the column is eluted with a suitable organic solvent or mixture of solvents. This moves A and B down the column. The rates at which they move depend on their distribution ratios, D_c or D_m. In partition chromatography, the distribution ratios are always defined so that the solute in the stationary phase is in the numerator:

$$D_m = \frac{\text{amount of solute in stationary phase}}{\text{amount of solute in mobile phase}}$$

$$D_c = \frac{\text{conc. of solute in stationary phase}}{\text{conc. of solute in mobile phase}}$$

Figure 21–2. Movement of bands during a chromatographic separation on a packed column.

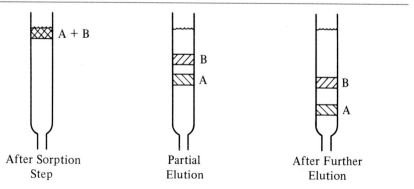

After Sorption Step   Partial Elution   After Further Elution

In these equations, D_m is the mass distribution ratio and D_c is the concentration distribution ratio (see Chapter 20). Although their nomenclature is perhaps illogical, chromatographers use the term "capacity ratio" or "capacity factor" instead of the mass distribution ratio, D_m. Capacity ratio is given the symbol k' or k and is defined exactly the same as D_m:

$$k = \frac{\text{amount of solute in stationary phase}}{\text{amount of solute in mobile phase}}$$

In keeping with actual practice, we will use capacity ratio in all discussions of chromatography and will denote it by the symbol k.

The smaller the capacity ratio of a solute, the faster the solute will move down the column. If the eluent is moving down the column at a linear flow rate of u(cm/sec), it can be shown that a sample chemical moves at the linear rate equal to $u/(1 + k)$ (cm/sec). Thus, if there is a reasonable difference in the k values for A and B, they will move at different rates and will be gradually resolved into separate, distinct bands. By collecting fractions of the column effluent, pure A and B can be collected as each comes off the column. In modern chromatography, a detector is placed just after the column to indicate automatically when A and B are eluted.

Displacement Chromatography.   In displacement chromatography, A and B are gradually resolved into separate bands, then move in tandem down the column at the same speed once equilibrium has been reached. This occurs when the eluting solvent is strongly taken up by the column. The solvent pushes B, and B pushes A ahead like a chemical piston. To maintain this situation, a rather large amount of each sample constituent is necessary. This technique is used for preparing pure substances. The profile curve, obtained by analysis of the fractions collected, is shown in Figure 21–3a.

The shift from A to B in the chromatogram can be seen because the detector will almost always have different sensitivities for the various sample chemicals. The

Figure 21–3.   Chromatographic elution curves: (a) displacement chromatography; (b) elution chromatography.

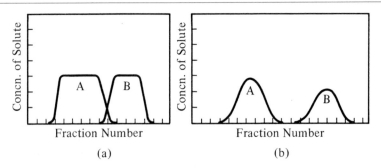

relative amounts of the sample components are indicated by their distances along the horizontal axis.

The obvious disadvantage of displacement chromatography is that there is a small area of mixed solutes at each interface between the pure solutes. However, displacement chromatography is a very efficient method for preparing relatively large amounts of pure compounds.

21–2. PLATE THEORY OF CHROMATOGRAPHY

Peak Retention Times and Peak Widths. The observed facts of any recorded chromatogram are these. The peaks are Gaussian in shape, and the widths of the peaks increase progressively as the retention time becomes larger. The retention time for any sample peak (abbreviated t_R or simply t) is the time (in minutes or seconds) from injection of the sample to the time the highest concentration of the sample (peak maximum) exits from the column and detector. The dead time (t_0) is the time (in minutes or seconds) for a nonsorbed substance ($k = 0$) to pass through the column and detector. The retention times for various sample peaks are designated as t_1, t_2, t_3, etc.

The retention time for any sample chemical is a function of both the column dead time and the capacity ratio of that chemical:

$$t = t_0(1 + k) \tag{21–1}$$

Later in the chapter we will show how this simple equation is derived. By experimentally measuring the values of t and t_0, we can calculate the value of the capacity ratio, k, with the aid of Equation 21–1.

Two parameters can be measured conveniently from recorded chromatograms. One is the retention time, and the other is the width of a peak at its base, w. The width, w, is obtained graphically by drawing tangents to the elution peak at the points of maximum slope and measuring w on the baseline, as shown in Figure 21–4; w is measured in the same time units as the retention time. It is convenient to convert peak width to standard deviation, σ, or to variance, σ^2. Values of σ are calculated from the fact that at the base of a peak, $w = 4\sigma$, or that at one-half peak height, $w_{1/2} = 2.35\sigma$. The latter is often easier to measure from recorded peaks.

Retention Volumes. Peak retention times and peak widths in time units can easily be converted to volume units provided that the volume flow rate of the eluent, f, is known. For a nonsorbed chemical ($k = 0$), the dead volume, V_0, is

$$V_0 \;=\; t_0 \;\times\; f \tag{21–2}$$
$$\text{(mL)} \quad \text{(min)} \quad \text{(mL/min)}$$

The *retention volume* for a sample peak (V_R or simply V) is obtained in a similar

Figure 21–4.   Elution curve with terms defined.

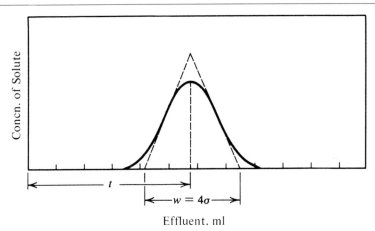

Effluent, ml

fashion from the eluent flow rate and the peak retention time:

$$V = t \times f \tag{21-3}$$

(mL)   (min)   (mL/min)

Plate Numbers.   In classical plate theory, the values of t and w are used to calculate a quantity, N, called the *theoretical plate number* or simply the *number of theoretical plates.*

$$N = 16\left(\frac{t}{w}\right)^2 = \frac{t^2}{\sigma^2} \tag{21-4}$$

(Some chromatographers use n in place of N.) Note that units of w or σ must match those in the numerator of the squared term. As mentioned above, it is easier to calculate σ from $w_{1/2}$. Thus, the right-hand term will be easier to use to calculate N.

A plate (or theoretical plate) is an imaginary segment of a column in which a sample chemical attains a partition equilibrium between the mobile and stationary phases (see Chapter 20). The number of theoretical plates in a column is commonly taken as a measure of its ability to separate various sample chemicals; separation power increases as N becomes larger.

N is supposed to be a constant for any given column operating under a fixed set of conditions (eluent, flow rate, etc.), but in practice this is not true. It is now well-documented that in any chromatogram, N becomes smaller as the retention time (or k) increases. N becomes *very large* as t (or k) approaches zero (see Figure 21–5).

> In some chromatograms, the value of N calculated from Equation 21–4 may be approximately constant for the various peaks, but this is probably because factors outside the column that also contribute to peak broadening have not been subtracted from the peak variance before calculating N. These additional peak-broadening effects will be discussed in Section 21–4.

Figure 21–5.   Plate numbers as a function of capacity ratio, k.

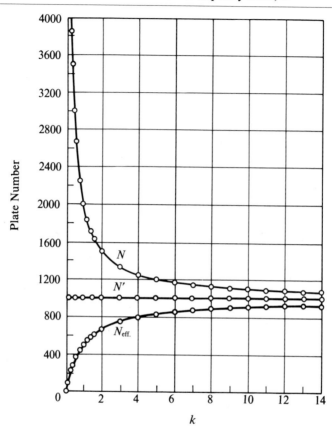

The actual separation of early-eluting peaks (those with a low t or k) is not nearly so good as would be predicted from the large N. This is especially true for very long columns (as used in gas chromatography) for which t_0 is large. It was largely because of this that a second plate number was introduced and is often used. This is N_{eff}, the number of *effective* theoretical plates in a column.

$$N_{eff} = 16\left(\frac{t - t_0}{w}\right)^2 = \frac{(t - t_0)^2}{\sigma^2} \tag{21–5}$$

(Some chromatographers use the symbol N to denote N_{eff}.)

If t_0 is rather large in comparison with t, the numerator in Equation 21–5 will be reduced significantly in comparison with the numerator in Equation 21–4, which does not include the t_0 term. In this case, N_{eff} will be much smaller than N. From Equation 21–5, we see that N_{eff} approaches 0 as t approaches t_0, which is in keeping with the poorer separation observed for peaks that elute very quickly. N_{eff} is not constant either but increases with increasing values of t or k, as shown in Figure 21–4.

A third plate number should also be considered. This number, which we shall designate by N', is defined as follows:

$$N' = \frac{t(t - t_0)}{\sigma^2} \tag{21-6}$$

The advantage of using N' (instead of N or N_{eff}) is that its value does not change as t increases; thus, N' is *constant* for all peaks in a chromatogram (see Figure 21–5).

> This is only true if N' is calculated by first subtracting other peak broadening effects from the experimentally measured value of σ^2 (see Section 21–4).
>
> A term equivalent to N' was proposed by Golay and is mentioned in a book by Ettre [2]. N' is also identical to an early but little-used term, $(N)(N_{eff})$, referred to by Nilsson [3]. More recently, Fritz and Scott [4] discussed the merits of using this plate number (which they called r, but which is identical to N').

Note from Figure 21–5 that all three plate numbers have about the same value when k is large. Because of its constant value, the plate number N' will be used in this book to designate the number of theoretical plates in a chromatographic column under given operating conditions. The well-known plate number N will be referred to as the *classical* number of theoretical plates. The relationship between N' and N is very simple:

$$N' = N\left(\frac{k}{1+k}\right); \qquad N = N'\left(\frac{1+k}{k}\right) \tag{21-7}$$

where k is the capacity ratio.

Chromatographers often use the term "plate height" as a measure of column efficiency. The classical plate height, H, is defined as

$$H = \frac{L}{N} \tag{21-8}$$

where L is the column length (cm) and N is the classical plate number. We will define plate height, H', as follows:

$$H' = \frac{L}{N'} \tag{21-9}$$

The smaller that H' or H is, the more efficient the column will be. Since N' is always smaller than N (Equation 21–7), it follows that H' will be somewhat larger than H. However, H' will be the same for all peaks in a chromatogram, while H should become larger as k increases.

It should be kept in mind that N', like N and N_{eff}, is strongly dependent on eluent flow rate. If the flow rate is doubled, the value of N' is cut in half. Cutting the flow rate in half should double the value of N', although the situation is complicated

[2] L. S. Ettre, *Open Tubular Columns in Gas Chromatography*, New York: Plenum Press, 1965, p. 19.
[3] O. Nilsson, *J. High Resolution Chromatogr.*, 5, 38, 143 (1982).
[4] J. S. Fritz and D. M. Scott, *J. Chromatogr.*, 271, 193 (1983).

by the increasing effect of diffusing peak broadening at slow flow rates (see Section 21–4). For our present purposes, it is sufficient to remember that a value of N' is valid only for a given, fixed flow rate.

21–3. MOVEMENT AND BROADENING OF PEAKS IN A COLUMN

In this section, elementary statistical methods will be used to account for the movement and broadening of chromatographic peaks. The behavior of a single molecule of a sample substance is calculated from statistical probability that is obtainable from its capacity ratio, k. Then the behavior of *all* molecules of the substance is summed with the aid of some simple statistical principles. The following information is of particular interest: (1) the rate of movement of a peak along a column and the time at which the peak leaves the column, (2) the width or variance of the peak, and (3) the resolution of two adjacent peaks, that is, how good the separation is.

According to this theory, the chromatographic column is considered to be divided into N' theoretical plates (abbreviated pl), each of equal length. The height of a theoretical plate is obtained by dividing the column length (in millimeters or centimeters) by the number of theoretical plates, N'. Eluent flows through the column at the rate of 1 pl per time period. In each time period, a given sample molecule in any plate makes a "decision" to stick in the stationary phase or to remain free and move on to the next plate. During any given time period, the probability of a molecule in the mobile phase sticking is $k/(1 + k)$. The probability of remaining free and moving to the next plate is $1/(1 + k)$. Similarly, the probability of a molecule's remaining stuck in any time period is $k/(1 + k)$, and the probability of breaking free and returning to the mobile phase is $1/(1 + k)$.

This leads to the relationship that a sample molecule moves down the column at a rate $1/(1 + k)$ with respect to that of the eluent. If the linear flow rate of the eluent is v plates/min, after t min the eluent will have moved vt plates and a sample chemical with capacity ratio, k, will have moved $vt/(1 + k)$ plates.

Not all molecules of a given sample chemical behave in exactly the same manner. The summation of the behavior of all molecules is a *statistical distribution*. After alternately sticking and breaking free a large number of times (which is the usual case in chromatography), the molecules assume approximately a normal, or Gaussian, distribution. Actual chromatograms show separated chemicals as Gaussian peaks.

In Chapter 3, the standard deviation, σ, or the variance, σ^2, was shown to be a measure of the width of a Gaussian, or normal, distribution. The broadening of a peak moving down a column can be predicted from simple statistical concepts applied to chromatographic plate theory. For a peak whose center has moved a distance of $vt/(1 + k)$ plates after t min (see above), the peak variance (in plates squared) is

$$\sigma^2 = \frac{vt}{1+k} \cdot \frac{k}{1+k} = \frac{vtk}{(1+k)^2} \qquad (21\text{--}10)$$

The objective of chromatography is, of course, to separate various sample substances from one another. Two compounds with different capacity ratios (k values) move at different rates down a column, and the peaks also broaden (spread out) as they move. A separation is achieved when the faster-moving peak has gotten sufficiently ahead of the lower peak that there is no overlap of the peaks, even at their base. This fundamental chromatographic process can be illustrated by a numerical example, using simple equations from statistics to calculate the movement and widths of two peaks.

Example: A column has 1500 theoretical plates; that is, $N' = 1500$. The eluent requires 1.83 min to traverse the entire column; that is, $t_0 = 1.83$ min. The linear flow rate, $v = 1500/1.83 = 820$ pl/min. Two sample compounds have the capacity ratios $k_1 = 2.0$ and $k_2 = 2.4$, respectively. Calculate the distance moved and the peak width of each compound after 1.0, 2.0, and 3.0 min. At which of these times is the separation of the two compounds complete (no band overlap at the 4σ level)?

$$\text{Distance (in plates)} = \frac{vt}{1+k}$$

$$\text{Variance (in plates}^2) = \frac{vtk}{(1+k)^2}$$

After 1.0 min	Cpd 1	Cpd 2	ΔPeak max
Peak max has moved	273 pl	241 pl	32 pl
Peak variance (σ^2)	182 pl^2	170 pl^2	
Peak width (4σ)	54 pl	52 pl	

After 2.0 min			
Peak max has moved	547 pl	482 pl	65 pl
σ^2	364 pl^2	340 pl^2	
4σ	76 pl	74 pl	

After 3.0 min			
Peak max has moved	820 pl	723 pl	96 pl
σ^2	547 pl^2	511 pl^2	
4σ	94 pl	90 pl	

Note that the peaks overlap after 1.0 and 2.0 min because Δpeak max is less than the average 4σ. However, the peaks are completely separated at 3.0 min.

Required Column Length. A calculation such as the one in the example gives a reasonable estimate of the number of plates that *must* be traversed for a separation of two substances to occur. The column length needed to provide the requisite number of theoretical plates depends on the column efficiency under the conditions used, as expressed by the height of a plate, H', given in Equation 21–9.

A very good packed column might have $H' = 0.002$ cm. If 800 pl is taken as an estimate of the needed theoretical plates, we calculate L from Equation 21–8 to be:

$$L = 800(\text{pl}) \times 0.002(\text{cm/pl}) = 1.6 \text{ cm}$$

Thus, with a column of high efficiency, a very short column indeed is sufficient for the separation in question. A column of the type used just a few years ago with $H' = 0.10$ cm would require $L = 800 \times 0.10$, or 80 cm. So the column length needed for a particular chromatographic separation will depend very much on the efficiency of the column.

21–4. FACTORS CAUSING PEAK BROADENING

In the previous section, we used plate theory to calculate the movement and broadening of sample compound zones (or peaks) *while they were still on the column*. One difficulty with this approach is the lack of experimental verification. A separated chemical peak can actually be measured only after it has left the column and passed through the detector. The measurements can be made from the peaks on the recorded chromatogram, which has time units (or sometimes volume units) on the horizontal axis. The retention time (t) and width of each peak (4σ) are measured in seconds or minutes. The variance (σ^2) can be easily calculated and has units of $\sec^2$ or $\min^2$.

While it is useful to estimate the number of plates needed for a separation, the division of a column into plates of equal length is artificial and, to some extent, inaccurate. Actually, several physical phenomena cause sample peaks to broaden as they move along the column and through the detector. The variance (σ^2) of any peak on the recorded chromatogram is the sum of the variances of these peak-broadening effects. Each of these effects will now be discussed briefly.

Eddy Diffusion (Variance Notation: σ_{ed}^2). There are many paths a molecule can take through a packed chromatographic column. Not only are these paths somewhat different in length, but the actual flow rate may change during different portions of a path. Swirling effects called "eddy diffusion" can also occur. The net result is a larger distribution of distances the molecules of a sample chemical have traveled and thus a broader peak. These effects are minimized by using small, spherical particles of uniform diameter as the column-packing material and packing the column carefully and evenly. The variance due to eddy diffusion in modern chromatographic columns is usually quite small and often negligible.

Axial Diffusion (Variance Notation: σ_d^2). When a sample chemical is in the mobile phase, the molecules tend to diffuse both in a forward and a backward direction. This causes a sample zone to broaden as it moves down the column. The amount of broadening depends on (1) the length of time the molecules spend free in the mobile phase and (2) the *rate* at which they diffuse, which is given by the diffusion coefficient, D. When the eluent flow rate is slow, there is more time for diffusion to occur and the zone broadening due to diffusion is greater. Smaller sample molecules tend to diffuse at a faster rate than the larger ones, although the diffusion coefficients of various sample components are often about the same. Peak broadening caused by axial diffusion is often negligibly small in liquid chromatography but is a major source of peak broadening in gas chromatography, in which diffusion coefficients typically are much larger. The variance of broadening due to axial diffusion is proportional to the diffusion coefficient, D, and is inversely proportional to the linear flow rate, u.

Resistance to Mass Transfer (Variance Notation: σ_{mt}^2).   This is the largest contributor to peak broadening. It stems from the fact that the various molecules of each sample chemical spend different periods of time in the stationary phase *each time they stick* (in the stationary phase) and different periods of time in the mobile phase each time they come back into that phase from the stationary phase. In packed columns, σ_{mt}^2 is inversely proportional to the particle size of the solid support. σ_{mt}^2 is smaller for sample compounds that have a high diffusion coefficient and thus move more quickly from somewhere in the mobile phase to an interface with the stationary phase. Finally, σ_{mt}^2 is a function of the eluent flow rate. At faster flow rates, the magnitude of σ_{mt}^2 (in sec^2 or min^2) is smaller but the efficiency of the column in achieving separations turns out to be lower.

The resistance to mass transfer term is also affected by the type and thickness of the stationary phase coating. If the coating is relatively thick, some molecules of a sample compound diffuse farther into the stationary phase than others, thus causing additional broadening of the peak. In capillary gas chromatographic columns with *thin films* (of stationary phase), this type of broadening is almost negligible.

Extra-Column Broadening (Variance Notation: σ_{ec}^2).   Ideally, a point injection of a sample should be made so that the spreading of the sample along the column is negligible. Usually, this is impossible, and the best that can be done is to inject the sample as a small "plug." As the volume of sample increases, this plug will become broader, as shown in the following diagram:

Small Sample   Larger Sample

Because of multipath and diffusion effects, the shape of the "plug" of even a nonsorbed substance becomes approximately Gaussian in shape as it moves down the column. Thus, sample injection will also contribute to the broadening of chromatographic peaks, especially if the sample volume is not extremely small.

Sample compounds can also spread out while passing through various connecting lines and the detector. When a liquid flows in an open tube, the walls produce a "drag" that impedes the flow of liquid in the vicinity of the wall. The result is a parabolic flow pattern (see the diagram) that broadens a chemical zone flowing through the tube.

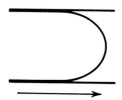

Small "dead volumes" in fittings and elsewhere in the system also cause mixing. This can be particularly bad if the flowing substances pass from a tube of small diameter to one of larger diameter. Extra-column peak broadening in liquid chromatography is apt to be substantial, even when commercial instruments of good quality are used. Better design of the various fittings and components that go into a liquid chromatograph should make it possible to reduce (but not eliminate entirely) the instrumental peak broadening.

Total Measured Broadening (Variance Notation: $\sigma_m{}^2$). To summarize, the measured peak variance is the sum of the variances just discussed:

$$\sigma_m{}^2 = \sigma_{ed}^2 + \sigma_d{}^2 + \sigma_{mt}^2 + \sigma_{ec}^2 \qquad (21\text{--}11)$$

Note that the *variances* are additive but *the peak widths are not* (4σ).

Some of the broadening variances are strongly affected by parameters such as capacity ratio (k) and dead time (t_0), while others are not. The largest variance is *almost always* σ_{mt}^2. According to the plate model, N', σ_{mt}^2 is a function of $t_0{}^2$, k, and $(1 + k)$:

$$\sigma_{mt}^2 = f[(t_0)^2(k)(1 + k)] \qquad (21\text{--}12)$$

According to the classical plate model,

$$\sigma_{mt}^2 = f[(t_0)^2(1 + k)^2] \qquad (21\text{--}13)$$

($t_0 = L/u$, where L is the column length and u the linear flow rate).

The axial diffusion variance $(\sigma_d{}^2)$ is also a function of $(1 + k)^2$ and t_0, as well as the diffusion coefficient (D) and the linear eluent flow rate (u):

$$\sigma_d{}^2 = f\left[\frac{2Dt_0(1 + k)^2}{u^2}\right] \qquad (21\text{--}14)$$

The magnitude of $\sigma_d{}^2$ may still be fairly small (if D is small and u is large), but it does increase with $(1 + k)^2$.

However, σ_{ed}^2 may be quite small, and it does not seem to increase with increasing k. σ_{ec}^2 is larger, but it tends to be almost the same for all peaks. The mathematics of adding peak variances is such that σ_{ec}^2 often has a very appreciable effect on the actual width (4σ) of early peaks but an almost negligible effect on the width of late peaks.

Example: A column operated under fixed conditions gives a variance for resistance to mass transfer (σ_{mt}^2) of 2.91 sec^2 for $k = 1.00$ and 39.20 sec^2 for $k = 4.00$. The value of σ_{ec}^2 is 1.45 sec^2 for both peaks. Calculate the total peak width (4σ) for each peak. Also calculate and compare the peak widths if there were not extra-column broadening.

For $k = 1.00$, total width $= 4(2.91 + 1.45)^{1/2} = 8.35$ sec. Neglecting extra-column broadening, width $= 4(2.91)^{1/2} = 6.8$ sec.

For $k = 4.00$, total width $= 4(39.20 + 1.45)^{1/2} = 25.5$ sec. Neglecting extra-column broadening, width $= 4(39.20)^{1/2} = 25.0$ sec.

Measurement of Theoretical Plates.   Under conditions used in practical chromatography, it often happens that both σ_{ed}^2 and $\sigma_d{}^2$ are small and σ_{ec}^2 affects only the early- to middle-eluting peaks. Thus, the widths of later eluting peaks result mainly from resistance to mass transfer (σ_{mt}^2). We can therefore estimate the number of theoretical plates in a column from σ_{mt}^2 of later-eluting peaks. Combining Equation 21–6 and Equation 21–1, we get

$$N' = \frac{(t_0)^2(k)(1 + k)}{\sigma_{mt}^2} \tag{21–15}$$

For the classical plate number, we get

$$N = \frac{(t_0)^2(1 + k)^2}{\sigma_{mt}^2} \tag{21–16}$$

As was stated earlier, N changes somewhat with k, but N' does not. Rearrangement of Equation 21–17 gives

$$\sigma_{mt}^2 = \frac{(t_0)^2(k)(1 + k)}{N'} \tag{21–17}$$

A plot of σ_{mt}^2 versus $(k)(1 + k)$ gives a straight line with slope $= (t_0)^2/N'$. For chromatographic conditions in which both σ_{ed}^2 and $\sigma_d{}^2$ are quite small, the intercept of this plot (at $k = 0$) is a good estimate of σ_{ec}^2.

This method of plotting data is illustrated in Figure 21–6, which is a chromatogram of some agricultural chemicals obtained with a commercial liquid

Figure 21–6.

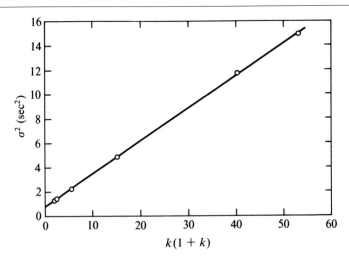

chromatograph using a 10-cm, high-performance column. The large value for N' confirms the excellent separating ability of the column, but the extra-column broadening is quite large. Much of this can be attributed to the use of a small "guard column" filled with a coarser packing than that used in the chromatographic column. Subsequent replacement of the guard column with one containing smaller (10-μm) particles reduced the extra-column variance by almost 50%.

21–5. RESOLUTION

The goal of chromatography is, of course, to obtain separations. But we need some way to describe how good a separation is. In particular, we would like to describe the resolution (the degree of separation) of two nearby peaks in a quantitative fashion. To do this, we define the resolution (R_s) of two peaks as the difference in their retention times, divided by their average peak width, $4\bar{\sigma}$:

$$R_s = \frac{t_2 - t_1}{4\bar{\sigma}} = \frac{\Delta t}{4\bar{\sigma}} \qquad (21\text{--}18)$$

A resolution of 1.0 means that the centers of two peaks (t_2 and t_1) are separated by $4\bar{\sigma}$, where $\bar{\sigma}$ is the average standard deviation of the peaks. At $R_s = 1.0$, there is still some overlap of the peaks near their base. Since chromatographic peaks are Gaussian, the Z values for a standard normal distribution curve in Table 3–1 enables us to calculate that 2.27% of the area of peak 1 is greater than ($t_1 + 2\sigma$) and therefore is mixed with peak 2. Likewise, 2.27% of peak 2 is outside the limit ($t_2 - 2\sigma$) and is mixed with peak 1.

If we move the two peaks farther apart so that the peak centers, t_2 and t_1, are separated by 5σ, the resolution is now 1.25, and the degree of peak overlap is less. From the standard normal distribution, we calculate that 0.62% of peak 1 lies beyond ($t_1 + 2.5\sigma$) and is mixed with peak 2; likewise, 0.62% of peak 2 is mixed with peak 1. When the peak centers are separated by 6σ, the resolution is 1.5 and each peak contains 0.13% of the other compound.

The difficulty of achieving good resolution of any two compounds is often expressed by the *separation factor*, α. By definition,

$$\alpha = \frac{t_2'}{t_1'} = \frac{k_2}{k_1} \qquad (21\text{--}19)$$

where the subscript 2 refers to the later-eluting peak and 1 to the earlier-eluting peak of the pair. (Recall that $t' = t - t_0$.) Thus, α always has a value greater than one.

The number of theoretical plates necessary to resolve any two peaks can be estimated for any given value of α. The following equation can be derived by combining Equations 21–6 and 21–19:

$$R_s = \frac{\alpha - 1}{\alpha + 1} \cdot \frac{\sqrt{N'}}{2} \cdot \left(\frac{\bar{k}}{1 + \bar{k}} \right)^{1/2} \qquad (21\text{--}20)$$

This equation ignores σ_{ec}^2, which can have an appreciable effect on peaks that elute quickly. A similar equation for resolution in terms of α, $\bar{k}$, and N_{eff} can be similarly derived:

$$R_s = \frac{\alpha - 1}{\alpha + 1} \cdot \frac{\sqrt{N_{eff}}}{2} \qquad (21-21)$$

Chromatography textbooks usually give the following equation for resolution:

$$R_s = \frac{\alpha - 1}{\alpha} \cdot \frac{\sqrt{N}}{4} \cdot \frac{k}{1 + k_2} \qquad (21-22)$$

which is equivalent to

$$R_s = \frac{\alpha - 1}{\alpha} \cdot \frac{\sqrt{N_{eff}}}{4} \qquad (21-23)$$

(k_2 is the capacity ratio of the later-eluting peak.) However, Equations 21–22 and 21–23 have been shown to be in error unless the widths of both peaks are exactly the same [5].

Equations 21–19 and 21–20 show that the resolution of any two peaks depends on three factors: (1) the column plate number, (2) the relative difference in their capacity factors, and (3) the average value of their capacity factors.

Example: Calculate the expected resolution of sample chemicals having $k_1 = 5.35$ and $k_2 = 5.66$ on a column in which $N' = 9860$ pl under the conditions used.
Calculating $\alpha = 5.66/5.35 = 1.058$ and $\bar{k} = 5.505$, and substituting into Equation 21–20,

$$R_s = \frac{0.058}{2.058} \cdot \frac{\sqrt{9860}}{2} \cdot \left(\frac{5.505}{6.505}\right)^{1/2} = 1.29$$

Example: A 30-m gas chromatographic column is found to have $N' = 84{,}000$ pl when operated at a certain flow rate. Calculate the minimum separation factor, α, that will give a resolution of 1.00 at that flow rate when $\bar{k} = 4.50$ and also when $\bar{k} = 0.45$. Substituting into Equation 21–20,

$$1.00 = \frac{\alpha - 1}{\alpha + 1} \cdot \frac{\sqrt{84{,}000}}{2} \cdot \left(\frac{4.50}{5.50}\right)^{1/2}$$

$$\frac{\alpha - 1}{\alpha + 1} = 0.00763; \qquad \alpha = 1.015_4 \quad (\text{for } \bar{k} = 4.5)$$

$$1.00 = \frac{\alpha - 1}{\alpha + 1} \cdot \frac{\sqrt{84{,}000}}{2} \cdot \left(\frac{0.45}{1.45}\right)^{1/2}$$

$$\frac{\alpha - 1}{\alpha + 1} = 0.01239; \qquad \alpha = 1.025 \quad (\text{for } \bar{k} = 0.45)$$

The column in the last example can resolve compounds that have a very small difference in their capacity factors. For two compounds having a given α, the

[5] A. S. Said, *J. High Resol. Chrom. & Chromatogr. Commun.*, 2, 193 (1979).

resolution is appreciably better if k is not too small. Parameters such as column temperature in gas chromatography and eluent composition in liquid chromatography can usually be adjusted to increase the capacity factors without affecting α.

Although Equations 21–20 and 21–21 are useful in relating resolution to the separation factor and the number of theoretical segments in the column, these equations do not include the effects of extra-column peak broadening. Because the variances contributing to peak broadening are additive (see Equation 21–15), it turns out that any extra-column variance can have a sizable effect on the widths of early-eluting peaks. The effect on later peaks is less and often is negligible. For example, extra-column broadening might seriously affect the resolution of two peaks when $\bar{k} = 0.5$ but have almost no effect when $\bar{k} = 5.0$.

21–6. PROGRAMMED CHROMATOGRAPHY

It often happens that some chemicals in a sample take a long time to pass through a column, while others may never make it. Compounds with a high k value move slowly and appear as broad, flattened-out peaks on the recorded chromatogram. If operating conditions are changed to decrease the k value of these peaks, the k values of early-eluting substances are reduced proportionately. This often means that the earlier-eluting substances are not separated from one another.

Programmed elution is often a good way out of this dilemma. In gas chromatography, the oven temperature is raised at a fixed rate throughout the chromatographic run. In liquid chromatography, the composition of the eluent is changed continuously. These changes cause the k values of *all* sample chemicals to decrease throughout the run. Substances with initially high k values move very slowly at first; but as the program continues, the k values decrease and their rate of movement becomes faster. Sample peaks tend to be moving relatively fast as they leave the column. This causes the peaks to be narrower and higher than they would be in an unprogrammed elution.

Sometimes a program is such that all peaks have about the same k as they leave the column. Since the rate at which a substance moves through the detector is proportional to $1/(1 + k)$, this means that all of the peaks have about the same width. The narrower peaks may give the impression that much better resolution is obtained in programmed chromatography than in ordinary chromatography. However, this is *not* the case because resolution really occurs while the sample chemicals are on the column. The faster rate of movement through the detector in programmed chromatography does not affect resolution, but it does improve the sensitivity detection because the exit peaks are more concentrated.

Programming will be discussed further in Chapter 22 on gas chromatography and in Chapter 23 on liquid chromatography.

Conclusions. Our ability to separate various substances by elution chromatography depends on the number of theoretical plates in the column used, on the difference in capacity ratios of the substances, and on the actual capacity ratios of the substances. The number of theoretical plates in the column can be increased by operating at a slower flow rate, although this advantage can be offset by greater peak broadening from diffusion if the flow rate becomes too slow. Also, separations can take a long time if the flow rate is too slow. Often nothing can be done to improve the difference in capacity ratios other than selecting an eluent that is appropriate for the compounds to be separated. However, elution conditions can be changed to increase or decrease the capacity ratios while maintaining about the same relative difference between any two of them. This can be advantageous because resolution is proportional to the quantity $[k/(1 + k)]^{1/2}$. If k is too low, the resolution will be poor because the substances move through the column without sticking very often. Extra-column peak broadening reduces the resolution of substances of low k much more than those having a higher capacity factor. When capacity factors are high (say above 15), resolution is good but the separations are slow and the peaks become very broad.

Programmed chromatography is a technique in which the capacity factors of all sample compounds are decreased throughout the chromatographic run. This decreases the time needed for a separation and sharpens up the late-eluting peaks, thereby improving the sensitivity of detection. Although the peaks are higher and more narrow, the resolution of any two peaks in programmed chromatography should be no better than when elution is carried out under fixed conditions.

QUESTIONS AND PROBLEMS

Definitions

1. Define each of the following terms: (a) stationary phase, (b) mobile phase, (c) eluent, (d) solid support, (e) multistage separation, (f) concentration distribution ratio, (g) mass distribution ratio, (h) adjusted retention volume, (i) retention time, (j) separation factor.
2. Distinguish between displacement and elution chromatography. Which is preferred for quantitative analytical separations?
3. Define each of the following terms: (a) holdup volume, (b) retention volume, (c) adjusted retention volume, (d) retention time, (e) adjusted retention time.
4. Scientists often refer to different types of chromatography by two or three letters. Give the name of each type of chromatography abbreviated.

   (a) LC
   (b) LLC
   (c) GC
   (d) TLC

5. Define each of the following plate numbers: (a) plate number, N', in terms of t and σ; (b) classical plate number, N, in terms of t and σ; (c) effective plate number, N_{eff}, in terms of t, t_0, and σ; (d) plate number, N', in terms of classical plate number, N.
6. Distinguish between plate height, H', and classical plate height, H. What advantage does H' have over H?

7. Tell how to experimentally determine the capacity ratio, k, for a sample peak.

8. Write an equation for retention time (t) in terms of t_0 and k. Also, define the theoretical plate number, N', in terms of t_0, k, and σ.

9. Briefly explain each of the following contributions to chromatographic peak broadening.

   (a) Interphase
   (b) Extra-column
   (c) Axial diffusion
   (d) Eddy diffusion

Movement and Broadening of Peaks

10. Briefly explain why chromatographic peaks resemble a normal distribution curve.

11. A column has $N' = 45,000$ when $t_0 = 30$ sec.

    (a) Calculate the eluent flow rate in plates/sec.
    (b) Calculate the distance (in plates) a peak of $k = 3.0$ has moved along the column after 10 sec and also after 20 sec.
    (c) Calculate the peak width, 4σ (in plates), of this same peak after 10 sec and after 20 sec.
    (d) Calculate the distance (in plates) a peak of $k = 3.3$ has moved after 10 sec. Also calculate the peak width, 4σ (in plates), after 10 sec. Is separation of the two peaks ($k = 3.0$ and $k = 3.3$) complete after 10 sec?

12. A column has $N' = 22,500$ when $t_0 = 30$ sec.

    (a) Calculate the eluent flow rate in plates/sec.
    (b) Calculate the distances moved (in plates) of peaks with $k = 3.0$ and $k = 3.3$ after 10 sec.
    (c) Calculate the peak widths, 4σ (in plates), of the two peaks in part (b). Is separation of the two peaks complete? How does this compare with part (d) of the preceding problem?

13. For a chromatographic column when $t_0 = 30$ sec, $t_1 = 120$ sec, and $t_2 = 124$ sec:

    (a) Calculate k_1 and k_2.
    (b) Calculate the width of each peak (4σ, in sec) when $N' = 2000$, $N' = 10,000$, $N' = 20,000$, $N' = 30,000$, $N' = 40,000$. (Assume no extra-column broadening.)
    (c) From the results in part (b), decide approximately how many theoretical plates are necessary for a good separation.

14. The following data were obtained on a 25.2-cm liquid chromatographic column at a flow rate such that $t_0 = 44$ sec.

Peak No.	t(sec)	k	σ^2(sec^2)
1	97	1.20	3.67
2	110	1.50	4.56
3	200	3.55	14.72
4	332	6.54	41.54
5	337	6.66	43.30

    (a) Use linear regression to measure N' and σ^2_{ec}.
    (b) Calculate the plate height, H'.

15. A column has $N' = 3050$ when operated at a flow rate such that $t_0 = 39.0$ sec. However, at this flow rate, the extra-column variance $(\sigma_{ec}^2) = 1.03$ sec^2.

   (a) Calculate the variance and width of each sample peak: $t_1 = 84$ sec, $t_2 = 131$ sec, $t_3 = 137$ sec, $t_4 = 315$ sec.
   (b) Calculate the width of each peak if there was no extra-column variance. Which peak widths are most affected by extra-column broadening?

16. The column in the previous problem still has $N' = 3050$ and $t_0 = 39.0$ sec, but the extra-column variance has been reduced to 0.31 sec^2 by better column fittings.

   (a) Calculate the peak variance and peak width of each sample peak: $t_1 = 84$ sec, $t_2 = 131$ sec, $t_3 = 137$ sec, $t_4 = 315$ sec.
   (b) Calculate the classical plate number for each peak (including σ_{ec}^2). $N = t^2/\sigma^2$.

Resolution

17. Calculate the resolution of two peaks, $t_1 = 131$ sec and $t_2 = 137$ sec, if the average peak width is 8.0 sec. Would the resolution of these two peaks be considered good?

18. Under given conditions, a column has $N' = 42,000$ theoretical plates. Calculate the expected resolution of two peaks, $k_1 = 5.35$ and $k_2 = 5.51$.

19. Two peaks have capacity ratios $k_1 = 8.15$ and $k_2 = 8.23$. Calculate the separation factor, α. Also calculate the number of theoretical plates (N') that will be required to give a resolution of 1.0 for these peaks.

20. Assuming an average capacity ratio $(\bar{k})$ of 5.00, calculate the necessary theoretical plate number (N') to obtain a resolution of 1.0 for each of the following values of α: (a) 1.005, (b) 1.01, (c) 1.02, (d) 1.05, (e) 1.10, (f) 1.20, (g) 1.40, (h) 1.80. Make a graph of N' versus α.

21. It is usually possible to vary the eluent composition in liquid chromatography or the temperature in gas chromatography to change capacity ratios drastically without changing the separation factor (α) of two peaks. Compare the resolution that will be obtained for $\alpha = 1.20$, $N' = 800$ when:

   (a) $\bar{k} = 0.80$ (b) $\bar{k} = 8.00$

22. Chromatograms were actually run to test the predictions made in the previous problem. The resolution obtained for two peaks where $\bar{k} = 8.0$ was about as predicted, but the resolution obtained for $\bar{k} = 0.80$ was much poorer than predicted. Explain the probable reason for this.

Programming

23. What is the purpose of using programmed chromatography? Tell how this is accomplished (a) in gas chromatography and (b) in liquid chromatography.

24. Assuming that the separation factor (α) remains the same, explain whether the resolution of two peaks will remain the same or will be better in part (a) or part (b):

   (a) Programmed run in which the *average* $\bar{k} = 1.5$
   (b) Programmed run in which the *average* $\bar{k} = 4.2$

   (Remember that all k values change throughout a programmed run; we are referring to the average for the run.)

25. A program is run such that the initial k for a peak is 15.0 and the final k is 3.0. How many times faster is the chemical moving through the column at the end of the run compared to the beginning?

22

Gas Chromatography

22-1. INTRODUCTION AND OVERVIEW

Gas chromatography (GC) is a type of chromatography in which the mobile phase is a gas, such as nitrogen or helium, and the stationary phase is either an inert liquid or a solid. The sample is usually a liquid at room temperature, but it is flash vaporized as it is injected into the instrument. The column is held at a high enough temperature to keep the sample components in the gaseous state during the entire separation. The separated sample chemicals are detected as they leave the column and pass through a detector cell. The electrical output is connected to a recorder, which records the entire chromatogram.

The resistance to gas flow is much less than that for a liquid, so the columns used in gas chromatography are much longer than can be used in liquid chromatography. This is especially true for the newer capillary columns used in GC, which contain no packing and are typically 30 m or longer. These columns are also very efficient, which means that samples containing a large number of chemical compounds can often be separated in minutes and the amount of each compound estimated from peak height. The impact of gas chromatography on chemical analysis has been tremendous, especially in analyzing organic samples. Sometimes gas chromatography is the only viable way to analyze complicated samples.

Components of a Gas Chromatograph. Instrumentation and columns used in GC have been changing and improving rapidly. However, all gas chromatographs have several things in common, and these are shown in the block diagram in Figure 22–1.

The *carrier gas* chosen depends on the detector to be employed; nitrogen is used with the most popular detector, the flame-ionization detector. The carrier gas is supplied at a reduced pressure from a large gas cylinder equipped with a pressure regulator. Often the carrier gas is filtered through a tube containing molecular sieves to remove impurities from the gas.

Figure 22–1. Block diagram of a gas chromatograph.

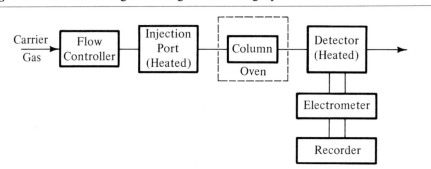

The *flow controller* is a needle valve or other device used to control the gas flow rate. In some instruments, a rotameter is used to measure the actual flow rate. A flow rate of around 75 mL/min is most often used for 1/4″ (6.4 mm) o.d. (outside diameter) columns; a flow rate of approximately 25 mL/min is used for 1/8″ (3.2 mm) o.d. columns.

For capillary columns, a much slower flow rate is used, although the *linear* flow rate is typically 30–60 cm/sec. As with packed columns, the flow rate is controlled by adjusting the carrier gas pressure.

A sample is introduced into the gas chromatograph through the *injection port*, a small heated chamber capped with a septum. The septum is pierced with a syringe needle and reseals when the needle is withdrawn. Liquid samples are injected by means of a small, calibrated syringe. Typically, 1–3 μL samples are used with packed columns, but smaller samples must be used with capillary columns. The injection port temperature should be high enough to vaporize the sample instantly. Sample injection should be rapid so that the vaporized sample is swept into the column as a discrete "plug." Gaseous samples can also be injected by using a gastight syringe or a gas-sampling valve attached to the chromatograph.

"On-column" injection is sometimes used for liquid samples. In such systems, there is no heated injection port. The liquid sample is simply deposited at the front end of the capillary column by means of a long silica capillary attached to a syringe.

The *column* where the actual separation occurs is enclosed in an oven that maintains the desired temperature. A conventional "packed" column is filled with a granular solid support coated with a thin layer of stationary phase; separation occurs by differences in the distribution of the various sample components between the carrier gas and the stationary phase. Stainless steel, 1/4″ or 1/8″ in o.d., is the most common construction material for a packed column. A column made of glass is used for samples containing pesticides or other materials that might react or be irreversibly adsorbed by a metal column. A column is normally conditioned before its first use by passing carrier gas at an elevated temperature through it for several hours to remove any volatile impurities.

Capillary columns are now widely used for GC separations. Typically, a capillary column is a long glass or silica tube only about 0.25 mm in i.d. (inside diameter) rolled into a coil. Although the coil is usually only about 11 cm in diameter, the tube itself may be 25–100 m in length. The capillary contains no packing, but the inside walls are coated with a thin layer of a stationary phase. Separation occurs by differences in the partitioning of sample components between this thin layer and the carrier gas. Capillary columns are quite efficient and, because of their great length, are capable of separating very complex samples.

The function of the *detector* is to sense when a compound is leaving the column and to provide a signal that is proportional to the concentration of the compound in the carrier gas stream. Several detectors are available (see Section 22–3), the choice of which depends on the sensitivity needed and on whether all compounds, or only a selected group (such as halogen compounds), are being sought. The detector is heated to the temperature needed to keep the sample compounds from condensing.

The FID, or flame-ionization detector, is the most commonly used detector; it will be described later in this chapter. The output of this detector, a very small electric current, is fed into an *electrometer*, which amplifies and converts the detector output to a voltage that is large enough to be recorded. The *recorder* records this voltage as a plot of voltage versus time, showing the separate sample components as chromatographic peaks. The recorder should have a fast pen speed and preferably a variable chart speed.

22–2. GC COLUMNS

Packed Columns.   A packed column is filled with a porous, granular support that is coated with a uniform thin film of the stationary phase. Packed columns are usually made of 1/4″ or 1/8″ o.d. stainless steel and are typically 4–6 ft (1.2–1.8 m) in length. The columns are usually coiled to fit conveniently into the oven of the gas chromatograph. Coiled glass columns are more inert than stainless steel but must be handled with care to prevent breakage. Columns must be carefully packed to obtain efficient separations. The packing is usually held by a small plug of glass wool at each end of the column.

Typically, a liquid sample of 1–3 μL can be injected onto a packed column. This is a larger sample than can be used with a capillary column. However, the separation ability of a packed column is *much less* than that of a capillary column. Typical plate numbers might be 2000 or less for a packed column and 50,000–100,000 for a capillary column.

For a packed column, the stationary phase is normally coated evenly on the surface of the solid support by slurrying the support with a solution of the liquid phase and then evaporating the solvent. The solid support must have a uniform pore diameter and a large surface area. These properties are needed to support an adequate coating of stationary liquid phase and to provide good contact with the mobile phase. The particles should be of regular shape with good mechanical strength to permit an efficient, well-packed column.

A number of solid supports are available from supply houses. One widely used support is made of diatomite, a largely silica material composed of the skeletons of microscopic algae. Other porous supports are made from silica and other materials.

A solid support should be inert, but silica supports contain silica-OH groups that interact by hydrogen bonding with polar solutes such as alcohols and amines, resulting in bad peak tailing. (Instead of being sharp and symmetrical, tailed peaks are drawn out on the trailing edge.) Tailing is reduced or eliminated by reacting the support with an organic silicon compound. This process, called *silanization*, is illustrated by a reaction to introduce TMS (trimethylsilyl) groups:

$$\text{Silica-OH} + (CH_3)_3\text{SiCl} \rightarrow \text{Silica-OSi}(CH_3)_3 + HCl$$

Porous organic polymers are also coming into use as column packings. (The Chromosorb Century Series, Poropak polymers, and Tenax are trade names for some popular types.) These polymers are available in a variety of different chemical structures; all serve as the stationary phase for sample solute partitioning and need no separate solid support. Porous polymers will not bleed from the column as liquid stationary phases can, although impurities sometimes come off slowly and cause a background problem. All packed columns are initially conditioned in the gas chromatograph for several hours at a relatively high temperature to remove impurities. Porous polymers often retain sample solutes quite strongly and require a higher temperature for separation than do conventional coated solid supports.

Capillary Columns. A column of very small inside diameter coated with a stationary phase on the inside walls has been called a WCOT (wall-coated open tubular) column, or simply a "capillary" column. The history, principles, and applications of capillary columns in gas chromatography have been reviewed [1].

Capillary columns must have a small inside diameter. To perform efficiently, molecules must be in close proximity to the walls for equilibration between the mobile and stationary phases to occur. The efficiency with which the molecules reach the walls is controlled by the diffusion coefficient, D_g, and by the square of the inside column radius, r^2. Typically, the inside diameter of a capillary column is about 0.25 mm.

Because there is no packing resistance, the gas flow is small and very long columns can be used. Commercial columns of 15–30 m give excellent separating ability, and samples containing many compounds can be separated on these columns. For even more complex samples, the column can be 100 m or even longer. Historically, the use of a mile-long capillary column (1.6 km) has been demonstrated.

Stainless-steel capillary columns were used a number of years ago, but they have the disadvantages that minute traces of sample compounds often sorb irreversibly on the metal walls. This small sorption might not be noticed on a packed column, but it becomes significant on a capillary column, where a much smaller sample is used. For this reason, glass or silica columns, which do not sorb impurities as strongly, are now used.

Glass capillary columns are also much less bulky than metal columns. A glass capillary 30 m or more in length comes in a coil only about 11 cm in diameter and

[1] M. Novotny, *Anal. Chem.*, *50*, 16A (1978).

2–3 cm thick. The outside diameter of the glass tubing may be 0.8 mm, and the inside diameter 0.25 mm.

More recently, capillary columns made of fused silica have been "taking over" in gas chromatography. Fused silica is more inert than glass and can be treated to give peaks with almost no tailing. A further advantage is that silica columns are actually quite flexible and can be bent more than glass when the column is being installed in the gas chromatograph. Silica columns have a brown polymeric coating on the outside to prevent damage to the column and thus preserve their flexibility and ruggedness.

Ordinary coated columns (glass or silica) have to be used with considerable care to avoid damage to the stationary phase, which is held on the surface only by relatively weak physical forces. Injection of too large a sample, or a sample dissolved in the wrong organic solvent, can wash off some of the stationary phase and ruin an expensive column. However, fused silica columns are now available with chemically bonded stationary phases. These are much more stable and can actually be "washed" with most organic solvents without danger. The exact nature of the chemical bonding tends to be a commercial secret. However, silica columns with various chemically bonded stationary phases can be purchased.

Solid supports used in packed columns have a very large surface area for holding stationary phase. Despite its great length, the surface area in a capillary column is much smaller than that of a packed column, so the amount of stationary phase is proportionately less. This means that the sample capacity is very limited and that only a very small sample can be injected without overloading. Often a liquid sample as small as $0.001–0.01$ μL is required. Samples this small cannot be injected with a syringe, so a somewhat larger sample is actually injected, and a *sample splitter* is used. This is a device that bleeds off most of the vaporized sample after injection and allows only a small, predetermined fraction to enter the column. The split ratio (the fraction of the injected sample that actually enters the column) can be varied and often ranges from 1:20 to 1:100.

Wide-Bore Capillary Columns.   Open-tubular columns with an inside diameter of approximately 0.5 mm are now available for gas chromatography. The length varies, although 15 m is fairly typical. The separation ability of such columns is less than that of capillary columns of smaller diameter but is much greater than that of packed columns. Comparison chromatograms with a wide-bore capillary column and a packed column are shown in Figure 22–2.

Wide-bore columns are very easy to install and require a very short equilibration time ($\sim$15 min) to establish a stable baseline. Liquid samples up to 1 μL or more can be injected without the use of a sample splitter.

Choosing a Column.   For samples containing many chemical compounds or for any sample that is difficult to separate, a capillary column is clearly the best choice. For samples that are relatively easy to separate, a packed column or a wide-bore unpacked column should do nicely. Packed columns are more rugged and permit the injection of larger samples than capillary columns. But wide-bore unpacked columns are also rugged and can handle relatively large sample volumes while

Figure 22–2. Comparison of results for EPA method No. 602 using a wide-bore capillary column (left chromatogram) and a packed column (right chromatogram). Peak identification: 1 = benzene, 2 = toluene, 3 = ethylbenzene, 4 = chlorobenzene, 5 = 1,3-dichlorobenzene, 6 = 1,4-dichlorobenzene, 7 = 1,2-dichlorobenzene. (Courtesy of J and W Scientific, Inc.)

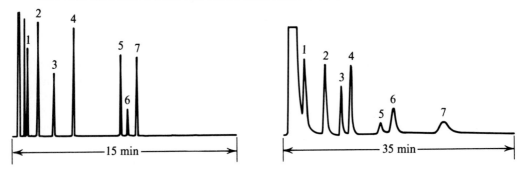

giving much better separations than packed columns. For these reasons, either a capillary column or a wide-bore open-tubular column is usually the best choice for gas chromatography.

A wide choice of stationary phases are available for GC columns. The selection of an appropriate stationary phase will be discussed in Section 22–4.

22–3. THEORY OF GAS-LIQUID CHROMATOGRAPHY

Capacity Ratio. In gas-liquid chromatography, the mobile phase is the carrier gas (usually nitrogen or helium) and the stationary phase is a liquid or soft solid coating on a solid support (packed columns) or on the walls of a capillary column. The partitioning ratio most used for sample chemicals between the two phases is the capacity ratio (or capacity factor), k, which is defined as follows:

$$k = \frac{\text{amount of solute in stationary phase}}{\text{amount of solute in mobile phase}} \qquad (22\text{–}1)$$

In gas chromatography, the retention time (t_R or simply t) and adjusted retention time (t_R' or t') are generally used rather than retention volume and adjusted retention volume, respectively:

$$t = t_0(1 + k) \qquad (22\text{–}2)$$

$$t' = t_0 k \qquad (22\text{–}3)$$

where t_0 (sometimes written as t_M) is the holdup time, the time required to displace the mobile phase between the point of injection and the detector.

The value of k can be calculated from Equation 22–2 after measuring t and t_0. To measure t_0, a gas (such as methane or air) that passes unimpeded through the column is injected, and its retention time is noted.

To achieve a separation, conditions must be adjusted so that each sample component has a large enough retention time to resolve all of the peaks. Since retention time is largely governed by k, we need to look at the factors that affect the values of k for the various sample components.

The value of k for a sample component is strongly temperature dependent. The concentration of a component in the mobile phase (carrier gas) is proportional to the partial pressure of the component in the carrier. A temperature increase will increase the volatility of the component in the stationary phase and, thus, the partial pressure of the component in the carrier. Therefore, a temperature increase will decrease k and decrease the retention time. The k values for solutes of the same chemical type (alkanes, alkenes, ketones, etc.) often vary with boiling point, the compound with the lowest boiling point being eluted first.

The k of a sample component also depends on the *chemical nature* and *amount* of stationary liquid phase in the column. Substantial changes in k can often be achieved by changing to a stationary phase with a different chemical structure. Selecting a suitable stationary phase will be discussed in a later section.

The amount of the stationary phase in a column of given length will vary with the thickness of the coating. The thickness of the stationary phase will vary with the concentration of the stationary phase in the solvent.

Gas chromatographic separations depend on the different rates at which sample chemicals move along the column and thus on the different times at which they arrive at the detector. As was stated in Chapter 21, the rate of movement along the column depends on the linear flow rate of carrier gas (u, in cm/sec) and on the capacity ratio (k):

$$\text{Rate} = u/(1 + k) \qquad \textbf{(22–4)}$$

The goodness of a separation depends on the width of peaks as well as the difference in retention times. Better separations are obtained when the peaks are very narrow.

Effect of Pressure on Flow Rate.   The linear flow rate of carrier gas through a column is easily obtained by measuring t_0 for a nonsorbed peak:

$$\underset{\text{(cm/sec)} \quad \text{(cm/sec)}}{u \;\; = \;\; L/t_0} \qquad \textbf{(22–5)}$$

(Here, L is the column length in centimeters.) This is really an *average* flow rate because the flow rate actually varies along the column, as we shall see. The average flow rate, $\bar{u}$, is adjusted by varying the pressure of the carrier gas at the column inlet. This is a linear relationship, as is demonstrated by the experimental plot in Figure 22–3. The applied pressure at the column inlet, p_i, can be measured with an appropriate pressure gauge.

The pressure tends to decrease linearly along a column, dropping to ambient pressure (approximately 1 atmosphere) at the column outlet. The carrier gas is compressed by the elevated pressure at the column inlet; however, the gas volume

Figure 22–3. Average flow rate ($\bar{u}$) as a function of applied carrier gas pressure ($P_i - P_0$).

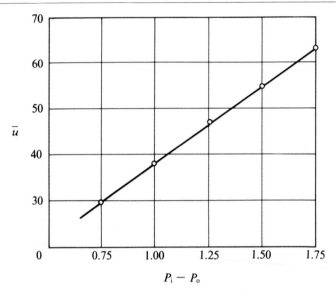

$P_i - P_o$

gradually increases as the pressure drops along the column. Since the product of pressure times volume is a constant, the increasing gas volume along the column means that the carrier gas flow rate will change appreciably from the column inlet to outlet. The term $\bar{u}$ simply represents the average of the various flow rates along the column.

The average linear flow rate ($\bar{u}$) and the linear flow rate at the colum outlet (u_0) are related as follows:

$$\bar{u} = ju_0 \tag{22–6}$$

where j is the gas compression factor. This j term is calculated by using the equation

$$j = \frac{3(P^2 - 1)}{2(P^3 - 1)} \tag{22–7}$$

where P is the ratio of the inlet to outlet pressures, $P = p_i/p_0$.

Chromatographic Efficiency. The width of chromatographic peaks is a measure of the efficiency of a chromatographic system. The system includes the entire instrument and not just the column. The ability to obtain sharp, narrow peaks is often expressed in terms of a plate number, such as N', N, or N_{eff} (see Section 21–2). A large value for a plate number indicates high chromatographic efficiency and excellent separation ability. However, it is difficult to express chromatographic efficiency by a single plate number because several effects contribute to peak broadening and these change according to experimental conditions. For capillary columns, the major contributors to measured peak variance (σ_m^2) are the variance due to diffusion (σ_d^2), the variance due to resistance to mass transfer (σ_{mt}^2), and the extra-column variance (σ_{ec}^2):

$$\sigma_m{}^2 = \sigma_d{}^2 + \sigma_{mt}^2 + \sigma_{ec}^2 \qquad\qquad (22\text{–}8)$$

All variances have the units sec^2 or min^2. Remember that peak width at the base of a peak $= 4\sigma$ and that variances, rather than peak widths, are additive.

In practical GC, σ_{mt}^2 is almost always the largest component of the measured peak variance. The variance due to axial diffusion ($\sigma_d{}^2$) becomes large at very slow flow rates but is relatively small (though usually not negligible) at the flow rates that are generally used. The mathematics of adding variances (and then calculating total peak width as 4σ) is such that σ_{ec}^2 tends to have a very small effect on the widths of late-eluting peaks. However, σ_{ec}^2 often has a *major* effect on the widths of early-eluting peaks.

Effect of Flow Rate. The Van Deemter equation is the classical statement of the effect of flow rate on chromatographic efficiency. This equation may be written

$$H = \sigma^2/L = A + B/\bar{u} + C\bar{u}$$

where H is the classical plate height in cm, σ^2 is the total peak variance in cm^2, L is the column length in cm, and $\bar{u}$ is the linear flow rate in cm/sec. The A term refers to multipath effects and is not needed in GC with a capillary column. The B term refers to broadening by axial diffusion (see Equation 22–9). The C term refers to "resistance to mass transfer."

> Often the C term is written (C_g or C_l)u, where the subscripts g and l refer to the gas and liquid phases, respectively. However, in modern thin-film capillary columns, the C_l term is usually small enough to be neglected.

Van Deemter plots, such as the one in Figure 22–4, are often published to show the relative effects of the B and C terms. At very slow flow rates, the $B/\bar{u}$ term is large, but it decreases rapidly with increasing flow rate, and the $C\bar{u}$ term predominates. The minimum of the curve is the most efficient flow rate for a separation. However, a somewhat faster flow rate is generally used because the separation is faster with only a small decrease in efficiency.

Golay developed a famous equation for capillary columns. It is similar in form to the Van Deemter equation but considers additional parameters. The Golay equation is now considered to give the most detailed and accurate description of the factors affecting peak broadening in capillary columns used in gas chromatography. Each of these factors will now be discussed.

Diffusion Broadening. In GC, molecules of a sample chemical tend to diffuse, causing the width of the band to increase while the chemical is in the gas phase in the column. This broadening is substantial when the carrier gas flow rate is slow and the sample chemicals spend a longer time in the column.

The B term in both the Van Deemter and Golay equations gives the diffusion contribution to plate height as

$$H = \frac{\sigma_d{}^2}{L} = \frac{2\bar{D}_g}{\bar{u}} \qquad\qquad (22\text{–}9)$$

Figure 22–4. Schematic plot of the van Deemter equation. The solid line represents the hyperbola obtained when H is plotted versus $\bar{u}$. The dashed lines represent the constant contribution of A and the increasing contribution of $C\bar{u}$ to H. The difference between the solid line and the sum of the dashed lines at any point represents the contribution of the $B/\bar{u}$ term.

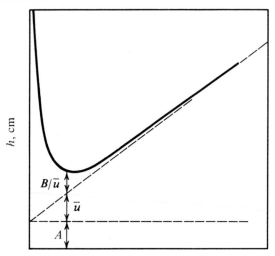

Average Linear Carrier Gas Velocity, $\bar{u}$, cm/sec

where H is the classical plate height in cm, $\sigma_d{}^2$ is the diffusion variance in cm² of a peak as it leaves the column, L is the column length in cm, $\bar{D}_g$ is the average axial diffusion coefficient of a sample chemical in cm²/sec, and $\bar{u}$ is the average linear flow rate of the carrier gas in cm/sec.

The amount of diffusion broadening can be calculated from measurable chromatographic parameters, provided that the diffusion coefficient (D_g) of a sample chemical is known. Giddings [2] has shown that D_g can be calculated to $\pm 5\%$ from constants related to the molecular weight of the carrier gas, the column pressure, and the absolute temperature. For our purposes, it should be noted that D_g is proportional to $T^{1.75}$, to $1/P$, and (approximately) to $1/\sqrt{M_c}$, where T is the absolute temperature in °K, P is the pressure in atmospheres, and M_c is the molecular weight of the carrier gas.

In Equation 22–9, the axial diffusion variance is for a peak that is still on the column. This cannot be checked experimentally. To calculate the peak variance due to diffusion after the peak has left the column and passed through the detector, it is necessary to multiply Equation 22–9 by the factor $(1 + k)^2/\bar{u}^2$. The resulting

[2] E. N. Fuller, K. Ensley, and J. C. Giddings, *Ind. Eng. Chem.*, *58* (No. 5), 18 (1966).

equation is

$$\sigma_{\text{d}}^2 = \frac{2D_{\text{g}}L}{\bar{u}} \cdot \frac{(1+k)^2}{\bar{u}^2} = \frac{2D_{\text{g}}L(1+k)^2}{\bar{u}^3} \tag{22–10}$$

where σ_{d}^2 is the diffusion variance in sec^2, D_{g} is the axial diffusion coefficient in cm^2/sec, L is the column length in cm, $\bar{u}$ is the average linear flow rate in cm/sec, and k is the capacity ratio of the sample chemical.

Calculation of σ_{d}^2 using Equation 22–10 is a reasonably accurate way to see the effect axial diffusion has on the width of peaks under a variety of experimental conditions.

Resistance to Mass Transfer. In both the Van Deemter and Golay equations, the C_{g} term describes the contribution to plate height due to resistance to mass transfer in the gas phase, and the C_1 the contribution from resistance to mass transfer in the stationary "liquid" phase. In modern capillary columns, which have a very thinly coated stationary phase, the C_1 term is very small and is often neglected.

The C_{g} term in the Golay equation is as follows:

$$H = \frac{a(1 + 6k + 11k^2)r^2\bar{u}}{24(1+k)^2\bar{D}_{\text{g}}} \tag{22–11}$$

Here, H is the plate height contribution in cm, a is a proportionality constant that varies from column to column, k is the capacity ratio, r is the column inner radius in cm, $\bar{u}$ is the average linear flow rate in cm/sec, and $\bar{D}_{\text{g}}$ is the average diffusion coefficient of a sample chemical in cm^2/sec.

The proportionality constant (a) was inserted in Equation 22–11 because the experimental value of H in the C_{g} term is usually somewhat larger than the calculated value of H. Equation 22–11 points out that, for any given column (fixed r^2 and a), the peak broadening will be affected by $\bar{u}/\bar{D}_{\text{g}}$ and not just by $\bar{u}$, as stated in the Van Deemter equation. This is an important concept because varying the pressure to change $\bar{u}$ also changes the value of $\bar{D}_{\text{g}}$. It must also be noted that the value of H in Equation 22–11 will be different for each peak in a chromatogram. This is because the value of the k terms varies from 1.0 at $k = 0$ to 11.0 at $k = \infty$.

The peak variance due to resistance to mass transfer is obtained by multiplying the right-hand side of Equation 22–12 by $L(1+k)^2/\bar{u}^2$:

$$\sigma_{\text{mt}}^2 = \frac{La(1 + 6k + 11k^2)r^2}{24\bar{D}_{\text{g}}\bar{u}} \tag{22–12}$$

where σ_{mt}^2 is the variance due to resistance to mass transfer in sec^2, L is the column length in cm, and the other parameters have the same units as in Equation 22–11.

Extra-Column Broadening. The extra-column variance represents the peak broadening that occurs outside the chromatographic column. This includes broadening that occurs in the injection port, the connecting tubing, and the detector. The *time* required for a sample to be injected also causes some broadening of the various peaks. In modern instruments, the time required for sample injection may be a major source of the extra-column variance. Desty [3] actually struck the syringe with a small hammer to get a very fast injection time and reduce the value of σ_{ec}^2.

The effect of extra-column peak broadening seems to have been ignored in the Van Deemter and Golay equations. However, the magnitude of σ_{ec}^2 can be estimated by a simple linear regression method. First, the axial diffusion variance (σ_d^2, in sec^2) is calculated for each chromatographic peak according to Equation 22–10 and subtracted from the measured peak variance (σ_m^2). (The latter is obtained by measuring peak width on the actual chromatogram.) Then a plot of ($\sigma_m^2 - \sigma_d^2$), which is really ($\sigma_{mt}^2 + \sigma_{ec}^2$), against a function of k gives a straight line with an intercept on the vertical axis (when $f(k) = 0$) equal to σ_{ec}^2. Similar (but not identical) values for σ_{ec}^2 are obtained by using $k(1 + k)$ as $f(k)$ (see Chapter 21) and $1 + 6k + 11k^2$ as $f(k)$ (Equation 22–12).

Effect of Column Length.   Suppose we double the length of the column but keep the column temperature and flow rate the same. Will the number of theoretical plates also double? Several published papers have provided the answer, which is "no." The plate number increases, but it is considerably less than doubled.

A likely explanation for this is that the plate number (N' or N) is proportional to $\bar{D}_g/\bar{u}$, as well as to L. (See Equation 22–11 and recall that N' or N is inversely proportional to H.) When the column length is doubled, a higher average pressure must be applied to keep the same average flow rate ($\bar{u}$). Since D_g is inversely proportional to pressure, the higher pressure means that $\bar{D}_g$ will be lower for the longer column and that N' per meter or N per meter will also be lower.

22–4. SELECTION OF CONDITIONS

Column.   In Section 22–2, we concluded that some type of an open-tubular column was preferable to a packed column for gas chromatography. Modern silica columns are rapidly replacing the older glass capillary columns. If possible, a silica column with a bonded stationary phase should be used. These columns are a lot more stable than columns in which the stationary phase is simply coated onto the silica surface.

The chemical nature of the stationary phase that is coated or bonded onto the silica surface will alter the capacity ratios of substances to be separated. In general, it is best to use a nonpolar stationary phase for the separation of compounds of low polarity (such as hydrocarbons). A phase of higher polarity is used to separate a sample containing more polar compounds (such as esters, alcohols, and amines). Table 22–1 lists some common stationary phases that are available in capillary (or wide-bore) bonded-phase columns. The phases are listed roughly in the order of increasing polarity.

[3] D. H. Desty in *Advances in Chromatography*, Vol. 1, J. C. Giddings and R. A. Keller, eds., New York: Dekker, 1965, p. 199.

Table 22-1. Some Popular Bonded-Phase Silica Columns

Phase	Older Commercial Designation	Temperature Range (°C)
Methylsilicone	OV-1, OV-101, SE-30	60–300°
5% Phenylmethyl-	SE-54, OV-3, etc.	60–300°
300°C 50% Phenylmethyl-	OV-17, etc.	40–260°
50% Trifluoropropyl-	OV-210, etc.	45–220°
75% Cyanopropylphenyl-	OV-225, XE-60	40–200°
Polyethyleneglycol	Carbowax 20M, SP-100	20–230°

Sample Injection. As was mentioned earlier, a liquid sample injected onto a capillary column is often so small that a sample splitter must be used. However, Grob and Grob [4] have developed a technique of "splitless" injection in which a dilute sample solution as large as 2–3 μL can be injected directly. The solvent for the sample must be carefully selected, volatile organic solvents such as pentane, hexane, or methylene chloride. As the solvent and sample components pass from the hot injection port into the relatively cool ($\sim 45°$C) capillary column, much of the solvent initially condenses in the first part of the column. The sample components are also held here while the solvent gradually evaporates and passes through the capillary column. Then the oven temperature is increased at a preset rate, moving the sample components through the column and separating them.

The peak height of a sample component separated on a capillary column is often about the same as that of a much larger volume of the same sample separated on a packed column. (The peak *area* will, of course, be greater on the packed column.) This happens because the very large number of theoretical plates on the capillary column causes peaks to be narrow, but high.

To prevent band spreading, all tubing within the gas chromatograph must be the same size. For example, if separated peaks pass from the end of the capillary column into a larger tube just before entering the detector, mixing will occur and the resolution of close peaks may be lost.

Temperature and Temperature Programming. Temperature is a major factor in adjusting conditions for a satisfactory separation. When sample components elute rapidly and are incompletely resolved, lowering the column temperature will slow down the elution and probably improve the peak resolution. On the other hand, when sample components are slow in coming off the column, raising the column temperature may be indicated.

If a mixture containing both high- and low-boiling components is to be separated, the temperature needed to separate the low-boiling compounds may slow the separation of the high-boiling components too much. Late-eluting peaks are always broader, and resolution is often poor. In such cases, *temperature programming* can be used. In this technique, the column temperature is increased linearly with time at a preset rate. The more volatile sample components are separated at the

[4] K. Grob and K. Grob, Jr., *J. Chromatogr.*, *94*, 53 (1974).

lower temperatures, while the higher-boiling compounds gradually move at a faster rate through the column so that their peaks appear earlier on the chromatogram.

The effect of temperature programming is illustrated in Figure 22–5. Isothermal separation of equal amounts of hydrocarbons shows that the later peaks become progressively broader and lower. When the sample is separated by using temperature programming, the peaks are higher and narrower, and the separation is completed more quickly. The higher peaks mean better sensitivity, that is, that a smaller quantity of a sample can be detected. Temperature programming is so useful that it has become almost a necessity in modern gas chromatography.

Figure 22–5. (a) Isothermal and (b) temperature-programmed separations of a 2-μL solution of C_{10}–C_{16} hydrocarbons (*n*-alkanes) in methylene chloride. Conditions: SE-30 glass capillary column, 28 m × 0.25 mm i.d.; He carrier gas, 3.3 mL/min; FID; chart speed, 0.1"/min. Isothermal run, 100°C; programmed run, 80°C for 4 min, then temperature increase at 2°C/min.

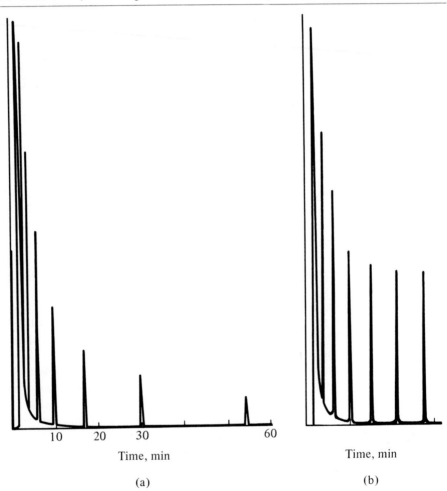

Time, min

(a)

Time, min

(b)

Compounds elute faster as the column temperature is increased because the capacity ratio (k) becomes smaller. Figure 22–6 shows a plot of k values for aliphatic hydrocarbons (n-alkanes) as a function of temperature. The rate of movement along a column is equal to $u/(1 + k)$ (see Section 21–3). Thus, at 60°C, the C_{11} compound in Figure 22–6 is moving at the rate $u/13.3$ while near the end of the column. Where k is perhaps 3.0, it is moving at the rate $u/4$, which is more than three times the initial rate. If we also consider the changes in u due to decompression of carrier gas, the increase in the rate at which the compound moves is even greater.

With temperature programming, the chemical peaks move off the column and through the detector faster than in isothermal GC, and the recorded peaks are therefore sharper and higher. If a plate number (N' or N) is calculated in the usual manner, it is usually several times larger when temperature programming is used. However, this is very misleading because the actual resolution of any two peaks is no better in a programmed run than in an isothermal run (at a suitable temperature for the peaks involved). In a programmed run in which the temperature is increased at too fast a rate, the compound may race through the latter part of the column so fast that very little separation occurs. In such a case, the resolution obtained in the programmed run will be *less* than in an isothermal run.

Detectors. After being separated on the chromatographic column, sample components pass through a detector that creates an electric signal that is proportional to the concentration of sample components in the carrier gas stream. This signal is

Figure 22–6. Capacity ratios of n-alkanes as a function of temperature.

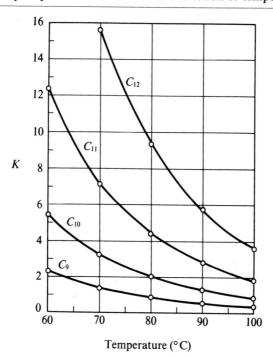

Temperature (°C)

amplified and fed to a recorded that traces the various chromatographic peaks. The amplification must be high enough that all of the peaks are measurable, but not so high that some peaks are off-scale. A device called an attenuator handles this situation. An input attenuator reduces the amplifier input voltage by various factors of 10 up to perhaps 10^5. An output attenuator reduces the amplifier output by definite fractions $1/2, 1/4, 1/8, \ldots, 1/1024$, as represented by settings of $2, 4, 8, \ldots,$ 1024. Thus, an input setting of 10^1 coupled with an output setting of 32 gives a total attenuation of 10×32, or 320. This means that the recorded signal is $1/320$ that of the unattenuated signal.

> The attenuation can be changed *during* a chromatographic separation. Low attenuation is used to obtain good sensitivity for minor peaks; then the attenuation is temporarily increased to keep major peaks on-scale.

Although a number of detectors are available, three of the most common types will be discussed here. Some characteristics of these detectors are summarized in Table 22–2.

Table 22–2. Comparison of Gas Chromatograph Detectors

Detector	Mechanism of Detector Response	Applications and Limitations
Thermal Conductivity Cell	Detector output is the difference in thermal conductivity between the pure carrier gas and carrier gas containing the sample. This is indicated by the electrical resistance of heated platinum filaments (or thermistors) in the gas stream.	Used for both inorganic and organic compounds. Sensitive to changes in temperature and gas flow.
Flame Ionization	Detector output is proportional to the number of ions produced when the sample compounds (in the carrier gas argon) are burned in H_2 and air. The ions are collected at an electrode, producing an electric current.	Used for organic compounds. No response to many inorganic gases: CO_2, H_2O, CO, SO_2, H_2S, NH_3. Useful for analyzing water extracts.
Electron Capture	Detector output is proportional to a continuous flow of "slow" electrons from the ionization of argon carrier gas by β particles. Reducible components in the carrier gas react with the electrons, reducing the flow of electrons to the anode and thus reducing the detector output.	Used for O, P, S, —NO_2, and halogen-containing compounds. Very weak response to ethers and hydrocarbons. Useful for analyzing insecticides and pesticides.

Thermal Conductivity (TC) Detector. This is the least sensitive of the common detectors. Its principle is that the rate of heat loss of a hot wire is proportional to the molecular weight of the surrounding gas. Hydrogen (mol wt = 2) or helium (mol wt = 4) is used as the carrier gas; gaseous sample components, which have a higher molecular weight, have a substantially lower thermal conductivity. Heater-metal filaments inside the TC cell form a Wheatstone bridge that measures the difference between the thermal conductivity of the pure carrier-gas stream and the gas stream containing the separated sample components. When an organic compound (such as hexane, which has a thermal conductivity 1/12 that of helium) flows past the detector, the filament (thermistor) temperature will increase. This heating increases the resistance of the filament, unbalancing the Wheatstone bridge. The resistance needed to restore the balance is measured by the recorder.

Flame-Ionization Detector (FID). This is now considered the standard detector for GC separations of organic compounds. Very sensitive and unusually stable, it will detect around 10^{-9} g (1 ng) or more of most organic compounds and remain unaffected by minor variations in flow rate.

The flame-ionization detector (Figure 22–7) is basically a microburner surrounded by an electrode used to collect ions formed by the burning sample. The

Figure 22–7. Diagram of a flame-ionization detector. (Courtesy of Hewlett-Packard Corp.)

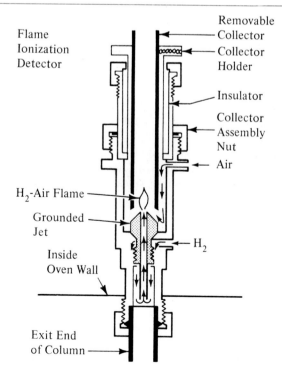

Flame
Ionization
Detector

Removable
Collector
Collector
Holder

Insulator

Collector
Assembly
Nut

Air

H_2-Air Flame

Grounded
Jet

H_2

Inside
Oven Wall

Exit End
of Column

argon gas stream coming from the GC column is mixed with hydrogen gas before entering the burner. Air or oxygen diffuses into the compartment through a separate inlet to support combustion. Combustion of organic sample components produces ions, which migrate to the electrode. The resulting electric current is then amplified and measured.

Electron-Capture Detector (ECD). The electron-capture detector has extraordinary sensitivity for the restricted class of compounds it can detect; in favorable cases, it will detect as little as 10^{-12} g (1 pg). It is particularly useful (see Table 22–2) for detecting organic halogen compounds such as chloroform ($CHCl_3$), which is found in chlorinated drinking water. It can easily pick minute traces of chlorinated pesticides out of the welter of organic compounds in biological samples.

The electron-capture detector contains a radioactive substance (usually ^{63}Ni) that gives off beta particles. Argon carrier gas (often mixed with a little methane) reacts with the beta particles to form electrons and other products. A voltage is applied to an anode and cathode in the detector; electrons are attracted to the anode giving a small electric current called the *standing current*. Organic solutes in the carrier gas stream react with ("capture") electrons, thus reducing the standing current and providing a signal for detection.

22–5. QUALITATIVE AND QUANTITATIVE

Identification of Peaks. Identifying the various peaks in a gas chromatogram is often a problem, particularly when the components are completely unknown. The simplest way is to match the retention times (or, preferably, the adjusted retention times) of the various sample peaks with those of known compounds separated under the same conditions. Some laboratories have libraries of known retention times available. This type of identification is rather uncertain because several organic compounds might have virtually the same retention time. A more positive identification is obtained if the match is again obtained when the sample peak and known compound are each chromatographed on a second column containing a *different type of stationary phase* from that used the first time.

The best method for the quantitative determination of peaks is to use a gas chromatograph interfaced with a mass spectrometer. The mass fragmentation pattern of a chromatographic peak can be matched with that in a library of known compounds. Sometimes the identity of a compound can be deduced from the mass fragmentation pattern without matching. Unfortunately, GC–mass spectrographic instruments are expensive and require an extensive computer system for efficient use.

Quantitative Analysis. The basis for quantitative analysis by gas chromatography is that the area under each peak is proportional to the concentration of the sample chemical, assuming constant conditions of column temperature, flow rate, etc. Areas of peaks are measured quite easily with a modern electronic integrator. However, the peaks obtained with a modern capillary column are so sharp and narrow that peak height can be conveniently used as a measure of relative concentration. Since the molar response of any detector varies for different compounds, each peak area (or height) has to be multiplied by an empirical factor to correct for this difference. The percentage of each component in the sample is calculated by dividing the corrected peak area for the component by the corrected sum of all peak areas.

This quantitative scheme requires careful control of all the experimental variables affecting the sensitivity and response of the detector; otherwise, the height and area of the chromatographic peaks will vary. Thus, quantitative analysis is not as straightforward as qualitative analysis. Most such instrumental errors can be eliminated by using the *internal standard method*. In this method, a constant amount of a pure compound (the internal standard) is added to a specified volume of the unknown sample and to several synthetic mixtures containing different amounts of the compound to be measured. The synthetic mixtures are chromatographed first, and a plot is made of percent compound versus ratio of the peak area of the compound to the peak area of the standard. Then the unknown is chromatographed, the peak areas of the compound and internal standard are measured, and their ratio is calculated. The percentage of the compound in the unknown sample is then read from the graph. The internal standard technique is used in Experiment 33, Chapter 33.

22-6. APPLICATIONS OF GAS CHROMATOGRAPHY

Gas chromatography is feasible only when the sample components to be separated are volatile at the operating temperature of the gas chromatograph. As a rule, compounds having a vapor pressure of 10 torr or more at oven temperatures can be separated. Since temperatures range up to about 300°C (and occasionally higher), a great many organic compounds (and some inorganic compounds) fulfill these requirements. Gas chromatography has been developed to the point at which almost any mixture of volatile chemical compounds can be separated, provided that favorable conditions are chosen. *Cis-* and *trans-* isomers can usually be separated. Even option isomers (enantiomers) of organic amino acids, alcohols, ketones, etc., can be separated on a capillary column coated with a special chiral phase. König has reviewed separations of enantiomers with chiral stationary phases [5].

A significant percentage of organic compounds are not volatile enough to be separated by gas chromatography. However, chemicals can often be added to form volatile derivatives from such compounds. Some of these compounds include the following:

1. High-molecular-weight fatty acids (RCOOH), which are converted to more volatile methyl esters ($RCOOCH_3$). (This process is particularly useful in food and biochemical analysis.)

2. High-molecular-weight alcohols (ROH) and steroids, which are converted to volatile trimethylsilyl ethers ($ROSi(CH_3)_3$).

[5] W. A. König, *J. HRC and CC*, 5, 588 (1982).

3. Amino acids $\left(\begin{array}{c}\text{RCHCOOH}\\|\\\text{NH}_2\end{array}\right)$, which, being nonvolatile, must be converted to amide esters by derivatizing both the amino and carboxyl groups. In one study, 17 amino acids were converted to amide esters $\left(\begin{array}{c}\text{RCHCOOCH(CH}_3)_2\\|\\\text{NHCOCF}_2\text{CF}_3\end{array}\right)$; in less than 30 min, two well-resolved peaks were obtained for each amino acid, one for the D form and one for the L form [6]. Gas chromatography is widely used to separate amino acid mixtures in medical or biochemical research.

4. Sugars can be successfully derivatized by first converting the carbonyl group to the oxime $\left(\text{>C=NOH}\right)$ and then reacting to form the trimethylsilyl ether $\left(\text{>C=NOSi(CH}_3)_3\right)$. A separation of several common sugars by this procedure is shown in Figure 22–8.

Mixtures of gases with very low boiling points are difficult to separate because such compounds ordinarily pass through a chromatographic column too quickly. To separate such mixtures, subambient column temperatures or an unsually retentive column packing must be used. One successful method uses a special

Figure 22–8. GC separation of sugar oxime-TMS derivatives. Peak identification: 1 = fructose, 2 = dextrose, 3 = internal standard, 4 = sucrose, 5 = lactose, 6 = maltose. (Courtesy of K. M. Brobst.)

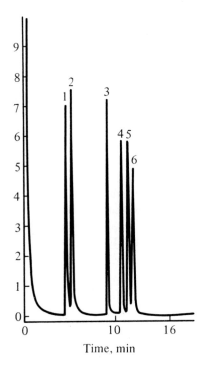

Time, min

[6] H. Frank, G. J. Nicholson, and E. Bayer, *J. Chromatogr. Sci.*, *15*, 174 (1977).

Figure 22–9. Separation of haloforms in pentane solution by gas chromatography using electron-capture detector. Peak identification: 1 = pentane, 2 = CHCl$_3$, 3 = CHBrCl$_2$, 4 = CHBr$_2$Cl, 5 = CHBr$_3$. Glass column, 6 ft × $\frac{1}{4}$ in, packed with 4% SE-30 + 6% OV-210 on Chromosorb; 75°C isothermal.

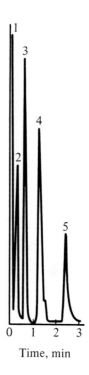

Time, min

graphitized carbon (Spherocarb) packing with a tremendous surface area (~1200 m^2/g). By this method, a mixture of hydrogen, oxygen, and nitrogen is separated with retention times of 1.1 min, 3.8 min, and 4.1 min, respectively, at 40°C on a 9 ft × $\frac{1}{8}$ in column with helium (30 mL/min) carrier gas [7]. A thermal conductivity detector is used.

Sometimes a concentration step is necessary before a GC separation can be carried out. It was discovered in 1974 that chlorinated drinking water usually contains ppb concentrations of chloroform (CHCl$_3$) and other haloforms (CHBrCl$_2$, CHBr$_2$Cl, and CHBr$_3$). It would be impossible to detect such small traces by direct injection of a water sample into a gas chromatograph, so the haloforms are first concentrated by extracting 10–100 mL of water with 1 mL of pentane. Then a portion of the pentane extract is injected, and the haloforms are separated [8]. A typical chromatogram is shown in Figure 22–9. An electron-capture detector is used for two reasons: (a) to achieve the necessary sensitivity and (b) to avoid interferences from other organic compounds extraced from water.

Solvent extraction is also used to isolate various compounds from urine before GC analysis. The presence (or changed concentration) of certain compounds in urine may be valuable in diagnosing various diseases. For example, the concentration of ketones, especially 4-heptanone, correlate with metabolic disorders related to diabetes. 4-Heptanone in urine can be determined by first extracting

[7] Analabs, Inc., 1976.
[8] J. J. Richard and G. A. Junk, *J. Am. Waterworks Assoc.,* **69,** 62 (1977).

Figure 22–10. Chromatogram of organics concentrated from drinking water. Conditions: capillary column 25 m × 0.25-mm i.d., coated with SP-1000; oven temperature 60°C at 3°/min to 220°C; FID attenuation of 10; 2 μL of ether solution using "splitless" injection. (Courtesy of L. D. Kissinger.)

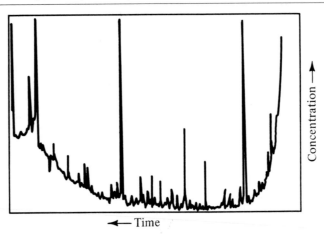

with cyclohexane and then chromatographing on a 100 m × 0.5 mm capillary column [9].

Organic substances can also be concentrated by extracting them from an aqueous solution onto a small column containing porous beads of organic resin. When drinking water containing trace concentrations of organic contaminants, for example, is passed through such a column, the contaminants are sorbed and retained by the beads. The sorbed organics are then eluted from the column by a small volume of ethyl ether. The volume of ether is reduced by evaporation, and a 2-μL aliquot is injected into a gas chromatograph to separate the individual organic solutes [10]. Figure 22–10 shows a chromatogram of the organic compounds concentrated from 55 gal of drinking water from a large U.S. city. Because of the large number of compounds present, the final separation was done on a glass capillary column.

QUESTIONS AND PROBLEMS

Principles and Practice

1. Explain why a liquid sample should be injected rapidly onto a gas chromatographic column.
2. Explain why a very low gas flow rate may result in an inefficient separation as measured by theoretical plate height, H. Why is a higher flow rate than the optimum, or minimum-H, flow rate sometimes used?

[9] H. M. Liebich and G. Huesgen, *J. Chromatogr.*, *126*, 465 (1976).
[10] G. A. Junk et al., *J. Chromatogr.*, *99*, 745 (1974).

3. List some desirable properties of solid support for gas chromatography. Explain how a thin, even coating of stationary phase can be applied to a solid support.

4. The amount of stationary phase applied to a solid support is doubled. What effect will this have on the t_R of sample peaks if all other conditions are held constant?

5. Why is a solid support sometimes "silanized" (treated with an organic silicon compound)?

6. List some desirable properties of a stationary phase used in gas chromatography. What simple, general rule governs the selection of a stationary phase for any given type of sample?

Capillary Columns

7. Briefly describe capillary columns with regard to dimensions, location of stationary phase, and approximate number of theoretical plates. What advantages do fused silica capillaries have over glass capillary columns?

8. A capillary column with a 0.25-mm inside diameter has a linear flow rate $\bar{u} = 45$ cm/sec. Calculate the average volume flow rate in cm^3/cm. How does this compare with a typical volume flow rate of 1.25 cm^3/sec for a 1/4" packed column?

9. For a packed column with 2000 theoretical plates, calculate the width of a chromatographic peak (in min) when $t_R = 2.00$ min. Taking this width as the average, calculate the maximum number of chromatographic peaks that can be resolved from $t_R = 1.50$ min to $t_R = 2.50$ min without overlap at the peak base.

10. Following Problem 9, calculate the width of a chromatographic peak eluted from a capillary column of $N' = 75,000$ when $t_R = 4.00$ min. Taking this width as the average, calculate the maximum number of peaks that can be resolved from $t_R = 3.50$ min to $t_R = 4.50$ min without overlap at the peak base.

11. Why must a smaller sample be injected onto a capillary column than onto a packed column? Explain how very small samples can be introduced onto a capillary column.

Column Efficiency, Flow Rate

12. The retention time (t_0) for methane on a capillary column is 1.45 min. Calculate k values for peaks having the following retention times: $t_1 = 3.05$ min, $t_2 = 3.94$ min, $t_3 = 6.17$ min.

13. Tell how the average linear flow rate $(\bar{u})$ can be measured. Calculate the ratio of u_0 to $\bar{u}$ for each of the following inlet pressures (assume an outlet pressure of 1.0 atm).

(a) 3.0 atm   (b) 2.0 atm   (c) 1.5 atm

14. For a column 2000 cm long, the average flow rate $\bar{u} = 43.6$ cm/sec. Two compounds to be separated have capacity ratios $k_1 = 6.05$ and $k_2 = 6.12$. Assuming that $\bar{u}$ is constant along the column, calculate how far (in cm) compound 2 will be along the column when the peak of compound 1 is just leaving the column. Also calculate the retention time (in sec) of each compound.

15. In a capillary column at 150°C, t_0 for methane is 47.0 sec, $t_1 = 91$ sec, $t_2 = 96$ sec, and $t_3 = 99$ sec. Resolution of the three sample peaks is very poor. Tell what might be done to improve the separation.

16. Explain how compression of carrier gas causes the flow rate to change along a column. Assuming that pressure times volume is a constant and that linear flow rate (u) is proportional to volume, calculate the linear flow rate at the locations indicated when $p_i = 3.0$ atm and $p_0 = 1.0$ atm if $u_0 = 80$ cm/sec. Assume a linear pressure drop along the column.

(a) At the column inlet and at 20%, 40%, 60%, 80%, and 90% of the distance along the column.

(b) Prepare a graph of flow rate as a function of the column location.

17. Explain why an increase in column length in GC does not give a proportionate increase in column efficiency, as indicated by N' or H (due only to σ_{mt}^2).

18. The D_g value for most sample compounds tends to be about 2.5 times higher with helium (mol wt = 4) than with nitrogen (mol wt = 28) as the carrier gas. Explain how this affects (a) the peak variance due to diffusion and (b) the peak variance due to resistance to mass transfer. Why is helium generally preferred?

19. (a) The D_g for a given chemical in helium is 0.163 at 80°C. Calculate the D_g at 120°C in helium assuming that the pressure and flow rate do not change.

(b) If at both 80°C and 120°C, k is 6.33, L is 1500 cm, $\bar{u}$ is 40.0 cm/sec, and u_0 is 64.2 cm/sec, calculate σ_d^2 (in sec^2) at 80°C and at 120°C.

20. At an average linear flow rate (u) of 30.0 cm/sec, a compound has $\bar{D}_g = 0.18$ cm^2/sec; at $\bar{u} = 48.0$ cm/sec, $\bar{D}_g = 0.108$ cm^2/sec; at $\bar{u} = 56.0$ cm/sec, $\bar{D}_g = 0.090$ cm^2/sec. Use Equation 22–9 to calculate the contribution to plate height (H) from axial diffusion at each of the flow rates listed.

21. Use Equation 22–11 to calculate the contribution of resistance to mass transfer to H for a compound with $k = 5.0$ at $\bar{u} = 30.0$ cm/sec, $\bar{u} = 48.0$ cm/sec, and $\bar{u} = 56.0$ cm/sec. Use the $\bar{D}_g$ values from the previous problem, $a = 1.43$, and $r = 0.0120$ cm.

22. A capillary column of $L = 1500$ cm and $r = 0.012$ cm was operated at an average gas flow rate $\bar{u} = 48.0$ cm/sec and a flow rate at the column outlet $u_0 = 70.0$ cm/sec. A sample compound had $k = 5.0$ and $D_g = 0.110$. Calculate the peak variance due to axial diffusion (σ_d^2) in sec^2.

23. With the same compound ($k = 5.0$) and conditions as in the previous problem, the measured peak variance was 0.740 sec^2. Assuming no extra-column contribution to peak broadening, calculate the value of the proportionality constant, a, in Equation 22–12.

Detectors and Instrumentation

24. Explain the principles of each of the following detectors used in gas chromatography: (a) thermal conductivity, (b) flame ionization, (c) electron capture.

25. Explain briefly which detector you would choose for each of the following GC suggestions: (a) traces of chlorinated pesticides in an extract from pheasant meat, (b) water impurities in organic solvents, and (c) benzene vapors concentrated from an industrial air sample.

26. Explain the function of an attenuator. Why is attenuation of chromatographic peaks often a practical necessity?

Programmed GC

27. Explain how temperature-programmed GC is able to give a faster separation of a complex sample with better sensitivity than an isothermal separation.

28. Why not just use isothermal conditions instead of temperature programming but choose a temperature that is high enough to give fairly rapid elution of the later peaks? Explain briefly.

29. Resolution of two peaks is a function of the average value of k, as well as the plate number (N') and separation factor (α) (see Equation 21–20). Compare the expected resolution for

an isothermal separation in which $\bar{k} = 6.5$ and for a programmed separation in which the average value of $\bar{k} = 3.3$. Assume that $N' = 56,000$ and that $\alpha = 1.008$ in both cases.

Quantitative Analysis

30. Modern integrators are available that give the area under each peak and print out the percentage that each peak area is of the total areas of all peaks. What, if anything, is wrong with using the printout values as the percentages of each compound in the sample?
31. It is difficult to reproduce accurately the very small volume of sample that must be used in gas chromatography. This being the case, explain how it is possible to perform an accurate quantitative analysis.

Methods

32. In a court case involving the possession of an illegal drug, the prosecution showed that GC analysis of the material in question gave a retention time that exactly matched that of a known illegal drug chromatographed under identical conditions. The defense argued that of over one million organic compounds, several innocuous compounds would be likely to have the same retention time as the drug. Suggest an analytical procedure for a more certain identification of the substance in question.
33. For each of the following, suggest one or more experimental conditions that might be changed to improve the gas chromatographic separation.

    (a) Retention times of 9.2, 10.3, and 12.5 min are obtained for the sample constituents on a 3-foot column, operated at $120°C$. The peaks are rather broad.
    (b) On a column containing Carbowax-600 as stationary phase, sample constituents A and B are nicely separated but resolution of constituents C and D is very poor. Doubling the length of the column improves the resolution of C and D only slightly.
    (c) On a column packed with OV-1 on Chromosorb-W at $80°C$, the following retention times are noted for sample constituents: 2.1, 2.3, 3.4, 7.8, 10.7, and 13.5 min. The later peaks are so broad that quantitative determination is difficult.
    (d) The peak of a sulfur-containing insecticide is almost impossible to pick out from numerous other peaks obtained when a vegetable extract is chromatographed using an FID.

34. In an industrial plant, various organic solvents are used to clean metal parts. Some of these solvent residues are washed into the water discharged from the plant into a nearby river. Outline an analytical procedure for determining whether the plant is meeting environmental regulations, which allow no more than 1 ppm of each solvent in the water discharged.
35. Corn syrup containing some fructose has a sweeter taste than "ordinary" corn syrup, which contains mostly glucose. Suggest a chromatographic procedure for analyzing syrup for fructose and glucose.
36. Vinyl chloride $(CH_2{=}CHCl)$ is volatile and has been identified as a hazardous substance. Very strict limits have been placed on its concentration in the air in and around factories. Outline an analytical procedure for determining trace quantities of vinyl chloride in air, given that a small column containing porous organic resin beads will effectively sorb all of the vinyl chloride from a measured volume of air passing through it.
37. Find one or more papers in a scientific journal in which gas chromatography has been used to solve some particular problem. Give the complete literature reference, outline the conditions used for the sample preparation and gas chromatographic separation, and summarize briefly the results obtained.

23

Liquid Chromatography

Just as gas chromatography (GC) includes chromatographic methods in which a gas is used as the eluent, liquid chromatography (LC) refers to those methods in which the eluent is a liquid. Methods in which an ion-exchange resin is the stationary phase are an exception; these are referred to as *ion-exchange chromatography* even though a liquid eluent is used. Ion-exchange methods will be discussed in Chapter 24.

Classical liquid chromatography will be discussed first to indicate the fundamental principles involved and to give a brief historical perspective. Then modern high-performance liquid chromatography (HPLC) will be covered in Section 23–2, and some modern examples will be given in Section 23–3. Finally, plane chromatography will be taken up in Section 23–4.

23–1. CLASSICAL LIQUID CHROMATOGRAPHY

General Principles. Chromatographic separations occur as a result of differences in the extent to which various solutes partition between a stationary phase and a mobile phase. In liquid chromatography, the stationary phase is usually a porous granular solid or is a liquid or solid phase coated onto the surface of a porous solid. The mobile phase, or eluent, can be a polar liquid such as water or an alcohol, or it can be a relatively nonpolar organic liquid.

Although conditions vary, a typical column for classical chromatography might be a tube of glass 1 cm or more in diameter and 10–30 cm in length. A glass frit or plug of glass wool holds the packing in place; eluent flow is controlled by a stopcock. The typical column packing is 100–200 mesh (140–74 μm diameter) granules, and the eluent flow rate for a 1-cm inside-diameter column is around 1 mL/min. A sample containing several milligrams of various solutes is added to the

Figure 23–1. Separation of lithium from sodium with DBM-TBP on Haloport F. [From D. A. Lee, *J. Chromatog.*, *26*, 342–45 (1967). By permission.]

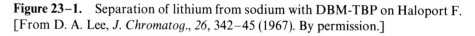

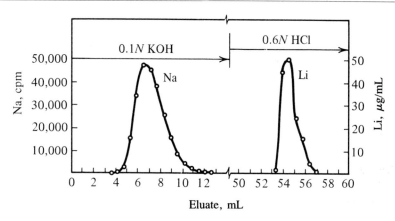

Eluate, mL

column, and the solutes are separated by elution with an appropriate mobile phase. Usually, fractions of several milliliters of column effluent are collected manually or with an automatic fraction collector. The amount of solute in each fraction is determined by titration, spectrophotometry, or some other analysis method. A plot of solute concentration versus fraction number shows peaks for the separated solutes (see Figure 23–1). Subsequent quantitative separations can be made somewhat more quickly by collecting all of each solute in a single fraction and determining the amount by a single titration (or some other analytical determination). Frequently, each solute requires 25 mL or more of eluent for elution, which means a separation time of one-half hour or more per sample constituent.

Classical Liquid-Liquid Chromatography (LLC)

Aqueous Stationary Phase–Organic Mobile Phase. Martin and Synge are credited with inventing liquid-liquid chromatography using this very system. In 1941, they published a paper entitled, "A New Form of Chromatography Employing Two Liquid Phases" [1]. They used granular silica gel to hold water as the stationary phase and separated mixtures of acetylated amino acids (derived from hydrolized protein) by eluting them with chloroform-butyl alcohol. Methyl orange was premixed with the silica gel to indicate the location of the organic acid bands on the column (pink bands on a yellow column). As each band was eluted off the column, the liquid effluent was collected and the amount of acid determined by titration with 0.01*N* barium hydroxide.

Many organic compounds have been separated on silica gel with various combinations of aqueous and organic phases, as have some inorganic substances. For example, uranium(VI) is separated from almost all other metal ions on a silica-gel column treated with aqueous 6*M* nitric acid [2]. After the aqueous samples are

[1] A. J. P. Martin and R. L. M. Synge, *Biochem, J.*, *35*, 1358 (1941).
[2] J. S. Fritz and D. H. Schmitt, *Talanta*, *13*, 123 (1966).

sorbed on the silica gel, uranium(VI) is selectively eluted from the column with methyl isobutyl ketone (MIBK). The nonextracted metal ions remain on the column but can be removed by washing with an aqueous solution.

Organic Stationary Phase–Inorganic Mobile Phase. Granular polymers of polystyrene, Teflon, and other materials can sorb and retain organic solvents for use in partition chromatography. These are special polymers of high porosity in which the aqueous solution and the sorbed organic phase come into contact throughout the bead, not just on the outer surface. LLC with this system is sometimes called "reversed-phase chromatography" because the stationary and mobile phases are the reverse of the systems developed earlier using an aqueous stationary phase (on silica gel) and organic mobile phase, which many scientists had come to regard as the normal arrangement. However, the reversed-phase system is now more widely used than the "normal" disposition.

Classical Liquid-Solid Chromatography (LSC). In LSC, the stationary phase consists of small, solid particles of a material such as activated alumina, silica, starch, or talc. The mobile phase is usually a pure organic liquid or a mixture of organic liquids. Most often, this type of chromatography is used to separate organic solutes. Separation depends on the relative affinity of the sample solute for the surface of the stationary phase and the eluting liquid. There is also competition between the eluent and the solutes for the adsorbing sites. The degree to which a solute is retained on the column depends on the *type* of solid absorbent, its *surface area*, and the type of eluent. Snyder has listed solvents in an "elutropic series" according to their ability to elute solutes [3].

Organic solids can also be used as a stationary phase in conjunction with an aqueous mobile phase. For example, a column filled with a special porous polystyrene solid is used to isolate various organic impurities from drinking water [4]. Neutral organic substances are strongly retained by the resin even at concentrations of less than 1 part per billion. Inorganic salts and some low molecular weight organic constituents are washed through the column by the water. After the water sample leaves the column, the sorbed organic compounds are eluted from the column with a small volume of an organic solvent such as ether. The organic compounds, which are now concentrated in the ether solution, are then separated by gas or liquid chromatography.

23–2. PRINCIPLES OF HIGH-PERFORMANCE LIQUID CHROMATOGRAPHY

Instrumentation and Equipment. In contrast to the relatively low efficiency of classical liquid chromatography, modern liquid chromatography is highly efficient

[3] L. R. Snyder, *J. Chromatogr., 8,* 178 (1962); *J. Chromatogr., 11,* 195 (1963).
[4] G. A. Junk *et al., J. Chromatogr., 99,* 745 (1974).

and is usually referred to as *high-performance liquid chromatography* (HPLC). It is described in books [5] and in numerous research articles in the literature.

HPLC is much faster, as well as more efficient, than classical liquid chromatography; it can resolve complicated multicomponent samples within a few minutes. It developed from classical liquid chromatography as a result of several major changes that occurred around 1968–1970 that made the great leap to HPLC possible. These changes included the following:

1. Much faster eluent flow rates were achieved by forcing the eluent through the column under pressure.
2. More efficient solid supports with generally smaller particle size were developed.
3. Much smaller samples were used, resulting in narrow peaks with better resolution.
4. Automatic detectors were devised and used for HPLC that could monitor the faster flow rates.

The various components that are used in a modern HPLC separation are shown in the following diagram.

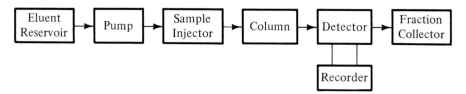

A pump is used to force the eluent through the column and detector. A pump provides constant flow regardless of the back-pressure of the column, which may be anything up to 3000 lb/sq in. The flow rate can be adjusted by changing the piston stroke of the pump. Unless damping is employed, however, pumps cause surges in liquid flow that may result in baseline noise on the chromatograph.

At low pressures, the *sample injector* can be simply a syringe that injects the sample into the eluent stream through a rubber septum. At higher pressures, a sample loop with a two-way valve is used. In one valve position, the loop can be filled with sample at atmospheric pressure. Then the value is switched, flushing the sample into the column along with the incoming eluent. Usually, only a few microliters of sample are injected.

The *column* is stainless steel or heavy-walled glass, built to withstand high pressures. The inside column diameter is typically 1–3 mm, and the length is up to 1 meter. The column is packed with small particles of porous silica, alumina, or organic resin (LSC) or with a porous support coated with a liquid (LLC). Column packing materials are discussed below in more detail.

The column and detector cell are often enclosed in a constant-temperature oven. Just after the column, the carrier stream enters a *detector*. Currently, the most

[5] L. R. Snyder and J. J. Kirkland, *Introduction to Modern Liquid Chromatography*, 2nd ed, New York: Wiley-Interscience, 1979.

commonly used detector is an ultraviolet detector with a fixed wavelength at 254 nm (or sometimes at 280 nm also). All aromatic organic compounds absorb intensely at this wavelength, as do a number of other organic compounds and many inorganic complexes. Variable-wavelength detectors covering the ultraviolet and visible spectral regions are also used. In the detector, the solution flows through a Z-shaped cell that provides a 1-cm light path through the solution, even though the tubing itself is much smaller in diameter (Figure 23–2). The solute's absorbance is converted into an electrical signal, which is recorded.

> When the separated components come off the column, it is vital not to allow the separated bands to remix. Passing the solution from 1-mm plastic tubing into larger tubing or into a cell with dead space causes turbulence and mixing. The Z-shaped flow-through cell in Figure 23–2 is excellent for allowing rapid sweep out of the cell without mixing.

A refractive index (RI) detector is sometimes used instead of the ultraviolet detector. This compares the refractive index of the exit stream with that of the pure carrier. When a sample solute is in the stream, the refractive index changes and the resulting signal is recorded. Although RI detectors are somewhat less sensitive than ultraviolet detectors and require better control of temperature and flow rate, they can be used to detect almost any organic solute as long as it is present in a high enough concentration.

A fluorescence detector is useful for detecting naturally fluorescent compounds and fluorescent derivatives of other compounds. The fluorescence detector is selective and finds extensive application in separating and analyzing vitamins, foods, and drugs.

Column Packings.   Column packings used to have a stationary phase that was physically adsorbed onto a porous silica support. However, the pressure and shearing effects of the mobile liquid flowing through the column often caused the stationary liquid to "bleed" off the porous support.

Figure 23–2.   Cross section of Z-flow cell. (From J. J. Kirkland, ed., *Modern Practice of Chromatography*, New York: Wiley-Interscience, 1971. By permission.)

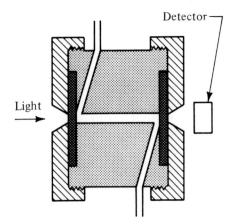

Detector

Light

Table 23–1. Some Commonly Available Silica Column Packings with a Chemically Bonded Organosilane

	Silane "R" Group	Chemical Structure
Increasing polarity ↓	Octadecyl	$C_{18}H_{37}$—
	Octyl	C_8H_{17}—
	Ethyl	C_2H_5—
	Benzyl	$C_6H_5CH_2$—
	Alkylcyano	$C{=}N(CH_2)_n$—
	Alkylamino	$NH_2(CH_2)_n$—

> One author described his experiment as follows: "The column bled profusely and soon expired."

Now the stationary phase is usually "chemically bonded" to a porous silica support. This is done by reacting the hydroxy groups on the surface of the silica support with an organo chlorosilane.

$$-\overset{|}{\underset{|}{Si}}-OH + ClSi(CH_3)_2R \;\rightarrow\; -\overset{|}{\underset{|}{Si}}-O-\overset{\overset{\displaystyle CH_3}{|}}{\underset{\underset{\displaystyle CH_3}{|}}{Si}}-R + HCl$$

The most common R group is $C_{18}H_{37}$; the silica so treated is often called ODS, for octadecylsilane. The long C_{18} groups stick out like tall trees from the silica, giving a very hydrophobic surface. Silica supports in which R is C_2H_5— or C_8H_{17}— are somewhat less hydrophobic. Coatings with still more polar (less hydrophobic) R groups are also used, as shown in Table 23–1.

Although many products are available, most modern column packings for liquid chromatography are small silica spheres that have one of the organosilanes listed above chemically bonded to the surface. The silica spheres are porous throughout and have a very large surface area, typically around 150–250 m²/g. For good chromatographic efficiency, the spheres should have a very small and uniform diameter.

Columns. As the technology of making column packings improves, the size gets smaller. The average diameter of state-of-the-art packings has decreased from 10 μm, to 5 μm, to 3 μm. Columns packed with these very small particles have a high resistance to liquid flow. Therefore, column lengths have become shorter as the size of column packings has become smaller. Columns are typically 10–30 cm in length, 10–15 cm being common for those packed with 3 μm particles. However, chromatographic efficiency is so much better with the very small particles that columns can be made shorter with little if any loss in separation ability. For example, a study [6] showed that a 4.6-mm (inside diameter) × 15-cm (length) column packed with 5-μm particles had a classical plate number, N, of 16,700 and

[6] N. H. C. Cooke, B. G. Archer, K. Olsen, and A. Berick, *Anal. Chem.*, *54*, 2277 (1982).

$H = 0.009$ mm. A similar column that was 7.5 cm in length but packed with 3-μm particles gave $N = 12,700$ and $H = 0.006$ mm.

Fast LC columns have become popular. These are very short (often 3 cm) and are packed with 3 μm spherical silica with a chemically bonded organic phase. These 3×3 columns have sufficient separation power for many types of samples. They find extensive use when rapid separations are required, as in quality-control applications.

Microbore columns, which have an inside diameter of only 1 or 2 mm, have very great separation power, although the separations often take much longer than with more conventional columns. The small diameter of microbore columns requires the use of special injectors and detectors with submicroliter volumes in order to minimize peak broadening outside the LC column.

The separation power of microbore columns in LC has been demonstrated by the experimental achievement of one million theoretical plates [7]. This was accomplished by separately packing and testing 23 microbore columns, then cutting off the ends and connecting them to form a single column 22 m in length!

Conditions for Chromatographic Separations.   The capacity ratio (see Chapter 21) of a sample component depends on the relative affinity of that component for the stationary and mobile phases. Usually, it is best to select a stationary phase of approximately the same polarity as the sample compound. For example, an octadecyl, octyl, or ethyl group on the stationary phase (see Table 23–1) would be appropriate for separating a mixture of nonpolar hydrocarbons, but a benzyl or alkylcyano stationary phase would be better for separating more polar compounds such as alcohols or ketones.

Often a mixture of methanol and water or acetonitrile and water is used as the eluent in liquid chromatography. The capacity ratio decreases as the percentage of organic solvent in the eluent is increased. Usually, a plot of log k versus eluent composition is linear. Thus, the elution of sample compounds can be speeded by increasing the organic content of the eluent or slowed down by decreasing the organic content.

Programmed Elution.   Changes in eluent composition *during* a chromatographic run are desirable for samples in which a given eluent quickly separates the early-eluting sample constituents but elutes the remaining constituents very slowly. Elution of the latter can be speeded up by increasing the organic solvent content of an organic-water eluent. By means of programmed pumps (or other devices that are sold commercially), the two eluent solvents can be mixed in continuously changing proportions throughout the chromatographic run. This creates a solvent gradient so that good resolution of the entire sample can be achieved in minimal elution time. This type of elution is called "gradient elution" or "programmed" liquid chromatography.

[7] H. G. Menet, P. C. Gareil, and R. H. Rosset, *Anal. Chem.*, *56*, 1770 (1984).

23–3. SOME APPLICATIONS OF LIQUID CHROMATOGRAPHY

Separation of Organic Compounds.   Modern liquid chromatography is widely used to analyze mixtures of organic compounds. Samples containing hormones, drugs, vitamins, nucleic acids, pesticides, synthetic organic chemicals, and others can be separated and the quantity of each component estimated. Figure 23–3 shows a liquid chromatogram from separating three common pesticides. Analyses of this type are useful in synthetic research, metabolic studies, residue analysis, and environmental monitoring.

Sample treatment is often needed to obtain a suitable solution for the chromatographic step. For example, to separate and determine the active ingredients in a contraceptive tablet (mestranol and norethindrone), the tablet is extracted with 8 mL of an organic solvent mixture (5% tetrahydrofurane in pentane) and the extract evaporated to 2 mL. Forty microliters (0.040 mL) of this solution are injected into a liquid chromatograph, and the active ingredients are separated from each other in about 12 min [8].

Figure 23–4 shows a separation of 7 nucleosides on a column of μ-Bondapak-C_{18} with a pH-5.8 phosphate eluent. A feature of this separation is the minute amount involved of each substance—only about 80 picomoles. An ultraviolet detector was used at 254 nm. To attain the required sensitivity, the detector was set at 0.02 absorbance unit full scale.

Separation of Metal Ions.   Liquid chromatography can be used for separation of inorganic ions if a suitable derivative is formed first by a chemical reaction. For example, several metal ions react with a sodium dialkyldithiocarbamate salt (Na^+DTC^-) to form complexes, as is illustrated for copper (II):

$$Cu^{2+} + 2R_2NC\overset{S}{\underset{S^-}{=}}Na^+ \rightarrow R_2N-C\overset{S}{\underset{S}{=}}Cu\overset{S}{\underset{S}{=}}C-NR_2 + 2Na^+$$

After reaction with the sodium DTC, the sample contains the DTC complexes of the various metal ions. In a published method [9], an aliquot of the sample is injected into a chromatograph fitted with a C_{18} column of 5-μm particle size. The mixture is then separated by elution with methanol-acetonitrile-water (40:35:25), and the eluted metal complexes are detected spectrophotometrically at 254 nm. A chromatogram of the separated metal complexes is shown in Figure 23–5.

Separation of BioPolymers.   The ability to separate proteins, peptides, polynucleotides, and other large molecules of biochemical interest is tremendously important. Until recently, the separation of these biopolymers by liquid chromatography has been difficult and very limited in scope. The pore size of conventional

[8] *Liquid Chromatography Application No. 2*, Chromatronix, Berkeley, Calif.
[9] S. R. Hutchins, P. R. Hadded, and S. Dilli, *J. Chromatogr.*, *252*, 185 (1982).

Figure 23–3. Separation of pesticides by liquid chromatography. Column is 2 × 500 mm, packed with ETH Permaphase. Eluent is isooctane; flow rate is 0.4 mL/min. (Courtesy of Chromatronix, Berkeley, Calif.)

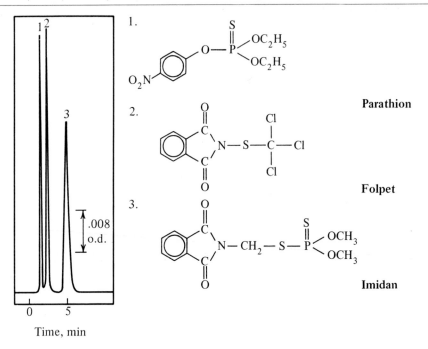

Parathion

Folpet

Imidan

Figure 23–4. Separation of nucleosides on μ-Bondapak. Peaks: (1) cytidine, (2) uridine, (3) xanthosine, (4) inosine, (5) guanosine, (6) thymidine, (7) adenosine. [From R. A. Hartwick and P. R. Brown, *J. Chromatog.*, *126*, 679 (1976). By permission.]

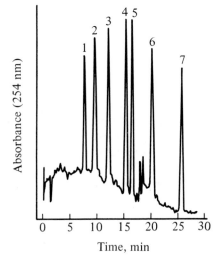

Figure 23–5. Separation of metal diethyldithiocarbamate complexes. Eluent is methanol-acetonitrile-water (40:35:25); detection at 254 nm. Peak identification: A = disulfiram, B = Cd, C = Pb, D = Ni, E = Co, F = Cr, G = Se, H = Cu, I = Hg, J = Te. [From S. R. Hutchins, P. R. Hadded, and S. Dilli, *J. Chromatogr.*, *252* 185 (1982).

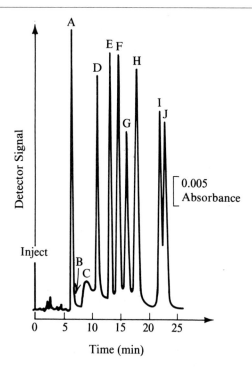

bonded-phase column packing is too small for effective partitioning of such large molecules. However, silica supports with very large pores are now available, and these have made possible some excellent separations of biopolymers. One such support, Vydac 218TP, is 5 μm in diameter and has an average pore size of 330Å; it accommodates molecules up to a molecular weight of 250,000. The spherical particles withstand high pressure so that high flow rates are possible.

The analytical separation of very large molecules is illustrated by the chromatogram of protein standards shown in Figure 23–6. A nice review on high-performance liquid chromatography of biopolymers is given by Regnier [10].

Size Exclusion Chromatography. Size exclusion chromatography (SEC) separates molecules by size on the basis of differences in their ability to permeate the column packing. Very large molecules have greater difficulty in entering the column packing and therefore elute more rapidly from the column. Polymers that have similar chemical structures but different molecular weights can be separated by SEC; the

[10] F. E. Regnier, *Science*, *222*, 245 (1983).

Figure 23–6. Separation of LDH iso-enzymes with a postcolumn enzyme detector. Column dimensions, 4 by 250 mm; packing, DEAE Glycophase/CPG (250-Å pore diameter, 5 to 10 μm particle size). Solvents: A, 0.025M tris pH 8.0; B, 0.025M tris, 0.2M sodium chloride, pH 8.0. Flow rate: 3 mL/min. Peak identification: a, LDH5; b, LDH4; c, LDH3; d, LDH2; e, LDH1. (From F. E. Regnier, *Science*, *222*, 245 (1983). Copyright 1983 by the AAAS.)

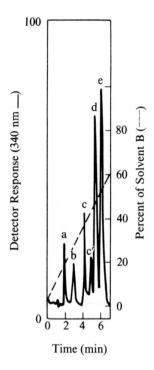

Time (min)

peaks elute in order of decreasing molecular weight. Size exclusion chromatography is particularly useful in biochemical research for separating very large molecules.

Two types of support materials have been used for high-performance SEC—organic gels and surface-modified silica particles. Gel beads are often rather soft and not well suited to rapid separations in which the eluent is pumped through the column under pressure. Porous silica supports are available in a variety of pore sizes so that a suitable material can be selected for the separation to be performed. These silica supports withstand high pressure and give good separations. However, they sometimes *adsorb* the substances to be separated, whereas the desired mechanism is separation according to the size of the sample molecules. Silica supports that are to be used for separation of biopolymers are often treated with a hydrophilic organosilane to cut down on adsorption effects.

Size exclusion chromatography is discussed in a review article [10] and a book [11].

23–4. PLANE CHROMATOGRAPHY

Chromatography can be carried out on a plane surface instead of a packed, cylindrical column. In *paper chromatography*, the separation is carried out on a

[11] W. W. Yau, J. J. Kirkland, and D. D. Bly, *Modern Size Exclusion Liquid Chromatography*, New York: Wiley-Interscience, 1984.

rectangular or circular piece of filter paper. The stationary phase is held by the paper or may be the paper itself. In *thin-layer chromatography*, the solid support is coated in a thin layer on a glass or plastic plate.

Both types of plane chromatography are simple, but effective, micro methods for separating very small quantities of substances. Small volumes of various samples are added near the edge of a rectangular plate, each forming a small spot. Then the edge of the plate is immersed in an eluting solvent, usually in a closed container so that the air becomes saturated with the vapors of the eluent. The eluent is drawn up (or down) the plate by capillary action. As the solvent front advances, the sample components are partitioned between the eluting solvent and the stationary phase and move at different rates along the plate. The final position of the sample components in ascertained by spraying the plate with a color-forming reagent or by using some other visualization technique. The components appear as a series of spots. If separation is complete, each spot is a single sample solute. The process of separating and visualizing the sample solutes sometimes is called "developing" the chromatogram.

The term "retardation factor," R_f, is commonly used to describe the chromatographic behavior of sample solutes under a given set of conditions. The R_f value for each substance is the distance it has moved divided by the distance the solvent front has moved. Usually, the center of each spot is the point taken for the measurement.

$$R_f = \frac{\text{Sample solute distance}}{\text{Solvent front distance}}$$

Paper Chromatography. Paper chromatography is remarkable in its simplicity. To illustrate, suppose a small ink spot is placed about $1''$ from the end of a strip of filter paper and the end immersed in $\frac{1}{2}''$ or less of water. As the water front moves up the filter strip, most inks will be separated into two or more colored constituents.

In typical scientific practice, several samples are spotted near the edge of a large sheet of filter paper, and the edge is immersed in a mixture of organic solvents or some other suitable eluent. Sometimes the paper is first impregnated with a solid or liquid organic phase and an aqueous eluent is used. After the separated spots are visualized, the amount of substance in each spot may be determined by reflectance spectrophotometry—that is, by measuring the intensity of light *reflected* by the spot. Alternatively, the spots are cut from the paper, and the solute is eluted and analyzed by microanalytical methods.

There are literally thousands of articles in chemical journals and other scientific publications dealing with specific examples of separations by paper chromatography. One book concerned exclusively with paper chromatography is nearly one thousand pages long [12].

Thin-Layer Chromatography. Resolution in paper chromatography is limited by fibers and other inhomogeneities in the paper. Thin-layer chromatography (TLC) gives sharper separations and is now preferred over paper chromatography. TLC

[12] I. M. Hais and K. Macek, eds., *Paper Chromatography*, New York: Academic Press, 1963.

employs a plastic sheet covered with a thin, dry layer of sorbent such as silica gel, alumina, cellulose powder, or polyamide. These are available commercially and are prepared by evenly spreading a thick slurry of sorbent plus a binder over the sheet. Heat is then applied to dry the applied layer and to set the binder to give a more adherent coating.

It is helpful to understand something about the two most common sorbents, silica gel and alumina. Silica gel contains —SiOH groups on its surface that can hydrogen-bond strongly to compounds such as phenols, amines, carboxylic acids, and even aromatic hydrocarbons (via the π-electron system). Separation can also occur by the solutes partitioning between water in the silica gel and the eluting solvent. Alumina functions by bonding sample components either to basic OH sites or to positive Al sites on its surface. Highly acidic molecules (such as carboxylic acids) are sorbed so strongly that they are not easily eluted, but aromatic hydrocarbons are more weakly sorbed and can be eluted easily.

A micropipet is used for spotting the sample solutions across the plate a short distance from the bottom. The sample size may range from several micrograms to a few milligrams. Frequently, a plastic template is employed for even spotting and as an aid in labeling the samples. Then the bottom of the plate is immersed in the eluting solvent in a closed chamber. The absorptive nature of the dried sorbent draws the solvent up the plate. If conditions have been chosen correctly, the sample components move up the plate at different rates and are separated from each other.

After the separation is complete, the individual components have usually formed invisible spots on the thin-layer plate or sheet. Numerous ways have been found to visualize these spots (many of these have been described by Bobbitt, Schwarting, and Gritter [13]). A common method for organic samples is to spray the entire plate with a universal reagent such as sulfuric acid, sulfuric acid–dichromate solution, or 1% iodine in methyl alcohol. (This method is not available in paper chromatography, since the sulfuric acid will destroy the paper—a definite advantage for thin-layer chromatography.) After spraying, the plate is heated in an oven. The hot sulfuric acid chars any organic compounds present, producing black carbon spots. The hot iodine reagent forms a brown spot but does not char the compound.

Other spray reagents, more selective, have also been used. For example, ferric chloride reagent can be used to locate phenols in the presence of compounds with no aromatic OH groups. The Fe^{3+} forms a red or violet color with many phenols.

Another visualization method involves the use of fluorescent plates or sheets. The fluorescence of the inorganic phosphor mixed with the sorbent is quenched or destroyed by most organic compounds; under ultraviolet radiation, the spots appear as dark shadows against a brilliant fluorescent background. (See Experiment 35 Chapter 33 for an example of this method.) A number of organic compounds fluoresce themselves; this fluorescence may be used to locate spots under the ultraviolet irradiation of nonfluorescent plates.

[13] J. M. Bobbitt, A. E. Schwarting, and R. J. Gritter, *Introduction to Chromatography*, New York: Reinhold, 1968, pp. 1–83.

Quantitative Thin-Layer Chromatography.   A number of methods may be used to measure the amount of an individual component in a spot [13]. In one general method, each spot can be carefully scraped off with a razor blade or washed off (after cutting out the appropriate section of a plastic sheet), and the sample constituent is analyzed by a spectrophotometric or colorimetric method. Alternatively, a micro-analytical method such as a microtitration may be used.

A second general method is to assay the compound directly on the thin-layer plate. This is done by a scanning instrument that uses a very narrow beam of light and has a photometric detector. The light reflected or absorbed is measured as the light beam moves along the plate. Colored spots of separated compounds show up as peaks, much the same as the recorded peaks in column chromatography. If the compounds are fluorescent, a special type of fluorometer can be used to measure the fluorescence emitted from a spot.

An Example of Thin-Layer Chromatographic Analysis.   A study of the metabolism of glyoxilic acid in bacteria [14] gives a fairly typical example of a practical use of thin-layer chromatography. It also illustrates some of the techniques employed in both paper and thin-layer chromatography. Nine intermediates associated with glyoxilic acid metabolism were to be separated using a glass microfiber support sheet impregnated with silica-gel sorbent. R_f values for the intermediates are shown in Table 23–2. It is impossible to obtain a complete separation of all nine com-pounds with any one eluting solvent mixture. However, the use of *two-dimensional* chromatography does permit a complete separation (Figure 23–7). First, the sample constituents are chromatographed using Solvent I; then the sheet is dried and chromatographed again at right angles to the first separation using Solvent II. The

Table 23–2.   Separation of Intermediates Related to Glyoxylate Metabolism

	R_f Values	
	Solvent I[a]	Solvent II[b]
Glyoxylate	0.17	0.55
Glycolate	0.59	0.50
α-Hydroxyglutarate	0.25	0.17
α-Hydroxyglutaryl lactone	0.33	0.73
Citramalate	0.41	0.21
Malate	0.17	0.11
α-Ketoglutarate	0.45	0.35
Succinate	0.80	0.65
Fumarate	0.97	0.70

Source: Data from A. S. Bleiweiss, H. C. Reeves, and S. J. Ajl, *Anal. Biochem.*, *20*, 335 (1967).

a. Petroleum ether (bp 30–60 °C), diethyl ether, and formic acid in the ratio 28:12:1.

b. Chloroform, methyl alcohol, and formic acid in the ratio 80:1:1.

[14] A. S. Bleiweis, H. C. Reeves, and S. J. Ajl, *Anal. Biochem.*, *20*, 335 (1967).

Figure 23–7.   Thin-layer chromatogram of intermediates associated with glyoxy-lic acid metabolism. The acids are numbered as follows: (1) fumarate, (2) succinate, (3) glycolate, (4) α-ketogluterate, (5) citramalate, (6) α-hydroxyglutarate, (7) malate, (8) α-hydroxyglutaryl lactone, and (9) glyoxylate. The origin is indicated by the circle at the baseline. [From *Anal. Biochem.*, **20**, 335 (1967). Copyright © 1967 by Academic Press. By permission.]

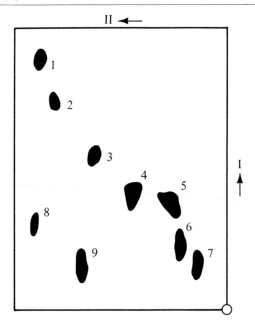

spot locations are shown by spraying with an acid-base indicator, bromophenol blue.

Some samples contained products obtained from the metabolism of carbon 14-labeled sodium glyoxylate. After spraying to visualize the separated spots, each substance is measured quantitatively by cutting its spot from the glass-fiber sheet and counting the radioactivity in the spot. This is a good way to measure very small amounts with reasonably good accuracy.

High-Performance TLC.   Although TLC is certainly a valuable technique, its separating ability is not as good as high-performance column chromatography. New techniques are now available that correct some of these shortcomings and make TLC also "high-performance" (HPTLC). The main changes are similar to those made in liquid chromatography a few years back. In HPTLC, the thin-layer support has a smaller particle size and a narrower particle-size distribution. The sample size, being significantly smaller, requires the use of a precision sample applicator.

The above changes result in a sharp reduction in solvent migration velocity. However, much shorter migration distances are needed for separation, so the net result is a decrease in separation times. HPTLC typically gives around 4000 theoretical plates for a migration distance of 3 cm and a time of 10 min [15]. Calculation of the height equivalent of a theoretical plate, H, shows this method to be unusually efficient:

$$H = \frac{L}{N} = \frac{30 \text{ mm}}{4000} = 0.0075 \text{ mm}$$

By comparison, ordinary TLC typically generates around 2000 theoretical plates for a migration distance of 12 cm and a time of 25 min. Calculation of H gives

$$H = \frac{120 \text{ mm}}{2000} = 0.06 \text{ mm}$$

    After spot visualization, each sample track on the plate is scanned by a microphotometer that detects how strongly each spot reflects or transmits a pinpoint of light traveling along the plate. The detector then plots reflectance or transmittance versus distance, generating a Gaussian-shaped peak for each spot. The entire chromatogram is not unlike that of a column-chromatographic separation.

    The theory and practice of HPTLC is discussed in a recent book [16].

QUESTIONS AND PROBLEMS

Principles of Column Chromatography

1. What is meant by LLC and by LSC? Explain the difference between "normal" and "reversed-phase" LLC.
2. Compare classical liquid chromatography with modern high-performance liquid chromatography.
3. Give the major advantages of using a chemically bonded stationary phase in LC over the older stationary phases.
4. A typical HPLC column is 4.6 mm × 15 cm and is packed with spherical particles that are 3 or 5 μm in diameter. Give typical column dimensions and packings employed in (a) "fast" LC columns and (b) microbore LC columns.
5. An LC column that is 10 cm in length and packed with 3-μm particles gives a plate height, H', of 0.008 mm.

    (a) Calculate N'.
    (b) If $t_0 = 30$ sec and 4 peaks have retention times $t_1 = 89$ sec, $t_2 = 92$ sec, $t_3 = 143$ sec, and $t_4 = 150$ sec, calculate the capacity ratio for each peak.
    (c) Calculate the resolution of peaks 1 and 2 and of peaks 3 and 4 assuming no extra-column broadening.
    (d) Calculate the resolution of peaks 1 and 2 and of peaks 3 and 4 assuming $\sigma_{ec}^2 = 0.210$ sec^2.

6. If $N' = 1,000,000$ for a 22-m microbore column and the extra-column peak broadening is

[15] T. H. Jupille and L. J. Glunz, *American Lab*, May 1977, p 85.
[16] W. Bertsch, S. Hara, R. E. Kaiser, and A. Zlatkis, *Instrumental HPTLC*, Heidelberg: Hüthig, 1980.

considered to be negligible, calculate the lowest value of separation factor (α) for two peaks of average $k = 5.00$ to be separated with $R_s = 1.00$.

7. Phenol has a capacity ratio $k = 19$ with a 50% methanol ($+50\%$ water) eluent and $k = 9.8$ with a 60% methanol eluent. 3,5-Dimethylphenol has $k = 52$ with 50% methanol and $k = 29$ with 60% methanol. Assuming a linear plot of $\log k$ versus percent methanol, calculate k for each phenol with 80% methanol eluent. If $t_0 = 40$ sec, calculate the retention time for each phenol with 80% methanol.

Gradient Elution

8. What is meant by gradient elution in LC? Explain the advantages of this type of elution over isocratic elution, which is elution with a fixed eluent.

9. A solute is eluted with an acetonitrile (AN)–water gradient starting with 40% AN (60% H_2O) and increasing the AN percentage by 5%/min. The following capacity ratios are measured: At 40% AN, $k = 20.5$; at 45% AN, $k = 15.2$; at 50% AN, $k = 11.3$.

   (a) Assuming a linear plot of $\log k$ versus percent AN, calculate k at 55%, 60%, 65%, 75%, and 85% AN.

   (b) If the linear eluent flow rate is 10.0 cm/min, estimate the distance the solute moves the first 2 min, also the distance it moves from 2 to 4 min, from 4 to 6 min, and from 6 to 8 min. (*Note:* An estimate of the distance is obtained by taking k at 55% AN as the average for the first 2 min, k at 65% AN as the average for the next 2 min, etc. See Chapter 21 for the relationship of peak movement to k.)

LC Methods

10. A laundry detergent contains a low percentage of a water-soluble polyacrylate (molecular weight is several thousand) as well as various organic and inorganic chemicals of much lower molecular weight. Suggest a way to separate the polyacrylate from the other chemicals so that the percentage of polyacrylate in the sample can be determined.

11. Sugars (glucose, fructose, lactose, etc.) are very polar compounds with several hydroxyl groups in the molecule. Suggest a reversed-phase support for separating sugars by LC. Since most sugars do not absorb in the visible or ultraviolet spectral region, suggest a suitable detector for the separation.

12. Polymerization reactions generally give products of a fairly broad range of molecular weights. Explain how a mixture of polymers could be separated chromatographically according to their molecular weights. What will be the order of elution of the polymers?

13. Some organic compounds have the ability to rotate polarized light to the right (D-isomers), and some rotate it to the left (L-isomers). Find an example of separation of D- and L-isomers by liquid chromatography in the recent chemical literature. Give the essential chromatographic conditions for the separation and explain what special type of stationary phase is needed for such a separation.

14. Consult recent scientific journals and find one paper dealing with the column-chromatographic separation of each of the following: (a) drugs or drug metabolites, (b) steroids, (c) metal-ion complexes, (d) pesticides, (e) vitamins or sugars. Give the complete journal reference and outline the conditions for the separation.

Thin-Layer Chromatography

15. List as many methods as possible for visualizing the location of organic substances separated using thin-layer chromatography.

16. Define retardation factor, R_f. Derive an equation that relates R_f to capacity ratio, k.

17. Explain what is meant by two-dimensional TLC and give the advantages of this technique.

18. Explain briefly how high-performance TLC differs from ordinary TLC in technique and performance. Explain how spots are quantified in HPTLC.

19. Briefly describe one example of HPTLC that you have found in a recent scientific journal. Give the complete journal reference.

24

Ion-Exchange Chromatography

24–1. INTRODUCTION

Ion exchange has many uses in analytical chemistry. Often only one ion (the cation or the anion) rather than both must be determined or separated. In such a case, that ion may be *exchanged* in a reaction for another ion, by using a substance called an *ion-exchange resin* that is available in the form of beads. The ion-exchange reaction is performed in a column packed with ion-exchange resin beads. The sample solution is added to the column and washed through with water or another liquid eluent. A cation-exchange resin is used for the exchange of cations; an anion-exchange resin is used for the exchange of anions.

Some Analytical Applications. A very useful analytical application of ion exchange is exchanging a sample ion that is difficult to determine, such as Na^+, for an exactly equivalent amount of an ion that is easily titrated, such as H^+. An ion-exchange column also may be used to exchange an anion or a cation that might interfere in an analysis for an anion or a cation that will cause no interference. It is also possible to adjust ion-exchange conditions so that several cations or anions in a sample will move down an ion-exchange column at different rates and thereby be separated *chromatographically*. Such ion-exchange chromatography, or *ion chromatography* (as the modern chromatographic separation of ions is often called), has become a very popular and important area of modern chromatography.

In this chapter, ion-exchange resins and the general principles of ion-exchange equilibria are discussed first. The final two sections deal with the principles and applications of modern anion chromatography and cation chromatography, respectively.

The principles and examples of ion exchange in analytical chemistry are covered in books [1, 2]. Modern ion chromatography is covered in more recent books [3, 4] and in various review articles [5].

24–2. ION-EXCHANGE RESINS

Classical Resins. Although certain glasses (Chapter 16), clays, minerals, and organic substances can exchange one ion for another, the most useful ion exchangers are organic polymers called resins. Those that exchange cations are called "cation-exchange resins," or *catex* for short. Those that exchange anions are called "anion-exchange resins," or *anex* for short.

The first step in making ion-exchange resins is usually the suspension polymerization of styrene to form small spherical particles of polystyrene. Some divinylbenzene (DVB) is added to the mixture to provide "cross-linking" between the linear polystyrene chains.

> The percentage of DVB in a polymerization mixture is often referred to as the cross-linking percentage of the resin.

Sulfonation of a styrene-DVB copolymer yields a cation-exchange resin (see Figure 24–1). The sulfonic acid group $—SO_3H^+$ are the active sites used for ion exchange. The $—SO_3^-H^+$ group is chemically bound to the resin, but the counter ion (H^+) is more free to move about and can be exchanged for another cation. For example, if a solution of sodium chloride is brought into contact with a catex resin in the hydrogen form, the following exchange reaction occurs:

$$R—SO_3^-H^+ + Na^+ \rightarrow R—SO_3^-Na^+ + H^+$$

If this reaction goes to completion, the resin is said to be in the sodium form.

If a polystyrene-DVB resin is chloromethylated and then reacted with an amine, the result is an anion-exchange resin. The most commonly used type has a quaternary ammonium group:

$$2R—CH_2N^+R_3Cl^- + SO_4^{2-} \rightleftharpoons (R—CH_2N^+R_3)_2SO_4^{2-} + 2Cl^-$$

In this example, two active sites are required for each sulfate because the sulfate anion has a double negative charge.

The *capacity* of an ion-exchange resin is the amount of exchangeable ion per gram of resin. The capacity of sulfonated cation-exchange resins is usually about

[1] W. Rieman and H. F. Walton, *Ion Exchange in Analytical Chemistry*, Oxford: Pergamon, 1970.

[2] O. Samuelson, *Ion-Exchange Separations in Analytical Chemistry*, New York: John Wiley, 1963.

[3] D. T. Gjerde and J. Fritz, *Ion Chromatography*, 2nd ed. Heidelberg: Hüthig, 1986.

[4] F. C. Smith and R. C. Change, *The Practice of Ion Chromatography*, New York: John Wiley, 1983.

[5] J. S. Fritz, *LC Magazine*, 2, 446 (1984).

Figure 24–1.   Chemical structure of a cation-exchange resin.

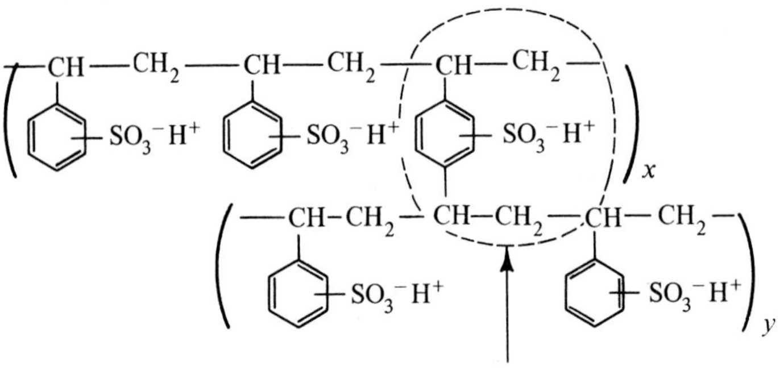

DVB Cross-Linking Unit

5 meq/g of dry hydrogen-form resin or 1.8 meq/mL of wet resin bed in a column. The capacity of quaternary ammonium types of anion-exchange resin is about 3–3.5 meq/g of dry chloride-form resin or 1.2 meq/mL of wet resin bed. *Particle size* is also important. Ion-exchange resins should be sieved to avoid large variations in particle size. Resin of 50–100 mesh or of 100 or 200 mesh is suitable for most analytical purposes.

   Ion-exchange resins exchange ions throughout the beads and not just on the surface. When a dry resin is placed in an aqueous solution, it takes up quite a bit of water and swells in the process; hence it is often called a gel-type resin. If it is a catex resin, cations from the solution can enter into the gel and exchange with the resin cation. Anions from the solution cannot penetrate the gel because of the electrostatic repulsion of the fixed $-SO_3^-$ ions. By the same token, anions can penetrate the gel of an anex resin, but cations are unable to do so because of the $-CH_2N^+R_3$ cation.

   Cross-linking affects the properties of either a catex or an anex resin. Resins with very low cross-linking (1–2%) swell excessively when they become wet and shrink when dried. Swelling or shrinking may also occur when the resin is converted from one form to another, making it difficult to maintain a properly packed column. A resin with around 8% cross-linking is less susceptible to volume changes and is commonly used in analytical and industrial ion-exchange columns.

Newer Resins.   A newer type of ion-exchange resin is finding increasing use. These are highly cross-linked "macroporous" polymers that do not form gels and hence undergo very little swelling or shrinking. The polymerization is conducted to form hundreds of hard "microspheres" inside each bead, connected by relatively large pores or channels. Macroporous polymers are useful sorbents for organic compounds (Chapter 23); ion-exchange resins made from them are especially useful for ion-exchange processes in nonaqueous solvents.

   Resins that are used in modern ion-exchange chromatography usually have a much lower exchange capacity than older resins. Sometimes the capacity of the newer resins is only 0.005–0.01 meq/g. The advantage of a low capacity is that sharper chromatographic peaks are obtained and that very dilute eluents can be used for chromatographic separations. Actually, the ion-exchange groups on these resins tend to be mostly near the outer part of the resin bead. It has been shown that sulfonation of spherical polystyrene beads under very mild conditions will position the sulfonic acid groups on the outer perimeter of the bead. Directions for preparing anion-exchange resins of very low capacity have been published [6].

Ion-exchange resins with a silica base are also used in modern ion-exchange chromatography. These consist of small, uniform, spherical beads of porous silica to which an organic silicon compound has been chemically or physically bonded. The organic bonding compound contains a sulfonic acid group to make the resin a catex or a quaternary ammonium group to make the resin an anex.

24–3. ION-EXCHANGE EQUILIBRIA

Selectivity Coefficient. Suppose we have a sample ion, A, and an eluent ion, E, which has the same ionic charge as A. If we start with the resin all in the E form and shake it up with a solution containing A, the following equilibrium will occur.

$$A_s + E_r \rightleftharpoons A_r + E_s \qquad (24-1)$$

Here the subscript "s" refers to the solution phase and "r" to the resin. This exchange can be expressed by an equilibrium constant, K, called the *selectivity coefficient*:

$$K = \frac{[A]_r[E]_s}{[A]_s[E]_r} \qquad (24-2)$$

Distribution Coefficient. The weight distribution coefficient, D_g, for the exchanged sample ion is given by:

$$D_g = \frac{[A]_r}{[A]_s} = \frac{\text{mmoles of } A_r/\text{g resin}}{\text{mmoles of } A_s/\text{mL solution}} \qquad (24-3)$$

Note that D_g is similar to the concentration distribution ratio, D_c, used in solvent extraction (Chapter 20).

Experimentally, the weight distribution coefficient of an ion, A, in equilibrium with an eluent ion is measured by equilibrating a known volume of a standard solution with a known weight of resin and then determining how much A has been taken up by the resin and how much remains in solution. The value of D_g is obtained by substituting these results into Equation 24–3.

Capacity Ratio. In ion-exchange chromatography, it is convenient to use the capacity ratio, k, instead of a distribution coefficient. The capacity ratio for an ion, A, that is being eluted from a column by an eluent ion, E, is the amount of A in the resin phase of the column, divided by the amount of A in the solution phase of the

[6] R. E. Barron and J. S. Fritz, *Reactive Polymers, 1,* 215 (1983).

column. Substituting k into Equation 24–3,

$$D_g = \left(k \cdot \frac{\text{mL solution}}{\text{g resin}} \right) \tag{24-4}$$

$$k = D_g \left(\frac{\text{g resin}}{\text{mL solution}} \right) \tag{24-5}$$

These equations show that k can be calculated from the D_g, obtained by batch equilibration of solution and resin, provided that both the grams of resin and the milliliters of solution in the ion-exchange column are known. However, it is usually much easier to simply elute the ion A from the column and calculate its capacity ratio from the retention time, t:

$$k = \frac{t - t_0}{t_0} \tag{24-6}$$

where t_0 is the dead time of the column (see Chapter 21). In brief, retention times and capacity ratios can be obtained and used in ion-exchange chromatography exactly as in ordinary liquid chromatography.

Effect of Chromatographic Conditions on Retention Time. Suppose the sample ion A has a charge of n ($n+$ for a cation, $n-$ for an anion) and that the eluent ion, E, has a charge of 1. Then Equations 24–1 and 24–2 can be rewritten as

$$A_s + nE_r \rightleftharpoons A_r + nE_s \tag{24-7}$$

$$K = \frac{[A]_r[E]_s^n}{[A]_s[E]_r^n} \tag{24-8}$$

Since D_g is proportional to the capacity ratio and retention time is proportional to the capacity ratio, let us replace $[A]_r/[A]_s$ in Equation 24–8 by $t'\alpha$. Here t' is the adjusted retention time ($t' = t - t_0$) and α is a proportionality factor. In a typical chromatographic run, only a small fraction of the total resin capacity of the column is occupied by a sample ion. Therefore, the $[E]_r$ term in Equation 24–8 is a reasonable measure of resin capacity. Making these substitutions and rearranging terms, we get

$$K = \frac{t'\alpha[E]^n}{(\text{Cap})^n} \tag{24-9}$$

$$t' = \frac{(\text{Cap})^n K}{[E]^n \alpha} = \frac{(\text{Cap})^n (\text{Constant})}{[E]^n} \tag{24-10}$$

Taking the log of both sides of this equation,

$$\log t' = n \log (\text{Cap}) - n \log [E] + \log (\text{Constant}) \tag{24-11}$$

This important equation states that *increasing* the concentration of the eluent ion will *decrease* the adjusted retention time. It also predicts that using a resin of lower capacity will make it possible to use an eluent of lower concentration and still obtain a reasonable value for t'.

Figure 24–2. Plot of log adjusted retention time versus log eluent concentration for divalent metal ions. (Courtesy of G. J. Sevenich.)

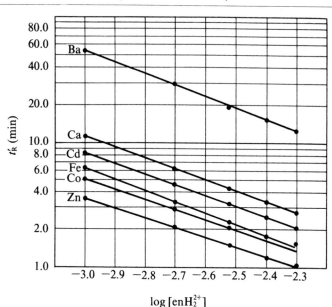

$$\log[\mathrm{enH}_2^{2+}]$$

Equation 24–11 further predicts that, for a given resin, a plot of $\log t'$ versus $\log[E]$ will be linear with a slope of $-n$. The validity of this prediction is demonstrated in Figure 24–2, in which a family of straight lines is obtained for $2+$ metal ions eluted from a cation-exchange column with H^+ (from perchloric acid). Each line has the expected slope of -2. The lines are displaced vertically because each metal cation has a different selectivity coefficient. (The selectivity coefficient is included in the log (Constant) term in Equation 24–11.)

24–4. SOME ANALYTICAL APPLICATIONS OF ION EXCHANGE

Determination of Salts. By using a cation-exchange resin in the hydrogen form, the cation content of most soluble salts may be determined. The reaction is

$$MA_z + z\,\mathrm{Res-H^+} \rightleftharpoons z\,HA + \mathrm{Res}_z\mathrm{-M^{z+}}$$

The cation content of the original sample is determined by titrating the acid formed, HA. One equivalent of hydrogen ion is formed for each equivalent of metal. Thus, a divalent metal cation gives two molecules of HA for titration.

In some cases, salt solutions also can be determined with an anion-exchange resin in the OH^- form. The reaction is

$$M_zA + z\,\mathrm{Res-OH^-} \rightleftharpoons z\,MOH + \mathrm{Res}_z\mathrm{-A^-}$$

The MOH is titrated with standard acid. This method is useful for alkali metal salts and a few other salts. It is limited by the fact that many metals form insoluble precipitates in the presence of strong bases.

Phosphates can be determined in the presence of other anions because phosphoric acid gives a double end point when titrated with a strong base. The exchange reaction is

$$\left.\begin{matrix} M^{z+} \\ H_2PO_4{}^-, HPO_4{}^{2-}, \text{etc.} \end{matrix}\right\} \xrightarrow{\text{Catex-H}^+} H_3PO_4$$

Other sample components in the column give other acids, such as hydrochloric, sulfuric, or nitric acid. These acids are titrated along with the first hydrogen of phosphoric acid. The amount of phosphate is calculated from the difference between the first and second end point breaks observed in titrating the mixed acids with sodium hydroxide.

Deionization.   Ion exchange is a useful means of deionizing water. Deionization is valuable to the analytical chemist, as it provides water of very high purity. The water to be purified is passed through an ion-exchange column containing a mixture of catex resin in the hydrogen form and anex resin in the hydroxyl form. The cation impurities in the water are exchanged for hydrogen and the anion impurities for hydroxyl. The net effect is that the impurities are converted into water.

Replacement of Interfering Ions.   It is frequently possible to exchange an unwanted ion for one that will not interfere in an analytical determination. For example, ferric iron and potassium ion interfere seriously in sulfate determination by coprecipitating with barium sulfate. This interference can be avoided if the sample solution is passed through a cation-exchange column in the hydrogen form, thus exchanging the metal impurities for hydrogen ions. Then the sulfate can be precipitated without cation coprecipitation.

Another example is the removal of phosphate, which interferes in the determination of calcium and certain other metal ions. In this case, the sample can be passed through an anion-exchange column in the chloride form to replace the phosphate with chloride, which does not interfere.

Separation of Metal Ions From Each Other.   This is one of the most important of all analytical applications of ion exchange. It will be discussed in the next sections.

24–5. ANION CHROMATOGRAPHY

The quantitative determination of fluoride, chloride, bromide, nitrate, sulfate, phosphate, and other common inorganic anions has always been a difficult analytical problem. This is especially true when the sample contains several different anions and when the concentration is quite low. However, in the middle to late

1970s, rapid chromatographic methods were developed that provided an answer to this long-standing problem.

Suppressed Systems. In 1975, a new method for separating and determining anions was invented [7] and sold commercially. A schematic diagram of the system is shown in Figure 24–3. The system consists of a separator column and a suppressor column. It is first equilibrated with a dilute, basic eluent such as sodium carbonate–sodium bicarbonate. Then the sample is injected, and the sample anions are separated on the *separator* column, which contains an anion-exchange resin of fairly low capacity. The separated anions then pass through a *catex suppressor column*, where the counter ion for each anion is changed from Na^+ to H^+. (H^+ is a very mobile ion and conducts electricity extremely well.) Then the H^+ and sample anion pass through the electrical conductivity detector, and the increase in conductance is recorded on a strip-chart recorder.

If there were no suppressor column, the conductance of the sodium carbonate eluent would be so high that any change in conductance due to the sample ions would be difficult to detect. However, the catex suppressor converts sodium carbonate to carbonic acid:

$$2Na^+ + CO_3^{2-} + 2\,Catex{-}H^+ \rightarrow 2\,Catex{-}Na^+ + H_2CO_3$$

Carbonic acid is very weakly ionized, and the background conductance is therefore quite low.

Figure 24–3. Dionex system.

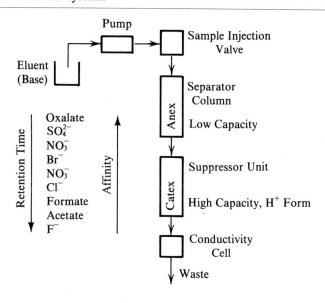

[7] H. Small, T. S. Stevens, and W. C. Bauman, *Anal. Chem.,* **47,** 1801 (1975).

In this system, the catex suppressor column requires frequent regeneration with strong acid to convert from the Na^+ form back to the H^+ form. More recently, this column has been replaced with a membrane suppressor unit that performs the same function but is continuously regenerated.

The performance of suppressed anion chromatography is sometimes demonstrated by separation of a mixture containing fluoride, chloride, nitrite, phosphate, bromide, nitrate, and sulfate in only 10–20 min. Concentrations in the low ppm range of these anions can be detected. Anions of very weak acids, such as borate and cyanide, cannot usually be separated because ionization of H_3BO_3, HCN, etc., is too slight.

Single-Column Methods. In 1979, a simple chromatographic method for rapid separation of common anions was developed at Iowa State University [8, 9]. This new system for anion chromatography uses an anion-exchange separator column connected directly to the conductivity detector. No suppressor column is required. This development was made possible by two principal innovations: (1) the use of an anion-exchange of very low capacity (mostly 0.007–0.09 meq/g) and (2) the use of an

Figure 24–4. The comparison of the separation of mixed monovalent and divalent anions on two resins with different functional groups. The resin on the left is a TMA resin with a capacity of 0.046 meq/g. The resin on the right is a THA resin with a capacity of 0.043 meq/g. The eluent is 0.4 mM KHP. (From R. E. Barron and J. S. Fritz, *J. Chromatogr.*, *316*, 201 (1984). With permission.)

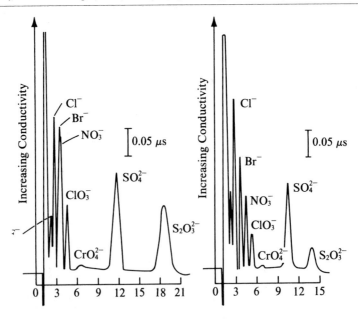

[8] D. T. Gjerde, J. S. Fritz, and G. Schmuckler, *J. Chromatogr.*, *186*, 509 (1979).
[9] D. T. Gjerde, G. Schmuckler, and J. S. Fritz, *J. Chromatogr.*, *187*, 35 (1980).

eluent in a $1.0 \times 10^{-4}M$ to $5.0 \times 10^{-4}M$ solution of sodium benzoate or potassium phthalate.

Very sharp separations of anions can be obtained as shown by the examples in Figure 24–4. The sample separated contained anions in the low ppm concentration range. Figure 24–4 also shows a better separation with an anex that contained hexyl groups (THA) than one with methyl groups (TMA) [10].

Single-column methods that use a conductivity detector require that the eluent anion have a lower equivalent conductance than the sample anions. The sodium benzoate eluent has an equivalent conductance of 82 (mhos cm²/equiv.) (Na = 50, benzoate = 32). Most inorganic anions have a much higher conductance than benzoate; $F^- = 54$, $Cl^- = 76$, $NO_3^- = 71$. Thus, the equivalent conductance of nitrate (with sodium as the accompanying cation) is 121, which is appreciably higher than that of the eluent.

Applications. Anions can be determined in drinking water, boiler water and condensate from steam turbines, clinical samples, beverages, and a whole host of industrial samples. Figure 24–5 shows the determination of fluoride in toothpaste.

Figure 24–5. Determination of fluoride in toothpaste. Eluent is $1.0 \times 10^{-3}M$ succinic acid. (Courtesy of Dean DuVal.)

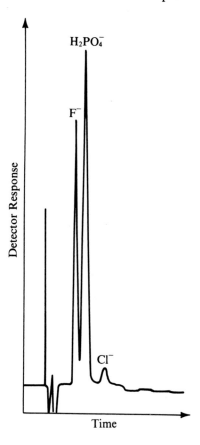

[10] R. E. Barron and J. S. Fritz, *J. Chromatogr., 316*, 201 (1984).

A weighed sample was mixed with water, and the solid abrasive was filtered off. The organic coloring material was removed by adsorption, and the remaining solutions was injected directly into an ion-chromatographic instrument.

24–6. CATION CHROMATOGRAPHY

Single-Column Methods with a Conductivity Detector. Metal cations and some organic cations can be separated very nicely by single-column ion chromatography with a conductivity detector. The separation is performed on a cation-exchange column of low capacity using a very dilute eluent. Instead of using an eluent of lower equivalent conductance than the sample ions (as in anion chromatography), cation chromatographic separations often use an eluent of *higher* equivalent conductance than the sample ions.

As in an example, the separation of alkali-metal and ammonium cations in Figure 24–6 uses approximately $1.3 \times 10^{-3} M$ nitric acid as the eluent. Nitric acid has an equivalent conductance of 421 ($H^+ = 350 + NO_3^- = 71$), while a typical

Figure 24–6. Separation of alkali-metal cations and ammonium cation on a low-capacity cation-exchange column using a conductivity detector. Eluent was 1.0mM nitric acid. Column was 350×2.0 mm packed with 0.059 meq/g cation-exchange resin. (Courtesy of G. J. Sevenich.)

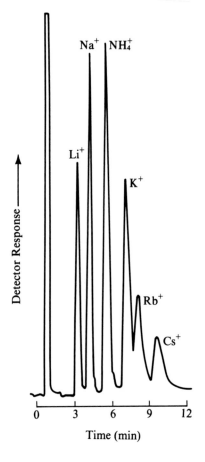

alkali metal–nitrate ion pair has an equivalent conductance of 145 ($K^+ = 74 +$ $NO_3^- = 71$). The conductance of the eluent is *decreased* when an alkali metal ion is eluted, and the "peaks" are actually in the direction of decreased conductance. In effect, the chromatogram in the figure has been turned upside down.

The peaks in Figure 24–6 can be explained as follows. The baseline is established at a relatively high conductance with nitric acid eluent flowing through the system. The sample is injected into the column as a short "plug," which has a lower conductance than the eluent. When this plug passes through the detector, a peak of decreased conductance called a "pseudo-peak" is recorded. After the sample plug has passed, the baseline is quickly restored to that obtained with the eluent alone. However, the alkali metal cations gradually move down the column, pushed by the mass-action effect of the eluent H^+ on the ion-exchange equilibrium. The total cation concentration in solution is fixed by the eluent cation concentration because an alkali metal cation can return to the solution from the resin phase only by

Figure 24–7. Separation of divalent metal ions using ethylenediammonium tartrate eluent. (Courtesy of G. J. Sevenich.)

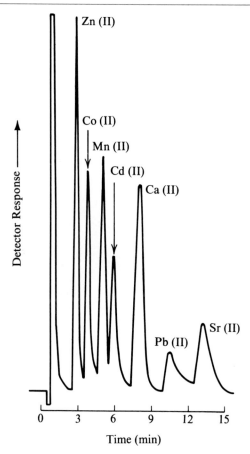

exchange of an equivalent number of eluent H^+. So when an alkali metal ion passes through the detector, the concentration of highly conductive H^+ is reduced by exactly the concentration of alkali metal ion, which has a much lower equivalent conductance. Therefore, the *decrease* in conductivity is directly proportional to the concentration of an alkali metal ion.

Divalent metal ions such as Mg^{2+}, Ca^{2+}, Sr^{2+}, and Ba^{2+} can also be separated by single-column ion chromatography using a conductivity detector. In this case, the divalent cation of ethylenediamine, $[H_3NCH_2CH_2NH_3]^{2+}$, is used in the eluent. This cation has a higher equivalent conductance than the sample cations, so peaks of decreasing conductance are again obtained. Even better separations can be obtained if an eluent containing a weakly complexing anion is employed. Figure 24–7 shows the separation of several divalent metal ions using a solution of ethylenediammonium tartrate $[H_3NCH_2CH_2NH_3]^{2+}[tartrate]^{2-}$, as the eluent.

Single-Column Methods with Spectrophotometric Detection. Direct spectrophotometric detection is used extensively for organic compounds separated by liquid chromatography (Chapter 23). However, many organic and inorganic cations do not absorb appreciably, in either the visible or ultraviolet spectral regions. One possibility for detecting such compounds is to mix the liquid coming from a chromatographic column with a reagent that will react rapidly with a sample cation to form a colored product. The reaction takes place in what is often called a "post-column reactor." A schematic diagram of this reaction and detection technique is shown in Figure 24–8.

Figure 24–8. Schematic of post-column reactor.

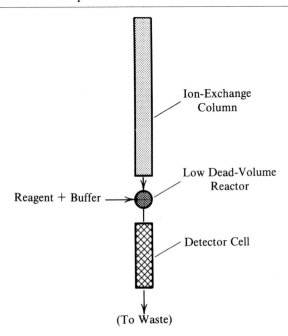

Ion-Exchange Column

Low Dead-Volume Reactor

Reagent + Buffer

Detector Cell

(To Waste)

Figure 24–9. Chromatographic determination of water hardness. A sample was injected every 30 sec to demonstrate reproducibility. The first peak is Mg^{2+}, and the second is Ca^{2+}. (Courtesy of G. J. Sevenich.)

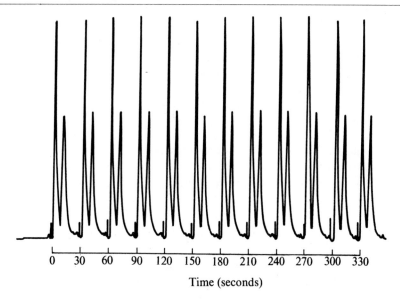

Figure 24–10. Separation of the lanthanides on 13-μm Aminex A-5. Eluent, gradient HIBA 0.17 m to 1 m; pH 4.6. Detection at 600 nm after post-column reaction with Arsenazo T. [Reprinted with permission from S. Elchuk and R. M. Cassidy, *Anal. Chem.*, *21*, 892 (1974). Copyright 1974 American Chemical Society.]

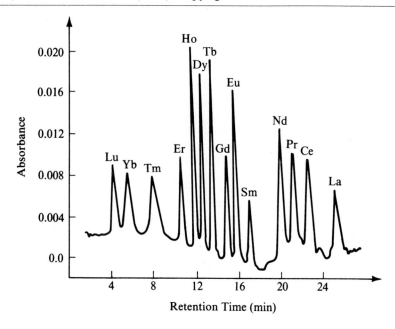

Magnesium and calcium cations in hard water and other aqueous samples can be separated very quickly on a modern cation-exchange column with an eluent such as dilute perchloric acid. The liquid coming off the column is mixed with reagent stream containing Arsenazo I (see Chapter 11), and a buffer to provide an alkaline pH for the mixed solutions. When either Ca^{2+} or Mg^{2+} is in the column stream, a reaction occurs with Arsenazo I to form a colored complex, which is detected as it passes through the detector cell. Figure 24–9 shows reproducible chromatograms for separations of calcium and magnesium in the same water sample.

More complicated samples can be separated by cation chromatography, and the metal ions can be detected spectrophotometrically after formation of colored complexes. Figure 24–10 shows a beautiful answer to a classical analytical problem—the separation of all the rare earth cations. This separation was done with a complexing eluent of gradually increasing concentration so that the separation could be completed more quickly.

QUESTIONS AND PROBLEMS

Resins

1. What is meant by the "capacity" of an ion-exchange resin? How do the more modern ion-exchange resins differ in capacity from the classical resins?
2. Explain what is meant by cross-linking and why it is desirable in a resin. What is meant by a macroporous resin, and what advantages do such resins have over gel resins?
3. Sulfonated cation-exchange resins are strong acids and therefore will be ionized and exchange cations even in solutions of very acidic pH (0–1). Another type of resin, known as a "weak-acid resin," has a carboxyl group (—COOH) that has a pK_a value of 4–5. Explain what pH limitations this type of cation-exchange resin might have.

Ion-Exchange Equilibria

4. Define each of the following terms: (a) selectivity coefficient, (b) distribution coefficient, (c) capacity ratio, (d) adjusted retention time.
5. Calculate the distribution coefficient, D_g, for calcium(II) on a cation-exchange column from the following data: 50 mL of a solution containing 1000 mg/L of dissolved calcium carbonate in $1.0M$ hydrochloric acid is equilibrated with 1.000 g of a sulfonic acid resin. After equilibration, 25.0 mL of the solution requires 13.6 mL of $0.0100M$ EDTA to titrate the calcium(II).
6. An anion-exchange column when packed contains 1.90 g of resin and 1.25 mL of aqueous solution. With a given eluent, NO_3^- has a retention time of 5.60 min. The dead time of the column at the flow rate used is 0.70 min. Calculate the values of k and D_g for NO_3^-.
7. Using the data from the previous problem, calculate the value of the selectivity coefficient for NO_3^- if a $0.001M$ solution of a 1 − anion is used as the eluent and if the resin used has a capacity of 0.010 mmole/g. Assume that most of the resin capacity is occupied by the eluent anion.
8. On a given column, Na^+ has an adjusted retention time of 1.06 min and K^+ an adjusted retention time of 1.23 min using an eluent that is $0.002M$ in H^+. Assuming theoretical

behavior, calculate the adjusted retention times for Na^+ and K^+ on the same column using an eluent $0.001 M$ in H^+.

9. Using data from F. W. E. Strelow, *Anal. Chem.*, *32*, 1185 (1960), plot log distribution coefficient versus log eluent concentration for Mg^{2+} and for Ag^+. Calculate the slope of each plot to the nearest whole number. What is the dependence of log D_g on the concentration of acid used to elute a metal cation? (*Note: D_c is proportional to D_g and hence to k and t'.*)

10. Calculate the theoretical slope of log adjusted retention time of a sample anion plotted against log phthalate ion concentration in the eluent for (a) SO_4^{2-} and (b) I^-. Assume that the eluent pH is such that all of the phthalate exists as the $2-$ anion.

11. Continuing the previous problem, suppose that iodide has an adjusted retention time of 5.4 min and sulfate an adjusted retention time of 4.4 min when a $5.0 \times 10^{-4} M$ phthalate eluent is used.

    (a) Calculate the adjusted retention times for both iodide and sulfate with an eluent concentration of $1.0 \times 10^{-4} M$ phthalate.
    (b) Calculate t' for iodide and sulfate with an eluent concentration of $1.0 \times 10^{-3} M$ phthalate.

Methods

12. A fertilizer contains ammonium phosphate, potassium phosphate, and ammonium nitrate. Outline a method for determining the phosphate content based on use of an ion-exchange and a subsequent acid-base titration.

13. The amount of free acid is aluminum salt solutions is difficult to determine because aluminum(III) hydrolyzes and eventually precipitates when a titration with sodium hydroxide is attempted. Suggest an ion-exchange method for determining free hydrochloric acid in an aluminum chloride solution of known aluminum content. Describe how to calculate the results.

14. An organic solvent such as acetone greatly increases the ability of metal cations to form complex anions at low concentrations of hydrochloric acid. By consulting F. W. E. Strelow, *Anal. Chem.*, *43*, 870 (1971), devise a cation-exchange scheme for separating at least 4 metal ions using acetone-HCl eluents.

15. Explain how a suppressor column (or hollow-fiber suppressor unit) is able to reduce the background conductance from a sodium carbonate eluent in anion chromatography. Propose a suitable eluent and suppressor column for use in cation chromatography with a conductivity detector.

16. Explain the changes in experimental conditions that make it possible to do anion chromatographic separations with a conductivity detector *without* the use of a suppressor column.

17. Tell how an eluent of high equivalent conductance, such as $H^+NO_3^-$, can be used successfully in single-column cation chromatography with a conductivity detector. Explain why negative peaks are obtained that are proportional to the concentration of a sample cation.

18. The pH at which a reagent reacts with a metal ion in a post-column reactor must be maintained within certain limits for the color-forming reaction to occur. Tell how the pH can be controlled.

19. Explain how a chromatographic peak will be affected if a post-column reactor does not have an extremely small volume relative to the column.

PART II

LABORATORY
TECHNIQUES AND
PROCEDURES

25

Introduction to Laboratory Work and Laboratory Safety

Before beginning laboratory work, the student should read this section carefully to acquire the proper foundation. Good habits in using a laboratory notebook, planning laboratory work, keeping equipment neat and clean, and using reagents properly are basic to successful measurement. Along with these factors, correct safety procedures are essential for your own protection.

25–1. LABORATORY NOTEBOOK

Importance of an Orderly, Tabular-Form Notebook. The key to your success in the analytical lab is an orderly notebook with data organized in tabular form. Neatness is secondary, but it is most important that you discipline yourself to organize your data as you record it rather than to scrawl numbers randomly across a page.

An orderly, tabular presentation of raw data and final results is important both to you and to the lab instructor. You will need to look at your raw data when calculating your results and deciding what value to report. It may not be easy to find or read your data if they are not organized. When your lab instructor grades your notebook, he or she may find that your reported value is far out of line with the true value. If your raw data and final results are organized, your instructor will be able to check it for an obvious mistake and perhaps give you a grade rather than rejecting your reported value. Therefore, from both points of view, organization pays!

Type of notebook. The type of notebook may be specified by your instructor. Usually, a bound notebook is preferred to a looseleaf one. A size that is popular with many instructors is a width of 5 or 6 inches and a length of 8 or 9 inches. A notebook of this size fits easily around the balance area and allows you to position several flasks or beakers near the balance as well. Larger notebooks, especially if opened, tend to get in the way of other students.

Using the Notebook.   Reserve the first two pages of your notebook for an index, in which you list the page number and title of each experiment. (If the pages in your notebook are not numbered, number each page in the upper corner.) Use a heading or title to identify data on each page. Write all data and words in ink to avoid smearing or erasures.

Recording Raw Data.   Your instructor may specify an arrangement for recording raw data for certain experiments. If not, you should proceed along the lines that are specified below. Whether or not your instructor does specify an arrangement, you should prepare your page(s) for recording data ahead of time— *before* you come to lab. Write down a title and draw up a table or other arrangement for recording data to enable you to record your data in an orderly manner.

If your instructor does not demand it, it is not necessary to be perfectly neat in recording raw data as long as they are legible and recorded in an orderly fashion. (For example, you may draw a line through a wrong word or number without detracting from your accuracy and orderliness.)

A convenient *modus operandi* is to record all raw data for an experiment on a left-hand page so that reports of results can appear opposite on a right-hand page. All raw data and any observations should be written in the notebook, whether or not they are to be used. *Nothing* should be erased; anything not needed should merely be crossed out with a single line, followed by an explanatory notation if needed.

Tables for Raw Data.   As stressed above, you should prepare a table or other arrangement for recording raw data *before* you come to lab. It is important to use a *vertical* table so as to organize all data for a particular sample *under* the sample number (see Table 25–1).

Note in Table 25–1 how easily each of the weights of an individual crucible may be compared because each successive weighing is written directly under the preceding weight. A vertical table is preferred over a horizontal table because of this and because there is more room in a vertical column for data.

Reporting Results.   Your instructor may specify an arrangement for reporting results. If not, you should observe all of the precautions given below. Regardless of what you do, your report should not only be organized and legible, but it should also be as neat as possible. It is better to prepare your report outside of the lab at your leisure rather than hastily writing it in the lab to meet a deadline.

Table 25–1.   Table of Raw Data for Crucible Weighings

Title: Sintered Glass Crucible Weighings for Exp. 3

	Crucible numbers		
	I	II	III
First weighing	31.0004 g	32.0103 g	33.0109 g
Second weighing	31.0001 g (constant weight)	32.0012 g	33.0104 g
Third weighing	——	32.0002 g	33.0100 g (constant weight)
Fourth weighing	——	32.0004 g	——

If raw data are limited only to the left-hand pages in your notebook, then you should place all report forms on right-hand pages. The report form itself should be given a title. It is also useful to include the chemical reaction and a page reference to experimental directions.

Gravimetric analysis report forms are fairly specialized, as you can see from Table 25–2. Note that the weight of the sample is given first. Then the weight (constant weight) of the crucibles containing the precipitate is given, followed by the weight (constant weight) of the empty crucibles. (Notice that the *last weight* of each

Table 25–2.   A Report Form for a Gravimetric Analysis

Title: Determination of Percent Chloride
Reaction: $Cl^- + AgNO_3 = AgCl(s) + NO_3^-$

	I	II	III
Weight Cl sample, g	0.5000	0.5001	0.5001
Weight crucible + AgCl	31.8002	32.8014	33.8250
Weight crucible, g	31.0001	32.0004	33.0100
Weight AgCl, g	0.8001	0.8010	0.8150
Gravimetric factor = Cl/AgCl	0.24737		
Weight Cl, g	0.1980	0.1984	0.2022
% Cl	39.60%	39.68%	40.44%
Application of Q test	——	——	$Q = 0.91$ (Retain)
Mean		39.91%	

Table 25–3.   A Typical Report Form

Determination ᵕof total basicity (as % Na_2O)
Chemical reaction: $2HCl + Na_2O = H_2O + 2NaCl$
Normality of HCl = 0.1011   Sample No. 101

	I	II	III
Container + Na_2O sample	9.3082	9.4120	9.8248
Container	9.1000	9.2000	9.6100
Na_2O sample weight	0.2082 g	0.2120 g	0.2148 g
Final buret reading	34.88	34.58	36.03
Initial buret reading	0.01	0.08	0.01
HCl titrant, mL	34.87	34.50	36.02
% Na_2O	47.80	4̶6̶.̶4̶9̶ (Dropped by Q test)	47.90
Average % Na_2O		47.85	

Calculations for sample I:

$$\% = \frac{34.87 \text{ mL} \mid 0.1011N \text{ HCl} \mid 31 \text{ (eq wt, } Na_2O) \mid \quad \mid 100}{\mid \mid \mid 208.2 \text{ mg sample} \mid}$$

crucible in Table 25–1 is used in Table 25–2, even though the third weight of crucible II is slightly lighter than the fourth weight.) Finally, the result is given for each column of data, and the *best value* (the mean in this case) is reported for all of the results.

Titrimetric analysis report forms are somewhat different (Table 25–3). A separate form should be used to show results for standardizing the titrant used; the final normality calculated should then be included in the report form following the chemical reaction used in the analysis. If the sample has been weighed by difference, the weights of the container before and after the sample has been removed should be given first in the table, followed by the weight of the sample. Then the final and initial buret readings should be given even if all of the initial readings are 0.00 mL. After the results are presented as shown, it is best to include a calculation setup for one of the sample to enable the lab instructor to check your setup as well as equivalent weight, etc.

25–2. PLANNING LABORATORY WORK

It is essential to plan laboratory work ahead of time, so as to use the laboratory time efficiently and avoid delays in using equipment that must be shared with other students. Drying the sample is often a time-consuming step that may be done outside of class. Ask what your instructor's policy is about entering the lab while other classes are in session so as to insert or remove a sample from the oven. If you may, follow the steps described below.

Drying the Sample.   The most important operation that can be done ahead of time is preparation of the sample. Not only should it be dried ahead of time, but it should also be allowed to cool ahead of time, since a hot sample cannot be weighed accurately on the balance. Always obtain the sample at least one lab period ahead of time. Put the sample in the oven close to the start of the lab so that it can be removed before the lab closes. If you must leave it in the oven overnight, obtain permission to remove it while another lab is in session.

Volumetric Analysis.   If a volumetric analysis experiment is to be performed as the first experiment of the term, obtain a portion of the primary standard needed for the titrant and dry it after checking in. For example, assume that the first experiment is the sodium hydroxide titration of an unknown acid. After checking in, obtain a portion of primary standard potassium acid phthalate (and the unknown if possible) and place it in the oven before learning how to use the analytical balance. Also bring distilled water to boil for preparing $0.1N$ sodium hydroxide titrant that is free of carbon dioxide. The water can then cool and the titrant be prepared without waiting for cooling in a subsequent lab.

Gravimetric Analysis.   A gravimetric method requires special planning. The keys to efficient work are first the crucibles and second the sample. Both must be heated before beginning the experiment. Since bringing the crucibles to constant weight requires two or more heating periods, the crucibles should be heated and weighed twice during the lab before the gravimetric experiment. If additional heating and weighing operations are necessary for constant weight for one or more crucibles, these may often be done while the precipitate is digesting or coagulating. Since you should always plan to start a precipitation at the *start* of a lab period, you should also dry and weigh out the sample in the previous lab.

25-3. CLEANLINESS AND ORDER

Since quantitative analysis involves measuring exact concentration or weight of a substance, an analyst must be neat and clean in all the *crucial* steps of the analysis. Although an experienced analyst knows which steps are crucial and may not be scrupulously clean during other steps, you will not have this advantage. You are advised to be neat and clean in all steps.

Volumetric Glassware.   Volumetric ware is the most important class of equipment that must be kept spotlessly clean. Burets and pipets should be cleaned at the start of the term and kept clean by rinsing with distilled water immediately after use. (Specific directions are given in Chapter 29.) Scrub other glassware, such as flasks and beakers, with hot detergent solution and brush. Rinse with tap water and then rinse twice with distilled water; drain upside down before storing. (Never waste distilled water by using it for all rinsings; use it only for the two final rinses.)

Storing Glassware.   Storing glassware properly in the drawer(s) is also important. Burets and pipets should be filled with distilled water and stored in a place where the tips will not easily be chipped when other equipment is moved in the drawer. Plug both ends of the pipets with bulbs from medicine droppers and store in a horizontal position. Plug the top of the buret with a rubber stopper only if it must be stored horizontally.

Analytical Balance Area.   It is obvious that your benchtop should be kept clean, but you should also recognize that the areas around the trip scale and analytical balance should be kept clean. Chemicals spilled in these areas may damage notebooks, clothing, or skin. If one of these areas needs cleaning, clean it before doing any weighing, and leave it clean. Discard all weighing papers and other papers to ensure that contaminated papers are not used again.

25-4. REAGENTS

The student should come to appreciate the purity of reagents and of the proper use of the various grades of reagents.

The most frequently used reagent-grade chemical is distilled water. However, its purity should not always be taken for granted. The quality of water delivered by the still may vary from day to day; older stills may occasionally contaminate distilled water with chloride ion. When chloride ion is to be determined, the water may have to be tested with silver nitrate to ensure that it is "chloride-free."

Distilled water is always used as the solvent for aqueous acid or base titrants, but it is usually boiled first to free it of carbon dioxide. Unknowns and primary-standard reagents are also dissolved in distilled water rather than tap water.

Distilled water is also used for the final rinse when washing glassware, but it should not be used for preliminary rinsing to remove detergents. It is extremely wasteful to rinse glassware by filling it with distilled water, instead of merely squirting the sides with a squeeze bottle.

Concentrated Acids and Bases. The next most frequently used reagents are concentrated acids and bases. In Table 25–4, their specific gravities and concentrations are listed. Of these, the only reagent not commercially available is 1:1 sodium hydroxide. This is prepared in the laboratory ahead of time so that the sodium carbonate impurity can be properly removed.

From these concentrated reagents, more dilute solutions can easily be prepared in the laboratory. Directions usually specify the parts of concentrated reagent first and then the parts of water used for dilution. For instance, directions for 1:3 hydrochloric acid indicate that 1 part of concentrated hydrochloric acid should be added to 3 parts of water.

Purity of Reagents. Chemical reagents of various grades, such as technical, USP, reagent-grade, and "meets ACS specifications," are commonly available. Unfortunately, these grades usually imply different levels of purity for different reagents. Neither technical-grade nor USP (United States Pharmacopoeia) reagents are generally suitable for good analytical work. Reagent-grade chemicals are often suitable for analytical work, but the percent purity and maximum limits of impurities listed on the label should be carefully inspected first.

Table 25–4. Concentrated Acids and Bases

Percent by Weight	Acid or Base	Specific Gravity	Molarity
99.5	CH_3CO_2H	1.05	17.4
37	HCl	1.19	12
72	HNO_3	1.42	15.7
70–2	$HClO_4$	1.68	11.6
85	H_3PO_4	1.69	14.7
95	H_2SO_4	1.83	18
28	NH_3	0.90	14.8
50	1:1 NaOH	——	16

The American Chemical Society's Committee on Analytical Reagents has established specifications for analytical reagents; if these are met by a particular reagent, then its label will state that the standards of purity for the reagent "meets ACS specifications." Reagents with such a label are usually of the highest purity commercially available and can be used with confidence. (An example of such a label is shown in Table 25–5.)

Primary-standard materials, of course, must be reagents of the highest purity, preferably within $\pm 0.05\%$ of 100.00% purity. It is of interest to examine a typical label for a primary-standard arsenious oxide (see Table 25–5). Although the purity is stated to be only 100.0% (implying an uncertainty of 0.1%), the impurities add up to less than 0.02%. The label also states that the purity "meets ACS specifications." Such substances as sulfide and antimony, which might interfere in the iodometric titration of arsenic, are each present at only the 0.001% level.

Not all primary-standard materials will analyze exactly at 100.0%. Potassium acid phthalate of 99.96% purity is frequently found; other primary-standard materials may analyze at slightly over 100%, such as 100.05%. Sometimes substances of only 99.0% purity will be used as primary standards, even though they are not. In determining water hardness of low levels, a 1% error is not significant. Hence, 99.0% calcium carbonate will be used as primary standard for EDTA.

Handling Reagents. Reagents may easily be contaminated when they are being used by a large number of students in the laboratory. Hence, it is important to handle all reagents so as to avoid contamination, even though some of the reagent must be wasted.

Chemicals should be transferred from their bottles by carefully rotating the bottle, which is tilted slightly downward, to give a fine flow of chemical onto a weighing paper or into a beaker. If a spatula must be used, it should be perfectly clean and dry, to avoid contamination. Any excess reagent should be thrown away, not put back into the bottle. Any spilled reagent should be instantly cleaned up; this is especially true for the area around the trip balance.

Table 25–5. The Label of a Primary-Standard Reagent

ARSENIC TRIOXIDE
Primary Standard
Meets ACS Specifications

Assay (As_2O_3)	100.0%
Residue after ignition	0.004%
Insoluble in HCl	0.005%
Chloride (as Cl)	0.005%
Sulfide (as S)	0.001%
Antimony (as Sb)	0.001%
Lead (as Pb)	0.0005%
Iron (as Fe)	0.0001%

Liquid reagents are best transferred by pouring a slight excess into a small beaker after the beaker has been rinsed with the reagent. A student's pipet should never be inserted into the reagent bottle, although the instructor may provide a common pipet for the transfer of that particular reagent alone. The excess reagent is then flushed down the drain with plenty of water. The stopper of a liquid reagent should never be put on a laboratory bench or in any place where the stopper could become contaminated. Hold the stopper between two fingers or, if necessary, place it on a clean piece of paper.

25–5. SAFETY IN LABORATORY WORK

Before starting laboratory work, it is important to understand the potential hazards in the laboratory and then to become familiar with the precautions and/or rules to be followed in lab work.

Some of the hazards in laboratory work are *outlined* below, after which precautions to avoid the hazards are discussed.

Some Hazards in the Chemical Laboratory

1. *Chemical hazards*
   a. From eye-skin contact
      (1) Spray from breakage, violent reactions, etc.
      (2) Spills
   b. Through the mouth
   c. From inhalation
      (1) Gases, fumes
      (2) Dust
2. *Fire hazards*
   a. From volatile solvents
   b. From electrical malfunctions
   c. From other sources
3. *Careless habits*
   a. And cuts (glass tubing, etc.)
   b. And falling objects
   c. And falling off a stool, etc.

Chemical Hazards.   The eyes are especially susceptible to injury from hazardous chemicals. The two most common types of accidents that threaten the eyes are breaking of glass containers of acids, bases, etc., and reactions that go out of control. Both types of accidents can cause chemicals to be sprayed on the eyes (as well as the skin). For this reason, safety glasses are essential. Since safety glasses make perfect vision difficult when reading a balance, buret, and the like, there is always the decision as to when to wear safety glasses. The rule is SAFETY GLASSES

SHOULD BE WORN AT ALL TIMES EXCEPT WHEN PERFECT VISION IS REQUIRED FOR MEASUREMENTS.

If any chemical is sprayed into the eyes, IMMEDIATELY FLOOD THE EYES WITH WATER. Some laboratories have special eye wash stations; but if one is not close at hand, quick flushing with water from the nearest tap will minimize the damage.

Injuries from contact of certain chemicals with the skin can occur:

1. by chemical burns, such as those from strong acids and bases, or
2. by absorption of certain chemicals through the skin.

To avoid injury, all chemicals should be handled with care, and even small spills should be cleaned up promptly. The hands, forearms, and elbows should be washed frequently when working with or around chemicals. Because help is often needed in such situations, it is important never to work alone in the laboratory, even if an instructor is present. One of the authors, while a student, was able to assist when acid was sprayed in the face of another student.

Hazards from Swallowing. To avoid any possibility of swallowing toxic chemicals, there should be NO FOOD OR DRINKS IN THE LABORATORY. It is too easy to contaminate food, especially with traces of chemicals on your hands.

To avoid any possibility of swallowing solutions containing chemicals, while using the pipet, all pipetting of chemical solutions should be done with a rubber bulb or other device. DO NOT FILL A PIPET WITH MOUTH SUCTION.

Hazards from Inhalation. Inhaling chemical fumes can irritate the lungs and mucous membranes. Some gases (hydrogen sulfide and HCN, for examples) are *extremely toxic* and must be handled with great care. Therefore, VOLATILE CHEMICALS OR CHEMICALS LIKELY TO EMIT IRRITATING FUMES SHOULD BE KEPT IN A LABORATORY HOOD. Some common examples include the heating of hydrochloric, nitric, perchloric, or sulfuric acids to dissolve samples; operations involving concentrated ammonium hydroxide; and cleaning of crucibles with nitric acid.

Chemical dust can also be a hazard. Although solid chemicals usually exist as crystal, finely powdered chemicals can be inhaled as dust. Such chemicals should also be handled in the hood.

Fire Hazards. A fire in the chemical laboratory can be dangerous and devastating. Accidental ignition of volatile organic solvents is perhaps the most common source of laboratory fires. Where possible, less flammable solvents should be substituted for highly flammable solvents like ethyl ether and low-boiling petroleum ether.

Most heating is now down with electric hot plates or heating mantles, but the following rule is still important: NEVER WORK WITH A VOLATILE SOLVENT AROUND AN OPEN FLAME. To avoid accidental spills and reduce fire hazards, KEEP VOLATILE SOLVENTS IN SMALL CONTAINERS AT THE BACK OF THE LAB BENCH. Volatile solvents should be stored in a special cabinet when not in use.

Hazards from Electrical Malfunctions. Frayed electrical cord or poor electrical connections can cause electric shock and/or fire. Therefore, it is important to periodically inspect any equipment involving electrical connections for possible hazards.

Hazards from Other Sources. Although volatile solvents and electrical malfunctions are the main fire hazards, there are other sources. Bad habits often can cause fires, such as failing to put out a match or smoking in the laboratory. Obviously, these are things that should never be done.

Careless Habits. Perhaps the most common personal injuries in the lab are cuts resulting from the improper use of glass tubing or other glassware. To insert glass tubing through a stopper, handle the tubing with a towel or gloves; do not *force* the tubing through a stopper *with bare hands.*

Hazards from Falling Objects. Falling objects can cause serious injuries. DO NOT PLACE HEAVY OBJECTS ON HIGH SHELVES. If a heavy container must be placed on a shelf, secure it with a belt or chain. Likewise, secure all gas cylinders with a belt. Finally, be careful in moving heavy instruments or heavy objects; use a lab cart where possible.

Hazards from Falling. AVOID STANDING ON LAB STOOLS. It is not uncommon for students to stand on a stool to read a buret; this risks a serious fall. Instead of using a stool, simply detach the buret and bring it down to eye level.

In conclusion, THINK ABOUT WHAT YOU ARE DOING BEFORE DOING IT. Most accidents in the lab are caused by doing something impulsively that later seems thoughtless. Think first, check with the instructor, and avoid accidents.

Toxicity of Chemicals. Virtually all chemicals are toxic to some extent, and care should be taken in handling them as well as avoiding contact with the skin, eyes, and cuts. It is important to read the label to be aware of whether the chemical is toxic, to what degree it is toxic, and what special handling precautions are necessary. When in doubt, consult a reference source such as the American Chemical Society monograph *Chemical Carcinogens* (Charles Searle, ed.). Table 25–6 lists chemicals that have been found to induce malignant tumors in animals or to induce benign tumors generally recognized as early stages of malignancies. Such chemicals are on the list of the Carcinogen Assessment Group (Environmental Protection Agency).

Use of Arsenic Compounds. Arsenic(III) oxide, As_2O_3, is a valuable primary standard compound for oxidation-reduction reactions, but it is also poisonous and carcinogenic. An examination of a typical label reads: "Danger! Contains inorganic arsenic. Cancer hazard. Poison: May be fatal if swallowed or inhaled. Use only with adequate ventilation."

Although this warning is accurate, it does not imply that arsenic(III) oxide cannot be used safely in the laboratory. It simply limits the operations that can be performed with it and the way in which this compound can be handled. For example, it should not be dried in the conventional laboratory oven; fortunately, it is not hygroscopic, so for most work it can be used without drying. It can be safely weighed as long as it does not contact the skin or in any way come in contact with the mouth. Gloves may be used for the weighing process, and the hood should be used to

Table 25–6. Some Carcinogenic Substances (EPA)

Substance	CAS No.
Arsenic compounds	——
Asbestos	1332-21-4
Benzene	71-43-2
Benzo[a]pyrene	50-32-8
Cadmium compounds	——
Carbon tetrachloride	56-23-5
Chloroform	67-66-3 ·
Chromium(VI) compounds	——
DDT	50-29-3
Formaldehyde	50-00-0
Hydrazine	302-01-2
Nickel compounds	——
Saccharin	81-07-2
Thioacetamide	62-55-5
Thiourea	62-56-6
Vinyl chloride	75-01-4

dissolve it. Once it is dissolved, it can be titrated safely in the open lab. After the analysis is finished, it should be disposed of according to the recommendations of the instructor. Some typical procedures for disposal are discussed below.

Disposal of Chemicals and Cleanup of Spills. The following rules should be followed when disposing of chemicals or cleaning up spilled chemicals.

1. Do not throw solid chemicals into the sink unless directed; use a waste jar or container provided in the lab.
2. Dispose of small amounts of acid or base by slowly pouring them into a stream of water and flushing them with a large amount of water.
3. Dispose of waste organic solvents and large amounts of acid by pouring them into marked containers in the lab. Do not mix acids with reagents such as bases. Pour into separate containers.
4. SPILLS. Clean chemical spills by first pouring large amounts of water on the chemicals to dilute them. Then neutralize if necessary; use sodium bicarbonate to neutralize acids and acetic acid (or sodium bicarbonate) to neutralize bases. After neutralization, sponge with water.
5. In general, keep bench surfaces clean at all times by sponging with a damp sponge and wiping with a paper towel.

26

The Analytical Balance

Measuring weight with the analytical balance is both the most basic and the most accurate operation in the chemical laboratory. Most analytical results are calculated as a percentage by weight, and even those reported differently normally depend on some weighing operation at some point in the analysis.

26–1. ACCURACY AND PRECISION

Two types of weighing are done in the laboratory: rough weighing and accurate weighing. Of course, the *analytical balance* should not be used for rough weighings, since *triple beam* or *toploading balances* give the two or three significant figures needed for rough weighings. The analytical balance is used for weighing objects with high accuracy and high precision (expressed as the uncertainty of single weighing). For example, the *absolute uncertainty* of a single weighing on the standard analytical balance is ± 0.0001 g (0.1 mg). (The absolute error varies but may be of that same order of magnitude.)

The relative uncertainty of a balance measurement is usually much smaller than that of analytical glassware or other instruments. The relative uncertainty of a single weighing of a 100.0-mg object may be calculated as parts per hundred (pph) or parts per thousand (ppt) as follows:

$$\frac{0.0001 \text{ g}}{0.1000 \text{ g}}(100) = 0.1 \text{ pph}$$

$$\frac{0.0001 \text{ g}}{0.1000 \text{ g}}(1000) = 1 \text{ ppt}$$

Since the relative deviation or relative uncertainty in careful analytical work should be no more than 0.1–0.2 pph, the minimum weight (min wt) desired for

Table 26–1. Measurements with a Low Relative Uncertainty

Uncertainty	Measurement
1 part in 10^{15}	Proposed hydrogen electronic clock in Harvard department of physics (error of 1 second in 30 million years)
2 parts in 10^{13}	National Bureau of Standards cesium clock based on the vibration of cesium atoms (error of 1 second in 150,000 years)[a]
~ 1 part in 10^9	Comparison of particular wavelengths of the spectra of cadmium and mercury[b]
1 part in 10^8	Determination of a particular wavelength in the spectrum of cadmium[b]
1 part in 10^7	Radio-frequency measurements[b]
1 part in 10^6	Determination of the weight of a 100-g object on the analytical balance

a. G. J. Whitrow, *The Nature of Time*, New York: Holt, Rinehart & Winston, 1972, p 88.
b. H. Diehl and G. F. Smith, *Quantitative Analysis*, New York: Wiley, 1952, p 47.

samples is at least 100 mg on the standard analytical balance:

$$\text{min wt} = \frac{\text{abs uncer}}{\text{rel uncer, pph}}(100) = \frac{0.0001 \text{ g}}{0.1 \text{ pph}}(100) = 0.1 \text{ g}$$

If the sample is weighed by difference, the possible absolute weighing error is doubled, and therefore the minimum weight for samples must be doubled to 200 mg to retain a relative uncertainty of 0.1 pph.

If an object of 100 g is weighed, the relative uncertainty of 1 part in $10^6 \cdot$ (0.0001 pph) in its weight is exceeded only by that of four or five kinds of measurements in the entire scientific community (Table 26–1).

26–2. PRINCIPLES OF THE EQUAL-ARM BALANCE

In principle, the analytical balance is a lever of the first class; the equal-arm, double-pan balance is, of course, a symmetrical lever, whereas the single-pan balance discussed later is an unsymmetrical lever. In either case, the fulcrum lies between the points of application of the forces. As shown in Figure 26–1, the fulcrum is the point at which the knife-edge K rests on the agate plate. The fulcrum lies halfway between the points of application of the forces F_L and F_R exerted by weights on the pans of the balance.

Suppose the force F_L just balances the force F_R (this is true if the balance comes to rest at its original rest point). Then

$$(F_L)(l_L) = (F_R)(l_R)$$

where l_L and l_R are the lengths of the lever arms from the knife-edges to the fulcrum. The forces F_L and F_R are proportional to the mass on each pan.

Figure 26–1. The pans and knife-edge as part of the lever system of the analytical balance. The fulcrum is at K, the points of application of the forces are at F_L and F_R, and the hypothetical lever arms are l_L and l_R.

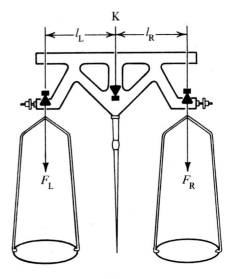

The origin of these forces is the attraction of the earth's gravity on an object of a given mass. Thus, *mass* is the quantity of matter in a body, and *weight* is the force exerted by gravity on that body. The relationship between force and mass is $F = Mg$, where M is the mass and g is the acceleration of gravity. Although the force of gravity varies somewhat throughout the world, it is constant for any given locality. Therefore, the force or weight of an object is proportional to its mass.

As it is customary to speak of the mass of an object as its weight, the following equation holds true for the analytical balance:

$$(W_L)(l_L) = (W_R)(l_R)$$

Since the balance is constructed so that l_L is equal to l_R within an uncertainty of one part in 10^5, then $W_L = W_R$. The weight of the object on the left pan is then known directly from the sum of weights on the right pan.

As emphasized by Schoonover [1], the balancing reading strictly speaking is not the mass of an object; it is really the "apparent mass versus the density (8.0 g/cm^3 currently) of the weights used." For weighings when air buoyancy causes a significant error (rare in student work), corrections are required [2].

Such double-pan balances are no longer in use because, first of all, they are too slow to use. A second reason involves *sensitivity*. Sensitivity is the displacement of the beam that is caused by a given change in force (apparent mass). On the two-pan balance, sensitivity changes significantly with a change in the weight on an object. On the single-pan balances to be discussed below, sensitivity is constant with changes in weight because the total weight on each side of the beam is always constant.

[1] R. M. Schoonover, *Anal. Chem. 54*, 973A (July 1982).
[2] R. M. Schoonover and F. E. Jones, *Anal. Chem. 53*, 900 (1981).

26–3. SINGLE-PAN BALANCES

The modern analytical balance is a single-pan, direct-reading balance that permits weighings within less than a minute. There are three types of such single-pan balances with two different weighing purposes:

1. Mechanical balance: Analytical (<1 g) and approximate (>1 g) weighing.
2. Hybrid balance: Analytical (<1 g) weighing.
3. Electronic (electromagnetic force) balance: Approximate (>1 g) weighing.

Formerly, mechanical balances were used for all weighing purposes, but now analytical weighings are done using the hybrid balances, a modification of the mechanical balance with electronics. Approximate weighings are done on the so-called toploading balances, which are fully electronic. All three types operate on the so-called substitution method of weighing.

Substitution Method for Single-Pan Balances. As shown in the schematic diagram of the side view of a single-pan balance in Figure 26–2, the empty pan and a set of removable weights suspended over the pan are balanced exactly by a large weight

Figure 26–2. Schematic diagram of substitution method of weighing on a single-pan balance.

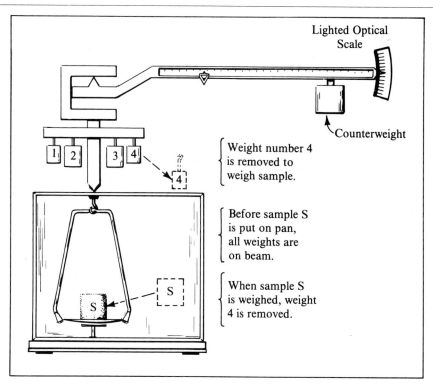

Lighted Optical Scale

Counterweight

Weight number 4 is removed to weigh sample.

Before sample S is put on pan, all weights are on beam.

When sample S is weighed, weight 4 is removed.

(counterweight) suspended at the rear of the beam. Next a sample (object marked "S") is placed on the pan after the balance beam is secured to avoid damage.

What follows differs in the electronic balance from either the hybrid or the mechanical balance. In the latter two, the beam is put into the semirelease position and a weight (no. 4), or several weights, is removed from the beam to achieve a close balance, to within 0.1 g. The remaining weight difference is determined on the mechanical balance by using a lighted optical scale calibrated to account for a limited range of imbalance. In the hybrid balance, the beam is returned to a predetermined null (reference) position by means of an electromagnetic servo system [1]. Unlike these two, the electronic balance uses no counterweight; instead, it has a much larger capacity electromagnetic servo system for opposing the gravitational (and buoyant) force imposed by the mass of the sample weighed. The beam is again brought to a predetermined null position by this system to give a weight.

In the case of the mechanical and hybrid single-pan balances, it is obvious that the balance operates with about the same weight on both ends of the beam at all times; that is, it operates at *constant load*. This has the advantage of maintaining a constant sensitivity.

Weight Readout. The weight readout may be different for all three types of single-pan balances. In the mechanical balances, the weight readout simply reflects the sum of the built-in weights plus the reading from the lighted optical scale. Usually, grams plus tenths of a gram are read directly, and the rest of the weights (0.0001–0.099 g) are read from the scale.

In the hybrid balances, the weight readout is a combination of the built-in weights plus the weight equivalent to the force exerted by the electromagnetic servo system to restore the beam to the null position. In many balances, the sum of the weights is combined into a single digital display, such as shown in Figure 26–3.

Note that in Figure 26–3, the number of grams appears on the left-hand portion of the horizontal scale. The first two decimals (0.73 g) are read from the middle of the vertical scale, and the last two decimals are read from the right side of the horizontal scale. This type of readout is available on some Mettler model balances such as that in Figure 26–4. Note that the readout is located near the floor of the balance, close to the location of a lab notebook.

In the electronic balance, there are no built-in weights, so the weight readout does not require a combination of such weights with the weight equivalent to the force exerted by the electromagnetic servo system. The readout is simply the weight equivalent referred to.

Analytical Weights. For both the mechanical balance and the hybrid balance, two features involving weights are important. One is that the knife-edges (Figure 26–1) are better protected because the heavier weights are manipulated in the semirelease position. The second is that the weights have more protection in the case from fumes and mishandling.

Although a knowledge of the different classes of analytical weights is of practical importance only for using double-pan balances, everyone should be aware

Figure 26–3. Optical scale with full-range weight readout found on the newer model single-pan, direct-reading balances. The weight readout is 132.7343 g. Note that the first two decimals are displayed on the vertical scale and that the number (73) in the middle of the vertical scale gives the correct first two decimals. (Courtesy of the Mettler Instrument Corporation.)

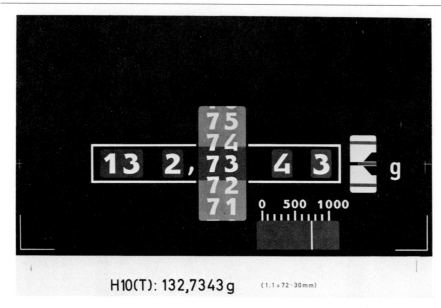

H10(T): 132,7343 g   (1:1 = 72 · 30 mm)

that there are at least four classes of analytical weights. The individual tolerances for these weights is given in Table 26–2. Class-M weights, for example, are used only for the most exact calibration work and presumably are still of use to chemists working on such measurements. Class P weights were formerly used for student work.

Table 26–2. Tolerance Limits for Analytical Weights

Denomination, g	Tolerances, mg			
	Class M	Class S	Class S-1	Class P
	Individual, Group[a]	Individual, Group[a]	Individual[b]	
50	0.25	0.12	0.60	1.2
30	0.15	0.074 ⎫	0.45	0.90
20	0.10	0.074 ⎬ 0.154	0.35	0.70
10	0.05	0.074 ⎭	0.25	0.50
5	0.034 ⎫	0.054 ⎫	0.18	0.36
3	0.034 ⎪ 0.065	0.054 ⎪ 0.105	0.15	0.30
2	0.034 ⎬	0.054 ⎬	0.13	0.26
1	0.034 ⎭	0.054 ⎭	0.10	0.20

a. Any combination of the bracketed weights must fall within the group tolerances.
b. Two-thirds of the weights in a set must be within one-half of the individual tolerances.

Figure 26–4.   Model Mettler H10T single-pan, direct-reading balance, with full-range weight readout and controls in the lower-front portion of the balance. (Courtesy of the Mettler Instrument Corporation.)

26–4. GENERAL RULES FOR THE USE OF ALL SINGLE-PAN BALANCES

Aspects of the Weighing Process Common to All Balances. There are four important aspects of the weighing process that are similar no matter what single-pan analytical balance is used. These are sample handling, preparation of the balance for weighing, recording weights, and securing the balance after weighing. Listed under each of these aspects are a number of important steps to be followed; each step is numbered for easy reference.

Sample Handling. Before the sample is placed on the pan of the balance, follow each of these steps:

1. Allow all heated samples to cool to room temperature in a desiccator to avoid causing the zero point to drift (see step 5). This also prevents convection currents from being set up within the balance.
2. Handle heavy objects to be weighed with a tongs. Handle light objects with finger gloves or a paper loop (Figure 26–5).
3. Weigh chemicals in a weighing bottle or scoop or on glazed paper; never place chemicals directly on the pan. Remove any spilled chemicals immediately with a soft brush.

Preparation of the Balance for Weighing

4. Brush any spilled chemicals off the pan and close the glass doors tightly.
5. Check the zero point, even if it has been checked recently. The zero point changes if anything is spilled on the pan or if the doors are left open. It drifts as a result of temperature changes as well.
6. Place the object to be weighed on the *center* of the pan.
7. Close the glass side door(s) tightly after placing object on the pan to avoid a draft of air.

Recording the Weights

8. Prepare the lab notebook for recording weights ahead of time—before coming

Figure 26–5. Handling a weighing bottle with a paper loop.

to the balance. Use a *vertical* table for recording weights and other data (see Tables 25–1, 25–2, and 25–3).

9. Record all weights *immediately* after weighing; write down four digits to the right of the decimal points if the last digit(s) is a zero.

Securing the Balance After Weighing

10. Secure the beam *before* removing objects from the pan.
11. Return all weights to zero and make sure that the pan is free from spilled chemicals, dust, or other material.

26–5. WEIGHING PROCEDURE FOR A TYPICAL SINGLE-PAN BALANCE

The operation of many single-pan balances will differ from those given below for the Mettler single-pan balance. Still, it will be worth reading these directions for general guidance and as a reminder of the zero point procedure.

Beam Controls for a Mettler Balance.   Follow these rules for controlling the beam before putting your sample on the pan. Locate the beam control and learn the three positions of the control: secured, semirelease, and complete release.

1. *Secured position.*   Place objects on the pan only when the beam control is in the secured position. (Remove them only at that same setting.)
2. *Semirelease position.*   Adjust weights (weight knobs) only when the beam control is in the semirelease position. In this position, a rough weight comparison is possible without damaging the beam.
3. *Complete release position.*   To obtain the weight to four decimal places, release the beam completely. On balances such as the Mettler H-15, this should be done only when the weight has been determined to the nearest 0.1 g.

Zero Point Setting.   As was noted above, the zero point should be checked before weighing even if you have adjusted it as recently as a half hour before.

4. *Zero point check.*   Make sure that the balance pan is clean and that all weights have been removed from the balance. Adjust the beam control to the complete release position. Turn the zero point control to zero in the balance if necessary. (Usually, this control is on the right-hand side of the balance.)

Weighing an Object.   This will vary; but on the Mettler H-15, the weight knobs should be adjusted first.

5. Adjust the weight knobs until the weight is known to the nearest 0.1 g; then adjust the beam control to the complete release position and read the weight to the fourth decimal.

26–6. BENEFITS AND ERRORS FROM USING SINGLE-PAN BALANCES

Schoonover [1] has summarized some of the benefits from using the electronic or hybrid single-pan balances. In the main, the electronic nature of either of these types of balances permits a direct electronic hookup with a microprocessor or a computer. This permits the user to adjust the time required for the balance to obtain an accurate weighing, for example, in the case of weighing an object that is at a different temperature than the balance. The weight readout may be fed into the computer and used to make any type of programmable computation.

Balance Errors. In general, the possible sources of error in any balance are balance-arm inequality, defective balance operation, defective weights, air buoyancy, temperature, electrification, humidity, and operator errors.

Balance arm inequality is only 1 part in 10^5. This is usually minimized in single-pan balances because the method of substitution is used for weighing. Defective balance operation is usually indicated by a fluctuating weight readout. The possibility of defective weights is assumed to be eliminated by the manufacturer's calibration of a single-pan balance, but in principle it would not be eliminated entirely. Buoyancy errors are quite small, but corrections for them can be made [2].

Temperature Effects. The most common source of error in analytical weighing is a temperature change. The temperature differential caused by a hot (or cold) object being weighed sets up convection currents, which in turn causes the zero point to drift. It is therefore important always to allow heated objects to cool to room temperature.

Avoid temperature changes of the balance by:

1. Never weighing warm or cool objects (allow to reach room temperature).
2. Never allowing a draft of cold or hot air to flow around the balance.

If you suspect a temperature change, check the zero point first and reweigh the previously weighed objects.

Operator Errors. Transcription errors are committed by everyone. Healy [3] has identified transcription errors as one of the two most common blunders (values that any worker would wish to correct if made aware of them). Healy estimates that some 0.5% of all items are copied incorrectly into the notebook, including weights. It is therefore important to double-check each balance readout written in the notebook.

Electrical Errors. Historically, nonconducting objects have caused weighing errors because of a static charge that caused two-pan balances to operate erratically. This often occurred because the object was wiped with a cloth, thus building up a charge of static electricity. A static charge of small magnitude is not a serious problem for most single-pan balances, although Schoonover [1] states that some

[3] M. J. R. Healy, *Clin. Chem. 25*, 675 (1979).

ferromagnetic materials may perturb the magnetic field in the electronic single-pan balances.

Experiment 1.   Weighing an Object

Check the balance rules and directions for weighing on a particular balance before proceeding with the weighing.

1. Find the zero point for the double-pan balance by the method chosen by the instructor. If a single-pan direct-reading balance is to be used, check the zero point and adjust if necessary.
2. The instructor may give you some unknown (such as a metal object) to weigh or direct you to weigh a porcelain crucible or some such object from your locker. Handle the object with the ivory-tipped forceps from your weight box or with appropriate tongs, not with your fingers. Weigh the object and record its weight. Weigh it a second time if directed to do so by the instructor, who may check your balance technique.
3. Next, perform the weighing by the difference technique, in the manner directed by the instructor. For instance, weigh a weighing bottle or waxed paper and then weigh the metal object together with the bottle or waxed paper. The *difference* between the two weights is the weight of the metal object. This weight should not differ by more than ± 0.0005 g from the weight in Step 2. Alternatively, the instructor may direct you to weigh a crucible cover and then to weigh the cover plus the crucible weighed in Step 2. The difference between these two weights should not differ by more than ± 0.0005 g from the crucible weight obtained in Step 2.

Experiment 2.   Statistical Evaluation of Weighing Data

This experiment illustrates the use of statistics in evaluating the manufacture of new pennies. Use ten new pennies for the experiment, and assume that all ten were made by the same machine at the same time. These ten new pennies will thus constitute a representative sample of the population of pennies made by this particular machine.

Each of the ten pennies will be weighed separately, and the weights will be used to calculate the mean weight. This will give an estimate of μ, the true mean weight of all the pennies made by this particular machine during one manufacturing run. The deviations of each weighing from the mean will also be calculated to give an idea of the machine's precision. The confidence limit for the mean will also be calculated. (Be sure to review all of the above concepts by rereading Chapter 3.)

Procedure

1. Obtain ten new pennies from your instructor (price, 10¢).
2. Before weighing, adjust the zero point carefully on the balance. Remember that all of the weighings are *absolute*; errors do not cancel out as they do in weighing by the difference.

3. Weigh each penny to the nearest 0.1 mg and record each weight in your notebook.

4. Calculations: Using the ten weighings, calculate the following: $\bar{X}$, the absolute $\bar{d}$, the absolute s, the relative s, and the confidence limits of $\bar{X}$ at the 90% probability level (use the t distribution, p. 42). Calculate s using all ten results, and also using only the results remaining after applying the Q test.

5. Report all data in tabular form and indicate any weighing(s) rejected by the Q test.

6. If the instructor directs, calculate the mean of all pennies weighed in the class using all ten of each student's weighings.

7. *Optional Work.*   Gather old pennies from a class collection and group them by years. Obtain at least ten for each year. Each student should weigh ten pennies from a different year and report the mean. Plot the means versus each year to see whether there is a trend toward an increase or a decrease in weight with the age of the penny.

QUESTIONS AND PROBLEMS

Definitions and Concepts

1. What is the absolute uncertainty on the usual analytical balance for:

   (a) A single weighing?   (b) Weighing by difference?

2. Explain weighing by difference.

3. Explain the difference between the operation of a double-pan balance and a single-pan balance.

4. Explain the advantage of the substitution method of weighing used on single-pan balances.

5. What is the purpose of the air-release device on single-pan balances?

6. What are some ways to handle samples without touching them during the weighing process?

7. Summarize the important rules for operation of a single-pan balance.

Balance Problems

8. Calculate the relative uncertainty of weighing each of the following objects on a standard analytical balance, using first a single weighing and then weighing by difference. Decide whether each relative uncertainty is within the usual desirable value (you should know the value).

   (a) A 3.3-g penny   (b) A 0.150-g salt sample   (c) A 100-mg salt sample

9. Suppose that a semimicro balance has an absolute uncertainty of 0.01 mg. Calculate the minimum weight for a sample using a relative uncertainty of 0.1 pph for:

   (a) A single weighing.   (b) Weighing by difference.

10. Calculate the minimum weight of a sample that is to be weighed under the conditions

specified below, assuming the usual desirable relative uncertainty (which you should know):

(a) Standard analytical balance, single weighing.
(b) Standard analytical balance, weighing by difference.
(c) Analytical balance with absolute uncertainty of 0.0002 g for each weighing, single weighing.
(d) Same balance as in part (c), weighing by difference.

11. During World War I, a clerk in a French intelligence bureau became curious about a suspected German agent who carried many pencils. He suspected that something was being carried inside the pencils. How could he have checked each pencil without breaking it open? (See A. A. Hoehling, *Women Who Spied*, Dodd, Mead & Co., New York, 1967, p. 58.)

12. Discover the maximum weight that can be weighed on the analytical balance in your laboratory and calculate the relative uncertainty (pph) of weighing that weight once.

13. Calculate the minimum weight in mg of a sample that is to be weighed by difference on a standard analytical balance, if the desirable relative uncertainty is 0.2 pph.

14. Calculate the minimum weight in mg of a sample that is to be weighed on an analytical balance with an absolute uncertainty of 0.0002 g, if the desirable relative uncertainty is 0.2 pph.

27

Sample Handling and Gravimetric Techniques

General directions are given below for preparing and weighing samples to analyzed by gravimetric, titrimetric, spectrophotometric, and other methods. *Specific* directions are then given for a complete gravimetric analysis, including weighing operations that are peculiar to a gravimetric determination.

27–1. SAMPLES

Student Samples.   Although sampling was discussed formally in Chapter 2, it will be helpful to you to read something about student samples before starting any experiments. It is helpful to consider both solid samples and liquid samples.

Solid Samples.   Solid samples are usually in a finely powdered form suitable for drying, weighing, and dissolution. Treat powdered solid samples according to the following rules:

1. As soon as you receive the sample, transfer it to a weighing bottle to protect it from the laboratory atmosphere.
2. Record the unknown number on the sample container in your notebook.
3. If the sample needs to be dried, dry it as soon as received and store it in a desiccator. If no drying is necessary, store the sample in a desiccator unless otherwise directed.

Guard against running out of your sample and the delay caused by having to dry more sample. The amount of sample is often depleted by spilling it during transfer operations, by weighing out more sample than is specified in the method, or by having to perform more than three or four analyses. If you discover that you will need more sample at the end of an analysis, the first thing you should do is obtain and dry it as quickly as possible.

Liquid Samples.   Liquid samples are usually homogeneous solutions. Treat liquid samples according to the following rules:

1. As soon as you receive the sample, transfer it to a place in your lab desk where it will not be easily knocked down. Make sure the cap is tight.
2. Record the unknown number on the sample container in your notebook.
3. Protect the liquid from contamination by sampling it only with a clean, dry pipet. Do not allow any extraneous liquid to enter the container and dilute the concentration of the substance to be measured.

If a heterogeneous sample such as milk of magnesia must be sampled, it should, of course, be shaken thoroughly and sampled immediately so that a pipetful will contain a representative amount of the suspended solid such as magnesium hydroxide.

Real Samples.   If you are analyzing a real sample such as Vitamin C tablets, a hard water sample of city water, or other such sample, you should consult with your lab instructor. In general, the rules given above apply, but you may not be required to dry a solid sample, for example. It is often best to dissolve several pieces or tablets of a real solid and dilute to volume in a volumetric flask. Then several pipetfuls can be analyzed instead analyzing just one piece or tablet; the results are more representative of the whole than the result for the one piece or tablet.

27–2. WEIGHING OPERATIONS

Drying the Sample.   The time and temperature for drying depend on the nature of the sample. Some samples need not be dried at all because they do not absorb a significant amount of water; arsenic(III) oxide is one of these. Samples such as alums are not usually dried either because drying will partially remove the water of hydration. Other samples, such as sodium carbonate, should properly be heated at high temperature (300°C) to decompose any sodium bicarbonate impurity to sodium carbonate:

$$2NaHCO_3(s) \rightarrow Na_2CO_3(s) + H_2O(g) + CO_2(g)$$

Dry all samples in a drying oven for 1–2 hours (unless directed otherwise) at 110°C. Place the sample in an open weighing bottle in a beaker covered with a watch glass on hooks (Figure 27–1). Samples that decompose or sublime at 110°C may be dried at room temperature in a desiccator containing a drying agent that is able to absorb water from the sample.

Desiccators.   A desiccator is typically a glass container used to equilibrate samples to a controlled but very low level of water. The desiccant, or drying agent, is stored in a layer on the bottom (Figure 27–2). After a sample has been dried and allowed to cool for five minutes, it is placed in the desiccator. You should leave the lid ajar for about 10 minutes so that a partial vacuum will not form in the desiccator after the air

Figure 27–1. Beaker covered with a watch glass and weighing hooks.

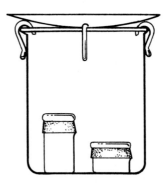

inside has cooled. Allow the sample to cool in the desiccator for at least 30 minutes before weighing.

For a tight seal, the surfaces joining the desiccator and lid should be greased lightly. To open or close, slide the lid sideways with a steady pressure; do not lift it upward.

Drying Agents (Desiccants). The efficiency of various chemical desiccants has been studied by Trusell and Diehl [1]. They found that magnesium perchlorate is the best means of drying. Because of cost, it is best to use only as good a drying agent as is necessary rather than always using magnesium perchlorate. Another difficulty is deciding when desiccant is ready to be replaced. An efficient drying agent that avoids this problem is the colored indicating form of Drierite ($CaSO_4 \cdot H_2O$). It is blue when fresh and pink when ready to be replaced. It can be reheated to restore its efficiency (and the blue color).

Coffee Can Desiccator. A low-cost, compact desiccator can be made from a 2- or 3-pound coffee can with a plastic lid. A wire mesh "hardware cloth" is folded, and holes are cut in the middle as shown in Figure 27–3. Desiccant is placed on the

Figure 27–2. A desiccator.

[1] F. Trusell and H. Diehl, *Anal. Chem. 35*, 674 (1963).

Figure 27–3. A coffee can desiccator: (a) Complete desiccator consisting of a 2- or 3-pound coffee can, folded wire mesh "hardware cloth" stand for crucibles, desiccant at the bottom of the can (or in a plastic tray), and a plastic lid. (b) Wire mesh stand before folding: cut out and discard the four corner sections marked "c," cut holes for the crucibles, and fold down on dotted lines. (Developed by Prof. Harvey Diehl and assistants at Iowa State University.)

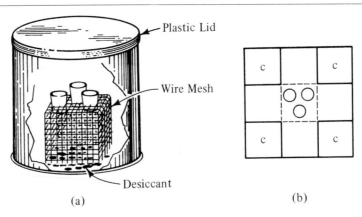

(a)   (b)

bottom of the can or in a plastic tray. The wire mesh is then placed over the desiccant, and the crucibles are placed in the holes. The can is sealed with the plastic lid. (Crucibles that are extremely hot should be cooled somewhat before being positioned in the mesh, and the lid should be left off for a time, since the heat may warp the plastic lid.)

Weighing Solid Samples. We will discuss three methods for weighing solid samples: the weighing bottle method, the use of coated paper, and the use of a scoop.
  Weighing Bottle for Solid Samples.   The bottom and top of a weighing bottle have standard tapered ground glass joints to retard absorption of water from the air by deliquescent samples. Throughout the weighing process, the weighing bottle should be handled with a paper loop (Figure 27–4) or with finger gloves on the thumb and first two fingers. If this is not done, moisture from the fingers will change the weight of the bottle and cause errors.

Figure 27–4. Handling a weighing bottle with a paper loop.

Figure 27–5. Lightweight scoop for handling solid samples.

Such weighings are best carried out by weighing by difference. In so doing, first weigh the bottle plus the sample and record the weight. Then carefully remove the estimated amount with a clean spatula, place in a marked beaker or flask, and weigh to find the sample weight by difference. Thus only four weighings are necessary to weigh three samples, etc. If the sample is deliquescent, then it is essential that the weighing bottle be weighed with the top on throughout. (If it is not, then the top may be left off during the weighings.)

Scoop for Solid Samples. Lightweight scoops (Figure 27–5), metal weighing dishes, and plastic weighing dishes are among the devices used instead of weighing bottles for nondeliquescent samples. The empty device is weighed accurately, sample is added, and then the sum of the two is weighed accurately. A small camel-hair brush should be used to brush the last traces of sample from the device.

Coated Weighing Paper for Solids. A polyethylene-coated paper square or waxed paper square is often used for weighing nondeliquescent samples. The paper alone is weighed first, sample is added, and the sum of the paper plus sample is weighed. Since the paper is thrown away, it can be touched with the fingers before the first weighing and after the second weighing.

Weighing Liquid Samples. Liquid samples may be weighed directly or by difference. Most liquid samples can be weighed by difference using a small bottle fitted with a medicine dropper. The bottle, sample, and dropper are weighed accurately, and part of the liquid is transferred from the bottle to a flask or beaker. The stopper is replaced, and the whole weighed accurately. Alternatively, the liquid sample is added to a container whose weight is known, the whole is weighed, and the increase in weight found is the weight of the liquid.

27–3. GRAVIMETRIC TECHNIQUES

Drying and Weighing the Sample. Follow the general directions given in the section on weighing operations. Some ores and metals need not be dried. Still other samples to be analyzed gravimetrically need only to be air dried, and the analysis is reported on this basis. This kind of report is often necessary for properly comparing analytical results, where product cost is involved.

Dissolving the Sample. The experimental procedure will specify the solvent used to dissolve the sample. (See Chapter 2 for a survey of various solvents.) Consult the instructor if the sample does not yield to the recommended solvent. If there is a fizzing action or if heat is required to dissolve the sample, cover the beaker or flask with a watch glass. When dissolution is complete, wash down the sides of the beaker or flask and wash the watch glass so that the washings fall into the vessel.

If a solution needs to be reduced in volume by evaporation, cover the beaker with a watch glass on hooks, or use a watch glass ribbed to allow space for evaporation. Evaporation is best carried out over a source of low, but constant, dry heat. A variable-current hot plate or a heated sand bath works well. Solutions evaporated over an open burner are apt to bump or spatter.

Time-Saving Preparations for Precipitation.   There are a number of details that can be done before lab starts, or in a previous lab, that can save precious lab time, especially if the precipitation process is time-consuming. Some of these are explained below.

Crucibles at Constant Weight.   Crucibles should be brought to constant weight preferably *before* precipitation is started; or at the very least, the heating and weighing process should be at the point at which only the last weighing needs to be done during the precipitation. The constant weight process is much more efficient if the instructor allows students to enter the lab during other hours to remove crucibles from the oven to save cooling time.

Planning When to Start Precipitation.   Read the experimental directions to see if the precipitate must be filtered immediately after precipitation is complete. If that is the case, then have all equipment and solutions ready to begin precipitation as close as possible to the beginning of the lab period. Be sure that any calculations are done ahead of time.

Calculating the Volume of Precipitating Solution.   Before lab starts, find the approximate volume of solution required by dividing the estimated number of mmoles of sample by the molarity of the precipitating solution. (If the sample is a solid, assume that it is 100% pure to find the number of mmoles.) If directed, add a 10% excess to this volume to ensure complete precipitation by the common ion effect.

Beakers and Stirring Rods.   Before lab, have ready marked beakers for each numbered sample and a clean stirring rod for each beaker. Use the stirring rod to mix the solution after the precipitating reagent has been added; do not transfer stirring rods from one beaker to another. Have ready a medicine dropper for adding a drop of reagent to each beaker to test for complete precipitation by observing cloudiness around the added drop.

Filtering Crucibles.   Filtering crucibles, such as the sintered-glass crucibles, are the most commonly used filtering devices because they filter crystalline and curdy precipitates such as silver chloride faster than filter paper. (However, gelatinous precipitates should not be filtered in these crucibles because they become absorbed in the porous bottom.)

Sintered-glass crucibles.   As shown in the schematic diagram of Figure 27-6, this type of filter has a *filtering disk* of sintered ground glass fused to the wall of the crucible. They are generally available in fine (F), medium (M), and coarse (C) porosities and may be used below 500°C. The fine porosity must be used for fine crystalline precipitates only, the medium porosity is suitable for easily reduced silver

Figure 27–6. A suction system for filtering crucibles.

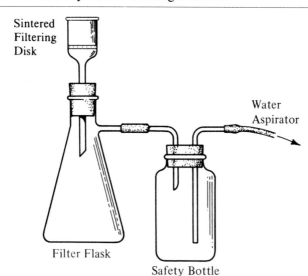

Sintered Filtering Disk

Water Aspirator

Filter Flask

Safety Bottle

halides and other such precipitates, and the coarse porosity is suitable for coarse crystals.

Sintered-glass crucibles are always dried in an oven, never over a flame. The reasons for this are that carbon particles may be deposited in the sintered-glass filtering disk, from which they are difficult to remove, and that the reducing portion of the flame may penetrate through the filtering disk and cause reduction of some ions in the precipitate.

Other Types of Crucibles. Two other types of crucibles that are not as much used are the *porcelain* crucible and the *Gooch* crucible. For heating precipitates above 500°C, porcelain crucibles with unglazed porous bottoms or Gooch crucibles with inexpensive glass filter disks may be used. Both are preferred to sintered-glass crucibles for temperatures much over 250°C. Porcelain crucibles are also satisfactory for filtering fine precipitates, such as calcium oxalate.

Cleaning and Auxiliary Apparatus. All three types of crucibles are usually cleaned by drawing a detergent solution and/or nitric acid through them. Alkaline cleaning solutions should not be used because strong alkali attacks the crucibles. After the crucibles are rinsed, they are ignited to constant weight.

All three types of filtering crucibles must be used with a crucible holder and suction system (Figure 27–6). Suction is supplied by a water aspirator powered by faucet pressure. A safety bottle like the one shown is recommended to keep tap water from being sucked back into the filter flask when the faucet is turned off.

Experiment 3 gives specific directions for using sintered-glass filtering crucibles and for cleaning them.

Filter Paper/Funnel Filtration. Although filter paper is not used much as a filtering medium for quantitative work, it is still useful for the gravimetric determination of sulfate, such as barium sulfate, and for the determination of silica. Since cellulose picks up water and cannot be brought easily to constant weight, filter paper must be burned off in a porcelain crucible. For quantitative work, ashless filter paper (0.1 mg of ash per paper after ignition) is used.

Very fine precipitates such as silica should be filtered through a highly retentive (slow) paper. Small crystals, such as barium sulfate, should be filtered through a moderately retentive (medium filtration rate) paper, such as Whatman no. 40 or Schleicher and Schuell no. 589 White Ribbon. Gelatinous or flocculent precipitates, such as hydrous ferric oxide, and large crystals should be filtered through a hardened (fast) paper.

To use filter paper, first fold it to form a cone and tear off about 1/4 inch from the corner (Figure 27–7). The vertices of the two quarters should miss coinciding by about 1/8 inch from the corners. Then place the cone in the funnel, add distilled water to wet the paper, and fill the stem of the funnel with distilled water. Press the fragile wet paper gently against the glass at the top to "seal" it to the glass. If the fit is proper, water will drain through the funnel without sucking air bubbles into the funnel; the suction caused by the weight of the water draining through the funnel will also increase the filtration rate.

If you are not ready to begin a filtration at this point, keep the funnel well filled with water by stoppering it with a bulb from a dropper. Regardless of when you start the filtration, be sure the funnel has some water in it and water is draining out when you start pouring the mixture of precipitate and solution into the filter. Do not overfill the funnel; fill it about three-quarters full, no more, to avoid the possibility of overflowing. Carry the filtration through to completion without stopping. At the finish, your precipitate should occupy no more than one-third to one-half of the filter paper. (See the next section for more details.)

Decantation, Filtration, and Washing.   Some of the techniques for decantation, filtration, and washing are shown in Figure 27–8. If a filtering crucible is to be used, apply gentle suction throughout the filtration step. The directions given are otherwise much the same, whether filter paper and funnel or filtering crucible is to be used.

1. *Decanting.*   Carefully decant the clear liquid above the precipitate through the filter or filtering crucible, keeping as much as possible of the precipitate in the beaker. Use a stirring rod (Figure 27–8, left) to direct the liquid into the filter or filtering crucible instead of down the outside of the beaker.

Figure 27–7.   A properly folded filter paper.

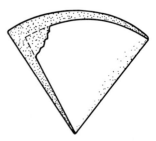

Figure 27–8. Decanting and transferring precipitate to filter paper. (Left) Decanting, keeping most of the precipitate in the beaker. (Center) Transferring the bulk of the precipitate down a stirring rod. (Right) Washing the remainder of the precipitate into the funnel with the wash bottle (a rubber policeman may also be needed in this step).

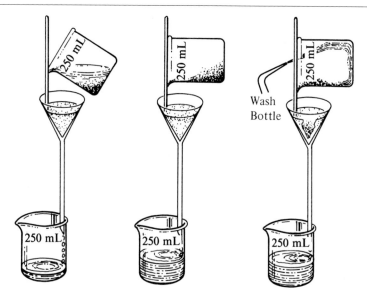

2. *Washing the Precipitate.* Direct the wash bottle stream around the upper sides of the beaker to wash the precipitate down from the walls of the beaker. Allow the precipitate to settle, and then pour the supernatant liquid into the filter as above. Repeat two or three times. Make sure that the stem of the funnel is well inside the beaker receiving the filtrate and that its tip touches the wall of the beaker if you are using filter paper.

3. *Transferring the Precipitate to the Filter.* Hold the beaker and stirring rod in one hand (Figure 27–8, right). Using the wash bottle with the other, wash the precipitate into the filter paper or filtering crucible. If some precipitate sticks to the beaker, moisten a rubber policeman with wash solution and scrub the walls to transfer the precipitate.

4. *Washing Last Traces of Impurities.* Direct the wash bottle stream around the upper sides of the filter to wash the precipitate to the bottom of the filter paper. Allow the level of the wash liquid to fall to the tip of the paper; then quickly add the next portion so that the suction effect is not broken. The object is to wash the impurities through the filter or filtering crucible, not to dilute the impurities. (A few washings, not more than six, effectively remove the last traces of impurities from the precipitate in the filter.) Test the last washing for traces of the precipitating agent.

Drying Using Filtering Crucibles. Silver halide precipitates in sintered-glass crucibles generally can be dried in an oven at 120°C. If any type of filtering crucible

Figure 27–9.   Heating filter paper in a crucible.

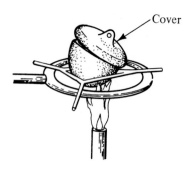

Cover

must be ignited over a flame, it should be heated in an ordinary porcelain crucible. (Expensive platinum crucibles are ideal for this.) Constant weight is achieved when successive weighings agree within ±0.4 mg.

Ignition of Filter Paper in a Crucible.   If filter paper is to be ignited in a porcelain crucible, remove the paper from the funnel by manipulating it where it is three layers thick. If time allows, fold the paper compactly and allow the water to evaporate overnight; otherwise, heat in a beaker in an oven. Thereafter, follow this procedure:

1. *Charring the Paper.*   Char the paper with gentle heat until the organic material gasifies and either escapes past the cover (left ajar as in Figure 27–9) or condenses as a tar on the inner surface of the cover.

2. *Burning off Carbon Residue.*   Remove the lid and gradually increase the flame temperature until the residue and the remaining tars on the crucible burn off completely. Avoid directing the reducing portion of the flame (just above the blue cone) into the crucible. (This could reduce a compound like barium sulfate to barium sulfide.) Heat so that the carbon residue burns with a glow but no flame. Turn the crucible occasionally for even heating. (If the covers have not been brought to constant weight, it is not necessary to ignite them.)

3. *Final Treatment.*   When the carbon is gone, ignite for the recommended time at the highest temperature of the burner with the bottom of each crucible just touching the top of the blue cone. Position the flame toward the back of the bottom of each crucible so that air can reach the precipitate.

4. *Cooling.*   Allow the crucibles to cool until the red glow disappears, and then place them in the desiccator. The lids need not be put on the crucibles at that point, but doing so reduces the possibility of mechanical loss of the precipitate during handling of the desiccator. Bring to constant weight.

28

Gravimetric Procedures

This chapter contains two classical experiments illustrating the use of inorganic precipitants for determining chloride and sulfate ion, and a modern experiment illustrating the use of organic precipitants for determining aluminum. The latter experiment utilizes homogeneous precipitation techniques to overcome some of the disadvantages of organic precipitants.

To perform and evaluate the experiments in this chapter properly, first read Chapter 27 on gravimetric techniques and Chapter 3 on treating experimental data.

Each of the experiments is introduced with a discussion of the chemistry involved. The actual experimental directions follow, with explanatory notes after the directions. The questions at the end of each experiment serve to review the chemistry and the directions and to stimulate further thinking.

Experiment 3.   Gravimetric Determination of Chloride

The directions for this method are written for either a solid sample or a liquid sample such as an aqueous solution of sodium chloride. As long as the solid samples consist of sodium chloride and an inert salt such as sodium sulfate, the directions may be followed without change.

Chemical Background

In this method, the chloride ion is precipitated as insoluble silver chloride. The precipitation does not yield filterable particles of silver chloride immediately; instead, a *colloid* (Section 4–2) is formed:

$$AgNO_3 + Ag^+ + Cl^- \rightarrow AgCl:Ag^+ \cdots\cdots NO_3^-$$
$$\text{(colloidal)}$$

The silver chloride molecules cannot approach one another because the positively charged adsorbed silver ions repel each other, the nitrate counter ions being too far from the silver ions to prevent the repulsion.

The colloidal silver chloride solution is then *digested* (Section 4–2) by heating until the white solution has turned crystal clear and the colloidal silver chloride has coagulated into a filterable curdy precipitate. The heat supplies energy to "shrink" the counter ion layer close enough to the primary silver ions to reduce the repulsion of the silver ions enough to allow the colloidal silver chloride to coagulate:

$$n\text{AgCl}:\text{Ag}^+ \cdots\cdots \text{NO}_3^- \xrightarrow{\text{digestion}} (\text{AgCl}:\text{Ag}^+\cdot\cdot\text{NO}_3^-)_n(s)$$
$$\text{(colloidal)}$$

After filtration, the coagulated silver chloride is washed with a dilute nitric acid solution to avoid peptizing the coagulated silver chloride back to the soluble colloidal form.

Planning Ahead

Sections 25–2 and 27–3 describe how to plan ahead in the laboratory to save time in gravimetric analysis. The most important points are summarized below; for details, read those sections.

1. *Solid Unknowns.* Obtain and dry a solid unknown at least one lab period ahead of time. If you must leave it in the oven to dry overnight, obtain permission to remove it for cooling before your lab meets again. Do not waste lab time waiting for a sample to cool.
2. *Crucibles.* Bring your crucibles to constant weight before starting precipitation. (The last weighing may be done during digestion.)
3. *Volume of Silver Nitrate.* Calculate the volume of silver nitrate solution needed before you come to lab to precipitate (see *Procedure*, Step 4).
4. *Filtering.* To complete filtering in one lab, start precipitation as soon as possible after the beginning of the lab. Put solid or liquid sample in beakers at the end of the previous lab and cover the beakers.
5. *Notebook.* Organize your notebook with a table for weighing data, etc. before coming to lab.

Crucibles

1. Use sintered-glass crucibles of medium (M) or fine (F) porosity for filtering silver chloride (do not use crucibles marked C for coarse). If your sample is a solid, dry it (*Sample Preparation*, Step 1) while cleaning the crucibles.
2. If the crucibles are not new, first remove any visible dirt with detergent solution and brush, and rinse. Assemble the filter flask, the rubber crucible holder, and the crucible. Connect the filter flask with rubber suction tubing to the aspirator.

Fill crucible about halfway with concentrated nitric acid and, using gentle suction, draw the acid slowly through the crucible. Fill halfway again, and interrupt the suction briefly to allow the acid to remain in contact with the crucible for a few minutes. Wash the crucible several times with distilled water. Number each crucible. Proceed to Step 3.

If silver chloride is apparently not removed by this procedure, draw $6M$ ammonium hydroxide through the crucible after the acid has been rinsed out with distilled water. Again rinse the ammonium hydroxide out with distilled water.

3. Dry the crucibles for 1 hour in an oven at 120°C using a beaker, glass hooks, and a cover glass. While the crucibles are drying, proceed with Step 1 of the procedure. Remove crucibles to the desiccator. Cool for 0.5 hour or longer. Weigh. Dry at 120°C for 0.5 hour and cool for 0.5 hour. Reweigh. Constant weight is attained if the second weighings agree with the first weighings within ± 0.4 mg. It is advisable to store crucibles in a desiccator until time to use.

Sample Preparation: Solid or Liquid

1. *Solid sample.* During crucible cleaning, put the solid sample in a weighing bottle, place the open weighing bottle and stopper in a marked beaker, cover the beaker with a watch glass on glass hooks, and dry in the oven at 120°C for 1–2 hours. Remove from the oven before the end of the lab, cool, and store in the desiccator. Weigh three samples of 0.4–0.7 g (see the instructor's directions) into clean 400-mL beakers numbered to correspond to your numbered crucibles.

2. *Liquid sample.* If the chloride sample is a solution of sodium chloride, pipet out *exactly* 25 mL (*volumetric pipet*) into each of three 400-mL numbered beakers. (See Section 29–3 for proper use of the pipet to obtain good accuracy.)

Procedure

1. Measure your sample into beakers numbered to correspond to your crucibles.

2. Add 150 mL of distilled chloride-free water to each beaker and acidify with 0.5 mL (10 drops) of concentrated nitric acid.

3. Insert a stirring rod into each beaker and stir. *Leave the rods in each of the beakers throughout the procedure to avoid cross-contamination.*

4. *Volume of $AgNO_3$ solution needed.* If your instructor has specified the volume of silver nitrate solution to be used, follow the instructor's directions. If not, calculate the volume needed according to whether your sample is a solid or liquid. A. *For a solid sample,* assume that the sample is pure sodium chloride and calculate the mmoles of silver nitrate needed as follows: for example, 410 mg of sample would be $410/58.5 = 7.0$ mmole NaCl $= 7.0$ mmole $AgNO_3$. If a $0.50M$ silver nitrate solution is available, use $7.0/0.50M = 14$ mL, etc. B. *For a liquid sample,* assume that the concentration of the sample is somewhere between 0.10 and $0.19M$. If 25 mL of liquid sample is used and if $0.50M$ silver nitrate is available, use $[(25)(0.14M)]/0.50M = 7.0$ mL for example.

5. *Precipitation from hot solution.* Heat the solutions to boiling on a hot plate, and with constant stirring, add most of the silver nitrate slowly to the first beaker. Continue heating and stirring until the precipitate has coagulated enough to test for complete precipitation. Then add silver nitrate a few drops at a time and look

for a cloud of precipitate. Continue heating, stirring, and adding silver nitrate dropwise until precipitation is complete. Then add a 10% excess of silver nitrate. Repeat with the other samples.

> *Alternative precipitation from cold solution*: If directed, add most of the silver nitrate slowly to a stirred solution of the unknown at room temperature until coagulation begins. Then heat just below boiling for ten minutes, test for complete precipitation, etc.

6. Continue heating with occasional stirring until the solutions are crystal-clear, indicating complete coagulation.
7. Remove beakers from heat, cover with watch glasses, and allow to cool for at least 1 hour in a hood or darkened area to reduce the solubility of silver chloride. (The beakers may be left to cool until the next lab if they are covered with watch glasses to protect from hydrochloric acid fumes.)

Filtration and Weighing

1. Place the weighed filter crucible in the suction filtration apparatus and apply gentle suction. Decant the clear supernatant from the corresponding beaker through the crucible, retaining the precipitate in the beaker.
2. Add about 10 mL of wash solution (0.6 mL of concentrated nitric acid per 200 mL of distilled water) to the beaker, and agitate the precipitate. Let it settle, and pour the wash solution through the crucible. Repeat twice.
3. Hold the stirring rod across the lip and top of the beaker with one hand, and direct the stream of wash solution from a wash bottle at the precipitate, washing it into the crucible (see Figure 27–8). If some precipitate sticks to the beaker, loosen it with a rubber policeman. Wash the precipitate completely into the crucible, and wash it in the crucible with several 5-mL portions of wash solution. Catch the last washing by itself, and test it for excess silver ion by adding a drop of 12M hydrochloric acid. Wash again if a precipitate is noted.
4. Completely drain the crucible with strong suction, place it in a marked beaker, cover it with a watch glass on hooks, and dry it for 2 hours in an oven at 120°C. Cool it in a desiccator for 0.5 hour or longer. Weigh. Dry it again for 45 minutes, and cool it for 0.5 hour or longer. Reweigh. Constant weight is attained if the second weighings agree with the first weighings within ±0.4 mg. Repeat the process if constant weight is not attained.

Calculations

1. *Solid sample.*   Calculate the percentage chloride in each solid sample, using 35.453 as the atomic weight of chloride and 143.32 as the formula weight of silver chloride. Report the percentage for each sample and the mean of the three percentages unless directed otherwise by the instructor. If one of the three results appears questionable, analyze additional samples and report the mean of all of the results. If there is no time to do this, apply the Q test to the questionable result

(Chapter 3). If the Q test does not reject the questionable result, then you must include it in your calculation of the mean.

2. *Liquid sample.* Calculate the molarity of the chloride ion (or sodium chloride) in your liquid sample by dividing the mg of silver chloride by the formula weight of 143.32 and by the volume of the sample. If one of the three results appears questionable, proceed according to the directions given in Step 1.

QUESTIONS

1. What is meant by weighing to constant weight? Give an example.
2. If a 400-mg sample is assumed to be pure sodium chloride, calculate the milliliters of $0.1\,M$ silver nitrate needed to precipitate it completely.
3. Why is an excess of silver nitrate needed? Explain in terms of K_{sp}.
4. If the solution is cloudy after precipitation and heating, what does this indicate? Can the solution be filtered at this point?
5. Why is the solution allowed to cool for an hour after coagulation?

Experiment 4. Gravimetric Determination of Aluminum

The directions for this experiment involve the use of sintered-glass filtering crucibles, just as for the previous experiment. Most of the techniques described there should be reviewed.

Chemical Background

Aluminum reacts with the organic precipitant 8-hydroxyquinoline to form an insoluble chelate, aluminum oxinate, in the pH range 4.5–9.5:

$$Al^{3+} + 3C_9H_6(OH)N: \rightleftharpoons Al[C_9H_6(O^-)N:]_3(s) + 3H^+ \qquad \textbf{(E4–1)}$$

The aluminum coordinates with a pair of electrons on nitrogen and on the phenoxide oxygen of each 8-hydroxyquinolinate anion. In doing so, it displaces three protons and assumes a coordination number of 6. Precipitation is complete as long as the pH does not fall below 4.5.

One advantage of using this organic precipitant is that a low temperature can be used to dry it. Unfortunately, the precipitate is not always easy to handle and filter; the excess 8-hydroxyquinoline frequently coprecipitates if the procedure outlined in Experiment 3 is used. The following modification overcomes these disadvantages [1].

Enough acetone is added to prevent precipitation when 8-hydroxyquinoline is added to the aluminum. The acetone is then gradually evaporated, and the aluminum oxinate precipitates slowly from a homogeneous solution in a readily

[1] L. C. Howick and J. L. Jones, *Talanta* 9, 1037 (1962).

filterable form. Enough acetone remains in the solution to keep the excess 8-hydroxyquinoline from coprecipitating.

Planning Ahead

Sections 25–2 and 27–3 describe how to plan ahead in the laboratory to save time in gravimetric analysis. The most important points are the following.

1. Obtain the solid unknown at least one lab period ahead of time and store in a weighing bottle. If the sample is an alum, do not dry; if it is not an alum and the instructor specifies that it should be dried, dry it at least one lab period ahead of time.
2. Bring your crucibles to constant weight before starting precipitation, or at least perform the last weighing while waiting for the acetone to evaporate.
3. To complete filtration in one lab, start precipitation as close to the beginning of lab as possible.
4. Organize your notebook with a table for data on weighings, etc., before lab.

Crucibles

1. Use sintered-glass crucibles of medium (M) porosity for filtering the precipitate formed (Equation E4–1). If your sample is not an alum and the *instructor directs that it should be dried, dry it while cleaning the crucibles.*
2. If the crucibles are not new, first remove any visible dirt with detergent solution and brush, and rinse. If this does not remove dirt and any discoloration, it may be that the crucibles were used to filter silver chloride. Clean them as directed under *Crucibles* in Experiment 3 and bring them to constant weight as directed in Step 3 under *Crucibles.*

Procedure

1. Obtain a sample. If it is an alum sample, $AlK(SO_4)_2 \cdot 12H_2O$, do not dry it, as this will remove water of hydration. If it is not an alum, follow the instructor's directions on drying. Weigh three 0.3–0.4 g samples (see the instructor's directions) into 250-mL beakers numbered to correspond to your crucibles (Note 1).
2. Steps 2 and 3 should be started and finished during one laboratory period. Turn on the hot plate or water bath to be used in Step 3 (Note 2). Add 50 mL of water, 60 mL of acetone (Note 3), 4 mL of 5% 8-hydroxyquinoline solution (Note 4), and 40 mL of $2M$ ammonium acetate to the sample.
3. Using a small hood, evaporate the acetone from the uncovered beaker for 2.5–3 hours on a hot plate or water bath at 70–75°C. Precipitation will be visible in about 15 minutes. It is essential that the hot plate be stabilized at a 70–75°C setting before putting the beaker on it, to evaporate enough acetone for complete precipitation. If any of the precipitate has adhered to the beaker walls because of evaporation, loosen it with a rubber policeman. After 2.5–3 hours, discontinue heating and allow the solution to cool to room temperature before filtering. The solution may stand overnight before filtering if it is covered with a watch glass.

Filtration and Weighing

1. Place the weighed filter crucible in the suction filtration apparatus and apply gentle suction. Decant the clear supernate from the beaker through the corresponding crucible, retaining the precipitate in the beaker. The pH of the filtrate should be about 5.5.

2. Transfer the precipitate to the crucible using distilled water from a wash bottle and a rubber policeman (see Step 3 under Filtration and Weighing in Experiment 3). Wash the precipitate in the crucible by filling the crucible three times with distilled water and draining.

3. Drain the crucible completely with strong suction, place it in a marked beaker, cover it with a watch glass on hooks, and dry it for 2.5 hours at 135°C. Cool it for 0.5 hour, weigh it, dry it again for 0.5 hour, etc., until constant weight is achieved.

4. Calculate the percentage of Al_2O_3 or Al in each sample. (The respective gravimetric factors are 0.11096 for Al_2O_3 and 0.05873 for Al.)

Notes

1. The procedure is designed to precipitate 2–10 mg of aluminum. Samples larger than 0.40 g should therefore be avoided. (No coprecipitation is found when calcium, magnesium, or cadmium ions are present.)

2. If a hot plate is to be used in Step 4 (Procedure), turn it on and adjust it to bring 100 mL of water in a beaker to not lower than 70–75°C. Measure the temperature with a thermometer.

3. Use reagent-grade acetone. This may need to be redistilled to remove a water-insoluble residue.

4. To make 5% 8-hydroxyquinoline, heat 5.7 mL of glacial acetic acid in a beaker to 70–75°C. Remove the acid from the heat. Add 2.5 g of 8-hydroxyquinoline, stir until it dissolves, and immediately dilute the solution to 50 mL with distilled water to cool it. No precipitate should form.

QUESTIONS

1. Why should the precipitate, but not the alum sample, be dried?
2. What is the purpose of the acetone, and why is it *slowly* evaporated?
3. What are two advantages of using an organic precipitant like 8-hydroxyquinoline? (*Hint:* Consider the magnitude of the gravimetric factor.)
4. What is meant by "constant weight"?
5. Write the structure of 8-hydroxyquinoline and its reaction with aluminum.
6. If the gravimetric factor (the formula weight of Al_2O_3 divided by twice the atomic weight of Al) is 1.8895, show how to calculate either of the gravimetric factors in Step 4 of Filtration and Weighing from the other.

Experiment 5.   Gravimetric Determination of Sulfate

The directions for this experiment involve the use of filter paper and porcelain crucibles. The techniques for using filter paper in Chapter 28 should be reviewed before starting.

Chemical Background

Sulfate is routinely determined by precipitating it with barium chloride to form insoluble barium sulfate (see Chapter 4). The barium sulfate particles precipitating initially are too small to filter and wash properly so they are digested to form larger crystals, which can be handled more easily. This process (Equation E5–1) produces extremely insoluble crystals:

$$Ba^{2+} + SO_4^{2-} \rightleftharpoons BaSO_4(s) \xrightarrow{\text{digestion}} BaSO_4(s) \qquad \text{(E5–1)}$$

$$\text{(BaCl}_2\text{)} \qquad\qquad \text{(small particles)} \qquad\qquad \text{(crystals)}$$

A serious source of error in this experiment is the coprecipitation of such cations as potassium(I) and iron(III). Since hydrogen ion does not cause a coprecipitation error, a cation exchange (catex) resin can be used to exchange such ions for hydrogen ion (see Chapter 24):

$$\text{Res-SO}_3^-\text{H}^+ + \text{K}^+ \rightarrow \text{Res-SO}_3^-\text{K}^+ + \text{H}^+ \qquad \text{(E5–2)}$$

It can be seen from this reaction (Equation E5–2) that the sulfate ion will not react with the resin and will pass through the catex column along with the hydrogen ion. A simpler alternative is to reverse the usual order of addition, and add the hot sample solution to the hot barium chloride solution. This minimizes the coprecipitation of cations during the precipitation because barium(II) is the primary adsorbed ion, and the counter ion will therefore be an anion. The coprecipitation of anions such as chloride will increase, but this will not cause as large an error as that caused by the coprecipitation of potassium.

Planning Ahead

Sections 25–2 and 27–3 describe how to plan ahead in the laboratory to save time in gravimetric analysis. The most important points are the following.

1. Obtain and dry the solid unknown at least one lab period ahead of time. If you must leave it in the oven overnight, obtain permission to remove it before your lab starts.

2. Bring your crucibles to constant weight before starting precipitation, or at least perform the last weighing while waiting for barium sulfate to digest.

3. To complete filtration in one lab, start precipitation as close to the beginning of lab as possible.

4. Organize your notebook with a table for data on weighings, etc., before lab.

Crucibles

1. Use porcelain crucibles and lids for igniting the barium sulfate to dryness. If

your sample is a solid, dry it while waiting for the crucibles to come to constant weight.

2. If your crucibles are not clean, clean them with a dilute detergent solution. After numbering them, heat the crucibles (without the lids) to constant weight (see Experiment 3) at the highest temperature of the Meker burner.

Procedure

1. If the sample is a solid, put it in a weighing bottle, place the bottle and stopper in a beaker, cover, and dry for 1 hour at 110°C. While the sample is drying, bring the crucibles to constant weight.

2. Weigh three 0.3–0.5 g samples of approximately equal size into clean 600-mL beakers numbered to correspond with your crucibles (see instructor's directions for any change in sample size).

3. Dilute each sample with 150 mL of distilled water and add 2 mL of concentrated hydrochloric acid. Heat nearly to boiling.

4. Assume that the largest sample is pure sodium sulfate and calculate the millimoles of barium chloride needed to precipitate it. For instance, 426 mg of sample would be $426/142 = 3$ mmoles $Na_2SO_4 = 3$ mmoles $BaCl_2$. If a $0.2M$ barium chloride stock solution is available, use 15 mL. If a 5% $(0.25M)$ stock solution is available, use 12 mL. Add to the calculated amount a 10% excess of the stock solution.

5. Steps 5–8 should be performed during the same laboratory period. Add about 50 mL of water to the calculated volume of barium chloride, and heat nearly to boiling. Stirring vigorously, pour the hot sample solution slowly into the hot barium chloride solution (see Step 3 of the Procedure in Experiment 3). Allow the precipitate to settle, and test for complete precipitation by adding a few drops of barium chloride solution. Repeat for the other two samples, using the calculated amount of barium chloride for the largest sample plus the 10% excess.

6. After precipitation is complete, cover the beaker with a watch glass and digest just below the boiling point for 1 hour. Longer periods are not harmful, but the solution should not be left to cool overnight.

7. Carefully decant the supernate through ashless filter paper (Whatman No. 40; see Note 1). Wash the precipitate in the beaker two or three times by stirring with warm water and decanting the wash water after the precipitate has settled. Empty the filtrate from the beaker below the funnel.

8. Using a stirring rod and wash bottle (see Figure 27–8), transfer the precipitate to the filter paper. Police the walls of the beaker to transfer the last traces of the precipitate. Wash the precipitate in the filter with 3–5 mL portions of warm water until the wash water, after acidification with 2 drops of concentrated nitric acid, gives only a faint turbidity when tested with silver nitrate stock solution.

9. Fold the filter paper compactly around the precipitate and place it in a porcelain crucible of known weight. If time allows, dry the filter paper overnight. If not, place the paper in a beaker and heat the beaker on a hot plate or in an oven until all of the water has been evaporated.

10. Cover the crucibles with lids, place them on triangles on ring stands, and heat them gradually with Meker burners until the escaping gases burn. Continue heating until gases are no longer evolved.

11. Remove the lids, which may have brown organic deposits on them, place the crucibles at an angle, and position the burners so that the flames touch the bottoms of the crucibles. The charred paper should burn with a glow but should not flame (see Note 2).

12. When the carbon is gone, ignite for 15 minutes at the highest temperature of the burner, the bottom of each crucible just touching the top of the blue cone of flame. Position the flame toward the bottom back of each crucible so that oxygen can reach the precipitate. Allow the crucibles to cool for 30 minutes in a desiccator, then weigh them. Repeat the 15-minute ignition, cooling, etc., until constant weight is achieved.

13. Calculate the results as percent SO_3. The gravimetric factor (the formula weight of SO_3 divided by the formula weight of $BaSO_4$) is 0.3430.

Notes

1. If, in the instructor's judgment, the precipitate is not well digested, the instructor may recommend a slower, finer filter paper, such as Whatman No. 42 or Schleicher and Schuell No. 598 White Ribbon. If precipitate passes through any of the filter papers, it may be useful to refilter the filtrate.

2. If some reduction of barium sulfate to barium sulfide is suspected, moisten the *cooled* precipitate with 2–4 mL of concentrated sulfuric acid and slowly increase the temperature to the highest temperature of the burner.

QUESTIONS

1. What is the chemical reaction when hydrochloric acid is added to the sample? Does this render the sample more, or less, soluble?
2. How many mL of 0.25M barium chloride is needed to precipitate 450 mg of sample? Assume that it is 100% sodium sulfate (form wt, 142).
3. Why is a 10% excess of barium chloride added? Answer in terms of K_{sp}.
4. What is the purpose of the digestion?
5. Why is the precipitate washed? Why is silver nitrate used to test for completeness of washing?
6. What could happen if the *wet* precipitate and filter paper were heated over the burner too rapidly?

29

Volumetric Glassware

Three basic items of precision volumetric glassware are used in chemical analysis: the buret, the volumetric pipet, and the volumetric flask. Burets and pipets of 1 mL or more are normally designed to *deliver* ("TD") the stated volume. Pipets are normally allowed to drain to deliver the stated volume, but pipets smaller than 1 mL (serological pipets) must be blown out to deliver their volume. Volumetric flasks are normally designed to *contain* ("TC") the stated volume.

29–1. CALIBRATION, ACCURACY, AND PRECISION

Calibration. Mass-produced glassware may not always be checked for accuracy by the manufacturer, so calibration is recommended for extremely accurate work. Most analytical measurements, however, do not require calibrated volumetric ware.

If it is necessary to calibrate glassware, this is done gravimetrically. The volume of water delivered or contained by the glassware is weighed and divided by the density (with a buoyancy correction if appropriate) to obtain the true volume. Experiment 6 gives directions for calibrating a 25-mL pipet.

Glassware should also be calibrated if it is to be used at a temperature that is markedly different from the calibration temperature of 20°C (specified by the National Bureau of Standards). Glass expands and contracts much less than does water, so the glassware will hold a smaller or larger weight of liquid than it should.

Accuracy and Precision. A careful distinction should be made between the accuracy and precision of volumetric ware. The *accuracy* of a given piece of volumetric ware is expressed as the maximum allowable error, or *tolerance*. The *precision* of a given piece of volumetric ware is expressed as the *uncertainty* of a buret reading or the *uncertainty* of filling a flask or pipet to the mark. By definition, there is no precision of a single buret reading, but (as was pointed out in Chapter 3) one can speak of the *uncertainty* of a single reading.

Table 29-1 gives some of the National Bureau of Standards tolerances for volumetric glassware. (The less-expensive equipment found in instructional analytical laboratories may have tolerances that are double those in the table.) These tolerances are *absolute* values of the maximum allowable error. For example, the tolerance of 0.05 mL for a 50-mL buret means that the absolute error in the volume delivered may be as large as 0.05 mL. If a volume of 40 mL were used from the buret, the *relative* value of the maximum allowable error, in parts per hundred (pph) and parts per thousand (ppt), would be

$$\frac{0.05 \text{ mL}}{40.00 \text{ mL}}(100) = 0.125 \text{ pph}$$

$$\frac{0.05 \text{ mL}}{40.00 \text{ mL}}(1000) = 1.25 \text{ ppt}$$

Uncertainty of Buret Measurements.   Since only single buret readings are to be characterized, the absolute uncertainty of a single buret reading is needed. This is somewhat arbitrary; the 50-mL buret is usually read to 0.01 mL, although a typical student absolute uncertainty might be ± 0.02 mL. Since two buret readings are subtracted to obtain the volume delivered, a student's maximum possible absolute uncertainty may be as much as ± 0.04 mL (see Section 3-7 on subtracting random variables).

For a buret, the absolute uncertainty is the same for any volume, but the relative uncertainty (pph or ppt) varies. The following calculation for typical student

Table 29-1.   Tolerances for Volumetric Glassware

	Maximum Error Allowable		
Capacity, mL	Volumetric Flasks	Volumetric Pipets	Burets
5	——	0.01	0.01
10	——	0.02	0.02
25	0.03	0.03	0.03
50	0.05	0.05	0.05
100	0.08	0.08	0.10
500	0.15	——	——
1000	0.30	——	——

Note: The Kimball brand, Kimax Class A, and the Corning brand, Pyrex, of glassware conform to these specifications (National Bureau of Standards). The less expensive brands may have tolerances twice as large.

measurements of 10.00 and 40.00 mL is instructive:

$$\frac{0.04 \text{ mL}}{10.00 \text{ mL}}(100) = 0.4 \text{ pph (4 ppt)}$$

$$\frac{0.04 \text{ mL}}{40.00 \text{ mL}}(100) = 0.1 \text{ pph (1 ppt)}$$

Note that the relative uncertainty of measuring the 10-mL volume in a 50-mL buret is far greater than the desirable relative precision of 0.1–0.2 pph (1–2 ppt) for a volumetric method. Obviously, titrations using a 50-mL buret should use 35–45 mL of titrant for good precision. To do a titration using only 10 mL of titrant, you should use a 10-mL buret with 0.02-mL subdivisions to obtain a lower relative uncertainty.

Significant Figures for Pipet and Flask Volumes. The number of significant figures allowable for the volume of a pipet or volumetric flask is an important consideration. Of course the volume delivered by a 5-mL or 10-mL pipet should be expressed to more than or two significant figures. But how many more? Since the tolerance of these pipets is ± 0.01 mL and ± 0.02 mL respectively (Table 29–1), it is clear that their volumes can be expressed as 5.00 ± 0.01 mL, or simply 5.00 mL, and 10.00 ± 0.02 mL, or simply 10.00 mL. It may not be as clear how to express the volume of a 50-mL pipet, or flask. Of course, such a volume can be expressed as 50.00 ± 0.05 mL, but would this simplify to 50.0 or 50.00 mL? First it would have to be recognized that 50.00 mL commonly implies an uncertainty of ± 0.01 mL. Then it would follow that 50.00 ± 0.05 mL would have to be simplified to 50.0 mL.

29–2. VOLUMETRIC FLASKS

Volumetric flasks used in laboratory work range from 5 to 5000 mL in capacity. They are designed to contain ("TC") the exact volume of liquid when the bottom of the meniscus just touches the etched line across the neck. Most volumetric flasks employ ground-glass stoppers, but some are equipped with screw caps lined with polyethylene.

Cleaning the Flask.   If distilled water leaves water breaks or droplets after draining from an inverted flask, it means that dirt or grease has contaminated the surface. The flask should be cleaned so that it will hold the stated volume. Because a flask has a tolerance of 0.1 mL or less (Table 29–1), it is important to remove even a small amount of dirt.

To clean a flask, use one of the approaches discussed in Section 29–5. It is recommended that a warm dilute (2%) detergent solution be tried first.

After cleaning, rinse several times with tap water, followed by three rinse portions of distilled water. Volumetric flasks may be used either wet or dry.

> *Dry volumetric flasks*: Use clean flasks without rinsing if desired. (Do not dry flasks by heating, only by air drying.)
>
> *Wet volumetric flasks*: Use without drying if they have been rinsed with distilled water (or the solvent to be used). Dry the neck above the mark with a tissue if solid is to be added.

Transferring Solutions into a Flask. The solution should be added to the flask first and the following steps followed.

1. If the solution is to be measured in a pipet and added, dry the outside of the pipet, touch the tip to the flask below the mark, and allow the pipet to drain according to procedure in Section 29–3. Remove the pipet from the side of the flask with a rotating motion to remove completely any fraction of a drop. Then fill as directed in Step 3.

2. If the solution is in a beaker, pour the solution from the beaker using a stirring rod (as in decantation in Figure 27–8) through a funnel placed in the neck of the flask. When the transfer is apparently complete, rinse the sides of the beaker and the trip of the stirring rod with distilled water. Fill as directed in Step 3.

3. Dilute the solution to close to the mark with distilled water, or as directed in the experiment. Swirl the flask briefly if desired. Before bringing the volume exactly to the mark, remove all water from above the mark (including the ground glass joint) with a clean tissue. Without contacting any surface above the mark, add distilled water dropwise to fill exactly to the mark. With flasks of 100 mL or less, it may be desirable to use a long-barreled medicine dropper to match the bottom of the meniscus to the mark. (Consider that one drop or 0.05 mL is exactly the tolerance of a 50-mL volumetric flask.)

4. Stopper the flask well, invert, and shake to mix thoroughly. Do not leave the solution in the flask but transfer to a *completely dry* reagent bottle. If the bottle cannot be dried, rinse four times with portions of the solution just large enough to wet all surfaces of the bottle when it is rotated upside down and shaken.

Transferring Solids into a Flask. This operation varies depending on whether or not heating is required to dissolve the solid.

1A. If no heating is required, insert a *powder funnel* into the flask and slowly pour the solid from a waxed paper or container as much as possible down the neck of the funnel. Rinse the funnel four times with distilled water before removing it. Then rinse the neck of the flask with distilled water to wash any solid into the body of the flask. Shake well and proceed to Step 3 under *Transferring Solutions into a Flask*.

2A. If heating is needed, transfer the solid quantitatively to a beaker and heat to dissolve. Cool to almost room temperature. Transfer the solution to the flask as directed in Steps 2–4 under *Transferring Solutions into a Flask*.

Storage. After using any volumetric flask, rinse it with tap water and then with three portions of distilled water. Invert and observe the drainage to see whether cleaning is necessary. If not, position in an inverted position for quick drainage before storing. If the flask is stored upright, cover the top with a beaker but do not stopper.

29–3. PIPETS AND THEIR USE

There seem to be a bewildering number of pipets available for delivering measured volumes of solution into containers. Mainly, pipets are of the automatic or one-hand dispensing type or of the two-hand all-glass type. It is more important to learn about the two-hand all-glass type, especially to understand the advantages and disadvantages of the former type. Therefore, only the two-hand all-glass pipets will be discussed, with the assumption that you will eventually use the other type.

The two common two-hand all-glass pipets are the measuring pipet and the volumetric pipet (Figure 29–1). Volumes from 0.5 to 100 mL may be handled with these pipets. For smaller volumes, the so-called clinical pipet and micropipets are used; these typically handle volumes from 1 μL to 500 μL.

Taking an Aliquot.   It is often necessary to dilute a solution by taking an *aliquot* of that solution. *Only volumetric pipets are accurate exact volume measurement such as taking an aliquot.*

> An *aliquot* is an accurately measured fractional volume of a standard solution. The fraction taken equals the aliquot volume/original volume.

Figure 29–1. Measuring pipet (left) and volumetric pipet (right) showing meniscus.

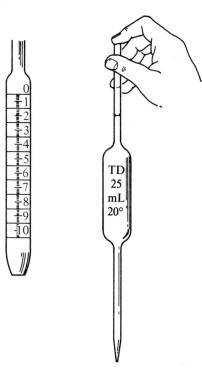

Cleaning the Pipet.   If distilled water leaves water breaks or droplets after a pipet is drained, it means that the pipet must be cleaned. Even a small amount of dirt is serious, since volumetric pipets must be accurate to $\pm 0.01-0.08$ mL (1 drop = 0.05 mL).

To clean a pipet, use one of the approaches discussed in Section 29–5. It is recommended that a warm dilute (2%) detergent solution be tried first.

After cleaning, rinse several times with tap water, followed by three rinse portions of distilled water. If a rack is available, place the rinsed pipet(s) in a draining rack until ready to use.

Use.   The lab instructor may check your technique while you are pipetting distilled water in Experiment 6 or pipetting a solution of potassium acid phthalate or acetic acid. In any use of the volumetric pipet, the following points are important:

1. *Use of Rubber Suction Bulb to Fill Pipet.*   Do not use the mouth for suction; use a rubber suction bulb to avoid swallowing hazardous chemicals (Chapter 25).
2. *Rinsing.*   Rinse the pipet with distilled water before using. Next, rinse the pipet with the solution to be pipetted, so that the water adhering to the inside of the pipet will not dilute the solution. To avoid inserting an unrinsed pipet into a container of solution, pour a little of the solution into a beaker and use that for rinsing. (If an alkaline solution is to be pipetted, it is preferable to take aliquots also from a beaker of the solution.) Rinse the pipet, not by filling completely, but by drawing in about one-fifth of the pipet's volume and twirling the pipet horizontally two or three times. Rinse above the mark by tipping the pipet slightly. Rinse at least twice in this manner with the solution to be pipetted. The pipet should then drain uniformly, or else it needs cleaning or further rinsing.
3. *Filling.*   Fill the pipet from the vessel to about an inch above the etched line (place the fleshy part of a forefinger over the top of the pipet to stop the outflow). Then place the tip of the pipet against the inside of the container of solution and rotate the pipet, allowing the solution to drain until the bottom of the meniscus just touches the etched line at eye level, as in Figure 29–1. There should be no air bubbles anywhere in the pipet.
4. *Carrying.*   The pipet may be conveniently carried by tilting it slightly, so that the solution flows back away from the tip slightly toward the other end.
5. *Draining.*   Wipe the outside of the pipet tip free of any liquid with a tissue before draining. Place the tip against the inside of the vessel to which the solution is to be transferred, and allow it to discharge. Keep the tip against the inside for 20 seconds after the pipet has emptied, for complete drainage. Remove the pipet from the side of the container with a rotating motion to completely remove any of the drop on the tip. NEVER BLOW OUT THE SMALL QUANTITY OF LIQUID INSIDE THE TIP, EVEN THOUGH IT APPEARS TO GROW LARGER WITH TIME. The pipet has been calibrated to account for this.

Storage.   The volumetric pipet should be immediately rinsed with distilled water after use, especially after transferring alkaline solutions. It is good practice to fill the pipet with distilled water and to cap both ends with rubber bulbs from medicine droppers. If this is not possible, the pipet should be thoroughly rinsed and stored in a part of the drawer where it will not easily be scratched or chipped.

29–4. BURETS AND THEIR USE

Burets are calibrated to deliver variable volumes of liquid. The essential parts of a buret are shown in Figure 29–2. Most modern burets are equipped with a glass stopcock lubricated with hydrocarbon greases or with a Teflon plastic stopcock, which requires no lubrication. The latter can be used for nonaqueous solvents and will not freeze after long contact with basic solutions.

Care of the Buret.   Observe the following points:

1. *Lubrication of Buret Stopcock.*   Plastic stopcocks do not need to be lubricated; proceed to Step 2. To clean a glass stopcock, remove all of the old grease. Apply a thin uniform layer of grease (not silicone lubricant); use less grease near the holes in the stopcock. Then insert the stopcock and rotate it several times. If too much grease has been applied, some of it will be forced from between the stopcock and barrel or may eventually work itself into the buret tip. If too little grease has been applied, the lubricant layer will not appear uniform and transparent.

2. *Cleaning the Buret.*   If distilled water does not drain uniformly from the buret but leaves drops of water clinging to the sides, the buret must be cleaned. To clean a buret, use one of the approaches in Section 29–5. It is recommended that a warm dilute (2%) detergent solution be used first.

Figure 29–2. Essential parts of a buret, showing the correct position of the meniscus illuminator (black portion just below meniscus).

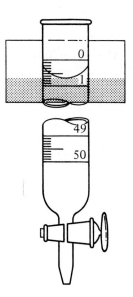

After cleaning, rinse several times with tap water, followed by three rinse portions of distilled water. Place in a buret stand and allow to drain before storage.

Use.   The lab instructor may check your buret technique while you are standardizing sodium hydroxide against potassium acid phthalate or while you are titrating acetic acid or potassium acid phthalate unknowns. In any titration, the following points are important:

1. *Filling.*   The buret must be first rinsed with titrant to remove water adhering to the inside of the buret. Do not rinse by filling the buret and draining, but by pouring about 10 mL of the titrant around the inside three times. Rotate the buret to wet the inside thoroughly, and leave the stopcock open while rinsing. ALLOW THE BURET TO DRAIN COMPLETELY BETWEEN RINSES. Close the stopcock and fill the buret to at least an inch above the zero mark.

2. *Cleanliness.*   During rinsing, check whether the titrant drains uniformly from the buret. If so, it is ready for use. IF DROPS OF TITRANT FORM ON THE INSIDE AFTER RINSING, THE BURET MUST BE CLEANED.

3. *Bubbles in the Tip.*   EXAMINE THE BURET TIP CAREFULLY FOR BUBBLES OR INCOMPLETE FILLING. Often bubbles are best observed by rapidly draining a milliliter or so of titrant when the buret is full. To remove bubbles, drain rapidly and refill buret. This is very important for dark solutions.

4. *Reading the Buret.*   Allow the titrant to drain slowly to the zero mark at the top. Using the meniscus illuminator shown in Figure 29–2, take an initial reading by estimating to 0.01 mL. The initial reading might be 0.00 mL or higher. Record the initial reading. Bring the meniscus illuminator up so that the black half is just below the meniscus. Be sure your eye is at the same level as the meniscus (seeing the meniscus from an angle causes parallax error).

5. *Titrating.*   Fold a white index card or a piece of paper and place it under and behind the sample flask for a white background. Position the tip of the buret within the neck of the flask. Swirl the flask with the right hand and manipulate the stopcock with the left hand from behind the buret. This maintains a slight pressure on the stopcock and avoids leakage. (For more efficient stirring, use a magnetic stirrer and stirring bar.) Add titrant rapidly at first. As the color of the indicator changes more slowly (signalling the approach of the end point) add the titrant dropwise.

6. *The End Point.*   Just before the end point, rinse down the sides of the flask with distilled water from a wash bottle. Split drops of titrant by allowing only a partial drop to form on the tip and washing it into the flask with distilled water. Since the buret can be read to 0.01 mL and a drop is 0.05 mL, splitting drops is essential for accuracy. Allow a minute for drainage before the final reading.

7. *The "Squirt" Technique.*   An alternative to dropwise titration and splitting drops near the end point is the "squirt" technique. Hold the barrel of the stopcock steady with the left hand and quickly twist the stopcock 180° with the right hand. This will deliver a 0.01–0.05 mL squirt of titrant.

8. *Vague End Points.* If the color change at the end point is uncertain (for example, with modified methyl orange), a useful general method is to record the volume of titrant added for successive addition (0.01–0.05 mL) of titrant, noting the color change with each addition. Usually, the point of maximum color change will be obvious after one or two further additions of titrant.

Storage. Burets filled with titrant should not be left standing long, especially if they are filled with a basic titrant such as sodium hydroxide. Such solutions can slowly dissolve glass and can freeze glass stopcocks. To prevent a stopcock from freezing, discard the titrant (but do not return it to the original container), and rinse the buret several times with distilled water. If possible, fill the buret with distilled water after rinsing and cap it to keep dust out. Otherwise, store the buret upside down. (Burets used in nonaqueous titrations may be rinsed with acetone and stored dry, upside down.)

29–5. CLEANING SOLUTIONS AND CLEANING GLASSWARE

Cleaning Solutions. The following cleaning solutions are listed in the order in which they should be tried for cleaning volumetric glassware.

1. A hot dilute (2%) detergent solution. Try a brief soaking first, and next use a brush if possible.
2A. A commercial substitute, such as Nochromix, for dichromate–sulfuric acid. Try a one-hour soaking, then try a two-hour soaking.
2B. A hot dilute (0.004M, pH 12) solution of EDTA to remove metal ions. Soak for 10–15 minutes first, then soak longer.

Commerical substitutes like Nochromix for dichromate–sulfuric acid usually consist of an oxidizing agent that is added to concentrated sulfuric acid. Such substitutes do not have the toxicity of chromium and hence are safer to use.

We do not recommend using dichromate–sulfuric acid cleaning solution because of the toxicity and because the mixture is so corrosive that it causes skin burns and clothing burns.

Cleaning Glassware. It is recommended that any dirty glassware be soaked first in hot dilute detergent. This usually floats most dirt and grease off of the surface in 10–15 minutes. A long soaking period is not recommended because the detergent may cause a rough surface to develop on the glassware that may prevent uniform wetting. If soaking fails for a buret, a careful brushing to avoid scratching the glass may remove dirt. This may also work for the necks of volumetric flasks.

Commercial dichromate–sulfuric acid substites should be used only if detergent solution fails to clean the glassware. Some of these substitutes, such as Nochromix, should not be used on glassware contaminated with silver salts unless the glassware has been thoroughly rinsed. After soaking for 1–2 hours, the substitute solution should be rinsed thoroughly to remove all traces of concentrated sulfuric acid.

After using any cleaning solution, the solution should be removed by adequate rinsing with tap water, followed by rinsing with three portions of distilled water.

Experiment 6.   Calibrating a 25-mL Pipet

1. Rinse a 25-mL volumetric pipet with distilled water. If drops cling to the inside of the pipet, clean the pipet as directed under *Cleaning the Pipet*.
2. Obtain and weigh a large weighing bottle to the nearest milligram (0.001 g).
3. Fill a beaker with distilled water and allow to come to room temperature; then measure the temperature of the water.
4. Rinse the pipet twice, then fill it to about an inch *above* the etched line. Dry the outside of the pipet with a clean towel or tissue. Place the tip of the pipet against the inside of the beaker and rotate the pipet, allowing the water to drain until the bottom of the meniscus just touches the etched line *at eye level*. There should be no air bubbles anywhere in the pipet.
5. The pipet can be conveniently carried by tilting it slightly so that the water flows back away from the tip and slightly toward the end covered by the right index finger.
6. Place the tip of the pipet against the inside of the large weighing bottle and allow the water to discharge. Keep the tip against the inside for 20 seconds after the pipet has emptied, to completely drain it.
7. Now stopper and reweigh the bottle and water to the nearest milligram. The difference between this weight and the weight in Step 2 is the weight of the water.
8. Enter the following data in your notebook:

Temperature of water_____	Vol of pipet calculated from	
Wt, empty bottle _____	Wt of water	_____
Wt, bottle and water _____	Error in mL	_____
Wt of water _____	Tolerance in mL	0.03

9. Corrected volume-weight relationship of water for calibration at 20°C:

Temp., °C	mL/g	Temp., °C	mL/g
19	1.0026	24	1.0036
20	1.0028	25	1.0039
21	1.0030	26	1.0041
22	1.0032	27	1.0043
23	1.0034	28	1.0046

30

Titrimetric Procedures

Titration continues to be the method of choice for analytical determinations when a combination of speed, high precision, and accuracy is desired. This chapter contains standard experiments illustrating acid-base titrations, precipitation titrations, complexometric titrations, and oxidation-reduction titrations. Special techniques are illustrated, such as titration in nonaqueous solvents (Experiment 9) and the use of ion exchange to remove interfering ions (Experiment 11). A new and improved method for total hardness in water is given (Experiment 13) as well as the older complexometric procedures (Experiment 12).

Before performing any of the experiments in this chapter, the student should read Chapter 29 on the use of volumetric glassware and Chapter 3 on the treatment of analytical data. As in Chapter 28, each experiment is introduced with a discussion of the chemistry involved, after which come experimental directions with explanatory notes. The questions at the end of each experiment are intended both to review the material and to stimulate further thinking.

Significant Figures in Final Result.   As long as the sample size is significantly larger than 200 mg, the number of significant figures justified for the final result of a titration depends on the relative uncertainties of the titrant concentration and titrant volume. A typical example is given in detail on p. 28 in Section 3–3. For a $0.1000M$ concentration and 40 mL volume, four significant figures were justified. Although this is common, this should always be checked for the particular experiment. For example, when a liquid sample is used, as in water hardness determinations, three significant figures may be the maximum number allowed because of the greater uncertainty of measuring the sample with a pipet.

Using the Proper Volume of Titrant.   Whatever the buret capacity, the size of the sample should be adjusted to avoid the two sources of larger uncertainty in volume measurement:

1. Using too small ($<40\%$) a percentage of the buret's capacity.
2. Using more than 100% of the buret's capacity and having to refill it.

In general, the sample size should be adjusted after the first titration so that at least 40% and preferably close to $80+\%$ of the buret capacity is used. If this is not done, the relative uncertainty of the volume may decrease the number of significant figures allowed for the final result.

    For example, for the common 50-mL buret, at least 20 mL and preferably close to 40 mL should be used in a titration. Suppose that the sample size were so small that only 4 mL of titrant were used. With the typical absolute uncertainty of 0.04 mL, the relative uncertainty would be

$$\frac{0.04 \text{ mL}}{4.00 \text{ mL}}(100) = 1 \text{ pph (10 ppt)}$$

If this volume were used to obtain a final result of 40.84%, the result would have to be expressed as 40.8% instead because the relative uncertainty of 0.024 pph of 40.84% would not have fallen between 0.2 and 2 times 1 pph.

Choosing the Proper Size Buret.   Although the 50-mL buret is commonly used, it is advantageous to use a 10- or 25-mL buret with a correspondingly smaller sample for some analyses. For titrations in nonaqueous solvents (Experiment 9), the use of a smaller buret saves a good deal of solvent with little reduction in accuracy. In determining water hardness (Experiments 12 and 13), the use of a 10- or 25-mL buret avoids pipetting an unduly large water sample. The student should consult with the instructor for the appropriate sample size or buret size for such experiments.

30–1. ACID-BASE TITRATIONS

Acid-base titration is a quick and accurate means of determining acidic or basic substances in analytical samples. The theory and practice of such titrations were discussed in Chapter 9.

Purpose of the Experiments in This Section.   The first experiment, *Experiment 7*, is designed to be the first titration experiment of the term. Thus, although it involves a sharp end point and is readily done, the directions are written in more detail to provide a basis for a routine for succeeding titrations.

    Experiment 8 is partly designed to illustrate the challenge of observing an end point color change that is not sharp (the methyl orange end point). It also illustrates the choices available for the standardization of hydrochloric acid titrant.

    Experiment 9 is designed to illustrate the use of nonaqueous titrants and solvents in acid-base titrations. The perchloric acid titrant may be used to titrate any number of different types of compounds: solutions of potassium acid phthalate, solutions of amines, pure liquid or solid amines, or solid metal acetates such as sodium acetate.

Experiment 27 in the chapter on electroanalytical procedures (Chapter 32) is also an acid-base titration in that the pK_a of an pure unknown acid is determined potentiometrically by titration with sodium hydroxide.

Experiment 7. Standardization of NaOH and Determination of Total Acidity

The directions for this experiment are written with the assumption that this is your first titration experiment. You should have read Chapter 29, especially Section 29–4 on the use of the buret. Do not assume that, because you have used a buret in earlier chemistry labs, you remember everything about titrations; reviewing Section 29–4 may enable you to avoid serious errors.

Chemical Background

In this method, potassium acid phthalate $(KHC_8H_4O_4$, or KHP) is used as the primary standard for sodium hydroxide titrant. The compound is a salt of phthalic acid $(H_2C_8H_4O_4$, or $H_2P)$; only one proton per molecule of the potassium salt is acidic. The two molecules are represented thus:

H₂P   KHP

Potassium acid phthalate reacts with sodium hydroxide according to the equation

$$OH^- + HP^- = H_2O + P^{2-}$$
$$\text{(NaOH)} \quad \text{(KHP)}$$

The divalent phthalate ion produced is responsible for the basicity of the solution at the equivalence point. Phenolphthalein is employed as indicator.

The same reaction is used for determining impure potassium acid phthalate unknowns; this is sometimes called "determining the total acidity" of a substance.

Planning Ahead

Section 25–2 describes how to plan ahead in the laboratory for any type of experiment. A few of the most important points are summarized below.

1. Check to see that saturated sodium hydroxide is available about two labs ahead; if not prepare it. Otherwise, the $0.1M$ sodium hydroxide titrant cannot be prepared ahead of time.

2. Prepare the $0.1M$ sodium hydroxide titrant during the lab *before* the one in which you will standardize it. This is necessary because it will be hot from the boiled water used and/or the heat of mixing saturated sodium hydroxide and water.

3. Obtain and dry both the primary standard potassium acid phthalate and the acid unknown at least one lab period ahead of time. If you must leave it in the oven overnight, obtain permission to remove it during the time another lab is in session, but do not wait until your lab meets again. You will just waste time waiting for the solids to cool before you can weigh them.

Preparation of 0.1M Sodium Hydroxide

1. Boil about 1 L of distilled water for 5 minutes to remove carbon dioxide. Allow it to cool covered with a watch glass, and transfer while still warm (40°C) to a 1-L bottle. Use either a polyethylene bottle or a Pyrex glass bottle with a rubber stopper, not a glass stopper or screw cap. (The instructor may direct that the boiling be omitted and that unboiled distilled water be used.)

2. Transfer about 7 mL of a clear saturated (1:1) solution of sodium hydroxide (see Note) to the bottle, using a transfer pipet and a rubber bulb. Mix thoroughly and keep covered.

Standardization of 0.1M Sodium Hydroxide

1. Dry 4–5 g of primary-saturated potassium acid phthalate in a weighing bottle at 110°C for 2 hours. Cool it for 30 minutes in a desiccator before weighing.

2A. *First Titration.* To achieve close to a 40-mL volume for the other titrations, it is best to run a rough titration at first to obtain an estimate of the concentration of the sodium hydroxide titrant. Weigh out no more than 800 mg of potassium acid phthalate into a 250-mL flask. Dissolve in 50 mL of distilled water, warming if necessary. Titrate as directed in Steps 3 and 4. Even if your titrant is as low as 0.081M, you should not need more than 50 mL to reach the end point.

2B. *Accurate Titrations.* Estimate the amount of potassium acid phthalate needed for at least 40 mL by multiplying the weight you used in the first titration by the ratio (40 mL/mL titrn.). For example if you used just 25 mL to titrate 790 mg of potassium acid phthalate, then you should weigh out 790 × (40 mL/25 mL) or 1260 mg for each of your other titrations. Weigh out this amount into each of three 250-mL flasks. Dissolve in 50 mL of distilled water, warming if necessary.

3. Fill the buret carefully to 0.00 mL or to a volume that you record before titrating Be sure that the buret drains cleanly and that no bubbles are in the tip. Keep the meniscus illuminator handy to read the buret at all times. After placing the first flask beneath the buret, check to make sure that you have a white background against which to observe the indicator color change accurately. Titrate as directed in Step 4, using only these results to calculate the concentration of sodium hydroxide. (Discard the result of the first titration.)

4. Add 3 or more drops of phenolphthalein indicator to the first flask, and titrate to the first faint pink end point that persists for 20 seconds. Titrate dropwise in the vicinity of the end point, and split drops at the end point. If necessary, adjust the volume of phenolphthalein indicator added to the other two flasks, and titrate them

also. Try to obtain the same intensity of pink color at the end point for all three titrations.

5. Calculate the normality of the sodium hydroxide by using an equivalent weight of 204.22 for potassium acid phthalate. If one of the three normalities is quite different from the other two and there is no experimental reason to discard it, apply the Q test. If this test rejects one of the three normalities, average the remaining two values.

Determination of Total Acidity

1. Dry the unknown sample of impure potassium acid phthalate or other acid for 2 hours at 110°C. Allow it to cool before weighing.

2. *First Titration.* Run a rough titration of the sample first to adjust the titrant volume to around 40 mL. Using dry sample, or the undried sample if it is still being dried, weigh out about 1 g of the sample. Dissolve in 50 mL of distilled water and titrate as in Steps 3 and 4 above.

3. *Accurate Titrations.* Estimate the amount of sample needed for at least 40 mL of titrant by multiplying the weight used in Step 2 by the ratio (40 mL/mL titrn.). Weigh this amount into each of three 250-mL flasks. Dissolve in 50 mL of distilled water, warming if necessary.

4. Follow Steps 3 and 4 under the standardization directions. Record all volumes to two decimal places: 20.00, not 20.0 or 20. mL, etc. If one volume appears to deviate from the others, run an additional titration or two to check this discrepancy.

5. Calculate the results as percentage of potassium acid phthalate (or see the instructor's directions), using the equivalent weight of 204.22.

6. *Data Handling.* Check the discussion at the beginning of this chapter to be sure you report the correct number of significant figures for the mean of your results. If necessary, apply the Q test to your results to check for a questionable value. Do not discard any results (except that of the rough first titration) unless justified by the Q test. (Some instructors may allow discarding a result where careful notebook records indicate a systematic (determinate) error. See Chapter 3.)

Note

A saturated (1:1) solution of sodium hydroxide is usually available in the laboratory. Such a solution may be prepared by mixing 50 g of sodium hydroxide pellets with 50 mL of water in a polyethylene bottle. Let stand until the insoluble sodium carbonate has settled.

QUESTIONS

1. Why is distilled water boiled before it is used in preparing sodium hydroxide titrant? Write the chemical reaction that occurs when unboiled distilled water is used.
2. If the chemical formula of phenolphthalein (a weak acid) is abbreviated as HIndic, write an

equation for its reaction with sodium hydroxide. Indicate which equilibrium form is colorless and which is pink.
3. Why should the titration end point be taken as a *faint* pink color of the phenolphthalein indicator?

Experiment 8.   Standardization of HCl and Determination of Sodium Carbonate

It is assumed that you have already performed your first titrations of the term in the previous experiment. If not, you should review Chapter 29, especially Section 29–4, as well the detailed directions on the first titration in the previous experiment (Step 2A).

Chemical Background

Hydrochloric acid solutions can be standardized by two titrimetric methods and one gravimetric method according to the following equations:

$$OH^- + H^+ \rightleftharpoons H_2O$$
$$\text{(NaOH)} \quad \text{(HCl)}$$

$$2H^+ + CO_3{}^{2-} \rightleftharpoons H_2CO_3$$
$$\text{(HCl)} \quad \text{(Na}_2\text{CO}_3)$$

$$Cl^- + Ag^+ \rightleftharpoons AgCl(s)$$

Each method has its own peculiar advantages and disadvantages. The gravimetric standardization by the precipitation of chloride as silver chloride is the slowest method; it is used only when high accuracy and precision are needed. The two titrimetric methods are given because of their convenience.

The titrimetric method with the better end point is the standardization of hydrochloric acid against previously standardized sodium hydroxide. If the latter is available, this method is also the most rapid. Unfortunately, any error incurred in standardizing the sodium hydroxide will also cause an error in the normality of the acid. However, the phenolphthalein end point is much sharper than the methyl orange end point in the sodium carbonate method.

The other titrimetric method is the standardization of hydrochloric acid against primary-standard sodium carbonate. If modified methyl orange indicator is used, the end point is not sharp and must be approached slowly. Even so, the titration is not especially accurate.

> Modified methyl orange is a mixed indicator and consists of xylene cyanole FF, a blue dye, added to methyl orange. The blue color is complementary to the orange color of methyl orange at a pH of about 3.8, so that the end point color actually observed is grey. The pH range over which the grey color is visible is smaller than that of methyl orange alone, making the modified methyl orange end point sharper than the methyl orange end point.

If methyl red indicator is used, the solution must be boiled to remove carbon dioxide. The end point, however, is sharper than the methyl orange end point.

Sodium carbonate samples are commonly analyzed by titration with standard hydrochloric acid and reported as percent sodium carbonate or as percent sodium oxide. The following equation indicates that the stoichiometry involving sodium oxide is the same as that involving sodium carbonate:

$$2HCl + Na_2O \rightleftharpoons 2NaCl + H_2O$$

The method involves the same steps as the standardization of hydrochloric acid against primary-standard sodium carbonate, which is discussed later in this experiment.

Standardization of Hydrochloric Acid versus Sodium Hydroxide

1. Standardize $0.1M$ sodium hydroxide as directed in Experiment 7. If two or more weeks have elapsed since standardization, recheck the normality of the sodium hydroxide against potassium acid phthalate while the hydrochloric acid is being titrated.

2. Transfer exactly 25 mL of $0.1M$ hydrochloric acid to each of three 250-mL flasks with a 25-mL volumetric pipet. (Alternatively, transfer 40 mL of $0.1M$ hydrochloric acid to the flasks with a 50-mL buret.)

3. Add 2–3 drops of phenolphthalein indicator and titrate with $0.1M$ sodium hydroxide to a light pink color that persists for at least 15 seconds. Three results within ± 0.05 mL should be obtained. Calculate the normality of the hydrochloric acid.

Standardization of Hydrochloric Acid versus Sodium Carbonate

1. Dry 1–1.5 g of primary-standard sodium carbonate at 110°C for 2 hours in a weighing bottle.

2. Weigh three 0.2-g samples by difference from the weighing bottle (to avoid absorption of water from the air) into 250-mL flasks and dissolve each of them in 50–100 mL of distilled water.

3A. This is the methyl orange procedure. Add 2–3 drops of modified methyl orange indicator to each flask, then titrate with $0.1M$ hydrochloric acid, taking the end point as the change from green to grey. If the end point is difficult to observe, record the volume and color and proceed dropwise past the grey to a purple. By comparison, the change to grey should now be evident. If the color change to purple appears to be sharper, take this to be the end point for standardization and for determining sodium oxide (total basicity) in the unknown. In this way, the errors will cancel out.

3B. This is the methyl red procedure. Add 2–3 drops of methyl red indicator. Titrate with $0.1M$ hydrochloric acid until the indicator has changed gradually from yellow to a definite red color. Then boil the solution gently for 2 minutes. The color of the indicator should revert to yellow. Cover the flask with a watch glass, cool to room temperature, and continue the titration to a sharp change to red at the end point.

4. Calculate the normality of the acid using an equivalent weight of 53.0 for sodium carbonate.

Procedure

1. Dry the impure carbonate sample at 110°C for 2 hours.
2. Weigh out four 0.25–0.35 g samples (see the instructor's directions) by difference from a weighing bottle into 250-mL flasks. Dissolve each sample in 50–100 mL of distilled water. Follow either the methyl orange or methyl red procedure given in Steps 3A and 3B in the section on standardizing HCl against Na_2CO_3. Use the first sample to practice the end point if the hydrochloric acid has been standardized against sodium hydroxide (see Note).
3. Calculate the sample as percent Na_2O or as percent Na_2CO_3 (see the instructor's directions) using an equivalent weight of 31.0 for sodium oxide or 53.0 for sodium carbonate.

Note

   If it is difficult to observe the end point using modified methyl orange, prepare a potassium acid phthalate solution in a ratio of 1 g/100 mL of solution, and add 2 drops of modified methyl orange indicator to it. This buffers the solution at pH 4 and should produce a grey indicator color. Titrate samples to match this color.

QUESTIONS

1. Compare each of the three standardization methods for their advantages and disadvantages.
2. Why is it preferable to titrate hydrochloric acid with sodium hydroxide rather than the reverse?
3. Explain why primary-standard sodium carbonate should be weighed by difference from a weighing bottle.
4. What effects will boiling off carbon dioxide have on the pH of the solution in the methyl red end-point method?
5. Why can't methyl orange be used as indicator if carbon dioxide has been removed by boiling?
6. Explain why the color change in the methyl orange end point is gradual. (*Hint:* Consider the first ionization constant of carbonic acid.)

Experiment 9.   Perchloric Acid Nonaqueous Titrations

Chemical Background

   In glacial acetic acid as solvent, perchloric acid has been found to be the strongest acid; that is, it has a higher degree of ionization than other acids (such as sulfuric acid). For this reason, it is the titrant of choice for weak bases in glacial acetic acid. Bases such as aniline and sodium acetate, which are too weakly basic to titrate in water, can be titrated to sharp end points.

The most suitable primary standard for perchloric acid is potassium acid phthalate, which behaves as a Brönsted base in this reaction (Equation E9–1). Two common types of weak bases that can be analyzed by perchloric acid titration are amines, such as aniline (Equation E9–2), and salts of carboxylic acids, such as sodium acetate (Equation E9–3). The reactions are written in molecular rather than ionic form because the acids dissociate so little in glacial acetic acid (even perchloric acid has an overall K_a of the order of 10^{-5}):

$$HClO_4 + KHP \rightleftharpoons H_2P(s) + KClO_4 \qquad \text{(E9–1)}$$

$$HClO_4 + C_6H_5NH_2 \rightleftharpoons C_6H_5NH_3ClO_4 \qquad \text{(E9–2)}$$

$$HClO_4 + CH_3CO_2Na \rightleftharpoons CH_3CO_2H + NaClO_4 \qquad \text{(E9–3)}$$

Procedure

1. Prepare $0.1M$ perchloric acid by adding 4.3 mL of 72% perchloric acid ($HClO_4 \cdot 2H_2O$) to 100 mL of acetic acid, mixing well, and add 10 mL of acetic anhydride (Note 1). The mixture will become quite warm. Allow it to stand for half an hour, so that the water in the perchloric acid can react with the acetic anhydride to form acetic acid (Note 2). Then dilute the solution to 500 mL with glacial acetic acid. Allow it to cool to room temperature.

2. Prepare the indicator by dissolving 0.2 g of methyl violet or crystal violet in 100 mL of glacial acetic acid or chlorobenzene. (This solution may already be available in the laboratory.)

3. Standardize the perchloric acid by weighing three 0.6–0.7 g exact samples of primary-standard potassium acid phthalate into 250-mL flasks. Dissolve the phthalate by adding 60 mL of glacial acetic acid to each flask and heating to boiling in a hood. If any of the solid creeps up the sides of the flask, wash it down with a little acetic acid.

4. Cool the solutions and add 2–3 drops of methyl violet indicator to each flask. Titrate each to the color change of violet to blue (not to blue-green or green or yellow), using $0.1M$ perchloric acid. Calculate the normality of the titrant, using an equivalent weight of 204.2 for potassium acid phthalate.

5. Weigh three 0.5-g samples (see the instructor's directions) of solid unknown into 250-mL flasks, or pipet three 25-mL aliquots (see the instructor's directions) of liquid unknown into 250-mL flasks (Note 3). Dilute each sample to 60 mL with glacial acetic acid, add 2–3 drops of indicator, and titrate with $0.1M$ perchloric acid to the color change of violet to blue.

6. For solid unknowns, calculate the equivalent weight of the unknown. For liquid samples, calculate the milliequivalents of base for each 25-mL aliquot of the liquid unknown.

Notes

1. If the solution is to be used only for titrating tertiary amines, acetates, or

phthalates, add 13 mL of acetic anhydride instead of 10 mL. This renders the titrant completely anhydrous and gives a slightly sharper end point.

2. The reaction of water with acetic anhydride is

$$(CH_3CO)_2O + H_2O \rightarrow 2CH_3CO_2H$$

This reaction is actually catalyzed by the perchloric acid; it is quite rapid if the solution is not diluted.

3. Liquid unknowns may be potassium acid phthalate dissolved in glacial acetic acid or a solution of an amine in chlorobenzene. Typical amines are n-butyl-amine, pyridine, or dimethylaniline. Solid unknowns may be acetates such as sodium acetate, barium acetate, and strontium acetate. (Potassium acetate is not very suitable because it is quite hygroscopic.)

30–2. PRECIPITATION AND COMPLEXATION TITRATIONS

The reactions of metal ions and ligands can often be used for analytical titrations. Precipitation of many -1 anions by silver(I) ion and of sulfate by barium(II) ion is an excellent, though limited, titration technique discussed in detail in Chapter 10. In contrast, complexation of cations by ethylenediaminetetraacetic acid (EDTA) is one of the most widely applicable titration methods (Chapter 11).

Purpose of the Experiments in This Section.   The first experiment, *Experiment 10*, is designed to illustrate the use of the adsorption indicator end point method for the precipitation titration of chloride. The end point chemistry is the most unusual chemistry of any titration method.

Experiment 11 is unique in that it involves the precipitation titration of the sulfate ion. It is important because there is no other convenient titration method for sulfate ion.

Experiments 12 and 13 are two similar experimental introductions to EDTA titrations for the determination of total calcium and magnesium concentration (water hardness). Experiment 12 uses the traditional Calmagite indicator, and Experiment 13 features Arsenazo indicator for a sharper color change.

Experiment 14 is an advanced EDTA method in which bismuth(III) and then cadmium(II) are successively titrated after pH adjustment.

Experiment 15 is another advanced EDTA method in which zinc(II) and/or cadmium(II) may be masked by cyanide and magnesium titrated without interference.

Experiment 10.   Determination of Chloride Using an Adsorption Indicator

In the adsorption indicator method, the chloride ion is titrated with silver nitrate while the end point is detected with a dye that imparts a distinctive color to the surface of the silver chloride precipitate. A weakly acidic organic dye, such as fluorescein or dichlorofluorescein (HIndic) is commonly used as indicator. The ionization of the dye molecule produces a yellow-green anion (Indic$^-$).

Chemical Background

Although silver chloride precipitates in colloidal form, it tends to coagulate when the solution is stirred during the titration. Because this reduces the surface area available for adsorption of the indicator and also reduces the sharpness of the end point, some dextrin or polyethylene glycol is added to stabilize the colloidal state of the precipitate.

As the equivalence point is approached and passed, chloride ion is no longer in excess and is replaced by the silver ion as the primary adsorbed layer on the surface of the precipitate (Equation E10–1). Nitrate ion replaces sodium ion as the counterion (Equation E10–1). The negatively charged dichlorofluorescein anion (Indic$^-$) then displaces the nitrate ion to become the counterion (Equation E10–2). On being adsorbed, its electronic structure changes so that it reflects pink light rather than yellow-green; this signals the end point.

$$3Ag^+ + 2NO_3^- + AgCl:Cl^- \vert Na^+(s) \rightleftharpoons 2AgCl:Ag^+ \vert NO_3^-(s) + Na^+ \quad \textbf{(E10–1)}$$

$$\underset{\text{(yellow)}}{AgCl:Ag^+ \vert NO_3^-(s) + HIndic} \rightleftharpoons \underset{\text{(pink)}}{AgCl:Ag^+ \vert Indic^-(s)} + H^+ + NO_3^- \quad \textbf{(E10–2)}$$

The pH should not drop below 4, or the point of equilibrium (Equation E10–2) will be unfavorably shifted in favor of the weak acid form of dichlorofluorescein. In the 6–10 pH range, the equilibrium concentration of the dichlorofluorescein anion is high enough to displace chloride ion as the primary adsorbed layer slightly before the equivalence point. This error is cancelled by standardizing the silver nitrate against primary-standard potassium chloride.

At pH 4, the end point exactly coincides with the equivalence point. Buffering the solution at pH 4 represses the ionization of the indicator enough to avoid the end point error, but it also modifies the end point color change. This change is from white to a difficult-to-observe light pink, instead of from a light pink to a dark pink.

Procedure

1. Dry the unknown and 1.5 g of primary-standard potassium chloride for 2 hours at 110°C.
2. Weigh about 8.5 g of silver nitrate into a dark-amber–colored glass-stoppered bottle, add 500 mL of chloride-free distilled water, and shake it to dissolve the silver nitrate. If the instructor directs, prepare a standard solution of 0.1M silver nitrate instead (Note 1).
3. Weigh three 0.25–0.30 g samples of primary-standard potassium chloride into 250-mL flasks, dilute them to 100 mL with distilled water, and check that the pH is 6–10 (Note 2).
4. Choose a place away from the sunlight for titration. Add 0.1 g of dextrin (or 5 mL of a 2% dextrin solution) and 10 drops of a 0.1% solution of dichlorofluorescein to the first flask (Note 3).

5. Titrate with 0.1M silver nitrate at a fairly rapid rate, using continuous shaking or magnetic stirring. The suspended precipitate will have a pink tinge because of some premature displacement of chloride ion by the dichlorofluorescein ion. To detect the end point, set the buret so that it delivers titrant at a slow dropwise rate, shake constantly, and continuously observe the color of the suspension until it changes from a light pink to a dark pink.

6. If the color change to dark pink is observed (Note 4), repeat Steps 4 and 5 for the remaining two samples. Calculate the molarity of the silver nitrate titrant.

7. Analyze the unknown by weighing three 0.28–0.32 g samples (see the instructor's directions) into 250-mL flasks, diluting them to 100 mL and checking to see that the pH is 6–10. Titrate as directed for pure potassium chloride according to Steps 4, 5, and 6.

8. Calculate the percentage of chloride in the unknown using the atomic weight of 35.45 for chloride.

Notes

1. To prepare standard 0.1M silver nitrate, dry primary-standard silver nitrate for no longer than 1 hour at 110°C, store it in the dark in a desiccator, and weigh out about 8.5 g, by difference, to the nearest milligram. Transfer the silver nitrate to a 500-mL volumetric flask and dilute it to volume with distilled water. Perform the titration under the conditions of Note 2 to avoid the slight error arising from a premature end point.

2. An alternative is to perform the titration at a pH of about 4 by adding 1 mL of a buffer made by mixing 0.4M acetic acid with 0.4M sodium acetate. This avoids the slightly premature end point and is recommended if the silver nitrate titrant is made by weighing out primary-standard silver nitrate and diluting it to a known volume.

3. If a 0.1% dichlorofluorescein solution in a mixture of 50% polyethylene glycol and 50% water is used, dextrin can be omitted. The recommended glycol is Polyethylene Glycol 400, manufactured by the Union Carbide Chemicals Company (see R. B. Dean et al., *Anal. Chem.*, *24*, 1638 (1952)). The usual indicator is a 0.1% solution of the acid form in 70% alcohol.

4. If a grey-pink or purple suspension is noted at the end point, the indicator has catalyzed the photoreduction of silver(I) to metallic silver. Avoid this by adding the indicator, but not the dextrin, about 0.5 mL before the calculated end point in the next two titrations.

QUESTIONS

1. Explain why a large surface area favors a sharp end point.
2. Write equations explaining the titration of potassium bromide with silver nitrate.
3. Write an equation to explain the premature-end-point error at pH 6–10 mentioned in Notes 1 and 2.
4. Use Equation E10–2 to explain why the pH should not be below 4.

Experiment 11.   Determination of Sulfate Using an Adsorption Indicator

Sulfate can be titrated with a standard solution of barium chloride or perchlorate (see Chapter 10). This is a fairly accurate procedure, and much quicker than the gravimetric method.

Chemical Background

The titration is carried out in a mixture of methyl alcohol and water using alizarin red S (sodium alizarin sulfonate) indicator. The indicator remains yellow as long as any sulfate remains unprecipitated. When the first excess of barium is added, a pink complex of barium alizarin red S forms on the surface of the $BaSO_4$ precipitate.

Many foreign cations and anions can coprecipitate during a barium sulfate error. One purpose of this experiment is to determine the magnitude and direction (toward high or low results) of the error. This is done by comparing the results of two titration procedures, one in which the ammonium ion is left in the solution, and another in which it is first removed by ion exchange.

In the ion-exchange procedure, cations are exchanged for hydrogen ions by passing the sample through a cation-exchange column (catex) in the hydrogen form:

$$2NH_4^+ + SO_4^{2-} + 2Res\text{-}SO_3^-H^+ \rightleftharpoons 2Res\text{-}SO_3^-NH_4^+ + 2H^+ + SO_4^{2-}$$

Before titration, the sulfuric acid is partially neutralized by magnesium acetate (magnesium is one of the *least* coprecipitated cations):

$$2H^+ + 2CH_3COO^- \rightarrow 2CH_3COOH$$
$$\text{(H}_2\text{SO}_4\text{)} \quad \text{[(CH}_3\text{COO)}_2\text{Mg]}$$

Procedure (Without Ion Exchange)

1. Pipet exactly 10 mL of 0.1M ammonium sulfate (Note 1) into a 250-mL flask.
   Add approximately 30 mL of distilled water and 45 mL of methanol. Add 2 drops of alizarin red S indicator (Note 2), then add dilute (1:10) hydrochloric acid dropwise until the indicator turns yellow.
2. Titrate rapidly with 0.05M barium chloride. (Note 3) until about 90% of the theoretical amount of titrant has been added. Then add 3 more drops of alizarin red S indicator.
3. Continue the titration in a dropwise manner while vigorously swirling the solution in the flask. When flashes of pink appear, allow 3–5 seconds between addition of each increment of titrant. Take the end point to be the first permanent color change to pale pink. Repeat the titration with a second 10-mL sample to give duplicate analyses.

Procedure With Ion Exchange

1. Obtain a column from the instructor (Note 4). If the column is not already filled with resin, add a very small wad of glass wool to the bottom fitting to hold the

resin in place. Pour some cation exchange resin (hydrogen form, 50–100 mesh) into water and stir (Note 5). Then pour the slurry into the column. Add enough resin to make a column 8–10 cm in height. (Never pour dry ion-exchange resin into a column and then add water. The rapid swelling of the resin may cause an explosion.) Wash the resin with a little distilled water. Allow the liquid to drain almost to the top of the resin, then shut off the flow—either with the stopcock or by quickly inserting a rubber stopper into the top of the column. The resin will serve for 3–4 runs; then it should be placed in the used resin container for regeneration (Note 6).

2. Pipet 10 mL of 0.1M ammonium sulfate directly into the ion-exchange column. Place a 250-mL flask under the column, and allow solution to flow from the column at a flow rate of about 2 mL/min.

3. When the liquid level is almost down to the top of the resin, wash the sample though the column with 30 mL of distilled water added in three or four separate portions. Add each portion when the liquid level is about 0.5 inch above the resin. (Never allow the liquid level to drop below the level of the resin.)

4. The effluent contains all of the sulfate, now present as sulfuric acid. Add 45 mL of methanol and 3.5 mL of 0.25M magnesium acetate solution to the flask containing the effluent. Then add 2 drops of alizarin red S indicator, and titrate as directed in Steps 2 and 3 of the Procedure. Repeat the ion-exchange procedure and the titration to give duplicate results.

Notes

1. The instructor may direct each student to prepare 100 mL of 0.1M ammonium sulfate (1.321 g per 100 mL), or it may be provided as a standard solution for common use.

2. The alizarin red S indicator solution is a 0.2% aqueous solution of sodium alizarin sulfonate.

3. The 0.05M aqueous barium chloride titrant must be adjusted to pH 3.0–3.5 with dilute HCl, using a pH meter.

4. An appropriate column is 1.6 cm i.d. and approximately 15 cm long. However, these dimensions are not very critical.

5. Dowex 50 or 50W or Amberlite IR-120 is a suitable resin.

6. Accumulated used resin is best regenerated in a large column. Pass 3 or 4 volumes of 3M hydrochloric acid through the column, then rinse it with several column volumes of distilled water added in several portions. (The column volume is the bulk volume of the resin, in milliliters.)

QUESTIONS

1. Explain why the same procedure gives different results when it is used with, and without, ion exchange.

2. Write a chemical reaction for the regeneration of the catex resin with hydrochloric acid.

3. Why must the sulfuric acid in the ion-exchange procedure be partially neutralized?
4. Does the indicator color change occur throughout the solution or on the surface of the precipitate? Does this explain why the solution must be stirred to make the end point easily visible?

Experiment 12.   EDTA Determination of Water Hardness Using Calmagite

Two methods for the EDTA determination of total water hardness are given in this experiment and the next. In this experiment, the recommended indicator is Calmagite, a more stable indicator than Eriochrome black T. In Experiment 13, Arsenazo indicator is recommended for a sharper end point.

Chemical Background

EDTA is usually standardized against a standard solution of calcium(II) ion prepared by dissolving reagent-grade calcium carbonate in hydrochloric acid and boiling to remove most of the carbonic acid as carbon dioxide:

$$CaCO_3(s) + 2H^+ \rightleftharpoons Ca^{2+} + H_2O + CO_2(g) \qquad \textbf{(E12–1)}$$

The titration (Equation E12–2) is performed by buffering the solution at pH 10. This permits the calcium-EDTA and magnesium-EDTA chelates to form stoichiometrically, along with a sharp indicator change. The breakup of the wine-red magnesium-Calmagite chelate (Equation E12–3) coincides with the equivalence point:

$$2H_2Y^{2-} + Ca^{2+} + Mg^{2+} \rightleftharpoons CaY^{2-} + MgY^{2-} + 4H^+ \qquad \textbf{(E12–2)}$$
$$\text{(Na}_2\text{H}_2\text{Y)}$$

$$H_2Y^{2-} + MgIndic^- \rightleftharpoons MgY^{2-} + HIndic^{2-} + H^+ \qquad \textbf{(E12–3)}$$
$$\text{(wine-red)} \qquad\qquad\qquad \text{(blue)}$$

Precautions are necessary to prevent traces of metal ions, such as iron(III), copper(II), and aluminum(III), from irreversibly forming complexes with the Calmagite and preventing a sharp color change. The ammonia pH 10 buffer is always added *before* the indicator to tie up iron(III) and aluminum(III) in unreactive forms. The buffer will not prevent the interference of *large* amounts of iron(III); cyanide ion is necessary to mask iron(III) in this case. The hydroxylamine hydrochloride added with the indicator reduces copper(II) to copper(I) and so prevents interference from copper(II).

The distilled water should be tested for appreciable concentrations of these ions, and deionized if necessary. Titration flasks should be scrupulously cleaned to remove traces of these ions also.

Reagents

1. Prepare 0.01*M* EDTA titrant by dissolving about 1.9 g of the reagent-grade disodium salt ($Na_2H_2Y \cdot 2H_2O$, form wt 372) in 500 mL of distilled water. Add about 0.5 g (6 pellets) of sodium hydroxide and about 0.1 g of magnesium chloride, $MgCl_2 \cdot 6H_2O$. Mix well. Store in a Pyrex bottle. If ACS grade EDTA (99.0% min

Na$_2$H$_2$Y · 2H$_2$O) is available, the instructor may direct you to use it as a primary standard. (In this case, omit the addition of magnesium chloride and follow Note 1.)

2. Prepare standard 0.0100M calcium(II) solution by weighing 0.500 g $\pm$ 0.2 mg of 99+ % calcium carbonate into 20 mL of distilled water in a 250-mL beaker. Pipet about 1 mL of concentrated hydrochloric acid down the side of the beaker, and place a watch glass directly on top of the beaker. When the calcium carbonate has dissolved, rinse the watch glass into the beaker, raise it on glass hooks, and evaporate the solution to a volume of about 2 mL to expel most of the carbon dioxide. Add 50 mL of distilled water, transfer the solution to a 500-mL volumetric flask, and dilute it to the mark.

3. Prepare the pH 10 ammonia buffer by diluting 32 g of ammonium chloride and 285 mL of concentrated ammonium hydroxide to 500 mL with distilled water. Store the buffer in a polyethylene bottle. The buffer is as concentrated as possible to avoid contaminating the solutions with interfering metal ions. It is not stored in glass because metal ions would be leached from the glass.

4. Prepare a dilute, aqueous solution of Calmagite. If Eriochrome black T indicator is to be used instead, prepare a solid mixture by grinding 50 mg of Eriochrome black T into 5 g of sodium chloride and 5 g of hydroxylamine hydrochloride.

Procedure

1. Pipet exactly 20 mL of 0.0100M calcium(II) solution into each of three clean 250-mL flasks (Note 2), and add about 1 mL of the pH 10 ammonia buffer to each flask. Add enough Calmagite indicator to the first flask to give a wine-red color (Note 3). Alternatively, add a small scoop of the solid Eriochrome black T indicator mixture.

2. Titrate with 0.01M EDTA from a 25-mL buret until a color change from wine-red through purple to clear blue is observed. The color change is somewhat slow, so titrant must be added slowly near the end point (Note 4). Repeat the titration for the remaining two samples. Calculate the molarity of the EDTA, using the 0.0100 molarity of the calcium(II) solution.

3. Determine the hardness of the unknown hard water by pipetting exactly 50 mL of the sample into each of three clean 250-mL flasks. Add 1 mL of ammonia buffer and Calmagite indicator to the first sample and titrate it with 0.01M EDTA from a 25-mL buret to a clear-blue end point (Notes 4 and 5).

4. Calculate the parts per million of calcium carbonate in the unknowns using the molarity of EDTA and 100.1 mg/mmole as the formula weight of calcium carbonate.

Notes

1. Magnesium(II) is added to the titrant as magnesium chloride to ensure a sharp end point, since calcium(II) does not form a strong enough chelate with the indicator. Instead of this, about 0.5 mL of 0.005M magnesium-EDTA chelate may be added to each flask. Prepare this by mixing equal volumes of 0.010M EDTA and 0.010M magnesium(II) solutions.

2. Remove traces of interfering metal ions from the sides of the titration flasks by rinsing each flask with about 10 mL of 1:1 nitric acid. Tilt and rotate the flask to contact the entire inside. Rinse well with distilled water, and allow it to drain upside down.

3. Possible air oxidation makes it advisable to add the indicator just before titrating.
   Traces of metal ions such as manganese(II) can catalyze this oxidation; this is prevented by adding a bit of ascorbic acid. Avoid adding too much indicator, because the end-point color change will be too gradual.

4. To avoid titrating slowly near the end point, the solutions may be warmed to about 60°C.

5. If the end-point color change is to a violet and not to a clear blue, a high level of iron in the water may be responsible. Avoid this in succeeding samples by adding a few crystals of potassium cyanide (as much as will fit on the tip of a spatula) *after* the buffer has been added. *CAUTION!* Potassium cyanide is a poison. On contact with acid, it forms hydrogen cyanide gas—which is especially dangerous.

Experiment 13. EDTA Determination of Water Hardness Using Arsenazo

The basic principle of this procedure is essentially the same as for Experiment 12. However, the use of Arsenazo as the indicator has several important advantages [1]. One advantage is that, since this indicator works well for both calcium (II) and magnesium(II), it is not necessary to add a magnesium salt to the titrant. The color change with Arsenazo is much faster than with Calmagite or Eriochrome black T, so there is less chance of overrunning the end point. Finally, small amounts of iron, copper, or aluminum in the water do not block the indicator or reduce the sharpness of the end point, as may happen with the other indicators.

A disadvantage of Arsenazo is that the color contrast between the metal complex (violet) and the free indicator (orange pink) is not as vivid as might be desired. To improve the sharpness of the color change, a blue dye and a yellow dye are mixed with the Arsenazo. These color screening dyes modify the Arsenazo colors so that a change from a pale violet to a greenish straw color is obtained. Both the violet and straw colors are grayed and intensified by the color-cancelling effects of the screening dyes.

Reagents

1. Prepare 0.01M EDTA titrant by dissolving about 1.9 g of reagent grade disodium salt (Na$_2$H$_2$Y · 2H$_2$O, form wt 372) in 500 mL of distilled water (Note 1).

2. Prepare standard 0.0100M calcium(II) solution from calcium carbonate as described in Experiment 12.

3. Prepare a pH 10 buffer solution by dissolving 13.1 g of THAM, tris-(hydroxy-

[1] J. S. Fritz, J. P. Sickafoose, and M. A. Schmitt, *Anal. Chem. 41*, 1954 (1969).

methyl)aminomethane, in 100 mL of distilled water and adding 4.0 mL of concentrated hydrochloric acid and 30 mL of concentrated ammonium hydroxide.

4. Prepare a mixed Arsenazo indicator by dissolving 100 mg of Arsenazo I (Aldrich Chemical Co.), 100 mg of martius yellow, 52.5 mg of xylene cyanol FF, and 1.0 g of THAM in 10 mL of isopropyl alcohol. Slowly add about 20 mL of distilled water, then transfer the solution to a 100-mL volumetric flask and dilute it to volume with distilled water.

Procedure

1. Standardize the EDTA by pipetting exactly 20 mL of 0.0100M calcium(II) solution into each of three clean 250-mL flasks. Add 2 mL of THAM buffer and two drops of mixed Arsenazo indicator, then titrate the solution with ~0.01M EDTA until the color changes from pale magenta to a golden or greenish yellow. Titrate the other two standards in similar fashion and calculate the molarity of the EDTA (Note 2).

2. Determine the total hardness-content of water of unknown hardness by pipetting 50-mL aliquots into each of three clean 250-mL flasks. Add 2 mL of THAM buffer and two drops of mixed Arsenazo indicator, and titrate the sample with the standardized 0.01M EDTA to a yellow end point as before (Note 3).

3. Calculate the total hardness (Ca + Mg) as ppm of calcium carbonate, using the molarity of the EDTA and 100.1 as the formula weight of calcium carbonate.

Notes

1. It is not necessary to add a magnesium salt to the EDTA (see Experiment 12), but magnesium-EDTA is not detrimental.

2. One of our graduate students recommends a "bias eliminator" for repeated titrations of aliquots. This is a short paper cylinder slipped around the buret to prevent the experimenter from observing the titrant level near the expected end point. Only after the end point has been firmly decided upon is the paper removed and the buret read.

Experiment 14.   Successive EDTA Titration of Bismuth(III) and Cadmium(II) Ions

This experiment involves the successive titration of first bismuth(III) ion at pH 1 and then cadmium(II) ion at pH 5. To understand how this is possible, consider the ratio of the stability constants of their EDTA complexes:

$$\frac{K_{BiY}}{K_{CdY}} = \frac{10^{22.8}}{10^{16.5}} = \frac{10^{6.3}}{1}$$

It can be shown that if such a ratio is greater than $10^6:1$, the titration of the metal ion with the stronger EDTA chelate can be performed quantitatively before the other metal ion reacts significantly with EDTA.

Chemical Background

Korbl and Pribil [2] have shown that bismuth(III) reacts quantitatively with EDTA at pH 1–3, and cadmium(II) does not react significantly. The reactions at the pH of 1 used in this experiment are

$$Bi(III) + H_2Y^{-2} \xrightarrow{pH \ 1} BiY^- + 2H^+$$

$$Cd(II) + H_2Y^{-2} \xrightarrow{pH \ 1} NR$$

After bismuth(III) has been titrated, the pH is increased to pH 5 with a buffer, and cadium(II) is titrated.

Reagents

1. Obtain $99.0 + \%$ ACS grade EDTA ($Na_2H_2Y \cdot 2H_2O$) and use it as a primary standard. Prepare $0.005M$ EDTA by weighing 0.93 g of the disodium salt to three significant figures and diluting to exactly 500 mL with distilled water. (If the salt does not dissolve readily, add no more than 3 pellets of sodium hydroxide.) Store in a Pyrex bottle and calculate the exact molarity, using 372.24 as the formula weight.

2. Prepare approximately $0.005M$ cadmium(II) ion by weighing out approximately 0.5 mmole of any 99% pure cadmium compound such as anhydrous cadmium chloride (form wt = 183.3) into a 100-mL volumetric flask. Add 10 mL of $1M$ nitric acid for a $0.1M$ final concentration and dilute with distilled water to the mark.

3. Prepare xylenol orange indicator by dissolving 0.1 g of the solid in 100 mL of 50% ethyl alcohol.

4. Obtain solid hexamethylenetetramine (Fisher Scientific) for buffering to pH 5.

Procedure

1. Determine the optimum amount of xylenol orange indicator needed by practicing the cadmium end point. Pipet 25 mL of $0.005M$ cadmium(II) into a clean flask. Add 5 mmole (700 mg) of hexamethylenetetramine to obtain a buffer mixture of equal amounts of the buffer acid and base, giving a pH of about 5. Try 5–6 drops of indicator at first, and work for a sharp color change from pink-purple to yellow.

2. Prepare for titration of the unknown by rinsing the sides and bottoms of three 250-mL flasks with 1:1 nitric acid. Rinse well with distilled water, and then pipet exactly 25 mL of the bismuth-cadmium unknown solution into each of the flasks.

3. If the unknown is not $0.1M$ in nitric acid, check with the instructor to adjust the pH to 1.0. (In any case, do not dilute the unknown with water; use only $0.1M$ nitric acid if dilution is needed.)

4. Add the optimum number of drops of xylenol orange indicator determined in Step 1 and titrate with standard $0.005M$ EDTA to a sharp color change from pink-purple to light yellow. Record the volume.

[2] J. Korbl and R. Pribil, *Chemist-Analyst 45*, 102 (1956) and *46*, 28 (1957). For bismuth(III) alone, see also J. S. Fritz, J. E. Abbink, and M. A. Payne, *Anal. Chem. 33*, 1381 (1961).

5. Adjust the pH to about 5 for the titration of cadmium(II) by adding 5.0 mmole
   (0.7 g) of hexamethylenetetramine. Check the pH with pH paper and add more
hexamethylenetetramine if needed. (If the unknown has been diluted with $0.1M$
nitric acid, about twice as many mmoles of hexamethylenetetramine as nitric acid
are needed.)

6. As the pH increases, the color should turn back to pink-purple from the complexation of cadium(II) ion. Titrate slowly with $0.005M$ EDTA to a light yellow
end point. (Near the end point, the color may at first appear more orange than
pink-purple, but it should still change to the same light yellow. The color change
may also be slow if too much buffer has been added.)

7. Calculate the molarity of each metal ion in the unknown for your report.

Experiment 15.   EDTA Titration of Magnesium(II) After Masking of Zinc(II) and/or Cadmium(II)

This experiment involves the use of cyanide ion to mask metal ions that
form stable cyanide complex ions after which an ion such as magnesium can be
titrated with EDTA. Magnesium and calcium do not form cyanide complexes and
thus are readily titrated with EDTA.

Chemical Background

The first chemical reaction involves addition to cyanide to the sample to
complex metal ions such as zinc(II) and cadmium(II):

$$Zn^{2+} + Cd^{2+} + 8CN^- \xrightarrow{pH\ 10} Zn(CN)_4^{2-} + Cd(CN)_4^{2-}$$

$$Mg^{2+} + CN^- \xrightarrow{pH\ 10} NR$$

It is important that the pH be adjusted to pH 10 before cyanide is added, since
poisonous HCN gas can form if the solution is not basic. Once the cyanide has
complexed zinc(II) and/or cadmium(II), then the magnesium can be titrated with
standard EDTA. The cyanide complexes of zinc(II) and cadmium(II) are too stable
to allow EDTA to react with the complexed metal ions.

Reagents

1. Obtain $99.0+\%$ ACS grade EDTA ($Na_2H_2Y \cdot 2H_2O$) and use it as a primary
   standard. Prepare $0.005M$ EDTA by weighing 0.93 g of the disodium salt to three
significant figures and diluting to exactly 500 mL with distilled water. (If the salt
does not dissolve readily, add no more than 3 pellets of sodium hydroxide.) Store
in a Pyrex bottle and calculate the exact molarity, using 372.24 as the formula
weight.

2. Prepare a pH 10 ammonia buffer by diluting 32 g of ammonium chloride and
   285 mL of concentrated ammonium hydroxide to 500 mL with distilled water.
Store in a polyethylene bottle.

3. Prepare a 0.05% Calmagite solution by dissolving 0.05 g of Calmagite in 100 mL of water and storing in any type of bottle. (It is stable indefinitely.)

Procedure

1. Prepare for the titration of the unknown by rinsing the sides and bottoms of three 250-mL flasks with 1:1 nitric acid. Rinse with distilled water at least five times to remove all acid. Then pipet exactly 10 mL of the magnesium-zinc-cadmium unknown into each of the flasks.

2. Dilute each solution with about 60 mL of distilled water, and add about 10 mL of the ammonia buffer.

3. Check the pH to be sure it is 10. *Put all three flasks in a hood* and add about 0.5 g of KCN to each flask *in the hood.* Again check that each pH is about 10.

4. Add a few drops of Calmagite indicator to the first flask, and shake well to obtain a pink solution.

5. Titrate with 0.005M EDTA to a clear blue end point, proceeding slowly near the end point to give the color change time to occur. Wash the contents of the flask down the drain with plenty of water, preferably using a drain in a hood. If a cyanide disposal vessel is available, pour the contents into that vessel instead of down the drain.

6. Repeat the titration of the other two flasks in the same manner.

7. If directed by the instructor, do a back-titration to determine the total mmoles of magnesium, zinc, and cadmium. Add a measured excess of 50–75 mL (instructor's direction) of 0.005M EDTA and back-titrate with a standard 0.005M magnesium solution prepared according to the instructor's directions.

8. Calculate the molarity of magnesium in the unknown and, if directed, the total molarity of all metal ions.

30–3. OXIDATION-REDUCTION TITRATIONS

A number of inorganic substances and a few types of organic compounds can be determined by oxidation-reduction titrations. The most important reactions are the iodometric reactions, especially since they may be used in fields related to chemistry, such as clinical analysis and drug analysis. The use of stronger oxidizing agents such as potassium dichromate is of classical interest, so such experiments are presented after the iodometric experiments.

Purpose of the Experiments in This Section.   The first experiment, *Experiment 16,* is a multipurpose experiment. It details the preparation of an iodine titrant as well as standardization of this titrant. Arsenic as arsenic(III) oxide is used for the standardization. Although this is a hazardous substance, it can be handled safely in small amounts. We recommend that a standard solution of arsenic(III) be prepared by the teaching assistant if individual handling is to be avoided.

Experiment 17 is recommended as a replacement for the determination of arsenic(III) in solid unknowns. The determination of ascorbic acid in vitamin tablets is a much safer experiment and also is of interest as an example of the type of analysis that a pharmaceutical manufacturer would have to perform for tablets. Iodine is used as titrant.

Experiment 18 is included as an example of the use of iodine to titrate arsenic(III) ion in solid arsenic(III) oxide unknowns. It requires that students handle these unknowns carefully in weighing and dissolving.

Experiment 19, the iodometric determination of copper, is included for those who prefer a more classical iodometric method instead of direct iodine titrations of ascorbic acid or arsenic(III).

Experiment 20 is included as an example of the determination of an organic compound, ethylene glycol, using iodometric methods.

Experiment 21 is a classic redox experiment involving the use of a strong oxidizing agent, potassium dichromate. Its use in the analysis of razor blades and in the analysis of an iron ore is illustrated by directions for both methods.

Experiment 16. Preparation and Standardization of Iodine

This experiment can be used for preparing and standardizing iodine for the determination of ascorbic acid in tablets (Experiment 17) or for the determination of arsenic (III) in impure arsenic(III) oxide (Experiment 18).

Chemical Background

A standard solution of iodine can be prepared by weighing the iodine exactly on an analytical balance or by standardizing it against primary standard arsenic(III) oxide. Since it is difficult to handle iodine without losing some of it, it is usually weighed out approximately and standardized. The arsenic(III) oxide is first dissolved in base and then neutralized to arsenious acid:

$$As_2O_3(s) + 2OH^- + H_2O \rightarrow 2H_2AsO_3^- + 2H^+ \rightarrow 2H_3AsO_3$$

The arsenious acid is then oxidized by the iodine titrant to arsenic acid in a solution buffered at about pH 8 with sodium bicarbonate:

$$I_2 + H_3AsO_3 + H_2O \rightarrow 2I^- + H_3AsO_4 + 2H^+$$

The end point is detected by the formation of the deep blue starch-triiodide color. Since iodine exists as the triiodide (I_3^-) ion in aqueous solution, the first excess of iodine titrant added will form the starch-triiodide complex.

Reagents

1. Prepare starch solution by making a paste of 2 g of soluble starch and 25 mL of water, and pour with stirring into 250 mL of boiling water. Boil for 2 min, add 1 g of boric acid as preservative, and allow to cool. Store in a glass-stoppered bottle.

2A. Preparation of 0.03N iodine for Vitamin C analysis (Experiment 17). Weigh
   about 4.5 g of reagent-grade iodine on a trip scale and transfer it to a 100-mL
beaker containing 20 g of potassium iodide dissolved in 25 mL of water. Stir
carefully to dissolve all the iodine, and pour the entire contents into a glass-
stoppered amber liter bottle. Rinse the beaker with 25 mL of distilled water and
pour this into the bottle. Dilute the mixture to 1 L using distilled water and shake
several times.

2B. Preparation of 0.1N iodine for arsenic analysis (Experiment 18). Weigh about
   12.7 g of reagent-grade iodine on a trip scale and transfer it to a 250-mL beaker
containing 40 g of potassium iodide dissolved into 25 mL of water. Stir carefully to
dissolve the iodine and pour into a glass-stoppered amber liter bottle. Rinse the
beaker with 50 mL of distilled water and pour this into the bottle. Dilute the mixture
to 1 L using distilled water and shake several times. (If there is any doubt that the
iodine has dissolved completely, allow to stand until the next lab.)

Standardization of 0.03N Iodine

1. If a 0.0300N arsenious acid solution is available in the lab, use it in the steps
   below. If not, prepare it by weighing exactly 0.370 g of primary standard
arsenic(III) oxide (do not dry) into a 250-mL beaker. (Avoid touching this chemical
or ingesting it; wash your hands if you have any reason to believe you have touched
it.) Add a solution of 1 g of sodium hydroxide pellets freshly dissolved in 20 mL of
distilled water. Swirl until the oxide dissolves completely, warming if necessary. Add
50 mL of water and 2 mL of 12M hydrochloric acid. Transfer quantitatively to a
250-mL volumetric flask and dilute to volume.

2. *Using a bulb*, pipet exactly 25 mL of 0.0300N arsenious acid into each of three
   flasks. Add 25 mL of water to each. Then add about 4 g of sodium bicarbonate to
each, and check the pH to be 7–8 with pH paper. If the pH is not in this range, add
more (1 + g) sodium bicarbonate.

3. To each flask, add about 5 mL of starch solution and titrate with 0.03N iodine
   to the first appearance of the blue starch-triiodide color. (For the first titration, it
is desirable to check that the pH is still 7–8 near the end point. Add more sodium
bicarbonate if needed.) Approach the first end point cautiously, since the color of
the indicator is so intense that you may obtain a very dark color with a very small
excess of titrant.

4. Calculate the normality of the iodine, using the average volume of iodine from
   the three titrations. (If two volumes, such as the second and third, vary from the
other, you should run at least one more titration and use all the titration volumes to
calculate the normality.)

Standardization of 0.1N Iodine

1. If a 0.1000N solution of arsenious acid is available in the lab, proceed to Step
   4 below. If not, prepare three individual solutions as follows. Put at least 1 g of
sodium hydroxide pellets into each of three 250-mL flasks and add no more than

20 mL of water. Swirl briefly, and quickly weigh exactly 0.2–0.25 g of arsenic(III) oxide into each flask.

2. Swirl to dissolve the arsenic(III) oxide, wash down the sides of the flask, and warm if necessary to dissolve undissolved particles.

3. To each flask, add 50 mL of water and not less than 2.5 mL of concentrated hydrochloric acid (measuring pipet). Add about 4–5 g of sodium bicarbonate to each flask, and check the pH to be 7–8 with pH paper.

4. At this point, you should have three flasks prepared from Steps 1–3, or three flasks into which you have pipetted a standard solution of arsenious acid available in the lab. To each flask, add about 5 mL of starch solution and titrate with 0.1N iodine to the first appearance of the deep blue starch-triiodide color. (For the first titration, it is desirable to check that the pH is still 7–8 near the end point. Add more sodium bicarbonate if needed.) Approach the first end point cautiously, since the color of the indicator is so intense that you may obtain a very dark color with a very small excess of titrant.

5. Calculate the normality of the iodine, using the average volume of iodine from the three titrations. (If two volumes vary from the other, you should run at least one more titration and use all the titration volumes to calculate the normality.) If you weighed out arsenic(III) oxide, you should use an equivalent weight of 197.8/4.

Experiment 17. Determination of Ascorbic Acid in Vitamin Tablets

A simple example of drug analysis is the determination of ascorbic acid (Vitamin C) in vitamin tablets. This illustrates the decision that a drug manufacturer must make for analysis of any tablet: How many tablets constitute a representative sample? If too few are used, the sample may not be representative, and the results may deviate widely from the true value. If too many are used, the sample may be difficult to handle or be too expensive. In the directions for your ascorbic acid analysis, several choices are suggested. You should consult with the instructor for the best choice.

Chemical Background

To analyze for ascorbic acid, iodine is used to oxidize the ascorbic acid quantitatively to dehydroascorbic acid:

$$I_2 + C_4H_6O_4(OH)C=COH \rightarrow 2I^- + 2H^+ + C_4H_6O_4C(=O)-C=O$$

Note that the essential reaction is the oxidation of the "enediol" group of ascorbic acid to an "alpha-dicarbonyl" group of dehydroascorbic acid. This reaction has long been used for analyzing for Vitamin C [3].

[3] J. W. Stevens, *Ind. Eng. Chem., Anal. Ed., 269* (1938); C. E. Moore, *J. Chem. Educ. 25,* 671 (1948).

Since oxygen is a stronger oxidizing agent ($E^0 = 1.23$ V) than iodine ($E^0 = 0.535$ V), dissolved oxygen oxidizes ascorbic acid also, but much more slowly. The samples should therefore be titrated immediately after being dissolved, and the flask should be covered to minimize absorption of oxygen from the air.

Preparation of Reagents

1. Prepare 0.03N iodine as directed in Experiment 16 and standardize it.

2. Prepare starch solution as directed in Experiment 16, or prepare a starch-urea solid complex as follows. Melt 40 g of urea in a small beaker, and add 10 g of soluble starch. Swirl or stir until liquid, then cool, grind in a mortar, and store in a bottle. This dissolves easily in water and is stable indefinitely.

Analysis of Vitamin C Tablets

1. In consultation with your instructor, decide how many tablets will be dissolved for the analysis. The 0.03N iodine titrant will allow you to titrate about 125 mg of ascorbic acid per aliquot. Here are some possible choices:

Number Tablets	Total mg	Vol. Flask	Aliquot
3 × 250 mg	750 mg	100 mL	10 mL
5 × 250 mg	1250 mg	250 mL	25 mL
10 × 250 mg	2500 mg	500 mL	25 mL

2. Because whole tablets may dissolve slowly, you should carefully cut each of the tablets into 5–6 pieces *before* weighing. (If the instructor wishes you to report the weight of vitamin C per tablet, be careful not to lose any of the tablet in cutting it. If you are required to report only the percentage, loss is not a problem.)

3. Before weighing the vitamin C tablets, be sure that your iodine titrant is prepared and in the buret. Do not dissolve the tablets unless you are able to titrate all samples during the lab period. Weigh the chosen number of vitamin C tablets (cut up), and pour into a *dry* volumetric flask of the size chosen.

4. When you are ready to titrate, dissolve the tablets by adding a volume of water that is equal to about half the volume of the volumetric flask and shake vigorously until the tablets have dissolved. Fill to the mark.

5. Pipet exactly 25 mL of the vitamin C solution into a 250-mL flask and add 5 mL of starch indicator. Cover the opening of the flask with a piece of cardboard with a small hole for the buret tip. Titrate rapidly to reduce air oxidation of the ascorbic acid, but proceed dropwise near the end point, a deep blue starch-triiodide color.

6. Repeat the titration for 2 or 3 additional samples (instructor's directions). Do not pipet a sample into a flask until you are ready to titrate it.

7. Calculate either the milligrams of ascorbic acid per tablet or the percent ascorbic acid in the tablets. Use an equivalent weight of 176.12/2 for ascorbic acid and a dilution factor equal to the volume of the flask divided by the volume of the pipet for your calculations.

QUESTIONS

1. A vitamin C tablet, claimed to contain 500 mg on the label, is dissolved and titrated with 28.54 mL of 0.2000N iodine. Calculate the mg of vitamin C (form wt = 176.12) in the tablet.
2. Two 250-mg vitamin C tablets are dissolved in a 100-mL volumetric flask. A 25.00-mL aliquot is removed and titrated with 28.44 mL of 0.0500N iodine. Calculate the average mg of vitamin C (form wt = 176.12) per tablet.
3. Four vitamin C tablets are dissolved in a 100.0-mL volumetric flask. A 10.00-mL aliquot is removed and titrated with 40.00 mL of 0.0300N iodine. Calculate the average mg of vitamin C (form wt = 176.12) per tablet.

Experiment 18.   Iodometric Determination of Arsenic(III)

This experiment is included for those who wish to continue to analyze impure arsenic(III) oxide samples for the content of arsenic(III). It requires careful handling of arsenic(III) oxide samples.

Chemical Background

Once arsenic(III) oxide has been dissolved, it can be converted to arsenious acid and oxidized with standard iodine titrant:

$$I_2 + H_3AsO_3 + H_2O \rightarrow 2I^- + 2H^+ + H_3AsO_4$$

Procedure (see Section 25—5 on "Use of Arsenic Compounds")

1. Do not dry arsenic(III) oxide samples; store in your desiccator until use.
2. Before weighing out the unknown, put at least 1 g of sodium hydroxide pellets into each of three 250-mL flasks. When you are ready to weigh the first portion of the unknown, add no more than 20 mL of water to the first flask. Then weigh exactly 0.3–0.35 g sample into the first flask, and swirl to use the heat released to dissolve the sample. (Warm if still insoluble.)
3. Repeat the process for the remaining samples.
4. To each flask, add 50 mL of water and not less than 2.5 mL of concentrated hydrochloric acid from a measuring pipet. Add 4 g of sodium bicarbonate to each flask and check that the pH is 7–8 with pH paper. (If not, add 1+ g of sodium bicarbonate.)
5. To each flask, add 5 mL of starch solution (Experiment 16) and titrate with standard 0.1N iodine (Experiment 16) to the first appearance of the deep blue starch-triiodide color. (For the first flask, it is desirable to check that the pH is 7–8 near the end point. If not, add more bicarbonate.)
6. Calculate the percentage of arsenic in the sample using an equivalent weight of 74.92/2.

Experiment 19. Iodometric Determination of Copper

Copper can be determined in an ore with the following series of reactions: (a) the ore is dissolved in nitric acid; (b) the pH of the solution is adjusted with ammonium bifluoride buffer; (c) the copper(II) in the solution is iodometrically reduced to copper(I) iodide; and (d) the iodine formed in the reduction reaction is titrated with sodium thiosulfate.

Chemical Background

The dissolution of copper or copper ore produces oxides of nitrogen (Equation E19–1), which are usually removed, together with any excess nitric acid, by fuming down with sulfuric acid. The pH is buffered at 3.5 by adding ammonium bifluoride, NH_4HF_2. This accomplishes three things:

1. It provides sufficient acid to repress hydrolysis of copper(II) ion (Equation E19–2). This is essential, because appreciable hydrolysis slows the iodometric reduction.

$$3Cu^0(s) + 2NO_3^- + 8H^+ \rightarrow 3Cu^{2+} + 2NO(\rightarrow HNO_2) + 4H_2O \quad \text{(E19–1)}$$

$$Cu^{2+} + H_2O \rightleftharpoons CuOH^+ + H^+ \quad \text{(E19–2)}$$

2. The buffer keeps the pH high enough to prevent any appreciable iodometric reduction of arsenic(V) to arsenic(III), if arsenic(V) is present. (At any pH below 3, the equilibrium conditions favor a more than 0.1% reduction of arsenic(V).)

3. Ammonium bifluoride prevents the iodometric reduction of iron(III) (Equation E19–3) by complexing iron(III) as an anionic fluoride complex ion. The complexation lowers the potential of iron(III) to oxidize iodide to iodine.

$$2Fe^{3+} + 2I^- \rightleftharpoons 2Fe^{2+} + I_2 \quad \textit{No Reduction if F}^- \textit{ Present} \quad \text{(E19–3)}$$

Quantitative reaction of iodide and copper(II) ions (Equation E19–4) is favored by the precipitation of copper(I) iodide, making the method useful only for macro amounts of copper. The iodine released is best titrated by standard sodium thiosulfate (Equation E19–5). Unfortunately, the precipitate absorbs iodine, which is only slowly released at the end point. Potassium thiocyanate will displace the iodine from the surface, giving a sharper end point (Equation E19–6). For best results, thiocyanate is added near the end point, since it forms a complex with iodine and is actually slowly oxidized by iodine to sulfate [4].

$$2Cu^{2+} + 4I^- \rightleftharpoons 2CuI(s) + I_2 \quad \text{(E19–4)}$$

$$\underset{(Na_2S_2O_3)}{2S_2O_3^{2-} + I_2} \rightarrow S_4O_6^{2-} + 2I^- \quad \text{(E19–5)}$$

$$CuI:I_2(s) + SCN^- \rightleftharpoons CuI:SCN^-(s) + I_2 \quad \text{(E19–6)}$$

By using pure copper wire as a primary standard, the indirect iodine titration may be

[4] C. Lewis and D. A. Skoog, *J. Am. Chem. Soc.*, *84*, 1101 (1962).

used to standardize sodium thiosulfate (Note 1). Although copper metal is oxidized to copper(II), the actual measurement step involves only a one-electron change in the iodometric reduction; hence, the equivalent weight of copper is its atomic weight.

Standardization of 0.1*M Sodium Thiosulfate*

1. Prepare 0.1*M* sodium thiosulfate by adding 12.5 g of $Na_2S_2O_3 \cdot 5H_2O$ to 500 mL of distilled water that has been boiled, then cooled. Add 0.1 g of sodium carbonate as a preservative.

2. It is preferable for beginning students to standardize sodium thiosulfate against iodine that has been standardized according to the directions of Experiment 16. If the instructor so directs, standardize sodium thiosulfate against primary standard copper metal as in Steps 3–8.

3. Weigh three 0.2–0.25 g pieces of shiny copper wire into 250-mL flasks. Add 5 mL of 1:1 nitric acid to each flask and heat it over a low flame in a hood to dissolve the copper. Do not boil.

4. Add 25 mL of water to each flask and bring the solutions to boil. Remove the flasks from the heat, add 0.4 g of urea, and boil them for 5 minutes to react the urea with any nitrous acid produced from the reaction of nitric acid and copper. (If not removed, nitrous acid will oxidize iodide ion to iodine.)

5. Cool the flask under tap water. Add 1:1 ammonium hydroxide dropwise from a measuring pipet until a light blue precipitate of copper(II) hydroxide forms in the flask. Do not add an excess; this will form a dark-blue copper-ammonia complex ion (Note 1). Add 4 mL of glacial acetic acid to each flask, and cool it if necessary.

6. At this point, treat each solution individually (Note 2). Add 2.5 g of potassium iodide to one flask and titrate immediately with 0.1*N* sodium thiosulfate until the brownish iodine color is almost gone. Observe the color by interrupting the titration and allowing the precipitate of copper(I) iodide to partially settle.

7. Add solid starch or 3 mL of starch indicator (see Experiment 12), and continue titrating dropwise until the blue starch-triiodide color just disappears with the addition of one drop of titrant. Then add 1–15 g of potassium thiocyanate, and titrate the solution dropwise until the blue color disappears permanently (Note 3). (In this case, "permanently" means "for 20–30 seconds," since more iodine is eventually produced by the air oxidation of iodide.)

8. If the color returns again and again before 20–30 seconds have elapsed, too much ammonia may have been added, or more acetic acid (or ammonium bifluoride for the ore) may be needed. Add a 20% excess of acetic acid (or ammonium bifluoride) to the next sample.

9. Calculate the normality of sodium thiosulfate using the atomic weight 63.54 of copper as its equivalent weight.

Analysis of Copper Ore

1. Weigh three 0.5–0.6 g samples of copper ore into 250-mL flasks (see the instructor's directions). To each flask, add 10 mL of concentrated hydrochloric

acid and evaporate the solution on a hot plate or burner until the volume is about 5 mL. Use a hood, and swirl to avoid bumping.

2. Cool each flask under tap water until it can be comfortably handled, then add 10 mL of concentrated nitric acid. Cover the flasks with watch glasses and heat them below the boiling point until the dark dense ore at the flask bottom has dissolved, leaving only light-colored floating silica particles or light-colored sulfur particles.

3. Remove the flasks from the heat, allow them to cool, and carefully pipet about 8 mL of 1:1 sulfuric acid down the side of each flask. Evaporate to white fumes of sulfur trioxide to expel nitric acid, which later might oxidize iodide to iodine.

4. Cool the solution, and carefully add 35 mL of water and 5 mL of saturated bromine water to each flask to oxidize any arsenic(III) to arsenic(V). Boil gently for 5 minutes in a hood to expel excess bromine, which later might oxidize iodide to iodine.

5. Let the solution cool, and carefully add 1:1 ammonium hydroxide down the side of each flask until a slight precipitate of brown ferric hydroxide begins to form. (If no iron(III) is present, watch for the blue copper-ammonia complex as a signal to stop adding ammonia.) Avoid adding an excess of ammonium hydroxide. Now add 3–4 mL of glacial acetic acid (Note 2) and 2 g of ammonium bifluoride (NH_4HF_2), to bring the pH to 3.5. Stir, to dissolve any ferric hydroxide.

6. Proceed to Step 6 under the standardization of thiosulfate. After reaching the end point, rinse the flask immediately to prevent extreme etching by the fluoride ion. Calculate the percentage copper in the ore, using 63.54 as the equivalent weight of copper.

Notes

1. If an excess of ammonia is formed, boil the solution to remove it. The precipitate of copper(II) hydroxide should reform. An excess of ammonia may interfere with the adjustment to pH 3.5 by addition of acetic acid, and may retard the reduction of copper(II).

2. The solutions may be left overnight after adding acetic acid, but do not add ammonium bifluoride or potassium iodide until the next laboratory period begins. Slow air oxidation of iodide to iodine occurs if potassium iodide is left standing in acid solution. This reaction is also catalyzed by copper(II).

3. Since iodine is removed from the surface of the precipitate by potassium thiocyanate, the precipitate should become lighter in color. If a slight excess of thiosulfate has been added before the potassium thiocyanate, no iodine color may be observed when the latter is added.

QUESTIONS

1. Write a balanced equation for the reaction between nitrous acid (HNO_2) and urea (NH_2CONH_2) to produce nitrogen gas and carbon dioxide.

2. Write a balanced equation for the air oxidation of iodide ion in acid solution.
3. Write a two-step reaction sequence to show how copper(II) might catalyze the reaction given in Question 2.
4. Why is it not necessary to add urea to the dissolved ore samples?
5. Why is it necessary to oxidize arsenic(III) to arsenic(V) if the former is present? Nitric acid is a good oxidizing agent; why is it also necessary to add bromine to oxidize arsenic(III)?

Experiment 20.   Periodate Determination of Ethylene Glycol

A specified method for determining the 1,2-diol, or α-glycol, group in organic compounds is cleaving the carbon-carbon bond with excess periodic acid. The periodate ion is reduced to iodate and the hydroxyl groups are oxidized to carbonyl groups

Chemical Background

The reaction of ethylene glycol and periodate is given in Equation E20–1:

$$\begin{array}{l} CH_2OH \\ | \qquad\qquad + IO_4^-(xs) \xrightarrow{pH\ 1} 2H_2C{=}O + IO_3^- + H_2O \\ CH_2OH \quad _{(HIO_4)} \end{array} \qquad (E20\text{--}1)$$

The analysis may be concluded either by reducing the unreacted periodate (and the resulting iodate) to iodine with iodide in acidic solution (Note 1) or by using the more accurate method that follows [5]. Here, the pH is adjusted to the neutral region with a buffer, and iodide reduces only the unreacted periodate to iodate and iodine (Equation E20–2). Since this reaction is somewhat slow, a measured excess of arsenic(III) is added to react with the iodine as it is produced (Equation E20–3). The unreacted arsenic(III) is then titrated with iodine (Equation E20–4).

$$IO_4^- + 2H^+ + 2I^- \xrightarrow{pH\ 7\text{--}8} I_2 + IO_3^- + H_2O \qquad (E20\text{--}2)$$

$$H_3AsO_3(xs) + I_2 + H_2O \underset{}{\overset{pH\ 7\text{--}8}{\rightleftharpoons}} HAsO_4{}^{2-} + 4H^+ + 2I^- \qquad (E20\text{--}3)$$

$$I_2 + H_3AsO_3 + H_2O \underset{}{\overset{pH\ 7\text{--}8}{\rightleftharpoons}} HAsO_4{}^{2-} + 4H^+ + 2I^- \qquad (E20\text{--}4)$$
$$_{(0.1N)}$$

A likely source of error is a failure to neutralize the solution properly; this will reduce the periodate irreversibly to 4 molecules of iodine per periodate ion (Note 1) instead of to one molecule of iodine per periodate ion. Hence, an excess of bicarbonate should always be added and the pH checked with pH paper. Iodate can also be irreversibly reduced to iodine (Note 1) by a region of localized acidity in an acidic arsenic(III) solution (Equation E20–3). This procedure avoids that error by using an arsenic(III) reagent buffered slightly with sodium bicarbonate. The reagent

[5] G. H. Schenk, *J. Chem. Educ.*, *39*, 32 (1962).

also contains enough iodide ion to reduce periodate as the arsenic(III) is being added.

Since periodic acid is not a primary standard, the determination is performed by running both a blank and a sample titration. Subtracting the two volumes gives the milliliters of iodine equivalent to the 1,2-diol. This method has been applied in determining pinacol and ethylene glycol [5].

Preparation of Reagents

1. Weigh exactly 2.500 g of primary-standard arsenic(III) oxide (As_2O_3) into a 250-mL beaker, add a solution of 5 g of sodium hydroxide pellets freshly dissolved in 20 mL of water, and swirl until the oxide dissolves. Add 50 mL of water, 10 mL of 12M hydrochloric acid, 3 g of sodium bicarbonate, and 4 g of sodium iodide; then transfer the solution to a 500-mL volumetric flask. Dilute to the mark and mix thoroughly. The pH should be 7–8. This is a 0.1011N solution of arsenic(III).

2. Dissolve 2.28 g of the dihydrate of periodic acid, H_5IO_6 (form wt = 227.96) in 100 mL of distilled water. This is a 0.1M solution.

3. Prepare and standardize 0.1N iodine as directed in Experiment 16. Prepare a starch indicator according to directions from the same experiment. Alternatively, pipet 25 or 50 mL of the 0.1011N arsenic(III) reagent into a flask and standardize the 0.1N iodine against this according to the directions in Experiment 16.

Assay of Blank

1. Pipet exactly 10 mL of 0.1M periodic acid into each of three 250-mL flasks. Add 25 mL of water and 5 g of sodium bicarbonate to each. Pipet exactly 50 mL of 0.1011N arsenic(III) into each flask and check that the pH is 7–8. Allow the solution to react for at least 5 minutes.

2. Add 5 mL of liquid starch indicator or a spatulaful of solid starch indicator, and titrate the excess arsenic(III) with 0.1N iodine, taking the first blue color as the end point. Label the milliliters of iodine consumed "B" (blank). Perform the blank assay until 2 or 3 values are obtained that agree within ±0.3 mL. Do not analyze the ethylene glycol sample until the blank values are reproducible.

Assay of Ethylene Glycol

1. Pipet exactly 10 mL of 0.1M periodic acid into four 250-mL flasks, add 10 mL of water to each, and mark one flask "pH 7–8." Add 5g of sodium bicarbonate to this flask (using more water if necessary) and set it aside.

2. Pipet exactly 10 mL of the unknown containing ethylene glycol (Note 2) into each of the three remaining flasks. After 12 minutes' reaction at room temperature, add about 3 g of sodium bicarbonate and exactly 50 mL of 0.1011N arsenic(III) into each flask. Check whether the pH is 7–8, and allow the solution to react for at least 5 minutes.

3. Add starch indicator and titrate the excess arsenic(III) with 0.1N iodine. Label

the milliliters of iodine consumed "S" (sample). Calculate the milligrams of ethylene glycol per milliliter of unknown solution.

$$\frac{\text{mg glycol}}{1 \text{ mL}} = \frac{(S - B) \text{ mL I}_2}{} \left| \frac{N_{I_2}}{x} \right| \frac{62}{10 \text{ mL}} \right|$$

Part of the assignment is to calculate the correct equivalent weight of ethylene glycol knowing that the formula weight is 62. Obviously, the equivalent weight is $62/x$, where x is the number of electrons lost by ethylene glycol in being oxidized (Note 3).

4. Repeat Steps 2 and 3 with the flask marked pH 7–8, but omit adding the sodium bicarbonate in Step 2. Compare the $(S - B)$ milliliters of iodine consumed at this pH with that consumed at the pH (about 1) of $0.1 M$ periodic acid. Answer Question 3 in your notebook.

Notes

1. In acid solution, periodate and iodate are reduced by iodide to iodine according to the following equations:

$$IO_4^- + 7I^- + 8H^+ \rightarrow 4I_2 + 4H_2O$$
$$(HIO_4)$$

$$IO_3^- + 5I^- + 6H^+ \rightarrow 3I_2 + 3H_2O$$

Iodate is not reduced by iodide above pH 7.

2. Pure ethylene glycol is a viscous liquid, inconvenient to weigh. A dilute solution containing 310–460 mL of glycol per 100 mL of water may be supplied as the unknown.

3. The number of electrons lost by ethylene glycol (and the equivalent weight) is easily found by one of the two following methods. (a) Using oxidation numbers, balance the half-reaction for the reduction of periodate to iodate, and use this electron change for the number of electrons lost by the glycol. (b) Using the ion-electron method, balance the half-reaction for the oxidation of ethylene glycol to 2 molecules of formaldehyde and protons. This gives the electron change directly.

QUESTIONS

1. Write a balanced equation for the periodate oxidation of glycerol,

$$CH_2OHCHOHCH_2OH$$

which reacts with 2 molecules of periodate to give 2 molecules of formaldehyde and 1 molecule of formic acid. Find the equivalent weight of glycerol.

2. Is the cleavage of ethylene glycol reversible or irreversible? If it is reversible, can pH affect the point of equilibrium?

3. If ethylene glycol reacts more slowly with periodate at pH 7–8 than at pH 1, the reaction may be said to be catalyzed by what species?

Experiment 21.   Dichromate Determination of Iron Ore; Analysis of a Razor Blade

Iron ore is analyzed by (a) dissolving the ore in hydrochloric acid solvent (Equation E21–1) to give iron(III), (b) prereducing the iron(III) (Equations E21–2 and E21–3) to give iron(II), and (c) titrating the iron(II) with a standard oxidizing agent such as potassium permanganate or potassium dichromate (Equation E21–5). In the simplest determination, the iron(III) is prereduced with stannous chloride, and the resulting iron(II) is titrated with potassium dichromate.

Chemical Background

Although reducing iron with stannous chloride is more convenient than using a column of a metallic reducing agent (such as amalgamated zinc), serious errors can result if a large excess of stannous chloride is added. When mercuric chloride is added to react with the excess stannous chloride, a high concentration of the latter tends to reduce mercury(II) to finely divided mercury metal (Equation E21–4). During the titration with dichromate, the mercury metal will interfere with the determination by being oxidized itself. If only a small excess of stannous chloride is present, mercuric chloride is reduced only to insoluble mercurous chloride.

$$Fe_2O_3 + 6H^+ \rightarrow 2Fe^{3+} + 3H_2O$$
$$\text{(HCl)}$$
$$\textbf{(E21–1)}$$

$$\underset{(xs)}{Sn(II)} + \underset{\text{(yellow)}}{2Fe^{3+}} \rightleftharpoons Sn(IV) + \underset{\text{(light green)}}{2Fe^{2+}}$$
$$\textbf{(E21–2)}$$

$$Sn(II) + 2Hg^{2+} + 2Cl^- \rightleftharpoons Hg_2Cl_2(s) + Sn(IV)$$
$$\text{(HgCl}_2\text{)}$$
$$\textbf{(E21–3)}$$

$$xs\ Sn(II) + \underset{\text{(HgCl}_2\text{)}}{Hg^{2+}} \rightleftharpoons \underset{\text{(black)}}{Hg^0(s)} + Sn(IV)$$
$$\textbf{(E21–4)}$$

$$\underset{\text{(K}_2\text{Cr}_2\text{O}_7\text{)}}{Cr_2O_7^{2-}} + 6Fe^{2+} + 14H^+ \rightarrow 2Cr^{3+} + 6Fe^{3+} + 7H_2O$$
$$\textbf{(E21–5)}$$

Because potassium dichromate is a primary-standard material, a standard solution of it is easily prepared by weight. Strictly speaking, potassium dichromate should be standardized against primary-standard iron wire because the end point does not coincide with the equivalence point when barium diphenylaminesulfonate indicator is used. However, this error is usually small enough to be disregarded in all but the most careful work.

Potassium dichromate has an additional advantage: it does not oxidize chloride ion from the hydrochloric acid solvent, as does potassium permanganate.

Preparation of Potassium Dichromate

Weigh a 2.452-g sample of dried primary-standard potassium dichromate, transfer it to a 500-mL volumetric flask, dissolve it in water, and dilute to the mark. This is a $0.1000N$ solution.

Dissolution of Ore and Prereduction of Iron(III)

1. Weigh three 0.5-g iron-ore samples into 500-mL conical flasks (see the instructor's directions). Add 10–15 mL of $12M$ hydrochloric acid to each flask (Note 1), and cover them with watch glasses.

2. Using a hood and a hot plate or wire gauze, heat the flasks below the boiling point until the ore dissolves (20–60 minutes). The solution will turn yellow as iron(III) dissolves and is complexed by chloride ion. A white flocculent residue of silica will also be observed. A heavy black residue may be insoluble sulfides or silicates (Note 2).

3. Evaporate the solution to about 5 mL and dilute it with water to 15 mL. Since the iron(II) produced in the reduction is easily oxidized by air, treat each sample individually at this point (Note 3) and titrate each reduced sample before reducing the next sample. Heat the solution to boiling, remove it from heat, and add $0.5M$ stannous chloride (Note 4) dropwise until the yellow iron(III) is completely reduced to light-green iron(II). (Most samples of iron(II) are so dilute that they appear virtually colorless.) Do not add more than 2 drops in excess (Note 5). Heat and a high concentration of chloride ion favor a high reaction rate.

4. Cool to room temperature, add 50 mL of water, stir rapidly, and add 10 mL of $0.25M$ mercuric chloride *all at once* to avoid reducing the mercuric ion to mercury metal. All excess tin(II) has now been oxidized to tin(IV).

5. Wait 3 minutes. If a white precipitate of mercurous chloride is observed, proceed with the titration as directed below. If no precipitate is observed, an excess of stannous chloride was not added. Discard the sample. If the precipitate is grey or black, mercury metal is present. Discard the sample.

Titration of Iron(II)

1. Immediately add 200 mL of water, 10 mL of 1:5 sulfuric acid, 5 mL of 85% phosphoric acid, and 8 drops of sodium or barium diphenylamine sulfonate indicator to the flask containing iron(II). Titrate slowly with potassium dichromate to the end point. As dichromate is reduced, it produces the blue-green chromium(III) ion; hence, the color change at the end point proceeds from a blue-green to a greyish tinge to a purple. The titration should be conducted dropwise when the grey tinge is noted, because the indicator oxidizes somewhat slowly.

2. Calculate the percentage of iron in the ore using an equivalent weight of 55.85.

Analysis of a Razor Blade

1. Obtain a double-edge razor blade from the instructor. Since different batches

of the same commercial brand may be given out, record the batch number on the razor blade given to you (Note 6).

2. There may be wax on the razor blade; this should be removed before weighing. Think of a way to remove the wax and use it. (*Hint:* Wax is an organic substance.)

3. Weigh the razor blade on the analytical balance. Add about 5 mL of water to a 250-mL Erlenmeyer flask, then carefully add about 25 mL of concentrated hydrochloric acid (Note 7). Using a hood, warm the solution for about 5 minutes on a hot plate to expel oxygen from the flask.

4. Add the razor blade to the flask. Allow the flask to stand for 10–15 minutes at room temperature in a hood (Note 8), then heat it to boiling to complete dissolution. The solution should be light green from iron(II). If a solution is yellow, the lab instructor may suggest the careful addition of 1–2 drops of stannous chloride to reduce any iron(III) to iron(II), followed by the addition of mercury(II) chloride.

5. Obtain a 250-mL volumetric flask and add about 190 mL of distilled water. Transfer the iron(II) solution quantitatively to the volumetric flask, then rinse the Erlenmeyer flask well with distilled water and add the rinse water to the volumetric flask. Dilute the solution to the mark (Note 9).

6. Using a 50-mL pipet (or a 25-mL pipet twice), transfer exactly 50 mL of the iron(II) solution to a 500-mL conical flask and titrate the iron(II) as directed in Step 1 under the Titration of Iron(II). Do not pipet more than one sample at a time, since iron(II) slowly oxidizes when it stands open to the air.

7. Calculate the percentage of iron in the razor blade using 55.85 as the equivalent weight of iron.

Notes

1. An alternative procedure at this point is to add 3 mL of stannous chloride along with the acid to speed the dissolution of the ore.

2. The instructor may direct you to add 5 mL of nitric acid at this point to dissolve the sulfides or silicates. The nitric acid is removed by fuming with 5 mL of concentrated sulfuric acid.

3. If there is not enough time to finish all the samples, they may be safely left in this condition until the next period.

4. The 0.5M stannous chloride solution is made by dissolving 115 g of $SnCl_2:2H_2O$ in 300 mL of 12M hydrochloric acid, adding 3 or 4 pieces of mossy tin, and diluting the resulting solution to 1 liter with water. The mossy tin prevents slow air oxidation of tin(II) to tin(IV).

5. If a large excess of stannous chloride has been added, oxidize the excess with a few drops of potassium permanganate solution.

6. Most double-edge razor blades have been found to be satisfactory for analysis, as long as they are *not* stainless steel. Some razor blades give some difficulty, since the printing on the blade does not dissolve.

7. It is recommended that this step be started near the beginning of the laboratory period.

8. The iron in the blade dissolves as iron(II) and releases explosive hydrogen gas.
    The hydrogen maintains a reducing atmosphere and prevents iron(II) from oxidizing to iron(III). Hence there is no need to reduce the iron with stannous chloride, etc.

9. It is not best practice to leave the sample overnight before analysis, although in some cases a tightly stoppered flask can be left overnight without appreciable error.

QUESTIONS

1. Why should a large excess of stannous chloride be avoided?
2. Show that 2.452 g of potassium dichromate gives a $0.1N$ solution when diluted to 500 mL.
3. Why should the mercuric chloride solution be added all at once to the solution containing excess stannous chloride?
4. Why does the chloride ion speed the reaction between iron(III) and tin(II)?
5. Why is phosphoric acid added before the iron(II) is titrated?

31

Spectrophotometric Instruments and Procedures

The experiments in this chapter may be performed with a standard filter colorimeter, an inexpensive grating spectrophotometer, or a more expensive precision spectrophotometer. We have also included an experiment in which the fluorescence of Vitamin D is measured with a simple fluorometer.

Spectrophotometer Readout. Inexpensive spectrophotometers and filter colorimeters are usually equipped with a scale readout rather than a digital readout (Chapter 17). Both absorbance (A) and percent transmittance ($100 \times T$) may be read from a scale readout. More modern spectrophotometers, such as the LKB Novaspec spectrophotometer, have a *digital* readout. Often there are two absorbance mode readouts, 0–1 (really 0.000 to 0.999) and 0–2 as shown in Figure 31–1.

To calibrate the spectrophotometer readout, the spectrophotometer must be set at one or two reference points manually. The $100\%T$ (zero absorbance) point is always set manually, using only the solvent in the sample cell. The $0\%T$ (infinite absorbance) point is set automatically by the microprocessor of microprocessor-spectrophotometers such as the LKB Novaspec; $0\%T$ must be set manually on inexpensive spectrophotometers. Once these points are set, then the sample is placed in the cell and its absorbance is read.

Table 31–1. Some Transmittance Values and Corresponding Absorbance Values

T	$\%T$	A	Relative Concentration Error per $0.5\%T$ Reading Error
0.89	89	0.050	5.2
0.75	75	0.125	2.3
0.35	35	0.460	1.36
0.10	10	1.000	2.17
0.01	1	2.000	10.8

Figure 31–1. The readout of the LKB Novaspec microprocessor-spectropho-tometer, showing the 0–1 absorbance mode readout ($A = 0.694$). Wavelength readout (631 nm) is also digital.

Often a sample reading will be obtained in percent transmittance instead of absorbance. It is important to review the conversion of percent transmittance to absorbance using the negative log of T. The values listed in Table 31–1 provide such an opportunity.

It is important to be aware that the errors in determining concentration at either end of the scale are higher than they are in the middle of the scale. Table 31–1 gives the relative concentration error per $0.5\%T$ scale reading for selected $\%T$ values. Note that a 10.8% relative concentration error is incurred when a sample reading of $1\%T$ is used for analysis!

31–1. OPERATING PRINCIPLES

Most inexpensive spectrophotometers have a schematic optical diagram much like that shown in Figure 31–2. Direct current from a transformer in the instrument enters the tungsten lamp, causing the lamp to emit radiation, which is focused by the field lens onto the objective lens. This lens focuses radiation on a reflection-type diffraction grating, which reflects a preselected bandwidth of radiation through the exit slit. (On the LKB Novaspec spectrophotometer, this bandwidth is 12 nm; on the Spectronic 20, it is 20 nm.)

The diffraction grating is a replica of a machine-ruled steel grating with 600 accurately spaced grooves per millimeter. The radiation falling on the grating is dispersed into a spectrum of radiation from about 320 nm (ultraviolet) to 900+ nm (near infrared). The dispersed radiation then passes through a light-control aperture

Figure 31–2. Optical diagram of the Spectronic 20, top view. (Courtesy of Bausch and Lomb, Inc., Rochester, N.Y.)

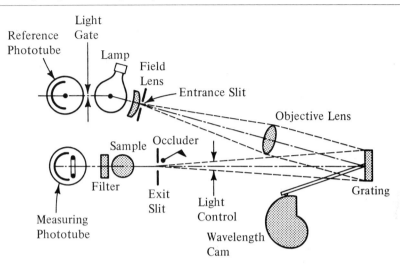

onto a screen with an exit slit. The desired wavelengths can be selected by means of a keypad control (see Figure 31–3) on microprocessor-spectrophotometers (Novaspec). By pressing the "minus" or the "plus" keypad, a lower or higher wavelength, respectively, can be selected. Where the bandwidth is 12 nm, a

Figure 31–3. Using the minus keypad to lower the wavelength on the LKB Novaspec microprocessor spectrophotometer.

wavelength setting of 506 nm passes radiation from 500 to 512 nm through the sample.

Radiation passing through the exit slit enters the sample cell, where part of it is absorbed. Radiation that is not absorbed passes through the cell and strikes the detector. Typical detectors are silicon photodiodes (Novaspec) or a phototube (inexpensive spectrophotometer). This radiation is converted to electrical current, which is amplified and displayed digitally or on a scale.

31–2. SPECTROPHOTOMETRIC OPERATING PROCEDURES

Each type of spectrophotometer has its own operating procedures. Directions are given below for both the LKB Novaspec and the Spectronic 20. Regardless of the instrument, read over Steps 1–4.

1. *Instrument startup.*   Turn on the spectrophotometer as soon as possible to allow it to warm up and stabilize. The LKB Novaspec will go through a three-step wavelength calibration procedure; when it is finished, the microprocess will set the wavelength automatically at 360 nm and select the $0–1A$ absorbance mode. Inexpensive spectrophotometers do not do this and will be at whatever wavelength setting was last set before shutdown.

2. *Cleaning Cells.*   If the cell is visibly contaminated, it should be cleaned with a warm dilute detergent solution. If not, then rinse with distilled water and then three times with the solution to be measured. Of course, the cell used to set $100\%T$ should be rinsed just three times with distilled water.

3. *Handling Cells.*   In most spectrophotometers, such as the Spectronic 20 spectrophotometer, radiation passes through the bottom half of the cell. Therefore you should handle only the top half of the cell in placing the cell in the cell holder. In any event, use a soft tissue to wipe off any liquid or smudges on as much of the outside of the cell as possible, including the entire bottom half.

4. *Recording Scale Readouts.*   Generally, absorbance should be recorded rather than $\%T$ from a scale readout. However, it may be desirable to record $\%T$ at very high absorbance values when the absorbance scale has few divisions. In any case, for inexpensive spectrophotometers, only two digits to the right of the decimal are significant for absorbance values from 0.00 to 0.99; for absorbance values at 1.0 and above, only one digit to the right of the decimal is significant.

Measurements on the Novaspec or Spectronic 20

5. *Setting Infinite Absorbance* $(0\%T)$.   The microprocessor in the Novaspec adjusts the instrument to $0\%T$ automatically after the wavelength calibration procedure. Go to the next step. If you are using the Spectronic 20, adjust the left knob so that the meter needle coincides with $0\%T$.

6. *Setting Zero Absorbance* $(100\%T)$.   Set the desired wavelength using the keypad

(Novaspec) or the top knob (Spectronic 20). Fill a test tube cell with distilled water and insert into the sample holder on the top of either instrument, matching the index line on the cell with that of the holder. Adjust to zero absorbance using the "set ref" key (Figure 31–3) on the Novaspec or the right-hand knob on the Spectronic 20. Remove the cell to avoid fatiguing the detector. On the Spectronic 20, check to make sure the needle returns to exactly $0\%T$; if not, repeat.

7. *Sample Measurement.*   After $100\%T$ and $0\%T$ have been set, fill the test tube cell with the sample solution (after rinsing, Step 2). Insert the test tube cell into the cell holder, matching the index lines. Record the absorbance immediately, and remove the test tube cell to avoid phototube fatigue. Check to make sure that the $0\%T$ setting has not changed. If desired, reinsert the cell for a check on the absorbance. Repeat Step 6 and the above directions for each wavelength at which measurements are needed.

8. *Care of the Instrument.*   After finishing, rinse all test tube cells with distilled water and allow to drain. Replace the cells in the rack or storage container. Check with the instructor to see whether the instrument is to be left on for other students; if not, turn it off.

Experiment 22.   Measurement of an Absorption Spectrum

The objective of this experiment will be to obtain an absorption spectrum manually using an inexpensive spectrophotometer such as the Spectronic 20 spectrophotometer. A secondary objective will be to identify the wavelength of maximum absorption and to calculate the molar absorptivity at this wavelength. Although sophisticated spectrophotometers can be used to record absorption spectra automatically, it is instructive to obtain one spectrum manually to understand what is involved in automatic recording.

Chemical Background

Directions are given for obtaining the absorption spectrum of either a copper(II) salt or a nickel(II) salt. When copper(II) sulfate is dissolved, it forms the light blue hexaaquo ion (also written less accurately with 4 waters):

$$CuSO_4 \cdot 5H_2O + H_2O \ \rightarrow \ Cu(H_2O)_6^{+2} + SO_4^{-2}$$

Thus the colored species contains only water molecules, not sulfate ions, so any copper(II) salt in principal should have the same color. (Copper halides are an exception.) The same is true for nickel(II) sulfate:

$$NiSO_4 \cdot 6H_2O \ \rightarrow \ Ni(H_2O)_6^{+2} + SO_4^{-2}$$

except that the light green hexaaquo ion is formed.

Preparation

1A. *Copper(II) spectrum:*   the instructor will assign one of the following weights of $CuSO_4 \cdot 5H_2O$ to be weighed to ± 0.001 g on the analytical balance: 0.499 g,

0.624 g, 0.684 g, 0.749 g, 0.811 g, 0.874 g, 0.936 g, 0.999 g, 1.061 g, 1.124 g, 1.186 g, 1.249 g, 1.311 g, 1.374 g, 1.436 g, 1.498 g, 1.556 g, or 1.622 g.

1B. *Nickel(II) spectrum:* the instructor will assign one of the following weights of $NiSO_4 \cdot 6H_2O$ (Note 1) to be weighed to ± 0.001 g on the analytical balance: 1.262 g, 1.420 g, 1.577 g, 1.735 g, 1.893 g, 2.050 g, 2.208 g, 2.366 g, 2.524 g, 2.682 g, 2.840 g, 2.997 g, 3.155 g, 3.313 g, 3.470 g, 3.628 g, 3.786 g, or 3.944 g.

2. Transfer the assigned amount of $CuSO_4 \cdot 5H_2O$ or $NiSO_4 \cdot 6H_2O$ to a 100-mL volumetric flask. Add distilled water carefully to the mark and shake until the salt is completely dissolved.

3. If the Spectronic 20 spectrophotometer is to be used to obtain the spectrum of $Cu(H_2O)_6{}^{+2}$, make sure it is equipped with a red measuring phototube and a red filter. If it is to be used to obtain the spectrum of $Ni(H_2O)_6{}^{+2}$, make sure it is equipped with a blue measuring phototube (Note 2). Regardless of the spectrophotometer to be used, read Steps 1–4 under the Spectrophotometric Operating Procedure section.

Procedure

1. Turn on the Spectronic 20 or other spectrophotometer, and allow it to warm up for 10 min (Spectronic 20) or longer (instructor's directions). Review Steps 5–8 under Spectrophotometric Operating Procedures if the LKB NOVASPEC or the Spectronic 20 spectrophotometer is to be used. Only brief reference will be made below to operation of the Spectronic 20.

2. Obtain two test tube cells (Spectronic 20) and clean or rinse as necessary. Wipe off the outside of each with a tissue.

3. Fill one cell with distilled water for the blank. Fill the other with the solution of copper sulfate or nickel sulfate after rinsing the cell with solution three times. Wipe the outside of each cell with a tissue.

4. Set $0\%T$ on the Spectronic 20 by adjusting the left-hand knob so that the meter needle reads $0\%T$. Check periodically. (This is unnecessary on the NOVASPEC.)

5. For $Cu(H_2O)_6{}^{+2}$, adjust the wavelength to 620 nm; for $Ni(H_2O)_6{}^{+2}$, adjust the wavelength to 330 nm. Set $100\%T$ by inserting the cell filled with distilled water into the cell holder, matching index lines. Close the top of the holder and adjust the right-hand control until the meter needle reads $100\%T$. After removing the cell, check the $0\%T$ setting. If it has changed, both the $0\%T$ and $100\%T$ settings must be readjusted.

6. Measure the sample absorbance by inserting the cell containing the metal ion solution into the cell holder, matching index lines. Record the absorbance to two significant figures to the right of the decimal.

7. Repeat Steps 5 and 6 for each wavelength used. In low-absorbance regions, measure the absorbance every 40 nm; near the region of the wavelength of maximum absorption, measure the absorbance every 5–10 nm, since you must report this as accurately as possible. (If directed for $Ni(H_2O)_6{}^{+2}$, change to the red phototube and continue measurements to 800 nm.)

8. After finishing measurements, empty the test tube cells and rinse with distilled

water and allow to drain upside down. Do not turn off the spectrophotometer unless directed to do so.

9. Using graph paper, plot absorbance versus wavelength for your solution of $Cu(H_2O)_6^{+2}$ or $Ni(H_2O)_6^{+2}$. Mark the wavelength(s) of maximum absorbance; there may be more than one for $Ni(H_2O)_6^{+2}$ if measurements were made above 600 nm.

10. Calculate the molarity of your sulfate solution from the weight taken. Use this molarity and $b = 1.17$ cm for the diameter of the Spectronic 20 test tube cell to calculate the molar absorptivity of $Cu(H_2O)_6^{+2}$ or $Ni(H_2O)_6^{+2}$ at the wavelength of maximum absorbance. Write it (2 significant figures) on your graph and turn in the graph.

Notes

1. Nickel(II) sulfate is the easiest nickel(II) salt to handle, particularly the J. T. Baker brand. Other nickel(II) salts, such as nickel(II) nitrate, often consist of large crystal lumps that are hard to pulverize.

2. The $Ni(H_2O)_6^{+2}$ ion has its most intense absorption in the visible region, but it also absorbs significantly beyond 600 nm. This absorption, of course, can be measured by switching to the red phototube and red filter.

QUESTIONS

1. In what region of the spectrum is the wavelength of maximum absorbance found for the aquo ion that you measured?

2. The usual *d–d* electronic transitions have molar absorptivities from 0 to 150, whereas charge transfer electronic transitions have molar absorptivities from 10^3 to 10^5. What kind of electronic transition is occurring in the solution you measured?

3. Solutions of copper(II) sulfate are colorless and solid copper(II) sulfate is white, yet copper(II) sulfate is known to absorb photons. In which two regions of the spectrum might it absorb? Which region is more likely because of the location of the absorption of hexaaquocopper(II) ion (mostly in the orange and red regions)?

Experiment 23.   Spectrophotometric Determination of Manganese in Steel

Manganese is determined in steel by dissolving the steel and oxidizing the resulting manganese(II) to manganese(VII), which is measured colorimetrically or spectrophotometrically. Three oxidizing agents are used.

Chemical Background

The steel is first dissolved in dilute nitric acid to form iron(III) and manganese(II) (Equation E23–1). During the dissolution, NO_2 and NO are produced. These may interfere later in the experiment by reducing periodic acid;

hence, they are partially removed by boiling. Ammonium persulfate (ammonium peroxydisulfate) is used as an auxiliary oxidant, to oxidize the remaining oxides of nitrogen (Equation E23–2) and any carbon or other organic material present. Unreacted peroxydisulfate is then decomposed by reducing it with boiling water (Equation E23–3).

$$3Mn^0 + 2NO_3^- + 8H^+ \rightarrow 3Mn^{2+} + 2NO(g) + 4H_2O \quad \textbf{(E23–1)}$$

$$\underset{[(NH_4)_2S_2O_3]}{2NO_2 + S_2O_8^{2-}} + 2H_2O \rightarrow 2NO_3^- + 2SO_4^{2-} + 4H^+ \quad \textbf{(E23–2)}$$

$$2S_2O_8^{2-} + 2H_2O \rightarrow 4SO_4^{2-} + O_2(g) + 4H^+ \quad \textbf{(E23–3)}$$

Theoretically, peroxydisulfate has the oxidizing potential to oxidize manganese(II) to manganese(VII), but the reaction is kinetically too slow to be useful. Silver(II)—added as silver(I)—can be used as catalyst, but the results are occasionally erratic. Potassium periodate oxidizes manganese(II) to manganese(VII) more rapidly and reproducibly and is preferred for the oxidation (Equation E23–4). The oxidation is carried out at the boiling point to speed up the reaction and increase the solubility of the sparingly soluble potassium periodate. Since periodate also decomposes slightly at these temperatures, it is a good technique to add it in two portions to maintain an excess (see Step 5).

$$\underset{(KIO_4)}{2Mn^{2+} + 5IO_4^-} + 3H_2O \rightarrow 2MnO_4^- + 6H^+ + 5IO_3^- \quad \textbf{(E23–4)}$$

Spectrophotometric Determination

This procedure will be performed with a standard steel containing a known percentage of manganese and a steel sample containing an unknown amount of manganese (Note 1). The final calculation of percentage manganese is greatly simplified if the sample weights of the two agree within ± 0.01 g.

1. Prepare the permanganate solutions as directed in Steps 2–6. If the Spectronic 20 spectrophotometer is to be used for this experiment, read Steps 1–8 of the Operating Procedures at the beginning of this chapter, and warm up the instrument before proceeding further. (If any other type of colorimeter or spectrophotometer is to be used, consult the instructor's directions.)

2. Weigh out equal weights (1.00–1.50 g) of standard steel and unknown steel to the nearest 0.01 g (see the instructor's directions). Do not dry the steels.

3. Place the steels in separate 400-mL beakers, add 50–60 mL of dilute (1:3) nitric acid, and heat in a hood to dissolve. Finally, cover and boil gently for 2 minutes to remove nitric oxide. A black residue of carbon may remain at this point.

4. Remove the beakers from the heat and *carefully* sprinkle 1 g of ammonium persulfate into each beaker. Boil gently for 10–15 minutes to oxidize carbon and destroy the excess ammonium persulfate (Note 2).

5. Dilute each solution to about 100 mL, add 15 mL of 85% phosphoric acid [or 40 mL of 6M phosphoric acid (Note 3)], and 0.5 g of potassium periodate. Boil gently for about 3 minutes to oxidize the manganese(II) to permanganate. Remove

the beakers from the heat and allow them to cool below boiling, then add an additional 0.2 g portion of potassium periodate. Boil them for another minute or two.

6. Transfer both solutions to 500-mL volumetric flasks and dilute them to volume.
[Allow them to cool first to avoid oxidizing any organic matter in the distilled water used for dilution (Note 4).]

7. Fill one test tube or cell with distilled water and a second with a solution of permanganate from the standard steel. Adjust the wavelength to 480 nm by turning the knob on top of the instrument. Set $0\%T$ according to Step 4 of the Operating Procedure. (This is unnecessary on the LKB NOVASPEC.)

8. Starting at 480 nm, set $100\%T$, and measure the absorbance of the standard solution at each of the following wavelengths: 480, 520, 560, and 600 nm. Follow Steps 5–7 in the Operating Procedure. This will establish the location of the absorbance peak. At each wavelength, be sure that the 0 and $100\%T$ settings are correct.

9. Select the wavelength of peak absorbance, and again establish the $100\%T$ setting at this wavelength. Rinse the test-tube cell used for the standard three times with the permanganate solution from the unknown steel, then fill it with this solution. Read the absorbance for this solution.

10. If the weights of the two samples are within ± 0.01 g of each other, calculate the percentage of manganese in the unknown steel by multiplying the percentage of manganese in the standard steel by the ratio between the absorbance of the unknown and the absorbance of the standard. If the weights are different, calculate the molar absorptivity ϵ of permanganate, using data from the standard steel. (Use 1.16 cm for the diameter of the test tube.) Using Beer's law, solve for the concentration of permanganate from the unknown steel, then use this concentration to calculate the percentage of manganese in the unknown steel.

Alternate Spectrophotometric Determination

The instructor may direct that a working (calibration) curve be prepared by diluting standardized $0.01N$ potassium permanganate to a known volume, measuring the absorbance of each solution, and plotting the absorbance versus concentration. Then the absorbance of the permanganate solution of the unknown steel is measured, and the concentration of the permanganate is read from the plot of absorbance versus concentration. To simulate the iron(III) concentration of the unknown, the standard solutions of permanganate should be made up to be $0.004M$ in ferric nitrate per 1 g of steel per 500 mL.

Notes

1. There is an interaction (probably of a charge-transfer nature) between the permanganate and iron(III) ions, which appears to reduce the absorbance of the permanganate ion somewhat. [See G. Schenk, *Rec. Chem. Progress*, *28*, 135 (1967).] To correct for this, a standard steel containing a known amount of manganese is used. If a standard solution of permanganate were used, ferric nitrate would have to be added.

2. A precipitate of manganese dioxide at this point may be removed by adding a few drops of a dilute sodium sulfite solution. Boil the sample to expel the excess sulfur dioxide.

3. Phosphoric acid complexes iron(III) so that it does not absorb visible light. In nitric or hydrochloric acid, iron(III) is yellow; large amounts actually absorb appreciably at 520 nm.

4. Permanganate solutions should be analyzed colorimetrically on the day of preparation, although the excess periodate will stabilize the permanganate against slow reduction. (See H. H. Willard and H. Diehl, *Advanced Quantitative Analysis*, New York: Van Nostrand, 1943, p 177.)

QUESTIONS

1. Explain the difference between a colorimetric and spectrophotometric determination of permanganate.
2. If the dichromate and permanganate ions were present together, which of these two methods would be more useful for determining the permanganate ion? Why? What preliminary experiment should be carried out?
3. Calculate the molar absorptivity of potassium permanganate from the following data on the Spectronic 20: the test-tube cell diameter (light path) is 11.6 mm, the $\%T$ is 40% at 500 nm, and the concentration of $KMnO_4$ is $1.7 \times 10^{-4}M$.
4. Why is phosphoric acid added to dissolved steel samples? If it were omitted, would the results affect a colorimetric determination more than a spectrophotometric determination?

Experiment 24. Spectrophotometric Analysis of Two-Component Mixture

In Chapter 17, the spectrophotometric analysis of a two-component mixture was briefly discussed. A two-component mixture of components A and B whose absorption spectra completely overlap each other cannot be analyzed by measuring for each component at any given wavelength. However, if the absorbance of the mixture is measured at two different wavelengths, say wavelength 1 and wavelength 2, then two Beer's law equations in two unknowns can be written and solved for the concentrations of A and B. The equations can be written in simplified form by assuming that $b = 1.00$ cm:

$$A_1 = \epsilon_{A(1)}c_A + \epsilon_{B(1)}c_B$$

$$A_2 = \epsilon_{A(2)}c_A + \epsilon_{B(2)}c_B$$

where A_1 and A_2 are the total absorbances at wavelengths 1 and 2, respectively, c_A and c_B are the concentrations of A and B in the unknown, and ϵ terms are the molar absorptivities of A and B at wavelengths 1 and 2.

As shown in Figure 17–11, wavelengths 1 and 2 must be chosen so that component A absorbs more strongly than component B at wavelength 1, and component B absorbs more strongly than component A at wavelength 2. A further requirement is that solutions of both components must obey Beer's law so that absorbances of both components are *additive* at both wavelengths. Furthermore, Beer's law should be obeyed for each component in the mixture; that is, neither component should interact chemically with the other.

Chemical Background

The two-component mixture used is a mixture of solutions of chromium(III) nitrate and cobalt(II) nitrate, which yield $Cr(H_2O)_6^{3+}$ and $Co(H_2O)_6^{2+}$, respectively. Although both solutions are stable and obey Beer's law, an optional procedure is included for testing for the additivity of the absorbances of both. The chromium(III) nitrate solution is purple because of the $Cr(H_2O)_6^{3+}$ ion. In contrast, a solution of chromium(III) chloride is green because of the $CrCl_2(H_2O)_4^{+}$ ion. The latter slowly reacts with water over a period of days to form the more stable $Cr(H_2O)_6^{3+}$ ion. Such slow reactions are characteristic of the chromium(III) ion.

Reagents

1. Prepare 0.1880M cobalt(II) nitrate by dissolving 5.47 g of $Co(NO_3)_2 \cdot 6H_2O$ in a 100-mL volumetric flask filled to the mark with distilled water.

2. Prepare 0.0500M chromium(III) nitrate by dissolving 2.00 g of $Cr(NO_3)_3 \cdot 9H_2O$ in a 100-mL volumetric flask filled to the mark with distilled water.

3. Prepare three solutions for a Beer's law plot for cobalt(II) ion by diluting 5 mL, 10 mL, and 20 mL of 0.1880M cobalt(II) nitrate to 25 mL in volumetric flasks. Calculate the three molarities.

4. Prepare three solutions for a Beer's law plot for chromium(III) ion by diluting 5 mL, 10 mL, and 20 mL of 0.0500M chromium(III) nitrate to 25 mL in volumetric flasks. Calculate the three molarities.

5. For the optional check on additivity below, add 10 mL of 0.1880M cobalt(II) nitrate and 10 mL of chromium(III) nitrate to a 25-mL volumetric flask and dilute to the mark, giving 0.0752M cobalt(II) ion and 0.0200M chromium(III) ion.

Spectrophotometric Determination, Including Optional Additivity Check

1. Turn on the spectrophotometer to be used (directions are given for the Spectronic 20 spectrophotometer) and allow it to stabilize for 10 min. Clean test-tube cells and rinse with distilled water.

2. From the solutions above, fill cells with 0.0752M cobalt(II) ion and 0.0200M chromium(III) ion. If the optional additivity check is to be done, fill a third cell with the 0.0752M cobalt(II)–0.0200M chromium(III) mixture.

3. Starting at 375 nm, measure the absorbance of all solutions at every 25 nm except for these additional measurements: 470, 480, 490, 500 nm and 570, 575, and 580 nm. When absorbance readings are over 0.7 absorbance units, read the $\%T$ and convert to A.

4. Plot the absorbance values on graph paper and choose the two wavelengths

needed for the multicomponent analysis. If the optional additivity check is to be done, also plot the absorbance values of the $0.0752M$ cobalt(II)–$0.0200M$ chromium(III) solution.

5. After choosing the two wavelengths for analysis, use the three cobalt(II) solutions and the three chromium(III) solutions prepared in *Reagent Steps* 3 and 4 to prepare a plot of absorbance wavelength for each wavelength for both cobalt(II) and chromium(III), giving four plots.

6. Using the slope of each plot in Step 5, calculate the molar absorptivity of cobalt(II) ion at both wavelengths and the molar absorptivity of chromium(III) at both wavelengths. Assume that the path length is 1.00 cm even though it is not, as long as the same test-tube cells are used throughout.

7. Fill a test-tube cell with the unknown solution of cobalt(II)–chromium(III) ions and measure the absorbance of the solution *twice* at each of the wavelengths chosen for analysis. (If the absorbance values differ significantly, measure a total of four times.) Average all of the absorbance values obtained at each wavelength for the final calculations (see Note).

8. Now set up two equations in two unknowns using the four molar absorptivities and the two absorbance values of the unknown. Assume that the path length is 1.00 cm as long as the same test-tube cells were used throughout for both unknown and knowns; this simplifies the calculations. Since the path length drops out as long as it is the same, the accuracy of the calculations is not affected.

9. Report the concentration of both cobalt(II) ion and chromium(III) ion in the unknown. Also report the results of your check of additivity of absorbance values if the instructor assigned you to do this. Turn in your plots as part of your report.

Note

Unknowns containing from $0.0.0100M$ chromium(III) nitrate–$0.125M$ cobalt(II) nitrate to $0.030M$ chromium(III) nitrate–$0.075M$ cobalt(II) nitrate were found to be usable.

Experiment 25.   Photometric Titrations of Copper(II) with EDTA

One result of the commercial availability of modern spectrophotometers is the development of the photometric titration [1]. In photometric titration, a spectrophotometer is used to detect an end point graphically. The absorbance of a titrated species is measured after each addition of titrant until well past the end point. A plot of absorbance versus milliliters of titrant consists of two straight lines that intersect at the end point of the titration.

[1] R. F. Goddu and D. N. Hume, *Anal. Chem.*, 26, 1740 (1965); P. B. Sweetser and C. E. Bricker, *Anal. Chem.*, 25, 253 (1953); A. L. Underwood, *Anal. Chem.*, 25, 1911 (1953).

Chemical Background

In this experiment, copper(II) ion will be titrated photometrically with the disodium salt of EDTA. The reaction is

$$H_2Y^{2-} + Cu^{2+} \rightleftharpoons CuY^{2-} + 2H^+$$
$$(Na_2H_2Y)$$

The titration is performed at 625 nm; both the copper-EDTA chelate and the copper(II) ion absorb at this wavelength, but the molar absorptivity of the chelate is much higher.

Figure 31–4a is a plot of absorbance (uncorrected for dilution) versus milliliters of titrant. The points fall below the extrapolated lines in the end-point region because the reaction is incomplete near the equivalence point. The excess EDTA titrant then forces the reaction to completion at some point after the equivalence point, as indicated by a maximum in absorbance. Because the addition of titrant causes dilution, the absorbance will then decrease slightly.

This dilution effect can be corrected by multiplying the absorbance by a dilution factor, giving a plot such as that shown in Figure 31–4b. In practice, a plot such as this can be obtained without correction by using a large volume (80–100 mL) of solution and a small volume (5 mL or less) of a relatively concentrated titrant. This will be the procedure used in this experiment.

The pH is critical for this titration, because a large change in pH changes the structure of the chelate and, thus, changes its absorbance. The mixture of the hydrochloric acid and acetic acid maintains the pH between 2.4 and 2.8 to avoid this error. It also permits the copper to be titrated in the presence of metal ions that form weaker chelates.

For example, copper(II) ion may be titrated in the presence of an equal concentration of zinc(II) ion, even though the copper-EDTA chelate is only slightly more stable (Cu-EDTA, $\log K = 18.8$; Zn-EDTA, $\log K = 16.5$). For the usual in-

Figure 31–4. Photometric titration of copper(II) with EDTA titrant at pH 2.4: (a) absorbance uncorrected for dilution; (b) absorbance corrected for dilution.

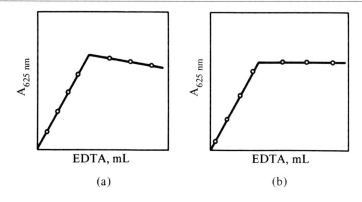

(a) (b)

dicator or potentiometric titration, the required difference in log K values is about 4. This difference ensures a favorable end-point equilibrium in titrating the ion that forms the more stable chelate. The difference is smaller for a photometric titration for two reasons: (a) the points taken at the beginning of the titration measure only the copper-EDTA chelate, and (b) the points taken after the end point show no change in absorbance even though the zinc-EDTA chelate is being formed; the zinc-EDTA chelate is colorless.

Photometric titrations are especially useful for determining low concentrations of substances, for analyzing substances whose reaction equilibrium is unfavorable at the end point, and for determining species for which there are no suitable indicators. In organic analysis, for example, phenols can be titrated with a strong base (such as tetraalkylammonium hydroxide) using ultraviolet light, even though the equilibrium is unfavorable at the end point.

Spectrophotometers used in photometric titrations are usually modified to accept a cell about 100-mm long provided with magnetic stirring. A filter colorimeter, such as the Fisher Electrophotometer II with built-in magnetic stirrer, is also convenient. The following procedure requires only the Bausch and Lomb Spectronic 20, a modified flask, and magnetic stirring.

Reagents and Equipment

1. Seal two 5–6 mm (o.d.) pieces of glass tubing to the side and bottom of a 125-mL Pyrex flask, as shown in Figure 31–5 [2]. (See Note 1.)
2. Prepare a solution of acetic acid and hydrochloric acid by adding 4.1 g of anhydrous sodium acetate (or an equivalent amount of the hydrate) to 50 mL of water. Add 1M hydrochloric acid to the buffer until the pH of the mixture is 2.2, as indicated by the pH meter.
3. Prepare 0.2M EDTA titrant by dissolving about 3.8 g of the reagent-grade disodium salt ($Na_2H_2Y \cdot 2H_2O$) in 50 mL of distilled water. If the solution is turbid, warm slightly, but do not add sodium hydroxide to remove the turbidity. Mix well.

Procedure

1. Turn on the spectrophotometer to allow it to warm up.
2. Assemble the titration flask and test tube cell as shown in Figure 31–5, but do not connect the outlet tube to the glass side-arm of the flask. Instead, connect the outlet tube to an aspirator, pour small portions of distilled water into the flask to rinse the apparatus, and aspirate the water from the flask. Repeat several times. Finally, aspirate all of the water from the flask and connecting tubes, but do not be concerned about the water left in the test tube cell. Connect the outlet tube to the flask.
3. Clamp the flask over the magnetic stirrer and place the stirring bar in the flask. Add 35 mL of distilled water, 20 mL of acetic acid–hydrochloric acid buffer

[2] C. Rehm, J. I. Bodin, K. A. Connors, and T. Higuchi, *Anal. Chem., 31*, 483 (1959).

Figure 31–5. Photometric titration assembly for use with the Bausch and Lomb Spectronic 20: (a) side view; (b) top view.

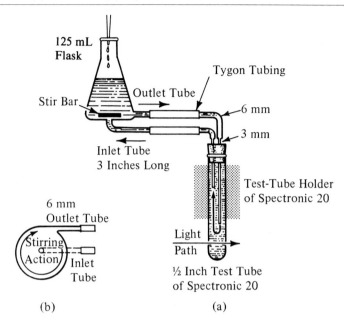

solution, and 25.00 mL of the unknown copper solution. (*Note:* The unknown copper solution contains about 1 mmole of copper per 25 mL). At this point, the pH of the solution should be 2.4–2.8.

4. Fill a 10-mL buret with standard 0.2*M* EDTA (Note 2).

5. Remove air bubbles from the connecting lines by lowering the cell below the level of the flask and flexing the tubes while the magnetic stirrer is spinning rapidly (Note 3). Leave the stirrer on for the rest of the experiment.

6. Set the wavelength at 625 nm (Note 4) and adjust the left-hand knob of the Spectronic 20 so that the meter reads 0%*T*.

7. Insert the test tube cell into the cell holder, ensuring that the rubber stopper completely covers the well of the cell to keep out all stray light (Note 5). Adjust the absorbance to read zero.

8. Titrate with 0.2*M* EDTA in 1-mL increments, allow the solution to circulate for 2 minutes (or until the absorbance is constant), and record the absorbance. Continue the titration until four or five points have been recorded past the end point. (In practice, it is desirable to use all 10 mL of EDTA.) Remove the test tube cell from the holder as soon as possible (Note 6).

9. Plot the absorbance against the volume of EDTA, determine the end point graphically, and determine the corresponding volume of EDTA. Calculate the molarity of the unknown copper solution (Note 7).

Notes

1. An inexpensive titration flask with sealed-on outlet and inlet tubes is commercially available from the Burrell Corp., Pittsburgh, Pa., for use with the Spectronic 20.

2. Standardize the EDTA as described in Experiment 12 or by photometrically titrating a standard copper(II) solution prepared by dissolving a known weight of pure copper wire in about 5 mL of 1:1 nitric acid, adding 25 mL of water, boiling, and diluting it to 500 mL to make a 0.04M solution.

3. An air bubble blocking the entire diameter of the tubing will prevent mixing, even with moderate magnetic stirring.

4. It is advantageous to titrate at 745 nm. To do this, substitute an infrared-sensitive phototube and a red filter for the phototube used in the 330–625 nm region. Use 0.1M EDTA and 25 mL of a 0.01M copper(II) solution.

5. Alternatively, reduce stray light by covering the top of the test tube and tubing with a black cloth or aluminum foil.

6. Removing the test tube automatically drops an occluder between the light source and the phototube and avoids fatiguing the phototube unnecessarily.

7. The unknown may contain copper(II) in the presence of equal quantities of zinc(II), cadmium(II), or aluminum(III). If no interferences are present, the titration may also be conducted at 580 nm using an ammonia buffer at pH 10.

QUESTIONS

1. Would it be difficult to perform a visual EDTA titration of *macro* amounts of copper(II) with an indicator?

2. List some advantages of photometric titrations over visual indicator titrations. List some disadvantages.

3. Could copper(II) be titrated photometrically if it formed a very weak chelate ($K_{CuY'(H)} = 10^6$) with EDTA?

Experiment 26. Fluorometric Determination of Vitamin D

If a molecule is to fluoresce, at least one benzene ring must be present in its structure. The D vitamins lack this feature and hence do not fluoresce.

Vitamin D₂ (Calciferol)

Chemical Background

In this experiment, the D vitamins are heated with acid in alcohol to transform them into fluorescing structures [3]. Both vitamins D_2 and D_3 yield a species that fluoresces primarily in the blue region, with a fluorescence emission maximum at 475 nm. This species is excited in the 300–450 nm region, with maximum excitation at 425 nm; thus, a *filter fluorometer* can be used (see Chapter 18) with glass, rather than quartz, optics and cells. The directions that follow are written for the Coleman Photofluorometer (fluorometer), although other comparable instruments can be used. The most convenient excitation wavelengths on the Coleman fluorometer are those around 366 nm, rather than those at the maximum excitation wavelength of 425 nm.

Calibration Curve Procedure for Vitamin D_2

This procedure yields a linear calibration curve over the range of 0.1–10 ppm [3]. The exact range obtained depends somewhat on the particular primary and secondary filters used. Your instructor will furnish specific volumes of stock solution to use with filters other than those given here.

1. Prepare a stock solution of vitamin D_2 (calciferol) containing 10 mg of vitamin D_2 in one liter of absolute ethanol.

2. Transfer 2.0, 4.0, 6.0, and 8.0 mL of the stock solution to 25-mL volumetric flasks using a 5- or 10-mL microburet. Mark the flasks 0.8 ppm, 1.6 ppm, 2.4 ppm, and 3.2 ppm. Dilute the solutions to the mark with absolute ethanol, and mix well.

3. Transfer 10-mL aliquots of each solution to 25-mL flasks, either Erlenmeyer or volumetric flasks. Pipet a 10-mL aliquot of a freshly prepared 20% sulfuric acid–ethanol (v/v) into each flask, then stopper them. Heat the flasks in a 75°C water bath for one hour (see Note).

4. Remove the flasks from the water bath and let them cool. After exactly 15 minutes (see Note), rinse each of four Coleman-fluorometer test-tube cells with solution from the corresponding flask, then fill it with about 15 mL of solution from the same flask. Wipe the outside of each cell with a clean tissue. Mark a fifth test-tube cell "blank" and fill it with 10 mL of absolute ethanol and 10 mL of the 20% sulfuric acid–ethanol reagent used in Step 3.

5. Turn on the Coleman fluorometer by turning the STD knob clockwise in the direction of the ON arrow. Insert the proper primary and secondary filters into the fluorometer. (A PC-6 narrow-pass primary filter and a PC-9A sharp-cut secondary filter are good choices.) Allow the fluorometer to warm up for 20 minutes to stabilize the excitation source.

6. After 20 minutes, set the STD knob full clockwise and the aperture control (APT CONT.) to the extreme right. Set the meter to about zero with the coarse BLK control.

7. The procedure that follows will electrically compensate for the solvent blank. Use the 0.8-ppm reaction mixture for the "DC solution" mentioned.

[3] A. J. Passannante and L. V. Avioli, *Anal. Biochem.*, *15*, 287 (1966).

a. Insert the tube containing the DC solution and cover it with a black cap. Depress the shutter button (SHUT. B) and slide the aperture control (APT CONT.) to the left till the meter reads 20 units. Allow the shutter button to return to normal position.

b. Remove the tube and insert the tube marked "Blank." Match the mark on the tube with the mark on the instrument. Depress the shutter button and set the meter to zero with coarse and fine BLK controls. Allow the shutter button to return to normal. (This tube, as well as all tubes, should be covered with a black cap.)

c. Reinsert the tube containing the DC solution and depress the shutter button. Adjust the aperture control to give a meter reading slightly greater than the recommended setting. Bring the meter exactly to the 20-unit setting with STD control. Allow the shutter button to return to normal.

d. Recheck the zero setting as directed in *b* and recheck the 20-unit setting as directed in *c.*

e. Measure the fluorescent intensities of the other tubes the same way. Insert the tube and match the marks. Cover the tube with a black cap. Depress the shutter button and record the intensity to the first decimal (26.5, for example). Allow the shutter button to return to normal, and remove the tube.

8. After you are finished, rinse the test tubes with absolute ethanol, wipe off their outsides, and allow them to drain upside down.

9. Plot the fluorescence intensity versus the concentration in ppm and use this graph to determine the concentration of the unknown.

Analysis of Liquid Vitamin D_2 Sample or Blood Sample

1. If the sample is a solution of Vitamin D_2 in ethanol, transfer 5.0 mL of the solution to a 25-mL volumetric flask using a volumetric pipet or microburet. Dilute it to the mark with absolute ethanol. Treat the sample solution as you treated the solutions used in plotting the calibration curve, starting with Step 3. Read its final concentration from the calibration curve and calculate the initial concentration of Vitamin D in the original solution.

2. If the sample is a blood-serum sample, measure out 1 mL of the fresh serum into a conical centrifuge tube and mix it with 4.0 mL of absolute ethanol. Centrifuge the mixture of precipitated protein and supernate for 10 minutes at 2000 rpm [3].

3. Decant the clear supernate into a 25-mL Erlenmeyer or volumetric flask. Add 5.0 mL of absolute ethanol and 10 mL of freshly prepared 20% sulfuric acid–ethanol (v/v). Heat the solution in a 75°C water bath for one hour; then treat it as in Steps 4 and 5 of the Calibration Curve Procedure for Vitamin D_2. Read the final concentration of Vitamin D from the calibration curve and calculate its initial concentration in the blood serum sample.

Note

If the calibration curve is to be reproducible, the solutions should not be heated for more than one hour. For the same reason, the fluorescence should be measured as soon as possible after the 15-minute cooling period. The flasks should be kept in the dark after the 15 minutes are past, since some form of photochemical decomposition sets in.

SUGGESTIONS FOR ADDITIONAL EXPERIMENTS

The following articles are suggested as material for additional work or research projects for advanced students:

J. S. Fritz and G. E. Wood, *Anal. Chem. 40*, 134 (1968). *Spectrophotometric titrations of olefins with bromine.*

M. Ozolins and G. H. Schenk, *Anal. Chem. 33*, 1035 (1961). *Spectrophotometric titration of Diels-Alder dienes with tetracyanoethylene.*

G. A. Parker and D. F. Boltz, *Anal. Chem. 40*, 420 (1968). *Ultraviolet spectrophotometric determination of chromium as the peroxychromic acid-2,2'-bipyridine complex.*

G. H. Schenk and W. E. Bazzelle, *Anal. Chem. 40*, 163 (1968). *Photometric titration of thallium(I) in mixtures of other metal ions with cerium(IV) titrant.*

Analytical Methods for Atomic Absorption Spectrophotometry, Perkin-Elmer Corp., Norwalk, Conn., 1966. *Gives exact working procedures for quantitative determination of most metallic elements by atomic absorption.*

H. Veening, *J. Chem. Ed. 43*, 319 (1966). *Infrared determination of* meta- and para-*xylene using* ortho-*xylene as an internal standard.*

C. D. West, R. L. Birke, and D. N. Hume, *Anal. Chem. 40*, 556 (1968). *Spectrophotometric study of the reaction of thorium chloranilate with fluoride as a basis for the spectrophotometric determination of traces of fluoride in water.*

32

Electroanalytical Procedures

In Chapter 15, we discussed various electroanalytical methods, such as polarography, amperometric titrations, coulometric titrations, and electrodepositions. In this chapter, potentiometric titrations are used in *Experiment 27* for determining equivalent weights and in *Experiment 28* for determining the formula and stability constant of a complex ion. Experiment 28 begins by discussing the basic fundamentals of potentiometric titrations and measurements.

In *Experiment 29*, copper ore is analyzed by electrodeposition on a platinum gauze cathode, without cathode potential control.

In *Experiment 30*, arsenic(III) is titrated coulometrically by electrolytic generation of iodine as oxidizing agent. In *Experiment 31*, mercaptans are amperometrically titrated with silver(I). The mercaptans are precipitated as silver mercaptides.

Suggestions for additional experiments are given at the end of this chapter.

Experiment 27.   Potentiometric Determination of the Equivalent Weight and K_a for a Pure Unknown Weak Acid

Properties of a newly synthesized organic acid, such as its equivalent weight and ionization constant, are often determined by titrating a purified sample of the acid with sodium hydroxide and determining the pH of the solution with a pH meter (a potentiometric titration). The pH is then plotted against the milliliters of titrant used.

The equivalent weight of the acid is obtained by dividing the milligrams of acid by the number of milliequivalents of titrant consumed (as determined from the number of milliliters of titrant at the equivalence point, the point where the curve changes in slope).

Chemical Background

The ionization constant of the acid is determined by reading the pH at the midpoint (50% neutralization) of the plot of pH versus milliliters of titrant. The pH at 50% neutralization is the pK_a of the acid, as shown in this derivation:

$$K_a = \frac{[H^+][A^-]}{[HA]} \tag{E27-1}$$

$$[H^+] = K_a \frac{[HA]}{[A^-]} \tag{E27-2}$$

$$-\log[H] = -\log K_a - \log \frac{[HA]}{[A^-]} \tag{E27-3}$$

$$pH = pK_a - \log \frac{[50\%]}{[50\%]} = pK_a - \log 1 = pK_a \tag{E27-4}$$

The graphical relationships are illustrated in Figure 32–1.

The object of this experiment is to perform a potentiometric sodium hydroxide titration of a pure unknown weak acid, plot the data, and report the equivalent weight and K_a of the weak acid. Before starting the experiment, review Chapter 16 for the operation of the pH meter and its electrodes.

Preliminary Work

1. Prepare and standardize $0.1M$ sodium hydroxide as directed in Experiment 7.
2. Obtain a sample of the unknown weak acid and dry it for 1 hour at 110°C if directed to do so by the instructor.

Figure 32–1. Hypothetical plot of pH versus milliliters of $0.1N$ sodium hydroxide for a potentiometric titration of 200 mg of a weak acid ($K_a = 1 \times 10^{-4}$) of molecular weight 100.

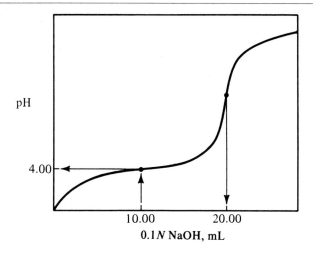

pH

4.00

10.00   20.00

$0.1N$ NaOH, mL

3. Assume that the approximate equivalent weight of the acid is 100 unless directed otherwise. Calculate the weight of sample that will be neutralized by not more than 40 mL of the standard sodium hydroxide.

4. Weigh out a sample of the size calculated in Step 3, and dissolve it in 100 mL of water in a beaker. If it is insoluble in water, dissolve it in about 50 mL of ethanol, and then add 50 mL of water (see Note).

Potentiometric Titration

1. Operating instructions for the pH meter are given on the instrument. The instructor may also give some further directions. The following precautions should be especially noted:

a. Plug the meter into the line, turn it on if necessary, and allow it to warm up (some meters operate on batteries and need only to be turned on).

b. Handle the electrodes carefully. They are of fairly rugged construction, but they will break if handled carelessly. Electrodes are expensive.

c. Always return the meter to the "reference" or "off" position before removing the electrodes from the solution.

d. After using the pH meter, rinse the electrodes using a wash bottle, place them in distilled water, and turn the meter to the "reference" or "off" position.

e. Know what you are doing and perform the titration as quickly as possible to allow others to use the pH meter.

2. If directed, standardize the pH meter by immersing the electrodes in a reference buffer solution [such as saturated potassium hydrogen tartrate (pH = 3.56)] and adjusting the meter to read the pH of the buffer.

3. Rinse the electrodes using a wash bottle, wipe them with a cleansing tissue, and insert them in the beaker containing the unknown weak acid. Measure the pH and record it opposite 0 mL of titrant in the record book.

4. Titrate potentiometrically with the standard $0.1M$ sodium hydroxide solution. Add rather large increments (2–5 mL) of base at first. Add smaller increments as the end point is approached, finally adding one-drop or two-drop portions between each reading in the immediate vicinity of the end point. After each increment, stir well and record the volume of titrant and the pH at this point. Continue adding titrant to about 5 mL beyond the end point.

5. Plot the titration curve on 20 div/in graph paper with pH on the ordinate (vertical axis) and milliliters of $0.1M$ sodium hydroxide on the abscissa (horizontal axis).

6. Calculate the equivalent weight of the acid. Locate the midpoint (50% neutralization) of the titration curve and calculate the ionization constant of the unknown acid from the pH at this point.

Note

If a 50% ethanol solvent must be used, the pH readings obtained will be "apparent pH" readings and may not agree with pH values for water alone. The ionization constant will also be an apparent constant and may not agree with a constant determined in water alone.

QUESTIONS

1. Calculate the equivalent weight of a pure weak acid, 100.0 mg of which consumes 50.00 mL of 0.04M sodium hydroxide. If the molecular weight of the acid is 200, how many protons per mole of the acid react?
2. Calculate the K_a of a weak acid with a pH of 5.33 at 50% neutralization.
3. Why must the weak acid be pure for this experiment?
4. Why should no more than 50.00 mL of sodium hydroxide titrant be used?

Experiment 28.   Determination of the Fomula and Stability Constant of a Silver Complex Ion

In this experiment, the formula and stability constant are determined for a silver complex ion of the type AgL_n^+.

Chemical Background

The ligand L is a basic nitrogen compound, such as ammonia, or an organic amine of the type RNH_2, R_2NH, or R_3N. The general reaction for the complex ion is

$$Ag^+ + nL \rightleftharpoons AgL_n^+ \qquad (E28-1)$$

The equilibrium constant (stability constant) of this reaction is

$$K = \frac{[AgL_n^+]}{[Ag^+][L]^n} \qquad (E28-2)$$

This equation rearranges to

$$[Ag^+] = \frac{[AgL_n^+]}{[L]^n K} \qquad (E28-3)$$

In logarithmic form, this equation becomes

$$\log[Ag^+] = -n\{\log[L]\} + \log[AgL_n^+] - \log K \qquad (E28-4)$$

In this experiment, the silver-ion concentration of a solution is measured potentiometrically as increasing amounts of ligand L are added. An excess of L is present at all times, so that almost all of the silver ion is present as the complex ion, AgL_n^+. Obviously, a small equilibrium concentration of Ag^+ is always present and can be measured.

A plot of $\log[Ag^+]$ versus $\log[L]$ will give a straight line with a slope of $-n$. The intercept of the curve on the vertical axis is equal to $\log[AgL_n^+] - \log K$. (This intercept occurs where the log of $[L]$ is zero, so $[L]$ is equal to 1 at the intercept.) Since $[AgL_n^+]$ is held virtually constant at 0.01M throughout the experiment, the vertical intercept is equal to $-2.0 - \log K$. Thus, the number of ligands combining with the silver ion can be obtained from the slope of the curve, and the stability constant of the complex ion can be determined from the intercept of the curve.

A silver-saturated calomel electrode system is used to measure the concentration of the silver ion. The saturated calomel electrode is a reference electrode at $+0.246$ V. The silver metal electrode responds to changes in silver-ion concentration; its potential is given by the Nernst equation for the Ag(I)-Ag couple:

$$E_{Ag} = E^0 + \frac{2.30RT}{F} \log [Ag^+] \qquad \text{(E28–5)}$$

Here, E^0 is $+0.800$ V, R is the gas constant (8.31 joules-deg^{-1}-mole^{-1}), T is the absolute temperature (273 + °C), and F is the Faraday constant (96,487 coul-equiv^{-1}). At 25°C, the ratio $RT/F = 0.0592$.

If E_{Ag} is known, $[Ag^+]$ can be calculated from the Nernst equation. However, the quantity measured is the *difference* in potential between the silver and saturated calomel electrodes. For the calomel reference electrode (Note 1), this is

$$E_{measured} = E_{Ag} - E_{calomel} = E_{Ag} - 0.246 \qquad \text{(E28–6)}$$

$$E_{Ag} = E_{measured} + 0.246$$

Reagents and Apparatus

1. Prepare $10M$ ammonia by diluting concentrated ($14.2M$) ammonium hydroxide. Pipet two aliquots of exactly 1 mL of $10M$ ammonia into two 250-mL flasks and titrate them to the methyl red end point with standard $0.5M$ hydrochloric acid. Calculate the molarity of the ammonia to three significant figures.

2. Prepare 10 mL of $0.2M$ silver perchlorate or silver nitrate by weighing the salt on a trip scale if such a solution is not available in the laboratory.

3. Obtain a student potentiometer or pH meter. (Read the instructor's directions and the discussion in Chapter 16 for information on its use.) Use the millivolt scale, not the pH scale, on the pH meter.

4. Obtain a silver-metal electrode and a saturated calomel electrode of the fiber type.

Procedure

1. Pour exactly 94 mL of distilled water into a dry 250-mL beaker. Add exactly 1 mL of standard $10M$ ammonia from a buret (Note 2).

2. Pipet exactly 5 mL of $0.2M$ silver perchlorate or nitrate into the beaker while stirring gently.

3. If the electrodes have been soaking in distilled water, remove them and dry them with a cleansing tissue. Insert the dried electrodes into the solution and take an initial potential reading. Add ten 1-mL increments of $10M$ ammonia from the buret, taking careful potential readings after each addition. Measure and record the temperature of the solution.

4. Tabulate the data under the following seven headings: mL of $10M$ ammonia, total volume of solution, $[NH_3]$, $\log [NH_3]$, $E_{measured}$, E_{Ag}, and $\log [Ag^+]$.

5. For each measurement, plot $\log[Ag^+]$ versus $\log[NH_3]$ on graph paper. Draw the best straight line through the points, then calculate its slope. This is n, the number of ligands in the complex ion. Write the formula of the silver complex.
6. From the curve, determine the value of $\log[Ag^+]$ when $\log[L] = 0$. From Equation E28–4, calculate the value of the stability constant K (assume that $\log[AgL_n^+]$ is -2.00).

Notes

1. In the following example, the silver-ion concentration is calculated in a solution where the measured voltage of the silver-saturated calomel electrode system is $+0.110$ volt at $23°C$.

$$E_{Ag} = +0.110 + 0.246 = +0.356 \text{ V} \qquad \textit{From Equation E28–6}$$

$$2.30\frac{RT}{F} = \frac{(2.30)(8.31)(296)}{96,500} = 0.0586$$

$$0.356 = 0.800 + 0.0586 \log[Ag^+] \qquad \textit{From Equation E28–5}$$

$$\log[Ag^+] = \frac{-0.444}{0.0586} = -7.58$$

2. A pure organic amine, such as pyridine, *n*-butylamine, dibutylamine, etc., may be used in place of ammonia.

QUESTIONS

1. Why must the temperature of the solution be measured?
2. What changes would be necessary in the experimental procedure if the stability constant of the tetraminecopper(II) ion were being measured?
3. Why must the volumes of the distilled water, silver nitrate, and ammonia be measured exactly?
4. Show that Equation 27–4 can also be expressed as $pAg = n\{\log[L]\} - \log[AgL_n^+] + pK$.

Experiment 29.   Electrodeposition of Copper

Chemical Background

The electrodeposition of copper is accomplished by plating copper metal on a previously weighed platinum gauze cathode (Equation E29–1). The deposit should be smooth (indicating that it is pure) and adherent (to avoid loss during handling and weighing). A second weighing of the cathode plus the copper gives the weight of the copper by difference. This is the most accurate method available for estimating macro amounts of copper.

$$Cu^{2+} + 2e^- \rightleftharpoons Cu^0(s) \qquad E^0 = +0.345 \text{ V} \qquad \textbf{(E29-1)}$$

Electrodeposition is a convenient method for separating copper from other less easily reduced metal ions, such as zinc(II) ($E^0 = -0.76$ V). Many ions that are reduced under the same conditions as copper interfere by codepositing with it on the cathode. Silver ($E^0 = +0.800$ V) is the most likely interference; it should be removed by precipitation before analysis. More than 0.2% silver roughens the copper deposit, making handling difficult. More than 0.4% silver may affect the accuracy of the analysis.

More than 0.5% iron may inhibit the quantitative deposition of copper if moderate amounts of nitric acid are present. (The nitric acid produces ferric nitrate, which dissolves copper from the cathode.) If the nitrate ion is kept at fairly low concentrations, good results are obtained.

Despite this possible interference, some nitric acid must be added to the electrolyte to depolarize the solution—that is, to prevent hydrogen ions from being reduced to hydrogen gas evolving from the cathode. Evolution of hydrogen gas will obviously roughen the copper deposit.

Another disadvantage of having nitrate ion present is that some of it may be reduced to the nitrite ion (Equation E29–2). The nitrite ion, in turn, prevents the quantitative reduction of copper. Fortunately, it is easily destroyed by adding urea (Equation E29–3).

$$NO_3^- + 2H^+ + 2e^- \rightleftharpoons NO_2^- + H_2O \qquad \textbf{(E29-2)}$$

$$2NO_2^- + (NH_2)_2CO + 2H^+ \rightleftharpoons 2N_2(g) + CO_2(g) + 3H_2O \quad \textbf{(E29-3)}$$

Finally, the chloride ion concentration should be very low; otherwise, the platinum anode will be oxidized to platinum(II), which will be codeposited on the cathode with copper.

The procedure described below uses electrolysis with mechanical stirring to analyze copper ores containing 2–25% copper with no interferences, such as silver. Commercial units, such as the Sargent Electroanalyzer, have a small cylindrical platinum gauze anode rotating inside a large platinum gauze cathode. Tall-form beakers covered with split watch glasses are used, to prevent loss by splattering.

Preparation of Electrodes

1. Clean the platinum gauze electrodes by immersing them in hot 1:2 nitric acid for about 5 minutes. This will remove any copper from the cathode, as well as grease and organic matter on both electrodes. Rinse them thoroughly with water and then distilled water, and dip them into a beaker of acetone or ethyl alcohol.
2. Handling the cathode by the stem (Note 1), place it on a watch glass and dry it at 110°C for 5–10 minutes. Cool the cathode in a desiccator and then weigh it carefully on the analytical balance (Note 1).

Dissolution of Ore

1. Weigh two 0.9–1.1 g samples of copper ore into 250-mL beakers (see the in-

structor's directions). Add 10 mL of nitric acid to each beaker and cover it with a watch glass. Heat the solutions in a hood until the ore dissolves.

2. Cool the beakers and add 10 mL of 1:1 sulfuric acid to each. Evaporate the solutions until dense sulfur trioxide fumes are evolved.

3. Dilute the cooled solutions to 90 mL with hot distilled water and filter them through medium-porosity filter paper into two 200-mL tall-form beakers. Wash the residue in the filter paper with three 5-mL portions of hot water.

Electrodeposition

1. The solutions being electrolyzed should contain no more than 1 mL of concentrated nitric acid for each 100 mL of solution. To each solution, add 1 mL of concentrated nitric acid per 100 mL of solution, 1 g of ammonium nitrate, and 0.4 g of urea.

2. Attach the electrodes to the electrodeposition apparatus (Note 1), placing the smaller anode inside the large cathode. (Do not let the two electrodes touch!)

3. Elevate the beaker around the electrodes until all but $\frac{1}{4}$ inch of the cathode is covered by the solution. Cover the beaker with a split watch glass.

4. Start the stirring motor and close the switch to start the electrolysis. Adjust the instrument until the current is 2–3 A and the voltage is less than 3 V (Note 2). Electrolyze at these conditions until the blue copper color has disappeared (about 45 minutes).

5. Add enough distilled water to cover the top of the cathode and lower the current to 0.5 A. Continue the electrolysis for another 15 minutes and note whether any copper is deposited on the top of the cathode. If not, the electrolysis is complete (Note 3).

6. Turn off the stirring motor, but do *not* turn off the current. Slowly lower the beaker from under the electrodes while continuously washing the exposed part of the cathode with distilled water. When the cathode is completely out of the solution, turn off the current (Note 4).

7. Immerse the electrodes in a beaker of distilled water and disconnect the cathode into this beaker. Lower the cathode and beaker away from the anode, remove the cathode from the beaker, and rinse it again with distilled water. Dip the cathode into a beaker of ethyl alcohol or acetone, place it on a watch glass, and dry it at 110°C for no longer than 5 minutes. (This period is short to avoid oxidizing the surface of the copper.)

8. Cool the cathode and weigh it carefully on the analytical balance. Calculate the percentage of copper in the ore using an atomic weight of 63.54. Clean the cathode as in Step 1 under Preparation of Electrodes and repeat the procedure for the second ore sample.

Notes

1. Avoid handling the gauze with the fingers; grease transferred to the gauze may prevent copper from adhering to the cathode. Handle the electrodes only by their stems.

2. The electrodeposition may be conducted overnight at 0.5 A, without stirring.
3. If the electrolysis is not complete, continue electrodeposition for another 15 minutes.
4. If the current is turned off before the cathode is completely out of the solution, the deposited copper may dissolve.

QUESTIONS

1. Compare the electrodeposition method for determining copper with the iodometric method.
2. The reduction of copper is a two-electron process. Compare the equivalent weight in the electrodeposition method with that in the iodometric method.
3. Why are the electrodes rinsed with acetone or ethyl alcohol?
4. Summarize the advantages and disadvantages of using nitric acid in this procedure.

Experiment 30. Coulometric Titration of Arsenic

The principles of coulometric titrations are discussed in Chapter 15. In this experiment, iodine is generated from a potassium iodide solution by a constant current at a platinum anode.

Chemical Background

The cathode reaction is the discharge of hydrogen gas:

$$2H_2O + 2e^- \rightleftharpoons H_2(g) + 2OH^-$$

The iodine generated reacts instantly with the arsenic(III) in the sample solution (see Experiment 18 for the reactions). The end point is indicated by the first appearance of a blue-violet color from the starch-triiodide complex. A constant known electrolysis current is employed, so the amount of arsenic can be calculated from the time required for the titration.

Apparatus and Reagents

1. If it is available, use a commercial coulometric power supply with a synchronous electric timer. If this is not available, assemble an apparatus similar to that described by Meloan and Kiser [1] (see Figure 32–2). Use squares of platinum foil for both electrodes and a small beaker for the titration cell. Stir the solution with a magnetic stirrer and record the time with an electric timer.
2. Weigh 200–250 mg of arsenic(III) oxide (As_2O_3) into a small beaker, add a solution of about 0.7 g (about 8 pellets) of sodium hydroxide freshly dissolved in

[1] C. E. Meloan and R. W. Kiser, *Problems and Experiments in Instrumental Analysis*, Columbus, Ohio: Merrill, 1963, p. 171.

Figure 32–2. Coulometric apparatus: R_1, 10,000-ohm variable resistor; R_2, 4700-ohm fixed resistor; R_3, 100.0-ohm wire-wound precision resistor, accurate to $\pm 0.05\%$, for measuring the iR drop by potentiometer.

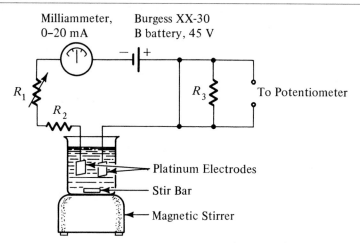

10 mL of water, and swirl to dissolve the sample. When it is completely dissolved, add about 50 mL of water and 1.9 mL of concentrated hydrochloric acid. Transfer the solution quantitatively to a 500-mL volumetric flask and dilute it to volume.

3. Prepare a 1M solution of potassium iodide.

4. Prepare a starch solution, or the solid starch-urea complex, as directed in Experiment 18.

Procedure

1. Warm up the coulometric generator briefly. If a commercial instrument is employed, set the current on 0.1 μeq/sec. If the other apparatus is used, adjust the variable resistor R_1 (Figure 32–2) to obtain a current of 9–10 mA. Calculate the exact current from Ohm's law, $E = iR$, by measuring the voltage across the 100.0-ohm precision resistor, R_3, with a potentiometer.

2. To a 150- or 250-mL beaker, add about 50 mL of water, about 3.5 g of reagent-grade sodium bicarbonate, about 5 mL of 1M potassium iodide, and about 5 mL of starch solution or a small scoop of solid starch-urea.

3. Immerse the two platinum electrodes in the solution and start the magnetic stirrer. Turn on the electrolysis current and simultaneously start the timer. Titrate until the very first darkening of the solution (starch-triiodide color) is observed. The time required for this constitutes a blank.

4. Reset the timer on zero. Pipet exactly 5 mL of the arsenic(III) solution into the blank that has just been titrated. Turn on the electrolysis current and simultaneously start the timer. Titrate until the very first darkening of the solution is observed.

5. Repeat the blank and sample titrations. Compute the milligrams of As_2O_3 in the sample aliquot taken, using an equivalent weight of $197.8/4$ for As_2O_3.

Experiment 31. Amperometric Titration of Mercaptans

The principles of amperometric titration are discussed in Chapter 15. In this experiment, $0.01M$ silver nitrate will be used to nitrate a dilute alcoholic solution of an organic mercaptan RSH to an amperometric end point.

Chemical Background

Mercaptans can be considered monosubstituted hydrogen sulfide. Thus, one acidic hydrogen is replaced by silver(I) to form an insoluble silver mercaptide in ammonia solution:

$$Ag(NH_3)_2^+ + RSH \rightleftharpoons AgSR(s) + NH_4^+ + NH_3$$
$$\text{(AgNO}_3\text{)}$$

Silver mercaptides, like silver sulfide, are far more insoluble than silver chloride or silver bromide. The ammonia will complex the silver(I) enough to keep it from precipitating silver chloride or silver bromide, but not enough to interfere with the quantitative precipitation of the silver mercaptide. Ammonia also neutralizes the proton released in the reaction.

The end point is signalled by the presence of a diffusion current and is easily found by graphical means. There is no diffusion current before the end point, because mercaptans, being in a reduced state, are not reduced at the platinum electrode used. After the end point, there is an excess of silver(I) ion, which is readily reduced at the platinum electrode. The voltage for this reduction is supplied by the reference electrode.

Apparatus and Reagents

1. Use either the apparatus of Kolthoff and Harris [2] or the apparatus of Grimes et al. [3]. (The apparatus of Kolthoff and Harris is shown in Figure 32–3.) Use a DC microammeter A with a sensitivity of at least $0.2 \mu A$ per scale division. Attach a rotating platinum wire electrode P 6–8 mm long and 0.5 mm wide (i.d.) to a stirring motor that will maintain a constant speed of about 1000 rpm. The instructor will have assembled the reference electrode R, the salt bridge S, and the rest of the apparatus (Note 1).

2. Prepare 100 mL of $0.010M$ silver nitrate solution, by exact weight, in a volumetric flask.

3. Prepare the supporting electrolyte solution by dissolving 5 g of reagent-grade ammonium nitrate in 25 mL of 1:1 ammonium hydroxide.

[2] I. M. Kolthoff and W. E. Harris, *Ind. Eng. Chem., Anal. Ed., 18*, 161 (1946).
[3] M. D. Grimes, J. E. Puckett, B. J. Newby, and B. J. Heinrich, *Anal. Chem., 27*, 152 (1955).

Figure 32–3. Amperometric titration apparatus. Shown are reference electrode (R), salt bridge (S), microammeter (A), mercury metal (Hg), glass tubing (T) with sintered-glass disk at the bottom, outer tube (E) containing potassium chloride electrolyte, and rotating platinum electrode (P).

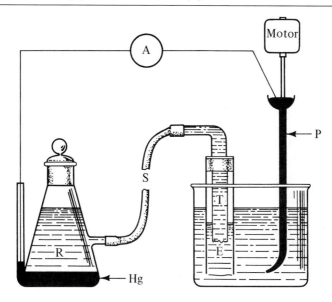

4. Prepare a solution of the unknown by weighing 0.2–0.3 mmoles of mercaptan (6–10 mg of mercaptan sulfur) into a 100-mL volumetric flask and diluting the solution to the mark with 95% ethyl alcohol.

Procedure

1. Pipet 25 mL of the unknown solution into a 250-mL beaker. Add about 75 mL of 95% ethyl alcohol and about 5 mL of the supporting electrolyte solution. This makes the solution about $0.1M$ in ammonium nitrate and $0.35M$ in ammonium hydroxide.

2. Fill a 10-mL buret with $0.01M$ silver nitrate.

3. Immerse the tips of the rotating platinum electrode and reference electrode at least 1 inch in the solution. Set the speed of the platinum electrode at about 1000 rpm. Close the electric circuit and observe the current on the microammeter. It may rise to 20 μA, but it will decrease to essentially zero after a few minutes (Note 2).

4. Add two to four 1-mL increments of $0.01M$ silver nitrate to the beaker and record the diffusion current after each addition. (The current should be almost zero unless the end point has been passed.) When the current increases, add 0.5-mL increments of titrant until at least five more points have been recorded.

5. Plot the microammeter readings in microamperes versus milliliters of $0.01M$ silver nitrate added. Determine the end point graphically and the volume of silver

nitrate corresponding to it. Calculate the percentage purity of the mercaptan or the percentage sulfur (atomic wt = 32.06).

Notes

1. The reference electrode R in Figure 32–3 has a potential of -0.23 V versus the saturated calomel electrode. Prepare its electrolyte by adding 1.3 g of mercuric iodide and 4.2 g of potassium iodide to 100 mL of saturated potassium chloride solution. Add a layer of mercury to cover the bottom of the flask and fill about 1 inch of the vertical side arm. To make electric contact with microammeter A, dip a wire into the mercury in the side arm. Obtain a glass tube T with a sintered-glass disk at the bottom and fill it with a gel of 3% agar and 30% potassium chloride. To protect the tube further, insert it into the outer tube E, which is filled with saturated potassium chloride and covered with an agar or sintered-glass plug. Fill a Tygon or rubber tube 2 feet long and 6 mm wide (i.d.) with saturated potassium chloride solution for a salt bridge S. Eliminate air bubbles and attach the horizontal side arm of the flask to the glass tube T.

2. Large currents (~ 20 μA) may be observed when a new or freshly cleaned platinum electrode is immersed in an ammoniacal silver solution. The platinum electrode should, however, be cleaned from time to time with concentrated nitric acid. It may also be necessary to wipe the platinum electrode after passing the end point, to remove silver mercaptide.

QUESTIONS

1. Name at least two different chemical methods for determining mercaptans.
2. Why will ammonium hydroxide prevent the interference of bromide and chloride ions, but not that of iodide ions?

SUGGESTIONS FOR ADDITIONAL EXPERIMENTS

The following articles are suggested as material for additional work or research projects for advanced students:

D. H. Evans, *J. Chem. Ed.*, *45*, 88 (1968). *Coulometric titration of cyclohexene with bromine using an amperometric end point with polarized electrodes.*

F. J. Feldman, *J. Chem. Ed.*, *43*, 378 (1966). *Simple experiments in amperometry.*

D. Jacques, *J. Chem. Ed.*, *42*, 429 (1965). *Potentiometric titration of mixed halide ions with silver(I).*

C. E. Meloan and R. W. Kiser, *Problems and Experiments in Instrumental Analysis*, Columbus, Ohio: Merrill Books, 1963.

33

Separation Procedures

If a chemical species cannot be determined in the presence of other constituents, then the desired species must be separated from the interfering species. Some of the common separation methods are precipitation, distillation, ion exchange, extraction, chromatographic separation, and electrodeposition.

A separation procedure must be carefully studied to find conditions for the quantitative transfer or removal of the desired species. *Experiment 32* illustrates how one systematically varies conditions to find a method for extracting zinc quantitatively from aqueous solution.

The gas chromatographic determination in *Experiment 33* illustrates the separation of simple organic compounds using an internal standard technique. The gas chromatograph is, of course, both a means of separation and an instrument for quantitative analysis.

Experiments 34 and 35 illustrate the burgeoning techniques of paper chromatography and thin-layer chromatography.

Complex mixtures of similar ionic species can also be separated by using ion-exchange resin. *Experiment 36* illustrates the separation of iron, cobalt, and nickel by taking advantage of differences in the stability of their respective chloro complexes.

Some of the experiments in other chapters also illustrate separation methods. The gravimetric determinations in Chapter 28 are basically separation methods; Experiment 29 might be thought of as the separation of copper by electrodeposition.

Experiment 32.   Solvent Extraction of Zinc

In a typical solvent extraction of metal ions, one of the ions is extracted from aqueous solution into an organic solvent. Conditions must be found in which the one ion is quantitatively transferred to the organic phase while, at the same time, all but traces of the other ions remain in the aqueous phase.

Chemical Background

A useful way of characterizing the degree of transfer of a metal ion is by its distribution ratio (Chapter 20), defined as

$$D_c = \frac{\text{concn. solute}_o}{\text{concn. solute}_w}$$

In this experiment, we will determine D_c when zinc(II) is extracted from aqueous hydrochloric acid into an organic solvent mixture of benzene and tributyl phosphate, TBP (Note 1). In the $4-10M$ hydrochloric acid used, the principal equilibrium species of zinc(II) are the trichlorozincate(II) and the tetra-chlorozincate(II) ions. Since zinc(II) is probably extracted in one of these forms (Equation E33–1), the hydrochloric acid concentration in the aqueous phase is critical. D_c will be determined for various concentrations between 4 and $10M$ acid.

$$\underset{\text{(aqueous phase)}}{ZnCl_4^{2-} + 2H^+ + 2Solv} \;\rightleftharpoons\; \underset{\text{(organic phase)}}{H_2S_2ZnCl_4} \tag{E32–1}$$

The benzene-TBP mixture interferes with the subsequent EDTA titration of zinc; to prevent this, zinc is stripped (back-extracted) from the organic phase with pure water. This back-extraction is quantitative because water favors the complete dissociation of the anionic zinc complexes into cationic forms of zinc(II), which are insoluble in the organic phase.

Reagents

1. Prepare 50 mL each of 4, 6, 7, 8.5, and $10M$ hydrochloric acid from concentrated ($12M$) hydrochloric acid.
2. Solutions of $35-50\%$ v/v TBP in benzene will be available. Use the solution assigned by the instructor.
3. If a $0.01M$ standard EDTA solution is not available, make a standard calcium(II) solution and standardize EDTA against it according to the directions of Experiment 12. (However, do not add magnesium chloride to the EDTA.)
4. A $0.5M$ standard zinc chloride solution will be available in the laboratory. If it is not standardized, standardize a 5- or 10-mL aliquot with $0.05M$ EDTA, as directed in Experiment 13.
5. Prepare a pH 10 buffer by dissolving 6.75 g of ammonium chloride and 57 mL of concentrated ammonium hydroxide in 100 mL of distilled water.
6. Prepare a 1% solution of Naphthyl Azoxine S indicator (NAS) in dimethylformamide (Note 2). Alternatively, prepare a 1% solution in dilute ammonia.

Procedure

1. Add 25 mL of $4M$ hydrochloric acid and 25 mL of the assigned TBP-benzene

solution to a 125-mL separatory funnel. Stopper the funnel, grasp it by stopper and stopcock, and shake it vigorously for a few seconds. Allow the layers to separate, then draw off and discard the lower aqueous layer. (Note 3).

2. Using a measuring pipet, measure 2 mL of $0.5M$ zinc chloride and about 23 mL of $4M$ hydrochloric acid into the separatory funnel (Note 4). Stopper the funnel, grasp as in Step 1, and shake vigorously for 1 minute. Allow the layers to separate and become clear. Carefully draw off the lower aqueous layer and discard it.

3. The TBP-benzene layer now contains much of the zinc(II) ion. Add 20 mL of distilled water to this layer and shake as before to back-extract zinc from the TBP-benzene phase. Place a 100-mL volumetric flask beneath the stem of the funnel and carefully drain the lower aqueous phase into the flask. Repeat two more times with fresh 20-mL portions of water, and drain both portions into the flask. Dilute to the mark with distilled water.

4. Label the flask No. 1. The zinc(II) content will be determined later.

5. Repeat Steps 1–3 with 6, 7, 8.5, and $10M$ hydrochloric acid. Label the 100-mL volumetric flasks containing the back extracts Nos. 2, 3, 4, and 5, respectively.

6. Pipet 25 mL from flask No. 1 into a 250-mL flask and dilute it to about 100 mL with distilled water. Buffer the pH at 6.5 by adding pyridine [check the pH with a pH meter (Note 5)]. Add enough NAS to give the solution a yellow color. Titrate with $0.01M$ EDTA to a color change of yellow to clear red. Repeat the procedure for flasks 2–5.

7. Calculate the millimoles of zinc(II) in the organic phase, remembering that only one-fourth of the zinc was titrated. Calculate, by difference, the millimoles of zinc(II) remaining in the aqueous hydrochloric acid layer, then calculate the distribution ratio D_c for each extraction. Plot D_c against the molarity of the hydrochloric acid (Note 4) on a piece of graph paper, and connect the points to form a smooth curve.

8. Calculate the percent extracted at the maximum of the extraction curve. Also calculate the total percent extracted if two identical extractions were made at the maximum of the extraction curve.

Notes

1. Tributyl phosphate, symbolized as $(BuO)_3PO$, is an organic ester of phosphoric acid.

2. Naphthyl Azoxine S is commercially available from the Eastman Kodak Company as No. 8643, 8-hydroxy-7-(6-sulfo-2-naphthylazo)-5-quinolinesulfonic acid, disodium salt. As an alternative indicator, use 50-mg of Calmagite ground into 10 g of potassium sulfate.

3. The purpose of this step is to saturate the TBP-benzene with hydrochloric acid to prevent the volume from changing during the extraction.

4. The aqueous zinc chloride solution dilutes the hydrochloric acid. Calculate the final concentration and use this value in plotting the graph.

5. If Calmagite is to be used, add about 3 mL of the pH 10 buffer instead of

pyridine. Add solid Calmagite to the solution until it is wine-red in color, then titrate it to a clear blue end point.

QUESTIONS

1. Sodium chloride also favors the formation of the $ZnCl_4^{2-}$ ion, but the extraction is not as efficient as when hydrochloric acid is used. Explain.
2. The densities of TBP and water are fairly close. Explain one purpose of benzene in the TBP-benzene mixture.
3. Review Chapter 20. Would the method used in the experiment be suitable for separating zinc(II) from iron(III)?

Experiment 33   Quantitative Gas Chromatographic Analysis of a Multicomponent Mixture

In this experiment, the gas chromatograph is used both for separation and quantitative analysis. The theory and practice of gas chromatography have already been discussed in Chapter 22. A schematic diagram of a chromatograph like that used in this experiment is shown in Figure 22–1; note that it uses a thermal-conductivity detector cell. A typical recorder output for a separated mixture is shown in Figure 22–2.

Chromatographic recordings are obtained by measuring the detector output, resulting in a series of *peaks*. The peak location, or *retention time*, identifies the component passing through the detector. If the signals produced by the components were all equal, then the area under each peak would be directly proportional to the mole percent concentration of the particular component, since thermal conductivity is directly proportional to the number of molecules present. The detector signal, however, is proportional to the *difference* between the thermal conductivity of the carrier gas and that of the component passing through the detector. When helium is used as carrier gas, the difference in thermal conductivity of the possible components yields signals that differ at most by 1–2%. This can be neglected, in most cases.

Internal Standard Technique

In this technique, a known amount of a selected compound (internal standard) is added to the sample mixture. Calibration curves are then established using the ratio of the peak areas of various concentrations of components to the area of the internal standard. The area can be measured by cutting out the peak and weighing it on the analytical balance, by using a planimeter, or by using the relation that the peak height times one-half the bandwidth is approximately equal to the area. The calibration curves should yield linear plots. (The concentration of one component can simply be calculated from the ratio of its area to that of a standard.)

The choice of internal standard depends on the nature of the components to be determined and also on the concentration range in which the components are present. The internal standard peak should be located close to those of the other components and its concentration should be approximately the same. In this manner, results can be satisfactorily reproduced even if operating conditions (such as flow rate and temperature) have varied slightly from run to run.

Apparatus and Reagents

1. A gas chromatograph with a thermal conductivity detector will be used for this experiment. Please become familiar with the operating instructions for the particular instrument available, including the directions for the power supply control unit and the recorder. Be sure that the column is connected to a cylinder of helium carrier gas through a pressure regulator and that the appropriate recorder paper is available for the recorder. (A good chromatographic column for this experiment uses 15% Carbowax stationary phase on a Chromosorb P solid support. The column is operated at approximately 175°C, giving retention times as long as 12 minutes.)

2. Obtain a watch with a digital readout to measure the rate at which paper comes off the recorder.

3. The diethyl ether solvent used should be anhydrous and should meet ACS specifications. Also obtains ACS grade benzene, cyclohexane, toluene, and ethylbenzene.

Preparation of Calibration Curve Solutions

1. Unknowns will contain varying amounts of benzene, cyclohexane, and ethyl-benzene dissolved in diethyl ether. They will not contain toluene, the internal standard, which must be added in Step 2. Bring a clean, *dry* 50-mL volumetric flask to the instructor to obtain a measured amount (such as 25 mL) of your unknown.

2. Using a *dry* pipet, add exactly 5.0 mL of toluene (internal standard) to your unknown. Dilute to the mark with diethyl ether, using an eyedropper for the last few drops.

3. Prepare the following calibration solutions in 50-mL volumetric flasks, and mix thoroughly before use.

Solution No.	Benzene	Cyclohexane	Ethylbenzene	Toluene	Diethyl Ether
1	5 mL	4 mL	3 mL	5 mL	33 mL
2	3	3	6	5	33
3	6	2	4	5	33
4	4	1	7	5	33
5	0	1	0	5	44

Gas Chromatographic Analysis

1. Turn on the chromatograph according to instructions with the instrument. Allow the proper warm-up time for the recorder, etc.

2. Using a watch with a digital readout, determine the rate, in divisions per minute, at which paper comes off the recorder. (Then, if directed, produce a chromatogram of solution no. 5 to determine the retention time of cyclohexane and to help identify one peak.)

3. Produce chromatograms of the calibration solutions prepared, labeling each chromatogram.

a. Inject a sample of solution no. 1 (see Note). When all of the components have passed through the column, you will probably find that some peaks are either off scale or too small. Adjust the sample size and/or attenuator setting so that the tallest peak is about 75% of full scale (except that of diethyl ether, which will go off scale). Clean the syringe with ether.

b. Use exactly the same sample size, attenuator setting, and other conditions for solutions 2–4 and the unknowns.

4. Now obtain from the instructor a solution that is equimolar in all components (except for the diethyl ether solvent). Produce a chromatogram of this solution, using the same conditions as in Step 3.

5. Check to be sure that you have ten good chromatograms: retention-time determinations (4), calibration curve solutions (4), your unknown (1), and the equimolar solution (1).

6. Turn the chromatograph off carefully according to instructions with the instrument.

7. Dispose of the calibration curve solutions and the unknown solution in a container kept in a hood.

Calculations and Report

1. From the chromatographic data obtained, determine the retention times of each component.

2. Determine the area under the peaks for each chromatogram (except those for retention time determination) by any method you choose.

3. Calculate the ratio between the area of the component and the area of the internal standard. For each component, plot this ratio against milliliters of component/50 mL (or total mmoles/50 mL). From these calibration curves, determine the concentration of each component in your unknown. Report your answers as total mmoles/50 mL.

4. From the ratios determined for the equimolar mixture, decide whether the thermal conductivities of these compounds are the same.

Note

If the 15% Carbowax on Chromosorb P column mentioned under the apparatus section is used, a reasonable sample size is 10 μL.

QUESTIONS

1. What are the advantages of using the internal standard technique?
2. Why did you choose your method for area determination?
3. Could you have predicted the exit order of the components without determining their retention times? What factors determine this?
4. Are the thermal conductivities of all components the same? Is this result what one would expect from the detector? Why? If a flame ionization detector had been used, would the conductivities have been the same? Why?
5. Do your chromatograms exhibit "tailing"? What causes it, and what conditions could be changed to prevent it?

Experiment 34. Separation of Metal Ions by Paper Chromatography

Paper chromatography is an important method for separating small amounts of metal ions and organic substances from each other. Its theory was discussed in Chapter 23. Two techniques will be illustrated in this experiment: (1) the separation of several metal ions using an organic solvent to elute (move) the metal ions selectively and (2) "reversed-phase" paper chromatography, in which the paper is first impregnated with a reagent that selectively complexes metal ions. The complexed metal ions are held by the fixed reagent and therefore do not move much; ions that are not complexed are free to move with the solvent front, which in this case is water.

Reagents

1. Prepare 0.1M solutions of bismuth nitrate, cobalt chloride or nitrate, copper chloride or nitrate, ferric chloride, and nickel chloride or nitrate. For experiments A and B, mix each solution with an equal portion of concentrated hydrochloric acid.
2. Prepare a DTC (dithiocarbamate) solution by dissolving 1 g of sodium diethyl-dithiocarbamate in 100 mL of water. This solution should be prepared freshly every few days.
3. For each day's work, prepare solutions of 5% (v/v) of concentrated hydrochloric acid in acetone and 1% (v/v) of concentrated hydrochloric acid in ethyl acetate. Shake the latter solution to mix it, and discard or ignore any trace of a lower aqueous phase which may remain.
4. Prepare a 10% (v/v) solution of IOTG (isooctylthioglycolate) in methyl alcohol. (IOTG may be obtained from Evans Chemetics, Inc., Waterloo, N.Y.)

General Instructions

1. Use pieces of chromatographic paper (such as Whatman No. 1) approximately $2\frac{1}{2} \times 3$ inches in size, cut square at the corners. Draw a straight pencil line across the paper approximately $\frac{1}{4}$ inch from the bottom. Using a small wooden applicator stick or a small glass stirring rod, apply a small spot of each sample solution (about $\frac{1}{16}$ to $\frac{3}{16}$ inch in diameter) along the pencil line.

2. Add 5 mL of the eluting solvent to a 250-mL beaker. Carefully insert the chromatography paper. (The liquid should come part way up to the spots but should not touch them immediately.) Place a watch glass over the beaker and allow the liquid to come up the paper until it is about $\frac{1}{4}$ inch from the top. Withdraw the paper and immediately mark the solvent front with a pencil.

3. Dry the paper over a hot plate, or let it air dry. Circle any visible colored spots with a pencil. Detect the remaining spots by spraying the paper with aqueous DTC solution or sprinkling DTC solution from a medicine dropper. Note the color of each spot and circle it with a pencil.

Experiment A

1. Apply a spot of each of the following metal ions along the pencil line at the lower edge of the paper: bismuth(III), cobalt(II), copper(II), iron(III), and nickel(II) (Note 1). Elute the paper with 5% hydrochloric acid in acetone in a covered 250-mL beaker.

2. Dry the paper and note the color and location of any visible spots. Detect the remaining spots using the DTC solution and circle the spots. Calculate and record the R_f value of each spot (Note 2).

Experiment B

Spot the same five elements and follow the procedure of Experiment A, but elute the paper with 1% hydrochloric acid in ethyl acetate instead of hydrochloric acid in acetone. Calculate and record the R_f values as before.

Experiment C

1. In this experiment, metal ions will be separated by using paper impregnated with an organic complexing agent, IOTG. First impregnate the paper by placing it in a 250-mL beaker containing 5 mL of IOTG in methyl alcohol (10% solution). Cover the beaker with a watch glass and allow the liquid to saturate the paper completely. Place the paper on some clean paper towel and let the methyl alcohol evaporate in a hood. (*CAUTION!* Avoid breathing methyl alcohol fumes.) The methyl alcohol will evaporate in a few minutes, leaving the nonvolatile IOTG in the paper.

2. Spot the dry IOTG paper with each of the five metal-ion solutions (Note 3). Elute the paper with 5 mL of $0.1M$ nitric acid in a covered, 250-mL beaker.

3. Dry the paper and circle the colored spots. Identify the other spot by spraying DTC solution. Calculate and record the R_f value of each spot.

Experiment D

You should now be able to separate a mixture of any two metal ions by a single paper-chromatographic method. Indicate clearly and concisely in your notebook how this might be done. Your instructor will now assign a mixture of two metal ions to separate in the lab. Prepare and separate such a mixture under the conditions you think best. Hand in the final, *dry* paper chromatogram with the separation conditions indicated and with the separated metal ions spotted and identified.

Notes

1. The spots should not be allowed to dry. Proceed with the separation without undue delay.
2. DTC reagent is very sensitive for some ions; circle only the most intense area of each spot and ignore any slight color fringes. Use the center of a spot to calculate the R_f value.
3. In this case, use metal solutions directly; do not dilute them with concentrated hydrochloric acid.

Experiment 35. Thin-Layer Chromatographic Separation of Nitroanilines on Fluorescent Sheets

Before discussing the aims of this experiment, we will first describe the ultraviolet lamp and fluorescent TLC sheets used and review the method used to locate separated compounds on the sheet.

Ultraviolet Lamps

To observe the fluorescence in this experiment, either an ultraviolet hand lamp or a view box containing an ultraviolet lamp can be used. A commercial view box is recommended because it is dark enough inside to make darkening the room unnecessary. The hand lamp and view-box lamp contain a mercury-arc ultraviolet source. Such lamps are available as either short-wave (SW) ultraviolet lamps, long-wave (LW) ultraviolet lamps, or *combination* ultraviolet lamps, which emit both short-wave and long-wave ultraviolet radiation. The short-wave lamp emits mainly a narrow *line* of ultraviolet radiation at 254 nm. The phosphor-coated tube in the long-wave lamp emits a more intense *band* from 313 to 400 nm, with most of the radiation centered at 345–365 nm, including an intense line at 365 nm.

Fluorescent TLC Sheets

As was mentioned in Chapter 23, plastic sheets precoated with adsorbent are now commercially available. Sheets are also available in which the adsorbent is mixed with an *inorganic phosphor*. Under ultraviolet radiation, the phosphor emits intense visible radiation over the entire length of the TLC chromatogram except where there are spots of separated organic compound. (This visible radiation is usually termed fluorescence, although it could be either phosphorescence or fluorescence, depending on the phosphor.) Most organic compounds *quench* the fluorescence of the phosphor; the spots appear as dark shadows against the brillant fluorescent background of unquenched phosphor.

The most common fluorescent color used on TLC sheets is green (522 nm). Typical green phosphors are pure zinc silicate or calcium silicate with a manganese-lead activator. Most such phosphors emit only under short wave (254 nm) ultraviolet excitation; therefore, the TLC sheet can also be examined under long-wave excitation for fluorescent organic compounds.

Separation of p-Nitroanilines

An excellent example of this procedure is the separation of a mixture of *o*-nitroaniline, *m*-nitroaniline, and *p*-nitroaniline. These isomers are readily separated on a *polyamide* adsorbent. In the experimental directions given below, the precoated TLC sheets used emit green light under short-wave ultraviolet excitation. The nitro group is an excellent quencher of luminescence, and each isomer is readily located after separation by a dark shadow against the brilliant green background of the phosphor emission.

Reagents

1. Obtain one or more 5 × 20 cm sheets of Baker-flex® polyamide 6-F TLC precoated sheets. These sheets contain a fluorescent indicator that is green under 254-nm ultraviolet radiation. (Do *not* activate the sheets by heating as is usually done with other TLC sheets.)
2. Obtain a beaker or jar deep enough to accommodate a 13–14 cm TLC sheet. (An 800- or 1000-mL beaker will do.) Also obtain enough plastic (sandwich) wrap and aluminum foil to cover the beaker.
3. Prepare about 70 mL of *developing solvent*, 90% carbon tetrachloride–10% glacial acetic acid by volume.

Procedure

1. Add the developing solvent to the beaker or jar until a 1-cm layer of solvent exists *around the walls*. (This will require roughly 50 mL of solvent.)
2. Cover the beaker or jar tightly—first with plastic wrap, then with aluminum foil. Let it stand at room temperature for 30–60 minutes before use to saturate the atmosphere inside the beaker with solvent.
3. Cut the 20-cm TLC sheet to 13–14 cm in length. Draw a light line at least 2 cm from one end with a pencil. Make four marks, evenly spaced, across the line as a guide for placing "spots" of your unknown solution and pure solutions of each of the three nitroanilines on the TLC sheet.
4. Obtain $0.1M$ solutions of the individual nitroanilines and a $0.03M$ solution of an unknown mixture. Using a micropipet, measure out 2-μL portions of the unknown and the three pure solutions. Use the capillary action of the pipet to draw up the solution, and let the capillary action of the TLC sheet draw it out. Apply the tip of the pipet several times rather than allowing all the solutions to flow out at once. Attempt to keep the spots small and evenly distributed.
5. Dry the spots 15–30 minutes in the desiccator. Place in the developing beaker, replace the cover, and allow it to develop to a 10-cm height (about 45 minutes).
6. Remove the sheet from the beaker, allow it to air dry, and examine it under

short-wave ultraviolet radiation. The nitroanilines should appear as shadows because they quench the green fluorescence.

7. Mark the *center* of each compound spot, both the single spots and the unknown mixture. Also mark the solvent front, if necessary.

8. Calculate the R_f value for each nitroaniline, using the definition

$$R_f = \frac{\text{distance moved, cm}}{\text{distance moved by solvent front, cm}}$$

(As a guide, the R_f value of *o*-nitroaniline should be about 0.4–0.5.) Report which nitroanilines are in your unknown mixture.

Experiment 36.   Anion-Exchange Separation of Iron, Cobalt, and Nickel

In this experiment, iron(III), cobalt(II), and nickel(II) are dissolved in $9M$ hydrochloric acid and separated on an anion-exchange column. Subsequently, the cobalt(II) and nickel(II) ions are titrated with EDTA. The separation depends on the selective formation of anionic complexes of cobalt(II) and iron(III).

Chemical Background

In $9M$ hydrochloric acid, the principal equilibrium forms of the three metal ions are $FeCl_4{}^-$, $CoCl_4{}^{2-}$, and Ni^{2+} (or $NiCl^+$). The two former ions are exchanged onto the resin (Equations 36–1 and 36–2) and all of the nickel(II) passes through the resin.

$$\text{Res-NR}_3{}^+\text{Cl}^- + \text{FeCl}_4{}^- \rightleftharpoons \text{Res-NR}_3{}^+\text{FeCl}_4{}^- + \text{Cl}^- \qquad \textbf{(E36–1)}$$

$$2\text{Res-NR}_3{}^+\text{Cl}^- + \text{CoCl}_4{}^{2-} \rightleftharpoons (\text{Res-NR}_3{}^+)_2\text{CoCl}_4{}^{2-} + 2\text{Cl}^- \qquad \textbf{(E36–2)}$$

In $4M$ hydrochloric acid, the tetrachlorocobaltate(II) ion breaks down to cationic forms of cobalt(II) which pass through the resin. The tetrachloroferrate(III) ion is more stable at this concentration and remains on the resin, although it moves down the column because of the mass-action effect of the acid. In $0.5M$ hydrochloric acid, this ion breaks down to cationic forms of iron(III), which pass through the resin. The $0.5M$ acid is used instead of water to keep iron(III) hydroxide from precipitating.

The effluents containing cobalt and nickel are then fumed down, to eliminate most of the hydrochloric acid. This limits the amount of pyridine needed to neutralize the excess acid (Equation E36–3). The pH is adjusted so that essentially only pyridine hydrochloride is present. EDTA is then used to titrate the separated cobalt and nickel solutions.

$$C_5H_5N + H^+ \rightleftharpoons C_5H_5NH^+ \qquad \textbf{(E36–3)}$$

Reagents

1. Prepare 9, 4, and $0.5M$ hydrochloric acid by diluting concentrated ($12M$) hydrochloric acid.

2. Prepare 0.01M EDTA by dissolving 1.9 g of the reagent disodium salt (Na$_2$H$_2$Y · 2H$_2$O, mol wt 372) in 500 mL of distilled water. Add 5 or 6 pellets of sodium hydroxide to remove any turbidity. Standardize the EDTA against a solution of standard 0.01M calcium, as directed in Experiment 12.

3. Prepare a 0.005M copper(II) solution by adding 150 mg of copper nitrate, Cu(NO$_3$)$_2$ · 6H$_2$O, to 100 mL of distilled water.

4. Prepare a 1% solution of NAS indicator in water (Note 1). Add enough ammonia to dissolve all of the indicator.

Procedure

1. Obtain a plastic column about 1.6 cm in diameter and 15 cm long. Place a slug of cotton or glass wool at the bottom to retain the ion-exchange resin. Slurry Dowex 1-X8 (100–200 mesh) anion-exchange resin in the chloride form in 9M hydrochloric acid, and add it to the column until the resin column is 3 inches high. Retain about 2 inches of liquid above the resin (never allow the liquid level to drop below the resin level).

2. Add about 10 mL of 9M hydrochloric acid in two portions to the column, the liquid flowing at 2–3 mL/min from the column (Note 2). Drain until about 1 inch of acid is left above the resin.

3. Pipet exactly 2 mL of the unknown solution (Note 3) into the column. Place a 250-mL beaker under the column. Using a flow rate of 2–3 mL/min, elute the nickel from the column with 75 mL of 9M hydrochloric acid added in 15-mL portions. Note that the blue cobalt band moves down, but not off, the resin during this operation.

4. Interrupt the flow of liquid. Remove the beaker and replace it with another 250-mL beaker to collect the cobalt. Using a flow rate of 2 mL/min, elute the cobalt with 50 mL of 4M hydrochloric acid added in 5 portions. Because the tetrachlorocobaltate(II) ion is breaking down, the intensity of the blue color will decrease somewhat.

5. Interrupt the flow of liquid. Place the beakers containing cobalt and nickel on a hot plate in a hood, and evaporate the solutions to near dryness (5 mL or less). Place a glass stirring rod in each beaker to prevent bumping.

6. During the evaporation, elute the iron with 100 mL of 0.5M hydrochloric acid added in 10 portions. Discard the eluate, because the iron content will not be determined. Rinse the resin with water and return it to the instructor.

7. Dilute the cooled samples of nickel and cobalt to 100 mL with distilled water. Adjust the pH to 5.5–6.0 with pyridine (Note 4), using a pH meter.

8. Add 2–3 drops of NAS indicator to the beaker containing cobalt and titrate with 0.01M EDTA to a color change from yellow (or yellow-orange) to red (Note 5).

9. Add 2–3 drops of NAS indicator to the beaker containing nickel and titrate with 0.01M EDTA to a color change from yellow to red. Add 0.005M copper(II) dropwise (counting the drops) until the yellow color returns. Titrate again with EDTA to the red color. Note the total volume of EDTA and also the volume of

EDTA needed to bring about the color change. Repeat the process (using 1 drop of copper) until the volume of EDTA required to restore the red color becomes *constant* (Note 6). This volume of EDTA will be equivalent to 1 drop of copper solution. To calculate the volume of EDTA equivalent to the nickel, multiply the volume of EDTA equivalent to 1 drop by the number of drops of copper solution added, and subtract this volume of EDTA from the total volume added.

10. Calculate the milligrams of metal ions per milliliter of unknown solution.

Notes

1. Naphthyl Azoxine S may be synthesized by the procedure of Fritz et al. [1].
   It is commercially available from the Eastman Kodak Company as No. 8643, 8-hydroxy-7-(6-sulfo-2-naphthylazo)-5-quinolinesulfonic acid, disodium salt.

2. The anion-exchange resin will shrink somewhat and become darker when fully saturated with $9M$ hydrochloric acid, but it does not decompose. It will return to normal when rinsed with distilled water.

3. The unknown is $9M$ hydrochloric acid solution, about $0.05-0.1M$ in cobalt(II) and nickel(II) and $0.05M$ in iron(III).

4. Dispense pyridine carefully; it is volatile, and injurious to health if inhaled for too long.

5. If not sufficiently dilute, the color of the Co-EDTA complex may decrease the color contrast of the NAS end point. In this case, the entire cobalt fraction should be diluted to volume in a volumetric flask and two or three aliquots taken for back-titration with EDTA as in Step 9.

6. This procedure is necessary because EDTA reacts so slowly with nickel(II). An alternative procedure is to add a measured excess of EDTA and back-titrate the unreacted EDTA with $0.005M$ copper(II) from a buret.

QUESTIONS

1. Explain why nickel or cobalt cannot be titrated with EDTA in the presence of iron.
2. Write chemical equations to represent the reactions in Step 9 of the Procedure.
3. Compare the reaction rates of copper(II) and nickel(II) with EDTA.
4. Why is copper(II) added in Step 9 of the Procedure?

SUGGESTIONS FOR ADDITIONAL EXPERIMENTS

The following articles are suggested as material for additional work or as research projects for advanced students:

S. Dal Nogare and L. W. Safranski, *J. Chem. Ed. 35*, 14 (1958). *Paper chromatographic separation of urea, biuret, guanyl urea, guanidine, etc.*

[1] J. S. Fritz, W. J. Lane, and A. S. Bystroff, *Anal. Chem., 29,* 821 (1957).

Eastman Kodak Co. (Rochester, N.Y.) "Analytical procedures for separations by thin-layer chromatography using the Eastman Chromagram System." *Typical examples of available experiments include separation of water-soluble vitamins, separation of some common dyes, separation of some diprotic acids, and separation of some ingredients from analgesic preparations.*

E. J. Goller, *J. Chem. Ed. 42*, 442 (1965). *Cation analysis (qualitative) using thin-layer chromatography.*

A. S. Ritchie, *J. Chem. Ed. 38*, 400 (1961). *A paper-chromatographic scheme for the identification of metallic ions.*

D. K. Sebera, *J. Chem. Ed. 40*, 476 (1963). *Preparation and analysis of a complex compound.*

H. F. Walton, *J. Chem. Ed. 42*, 477 (1965). *Experiments in inorganic paper chromatography.*

1

Literature of Analytical Chemistry

Analytical chemistry is not a static subject; new procedures and principles of analysis are constantly being developed by researchers. The original source of most new information is research papers and review articles published in chemical journals. Shortly after publication, brief abstracts of original papers are published periodically in *Chemical Abstracts* (published by the American Chemical Society) and in certain foreign abstracts journals such as the British journal *Analytical Abstracts*. Finally, new information becomes incorporated in various books, including general and specialized textbooks on different phases of analytical chemistry, treatises and extensive collections of analytical procedures, and collections of official analytical methods published by various technical societies.

A list of journals and selected books devoted to analytical chemistry follows; this is designed to acquaint the student with the literature available and to provide some suggestions for further reading. (Many excellent and useful books were omitted from this list for lack of space.)

ANALYTICAL CHEMISTRY JOURNALS

The abbreviations used in Chemical Abstracts are given in italics.

American Laboratory
Analusis
Analyst, The
Analytical Biochemistry
Analytical Chemistry
Analytica Chimica Acta
Analytical Letters
Bunseki Kagaku (Japan Analyst)
Fresenius' Zeitschrift für Analytical Chemistry

Journal of *Chromatographic Science*
Journal of *Chromatography*
Journal of *Electroanalytical Chemistry*
Journal of *High Resolution Chromatography*
Microkimica Acta
Talanta
Zhurnal Analiticheskoi Khimii (English translation available)

OFFICIAL METHODS

American Society for Testing and Materials (ASTM), *Chemical Analysis of Metals, Sampling and Analysis of Metal Bearing Ores*, Philadelphia: ASTM, published annually).
W. Horwitz, ed, *Official Methods of Analysis of the Association of Official Analytical Chemists*, 11th ed., Washington: AOAC, 1970.
Standard Methods for Testing Petroleum and Its Products, London: Institute of Petroleum, 1952.
M. J. Taras et al., *Standard Methods for the Examination of Water and Wastewater*, 14th ed., New York: American Public Health Association, 1975.
E. Wichers et al., *Reagent Chemicals*, Washington, D.C.: American Chemical Society, 1955.

TREATISES AND EXTENSIVE COLLECTIONS OF METHODS

N. H. Furman, ed. (Vol. I); F. J. Welcher, ed. (Vol. II), *Standard Methods of Chemical Analysis*, 6th ed., Princeton, N.J.: Van Nostrand, 1963.
I. M. Kolthoff and P. J. Elving, eds., *Treatise on Analytical Chemistry*, New York; Wiley-Interscience. A multivolume work, begun in 1959. New and revised volumes are published every year or two.

ABSTRACTS

Chemical Abstracts
Analytical Abstracts
Fresenius' Zeitschrift für Analytische Chemie

REVIEWS

Although review articles appear frequently in various journals, the annual reviews appearing in *Analytical Chemistry* are unusually comprehensive. In even-numbered years, *Fundamental Reviews* is published with references to virtually all significant research published during the two-year period in some 39 different areas of analytical chemistry. In the odd-numbered years, reviews of new work are published in *Applications*. These include fields such as air pollution, clinical chemistry, fuels, metallurgy, pharmaceuticals, and water analysis.

Critical Reviews in Analytical Chemistry publishes quarterly in-depth reviews on various phases of analytical chemistry.

TEXTBOOKS

Elementary Quantitative Analysis

A large number of texts are available.

Advanced Quantitative Analysis

L. Erdey, *Gravimetric Analysis*, Oxford: Pergamon, 1965.
H. A. Laitinen and W. E. Harris, *Chemical Analysis*, 2nd ed., New York: McGraw-Hill, 1975.
R. L. Pecsok and L. D. Shields, *Modern Methods of Chemical Analysis*, 2nd ed., New York: Wiley, 1976.

Instrumental Analysis

H. H. Bauer, G. D. Christian, and J. E. O'Reilly, eds., *Instrumental Analysis*, Boston: Allyn and Bacon, 1978.
C. K. Mann, T. J. Vickers, and W. M. Gulick, *Instrumental Analysis*, New York: Harper and Row, 1974.
D. A. Skoog and D. M. West, *Principles of Instrumental Analysis*, New York: Holt Rinehart, and Winston, 1971.
H. Strobel, *Chemical Instrumentation*, 2nd ed., Boston: Addison-Wesley, 1973.
H. H. Willard, L. L. Merritt, Jr., J. A. Dean, and F. A. Settle, *Instrumental Methods of Analysis*, 6th ed., Princeton, N.J.: Van Nostrand, 1981.

Books on Special Topics

M. R. F. Ashworth, *Titrimetric Organic Analysis*, Part I, "*Direct Methods*"; Part II, "*Indirect Methods*," New York: Wiley-Interscience, 1965.
R. G. Bates, *Determination of pH: Theory and Practice*, New York: Wiley, 1964.
F. E. Beamish and J. C. Van Loon, *Recent Advances in the Analytical Chemistry of the Noble Metals*, Oxford: Pergamon, 1972.
W. Bertsch, S. Hara, R. E. Kaiser, and A. Zlatkis, *Instrumental HPTLC*, Heidelberg: Hüthig, 1980.
E. Bishop, ed., *Indicators*, Oxford: Pergamon, 1972.
D. F. Boltz and J. A. Howell, *Colorimetric Determination of Nonmetals*, 2nd ed., New York: John Wiley, 1978.
R. A. Durst, ed., *Ion Selective Electrodes*, NBS special publication 314, Washington, D.C.: National Bureau of Standards, 1969.
D. T. Gjerde and J. S. Fritz, *Ion Chromatography*, 2nd ed. Heidelberg: Hüthig, 1986.

R. L. Grob, ed., *Modern Practice of Gas Chromatography*, New York: Wiley-Interscience, 1977.

E. Heftman, *Chromatography*, 2nd ed., New York: Reinhold, 1967.

W. Jennings, *Gas Chromatography with Glass Capillary Columns*, 2nd ed., New York: Academic Press, 1980.

B. L. Karger, *An Introduction to Separation Science*, New York: John Wiley, 1974.

B. L. Karger, R. Snyder, and C. Horvath, *An Introduction to Separation Science*, New York: Wiley, 1974.

L. H. Keith, *Identification and Analysis of Organic Pollutants in Water*, Ann Arbor, Mich.: Ann Arbor Science, 1976.

J. J. Kirkland, ed., *Modern Practice of Liquid Chromatography*, New York: Wiley-Interscience, 1971.

J. Korkisch, *Modern Methods for the Separation of Rarer Metal Ions*, London: Pergamon, 1969.

A. K. Majumdar, *N-Benzoylphenylhdyroxylamine and Its Analogues*, Oxford: Pergamon Press, 1972.

J. M. Miller, *Separation Methods in Chemical Analysis*, New York: Wiley-Interscience, 1975.

G. H. Morrison and H. Freiser, *Solvent Extraction in Analytical Chemistry*, New York: Wiley, 1957.

J. D. Mulik, E. Sawicki, *Ion Chromatographic Analysis of Environmental Pollutants*, Vol. 2, Ann Arbor, Mich.: Ann Arbor Science, 1979.

W. J. Price, *Analytical Absorption Spectrometry*, London: Heyden, 1972.

W. Rieman and H. F. Walton, *Ion Exchange in Analytical Chemistry*, Oxford: Pergamon, 1970.

A. Ringbom, *Complexation in Analytical Chemistry*, New York: Wiley-Interscience, 1963.

J. Ruzicka and E. H. Hansen, *Flow Injection Analysis*, New York: John Wiley, 1981.

O. Samuelson, *Ion Exchange Separations in Analytical Chemistry*, New York: Wiley, 1963.

G. H. Schenk, *Absorption of Light and Ultraviolet Radiation*, Boston: Allyn and Bacon, 1973.

G. Schwarzenbach and H. Flaschka, *Complexometric Titrations*, 2nd English ed., trans. H. M. N. H. Irving, London: Methuen, 1969.

L. G. Sillen and A. E. Martell, *Stability Constants of Metal-Ion Complexes*, Vol. I, "*Inorganic Ligands*"; Vol. II, "*Organic Ligands*," London: The Chemical Society, 1964.

S. Siggia, *Quantitative Organic Analysis via Functional Groups*, 3rd ed., New York: Wiley, 1963.

L. R. Snyder and J. J. Kirkland, *Introduction to Modern Liquid Chromatography*, 2nd ed., New York: Wiley-Interscience, 1979.

N. W. Tietz, ed., *Fundamentals of Clinical Chemistry*, Philadelphia: W. B. Saunders, 1976.

W. W. Yaw, *Modern Size-Exclusion Liquid Chromatography*, New York: John Wiley, 1979.

M. Zief and J. W. Mitchell, *Contamination Control in Trace Element Analysis*, New York: Wiley-Interscience, 1977.

A. Zlatkis and L. S. Ettre, *Advances in Chromatography*, Amsterdam: Elsevier (published annually).

Equilibrium Constants

The data given below are taken with permission from A. Ringbom, *Complexation in Analytical Chemistry*, New York: Wiley-Interscience, 1963. In general, the values given are for 25°C; μ refers to the ionic strength of the solution (see Chapter 1, Section 1–2).

Solubility Product Constants

Compound	K_{sp} ($\mu = 0$)	K_{sp} ($\mu = 0.1$)
AgBr	4.9×10^{-13}	8.7×10^{-13}
AgCl	1.8×10^{-10}	3.2×10^{-10}
$Ag_2C_2O_4$	1×10^{-11}	4.0×10^{-11}
Ag_2CrO_4	1.1×10^{-12}	5.0×10^{-12}
AgI	8.3×10^{-17}	1.5×10^{-16}
AgOH	1.95×10^{-8}	2.5×10^{-8}
AgSCN	1.1×10^{-12}	2.0×10^{-12}
$Al(OH)_3$	4.6×10^{-33}	2.5×10^{-32}
$BaCO_3$	4.9×10^{-9}	3.2×10^{-8}
$BaCrO_4$	1.2×10^{-10}	8.0×10^{-10}
$Ba(IO_3)_2$	1.5×10^{-9}	6.3×10^{-9}
$Ba(Oxinate)_2$	5.0×10^{-9}	2.0×10^{-8}
$BaSO_4$	1.1×10^{-10}	6.3×10^{-10}
CaC_2O_4	2.3×10^{-9}	1.6×10^{-8}
CaF_2	3.4×10^{-11}	1.6×10^{-10}
$Ca(Oxinate)_2$	1.0×10^{-11}	4.0×10^{-11}
$Ca_3(PO_4)_2$	1×10^{-26}	1×10^{-23}
$CaSO_4$	2.4×10^{-5}	1.6×10^{-4}
CdS	8×10^{-27}	5×10^{-26}
$Fe(OH)_3$	2.5×10^{-39}	1.3×10^{-38}
$MgNH_4PO_4$	2.5×10^{-13}	——
$Mg(Oxinate)_2$	4.0×10^{-16}	1.6×10^{-15}
$PbBr_2$	3.9×10^{-5}	2.0×10^{-4}

Solubility Product Constants—*continued*

Compound	K_{sp} ($\mu = 0$)	K_{sp} ($\mu = 0.1$)
PbCl$_2$	1.6×10^{-5}	8.0×10^{-5}
PbCrO$_4$	1.8×10^{-14}	1.3×10^{-13}
PbF$_2$	2.7×10^{-8}	1.3×10^{-7}
PbI$_2$	6.45×10^{-9}	3.2×10^{-8}
PbS	2.5×10^{-27}	1.6×10^{-26}
PbSO$_4$	1.66×10^{-8}	1.0×10^{-7}
SrC$_2$O$_4$	5.6×10^{-8}	3.2×10^{-7}
Sr(Oxinate)$_2$	5.0×10^{-10}	2.0×10^{-9}
SrSO$_4$	2.5×10^{-7}	1.6×10^{-6}
Zn(Oxinate)$_2$	5×10^{-25}	2×10^{-24}
ZnS	1.6×10^{-24}	——

Ionization Constants of Acids

Acid	K_a	pK_a
Acetic, CH$_3$COOH	1.75×10^{-5}	4.76
Acrylic, CH$_2$=CHCOOH	5.56×10^{-5}	4.25
Adipic, (CH$_2$)$_4$(COOH)$_2$		
K_1	3.89×10^{-5}	4.41
K_2	5.25×10^{-6}	5.28
Arsenic, H$_3$AsO$_4$		
K_1	6.5×10^{-3}	2.2
K_2	1.3×10^{-7}	6.9
K_3	3.2×10^{-12}	11.5
Benzoic, C$_6$H$_5$COOH	6.25×10^{-5}	4.20
Carbonic, H$_2$CO$_3$		
K_1	4.3×10^{-7}	6.4
K_2	4.8×10^{-11}	10.3
Chloroacetic, CH$_2$ClCOOH	1.38×10^{-3}	2.86
o-Chlorobenzoic, C$_6$H$_4$ClCOOH	1.14×10^{-3}	2.94
m-Chlorobenzoic, C$_6$H$_4$ClCOOH	1.50×10^{-4}	3.82
p-Chlorobenzoic, C$_6$H$_4$ClCOOH	1.03×10^{-4}	3.99
o-Chlorophenol, C$_6$H$_4$ClOH	3.33×10^{-9}	8.48
m-Chlorophenol, C$_6$H$_4$ClCOOH	9.48×10^{-10}	9.02
p-Chlorophenol, C$_6$H$_4$ClCOOH	4.19×10^{-10}	9.38
Dichloroacetic, CHCl$_2$COOH	5.5×10^{-2}	1.3
EDTA, H$_4$Y		
K_1	8.5×10^{-3}	2.1
K_2	1.8×10^{-3}	2.7
K_3	5.8×10^{-7}	6.2
K$_4$	4.6×10^{-11}	10.3
Formic, HCOOH	1.77×10^{-4}	3.75
Fumaric, C$_2$H$_2$(COOH)$_2$		
K_1	9.6×10^{-4}	3.0
K_2	4.1×10^{-5}	4.4

Acid	K_a	pK_a
2-Furoic, C_4H_3OCOOH	8.63×10^{-4}	3.064
Hydrazoic, HN_3	1.9×10^{-5}	4.8
Hydrogen Cyanide, HCN	4.9×10^{-10}	9.3
Hydrogen Fluoride, HF	6.75×10^{-4}	3.17
Iodic, HIO_3	1.67×10^{-1}	0.78
Lactic, $CH_3CH(OH)COOH$	1.32×10^{-4}	3.88
Maleic, $C_2H_2(COOH)_2$		
K_1	1.00×10^{-2}	2.00
K_2	5.50×10^{-7}	6.26
Malonic, $CH_2(COOH)_2$		
K_1	1.43×10^{-3}	2.84
K_2	2.2×10^{-6}	5.7
Nitrilotriacetic, $N(CH_2COOH)_3$		
K_1	1.0×10^{-2}	2.0
K_2	2.5×10^{-3}	2.6
K_3	1.6×10^{-10}	9.8
o-Nitrophenol, $C_6H_4(NO_2)OH$	6.2×10^{-8}	7.2
m-Nitrophenol, $C_6H_4(NO_2)OH$	4.0×10^{-9}	8.4
p-Nitrophenol, $C_6H_4(NO_2)OH$	5.2×10^{-8}	7.3
Nitrous, HNO_2	5.1×10^{-4}	3.3
Oxalic, $(COOH)_2$		
K_1	8.8×10^{-2}	1.06
K_2	5.1×10^{-5}	4.3
Periodic, HIO_4	2.3×10^{-2}	1.6
Phenol, C_6H_5OH	1.4×10^{-10}	9.9
Phosphoric H_3PO_4		
K_1	7.5×10^{-3}	2.1
K_2	6.2×10^{-8}	7.2
K_3	4.8×10^{-13}	12.3
o-Phthalic, $C_6H_4(COOH)_2$		
K_1	1.20×10^{-3}	2.92
K_2	3.9×10^{-6}	5.4
Propionic, CH_3CH_2COOH	1.34×10^{-5}	4.87
Pyridinecarboxylic, $C_5H_4N(COOH)$	5.0×10^{-6}	5.3
Pyruvic, $CH_3COCOOH$	3.24×10^{-3}	2.49
Salicylic, $C_6H_4(OH)COOH$		
K_1	1.05×10^{-3}	2.98
K_2	4.0×10^{-14}	13.4
Sulfurous, H_2SO_3		
K_1	1.3×10^{-2}	1.9
K_2	6.3×10^{-8}	7.2
Tartaric, $[CH(OH)COOH]_2$		
K_1	9.20×10^{-4}	3.04
K_2	4.31×10^{-5}	4.37
Trichloroacetic, CCl_3COOH	2.2×10^{-1}	0.66

Ionization Constants of Bases

Base	K_a of Conjugate Acid	pK_a	pK_b of Base
4-Aminopyridine, $C_5H_4N(NH_2)$	4.27×10^{-10}	9.37	4.63
Ammonia, NH_3	5.62×10^{-10}	9.25	4.75
Aniline, $C_6H_5NH_2$	2.38×10^{-5}	4.62	9.38
Butylamine, $C_4H_9NH_2$	2.44×10^{-11}	10.61	3.39
Ethanolamine, $HOCH_2CH_2NH_2$	3.57×10^{-10}	9.45	4.55
Ethylamine, $C_2H_5NH_2$	2.13×10^{-11}	10.67	3.33
Ethylenediamine, $H_2NCH_2CH_2NH_2$			
K_{1a}	5.0×10^{-8}	7.3	6.7
K_{2a}	7.8×10^{-11}	10.1	3.9
Glycine, H_2NCH_2COOH			
K_{1a}	4.45×10^{-3}	2.35	11.65
K_{2a}	1.66×10^{-10}	9.78	4.22
Hydrazine, NH_2NH_2	1.0×10^{-8}	8.0	6.0
Methylamine, CH_3NH_2	2.3×10^{-11}	10.6	3.35
Pyridine, C_5H_5N	6.7×10^{-6}	5.2	8.8
THAM, $(HOCH_2)_3CNH_2$	7.94×10^{-9}	8.10	5.90
Triethanolamine, $(HOCH_2CH_2)_3N$	1.5×10^{-8}	7.8	6.2
Trien, $(—CH_2NHCH_2CH_2NH_2)_2$			
K_{1a}	4.0×10^{-4}	3.4	10.6
K_{2a}	1.8×10^{-7}	6.7	7.25
K_{3a}	5.3×10^{-10}	9.3	4.7
K_{4a}	1.0×10^{-10}	10.0	4.0
Urea, NH_2CONH_2	7.1×10^{-1}	0.15	13.85

Stepwise Formation Constants for Selected Complexes

Metal Ion	Ligand	$\log K_1$	$\log K_2$	$\log K_3$	$\log K_4$	$\log K_5$	$\log K_6$
Ag^+	NH_3	3.4	4.0	—	—	—	—
Cu^{2+}	NH_3	4.1	3.5	2.9	2.1	—	
Zn^{2+}	NH_3	2.3	2.3	2.4	2.05	—	
Ag^+	Ethylenediamine	4.7	3.0	—			
Cu^{2+}	Ethylenediamine	10.55	9.05	—			
Zn^{2+}	Ethylenediamine	5.7	4.7	—			
Ag^+	Trien (see Chapter 11, Section 11–2)	7.7	—				
Cu^{2+}	Trien	20.4	—				
Zn^{2+}	Trien	12.1	—				
Ag^+	CN^-	—	$\log(K_1K_2)$ 21.1				
Ni^{2+}	CN^-	—	—	—	—	$\log(K_1K_2K_3K_4)$ 31.3	
Zn^{2+}	CN^-	—	—	—	—	$\log(K_1K_2K_3K_4)$ 16.7	
Fe^{3+}	SCN^-	2.3	1.9	1.4	0.8	—	

Stepwise Formation Constants for Selected Complexes—*continued*

Metal Ion	Ligand	$\log K_1$	$\log K_2$	$\log K_3$	$\log K_4$	$\log K_5$	$\log K_6$
Al^{3+}	F^-	6.1	5.1	3.8	2.7	1.7	0.3
Fe^{3+}	F^-	5.2	4.0	2.7	——		
TiO^{2+}	F^-	5.4	4.4	3.9	3.7	——	
Cd^{2+}	I^-	2.4	1.0	1.6	1.15	——	
Cu^{2+}	$C_2O_4{}^{2-}$	4.5	4.4	——			
Mg^{2+}	$C_2O_4{}^{2-}$	2.4	——				
Zn^{2+}	$C_2O_4{}^{2-}$	3.7	2.3	——			
Cu^{2+}	Tartrate (-2)	3.2	1.9	-0.3	1.7	——	
Various	EDTA	(see Table 12–2)					

Standard Electrode Potentials

Half-Reaction	E^0 (V)
$F_2 + 2H^+ + 2e^- = 2HF$	3.06
$S_2O_8^{2-} + 2e^- = 2SO_4^{2-}$	2.01
$Ag^{2+} + e^- = Ag^+$	1.98
$Co^{3+} + e^- = Co^{2+}$	1.82
$H_2O_2 + 2H^+ + 2e^- = 2H_2O$	1.776
$Ce(IV) + e^- = Ce^{3+}$ (in $1M$ $HClO_4$)	1.70
$Ce(IV) + e^- = Ce^{3+}$ (in $1M$ HNO_3)	1.61
$H_5IO_6 + H^+ + 2e^- = IO_3^- + 3H_2O$	1.60
$MnO_4^- + 8H^+ + 5e^- = Mn^{2+} + 4H_2O$	1.51
$Mn^{3+} + e^- = Mn^{2+}$	1.51
$PbO_2 + 4H^+ + 2e^- = Pb^{2+} + 2H_2O$	1.455
$Ce(IV) + e^- = Ce^{3+}$ (in $1N$ H_2SO_4)	1.44
$Cl_2 + 2e^- = 2Cl^-$	1.360
$Cr_2O_7^{2-} + 14H^+ + 6e^- = 2Cr^{3+} + 7H_2O$	1.33
$Ce(IV) + e^- = Ce^{3+}$ (in $1N$ HCl)	1.28
$MnO_2 + 4H^+ + 2e^- = Mn^{2+} + 2H_2O$	1.23
$O_2 + 4H^+ + 4e^- = 2H_2O$	1.229
$IO_3^- + 6H^+ + 5e^- = \frac{1}{2}I_2 + 3H_2O$	1.195
$Br_2 + 2e^- = 2Br^-$	1.065
$Fe(phen)_3^{3+} + e^- = Fe(phen)_3^{2+}$	1.06
$Br_3^- + 2e^- = 3Br^-$	1.05
$VO_2^+ + 2H^+ + e^- = VO^{2+} + H_2O$	1.00
$HNO_2 + H^+ + e^- = NO + H_2$	1.00
$2Hg_2 + 2e^- = Hg_2^{2+}$	0.920
$SO_4^{2-} + 2H^+ + 2e^- = SO_3^{2-} + H_2O$	0.868
$Cu^{2+} + I^- + e^- = CuI$	0.86
$Hg^{2+} + 2e^- = Hg^0$	0.854
$Ag^+ + e^- = Ag$	0.800
$Hg_2^{2+} + 2e^- = 2Hg$	0.789
$Fe^{3+} + e^- = Fe^{2+}$	0.771

Half-Reaction	E^0 (V)
Quinone + $2H^+$ + $2e^-$ = Hydroquinone	0.699
O_2 + $2H^+$ + $2e^-$ = H_2O_2	0.682
H_3AsO_4 + $2H^+$ + $2e^-$ = H_3AsO_3 + H_2O	0.559
I_2 + $2e^-$ = $2I^-$	0.535
Cu^+ + e^- = Cu	0.521
H_2SO_3 + $4H^+$ + $4e^-$ = S + $3H_2O$	0.45
Dehydroascorbic acid + $2H^+$ + $2e^-$ = Ascorbic acid	0.390
VO^{2+} + $2H^+$ + e^- = V^{3+}	0.361
$Fe(CN)_6^{3-}$ + e^- = $Fe(CN)_6^{4-}$	0.36
Cu^{2+} + $2e^-$ = Cu	0.337
UO_2^{2+} + $4H^+$ + $2e^-$ = U^{4+} + $2H_2O$	0.334
Hg_2Cl_2 + $2e^-$ = 2Hg + $2Cl^-$	0.268
Pyruvate + $2H^+$ + $2e^-$ = Lactate	0.224
AgCl + e^- = Ag + Cl^-	0.222
Cu^{2+} + e^- = Cu^+	0.153
Sn^{4+} + $2e^-$ = Sn^{2+}	0.154
S + $2H^+$ + $2e^-$ = H_2S	0.141
TiO^{2+} + $2H^+$ + e^- = Ti^{3+} + H_2O	0.10
$S_4O_6^{2-}$ + $2e^-$ = $2S_2O_3^{2-}$	0.08
AgBr(s) + e^- = Ag(s) + Br^-	0.071
$2H^+$ + $2e^-$ = H_2	0.000 (Exact no.)
U^{4+} + e^- = U^{3+}	-0.61
AgI + e^- = Ag + I^-	-0.151
Pb^{2+} + $2e^-$ = Pb	-0.126
Sn^{2+} + $2e^-$ = Sn	-0.136
Ni^{2+} + $2e^-$ = Ni	-0.250
V^{3+} + e^- = V^{2+}	-0.255
Co^{2+} + $2e^-$ = Co	-0.277
$Ag(CN)_2^-$ + e^- = Ag + $2CN^-$	-0.31
Cd^{2+} + $2e^-$ = Cd	-0.403
Cr^{3+} + e^- = Cr^{2+}	-0.41
Fe^{2+} + $2e^-$ = Fe	-0.440
Zn^{2+} + $2e^-$ = Zn	-0.763
Mn^{2+} + $2e^-$ = Mn	-1.18
V^{2+} + $2e^-$ = V^0	-1.18
Al^{3+} + $3e^-$ = Al	-1.66
Mg^{2+} + $2e^-$ = Mg	-2.37
Na^+ + e^- = Na	-2.71
Ca^{2+} + $2e^-$ = Ca	-2.87
Li^+ + e^- = Li	-3.04

Balancing Oxidation-Reduction Equations

In all written oxidation-reduction reactions, the coefficients preceding substances undergoing oxidation and reduction must be such that the net electron gain by reduced substances is equal to the net electron loss by oxidized substances. This should always be remembered when writing and balancing redox equations. When the proper coefficients for the oxidized or reduced elements have been established, then the remaining elements can be balanced.

The electron change can be calculated from the change in *oxidation number*. For simple ions such as Fe^{2+} and Fe^{3+}, the oxidation number is simply the valence of the element. The oxidation number of elements in their elemental state is 0 (for example, H_2, Cl_2, S, Na, Pb). The oxidation number of hydrogen in compounds is usually $+1$. Oxygen in compounds is usually -2 except in peroxides where it is -1.

The sum of oxidation numbers in a neutral *compound* must equal zero. The sum of oxidation numbers in an *ion* must equal the charge on the ion. These facts usually make it possible to calculate the oxidation number of an element of unknown oxidation state when that element is present in a compound or ion. For example,

nitrate (NO_3^-)	nitrite (NO_2^-)	ammonia (NH_3)
$3(O) = -6$	$2(O) = -4$	$3(H) = +3$
$N + (-6) = -1$	$N + (-4) = -1$	$N + (+3) = 0$
$N = +5$	$N = +3$	$N = -3$

To balance an oxidation-reduction reaction, proceed as follows:

1. Write the reaction products.
2. Ascertain what is oxidized and calculate the total number of electrons lost.
3. Ascertain what is reduced and calculate the total number of electrons gained.
4. Multiply the oxidized and reduced species by numbers that will make the electron gain equal to the loss.
5. Balance the other elements involved in the reaction.

Example:

$$\text{gain} = 5e^-$$

$$MnO_4^- + Fe^{2+} + H^+ \rightarrow Mn^{2+} + Fe^{3+} + H_2O$$

$$\text{loss} = 1e^-$$

Manganese in permanganate has an oxidation number of $+7$. This can be determined by recalling that oxygen in most chemical compounds has an oxidation number of -2. In permanganate, four atoms of oxygen, each with an oxidation number of -2, give a total of -8. Since the ion must have net charge of $1-$, the manganese must have an oxidation number of $+7$. Multiplying the iron species by 5 makes the electron gain equal to the electron loss in the reaction shown. Balancing is completed by noting that permanganate carries 4 oxygens. Hence, 4 water molecules are required on the right to balance the oxygen, and 8 hydrogen ions are required on the left to balance the hydrogen atoms in the 4 water molecules. The completely balanced equation is

$$MnO_4^- + 5Fe^{2+} + 8H^+ \rightarrow Mn^{2+} + 5Fe^{3+} + 4H_2O$$

Example:

$$\text{gain} = 6e^-$$

$$Cr_2O_7^{2-} + Sn^{2+} + H^+ \rightarrow 2Cr^{3+} + Sn^{4+} + H_2O$$

$$\text{loss} = 2e^-$$

Each chromium in dichromate is reduced from $+6$ to $+3$, but the net change for 1 dichromate molecule is 6 electrons. Multiplying the tin species by 3 balances the electrons. The completely balanced equation is

$$Cr_2O_7^{2-} + 3Sn^{2+} + 14H^+ \rightarrow 2Cr^{3+} + 3Sn^{4+} + 7H_2O$$

Example: More than one element can be oxidized or reduced in the same reaction. In this example, nitrogen is reduced from -5 to -2 and both iron and sulfur are oxidized:

$$\text{gain} = 3e^-$$

$$\overset{+5}{H}NO_3 + \overset{+2}{Fe}\!\!-\!\!\overset{-2}{S} + H^+ \rightarrow \overset{+2}{N}O + Fe^{3+} + \overset{+6}{S}O_4^{2-} + H_2O$$

$$\text{loss} = 1e^-$$

$$\text{loss} = 8e^-$$

The total electron loss is 9 electrons for each FeS molecule. Multiplying the nitrogen-containing species by 3 balances the electrons. The completely balanced equation is

$$3HNO_3 + FeS + H^+ \rightarrow 3NO + Fe^{3+} + SO_4^{2-} + 2H_2O$$

Least Squares Method
for Linear Plots

This is a method for drawing the best straight line through a set of data points. By using this method, a better fit of the straight line and a more accurate slope and intercept can be obtained that when a graphical method is used. The formula for a straight line is

$$y = mx + b$$

where y is the vertical axis; x is the horizontal axis; m is the straight line slope; and b is the straight line intercept (the value of y when $x = 0$). For n points with x, y coordinates, the slope m by the least squares method is:

$$m = \frac{n\sum x_i y_i - \sum x_i \sum y_i}{n\sum x_i^2 - (\sum x_i)^2} \qquad \text{(A5–1)}$$

$$b = \bar{y} - m\bar{x} \qquad \text{(A5–2)}$$

where
$$\bar{y} = \frac{y_1 + y_2 + \cdots + y_n}{n}$$

and
$$\bar{x} = \frac{x_1 + x_2 + \cdots + x_n}{n}$$

Example: Calculate m and b for the following data points:

$$x = 1.0, y = 2.0$$
$$x = 2.0, y = 3.8$$
$$x = 3.0, y = 6.2$$

Substituting into Equation A5–1,

$$m = \frac{3[(1 \times 2) + (2 \times 3.8) + (3 \times 6.2)] - [(1 + 2 + 3)(2 + 3.8 + 6.2)]}{3(1 + 4 + 9) - (1 + 2 + 3)^2}$$

$$m = \frac{84.6 - 72.0}{42.0 - 36.0} = 2.10$$

Substituting into Equation A5–2,

$$b = \frac{2 + 3.8 + 6.2}{3} - 2.10\frac{(1 + 2 + 3)}{3} = -0.20$$

Programmable calculators are available that can make a least squares calculation very quickly.

Solving Quadratic Equations

QUADRATIC FORMULA

When an equilibrium constant is larger than about 1×10^{-4}, the concentration of the desired substance usually must be calculated by solving a quadratic equation. In such cases, the equilibrium-constant expression can be arranged into the general form of the quadratic equation,

$$ax^2 + bx + c = 0$$

for which the solution is

$$x = \frac{-b \pm \sqrt{b^2 - 4ac}}{2a}$$

Only the positive root is meaningful in equilibrium problems.

As an example, consider the calculation of $[H^+]$ of $0.1000M$ picric acid, HPi, $K_a = 5 \times 10^{-3}$. The equilibrium concentration of picric acid will differ significantly from the analytical concentration of $0.1000M$ acid, and is equal to $0.1000M - [H^+]$.

The ionization-constant expression is

$$K_a = \frac{[H^+][Pi^-]}{0.1000M - [H^+]} = \frac{[H^+]^2}{0.1000M - [H^+]}$$

This rearranges to the form of a quadratic equation:

$$[H^+]^2 + K_a[H^+] - 0.1000M(K_a) = 0$$

In this example, $x = [H^+]$, $b = K_a = 0.005$, and $c = -0.0005$. Substituting these values into the solution equation for a quadratic equation gives $x = 2.0 \times 10^{-2}$.

Solving quadratic equations is easy with a programmable calculator. With Hewlett-Packard calculators, the following program will give the *positive* root: RCl 2, X^2, RCl 1, RCl 3, $\times$, y, $\times$, $-$, $\sqrt{}$, RCl 2, $-$, RCl 1, $\div$, 2, $\div$. Simply store the value for a in memory 1, store b in memory 2, store c in memory 3 (use the proper sign), and press the R/S key.

Scientific Notation, Logarithms, Antilogarithms, and Roots

SCIENTIFIC NOTATION

Numbers written in scientific notation have the form 1.234×10^n, where the decimal comes after the first digit. For example,

$$0.020 = 2.0 \times 10^{-2}$$

$$31.5 = 3.15 \times 10^1$$

$$0.000315 = 3.15 \times 10^{-4}$$

Large numbers are often written in scientific notation to indicate the number of significant figures. Thus, one million, written with three significant figures is 1.00×10^6.

To enter a number such as 3.15×10^{-4} into a calculator, enter 3.15, press the "enter exponent" key, then 4, then the "change sign" key. In some calculators, a number can be entered and then converted to scientific notation simply by pressing a "scientific notation" key.

LOGARITHMS

The logarithm (log) of a number is the exponent of the power to which 10 must be raised to give that number. For example,

$$\log 100 = \log 10^2 = 2$$

$$\log 0.01 = \log 10^{-2} = -2$$

$$\log 2.0 = 0.30$$

$$\log 0.020 = \log \times 10^{-2} = -1.70$$

It is very easy to obtain the log of a number using an electronic calculator; simply enter the number and press the log key. Note that the log of a negative number is not defined.

The natural logarithm (ln) of a number is the exponent of the power to which e must be raised to give that number. (The value of e is 2.7183.) For example,

$$\ln 7.389 = \ln (e^2) = 2$$

$$\ln 2 = \ln (e^{0.693}) = 0.693$$

$$\ln 0.02 = \ln (e^{-3.912}) = -3.912$$

To obtain the natural logarithm of a number using a calculator, simply enter the number and press the ln key.

The natural logarithm of a number is 2.303 times as large as the logarithm with base 10. If an equation has a ln term (such as the Nernst equation in Chapter 12), we can replace it with 2.303 times the log.

ANTILOGARITHMS

The process of obtaining an antilog is the inverse of taking the log—one goes from the log back to an ordinary number. For example,

$$\text{antilog } 2.0 = 10^{2.0} = 100$$

$$\text{antilog } 0.301 = 10^{0.301} = 2.0$$

To obtain the antilog, simply enter the number (which is really a log) and press the 10^x key. To obtain the antiln, enter the number and press the e^x key. (On some calculators, antilog or antiln can be taken by entering the number and pressing an "inverse" key and then the log key or the ln key.)

$$\text{antilog } 0.301 = 2.0$$

$$\text{antilog } 0.477 = 3.00$$

$$\text{antilog } -0.477 = 0.333$$

In the last example, enter 0.477, then press the "change sign" key, and then the 10^x key.

We often write the antilog function as antilog $(x) = 10^x$; similarly, antiln $(x) = e^x$. Some basic properties of these functions are

$$\log (10^x) = \log (\text{antilog } (x)) = x$$

$$10^{\log x} = \text{antilog} (\log (x)) = x$$

(log and antilog are inverse functions)

$$\log (xy) = \log x + \log y \qquad 10^{x+y} = (10^x)(10^y)$$

$$\log (x/y) = \log x - \log y \qquad 10^{x-y} = 10^x/10^y$$

$$\log (1/x) = -\log x \qquad 10^{-x} = 1/10^x$$

$$\log (y^x) = x \log y \qquad (10^x)^y = 10^{x \cdot y}$$

The log function is defined only for positive numbers, while the antilog function is defined for all numbers.

ROOT OF A NUMBER

Virtually all calculators have a square root key (usually labeled $\sqrt{x}$). Simply enter the number and press the $\sqrt{x}$ key to obtain its square root. Recall that $\sqrt{x}$ means $x^{1/2}$ or $x^{0.5}$. The nth root of a number x is $x^{1/n}$. On many calculators the nth root can be obtained by entering the number, then entering the exponent $(1/n)$ and pressing the y^x key. (The order of the last two steps may be reversed on some calculators.) For example:

$$2^{1/2} = 1.414 \text{ (square root)}$$

$$2^{1/3} = 1.260 \text{ (cube root)}$$

$$27^{1/3} = 3.000$$

$$27^{0.48} = 4.865$$

If the calculator lacks a y^x key, the nth root of x can be obtained by taking the log of x, dividing by n, and then taking the antilog. For example, take the cube root of 2:

$$\log 2 = 0.3010; \qquad 0.301/3 = 0.1003; \qquad \text{antilog } 0.1003 = 1.260$$

APPENDIX *8*

Four-place Logarithms

No.	0	1	2	3	4	5	6	7	8	9
10	0000	0043	0086	0128	0170	0212	0253	0294	0334	0374
11	0414	0453	0492	0531	0569	0607	0645	0682	0719	0755
12	0792	0828	0864	0899	0934	0969	1004	1038	1072	1106
13	1139	1173	1206	1239	1271	1303	1335	1367	1399	1430
14	1461	1492	1523	1553	1584	1614	1644	1673	1703	1732
15	1761	1790	1818	1847	1875	1903	1931	1959	1987	2014
16	2041	2068	2095	2122	2148	2175	2201	2227	2253	2279
17	2304	2330	2355	2380	2405	2430	2455	2480	2504	2529
18	2553	2577	2601	2625	2648	2672	2695	2718	2742	2765
19	2788	2810	2833	2856	2878	2900	2923	2945	2967	2989
20	3010	3032	3054	3075	3096	3118	3139	3160	3181	3201
21	3222	3243	3263	3284	3304	3324	3345	3365	3385	3404
22	3424	3444	3464	3483	3502	3522	3541	3560	3579	3598
23	3617	3636	3655	3674	3692	3711	3729	3747	3766	3784
24	3802	3820	3838	3856	3874	3892	3909	3927	3945	3962
25	3979	3997	4014	4031	4048	4065	4082	4099	4116	4133
26	4150	4166	4183	4200	4216	4232	4249	4265	4281	4298
27	4314	4330	4346	4362	4378	4393	4409	4425	4440	4456
28	4472	4487	4502	4518	4533	4548	4564	4579	4594	4609
29	4624	4639	4654	4669	4683	4698	4713	4728	4742	4757
30	4771	4786	4800	4814	4829	4843	4857	4871	4886	4900
31	4914	4928	4942	4955	4969	4983	4997	5011	5024	5038
32	5051	5065	5079	5092	5105	5119	5132	5145	5159	5172
33	5185	5198	5211	5224	5237	5250	5263	5276	5289	5302
34	5315	5328	5340	5353	5366	5378	5391	5403	5416	5428
35	5441	5453	5465	5478	5490	5502	5514	5527	5539	5551
36	5563	5575	5587	5599	5611	5623	5635	5647	5658	5670
37	5682	5694	5705	5717	5729	5740	5752	5763	5775	5786
38	5798	5809	5821	5832	5843	5855	5866	5877	5888	5899
39	5911	5922	5933	5944	5955	5966	5977	5988	5999	6010
40	6021	6031	6042	6053	6064	6075	6085	6096	6107	6117
41	6128	6138	6149	6160	6170	6180	6191	6201	6212	6222
42	6232	6243	6253	6263	6274	6284	6294	6304	6314	6325
43	6335	6345	6355	6365	6375	6386	6395	6405	6415	6425
44	6435	6444	6454	6464	6474	6484	6493	6503	6513	6522
45	6532	6542	6551	6561	6571	6580	6590	6599	6609	6618
46	6628	6637	6646	6656	6665	6675	6684	6693	6702	6712
47	6721	6730	6739	6749	6758	6767	6776	6785	6794	6803
48	6812	6821	6830	6839	6848	6857	6866	6875	6884	6893
49	6902	6911	6920	6928	6937	6946	6955	6964	6972	6981
50	6990	6998	7007	7016	7024	7033	7042	7050	7059	7067
51	7076	7084	7093	7101	7110	7118	7126	7135	7143	7152
52	7160	7168	7177	7185	7193	7202	7210	7218	7226	7235
53	7243	7251	7259	7267	7275	7284	7292	7300	7308	7316
54	7324	7332	7340	7348	7356	7364	7372	7380	7388	7396
	0	1	2	3	4	5	6	7	8	9

No.	0	1	2	3	4	5	6	7	8	9
55	7404	7412	7419	7427	7435	7443	7451	7459	7466	7474
56	7482	7490	7497	7505	7513	7520	7528	7536	7543	7551
57	7559	7566	7574	7582	7589	7597	7604	7612	7619	7627
58	7634	7642	7649	7657	7664	7672	7679	7686	7694	7701
59	7709	7716	7723	7731	7738	7745	7752	7760	7767	7774
60	7782	7789	7796	7803	7810	7818	7825	7832	7839	7846
61	7853	7860	7868	7875	7882	7889	7896	7903	7910	7917
62	7924	7931	7938	7945	7952	7959	7966	7973	7980	7987
63	7992	8000	8007	8014	8021	8028	8035	8041	8048	8055
64	8062	8069	8075	8082	8089	8096	8102	8109	8116	8122
65	8129	8136	8142	8149	8156	8162	8169	8176	8182	8189
66	8195	8202	8209	8215	8222	8228	8235	8241	8248	8254
67	8261	8267	8274	8280	8287	8293	8299	8306	8312	8319
68	8325	8331	8338	8344	8351	8357	8363	8370	8376	8382
69	8388	8395	8401	8407	8414	8420	8426	8432	8439	8445
70	8451	8457	8463	8470	8476	8482	8488	8494	8500	8506
71	8513	8519	9525	8531	8537	8543	8549	8555	8561	8567
72	8573	8579	8585	8591	8597	8603	8609	8615	8621	8627
73	8633	8639	8645	8651	8657	8663	8669	8675	8681	8686
74	8692	8698	8704	8710	8716	8722	8727	8733	8739	8745
75	8751	8756	8762	8768	8774	8779	8785	8791	8797	8802
76	8808	8814	8820	8825	8831	8837	8842	8848	8854	8859
77	8865	8871	8876	8882	8887	8893	8899	8904	8910	8915
78	8921	8927	8932	8938	8943	8949	8954	8960	8965	8971
79	8976	8982	8987	8993	8998	9004	9009	9015	9020	9025
80	9031	9036	9042	9047	9053	9058	9063	9069	9074	9079
81	9085	9090	9096	9101	9106	9112	9117	9122	9128	9133
82	9138	9143	9149	9154	9159	9165	9170	9175	9180	9186
83	9191	9196	9201	9206	9212	9217	9222	9227	9232	9238
84	9243	9248	9253	9258	9263	9269	9274	9279	9284	9289
85	9294	9299	9304	9309	9315	9320	9325	9330	9335	9340
86	9345	9350	9355	9360	9365	9370	9375	9380	9385	9390
87	9395	9400	9405	9410	9415	9420	9425	9430	9435	9440
88	9445	9450	9455	9460	9465	9469	9474	9479	9484	9489
89	9494	9499	9504	9509	9513	9518	9523	9528	9533	9538
90	9542	9547	9552	9557	9562	9566	9571	9576	9581	9586
91	9590	9595	9600	9605	9609	9614	9619	9624	9628	9633
92	9638	9643	9647	9652	9657	9661	9666	0671	9675	9680
93	9685	9689	9694	9699	9703	9708	9713	9717	9722	9727
94	9731	9736	9741	9745	9750	9754	9759	9763	9768	9773
95	9777	9782	9786	9791	9795	9800	9805	9809	9814	9818
96	9823	9827	9832	9836	9841	9845	9850	9854	9859	9863
97	9868	9872	9877	9881	9886	9890	9894	9899	9903	9908
98	9912	9917	9921	9926	9930	9934	9939	9943	9948	9952
99	9956	9961	9965	9969	9974	9978	9983	9987	9991	9996
	0	1	2	3	4	5	6	7	8	9

Answers to Selected Questions and Problems

Answers are given to all odd-numbered problems involving calculations. Answers are also given to most of the other odd-numbered questions and problems, except where the answer is obvious from the text material or involves a somewhat lengthy discussion.

CHAPTER 1

1. **(a)** 1:1, 1.39:1 **(c)** 1:2, 2.88:1
3. $0.01M$; $[K^+] = 0.02M$; $[SO_4{}^{-2}] = 0.01M$
5. 260 mL
7. $[H^+] = [\text{Acetate}] = 0.004M$,
   $[\text{Acetic acid}] = 0.996M$
9. **(a)** $\mu = 0.5$ **(c)** $\mu = 3.0$
11. Ionic strength will increase.
13. Activity of Li^+ will decrease because addition of KCl increases the ionic strength of the solution.

CHAPTER 2

(No problems)

CHAPTER 3

1. **(a)** 0.12 ppt **(b)** 0.09 ppt **(c)** 50 ppt
   (d) 2.0 ppt
3. **(a)** 226.4 **(b)** 0.427 **(c)** 12.0_6
   (d) 1.9×10^{-4}
5. **(a)** 1.0×10^{-7} **(b)** $1.00_4 \times 10^{-7}$
   (c) $1.003_9 \times 10^{-7}$
7. **(a)** $10^{-2}M$ (no sig figs) **(b)** $1._9 \times 10^{-2}M$
   (c) $2.00 \times 10^{-2}M$
9. Relative error = 2 ppt
13. The likelihood of a mode existing for the three conditions indicated decreases from (a) to (c).

15. **(a)** $m = 1.3$ **(b)** $m = 1.4$
17. Delete the 10.5% result; area $\cong 1.0$.
19. The probability that a result is between points a and b is obtained from the integrated area under the density function curve from point a to point b.
21. $\mu = 12.04$, $\sigma^2 = 0.70$
23. $\bar{X} = 0.496$, $S = 0.005_4$, $S^2 = 2.91 \times 10^{-5}$
25. **(a)** 0.24 **(b)** 7.9 ppt (not satisfactory)
27. $\sigma = 0.020$
29. **(a)** 12.09 is rejected, and $\bar{X} = 11.13\%$
   (b) 4.97 is rejected, and $\bar{X} = 5.20$
31. Retain 11.00%; median = 10.20%
33. **(a)** $Q = 0.921$; can't reject 7.93
   (b) No
   (c) 3 and 9 were mixed up; pH was really 7.39.
35. A standard normal distribution curve has a mean of 0 and a variance of 1.
37. $0.496 \pm 0.005_7$
39. For $n = 3$, $\sigma = 0.1_2$; for $n = 6$, $\sigma = 0.0_8$
41. C.I. $= \bar{X} \pm 0.0950$
43. Rel $S = 9.12$ ppt
47. $\epsilon = 1.069 \times 10^4$
49. **(a)** $R_{0.05} = 0.941 = Q_{0.90}$

CHAPTER 4

1–19. See text
21. $100 \times \dfrac{10 \times \text{f.w. CaO}}{Ca_5(PO_4)_3OH} = \dfrac{100 \times 560.8}{502.3}$
   $= 11.6$ mg
23. $1.000 \times \dfrac{\text{f.w. AgCl}}{\text{f.w. KCl}} = 1.922$ g AgCl

25. $1000 \times \dfrac{\text{f.w. BaCl}_2 \cdot 2H_2O}{\text{f.w. K}_2SO_4} = \dfrac{1405 \text{ mg}}{60 \text{ mg/mL}}$

$= 23.4$

27. $1.000 \times \dfrac{2\,\text{f.w. SiO}_2}{\text{f.w. NaAl(SiO}_3)_2} = 0.594 \text{ g}$

29. $\dfrac{10 \times \text{f.w. PO}_4}{\text{f.w. (NH}_4)_3\text{PMo}_{12}O_{40}} = \dfrac{10 \times 94.97}{1873.3}$

$= 0.506 \text{ mg}$

31. (a) $\dfrac{0.00011}{113.0} \times 100 = 9.7 \times 10^{-5}\% \text{ Hg}$

  (b) $\dfrac{0.11}{113.0} \times 1000 = 0.97 \text{ ppm}$

33. (a) $3.5 \times \dfrac{\text{f.w. CH}_3\text{Hg}}{\text{f.w. Hg}} = 3.8 \text{ ppm}$

  (b) $3.5 \times \dfrac{\text{f.w. (CH}_3)_2\text{Hg}}{\text{f.w. Hg}} = 4.0 \text{ ppm}$

35. $\dfrac{0.7715 \times \dfrac{\text{f.w. Cl}}{\text{f.w. AgCl}} \times 100}{0.500} = 38.17\% \text{ Cl}$

37. $\dfrac{0.1262 - 0.0012}{1.000} \times 100 = 12.5\% \text{ SiO}_2$

39. $\text{Ca} = \dfrac{16.2}{40} = 0.405, \quad \text{Fe} = \dfrac{22.5}{55.85} = 0.403, \quad \text{Si} =$

$\dfrac{22.6}{28} = 0.807, \quad O = \dfrac{38.7}{16} = 2.419.$ Divide each

by the smallest ratio (0.403), giving a relative
ratio of Ca:Fe:Si:O of 1:1:2:6. The formula
is $CaFe(SiO_3)_2$.

41. $0.3999 \times \dfrac{2\,\text{f.w. Al(C}_9H_6NO)_3}{\text{f.w. Al}_2(SO_4)_3 \cdot xH_2O} = 0.4185$

f.w. $\text{Al}_2(SO_4)_3 \cdot xH_2O = 877.2$
$x = 29.7(\sim 30)$

43. $0.1730 \times \dfrac{60 + 14 + 79.9x}{187.8x} = 0.0962$

$x = 3.02, (CH_3)_4NBr_3$

45. 135.9 mg/135.9 = 1.0 mmole KHSO$_4$;
9.0008 mg/18.0016 = 0.5 mmole H$_2$O. The
formula is $K_2S_2O_7$

47. (a) $Na^+Br^- + Ag^+NO_3^- \rightarrow$
$\qquad\qquad -Ag\text{-}BrAg:Br^- \cdot\cdot Na^+$

  (b) $K^+Br^- + Ag^+ClO_4^- \rightarrow$
$\qquad\qquad -Ag\text{-}BrAg:Br^- \cdot\cdot K^+$

49. (b), (c), and **(d)** would not be good

9. mL HCl $\times 12.0 = 1000 \times 0.25$
  mL HCl $= 20.8$

11. $\dfrac{1690 \text{ mg/mL} \times 0.85}{98.00} = 14.7M$

13. mL HCl $\times 1.183 = 1000 \times 0.1000$
  mL HCl $= 84.5$
15. $M = 0.0929_5$
17. $M = 0.00216$; original solution $= 0.01079M$
19. (a) $H_3AsO_4 = 0.1600M$ **(b)** $0.3200N$
21. (a) $M = 0.1022$ **(b)** Normality is also 0.1022
23. 712.2 mg Sn
25. 34.30% H_3PO_4, 83.99% NaH_2PO_4
27. (a) 53.18% Cl **(b)** 71.42% $MgCl_2$
29. 11.73% Cl; the difference is more than 3%
31. 41.46% As_2O_3
33. (b) 98.77% diffused
35. 31.83% NaF
37. 45.67% Na_2CO_3
39. form. wt. $= 299.4$
41. Ratio I:Bi $= 4.005$; complex is BiI_4^-
43. (form. wt.) $= 125.9$

CHAPTER 6

3. $A = 0.06085$; $B = 0.1609$; $AB = 0.0392$
5. (a) $[H^+] = 0.0742M$ **(b)** $[H^+] = 0.0516M$
7. $[H^+] = 0.0011M$
9. $\dfrac{[A^-]}{[HA]} = 1.77$

11. Mg $= 0.0238$, tart $= 0.324$, Mg tart $= 0.176$
13. (a) $3.2 \times 10^{-7}M$ **(b)** $x = 1.89 \times 10^{-3}M$
15. (a) $(Ag^+) = 1.34 \times 10^{-5}M$
  (b) $(Ag^+) = 2x = 13.0 \times 10^{-5}M$ (more soluble)
17. $[Mg^{2+}] = 1.82 \times 10^{-5}M$
19. $\mu = \frac{1}{2}[(7.2 \times 10^{-4})(4) + (14.4 \times 10^{-4})]$
  $= 2.16 \times 10^{-3}$
21. (a) $K' = 1.89 \times 10^{-10}$ **(b)** $K' = 2.11 \times 10^{-10}$
  (c) $K' = 2.80 \times 10^{-10}$
23. (a) $\alpha_F = 0.063$ **(b)** $\alpha_F = 0.403$
  (c) $\alpha_F = 0.871$ **(d)** $\alpha_F = 0.985$
25. $[Pb^{2+}] = 3.16 \times 10^{-6}M$
27. $[Ba'] = 7.9 \times 10^{-8}$ (quantitative), $[Pb'] = 10^{4.4}$ (no precipitate)
29. (a) Straight line; slope is the number of dye
  molecules per silver in the precipitate
  (b) Intercept $= \log K_{sp}$

CHAPTER 5

7. (a) $117.4/169.9 = 0.691M$
  (b) $9.72 \text{ g/L}/97.18 = 0.100M$
  (c) $0.200 \times 5/244.3 = 0.0004M$
  (d) $72.0/72 \times 142.0 = 0.007M$

CHAPTER 7

1. (a) NO_2^- **(b)** $C_2H_2(COO)_2^{2-}$ **(c)** NH_3
  (d) $\underset{\underset{NH_2}{|}}{CH_2COOH}, \underset{\underset{NH_2}{|}}{CH_2COO^-}$

3. **(a)** $HA + C_2H_5OH \rightleftharpoons C_2H_5OH_2^+ + A^-$
$B + C_2H_5OH \rightleftharpoons BH^+ + C_2H_5O^-$
5. $K_b = 1.26 \times 10^{-15}$
7. **(a)** $H^+ = 0.707$, $H^+ = 0.683$
(b) $H^+ = 0.092$, $H^+ = 0.070$
(c) $H^+ = 0.050$, $H^+ = 0.031$
9. **(a)** $pOH = 2.85$, $pH = 11.15$
(b) $pOH = 5.14$, $pH = 8.86$
(c) $pOH = 2.87$, $pH = 11.13$
(d) $pOH = 2.18$, $pH = 11.82$
11. $pH = 2.879 \, (25°)$, $pH = 2.907 \, (60°)$
13. $pH = 6.57$
15. **(a)** Fraction ionized = 0.025
(b) Fraction = 0.076 **(c)** Fraction = 0.221
(d) Fraction = 0.538
17. $pH = 8.55$
19. $pH = 4.49 \, (\mu = 0)$
21. $pH = 8.35$
23. **(a)** $pH = 3.83$ **(b)** $pH = 8.41$
25. $\alpha_Y = 10^{-5.58}$
27. Buffer capacity is greatest at a 1:1 ratio.
29. The acid chosen should have a $pK_a \cong pH$ of the buffer.
31. **(a)** $pH = 10.08$
(b) $x = 0.086$ (buffer capacity)
(c) $x = 0.082$ (buffer capacity)
33. $\dfrac{[BH^+]}{[B]} = 0.15$ (Mix 0.15 mole of HCl with

1.15 mole of pyridine.)

CHAPTER 8

7. 50% titration: $pH = pK_a = 9.37$
100% titration: $pH = 5.69$; use methyl red indicator
9. **(a)** $pH = 5.00$ **(b)** $pH = 6.34$
(c) $pH = 7.23$ **(d)** $pH = 9.60$
13. **(a)** $pH = 4.08$ **(b)** $pH = 2.77$
(c) $pH = 4.85$
15. Titrate NaH_2PO_4 with standard NaOH. On another aliquot, titrate Na_2HPO_4 with standard HCl.
17. **(a)** $pH = 4.65$, $pH = 9.75$
(b) $pH = 2.70$
(c) $pH = 2.30$, $pH = 6.20$
(d) $pH = 4.45$, $pH = 8.25$
19. There will be a sharp break around $pH = 9$ when the sodium hydroxide has been titrated and a second, less sharp break around $pH = 4.3$ when the malonate has been titrated to hydrogen malonate.
21. 60.9% malonic acid, 28.93% Na_2Malonate
23. Eq. wt. = 274.8
25. 18.54% Na_2Co_3, 4.28% $NaHCO_3$
27. Eq. wt. = 40
29. **(1)** $NaHCO_3$ **(2)** $NaOH$-Na_2CO_3
(3) Na_2CO_3 **(4)** Na_2CO_3-$NaHCO_3$

31. Titration of oxalic acid with standard NaOH gives two end points (although the first is not as sharp as could be desired). Titrate the mixture with standard NaOH, following the titration potentiometrically. The difference between the first and second end points represents the titration of the second H^+ of the oxalic acid. The second (and sharper) end point represents the titration of oxalic acid (2 hydrogens) plus sulfuric acid (2 hydrogens).
33. Titrate with standard NaOH; mmoles SO_2 = mmoles of NaOH to titrate from first to second potentiometric end point; mmoles SO_3 = (mmoles NaOH to first end point − mmoles SO_2)/2.
35. Titrate sodium borate with HCl. On another aliquot, titrate boric acid with NaOH after adding sorbitol.
39. 80.04% $(NH_y)_2SO_y$
41. **(a)** By Kjeldahl
(b) Hydrolyze with urease, then distill and titrate the ammonia.

CHAPTER 9

5. Sodium acetate
7. It is a weaker base and might permit the titration of weaker bases. However, it is more acidic and could compete with the acid titrant. Many samples bases will be leveled in dichloroacetic acid.
9. $NH_4^+Ac^- + OH^- \rightarrow NH_3 + H_2O + Ac^-$
    $(Bu_4N^+OH$ in 2-PrOH)
$NH_4^+Ac^- + H^+ \rightarrow NH_4 + HAc$
    ($HClO_4$ in dioxane)
11. **(a)** No; acetic acid, $HClO_4$ in HAc **(b)** Yes
(c) Yes **(d)** No; acetic acid, $HClO_4$ in HAc

CHAPTER 10

1. No
3. Potentiometric measurement of $[Ag^+]$ using a silver indicator electrode. Would need to consider the equilibrium, $RS^- + H^+ \rightleftharpoons RSH$.
5. See text; yes
7. **(a)** From solubility product, $pAg = 8.1$
(b) From solubility product, $pAg = 6.2$
(c) $\Delta[Ag^+] = 10^{-1.9}$; $\Delta E = 0.059 \log 10^{-1.9}$
    $= -0.112$ V
9. $\dfrac{[(40.00)90.1234] - (12.20}{303.0}$
    $= 44.47\%$ Cl
11. See text.

CHAPTER 11

3. No, there is a series of complexes.
7. (a) Halide ion (b) SCN^- (c) EDTA
   (d) SCN^-
9. $K_{MY'} = 1 \times 10^8$
11. (a) $\alpha_Y = 10^{-8.7}$, pH $\lesssim 3.8$
    (b) $\alpha_Y = 10^{-6.7}$, pH $\lesssim 4.8$
    (c) $\alpha_Y = 10^{-12.0}$, pH $\lesssim 2.5$
    (d) $\alpha_Y = 10^{-14.1}$, pH $\lesssim 1.8$
13. (a) $\alpha_Y = 10^{-8.1}$, pH ≥ 4.0
    (b) p[OH] $= 9.87$, pH ≤ 4.1
15. $\alpha_M = 4.9 \times 10^{-4}$, pM $= 6.3$
17. $K_{Cd'Y'} = K_{CdY}\alpha_Y\alpha_{Cd} = 10^{16.5}10^{-4.7}10^{-2.5}$
    $= 10^{9.3}$
19. (a) $\alpha_{Cu} = 10^{-11.1}$, pCu $= 13.1$
    (b) pCu $= 18.3$
    (c) There is some break at the end point, but it is not sharp.
21. (a) 0%, pCa $= 2.0$; 200%, pCa $= 8.5$
    (b) 0%, pMg $= 2.0$; 200%, pMg $= 2.7$
    (c) Titration should be possible.
23. Titration must be done at a pH where the indicator is an acid.
25. $K = \dfrac{[MgAr]}{[Mg][Ar]} = \dfrac{(5 \times 10^{-6})}{(5 \times 10^{-6})^2}$
    $K = 1/(5 \times 10^{-6}) = 10^{5.3}$
27–29. See text.
31. (a) $(10.00)(0.0120)/50 = 0.00240M$ Ca
    (b) $(15.00)(0.0100)/50 - 0.00240)$
       $= 6.0 \times 10^{-4}M$ Mg
    (c) $\dfrac{(15.00)(0.0100)(100.1)}{0.050}$
       $= 300$ ppm hardness (as $CaCO_3$)

CHAPTER 12

5. (a) An oxidant (or reductant) usually does not react with the solvent to form a weaker oxidant (or reductant). (b) Formation of an OH radical could be considered a form of leveling.
7. (a) Yes, because I^- from reduction of I_2 will precipitate AgI (b) Yes (c) No (d) Yes
9. The Co(III) complex is more stable than the Co(II) complex.
11. (a) $+0.751$ V (b) $+0.771$ V
13. (a) $E = 0.699 = 0.708$ V
    (b) $E = 0.699 = 0.412$ V
15. Measure the potential of a platinum wire dipping into a solution containing equal concentrations of the HQ and the Q at fixed pH. Changes in pH will affect the potential.

17. (a) $\log[Ag^+] = -6.19$
    (b) $[Ag^+] = 6.5 \times 10^{-7}M$;
       $K_{sp} = 4.17 \times 10^{-13}$
19. $[O_2] = 2.34 \times 10^{-4}M$
21. Minimum $K = 10^8$; $E_A{}^0 - E_B{}^0 = 8(0.05915)$
    $= 0.473$ V
23. $K = 7.4 \times 10^3$
25. $K = 1.30 \times 10^8$; the second half-reaction
27. $K = 4.00 \times 10^9; \dfrac{[Cr^{2+}]}{[Cr^{3+}]} = 0.0158$
29. $E^0 = 0.074$ V
31. (a) $[AsIII] = 9.1 \times 10^{-7}M$
    (b) $[AsIII] = 9.09 \times 10^{-3}M$; reaction is *not* quantitative
33. $E_{ind} = E_{meas} + E_{ref} = -0.590 + 0.246$
    $= -0.344$ V
35. (a) 0.10 V (b) 0.771 V
    (c) 1.44 V, titration curve will show two breaks
37. $E_{eq} = 0.556$; methylene blue
41. Dif. $= +0.88$ V in sulfuric acid. Activation energy is small.
43. Iodide in neutral or slightly basic solutions:
    $IO_4{}^- + 2I^- \rightarrow IO_3{}^- + I_2$
45. Reaction is catalyzed by OH^-; a higher concentration of OH^- is present in dilute $HClO_4$ than in concentrated $HClO_4$.
47. Fe(II) was used in the paper cited.

CHAPTER 13

1. (a) $+2.5$ (b) $+7$ (c) $+6$ (d) $+2$
3. Equivalent weights: (a) $KMnO_4/5$
   (b) $H_3ASO_3/2$
   (c) $K_2Cr_2O_7/6$, $Fe(CN)_6{}^{4-}/1$
   (d) $Mn^{+2}/5$ (e) $KBrO_3/5$, $NaBr/1$.
5. (a) $HIO_4{}^+CH_2OHCH_2OH \rightarrow$
       $HIO_3 + 2H_2CO + H_2O$
   (b) $2HIO_4 + CH_2OHCHOHCH_2OH \rightarrow$
       $2HIO_3 + 2H_2CO + HCO_2H + H_2O$
   (c) $2Ce(IV) + CH_3CHO + H_2O \rightarrow$
       $2Ce(III) + CH_3CO_2H + 2H^+$
   (d) $2Ce(VI) + CH_3CHOHCH_3 \rightarrow$
       $2Ce(III) + CH_3COCH_3{}^+2H^+$
7. $\dfrac{558.5 \text{ mg}}{(55.85 \text{ mg/meq}) \, 33.00 \text{ mL}} = 0.3030$
9. $\dfrac{(30.0 \text{ mL})(0.1N)(55.85)(100)}{55.85\%} = 300$ mg
11. $\dfrac{198.0 \text{ mg}}{(197.8/4)(20.00 \text{ mL})} = 0.2002$
13. $\dfrac{(15.50 \text{ mL})(0.1N)(207.2/3)(100)}{200.0 \text{ mg}} = 53.5\%$ Pb

15. 1.1794 meq, 0.377 mg/mL

17. $\dfrac{(20.00 \text{ mL})(0.2000 \ N)(197.8/4)(100)}{1000 \text{ mg}}$

= 19.78% As_2O_3; % As_2O_5 = 11.49%

19. $\dfrac{(24.0 - 3.00) \text{ meq}(94.1/6)(100)}{470.0 \text{ mg}}$

= 70.07% phenol

21. According to the Nernst equation, a *small portion* of Ag^+ can be oxidized to Ag^{2+} at a lower potential.

23. Use a new reductor column that has a good amalgam coating on the surface. Possibly the acidity of the iron solution is too great.

25. Potentiometric titration using Pt and calomel electrodes. The Pt electrode will respond to the iron(III)–iron(II) couple. A photometric titration might also work.

27. See text.

29. See text.

31. Buffer and titrate the sulfite to sulfate with standard iodine solution, starch indicator. The SO_2 concentration can be calculated from the titration and from the volume of air passed through the scrubber.

CHAPTER 14

1. (a) $t = 6.9/10^0 \ \text{sec}^{-1} = 6.9$ sec
(b) $t = 4.6/10^0 \ \text{sec}^{-1} = 4.6$ sec

3. (a) $t = 1.14_1 \ \text{hr} \ (> 1 \ \text{hr})$ (b) $t = 0.992$ hr
(c) $t = 0.843$ hr

5. $t = 3 \times 10^{-3}$ sec (slower than the first-order time of $7._6 \times 10^{-69}$ min)

7. $t = 7._1$ sec (time is short enough)

9. $[A] + [C] = 0.0800M$,
$[C] = 0.0800M - 0.0300M = 0.0500M$

CHAPTER 15

1. See text.

3. Stirring reduces concentration polarization.

5. grams Ag = 0.00252

7. It should be possible to plate out Ag^0 and Cu^0 selectively at around 0V. Then dissolve the plated metals in HNO_3, dilute, and measure by selective spectrophotometric methods.

9. To prevent secondary reactions at low Fe^{+2} concentration. Instead, Ce^{3+} oxidized to Ce^{IV}, which oxidizes Fe^{2+}.

11. Use equation in Problem 5.

$154.1 = \dfrac{(96{,}487)(mg)}{(19.3)(90)}$ mg = 2.77

13. 0.25449. Purity = 99.8%

15. Can be used at a very negative applied potential or in acidic solution. Pt is used for reactions requiring a positive applied potential.

17. $HgCL(s) + Cl^- \rightarrow HgCl_2 + e^-$; none

19. Base the method on 2-electron reduction of oxygen, first to H_2O_2 and then to OH^-.

21. The diffusion current increases as the area of each mercury drop grows, then decreases abruptly as the drop falls and a new drop starts growing.

23. It should increase the current over that obtained with no stirring.

27. See text.

CHAPTER 16

3. At the glass electrode, $AgCl(s) + e^- \rightarrow Ag^0(s) + Cl^-$. At the calomel electrode, $Cl^- + Hg^0(s) \rightarrow HgCl(s) + e^-$. The current flows from calomel to the glass electrode.

5. See text.

7. See text.

9. Measurement gives the activity of H^+.

11. (a) 0.011 (b) 0.1150 (c) 1.096 (d) 0.000
(e) 1.0093×10^{-5} (f) 1.26×10^{-7}

13. (a) 0.0589 (b) 1.096 (c) 0.631 (d) 2.000

15. Error = +6.0%

17. Error = +10%

19. (a) Glass electrode; no interference by H^+
(b) Glass electrode; $10^{-6}M \ H^+$ interferes, so use a buffer
(c) Glass electrode; $10^{-5}M \ H^+$ interferes, so use a buffer
(d) Solid state electrode; OH^- interferes, so use a buffer

21. (a) Liquid membrane electrode; measure Ca^{+2}
(b) Glass electrode; measure Na^+
(c) Glass electrode; measure Na^+

CHAPTER 17

1. (a) $A = -\log(7.2 \times 10^{-1}) = 0.14$ or 0.14_3
(b) $A = -\log(1.7 \times 10^{-1}) = 0.77$ or 0.76_9

3. (a) $A = (2.0 \text{ cm}/1.0 \text{ cm})(0.20) = 0.40$
(b) $A = (2.0 \text{ cm})/5.0 \text{ cm})(0.55_3) = 0.22_1$

5. The correct $T = (P/P_0) = 35.2/95 = 0.37$ or 0.37_{05}

7. If it were used for a known or standard, the A standard would be too high and the calculated concentration of the unknown would be too low. If it were used for the unknown, its calculated concentration would be too high.

9. (a) 4.30 (b) -1.70 (c) 1.04 (d) 0.009

11. (a) 2.0×10^3 (b) 0.035
    (c) 1.4×10^3 (280), 1.1×10^3 (235)

13. (a) $\epsilon = 197$ (b) $\epsilon = 610$

15. (a) $0.021M$ (b) 136.0 mg/dL

17. (a) $3.5 \times 10^{-4}M$ (b) $1.25 \times 10^{-4}M$

27. (b) $\epsilon = 2.0 \times 10^3$

29. (a) 0.185% (b) 0.740%

31. Acidic form $= 2.0 \times 10^{-4}M$,
    basic form $= 8.0 \times 10^{-4}M$

33. (a) 0.90 mL, 1 to 1 (b) $1.38 \times 10_4$
    (c) $K = 3.17 \times 10^3$

35. (a) $\epsilon = 3.73 \times 10^4$ (MR$_2$)
    (b) $\epsilon = 2.15 \times 10^3$ (R)

39. Measure ϵ for Br$_2$, then add excess HBr and measure ϵ for Br$_3^-$. Measure absorbance of the sample mixture. Knowing total concentration of Br$_2$ + Br$_3^-$, calculate the concentration of each from this equation:

$$A_{mix} = \epsilon_{Br_2} C_{Br_2} + \epsilon_{Br_3^-} C_{Br_3^-}$$

41. (a) At about 440 nm
    (b) Iron(III) might interfere
    (c) Complex iron(III) with phosphoric acid

CHAPTER 18

5. Ag(I) ion will complex with the pi orbital of the olefin.

7. The lowest energy transition of butadiene occurs at lower energy and longer wavelength (217 nm) than that of ethylene.

9. Two approaches might be tried: (a) Add a large (10–100-fold excess) to overcome the small formation constants so as to form all HgCl$_4^{2-}$ (b) Add slightly less chloride than mercury(II) so as to form only HgCl$^+$.

11. (a) At 200 nm: $n \rightarrow \pi^*$; at 208 nm: $n \rightarrow \pi^*$
    (b) The acetic acid $n - \pi^*$ transition of forbidden (low probability)

13. FIA detection limit $= 0.00005/13 = 3._8 \times 10^{-6}M$; spectrophotometer detection limit $= 0.010/13 = 7.6_9 \times 10^{-4}$

15. Since high concentrations of benzene could overlap the 175-nm band of naphthalene, measure naphthalene at 314 nm, where benzene does not absorb.

17. (a) Use Beer's law: $c_{anion} = A/b\epsilon$.
    (b) As shown in Figure 14–1, plot the reaction rate against [ester] or [anion]. Calculate the change in concentration of ester, or anion, and divide it by the time interval corresponding to this change. Plot this against [ester] or [anion]. The ester plot gives a slope $= -k$; the anion plot gives a slope $= k$.

19. (a) Selectively excite only A from 200 to 250 nm.
    (b) Measure only the fluorescence of A from 425 to 460 nm.
    (c) Separate before determining A.

21. $(10^{-8} \mu mole/\mu L)(1 \times 10^5 \mu L)(300 \mu g/\mu mole) = 0.3 \mu g$

23. One: excite only phenanthrene at 331 nm; measure at 350 nm. Two: excite both at a wavelength such as 280 nm and measure only phenanthrene at 380 nm.

27. The energy of most IR radiation is too weak to eject electrons from the cathode of a vacuum photodiode (phototube).

29. (a)
$$\frac{C—D \text{ freq}}{C—D \text{ freq}} = \sqrt{\frac{(12 + 2.0)/(12)(2.0)}{(12 + 1.0)/(12)(1.0)}}$$
$$= \sqrt{\frac{(14)(1)}{(13)(2)}} \cong \sqrt{\frac{1}{2}}$$
$$C—D \text{ freq} = 2980 \text{ cm}^{-1} \times \sqrt{\frac{1}{2}}$$
$$= 2107 \text{ cm}^{-1}$$

(b) $C{=}O^{18}$ freq $= 1635$ cm^{-1}

31. (a) The 3,3-μm band is the O—H intramolecular hydrogen-bonded stretching occurring in the enol form of the pentanedione.
    (b) Two carbonyl bands are observed because the enol has a conjugated carbonyl at 6.1 μm and the keto has two unconjugated carbonyls at 5.8 μm.

33. $C_6H_5NO_2$ (nitrobenzene)

35. Groups aromatic ring and primary amine.

CHAPTER 19

1. 3220 Å, 4590 Å

3. $E_{eV} = \dfrac{1.240 \times 10^3}{330.0} = 3.76$ eV

9. Use a fuel-rich flame. An ICP should be possible because the plasma contains no oxygen.

15. Add a salt of lower ionization potential than that of the elements to be measured. The electrons released by ionization of the added salt will repress the ionization of the sample atoms.

19. Atomic absorption

CHAPTER 20

1. Extract aniline into an acidic aqueous solution; the aniline forms a water-soluble salt, $C_6H_5NH_3^+$.
3. (1) Try a solvent other than diethyl ether.
   (2) Try other concentrations of HBr.
   (3) Add a larger cation than H_3O^+, such as R_4N^+.
   (4) Add a salting-out agent.
5. $x = 1.91$ (or approximately 2.0). Yes, this could account for the large difference.
7. $D_m = 11.3$; percentage $= 91.9\%$
9. For A, $f = 0.0023$ (99.8%); for B, $f = 0.961$ (3.9%). No, too much B is extracted.
11. (1) Hold $[UO_2^{2+}]$ and $[NO_3^-]$ constant and vary concentration of L in toluene. Plot $\log D$ versus $\log [L]$; slope $= x$.
    (2) Hold $[UO_2^{2+}]$ and $[L]$ constant and vary concentration of NO_3^-. Plot $\log D$ versus $\log [NO_3^-]$; slope $= y$.
13. Measure D at different concentrations of the liquid amide in toluene. The slope of a plot of $\log D$ versus $\log [\text{Amide}] = x$.
15. Acidify urine and extract neutral organic compounds. Then make the urine basic and again extract. Only organic bases should extract.

CHAPTER 21

7. Experimentally measure t and t_0; use $t = t_0 (1 + k)$ to calculate k.
11. (a) $u = 1500$ plates/sec
    (b) 3750 distance in plates, 7500 plates after 20 sec, 3488 plates in 10 sec, 6977 plates in 20 sec
    (c) $4\sigma = 212$ plates, 300 plates
    (d) $4\sigma = 207$ plates, 293 plates; $1.25(= R_s)$; yes, separation is complete.
13. (a) $k_1 = 3.00$, $k_2 = 3.13$
    (b)

N'	$4\sigma(k = 3.00)$	$4\sigma(k = 3.13)$
10,000	4.16 sec	4.31 sec
20,000	2.94	3.05
30,000	2.40	2.49
40,000	2.08	2.16

   (c) If $R_s = 1$, $4\bar{\sigma} = 4.0$, which occurs at a little more than $N' = 10,000$ would be safe with $N' = 20,000$.

15.

t	σ^2	4σ	σ^2(w/o e.c.)	4σ(w/o e.c.)
84	2.27	6.03	1.24	4.45
131	4.98	8.93	3.95	7.95
137	5.43	9.32	4.40	8.39
315	29.53	21.74	28.50	21.35

The earliest peaks are most affected.

17. $R_s = 0.75$; no
19. $\alpha = 1.010$, $\bar{k} = 8.19$; $N' = 181,300$
21. (a) $\bar{k} = 0.80$, $R_s = 0.86$; $\bar{k} = 8.00$, $R_s = 1.21$
23. To speed up elution of later peaks without losing resolution of earlier peaks:
    (a) raise oven temperature at a fixed rate throughout chromatographic run
    (b) continuously change eluent composition
25. Initial rate $= 4$

CHAPTER 22

1. To reduce extra-column broadening
5. To reduce tailing
9. 0.18 min/peak $= 5$ peaks
13. (a) $j = 0.46$ (b) $j = 0.64$ (c) $j = 0.79$
15. Use a lower temperature so that the retention times will be longer and the resolution better.
17. N' is proportional to $LD_g/\bar{u}$. As L increases a higher pressure is needed to keep the same $\bar{u}$; this decreases $\bar{D}_g$.
19. (a) $D_g = 0.198$
    (b) At $T = 80°C$, $\sigma_d^2 = 0.41$ sec^2; at $T = 120°C$, $\sigma_d^2 = 0.50$ sec^2
21. $H = 0.0122$ cm $(\bar{u} = 30)$; $H = 0.032$ cm $(\bar{u} = 48)$; $H = 0.045_4$ cm $(\bar{u} = 56)$
23. $a = 1.21$
25. (a) Electron capture, since halogens are involved
    (b) Thermal conductivity, since other available detectors do not give a signal for water
    (c) Flame ionization, since electron capture will not work and thermal conductivity is not as sensitive
29. $R_{s2} = 1.06$; the resolution is slightly better for the isothermal run.
31. Use an internal standard, which is added to the calibration samples and the unknown sample. The *ratio* of compound of interest to internal standard is used both to plot the calibration curve and to determine the unknown amount of compound.
33. (a) Increase the temperature; use a capillary instead of a packed column.
    (b) Use a different stationary phase; change the temperature.
    (c) Use temperature programming, starting at a temperature that is much lower than $80°C$ to improve the resolution of the first two peaks and ending at a temperature that is higher than $80°C$ to more quickly elute the later peaks.
35. Derivatize the sugars by oximating the carbonyl group and then forming the trimethylsilyl ether. The sugars will then be volatile

enough to analyze with gas chromatography. The presence of fructose and/or glucose may be determined by comparing chromatograms of corn syrup with those of solutions of pure derivatized fructose and glucose.

CHAPTER 23

5. (a) $N' = 12,500$
   (b) $k_1 = 1.97, k_2 = 2.07, k_3 = 3.77, k_4 = 4.00$
   (c) $R_s = 1.12, R_s = 1.45$
   (d) For peaks 1 and 2, $R_s = 0.93$; for peaks 3 and 4, $R_s = 1.40$
7. $t_{phenol} = 144$ sec; $t_{3,5\text{-dimethyl/phenol}} = 401$ sec
9. (a) $55\%, k = 8.39; 60\%, k = 6.23; 65\%, k = 4.62; 75\%, k = 2.55; 85\%, k = 1.40$
   (b) $0-2$ min, 1.23 cm; $2-4$ min, 2.13 cm; $4-6$ min, 3.56 cm; $6-8$ min, 5.63 cm
11. Since polar solutes, use polar stationary phase such as an alkylamino stationary phase. Detect with a refractive index detector.

CHAPTER 24

3. At low pH, the resin will lose its capacity owing to nonionized exchange sites
5. $\dfrac{1.000}{100}$ mg/mL $= 0.01M$ Ca^{2+} 50×0.01

   $= $ mmole Ca taken;
   $2 \times 13.6 \times 0.01$
   $= 0.272$ mmole Ca aqueous
7. $K = 0.46$
9. 1.0 for Ag^+, 2.0 for Mg^{2+}. As $[H^+]$ increases, D_g decreases.
11. (a) $t'_1 = 12.1$ min, $t'^{2-}_{SO_4} = 22.0$ min
    (b) $t'_1 = 3.8_2$ min, $t'^{2-}_{SO_4} = 2.2$ min
13. The sample can be passed through a catex column in the H^+ form. Since Al^{3+} is known, the original H^+ can be calculated.
19. Resolution will be lost owing to broadening of the eluted peaks.

Index

FORMULA WEIGHTS

(Arranged alphabetically according to atomic symbol. All weights rounded to relative uncertainties between 0.2 and 0.02 ppt.)

$AgBr$	187.78
$AgCl$	143.32
AgI	234.77
$AgNO_3$	169.88
$Al(C_9H_6NO)_3$ (aluminum oxinate)	459.46
Al_2O_3	101.96
As_2O_3	197.84
As_2O_5	229.84
$BaCl_2$	208.25
$BaCl_2 \cdot 2H_2O$	244.27
$BaCO_3$	197.35
$BaCrO_4$	253.33
BaO	153.34
$BaSO_4$	233.40
$CHCl_3$	119.38
CO_2	44.011
$C_2H_4(OH)_2$ (ethylene glycol)	62.07
$(CH_3)_2Hg$	230.66
C_4H_9OOH (t-butyl hydroperoxide)	90.13
$(CH_2OH)_3CNH_2$ (THAM)	121.14
C_6H_5Br (bromobenzene)	157.02
$CaCO_3$	100.09
CaC_2O_4	128.10
$CaMg(CO_3)_2$	184.41
CaO	56.08
$Ca(OH)_2$	74.10
$CaSO_4$	136.14
$Ca_3(PO_4)_2$	310.18
$Ce_2(C_2O_4)_3$	544.47
CeF_3	197.12
Cr_2O_3	152.00
CuO	79.54
CuS	95.60
$CuSO_4 \cdot 5H_2O$	249.68
FeO	71.85
FeS	87.91
$FeSO_4 \cdot C_2H_4(NH_3)_2SO_4 \cdot 4H_2O$	382.18
$FeSO_4 \cdot (NH_4)_2SO_4 \cdot 6H_2O$	392.15
FeS_2	119.97
Fe_2O_3	159.69
Fe_3O_4	231.54
$2Fe_2O_3 \cdot 3H_2O$	373.43
HCO_2CH_3 (acetic acid)	60.05
$HCO_2C_6H_5$ (benzoic acid)	122.12